Lecture Notes in Artificial Intelligence 5246

Edited by R. Goebel, J. Siekmann, and W. Wahlster

Subseries of Lecture Notes in Computer Science

Petr Sojka Aleš Horák
Ivan Kopeček Karel Pala (Eds.)

Text, Speech and Dialogue

11th International Conference, TSD 2008
Brno, Czech Republic, September 8-12, 2008
Proceedings

Series Editors

Randy Goebel, University of Alberta, Edmonton, Canada
Jörg Siekmann, University of Saarland, Saarbrücken, Germany
Wolfgang Wahlster, DFKI and University of Saarland, Saarbrücken, Germany

Volume Editors

Petr Sojka
Masaryk University
Faculty of Informatics
Department of Computer Graphics and Design
Botanická 68a, 602 00 Brno, Czech Republic
E-mail: sojka@fi.muni.cz

Aleš Horák
Ivan Kopeček
Karel Pala
Masaryk University
Faculty of Informatics
Department of Information Technologies
Botanická 68a, 602 00 Brno, Czech Republic
E-mail: {hales,kopecek,pala}@fi.muni.cz

Library of Congress Control Number: 2008934474

CR Subject Classification (1998): H.3.1, J.5, H.5.2, I.2.6, I.2.7

LNCS Sublibrary: SL 7 – Artificial Intelligence

ISSN 0302-9743
ISBN-10 3-540-87390-2 Springer Berlin Heidelberg New York
ISBN-13 978-3-540-87390-7 Springer Berlin Heidelberg New York

This work is subject to copyright. All rights are reserved, whether the whole or part of the material is concerned, specifically the rights of translation, reprinting, re-use of illustrations, recitation, broadcasting, reproduction on microfilms or in any other way, and storage in data banks. Duplication of this publication or parts thereof is permitted only under the provisions of the German Copyright Law of September 9, 1965, in its current version, and permission for use must always be obtained from Springer. Violations are liable to prosecution under the German Copyright Law.

Springer is a part of Springer Science+Business Media

springer.com

© Springer-Verlag Berlin Heidelberg 2008

Typesetting: Camera-ready by author, data conversion by Scientific Publishing Services, Chennai, India
Printed on acid-free paper SPIN: 12525645 06/3180 5 4 3 2 1 0

Preface

The annual Text, Speech and Dialogue Conference (TSD), which originated in 1998, is now starting its second decade. So far almost 900 authors from 45 countries have contributed to the proceedings. TSD constitutes a recognized platform for the presentation and discussion of state-of-the-art technology and recent achievements in the field of natural language processing. It has become an interdisciplinary forum, interweaving the themes of speech technology and language processing. The conference attracts researchers not only from Central and Eastern Europe, but also from other parts of the world. Indeed, one of its goals has always been to bring together NLP researchers with different interests from different parts of the world and to promote their mutual cooperation. One of the ambitions of the conference is, as its title says, not only to deal with dialogue systems as such, but also to contribute to improving dialogue between researchers in the two areas of NLP, i. e., between text and speech people. In our view, the TSD conference was successful in this respect in 2008 as well.

This volume contains the proceedings of the 11th TSD conference, held in Brno, Czech Republic in September 2008. Following the review process, 79 papers were accepted out of 173 submitted, an acceptance rate of 45.7%. We would like to thank all the authors for the efforts they put into their submissions and the members of the Program Committee and the reviewers, who did a wonderful job helping us to select the most appropriate papers. We are also grateful to the invited speakers for their contribution. Their talks provided insight into important current issues, applications and techniques related to the conference topics.

Special thanks are due to the members of the Local Organizing Committee for their tireless effort in organizing the conference. Dagmar Janoušková and Dana Komárková carried the main administrative burden and contributed in many other ways to the preparation of the conference. The TEXpertise of Petr Sojka resulted in the production of the volume that you are holding in your hands.

We hope that you benefitted from the event and that you also enjoyed the social program prepared by the Organizing Committee.

July 2008

Aleš Horák
Ivan Kopeček
Karel Pala
Petr Sojka

Organization

TSD 2008 was organized by the Faculty of Informatics, Masaryk University, in cooperation with the Faculty of Applied Sciences, University of West Bohemia in Plzeň. The conference webpage is located at `http://www.tsdconferences.org/tsd2008/`

Program Committee

Jelinek, Frederick (USA),
General Chair
Hermansky, Hynek (USA),
Executive Chair
Agirre, Eneko (Spain)
Baudoin, Geneviève (France)
Černocký, Jan (Czech Republic)
Ferencz, Attila (Romania)
Gelbukh, Alexander (Mexico)
Guthrie, Louise, (UK)
Hajič, Jan (Czech Republic)
Hajičová, Eva (Czech Republic)
Hanks, Patrick (Czech Republic)
Hitzenberger, Ludwig (Germany)
Hlaváčová, Jaroslava (Czech Republic)
Horák, Aleš (Czech Republic)
Hovy, Eduard (USA)
Kopeček, Ivan (Czech Republic)
Krauwer, Steven (The Netherlands)
Kunzmann, Siegfried (Germany)
Loukachevitch, Natalija (Russia)
Matoušek, Václav (Czech Republic)
Nöth, Elmar (Germany)
Ney, Hermann (Germany)
Oliva, Karel (Czech Republic)
Pala, Karel (Czech Republic)
Pavesić, Nikola (Slovenia)
Petkevič, Vladimír (Czech Republic)
Pianesi, Fabio (Italy)
Psutka, Josef (Czech Republic)
Pustejovsky, James (USA)
Rothkrantz, Leon (The Netherlands)
Schukat-Talamazzini, E. Günter (Germany)
Skrelin, Pavel (Russia)
Smrž Pavel (Czech Republic)
Tadić, Marko (Croatia)
Varadi, Tamas (Hungary)
Vetulani, Zygmunt (Poland)
Vintsiuk, Taras (Ukraine)
Wilks, Yorick (UK)
Zakharov, Victor (Russia)

Referees

Aitor Soroa, Arantza Diaz de Ilarraza Sánchez, Daniel Zeman, David Guthrie, Diego Giuliani, Dino Seppi, Drahomíra Johanka Spoustová, Emanuele Pianta, Eva Navas, Fabio Valente, František Grézl, Gerard Ligozat, Hamed Ketabdar, Hari Parthasarathi, Christian Zieger, Christopher Brewster, Inaki Alegria, Inmaculada Hernaez Rioja, Izaskun Aldezabal, Jacek Marciniak, Jarmila Panevová, Joel Pinto, Kepa Sarasola, Luděk Bártek, Maria Khokhlova, Markéta Lopatková, Olga Mitrofanová, Ondřej Glembek, Pavel Rychlý, Petr Motlíček, Petr Sgall, Petr Sojka, Phil Garner, Samuel Thomas, Sanaz Jabbari, Silvie Cinková, Sivaram Garimella, Sriram Ganapathy, Stefan Grocholewski, Tomasz Obręski, Sazhok Vintsiuk, Weifeng Li, Wei Liu, Yulia Ledeneva.

Organizing Committee

Dana Hlaváčková *(administrative contact)*, Aleš Horák, Dagmar Janoušková *(accounting)*, Dana Komárková *(secretary)*, Ivan Kopeček *(co-chair)*, Karel Pala *(co-chair)*, Adam Rambousek *(web system)*, Pavel Rychlý, Petr Sojka *(proceedings)*

Sponsors and Support

The TSD conference is regularly supported by the International Speech Communication Association (ISCA). We would like to express our thanks to Lexical Computing Ltd., IBM–Česká republika, spol. s r. o. and ASEC–elektrosystémy s .r. o. for their kind sponsoring contributions to TSD 2008.

Table of Contents

I Invited Papers

II Text

III Speech

IV Dialogue

Part I

Invited Papers

The Future of Text-Meaning in Computational Linguistics

Graeme Hirst

Department of Computer Science
University of Toronto
Toronto, Ontario, Canada M5s 3G4
gh@cs.toronto.edu

Abstract. Writer-based and reader-based views of text-meaning are reflected by the respective questions "What is the author trying to tell me?" and "What does this text mean to me personally?" Contemporary computational linguistics, however, generally takes neither view. But this is not adequate for the development of sophisticated applications such as intelligence gathering and question answering. I discuss different views of text-meaning from the perspective of the needs of computational text analysis and the collaborative repair of misunderstanding.

1 Introduction: Text and Text-Meaning in Computational Linguistics

In this paper, I will describe how new applications in computational linguistics and natural language processing are leading to changes in how the field views the idea of the meaning of a text. By a *text*, I mean a document or a dialogue, or a structurally complete fragment, such as a paragraph or section of a document or a turn or sequence of turns in a dialogue. Although a text could be just a single sentence, when we call it a text we add the idea that we are regarding it as complete by itself. The term *text-meaning*, then, denotes the complete in-context meaning or message of the text. It is thus potentially much more than just the sum of the sentence-meanings of the individual sentences of the text; it includes the broader "message", and possibly even a subtext as well. It will include implicatures and pragmatic and logical inferences that follow immediately from the text. A computer system (or person) might be able to understand each sentence of a text at the sentence-meaning level and yet fail to understand the text itself.

2 Where Is the Meaning of a Text?

Given this definition, we can now ask: Who decides what a text means? Where does text-meaning lie, what is its *locus*? There have been three traditional answers to this: It is the writer or speaker; it is the reader or listener; or it is the text itself.

In the first answer, that the locus of meaning is the writer or speaker, the view is that meaning is the *intent* of the agent who is communicating or, at least, is derived from that intent. A text means whatever its author intended it to mean, which might be quite different from the "literal" (sentence-level) meaning, regardless of whether the reader

P. Sojka et al. (Eds.): TSD 2008, LNAI 5246, pp. 3–11, 2008.
© Springer-Verlag Berlin Heidelberg 2008

or listener is able to determine that intent correctly or not. This view is associated with philosophers such as Grice [8]. In the second answer, that the locus is the reader or listener, the view is that a text-meaning is each reader's individual response to, or experience of, reading the text; a reader, after all, cannot ever know a writer's intentions for certain, but can know only their own response to the text. This view, in various forms, is associated with postmodernists such as Stanley Fish and Roland Barthes, among others. In particular, Fish [6] claims that when readers agree on a meaning – that is, have the same response – it is because they are members of the same "interpretive community".

The third answer, that meaning is in the text itself, is what Reddy [23] has called the *conduit* view: that the text is a conduit through which meaning is sent from writer to reader. While Reddy argued against this as simplistic, a view of text as the locus of meaning was central to theories of literary criticism (such as that of Wimsatt and Beardsley; see Fish ([6]: 2)) that regarded the reader as too variable and the writer's intents as unknowable; only the text itself is fixed and knowable. Olson [21,22] has argued that, historically, the gradual emergence of the concept of the written *autonomous text* as a bearer of meaning was a development from the more-natural and more-fundamental concept of meaning as something that a listener adds to a spoken utterance in its context.

The question of where the meaning lies can, of course, be asked not only of texts but equally of linguistic elements at a lower level – words, sentences, semantic roles, lexical relations, and so on – with the same three potential answers. It does not follow, however, that the same answer need be chosen for each of these elements, nor that it be the same answer chosen for the text-meaning level. For example, one could argue that the effects of individual readers or of writers' intents are apparent only at the text-meaning level and not below. But equally, one could argue conversely that the idiosyncrasies of individual readers that are observed at lower levels (for example, in interpretations of lexical relations (Klebanov [14], Morris and Hirst [19], Hollingsworth and Teufel [12])) are *dampened* by constraints that arise from consideration of a text as an integrated whole, and this dampening effect is what enables texts to be autonomous bearers of meaning.

Each of the three views has its passionate defenders, but my goal here is not to argue for one over the other, but rather to regard each one as a view of meaning that is helpful in some situations in NLP.

Nonetheless, I take the view that the text itself must always be taken as a locus of meaning for all elements at all levels. (*A fortiori*, I would argue that the meaning of closed-class words is always *solely* textual, constant for all competent users of a language.) The text, after all, plays a central role in the meaning business. This is not to say that meaning cannot exist in the absence of text. Any kind of information transfer can entail meaning; one can say that a person's (non-verbal) actions have meaning and that events in the world have meaning. For example, Nadia putting the garbage bin out means that (she thinks that) it's Wednesday; or that smoke means fire (a famous example of Barwise and Perry [2]). But in the case where text is a medium of information transfer from which meaning somehow arises, one cannot overlook its causal role even if one does not accept the conduit metaphor. The question then remains as to whether the reader (listener) and/or the writer (speaker) are to be considered loci *in addition* to the text.

3 Computational Views of Text-Meaning: A Very Short History

Computational linguists are not philosophers or literary theorists, and generally don't think very much about the issues raised in the previous section. Nonetheless, any research in computational linguistics and natural language processing that involves semantics must at least implicitly choose a view of meaning. In Hirst [11], I pointed out that each of the three views of text-meaning has dominated one of the last three decades.

Ironically, in the traditional, more logic-based approaches of the mid-1970s to the mid-1980s, the dominant view was the most postmodern one, the reader-based view: language understanding was viewed as a knowledge-based enterprise, and so a text was understood by a system in the light of its own particular knowledge. This was so at both the sentence-meaning level and the text-meaning level; in the latter case, exemplified by the work of Schank and his colleagues (e.g., Schank [24], Schank and Abelson [25]), to understand language was to use knowledge and inference to "fill in the gaps" in what was said. The reader-based view became most explicit in Corriveau's computational model of time-constrained text comprehension (published [4], but developed in the late 1980s).

In the subsequent decade, with a greater interest in interactive dialogues, attention became focused on the speaker and their intentions. Again, this was so at both levels. At the text-meaning level, the work was typified by Carberry's book *Plan Recognition in Natural Language Dialogue* [3]. And below the text-meaning level, Grosz and Sidner [9], for example, took a speaker- or writer-based view of discourse structure.[1]

From the mid-1990s into the 2000s, with the increasing availability and importance of autonomous text such as newswire and Web pages and the rise of statistical and machine-learning methods, attention became focused solely on the text as the locus of meaning, with no thought or individuation of the writer or the reader. Tasks that implicitly entailed a regard for text-meaning, such as summarization, machine translation, and topic tracking, became seen just as meaning-preserving statistical transformations of the text. Text was to be "processed", not understood. In fact, in this paradigm, text-meaning was taken to be little or nothing more than the sum of sentence-meanings; for example, a text could be summarized, in this view, just by finding and displaying its "most important" sentences.

4 Computational Views of Text-Meaning: The Future

With this as background, I want to now turn to future views of text-meaning in computational linguistics and the future role of *the linguistic computer*. We are now starting to again see applications in natural language processing in which writer- and reader-based views of text-meaning are more explicit in the task, and I believe that, as NLP methods advance, these kinds of application will become increasingly important. I will give examples below.

In fact, the word *processing* in *natural language processing* will become less appropriate, as these developments will bring us closer to what early researchers optimistically

[1] Contrary to remarks by Morris [18], the listener or reader played no role in determining the discourse structure or focus of attention in Grosz and Sidner's theory; see Section 4.3 below.

called *natural language understanding*. But *understanding* is not the right word either; it's too loaded, and tends to imply a single "correct" meaning. A better word would be *interpretation*, implying one of many possible kinds of understandings. And it's exactly in the distinctions between different kinds of understanding or interpretations that these applications have their value.

In the discussion below, I will divide applications into two classes: Those in which the computer is an **observer** of external text, playing only the role of reader; and those involving interactive dialogue in which the computer is an **active participant** and hence plays alternating roles of reader and writer or speaker and listener. In the first case, the text (which could be a dialogue between others) has been created for some other reader in some other context; typically, the computer will be looking at it on behalf of a user in order to achieve some goal for that user. In the second case, the text that the computer reads is created by the writer expressly for the computer in the context of the dialogue; typically, this is a user interacting with a computer in natural language in order to achieve some goal.[2] In the space available here, I'll concentrate mostly on the first case, observing external text.

4.1 What Does This Text Mean to Me?

One of the profound changes to everyday life that has resulted from the rise of the search engine and the World Wide Web over the last decade is that the average person has become a researcher. In addition to students using Web research in their school assignments, people are now turning to Google and the Web to answer questions arising in their daily life that they previously would not have bothered about, and usually their starting point is a search engine.[3] Questions include those about health matters, finding a place to live, hobbies, government services, and matters of general curiosity. But of course, such questions must be re-expressed by the user as search terms, and the resulting documents must be read to extract the answer – assuming that they do, in fact, contain the answer.

Now, observe that a document might happen to provide an answer to a user's particular question without the writer having had any intent of doing so. But when search is based on keyword matching, this will be relatively infrequent; in the canonical case, a relevant document contains the search terms from the question[4] (and probably in close proximity if it is highly ranked by Google) exactly because the writer has made an explicit statement that answers the question. However, the more abstract, wide-ranging, or unusual the question, the less likely this is, and indeed the more likely it is that a relevant document will not be a good keyword match at all. These characteristics can be seen, for example, in the test topics used in the "HARD" task of the Text Retrieval Conference (TREC; `trec.nist.gov`) (Allan [1]); for example, "What factors contributed to the

---

[2] Of course, the two classes might occur together in a single application; the user might have a dialogue with the computer in order to achieve a goal pertaining to some observed texts.

[3] This is the general trend in the United States shown by recent surveys by the Pew Internet & American Life Project (`http://www.pewinternet.org`). There is no reason to think that things are significantly different in other developed countries.

[4] Google's PageRank algorithm may return documents that do not contain the search terms, but this is a minor effect for this argument.

growth of consumer on-line shopping?" In TREC, the requirements of this task were merely document or passage retrieval; but the task more generally can be thought of as the evolution and confluence of several streams of research in NLP and information retrieval – document retrieval, question-answering, and multi-document summarization – and future systems (we anticipate) will construct a full and relevant answer to the user's question by selecting relevant information from many documents. To do so, such a system must consider each document or passage from the point of view of the user's question (and anything else known about the user);[5] that is, it must take a reader-based view of the meaning of the text: "What does this text mean to me?"

Another particular research direction of relevance here is that known as "learning by reading" (LbR) (Hovy [13]). Much as in the research of the 1970s–80s mentioned in Section 3 above, the goal is for the system to achieve a full understanding of a (possibly lengthy) text, beginning from an initial knowledge base (which might be quite large), making inferences and interpretations, and then answering questions about what it has read. And like that earlier work, because of its dependence on prior knowledge for the interpretation of new text, it takes a reader-based view of text-meaning (but see also Section 4.2 below); and if its prior knowledge is deficient, it may misinterpret text (Forbus et al. [7]). Moreover, in the context of LbR, Ureel *et al* [28] and Forbus et al. [7] present a prototype of a "ruminator" process that "reflects" on what it has read by using its prior knowledge to generate questions that, if answered, would fill gaps in the system's understanding of the text.[6]

4.2 What Are They Trying to Tell Me?

The complement of applications that take a reader-based view of text is applications that try to determine the intent of the author of the text, and, to the extent that it's possible, without the bias or influence of any reader-based information. That is, the goal is to figure out "what is the writer trying to say?" Such applications can be broadly characterized as a kind of *intelligence gathering* – not necessarily in the military or espionage sense, though those domains are canonical examples, but in the broader sense of trying to find out, for any reason, what other people believe, or are thinking, or are planning.

A good example is the development in the last few years of computational methods for *sentiment analysis*, especially for blogs and similar online texts, and the more general idea of *opinion analysis, extraction,* or *mining* that has emerged in the NLP literature. The goal is to discover, from evaluative opinions expressed online, what consumers really think about various commercial products and services; companies can use this information in planning their market strategies. Clearly, if such information is to be maximally useful, it must be a close reflection of the writer's opinion, uncolored by any individualities or preconceptions of the reader. Work on tasks such as multi-perspective question answering (Stoyanov, Cardie, and Wiebe [26]) has extended this research into

[5] This could be facilitated by a personalized lexicon or thesaurus based on the user's past reading, as proposed by Yoshida et al. [29].

[6] Ureel et al. describe rumination as an "offline" process; hence the reflections or questions themselves can't be considered part of the text-meaning. But by its nature the process is nonetheless closely related to a reader-based view of text-meaning.

finding and summarizing reported or expressed opinions in newswire text. But beyond this, we can expect sophisticated applications of the future to be able to find implicitly stated opinion and sentiment without restriction to any pre-defined topic. Eventually, this may use (possibly inferred) knowledge of the writer and the writer's prior beliefs to determine just what the writer intends to say; that is, systems may be able to classify texts by the ideological background of the writer and interpret and analyze them accordingly. (This goes well beyond Malrieu's [17] description of a computational method of checking texts for *ideological consistency*.)[7]

Learning by reading will eventually come under this heading too, because as defined by Hovy [13], its goal of being able to answer questions about the text is construed as *test-taking*: in the prototypes, the text is taken from a high-school book on chemistry and the questions are from an advanced-placement exam on the material. Thus there is a "right answer" to the questions, and despite the reader-based view of understanding that the LbR architectures entail (see Section 4.1 above), LbR nonetheless has the writer-based goal of figuring out "what are they trying to tell me" in order to learn enough to pass the exam. However, this is for the future; there is no recognition of the writer or of their intent in the LbR architectures presently proposed (Hovy [13], Forbus et al. [7]).

Last, it may be recalled that the goal of high-quality machine translation is, by definition, determining, and preserving across languages, an author's intent – a fact that is largely obscured by the still-dominant purely statistical text-transformation approach. As the renewed interest in interlingual, and hence semantic, methods in MT develops further (e.g., Nirenburg and Raskin [20], Farwell et al. [5]), MT too will take on a more-explicit writer-based view of text-meaning.

4.3 Recovering from Misunderstanding

Despite any latitude a reader has in interpretation in all that we have said above, an interpretation can nonetheless count as a *misunderstanding* or *misinterpretation* if it does not fully respect the given text. That could result from a simple mishearing or misreading of the text, or from inattention or mental confusion; but it could also result from linguistic processing that is at odds with what the writer or speaker expected of the reader – incorrect resolution of an ambiguity. Examples of this include resolving an anaphor to a different antecedent than intended, choosing a different sense of a homonym, and attaching a phrase to a different point in the parse tree.

Although there is no guarantee that a misunderstanding will be detected as such, a listener or reader might hypothesize a misunderstanding, either at the present point in the text or at some earlier point, if he or she cannot interpret the present text. A speaker might hypothesize a misunderstanding by the other if they respond in an unexpected way. In either case, recognition that a misunderstanding has occurred might be followed

[7] Some of what I earlier suggested were purely text – view-based applications might also be considered to be implicitly about opinion analysis. For example, in processing newswire text about current objective events, while there might be said to be no overt authorial intent other than to comprehensively communicate facts, often, even in "objective" news reporting, an authorial bias or intent or ideology may nonetheless be implied in what facts are communicated and what words are used to describe them; such an intent might be discovered by comparing two authors' reports of the same event.

by recognition of the likely misunderstanding itself, leading to a reinterpretation by the reader or listener in the first scenario, or to a clarification by the speaker in the second scenario.

Given that computers will remain relatively poor understanders for some time yet, it is important that they be able to detect misunderstanding in dialogue, both their own and that of their conversant. Some years ago, McRoy ([15], McRoy and Hirst [16]) developed computational models of negotiation in the collaborative, constructive repair of misunderstanding in dialogues. Reflecting the inherent symmetry of the negotiation of meaning, the model could play both the role of the conversant who is misunderstood and the role of the conversant who fails to understand.

For example, the model could account for the text-meaning-level misunderstanding in this fragment of a conversation between a mother and her child Russ about a forthcoming parent-teacher meeting (Terasaki [27]):

1. MOTHER: Do you know who's going to that meeting?
2. RUSS: Who?
3. MOTHER: I don't know.
4. RUSS: Oh. Probably Mrs McOwen and some of the teachers.

Russ initially interprets line 1 as expressing Mother's desire to tell, that is, as a *pretelling* or *pre-announcement* as if Mother intends to surprise him (cf *Guess who's going to that meeting!*). But Russ finds this interpretation inconsistent with her next utterance; in line 3, instead of telling him who's going, as he would expect after a pretelling, Mother claims that she does not know. Russ recovers by reinterpreting line 1 as an indirect request for information, which his line 4 then responds to. In this example, we see that both the speaker and listener negotiate and refine the meaning of a prior utterance. (They swap roles of speaker and listener in the dialogue as they do, but not their roles with respect to the original utterance.) McRoy's model is thus a rare example in computational linguistics of one in which both speaker and hearer are loci of meaning. It thus contrasts with, for example, the earlier work of Grosz and Sidner [9] on discourse segmentation in which, even in the case of interactive dialogue, nothing is negotiated or refined and only the speaker, not the listener, is a locus of meaning; each participant may modify the structure or focus of the discourse only when taking their turn to act as speaker; and when the other takes their turn as speaker, they cannot change or undo what the first has done.

5 Conclusion

There are many other aspects of how we view meaning in computational linguistics and natural language processing and how that will change in the future that I haven't had space to touch on in this paper. These include the negotiation or collaborative construction of meaning in interactive dialogue, particularly in the interactive elicitation of knowledge; methodological issues that arise within annotation-based learning paradigms; the idea of current research in textual entailment as a preliminary to the more-general task of searching for differing interpretations of a text. This in turn could lead to automatic processes for reconciling seemingly incompatible interpretations of the same text or situation: the linguistic computer as an aid to mediation and reconciliation.

Acknowledgments. Some of the ideas in this paper are elaborations or improvements on points made in Hirst [10] and Hirst [11]. I am grateful to Stephen Regoczei and Jean-Pierre Corriveau for first introducing me to the ideas of reader-based models of meaning; to Susan McRoy, Peter Heeman, Phil Edmonds, and David Traum for discussions of models of misunderstanding; and to Jane Morris for drawing my attention to how the view in CL of the locus of text-meaning differed in the early 2000s from that of the 1980s. I thank Nadia Talent for helpful comments on an earlier draft of this paper. The preparation of this paper was supported by a grant from the Natural Sciences and Engineering Research Council of Canada.

References

1. Allan, J.: HARD track overview in TREC 2005. In: The 14th Text REtrieval Conference (TREC 2005) Proceedings, NIST (2006)
2. Barwise, J., Perry, J.: Situations and Attitudes. MIT Press, Cambridge (1983)
3. Carberry, S.: Plan Recognition in Natural Language Dialogue. MIT Press, Cambridge (1990)
4. Corriveau, J.-P.: Time-constrained Memory: A reader-based approach to text comprehension. Lawrence Erlbaum Associates, Mahwah (1995)
5. Farwell, D., et al.: Interlingual annotation of multilingual text corpora and FrameNet. In: Boas, H. (ed.) Multilingual FrameNets in Computational Lexicography, Mouton de Gruyter (to appear)
6. Fish, S.: Is there a text in this class? The authority of interpretive communities. Harvard University Press (1980)
7. Forbus, K.D., et al.: Integrating natural language, knowledge representation and reasoning, and analogical processing to learn by reading. In: Proceedings, 22nd AAAI Conference on Artificial Intelligence (AAAI-2007), Vancouver, pp. 1542–1547 (2007)
8. Grice, H.P.: Utterer's meaning, sentence-meaning, and word-meaning. Foundations of Language 4, 225–242 (1968)
9. Grosz, B.J., Sidner, C.L.: Attention, intentions, and the structure of discourse. Computational Linguistics 12(3), 175–204 (1986)
10. Hirst, G.: Negotiation, compromise, and collaboration in interpersonal and human-computer conversations. In: Proceedings, Workshop on Meaning Negotiation, 18th National Conference on Artificial Intelligence (AAAI-2002), Edmonton, pp. 1–4 (2002)
11. Hirst, G.: Views of text-meaning in computational linguistics: Past, present, and future. In: Dodig-Crnkovic, G., Stuart, S. (eds.) Computation, Information, Cognition – The Nexus and the Liminal, pp. 270–279. Cambridge Scholars Publishing (2007)
12. Hollingsworth, B., Teufel, S.: Human annotation of lexical chains: Coverage and agreement measures. In: Workshop on Methodologies and Evaluation of Lexical Cohesion Techniques in Real-world Applications, Salvador, Brazil (2005)
13. Hovy, E.: Learning by reading: An experiment in text analysis. In: Sojka, P., Kopeček, I., Pala, K. (eds.) Text, Speech and Dialogue. LNCS (LNAI), vol. 4188, pp. 3–12. Springer, Heidelberg (2006)
14. Klebanov, B.B.: Using readers to identify lexical cohesive structures in texts. In: Proceedings, Student Research Workshop, 43rd Annual Meeting of the Association for Computational Linguistics, Ann Arbor, pp. 55–60 (2005)
15. McRoy, S.: Abductive interpretation and reinterpretation of natural language utterances. Ph.D. thesis, Department of Computer Science, University of Toronto (1993)
16. McRoy, S., Hirst, G.: The repair of speech act misunderstandings by abductive inference. Computational Linguistics 21(4), 435–478 (1995)

17. Malrieu, J.P.: Evaluative Semantics. Routledge (1999)
18. Morris, J.: Readers perceptions of lexical cohesion in text. In: Proceedings of the 32nd annual conference of the Canadian Association for Information Science, Winnipeg (2004)
19. Morris, J., Hirst, G.: The subjectivity of lexical cohesion in text. In: Shanahan, J.G., Qu, Y., Wiebe, J. (eds.) Computing attitude and affect in text, pp. 41–48. Springer, Heidelberg (2005)
20. Nirenburg, S., Raskin, V.: Ontological Semantics. MIT Press, Cambridge (2004)
21. Olson, D.R.: From utterance to text: The bias of language in speech and writing. Harvard Educational Review 47(3), 257–281 (1977)
22. Olson, D.R.: The World on Paper. Cambridge University Press, Cambridge (1994)
23. Reddy, M.J.: The conduit metaphor: A case of frame conflict in our language about language. In: Ortony, A. (ed.) Metaphor and Thought, pp. 284–324. Oxford University Press, Oxford (1979)
24. Schank, R.C. (ed.): Conceptual Information Processing. North-Holland, Amsterdam (1975)
25. Schank, R.C., Abelson, R.P.: Scripts, Plans, Goals and Understanding. Lawrence Erlbaum Associates, Mahwah (1977)
26. Stoyanov, V., Cardie, C., Wiebe, J.: Multi-perspective question answering using the OpQA corpus. In: Proceedings, Human Language Technology Conference and Conference on Empirical Methods in Natural Language Processing (HLT/EMNLP), Vancouver, pp. 923–930 (2005)
27. Terasaki, A.: Pre-announcement sequences in conversation. Social Science Working Paper 99. University of California, Irvine (1976)
28. Ureel II, L., et al.: Question generation for learning by reading. In: Proceedings of the AAAI Workshop on Inference for Textual Question Answering, Pittsburgh, pp. 22–26 (2005)
29. Yoshida, S., et al.: Constructing and examining personalized cooccurrence-based thesauri on Web pages. In: Proceedings, 12th International World Wide Web Conference, Budapest (2003)

Deep Lexical Semantics

Jerry Hobbs

University of Southern California / Information Sciences Institute
4676 Admiralty Way, CA 90292-6695 Marina del Rey, USA
hobbs@isi.edu

Abstract. The link between words and the world is made easier if we have conceptualized the world in a way that language indicates. In the effort I will describe, we have constructed a number of core formal theories, trying to capture the abstract structure that underlies language and enable literal and metaphorical readings to be seen as specializations of the abstract structures. In the core theories, we have axiomatized composite entities (or things made out of other things), the figure-ground relation, scalar notions (of which space, time and number are specializations), change of state, causality, and the structure of complex events and processes. These theories explicate the basic predicates in terms of which the most common word senses need to be defined or characterized. We are now encoding axioms that link the word senses to the core theories, focusing on 450 of words senses in Core WordNet that are primarily concerned with events and their structure. This may be thought of as a kind of "advanced lexical decomposition", where the "primitives" into which words are "decomposed" are elements in coherently worked-out theories.

P. Sojka et al. (Eds.): TSD 2008, LNAI 5246, p. 13, 2008.
© Springer-Verlag Berlin Heidelberg 2008

Toward the Ultimate ASR Language Model

Frederick Jelinek and Carolina Parada

The Center for Language and Speech Processing
John Hopkins University, Baltimore, MD 21218, United States
carolinap@jhu.edu

Abstract. The n-gram model is standard for large vocabulary speech recognizers. Many attempts were made to improve on it. Language models were proposed based on grammatical analysis, artificial neural networks, random forests, etc. While the latter give somewhat better recognition results than the n-gram model, they are not practical, particularly when large training data bases (e.g., from world wide web) are available. So should language model research be abandoned as a hopeless endeavor? This talk will discuss a plan to determine how large a decrease in recognition error rate is conceivable, and propose a game-based method to determine what parameters the ultimate language model should depend on.

P. Sojka et al. (Eds.): TSD 2008, LNAI 5246, p. 15, 2008.
© Springer-Verlag Berlin Heidelberg 2008

Practical Prosody

Modeling Language Beyond the Words

Elizabeth Shriberg[1,2]

[1] Speech Technology & Research Laboratory, SRI International
333 Ravenswood Avenue, Menlo Park, CA 94025, USA
[2] International Computer Science Institute
1947 Center Street, Suite 600, Berkeley, CA 94704
ees@speech.sri.com, ees@icsi.berkeley.edu

Abstract. Prosody is clearly valuable for human understanding, but can be difficult to model in spoken language technology. This talk describes a "direct modeling" approach, which does not require any hand-labeling of prosodic events. Instead, prosodic features are extracted directly from the speech signal, based on time alignments from automatic speech recognition. Machine learning techniques then determine a prosodic model, and the model is integrated with lexical and other information to predict the target classes of interest. The talk presents a general method for prosodic feature extraction and design (including a special-purpose tool developed at SRI), and illustrates how it can be successfully applied in three different types of tasks: (1) detection of sentence or dialog act boundaries; (2) classification of emotion and affect, and (3) speaker classification.

P. Sojka et al. (Eds.): TSD 2008, LNAI 5246, p. 17, 2008.
© Springer-Verlag Berlin Heidelberg 2008

Part II

Text

"**Text**: a book or other written or printed work, regarded in terms of its content rather than its physical form: *a text which explores pain and grief.*"

NODE (The New Oxford Dictionary of English), Oxford, OUP, 1998, page 1998, meaning 1.

Sentiment Detection Using Lexically-Based Classifiers

Ben Allison

Natural Language Processing Group, Department of Computer Science
University of Sheffield, UK
b.allison@dcs.shef.ac.uk

Abstract. This paper addresses the problem of supervised sentiment detection using classifiers which are derived from word features. We argue that, while the literature has suggested the use of lexical features is inappropriate for sentiment detection, a careful and thorough evaluation reveals a less clear-cut state of affairs. We present results from five classifiers using word-based features on three tasks, and show that the variation between classifiers can often be as great as has been reported between different feature sets with a fixed classifier. We are thus led to conclude that classifier choice plays at least as important a role as feature choice, and that in many cases word-based classifiers perform well on the sentiment detection task.

Keywords: Sentiment Detection, Machine Learning, Bayesian Methods, Text Classification.

1 Introduction

Sentiment detection as we approach it in this paper is the task of ascribing one of a pre– (and well–) defined set of non-overlapping sentiment labels to a document. Approached in this way, the problem has received some considerable attention in recent computational linguistics literature, and early references are [1,2].

Whilst it is by no means obligatory, posed in such a way the problem can easily be approached as one of classification. Within the scope of the supervised classification problem, to use standard machine learning techniques one must make a decision about the features one wishes to use. On this point, several authors have remarked that for sentiment classification, standard text classification techniques using lexically-based features (that is, features which describe the frequency of use of some word or combination of words) are generally unsuitable for the purposes of sentiment classification. For example, [3] bemoans the "initially dismal word-based performance", and [4] conclude their work by saying that "traditional word-based text classification methods (are) inadequate" for the variant of sentiment detection they approach.

This paper revisits the problem of supervised sentiment detection, and whether lexically-based features are adequate for the task in hand. We conclude that, far from providing overwhelming evidence supporting the previous position, an extensive and careful evaluation leads to generally good performance on a range of tasks. However, it emerges that the choice of method plays at least as large a role in the eventual performance as is often claimed for differing representations and feature sets.

P. Sojka et al. (Eds.): TSD 2008, LNAI 5246, pp. 21–28, 2008.
© Springer-Verlag Berlin Heidelberg 2008

The rest of this paper is organised as follows: §2 describes our evaluation in detail; §3 describes the classifiers we use for these experiments; §4 presents results and informally describes trends. Finally, §5 ends with some brief concluding remarks.

2 Experimental Setup

The evaluation presented in this work is on the basis of three tasks: the first two are the movie review collection first presented in [1] which has received a great deal of attention in the literature since, and the collection of political speeches presented in [5]. Since both of these data sets are binary (i.e. two-way) classification problems, we also consider third problem, using a new corpus which continues the political theme but includes five classes. Each of them is described separately below.

The movie review task is to determine the sentiment of the author of a review towards the film he is reviewing – a review is either positive or negative. We use version 2.0 of the movie review data.[1] The corpus contains approximately 2000 reviews equally split between the two categories, with mean length 653 words and median 613.

The task for the political speech data is to determine whether an utterance is in support of a motion, or in opposition to it, and the source of the data is automatically transcribed political debates. For this work, we use version 1.1 of the political data.[2] We use the full corpus (i.e. training, tuning and testing data) to create random splits: thus the corpus contains approximately 3850 documents with mean length 287 words and median 168.

The new collection consists of text taken from the election manifestos of five UK political parties for the last three general elections (that is, for the elections in 1997, 2001 and 2005). The parties used were: Labour, Conservative, Liberal Democrat, the British National Party and the UK Independence Party. The corpus is approximately 250,000 words in total, and we divide the manifestos into "documents" by selecting non-overlapping twenty-sentence sections. This results in a corpus of approximately 650 documents, each of which is roughly 300–400 words in length.

We also wished to test the impact of the amount of training data; various studies have shown this to be an important consideration when evaluating classification methods. Of particular relevance to our work and results is that of [6], who show that the relative performances of different methods change as the amount of training data increases. Thus we vary the percentage of documents used as training between 10% and 90% at 10% increments. For a fixed percentage level, we select that percentage of documents from each class (thus maintaining class distribution) randomly as training, and use all remaining as testing. We repeat this procedure five times for each percentage level. All results are in terms of the simplest performance measure, and that most frequently used for non-overlapping classification problems, accuracy.

Otherwise, all "words" are identified as contiguous alpha-numeric strings. We use no stemming, no stoplisting, no feature selection and no minimum-frequency cutoff.

We were also interested to observe the effects of restricting the vocabulary of texts to contain only words with some emotional significance, since this in some ways seems

[1] `http://www.cs.cornell.edu/people/pabo/movie-review-data/`

[2] `http://www.cs.cornell.edu/home/llee/data/convote.html`

a natural strategy, ignoring words with specific topical and authorial associations. We thus perform experiments on the movie review collection, but using only words which are marked as *Positive* or *Negative* in the General Inquirer Dictionary [7].

3 Methods

This section describes the methods we evaluate in detail. To test the applicability of both word-presence features and word-count features, we include standard probabilistic methods designed specifically for these representations. We also include a more advanced probabilistic method with two possibilities for parameter estimation, and finally we test an SVM classifier, which is something of a standard in the literature.

3.1 Probabilistic Methods

In this section, we briefly describe the use of a model of language as applied to the problem of document classification, and also how we estimate all relevant parameters for the work which follows.

In terms of notation, we use $\tilde{c}$ to represent a random variable and c to represent an outcome. We use roman letters for observed or observable quantities and greek letters for unobservables (i.e. parameters). We write $\tilde{c} \sim \varphi(c)$ to mean that $\tilde{c}$ has probability density (discrete or continuous) $\varphi(c)$, and write $p(c)$ as shorthand for $p(\tilde{c} = c)$. Finally, we make no explicit distinction in notation between univariate and multivariate quantities; however, we use θ_j to refer to the j-th component of the vector θ.

We consider cases where documents are represented as vectors of count-valued (possibly only zero or one, in the case of binary features) random variables such that $d = \{d_1 \ldots d_v\}$. As with most other work, we further assume that words in a document are exchangeable and thus a document can be represented simply by the number of times each word occurs.

In classification, interest centres on the conditional distribution of the class variable, given a document. Where documents are to be assigned to one class only (as in the case of this paper), this class is judged to be the most probable class. Classifiers such as the probabilistic classifiers considered here model the posterior distribution of interest from the joint distribution of class and document: this means incorporating a *sampling model* $p(d|c)$, which encodes assumptions about how documents are sampled. Thus letting $\tilde{c}$ be a random variable representing class and $\tilde{d}$ be a random variable representing a document, by Bayes' theorem:

$$p(c|d) \propto p(c) \cdot p(d|c) \tag{1}$$

For the purposes of this work we also assume a uniform prior on c, meaning the ultimate decision is on the basis of the document alone.

For each of the probabilistic methods, what sets them apart is the sampling model $p(d|c)$; as such, for each method we describe the form of this distribution and how parameters are estimated for a *fixed* class. We estimate a single model of the types described below for each possible class, and combine estimates to make a decision as above, and as such we will not use subscripts referring to a particular class for clarity in notation. Where training documents and/or training counts are mentioned, these relate only to the class in question.

Binary Independence Sampling Model. For a vocabulary with v distinct types, the simplest representation of a document is as a vector of length v, where each element of the vector corresponds to a particular word and may take on either of two values: 1, indicating that the word appears in the document, and 0, indicating that it does not. Such a scheme a long heritage in information retrieval: see e.g. [8] for a survey, and [9,10] for applications in information retrieval and classification respectively. This model depends upon parameter θ, which is a vector also of length v, representing the probabilities that each of the v words is used in a document.

Given these parameters (and further assuming independence between components of d), the term $p(d|c)$ is simply the product of the probabilities of each of the random variables taking on the value that they do. Thus the probability that the j-th component of $\tilde{d}$, $\tilde{d}_j$ is one is simply θ_j (the probability that it is zero is just $1-\theta_j$) and the probability of the whole vector is:

$$p_{bin-indep}(d|\theta) = \prod_j p_{bi}(d_j|\theta_j) \tag{2}$$

Given training data for some particular class, we estimate the θ_j as their posterior means, assuming a uniform prior.

Multinomial Sampling Model. A natural way to model the distribution of word *counts* (rather than the presence or absence of words) is to let $p(d|c)$ be distributed multinomially, as proposed in [11,10] amongst others. The multinomial model assumes that documents are the result of repeated trials, where on each trial a word is selected at random, and the probability of selecting the j-th word is θ_j.

Under multinomial sampling, the term $p(d|c)$ has distribution:

$$p_{multinomial}(d|\theta) = \frac{\left(\sum_j d_j\right)!}{\prod_j (d_j!)} \prod_j \theta_j^{d_j} \tag{3}$$

Once again, as is usual, given training data we esimate the vector θ as its posterior mean assuming a uniform Dirichlet prior.

A Joint Beta-Binomial Sampling Model. The final classifier decomposes the term $p(d|c)$ into a sequence of independent terms of the form $p(d_j|c)$, and hypothesises that conditional on known class (i.e. c) $\tilde{d}_j \sim Binomial(\theta_j, n)$. However, unlike before, we also assume that $\tilde{\theta}_j \sim Beta(\alpha_j, \beta_j)$, that is $\tilde{\theta}_j$ is allowed to vary between documents subject only to the restriction that $\tilde{\theta}_j \sim Beta(\alpha_j, \beta_j)$. Integrating over the unknown θ_j in the new document gives the distribution of d_j as:

$$p_{bb}(d_j|\alpha_j, \beta_j) = \frac{n!}{d_j!(n-d_j)!} \times \frac{\mathrm{B}(d_j+\alpha_j,\, n-d_j+\beta_j)}{\mathrm{B}(\alpha_j, \beta_j)} \tag{4}$$

where $\mathrm{B}(\bullet)$ is the Beta function. The term $p(d|c)$ is then simply:

$$p_{beta-binomial}(d|\alpha, \beta) = \prod_j p(d_j|\alpha_j, \beta_j) \tag{5}$$

As with most previous work, our first estimate of parameters of the beta-binomial model are in closed form, using the method-of-moments estimate proposed in [12]. We also experiment with an alternate estimate, corrected so that documents have the same impact upon parameter estimates regardless of their length. We refer to the original as the Beta-Binomial model, and the modified version as the Alternate Beta-Binomial.

3.2 A Support Vector Machine Classifier

We also experiment with a linear Support Vector Machine, shown in several comparative studies to be the best performing classifier for document categorization [13,14]. Briefly, the support vector machine seeks the hyperplane which maximises the separation between two classes while minimising the magnitude of errors committed by this hyperplane. The preceding goal is posed as an optimisation problem, evaluated purely in terms of dot products between the vectors representing individual instances. The flexibility of the machine arises from the possibility to use a whole range of kernel functions, $\phi(x_1, x_2)$ which is the dot product between instance vectors x_1 and x_2 in some transformed space.

Despite the apparent flexibility, the majority of NLP work uses the linear kernel such that $\phi(x_1, x_2) = x_1 \cdot x_2$. Nevertheless, the linear SVM has been shown to perform extremely well, and so we present results using the the linear kernel from the SVM^{light} toolkit [15] (we note that experimentation with non-linear kernels made little difference, with no consistent trends in performance). We use the most typical method for transforming the SVM into a multi-class classifier, the One-Vs-All method, shown to perform extremely competitively [16]. All vectors are also normed to unit length.

4 Results

This section presents the results of our experiments on the collections described in §2.

Figure 1 shows performance on [1]'s movie reviews collection. Several trends are obvious; the first is that, reassuringly, performance generally increases as the amount of training data increases. Note, however, that this is not always the case – a product of the random nature of the training/testing selection process, despite performing the procedure multiple times for each data point. Note also that individual classifiers experience difficulties with particular splits of the data which are not experienced by all. The most telling example of this is the pronounced dip in the performance of the SVM at 40% training not reflected in other classifiers' performance. Also, we note that the classifier specifically designed to model binary representations fails to perform as well as the multinomial and Beta-Binomial models, in contradiction to [1], who observed superior performance using binary features, but inkeeping with results on more standard text classification tasks [10,12].

Figure 2 shows results on the same data using only words deemed *Positive* or *Negative*. Note here that relative performance trends are markedly different, with the SVM experiencing a particular reversal of fortunes. Otherwise, the same idiosyncrasies are evident – occasional dips in one classifier's performance not observed with others, and crossing of lines in the graphs.

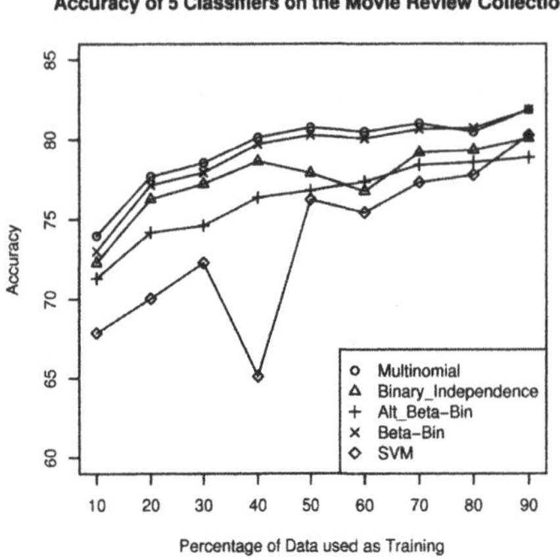

Fig. 1. Results for [1]'s Movie Review Collection

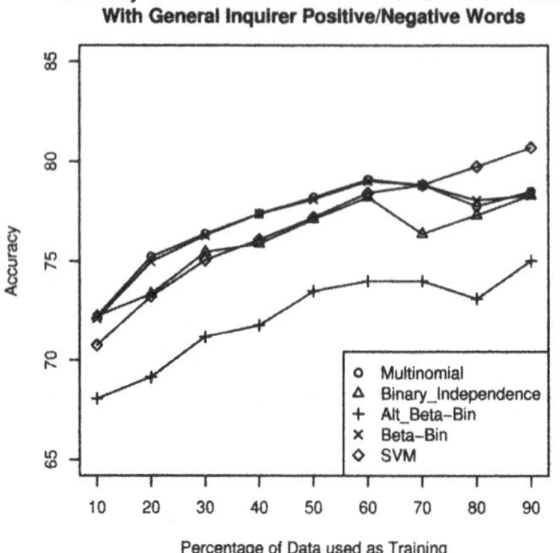

Fig. 2. Results for [1]'s Movie Review Collection, using only words marked as *Positive* or *Negative* in the General Inquirer Dictionary

Figure 3 presents a slightly less changeable picture, although what is apparent is the complete reversal in fortunes of the methods when compared to the previous collection. The binary classifier performs worst by some margin, and the alternate Beta-Binomial classifier is superior by a similar margin. Also, note that at certain points performance for some classifiers dips, while for others it merely plateaus.

Finally, Figure 4 displays results from [5]'s collection of political debates. The results here are perhaps the most volatile of all – the impact of using any particular classifier over others is quite pronounced, and the SVM is inferior to the best method by up to 7% in some places. Furthermore, the binary classifier is even worse, and this is exactly the

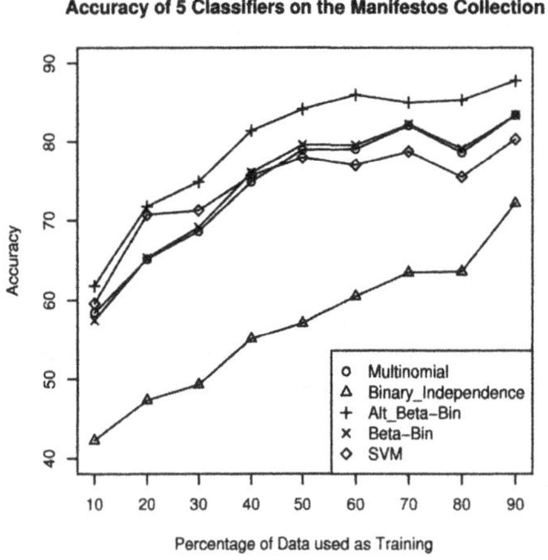

Fig. 3. Results for the Manifestos Collection

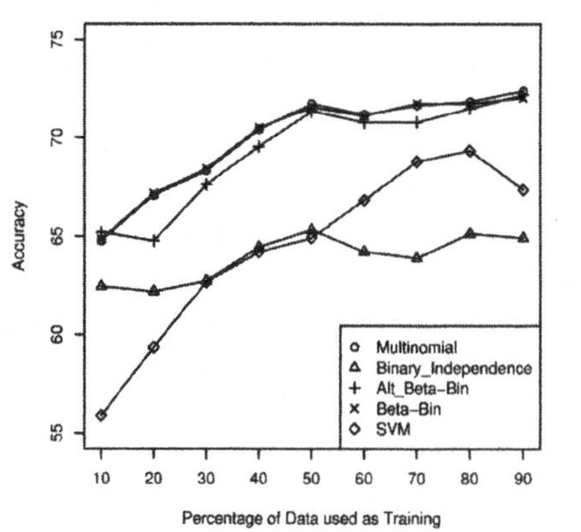

Fig. 4. Results for [5]'s Political Speeches Collection

combination used in the original study. The difference between classifiers is in many cases the same as the difference between the general document-based classifier and the modified scheme presented in that paper.

5 Conclusion

In terms of a conclusion, we revisit the initial question. Is it fair to say that the use of lexically-based features leads to classifiers which do not perform acceptably? Of course, this question glosses over the difficulty of defining "acceptable" performance; however, the only sound answer can be that it depends upon the classifier in question, the amount of training data, and so on. While it would be easier if sweeping generalisations could be made, clearly they are not justified.

References

1. Pang, B., Lee, L., Vaithyanathan, S.: Thumbs up? sentiment classification using machine learning techniques. In: Proceedings of the 2002 Conference on Empirical Methods in Natural Language Processing (EMNLP) (2002)
2. Turney, P., Littman, M.: Measuring praise and criticism: Inference of semantic orientation from association. ACM Transactions on Information Systems 21(4), 315–346 (2003)
3. Efron, M.: Cultural orientation: Classifying subjective documents by cociation (sic) analysis. In: Proceedings of the AAAI Fall Symposium on Style and Meaning in Language, Art, Music, and Design, pp. 41–48 (2004)
4. Mullen, T., Malouf, R.: A preliminary investigation into sentiment analysis for informal political discourse. In: Proceedings of the AAAI Workshop on Analysis of Weblogs (2006)
5. Thomas, M., Pang, B., Lee, L.: Get out the vote: Determining support or opposition from Congressional floor-debate transcripts. In: Proceedings of EMNLP, pp. 327–335 (2006)
6. Banko, M., Brill, E.: Mitigating the paucity of data problem: Exploring the effect of training corpus size on classifier performance for nlp. In: Proceedings of the Conference on Human Language Technology (2001)
7. Stone, P.J., Dunphy, D.C., Smith, M.S., Ogilvie, D.M.: Associates: The General Inquirer: A Computer Approach to Content Analysis. MIT Press, Cambridge (1966)
8. Lewis, D.D.: Naïve (Bayes) at forty: The independence assumption in information retrieval. In: Proceedings of ECML 1998, pp. 4–15 (1998)
9. Robertson, S.E., Jones, K.S.: Relevance weighting of search terms. Document retrieval systems, 143–160 (1988)
10. McCallum, A., Nigam, K.: A comparison of event models for naïve bayes text classification. In: Proceedings AAAI 1998 Workshop on Learning for Text Categorization (1998)
11. Guthrie, L., Walker, E., Guthrie, J.: Document classification by machine: theory and practice. In: Proceedings COLING 1994, pp. 1059–1063 (1994)
12. Jansche, M.: Parametric models of linguistic count data. In: ACL 2003, pp. 288–295 (2003)
13. Dumais, S., Platt, J., Heckerman, D., Sahami, M.: Inductive learning algorithms and representations for text categorization. In: CIKM 1998, pp. 148–155 (1998)
14. Yang, Y., Liu, X.: A re-examination of text categorization methods. In: 22nd Annual International SIGIR, Berkley, pp. 42–49 (August 1999)

15. Joachims, T.: Making large-scale svm learning practical. Advances in Kernel Methods– Support Vector Learning (1999)
16. Rennie, J.D.M., Rifkin, R.: Improving multiclass text classification with the Support Vector Machine. Technical report, Massachusetts Insititute of Technology, Artificial Intelligence Laboratory (2001)

First Steps in Building a Verb Valency Lexicon for Romanian

Ana-Maria Barbu

Romanian Academy, Institute of Linguistics
13 Calea 13 Sepetembrie, 050711, Bucharest, Romania
anabarbu@unibuc.eu

Abstract. This paper presents some steps in manually building a verb valency lexicon for Romanian. We refer to some major previous works by focusing on their information representation. We select that information for different stages of our project and we show the conceptual problems encountered during the first phase. Finally we present the gradually building procedure of the lexicon and we exemplify the manner in which the information is represented in a lexicon entry.

Keywords: Valency, verb, semantic preferences, sense disambiguation, Romanian.

1 Introduction

The goal of our project is to build a lexicon of the verb valencies for Romanian. Valences are sets of elements required by a predicate. Such a valuable resource does not exist yet for this language. Unfortunately, we do not dispose of a large enough corpus of Romanian texts, or a syntactically annotated one, as a starting point. In these conditions, we have to build this lexicon in several steps, by starting from the printed dictionaries, going to the corpus and coming back to improve the lexicon.

The main concern for achieving this task is the nature of the information we have to gradually gather in entry descriptions in order to use this resource first for syntactical dependency annotation, then for sense disambiguation, and finally for knowledge representation. At the first glance there is a rich literature on (automated or not) building subcategorisation or semantic frames from corpora. Nevertheless there are few works dealing with our problem which is building such a resource from scratch. Therefore, we insist on the conceptual part of the project by highlighting aspects which seem not to be referred in the literature.

In the first section of this paper, we review previous researches on this topic. The main section is devoted to the analysis of the points we take over from these approaches and the information necessary to be encoded. We also discuss the problems we encountered. Another section describes the present stage of our lexicon: the building steps and the entry structure. Finally, we show the further steps in aiming at our goal.

2 Previous Work Analysis

In this section we briefly review the previous approaches we used as a starting point for our project, namely FrameNet, VerbNet, VALLEX and CPA. We especially pay

P. Sojka et al. (Eds.): TSD 2008, LNAI 5246, pp. 29–36, 2008.
© Springer-Verlag Berlin Heidelberg 2008

attention to the information encoded in each studied framework. Then we analyze what information is proper to take over for our lexicon, at this stage and in the future, and also whether some other information is needed.

FrameNet. We confine ourselves to point out only some of the FrameNet project characteristics described in [1].

- Parts in British National Corpus (BNC) are annotated with frame semantic information.
- Semantic frames mainly use five types of information: *domains*: Body, Cognition, etc., *frame elements* (FE) expressed by semantic roles: Buyer, Seller, etc., *phrase types* (PT): Noun Phrase, Verb Phrase etc., *grammatical functions* (GF): External Argument, Complement etc., and cases of *nonexpressed frame elements*: Null Instantiation of Indefinite (INI), Definite (DNI) and Constructionally licensed (CNI) types.
- The phrase types and grammatical functions are automatically inferred from BNC, which is a syntactically annotated corpus.

An attempt to use FrameNet for building a Romanian valency lexicon has been done within the project described in [2]. This supposes a parallel Romanian-English FrameNet using annotation import. The XML files of the annotated English sentences are aligned word-by-word with its corresponding Romanian translation, then it is automatically created a set of XML files containing a corpus of FE-annotated sentences for Romanian language. In our opinion this procedure suffers from the following drawbacks. First, it needs a translation of English texts or a huge amount of Romanian-English parallel texts which, so far, are lacking. Second, the translation introduces an artificial intervention of translators by choosing preferred words and structures. Third, the procedure inherits all the flaws of each automatic task it uses. In general, we plead for avoiding the translation of foreign resources for Romanian, especially those which are language-specific to a large extent and the final quality of which is as important as a valency lexicon.

VerbNet. VerbNet project described in [3] has the following characteristics.

- It use a variant of Levin verb classes to systematically construct lexical entries.
- Each verb refers to a set of classes corresponding to the different senses of the verb (e.g. “run” refers to *Manner of Motion* class for “John runs”, and to *Meander* class for “The street runs through the district”).
- For each verb sense specific selectional restrictions and semantic characteristics are added, if they can not be captured by the class membership.
- Each verbal class lists the thematic roles and the selectional restrictions for each argument in each frame, as well as the syntactic frames corresponding to licensed constructions, which are linked to syntactic trees in Lexicalized Tree Adjoining Grammar representation. Besides, each member of a verbal class is mapped to the appropriate WordNet synset.
- Each frame also includes semantic predicates describing the participants during various stages of the event (namely *during*(E), *end*(E) and *result*(E)).

VALLEX. The Valency Lexicon of Czech Verbs (VALLEX 2.0) uses the information described in [4] and refers to the following aspects.

- A dictionary entry represents a lexeme structured on lexical forms and lexical units.
- A lexical form specifies the "base" infinitive form (eventually with its variants), its morphological aspect (imperfective, perfective, etc.), aspectual counterparts, reflexive particle, if it is the case, and numbers for homographs, if they exist.
- Lexical units correspond to the meanings of the lexeme. Each lexical unit displays obligatory information: valency frame, gloss (a synonym or a paraphrase) and example, as well as optional information: flag for idiom, information on control, possible type(s) of reflexive constructions, possible type(s) of reciprocal constructions and affiliation to a syntactico-semantic class.
- Valency frames contain the following types of information assigned to frame slots:
  - Functors (Actor, Addressee etc.) structured on inner participants, quasi-valency complementation and free modification.
  - Morphemic forms (case, prepositional case, infinitive construction etc.)
  - Types of complementations (obligatory or optional for inner participants and quasi-valency complementations, and obligatory or typical for free modifications).
  - Marks for slot expansion.

CPA. "Corpus Pattern Analysis" (CPA) project, see [5] and [6], has the goal to extract, from BNC, all normal patterns of use of verbs and to assign a meaning to each pattern. The relationship between a pattern and its meaning seems to be of type one-to-one, that is, unambiguous. An entry of CPA lexicon displays the following characteristics.

- Each valency contains the following types of information:
  - A pattern described in terms of semantic values (*semantic types* and *semantic roles*), lexical sets, words or/and morpho-syntactic characteristics (e.g. [[Person]] grasp {[[Abstract]]|[N-clause]}).
  - One or more implicatures describing the meaning in terms of synonyms or paraphrases of the entry verb, and semantic values for its arguments (e.g. [[Person=Cognitive]] understands {[[Abstract=Concept]]|[N-clause]}).
  - Lexical alternation (e.g. [[Person]] <-> {hand, finger}).
  - Lexical set (e.g. [[Oportunity<Abstract]]: opportunity, chance, offer, moment....
  - Clues, that is, specific elements (words, phrases) usually used in the context of the respective meaning able to disambiguate that meaning.
  - Comment, especially for marking idioms.
- A *semantic type* is a class to which a term is assigned (e.g. Person, BodyPart).
- A *semantic role* is the context-specific role of the related semantic type. For instance, a semantic type Person can play in a health context the semantic role Doctor: [[Person=Doctor]] or Patient: [[Person=Patient]].

3 Analysis of Required Information

We are now trying to analyze the types of information used in the above mentioned approaches, in order to select them in accordance with the needs of the different stages of our project. We also discuss some problems they raised in our representation. Information generally used is of the following kinds.

Phrase types. Information concerning the morpho-syntactic nature of valency elements is present in every approach either extracted from corpora (CPA, FrameNet) or built manually (VerbNet, VALLEX). We also need to represent it in our lexicon especially because there is no syntactic frames lexicon for Romanian and such a resource is basic for almost any task in NLP.

It is worth mentioning here an aspect concerning free alternation. Alternation preserving the meaning can be done either locally, for instance, one of type AdvP/PP: *He behaves well / in snobbish way*, or by restructuring the whole frame: *Imi place de Ion* ("Me like DE John") / *Eu il plac pe Ion* ("I him like PE John"). The local alternation can be regular, such as AdvP/PP, or verb-specific (NP/PP): *Ion priveste ceva / la ceva* ("John watches something / at something"). We tried to capture all these cases in our representation.

Lemma characteristics. Even if there is an obvious relationship between some morphological characteristics of a lemma and its valency frames, as it is also pointed out in [7], only VALLEX takes this into account. Some meanings can be appropriate, for instance, only for a verb at third person singular form (e.g. with impersonal use), or only for its negated form. On the other hand, only some meanings allow certain diathesis alternations and they have to be marked as such. Therefore we have assigned a slot for describing lemma characteristics even at this stage of our project.

Grammatical functions. Only FrameNet provides explicit information about grammatical functions of the frame elements. A Romanian approach on valence frames [8] also proposes a description in terms of grammatical functions based on the inventory in traditional Romanian grammar. However, we set aside this kind of information because it is too dependent on the theoretical background and because it is not reflected, in any way, in what we can extract from corpora. Besides, for a rich inflected language like Romanian functions can be often inferred from grammatical cases of arguments.

Selectional preferences. Surprisingly, this information is only captured in VerbNet and in CPA (as semantic types), despite its importance, in our opinion, for distinguishing valencies and for word sense disambiguation. Therefore we pay special attention to them.

For our description we set up the following criterion: *two valency frames are distinct if their elements display different selectional preferences.* For instance, the intransitive usage of the Romanian verb *a toca* ("to hash") roughly means to make repeated short noises, to knock, to rattle. This valency (of intransitive form) has three different meanings depending on the semantic type of the subject. If the subject is a Person, the verb means "to hammer on a board (for asking people to church)", if the subject is a Gun, the verb means that the (machine) gun is fired and if it is a Stork, the meaning is that the stork rattles its beak. Therefore we actually record three valency frames: NP[Person], NP[Gun], NP[Stork], corresponding to the three meanings, instead of one general frame such as NP[Concrete]. Note that the intransitive form of the verb *a toca* cannot have other subject type than a person, a machine gun or a stork with the corresponding meanings. For instance, in a sentence such as "The house shook and the doors and windows rattled" Romanian does not translate "to rattle" by *a toca*, even if the common feature of the three above mentioned meanings is "to make repeated short noises".

A consequence of our criterion (also mentioned above) is that, if given two meanings (registered in printed dictionaries) about which we cannot specify different semantic types for the elements of their corresponding valencies, then we assign both meanings to one and the same valency frame. For instance, the verb *a omori* ("to kill") means either "to cause to die" or "to torture" someone. For both senses, the subject can be practically anything and the object has to be Animate. At this stage of our project, we can not distinguish two different valency frames for the two meanings, but a future solution could be to implement semantic predicates, in VerbNet fashion, for marking the result aspect of the first meaning and the durative one of the second or/and to introduce Clues, in CPA style, because it is obvious that only external elements in context can disambiguate them.

Using semantic types for disambiguating meanings raises the following problem, as well. If there are two semantic types, one more general than the other, the problem is to determine to what extent the particular semantic type defines a new meaning. So for example, the verb *a lepada* has the meaning "[Person] sheds [Concrete]". If Concrete=Clothes, its meaning is "[Person] sheds/takes off [Clothes]". The question is, on the one hand, to what extent the meaning changed so that two valency frames are needed, i.e. "[Person] sheds [Concrete]" and "[Person] sheds [Clothes]". On the other hand, if there are two meanings, can one assume that Clothes do not ever participate at the first meaning, so that no ambiguity exists? So far, in cases of subordinated semantics types without major difference of meanings we retain only the more general valency frame.

We do not use semantic types in a pre-determined inventory. We implement them in as much as they are needed. A problem we encounter is to establish the most general semantic type, able to include all the real situations a verb can imply. For instance, a verb like *a exprima* ("to express") (or any other denotation verb) can have as its subject a person, a text, an image, an attitude, a situation, a sign etc. Which is the appropriate semantic type for all these cases? In these situations we preferred not to specify any semantic type on the subject. We think not even *lexical set* in CPA could solve the problem, but at the most we can choose a semantic type like Entity.

As it is mentioned in [6], sometimes it is worth putting down the axiological aspect of an element. *A oferi* ("to offer") has a complement which normally has a positive connotation. One offers good things, unlike "to give" which is neutral. So, axiological semantic types have also to be introduced in our inventory. Besides, it turns out that a valency element may be described by more than one semantic restriction. It is likely that *sets* of semantic types are needed for describing one argument.

Obligatory vs. optional elements. This kind of information is explicit in FrameNet and VALLEX and implicit in VerbNet and CPA, where it is covered by enumerating all possible valency frames. We have chosen the method adopted for Czech lexicon in that we mark each argument as obligatory or optional. For a manually built lexicon this way is more efficient. However there is a problem here too. For a verb like *a se desfasura* ("to take place", "to proceed") arguments indicating the place, the time and/or the manner in which an event can take place are needed. The problem is how to express the fact that any of these arguments can miss but not all of them. It is not clear how the mentioned approaches grasp, on the one hand, the fact that in the sentences: "The show takes place tomorrow / here / somehow", "The show takes place here tomorrow" etc. the verb "take place" has the same meaning and, on the other hand, "The show takes place" is ungrammatical. In other words, we need a mean of marking that, in this case, any argument is optional but at least one is obligatory.

Modifiers. FrameNet, VALLEX and CPA include modifiers in valency structures. At the first stage, we confine ourselves to represent the "minimal" valencies because of the lack of accessible corpus evidence. Furthermore, this kind of information has to be added in order to reflect the typical use of verbs and to distinguish meanings more precisely (see Clues in CPA).

Meanings. Sense descriptions are made by means of synonyms and paraphrases (VALLEX), WordNet synsets (VerbNet) or "primary implicature" (CPA). We join VALLEX by using synonyms and paraphrases for describing one or more meanings of a valency. In general, we follow the meanings displayed in a verb entry of printed dictionaries. However, there are cases in which we have to split one dictionary definition because it corresponds to more than one valency structure and meaning. On the other hand, there could be, for instance, two definitions for which we are not able to say precisely what distinguishes the another and thus we describe one valency frame with two meanings. For instance, with the means we use to describe valencies at this stage we are not able to assign different valency frames to the verb *a trai* "to live" for the sentences *Ion isi traieste tineretea (intens)* "John lives his youth (intensively)" and *Spectatorii au trait momentul (cu entuziasm)* "The public lived the moment (enthusiastically)" (see Figure 1 below) even if they represent two different meanings in the printed dictionary.

Semantic roles. This information is referred in all studied approaches but always in different ways. The common point is that they reflect the role which an argument plays *in context* unlike the semantic types or selectional preferences which reflect lexical features of a given lexeme, able to be organized into a semantic ontology. We postponed specifying semantic roles in our valencies description because there is not an unanimously accepted inventory, so that we have to reflect on criteria for building or adopting it and because semantic roles are less informative for our immediate purposes.

Verb classes. Classes are useful for generalizations and systematic work. They are extensively used in VerbNet, but also in VALLEX. An approach which mainly distributes valencies in classes is [10], as well. We have postponed any classification for the moment when we get a significant inventory of verb valencies.

Semantic predicates. This information described in VerbNet or [9] is very important for knowledge representation. However it goes beyond the immediate aims of our project.

4 Building Steps and Results

Building the lexicon means four main steps.

I. Describing subcategorization frames which provide morpho-syntactic information on verb arguments and their "primary" semantic types.
II. Extended search on corpus, by using the lexicon obtained at the previous step, for adding information on modifiers and for refining semantic types.

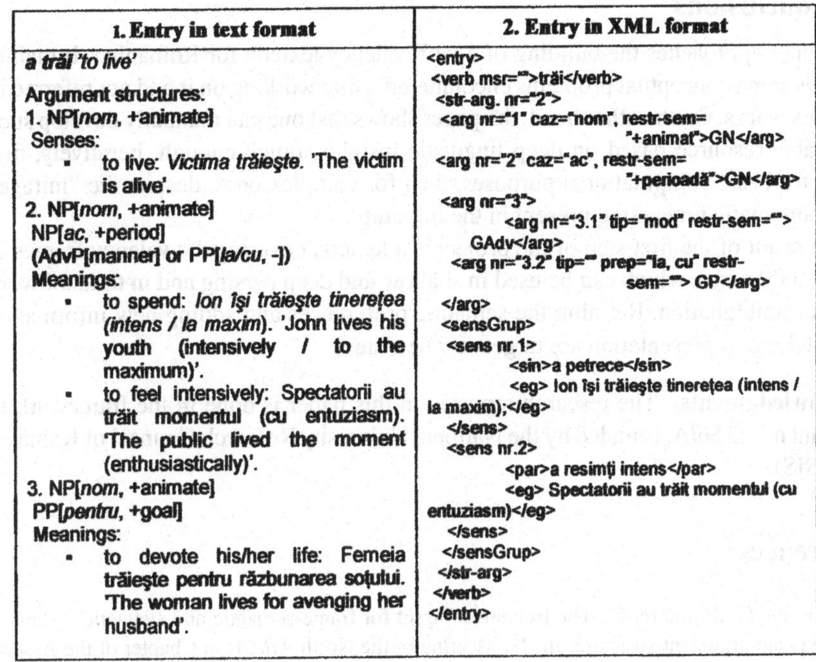

1. Entry in text format	2. Entry in XML format
a trăi 'to live' Argument structures: 1. NP[*nom*, +animate] Senses: ▪ to live: *Victima trăieşte.* 'The victim is alive'. 2. NP[*nom*, +animate] NP[*ac*, +period] (AdvP[manner] or PP[*la/cu*, -]) Meanings: ▪ to spend: *Ion îşi trăieşte tinereţea (intens / la maxim).* 'John lives his youth (intensively / to the maximum)'. ▪ to feel intensively: Spectatorii au trăit momentul (cu entuziasm). 'The public lived the moment (enthusiastically)'. 3. NP[*nom*, +animate] PP[*pentru*, +goal] Meanings: ▪ to devote his/her life: Femeia trăieşte pentru răzbunarea soţului. 'The woman lives for avenging her husband'.	<entry> <verb msr="">trăi</verb> <str-arg. nr="2"> <arg nr="1" caz="nom", restr-sem= "+animat">GN</arg> <arg nr="2" caz="ac", restr-sem= "+perioadă">GN</arg> <arg nr="3"> <arg nr="3.1" tip="mod" restr-sem=""> GAdv</arg> <arg nr="3.2" tip="" prep="la, cu" restr-sem=""> GP</arg> </arg> <sensGrup> <sens nr.1> <sin>a petrece</sin> <eg> Ion îşi trăieşte tinereţea (intens / la maxim);</eg> </sens> <sens nr.2> <par>a resimţi intens</par> <eg> Spectatorii au trăit momentul (cu entuziasm)</eg> </sens> </sensGrup> </str-arg> </verb> </entry>

Fig. 1. Examples of an entry in Romanian Verb Valency Lexicon

III. Adding information about control and passivization and implementing rules for expanding structural regularities and alternations in valence representation.

IV. Adding information about semantic roles and semantic predicates.

The tasks of the first stage of the project have been fulfilled. First, we have chosen about 3,000 verbs from the Romanian core lexicon. The main resource for getting the senses of the verbs was a Romanian Explanatory Dictionary (DEX). The information in the printed dictionary was confronted with Romanian texts on Internet and a corpus of about 14 millions words from newspapers. This task aimed at actualizing senses of verbs, getting primary semantic types and examples for valency lexicon. Entries were built manually, during about three years, by a team of five linguists. They followed the meta-language described in [11], where the significance of the formal representation and the grammatical characteristics appropriate for describing verb valencies for Romanian are fully detailed.

The result of the first phase is a lexicon which, this year, reaches a number of about 3,000 verbs, in a text format and XML format. Figure 1 shows three (from ten) argument structures of the verb *a trai* ("to live") in text format and the corresponding XML representation of the second argument structure. The XML format is automatically obtained from the text description which is meant to follow the meta-language strictly. Any enhancement on lexicon is done on the text format, which is afterwards translated into the XML one.

5 Conclusions

This paper approaches the building of a verb valency lexicon for Romanian. It mainly presents some conceptual problems encountered while working on it and not referred in previous works. On the other hand, the paper shows that one can manually develop such a valuable resource based on deep linguistic insights, quick enough, iteratively, first meant for basic computational purposes, then for complex ones, despite the "mirage" of the automatic processing present in the literature.

The result of the first step of the project is a lexicon covering the valency frames of about 3,000 verbs, which can be used in shallow and deep parsing and in tasks of word sense disambiguation. Refining the semantic preferences and adding new information of knowledge representation are targets for new steps.

Acknowledgments. The research reported in this paper is done in the framework of the grant no. 1156/A, founded by the National University Research Counsel of Romania (CNCSIS).

References

1. Johnson, C., Fillmore, C.: The FrameNet tagset for frame-semantic and syntactic coding of predicate-argument structure. In: 1st Meeting of the North American Chapter of the Association for Computational Linguistics (ANLP-NAACL 2000), Seattle, WA, pp. 56–62 (2000)
2. Trandabat, D.: Semantic Frames in Romanian Natural Language Processing. In: Consortium, D. (ed.) Proceedings of the North American Chapter of the Association for Computational Linguistics NAACL-HLT 2007, Companion Volume, April 2007, pp. 29–32. Association for Computational Linguistics, Rochester (2007)
3. Kipper, K., Dang, H.T., Palmer, M.: Class-based Construction of a Verb Lexicon. In: AAAII-2000 Seventeenth National Conference on Artificial Intelligence, Austin, TX (2000)
4. Žabokrtský, Z., Lopatková, M.: Valency Information in VALLEX 2.0: Logical Structure of the Lexicon. The Prague Bulletin of Mathematical Linguistics 87, 41–60 (2007)
5. Hanks, P., Pustejovsky, J.: A Pattern Dictionary for Natural Language Processing. Revue française de linguistique appliquée X-2, 63–82 (2005)
6. Hanks, P.: The Organization of the Lexicon: Semantic Types and Lexical Sets (2006), `www.cs.cas.cz/semweb/download/06-11-hanks.doc`
7. Przepiorkowski, A.: Towards the Design of a Syntactico-Semantic Lexicon for Polish. In: New Trends in Intelligent Information Pocessing and Web Mining, Zakopane. Springer, Heidelberg (2004)
8. Serbanescu, A.: Pentru un dictionar sintactic al verbelor românesti, Studii si Cercetari Lingvistice, XLV, nr. 3–4, Bucuresti, pp.133–150, (1994)
9. Pustejovsky, J.: The Generative Lexicon. MIT Press, Cambridge (2001)
10. Leclere, C.: The Lexicon-Grammar of French Verbs: a syntactic database. In: Kawaguchi, Y. (ed.) Linguistic Informatics - State of the Art and the Future, Tokyo University of Foreign Studies, UBLI 1, pp. 29–45. Benjamins, Amsterdam / Philadelphia (2003)
11. Barbu, A.M., Ionescu, E.: Designing a Valence Dictionary for Romanian. In: International Conference Recent Advances in Natural Language Processing (RANLP), Borovets, Bulgaria, pp. 41–45 (2007)

Hyponymy Patterns
Semi-automatic Extraction, Evaluation and Inter-lingual Comparison

Verginica Barbu Mititelu

Romanian Academy, Research Institute for Artificial Intelligence
13, Calea 13 Sepetembrie, 050711, Bucharest, Romania
vergi@racai.ro

Abstract. We present below an experiment in which we identified in corpora hyponymy patterns for English and for Romanian in two different ways. Such an experiment is interesting both from a computational linguistic perspective and from a theoretical linguistic one. On the one hand, a set of hyponymy patterns is useful for the automatic creation or enrichment of an ontology, for tasks such as document indexing, information retrieval, question answering. On the other hand, these patterns can be used in papers concerned with this semantic relation (hyponymy) as they are more numerous and are evaluated as opposed to those "discovered" through observation of text or, rather, introspection. One can see how hyponymy is realized in text, according to the stylistic register to which this belongs, and also a comparison between such patterns in two different languages is made possible.

Keywords: Hyponymy patterns, ontology.

1 Introduction and Related Work

Pattern extraction from corpora is quite widely used for question answering. It seems that patterns are very useful to extract various kinds of information. Etzioni et al. (2004) extract names of towns, states, countries, actors and films; Ravichandran and Hovy (2002) recognize in text the birth year of different personalities, inventors, discoverers, etc.; Szpektor et al. (2004) find entailment in text, and Chklovski and Pantel (2004) semantic relations between verbs.

The idea to extract hyponymy patterns from corpora is not new. Hearst (1992 and 1998) identifies six such patterns by mere observation of text and by analyzing the context between a hyponym and its superordinate found in the corpus: NP_0 *such as* {NP_1, NP_2..., (and|or)}NP_n; *such* NP *as* {NP,}*{(or|and)}NP; NP{, NP}*{,} *or other* NP; NP{, NP}*{,} *and other* NP; NP{,} *including* {NP,}*{(or|and)}NP; NP{,} *especially* {NP,}*{(or|and)}NP. The only pattern evaluated by Hearst (1998) is that containing *or other*: 63% of the contexts in which it occurs contain hyponym-hypernym pairs.

Alfonseca and Manandhar (2001) use Hearst's algorithm to enrich WordNet with new concepts. They use the first order predicate logic to identify hyponym-hypernym pairs: *Shakespeare* was a first-class *poet*; *Shakespeare*, the *poet*,...; The English *dramatist*, *Shakespeare*,...;... the *city* of *Seville*...

P. Sojka et al. (Eds.): TSD 2008, LNAI 5246, pp. 37–44, 2008.
© Springer-Verlag Berlin Heidelberg 2008

Oakes (2005) uses Hearst's patterns to extract hyponym-hypernym pairs from a pharmaceutical text and remarks their high effectiveness. Then, he identifies in text patterns in which these pairs occur and calculates their frequency. He is able to find Hearst's patterns and a few others (*bt and nt, bt in nt, bt outside the nt, nt bt*[1]) which are not specific for the syntagmatic realization of this relation.

In their studies dedicated to semantic relations, Lyons (1977) and Cruse (1986) also tackle the hyponymy relation and mention some structures in which hyponyms and hypernyms tend to co-occur at short distance in a sentence. Lyons's structures (1977: 292-4) are the following: analytic sentences (*A cow is a kind of animal.*), *cows and other (kind of) animals, some other, a certain, some other kind, of a certain kind.* Cruse (1986: 91) presents the following structures, whitout evaluating them: *dogs and other animals, There's no flower more beautiful than a rose, He likes all fruits except bananas, She reads books all day – mostly novels*. So, the interest for the patterns in which hypynym-hypernym pairs co-occur in context exists in the semantic literature, but we were not able to find any study dedicated to this matter.

2 Semi-automatic Extraction of Hyponymy Patterns from Corpora

The experiment we present below focuses on two languages: English and Romanian. For the former, we semi-automatically identify hyponymy patterns in a corpus, we compare them with patterns identified by other researchers and evaluate their precision. For the latter, we identify patterns in two ways: on the one hand, we translate the English patterns and check the occurrence and precision of these translated patterns against a Romanian corpus; on the other hand, we run on a Romanian corpus the same experiment as in English. The patterns obtained via the two methods are compared with each other and then with the English ones.

For English, we used five .xml converted files from British National Corpus (BNC). Four of them (i.e. more than 38 million words) were used for training and the fifth (almost 7 million words) for testing. From the training files we automatically extracted, using a Perl script, the sentences containing nouns between which there is a (direct or indirect) hyponymy relation, which we recognized using Princeton WordNet (PWN) 2.1. We made use of the lemmatization in BNC and automatically grouped the extracted sentences according to the similarity of the lexical material (i.e. lemmas) between the hyponym and its co-occurring hypernym, thus resulting groups of examples with identical lexical material between the hyponym and its hypernym. These groups of examples were then manually inspected to extract the patterns specific to the relation of interest.

Beside the patterns in Table 1, we also found what we could call a zero pattern: the hyponym and the hypernym follow immediately after each other in the sentence: "In many subtle ways our care demonstrates that we believe the *ageing* [hyponym] *process* [hypernym] makes independent action impossible."

The next step in our experiment was to test the identified patterns, i.e. to establish their precision: from the total number of occurrences, how many contain hyponym-hypernym pairs. We automatically extracted the sentences containing these patterns

[1] *Bt* stands for broad term, so the hypernym, and *nt* for narrow term, so the hyponym.

Table 1. The relevance of the English hyponymy patterns

No.	Pattern	Number of occurrences	Number of relevant occurrences	Intermediary precision (%)	Number of relevant occurrences after parsing	Final precision (%)
1.	other than	168	164	97.6	168	100
2.	especially	120	90	75	120	100
3.	principally	11	6	54.5	11	100
4.	usually	18	14	77.8	18	100
5.	such as	2470	1950	78.9	2450	99.2
6.	in particular	78	48	61.5	72	92.3
7.	e(.)g(.)	280	216	77.1	256	91.4
8.	become	780	510	66.7	710	91
9.	another	92	72	78.3	80	87
10.	notably	76	42	55.3	66	86.8
11.	particularly	130	80	61.5	110	84.6
12.	except	13	4	30.8	11	84.6
13.	called	270	220	81.5	220	81.5
14.	like	1600	1300	81.3	1300	81.3
15.	including	670	430	64.2	540	80.6
16.	mainly	40	25	62.5	30	75
17.	mostly	72	48	66.7	51	70.8
18.	for example	300	140	46.7	-	-
19.	that is	195	75	38.5	-	-
20.	apart from	120	44	36.7	-	-
21.	even	78	26	33	-	-
22.	be (are)	190	60	31.6	-	-
23.	i.e.	13	4	30.8	-	-
24.	for instance	150	36	24	-	-
25.	as	15618	3288	21	-	-
26.	either	30	6	20	-	-
27.	as well as	252	14	5.5	-	-

from the test corpus. We manually[2] went through all these sentences and checked if the pattern/s they contain is/are used for the lexicalization of the hyponymy relation.

In Table 1 there are the identified patterns, the number of occurrences for each of them, the number of relevant occurrences, and their precision (see the column *Intermediary precision*). We mention that some of the patterns identified in the training corpus were not found in the testing one: *be another, namely, and other, or other, a form of, or another, and similar, or similar, not least, but not, a kind of, like other, in common with*

[2] We preferred the manual verification instead of an automatic one, in order not to lose the possible examples in which the respective words were not marked as hyponyms in WordNet (we also include here the contextually created hyponyms) and also the situations in which the syntactic analysis of the context can help to the correct identification of the hyponymic pair.

other, i.e., and sometimes other, and many other, and in other, or any other, which be. Other patterns are far less specific to hyponymy, so we disregarded them: *a/an, of.*

We noticed that sometimes the patterns are relevant with the condition of establishing the dependences between the phrases (i.e. syntactically analyzing or parsing the sentences). For instance, in the context "in **areas** with a long history of mining such as **South-west England**", the pattern *such as* occurs in the context "N_1 such as N_2", but N_2 (*South-west England*) is not the hyponym of N_1 (*mining*); looking further to the left context of N_1 there is another noun (*areas*), N_3, that is N_2's hypernym. The prepositional phrase *with a long history of mining* is a(n optional) adjunct of the noun *areas*. Once this determination relation identified, the pattern *N such as N* refers to the correct pair: *areas – South-west England*. Due to the existence of such examples, we also included in Table 1 the number of cases when the patterns are relevant after parsing the text[3]; the precision was recalculated (see the column *Final precision*) only for those patterns with an intermediary precision above 50%.

Four of these patterns have the highest precision in our test file: *other than, especially, principally, usually*. Others have a very good precision: *such as, in particular, e(.)g(.), become, another, notably, particularly, except, called, like, including, mainly, mostly, i.e.*. The last category are those patterns that allow for the occurrence of hyponymy pairs in a very low degree. Some patterns are very frequent (*such as, like*), other less frequent. So, besides precision, one needs to consider frequency, as well. A high frequency and a high precision are the ideal characteristics of a pattern specific to hyponymy.

Comparing our results with those of other researchers, we notice that our patterns include those identified by Hearst (1992 and 1998), by Cederberg and Widdows (2003), by Oakes (2005), by Lyons (1977) and Cruse (1986). We also identified patterns that had not been identified by other researchers: *other than, principally, usually, in particular, notably, particularly, except, called, another, mainly, mostly, either, that is, i(.)e(.), for instance/example, rather than, as well as, apart from, even*. The first three appear with the highest precision in the table.

The method we used to identify the patterns is lexical, with a few syntactic pieces of information added, for testing. Snow et al. (2005) uses a corpus parsed with dependency grammars and learns hyponymy patterns from it; they resemble some subcategorization frames used to identify new hyponyms in text. Beside Hearst's patterns, Snow et al. also found: *NP like NP, NP called NP, NP is NP, NP, a/an NP*. The most frequent pattern found by them is *NP such as NP* (almost 70%). It is followed by *NP including NP* (approximately 30%), *NP like NP* (around 25%), *NP called NP* (some 20%), *NP especially NP* and *NP and/or other NP* (around 10%), *NP is NP* and *NP, a/an NP* (below 10%).

If for English we had quite a big corpus from which to learn the hyponymy patterns, this is not the case for Romanian. That is why we decided to use two different methods. Firstly, we translated the English patterns and then searched for them in a Romanian corpus (of 881817 lexical units) in order to establish their relevance. The results are in Table 2.

The translations that were not found in the Romanian corpus were not included in this table: *îndeosebi, aşa cum este/sunt, de talia, de ex., minus, în mod obişnuit, de cele mai*

---

3 These cases are identified during the manual evaluation of the sentences containing the patterns subject to testing.

Table 2. The relevance of the Romanian patterns identified by translating the English patterns

No.	Pattern	Number of occurrences	Number of relevant occurrences	Intermediary precision (%)	Number of relevant occurrences after parsing	Final precision (%)
1.	de exemplu	1	1	100	1	100
2.	ca de pildă	1	1	100	1	100
3.	de pildă	2	2	100	2	100
4.	cum ar fi	7	5	71.4	7	100
5.	, mai puţin	1	1	100	1	100
6.	de obicei	1	1	100	1	100
7.	altul/alta/alţii /altele decât	2	2	100	2	100
8.	sau alt/altă /alţi/alte	9	9	100	9	100
9.	sau orice alt /altă/alţi/alte	2	2	100	2	100
10.	şi anume	2	2	100	2	100
11.	numit	51	43	84.3	50	98
12.	deveni	114	56	49.1	101	88.6
13.	mai ales	8	5	62.5	7	87.5
14.	şi alt/altă /alţi/alte	53	42	79.2	44	83
15.	adică	30	14	46.7	19	63.3
16.	cu excepţia	16	8	50	10	62.5
17.	care fi	8	1	12.5	5	62.5
18.	în special	5	3	60	3	60
19.	inclusiv	53	19	35.8	29	54.7
20.	afară de	11	4	36.4	5	45.5
21.	precum	6	2	33.3	2	33.3
22.	în afară de	9	2	22.2	2	22.2
23.	chiar şi	29	3	10.3	6	20.7
24.	un fel de	27	4	14.8	5	18.5
25.	alt/altă/alţi /alte	8	1	12.5	1	12.5
26.	ca	700	40	5.7	40	5.7
27.	până şi	11	0	0	0	0
28.	dar nu	2	0	0	0	0

multe ori, în general, mai cu seamă, ca (ş) alt/altă/alţi/alte, la fel ca alt/altă/alţi/alte, un tip de, cu alte cuvinte, în primul rând, fi alt/altă/alţi /alte, exceptând, sau... asemănători/asemăn(a)toare. One pattern was specific to the co-occurrence of cohyponyms: *precum şi*.

The second method we used to identify Romanian hyponymy patterns was identical to the one used for English. However, Romanian WordNet (RoWN) is not as rich as PWN 2.1. The version we used (dated January 2008) contains 46.269 synsets. Running the experiment, we identified the patterns in Table 3, which also contains the evaluation

Table 3. The relevance of Romanian hyponymy patterns identified by running the algorithm

No.	Pattern	Number of occurrences	Number of relevant occurrences	Intermediary precision (%)	Number of relevant occurrences after parsing	Final precision (%)
1.	şi orice alt	2	1	50	2	100
2.	şi celălalt	4	3	75	4	100
3.	mai ales	1	1	100	1	100
4.	fi considerat	3	0	0	3	100
5.	sine numi	6	2	33.3	6	100
6.	fi un	36	24	66.7	36	100
7.	sau alt	2	1	50	2	100
8.	ci (şi/doar) un	3	2	66.7	3	100
9.	deveni	15	9	60	14	93.3
10.	şi anume	11	4	36.4	10	90.1
11.	şi alt	7	6	85.7	6	85.7
12.	inclusiv	31	9	29	23	74.2
13.	adică	8	3	37.5	5	62.5
14.	nu ca un	1	0	0	0	0

of these patterns against a corpus of 900.000 lexical units. Two of the patterns found in the training corpus were not found in the testing one: *fi un fel de*, *care avea fi*.

The quantitative difference between the patterns found by the two methods can be accounted for by three factors: the quantitative difference between the two wordnets used: the English (PWN 2.1) contains 117.597 synsets and the Romanian 46.269; the quantitative difference between the corpora we used: the BNC fragments we used have altogether 1863MB, and the Romanian one 25,6MB; when we evaluated the occurrences of the translated patterns we also accepted as relevant those situations when contextual hyponymy occurred, i.e. words that are not registered as hoponyms in the wordnets, but that function as hyponyms in the respective context or can be considered hyponyms after the metonymic interpretation of the context (see also Hearst 1992).

Comparing the set of Romanian patterns identified by translating the English patterns with the ones identified by applying the algorithm, we notice that the intersection of the two sets contains: *sau alt*, *şi anume*, *sine numi*, *deveni*, *mai ales*, *şi alt*, *adică*, *inclusiv*. This means that almost 30% of the translated patterns were also identified by applying the algorithm and that almost 60% of the ones identified by applying the algorithm were also identified by translating the English patterns. These common patterns, as one could have expected, have quite a high precision: over 80%, exceptions being *adică* and *inclusiv*. If we calculate the arithmetic mean of the final precisions for these common patterns, *sau alt* has maximal precision; the others have a mean precision over 90%: *numit/sine numi* 99%, *şi anume* 95%, *mai ales* 93.8%, *deveni* 91%; then we have *şi alt* with 84.4%, *inclusiv* with 64% and *adică* with almost 63%. In linguistic terms, the lower the precision, the more syntactically polysemous the respective pattern, that is, it occurs in more syntactic structures, with different values. The common patterns are, as expected, not necessarily the most frequent ones: the more polysemantic a word, the more frequent it is.

When we compare the English patterns with maximal precision with the Romanian patterns with maximal precision, we notice that the equivalents of three English patterns are among the patterns with maximal precision in Romanian: *other than – altul decât*, *especially – mai ales*, *usually – de obicei*.

3 Conclusions and Further Work

We presented an experiment in which we identified and evaluated quite a big number of hyponymy patterns, we checked the possibility to export these patterns into another language via translation. Former experiments on this topic did not present a complete evaluation of the patterns identified. Moreover, semantic studies only mention some patterns (probably identified via introspection), without any remark on their frequency or precision. It is worth mentioning that automatic extraction of such patterns is more reliable than introspection, especially in the case of large data.

We are aware of the unreliable results of the experiment with Romanian, due to the small size of the corpora used. However, the next step in our work is to enlarge the corpus until it reaches a size comparable to the English one and re-run the experiments.

References

1. Alfonseca, E., Manandhar, S.: Improving an Ontology Refinement Method with Hyponymy Patterns. In: Third International Conference on Language Resources and Evaluation, Las Palmas, Spain, pp. 235–239 (2001)
2. Chklovski, T., Pantel, P.: VERBOCEAN: Mining the Web for Fine-Grained Semantic Relations. In: Proceedings of Conference on Empirical Methods in Natural Language Processing (EMNLP-2004), Barcelona, Spain, pp. 33–40 (2004)
3. Cruse, D.A.: Lexical Semantics. Cambridge University Press, Cambridge (1986)
4. Etzioni, O., Cafarella, M.J., Downey, D., Popescu, A.M., Shaked, T., Soderland, S., Weld, D.S., Yates, A.: Methods for domain-independent information extraction from the Web: An experimental comparison. In: Proceedings of the 19th AAAI, pp. 391–398. AAAI Press/MIT Press (2004)
5. Hearst, M.A.: Automated Acquisition of Hyponyms from Large text Corpora. In: Proceedings in the Fourteenth International Conference on Computational Linguistics, Nantes, France, pp. 539–545 (1992)
6. Hearst, M.A.: Automated Discovery of WordNet relations. In: Fellbaum, C. (ed.) WordNet: An Electronic Lexical Database, pp. 131–151. MIT Press, Cambridge (1998)
7. Chun, H.W., Tsuruoka, Y., Kim, J.D., Shiba, R., Nagata, N., Hishiki, T., Tsujii, J.I.: Extraction of Gene-Disease Relations from Medline Using Domain Dictionaries and Machine Learning. In: Pacific Symposium on Biocomputing, pp. 4–15 (2006)
8. Huang, M., Zhu, X., Hao, Y., Payan, D.G., Qu, K., Li, M.: Discovering patterns to extract protein-protein interactions from full texts. Bioinformatics 20, 3604–3612 (2004)
9. Lyons, J.: Semantics, vol. 1. Cambridge University Press, Cambridge (1977)
10. Malaisé, V., et al.: Detecting semantic relations between terms in definitions. In: Proceedings of CompuTerm 2004 – 3rd International Workshop on Computational Terminology, Geneve, pp. 55–62 (2004)
11. Oakes, M.P.: Using Hearst's Rules for the Automatic Acquisition of Hyponyms for Mining a Pharmaceutical Corpus. In: Proceedings of the Workshop Text Mining Research, Practice and Opportunities, Borovets, Bulgaria, pp. 63–67 (2005)

12. Ravichandran, D., Hovy, E.: Learning surface text patterns for a question answering system. In: Proceedings of 40th ACL Conference, pp. 41–47 (2002)
13. Snow, R., Jurafsky, D., Ng, A.Y.: Learning syntactic patterns for automatic hypernym discovery. In: NIPS, Vancouver, Canada, vol. 17, pp. 1297–1304 (2005)
14. Szpektor, I., Tanev, H., Dagan, I., Coppola, B.: Scaling Web-based Acquisition of Entailment Relations. In: Empirical Methods in Natural Language Processing (EMNLP), Barcelona, Spain, pp. 49–56 (2004)

Temporal Issues and Recognition Errors on the Capitalization of Speech Transcriptions*

Fernando Batista[1,2], Nuno Mamede[1,3], and Isabel Trancoso[1,3]

[1] L[2]F – Spoken Language Systems Laboratory – INESC ID Lisboa
R. Alves Redol, 9, 1000-029 Lisboa, Portugal
http://www.l2f.inesc-id.pt/

[2] ISCTE – Instituto de Ciências do Trabalho e da Empresa, Portugal

[3] IST – Instituto Superior Técnico, Portugal

Abstract. This paper investigates the capitalization task over Broadcast News speech transcriptions. Most of the capitalization information is provided by two large newspaper corpora, and the spoken language model is produced by retraining the newspaper language models with spoken data. Three different corpora subsets from different time periods are used for evaluation, revealing the importance of available training data in nearby time periods. Results are provided both for manual and automatic transcriptions, showing also the impact of the recognition errors in the capitalization task. Our approach is based on maximum entropy models and uses unlimited vocabulary. The language model produced with this approach can be sorted and then pruned, in order to reduce computational resources, without much impact in the final results.

Keywords: Capitalization maximum entropy, discriminative methods, speech transcriptions, language dynamics.

1 Introduction

The capitalization task consists of rewriting each word of an input text with its proper case information. The intelligibility of texts is strongly influenced by this information, and different practical applications benefit from automatic capitalization as a preprocessing step. It can be applied to the speech recognition output, which usually consists of raw text, in order to provide relevant information for automatic content extraction, Named Entity Recognition (NER), and machine translation.

This paper addresses the capitalization task when performed over Broadcast News (BN) orthographic transcriptions. Written newspaper corpora are used as sources of capitalization information. The evaluation is conducted in three different subsets of speech transcriptions, collected from different time periods. The importance of training data collected in nearby testing periods is also evaluated. The paper is structured as follows: Section 2 presents an overview on the related work. Section 3 describes the approach. Section 4 provides the upper-bound results by performing the evaluation over written corpora. Section 5 shows results concerning speech transcriptions. Section 6 concludes and presents future plans.

* This work was funded by PRIME National Project TECNOVOZ number 03/165 and supported by ISCTE.

P. Sojka et al. (Eds.): TSD 2008, LNAI 5246, pp. 45–52, 2008.
© Springer-Verlag Berlin Heidelberg 2008

2 Related Work

The capitalization problem can either be seen as a disambiguation problem or as a sequence tagging problem [1,2,3], where each lower-case word is associated to a tag that describes its capitalization form. The impact of using increasing amounts of training data as well as a small amount of adaptation data is studied by [1]. This work uses a Maximum Entropy Markov Model (MEMM) based approach, which allows the combination of different features. A large written newspaper corpora is used for training and the test data consists of BN data. The work of [2] describes a trigram language model (LM) with pairs (word, tag) estimated from a corpus with case information, and then uses dynamic programming to disambiguate over all possible tag assignments on a sentence. Other related work includes a bilingual capitalization model for capitalizing machine translation (MT) outputs, using conditional random fields (CRFs) reported by [4]. This work exploits case information both from source and target sentences of the MT system, producing better performance than a baseline capitalizer using a trigram language model. A previous study on the capitalization of Portuguese BN can be found in [5]. The paper makes use of generative and discriminative methods to perform capitalization of manual orthographic transcriptions.

The language dynamics is an important issue in different areas of Natural Language Processing (NLP): new words are introduced everyday and the usage of some other words decays with time. Concerning this subject, [6] conducted a study on NER over written corpora, showing that, as the time gap between training and test data increases, the performance of a named tagger based on co-training [7] decreases.

3 Approach Description

This paper assumes that the capitalization of the first word of each sentence is performed in a separated processing stage (after punctuation for instance), since its correct graphical form depends on its position in the sentence. Evaluation results may be influenced when taking such words into account [3]. Only three ways of writing a word will be considered here: lower-case, first-capitalized, and all-upper. Mixed-case words, such as "McLaren" and "SuSE", are also treated by means of a small lexicon, but they are not evaluated in the scope of this paper.

The evaluation is performed using the metrics: Precision, Recall and SER (Slot Error Rate) [8]. Only capitalized words (not lowercase) are considered as slots and used by these metrics. For example: Precision is calculated by dividing the number of correctly capitalized words by the number of capitalized words in the test data.

3.1 The Method

The modeling approach used is discriminative, and is based on maximum entropy (ME) models, firstly applied to natural language problems in [9]. An ME model estimates the conditional probability of the events given the corresponding features. Figure 1 illustrates the ME approach for the capitalization task, where the top rectangle represents the training process using a predefined set of features, and the bottom rectangle illustrates the classification using previously trained models. This framework provides

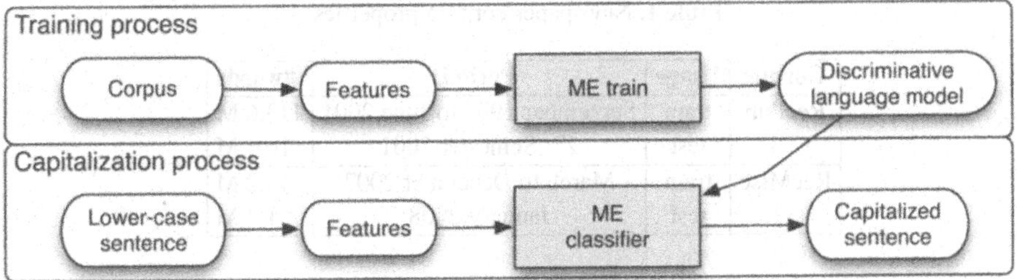

Fig. 1. Outline of the maximum entropy approach

a very clean way of expressing and combining several knowledge sources and different properties of events, such as word identification and POS tagging information. This approach requires all information to be expressed in terms of features, causing the resultant data file to become several times larger than the original. This constitutes a training problem, making it difficult to train with large corpora. The classification however, is straightforward, making it interesting for on-the-fly usage.

The memory problem can be mitigated by splitting the corpus into several subsets. The first subset is used for training the first language model (LM), which is then used to provide initialized models for the next iteration over the next subset. This goes on until all subsets are used. The final LM contains information from all corpora subsets, but, events occurring in the latest training sets gain more importance in the final LM. As the training is performed with the new data, the old models are iteratively adjusted to the new data. This approach provides a clean framework for language dynamics adaptation, offering a number of advantages: (1) new events are automatically considered in the new models; (2) with time, unused events slowly decrease in weight; (3) by sorting the trained models by their relevance, the amount of data used in next training stage can be limited without much impact on the results.

These experiments use only features comprising word identification, combined as unigrams or bigrams: w_i (current word); $\langle w_{i-1}, w_i \rangle$, $\langle w_i, w_{i+1} \rangle$. All the experiments used the `MegaM` tool [10], which uses conjugate gradient and a limited memory optimization of logistic regression.

4 Upper-Bound Performance Using Written Corpora

This section presents results achieved for written corpora. Two different newspaper corpora are used, collected in separate time periods. The oldest and largest corpus is named RecPub and consists of collected editions of the Portuguese "Público" newspaper. RecMisc is a recent corpus and combines information from six different Portuguese newspapers, found on the web. Table 1 shows corpora properties and the corresponding training and testing subsets.

All the punctuation marks were removed from the texts, making them close to speech transcriptions, but without recognition errors. Only events occurring more than once were included for training, thus reducing the influence of misspelled words and memory limitations. The approach described in Section 3 is followed, where the training corpus

Table 1. Newspaper corpora properties

Corpus	Usage	Period	#words
RecPub	train	September 1995 to June 2001	113.6 M
	test	2nd Semester 2001	16.4 M
RecMisc	train	March to December 2007	19.2 M
	test	January 2008	1.3 M

Table 2. Forward and Backward training using unigrams and bigram features

	Training			LM	RecPub test			RecMisc test		
Exp	Corpus	Type	Last month	#Lines	Prec	Rec	SER	Prec	Rec	SER
1	RecPub	Back	1995-09	10.6 Million	92%	81%	0.258	93%	80%	0.250
2		Forw	2001-06	10.8 Million	**94%**	**82%**	**0.229**	94%	80%	0.238
3	RecMisc	Back	2007-03	5.2 Million	89%	75%	0.342	94%	85%	0.205
4		Forw	2007-12	5.2 Million	89%	75%	0.344	93%	85%	0.201
5	All	Back	1995-09	12.7 Million	91%	82%	0.256	92%	83%	0.228
6		Forw	2007-12	12.9 Million	90%	82%	0.268	**93%**	**87%**	**0.186**

is split into groups containing a month of data. Table 2 shows the corresponding results. Each pair of lines in the table corresponds to using a given corpus, either by performing a normal training or training backwards. For example, both experiments 1 and 2 use RecPub training corpus, but while the training process of experiment 1 started at 2001-06 and finished at 1995-09, experiment 2 started at 1995-09 and finished at 2001-06. These two experiments use the same training data, but with a different training order. Results reveal that higher performances are achieved when the temporal difference between the time period of the last portion of training data and the time period of the testing data is smaller. The last month used in the training process seems to establish the period for which the LM is more adequate. The SER achieved in experiment 2 is better for both testing sets, given that their time period is closer to 2001-06 than to 1995-09. Notice, however, that experiments using different training sets can not be directly compared.

5 Speech Transcription Results

The following experiments use the Speech Recognition corpus (SR) – an European Portuguese broadcast news corpus – collected in the scope of the ALERT European project [11]. Table 3 presents details for each part of the corpus. The original corpus included two different evaluation sets (Eval and JEval), and it was recently complemented with a collection of six BN shows, from the same public broadcaster (RTP07).

The manual orthographic transcription of this corpus constitutes the reference corpus, and includes information such as punctuation marks, capital letters and special marks for proper nouns, acronyms and abbreviations. Each file in the corpus is divided into segments, with information about their start and end locations in the signal file, speaker id, speaker gender, and focus conditions. Most of the corpus consists of planned speech. Nevertheless, 34% is still a large percentage of spontaneous speech.

Table 3. Different parts of the Speech Recognition (SR) corpus

Sub-corpus	Recording period	Duration	Words
Train	2000–October and November	61h	449k
Eval	2001–January	6h	45k
JEval	2001–October	13h	128k
RTP07	2007–May, June, September, October	6h	45k

Table 4. Alignment report, where: *Cor, Ins, Del,* and *Sub* corresponds to the proportion of correct, insertions, deletions, and substitutions in terms of word alignments

Corpus part	Alignment errors		Cor	Ins	Del	Sub				WER
	c. words	sclite				lower	firstcap	allcaps	fail	
Train	282	2138	419872	25687	10193	25841	2630	637	1920	14.5%
Eval	17	283	38162	3122	1701	5291	471	99	338	23.9%
JEval	98	781	103365	6328	5647	12745	1455	212	1002	22.0%
RTP07	23	287	38983	2776	1493	4934	547	106	341	22.0%

Besides the manual orthographic transcription, we also have available the automatic transcription produced by the Automatic Speech Recognition (ASR) module, and other information automatically produced by the Audio Preprocessor (APP) module namely, the speaker id, gender and background speech conditions (Noise/Clean). Each word has a reference for its location in the audio signal, and includes a confidence score given by the ASR module.

5.1 Corpus Alignment

Whereas the reference capitalization already exists in the manual transcriptions, this is not the case of the automatic transcriptions. Therefore, in order to evaluate the capitalization task over this data, a reference capitalization must be provided. In order to do so, we have performed an alignment between the manual and automatic transcriptions, which is a non-trivial task mainly because of the recognition errors. Table 4 presents some issues concerning the word alignment. The alignment was performed using the NIST SCLite tool[1], but it was further improved in a post-processing step, either by aligning words which can be written differently or by correcting some SCLite basic errors. For example: the word "primeiro-ministro" (head of government) is sometimes written and recognized as two isolated words "primeiro" (first) and "ministro" (minister). The second and third columns present the number of corrected alignment errors.

When in the presence of a correct word, the capitalization can be assigned directly, but insertions and deletions do not constitute a problem either. Moreover, most of the insertions and deletions consist of functional words which usually appear in lowercase. The problem comes from the substitutions where the reference word appears capitalized (not lowercase). In this case, three different situations may occur: (1) the two words have different graphical forms, for example: "Menezes" and "Meneses" (proper nouns);

[1] Available from `http://www.nist.gov/speech`

(2) the two words are different but share the same capitalization, for example: "Andreia" and "André" (proper nouns); and (3) the two words have different capitalization forms, for example "Silva" (proper noun) and "de" (of, from). We concluded, by observation, that most of the words in these conditions share the same capitalization if their lengths are similar. As a consequence, we decided to assign the same capitalization when the number of letters do not differ by more than 2 letters. The column "fail" shows the number of unsolved alignments (kept lowercase).

5.2 Results over Manual Transcriptions

The initial capitalization experiments with speech transcriptions were performed with the LMs also used for Table 2 results. Nevertheless, subsequent experiments have shown that the overall performance can be increased by retraining such models with speech transcription training data. By doing so, the SER performance increased about 3% to 5%. Table 5 shows the results concerning manual transcriptions, after retraining also with manual transcriptions. As expected, an overall lower performance is achieved when compared to written corpora, nonetheless, only a marginal difference is obtained for *RTP07*. The biggest difference is observed for the *Eval* and *JEval* test sets, however, *JEval* may be more representative, given that its size is almost three times the size of *Eval*. The smaller size of the RecMisc training data justifies the lower results achieved. The last two lines show the results when all the training is used. The relation between temporal issues and the performance can still be observed speech transcriptions but the differences are now much smaller, in part because manual transcriptions were used for retraining the final discriminative language models.

Table 5. Retraining and evaluating with manual transcriptions

Training			**Eval**			**JEval**			**RTP07**		
Corpus	**Type**	**Last month**	**Prec**	**Rec**	**SER**	**Prec**	**Rec**	**SER**	**Prec**	**Rec**	**SER**
RecPub	Back	1995-09	84%	81%	0.35	86%	85%	0.29	92%	83%	0.24
	Forw	2001-06	**83%**	**81%**	**0.35**	**87%**	**86%**	**0.27**	93%	83%	0.23
RecMisc	Back	2007-03	82%	78%	0.39	85%	84%	0.31	91%	86%	0.22
	Forw	2007-12	81%	78%	0.40	84%	84%	0.31	**91%**	**87%**	**0.21**
All	Back	1995-09	83%	81%	0.35	86%	85%	0,29	92%	85%	0.22
	Forw	2007-12	82%	80%	0.38	84%	87%	0.29	**91%**	**88%**	**0.21**

5.3 Results over Automatic Transcriptions

Table 6 shows the results of capitalizing automatic transcriptions. These experiments share the LMs also used for Table 5 results. Other tests were conducted, for example, by retraining with automatic transcriptions, but only small differences were achieved. The SER decreased about 15% to 20%, however, these results are influenced by alignment problems and more accurate results can be achieved by manually correcting this capitalization alignment.

Table 6. Retraining with manual and evaluating with automatic transcriptions

Training			Eval			JEval			RTP07		
Corpus	**Type**	**Last month**	**Prec**	**Rec**	**SER**	**Prec**	**Rec**	**SER**	**Prec**	**Rec**	**SER**
RecPub	Back	1995-09	72%	73%	0.55	74%	77%	0.50	79%	74%	0.46
	Forw	2001-06	70%	73%	0.58	72%	79%	0.51	77%	77%	0.45
RecMisc	Back	2007-03	72%	74%	0.55	74%	77%	0.50	79%	75%	0.45
	Forw	2007-12	72%	73%	0.56	73%	76%	0.52	79%	76%	0.44
All	Back	1995-09	72%	74%	0.55	74%	78%	0.48	79%	73%	0.45
	Forw	2007-12	71%	72%	0.58	73%	76%	0.52	79%	76%	0.44

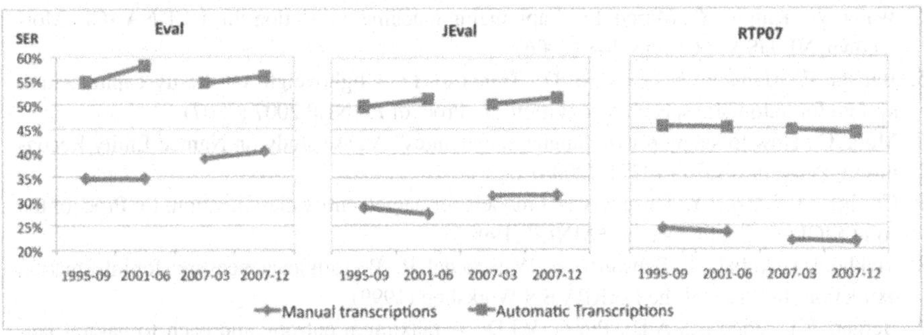

Fig. 2. Comparing the capitalization results of manual and automatic transcriptions

Figure 2 illustrates the differences between manual and automatic transcriptions. Results show that the RTP07 test subset consistently presents best performances in opposition to the *Eval* subset. The worse performance of *Eval* and *JEval* is closely related with the main topics covered in the news by the time the data was collected (US presidentials and War on Terrorism). Results of the capitalization performed with RecPub for manual transcriptions suggest a relation between the performance and the training direction, but this relation cannot be found the same way in the automatic transcriptions.

6 Conclusions and Future Work

This paper have presented capitalization results, both on written newspaper corpora and broadcast news speech corpora. Capitalization results of manual and automatic transcriptions are compared, revealing the impact of the recognition errors on this task. Results show evidence that the performance is affected by the temporal distance between training and testing sets. Our approach is based on maximum entropy models, which provide a clean framework for language dynamics adaptation.

The use of generative methods in the capitalization of newspaper corpora is reported by [5]. Using WFSTs (Weighted Finite State Transducers) and a bigram language model, the paper reports about 94% precision, 88% recall and 0.176 SER. Using similar conditions, our approach achieves only about 94% precision, 82% recall and 0.229

SER. In the near future other features will be explored for improving the discriminative approach results. For example, the word confidence score given by the recognition system will be used in future experiments.

References

1. Chelba, C., Acero, A.: Adaptation of maximum entropy capitalizer: Little data can help a lot. In: EMNLP 2004 (2004)
2. Lita, L.V., Ittycheriah, A., Roukos, S., Kambhatla, N.: tRuEcasIng. In: Proc. of the 41st annual meeting on ACL, Morristown, NJ, USA, pp. 152–159 (2003)
3. Kim, J., Woodland, P.C.: Automatic capitalisation generation for speech input. Computer Speech & Language 18, 67–90 (2004)
4. Wang, W., Knight, K., Marcu, D.: Capitalizing machine translation. In: HLT-NAACL, Morristown, NJ, USA, ACL, pp. 1–8 (2006)
5. Batista, F., Mamede, N., Caseiro, D., Trancoso, I.: A lightweight on-the-fly capitalization system for automatic speech recognition. In: Proc. of RANLP 2007 (2007)
6. Mota, C.: How to keep up with language dynamics? A case study on Named Entity Recognition. Ph.D. thesis, IST / UTL (2008)
7. Collins, M., Singer, Y.: Unsupervised models for named entity classification. In: Proc. of the Joint SIGDAT Conference on EMNLP (1999)
8. Makhoul, J., Kubala, F., Schwartz, R., Weischedel, R.: Performance measures for information extraction. In: Proc. of the DARPA BN Workshop (1999)
9. Berger, A.L., Pietra, S.A.D., Pietra, V.J.D.: A maximum entropy approach to natural language processing. Computational Linguistics 22, 39–71 (1996)
10. Daumé III, H.: Notes on CG and LM-BFGS optimization of logistic regression (2004)
11. Meinedo, H., Caseiro, D., Neto, J.P., Trancoso, I.: Audimus.media: A broadcast news speech recognition system for the european portuguese language. In: Mamede, N.J., Baptista, J., Trancoso, I., Nunes, M.d.G.V. (eds.) PROPOR 2003. LNCS, vol. 2721, pp. 9–17. Springer, Heidelberg (2003)

Web-Based Lemmatisation of Named Entities

Richárd Farkas[1], Veronika Vincze[2], István Nagy[2], Róbert Ormándi[1], György Szarvas[1], and Attila Almási[2]

MTA-SZTE, Research Group on Artificial Intelligence,
6720 Szeged, Aradi Vértanúk tere 1., Hungary,
University of Szeged, Department of Informatics,
6720 Szeged, Árpád tér 2., Hungary
{rfarkas,vinczev,ormandi,szarvas}@inf.u-szeged.hu,
Istvan.Nagy@stud.u-szeged.hu, vizipal@gmail.com

Abstract. Identifying the lemma of a Named Entity is important for many Natural Language Processing applications like Information Retrieval. Here we introduce a novel approach for Named Entity lemmatisation which utilises the occurrence frequencies of each possible lemma. We constructed four corpora in English and Hungarian and trained machine learning methods using them to obtain simple decision rules based on the web frequencies of the lemmas. In experiments our web-based heuristic achieved an average accuracy of nearly 91%.

Keywords: Lemmatisation, web-based techniques, named entity recognition.

1 Introduction

This paper seeks to lemmatise Named Entities (NEs) in English and in Hungarian. Finding the lemma (and inflectional affixes) of a proper noun can be useful for several reasons: the proper name can be stored in a normalised form (e.g. for indexing) and it may prove to be easier to classify a proper name in the corresponding NE category using its lemma instead of the affixed form. Next, the inflectional affixes themselves can contain useful information for some specific tasks and thus can be used as features for classification (e.g. the plural form of an organisation name is indicative of org-for-product metonymies and is a strong feature for proper name metonymy resolution [1]). It is useful to identify the phrase boundary of a certain NE to be able to cut off inflections. This is not a trivial task in free word order languages where NEs of the same type can follow each other without a punctuation mark.

The problem is difficult to solve because, unlike common nouns, NEs cannot be listed. Consider, for instance, the following problematic cases of finding the lemma and the affix of a NE:

1. the NE ends in an apparent suffix,
2. two (or more) NEs of the same type follow each other and they are not separated by punctuation marks,
3. one NE contains punctuation marks within.

P. Sojka et al. (Eds.): TSD 2008, LNAI 5246, pp. 53–60, 2008.
© Springer-Verlag Berlin Heidelberg 2008

In morphologically rich languages such as Hungarian, nouns (hence NEs as well) can have hundreds of different forms owing to grammatical number, possession marking and grammatical cases. When looking for the lemmas of NEs (case 1), the word form being investigated is deprived of all the suffixes it may bear. However, there are some NEs that end in an apparent suffix (such as *McDonald's* or *Philips* in English or *Pannon* in Hungarian, with *-on* meaning "on"), but this pseudo-suffix belongs to the lemma of the NE and should not to be removed. Such proper names make the lemmatisation task non-trivial.

In the second and third cases, when two NEs of the same type follow each other, they are usually separated by a punctuation mark (e.g. a comma). Thus, if present, the punctuation mark signals the boundary between the two NEs. However, the assumption that punctuation marks are constant markers of boundaries between consecutive NEs and that the absence of punctuation marks indicates a single (longer) name phrase often fails, and thus a more sophisticated solution is necessary to locate NE phrase boundaries. Counterexamples for this naive assumption are NEs such as *Stratford-upon-Avon*, where two hyphens occur within one single NE.

In order to be able to select the appropriate lemma for each problematic NE, we applied the following strategy. Each ending that seems to be a possible suffix is cut off the NE. With those NEs that consist of several tokens either separated by punctuations marks or not, every possible cut is performed. Then Google and Yahoo searches are carried out on all the possible lemmas. We trained several machine learning classifiers to find the decision boundary for appropriate lemmas based on the frequency results of the search engines.

2 Related Work

Lemmatisation of common nouns. Lemmatisation, that is, dividing the word form into its root and suffixes, is not a trivial issue in morphologically rich languages. In agglutinative languages such as Hungarian, a noun can have hundreds of different forms owing to grammatical number, possession marking and a score of grammatical cases: e.g. a Hungarian noun has 268 different possible forms [2]. On the other hand, there are lemmas that end in an apparent suffix (which is obviously part of the lemma), thus sometimes it is not clear what belongs to the lemma and what functions as a suffix. However, the lemmatisation of common nouns can be made easier by relying on a good dictionary of lemmas [3].

Lemmatisation of NEs. The problem of proper name lemmatisation is more complicated since NEs cannot be listed exhaustively, unlike common nouns, due to their diversity and increasing number. Moreover, NEs can consist of several tokens, in contrast to common nouns, and the whole phrase must be taken into account. Lots of suffixes can be added to them in Hungarian (e.g. *Invitelben*, where *Invitel* is the lemma and *-ben* means "in"), and they can bear the plural or genitive marker *-s* or *'s* in English (e.g. *Toyotas*). What is more, there are NEs that end in an apparent suffix (such as *McDonald's*), but this pseudo-suffix belongs to the lemma of the NE.

NE lemmatisation has not attracted much attention so far because it is not such a serious problem in major languages like English and Spanish as it is in agglutinative languages. An expert rule-based and several string distance-based methods for Polish person name inflection removal were introduced in [4]. And a corpus-based rule induction method was studied for every kind of unknown word in Slovene in [5]. The scope of our study lies between these two as we deal with different kinds of NEs. On the other hand, these studies focused on removing inflection suffixes, while our approach handles the separation of consecutive NEs as well.

Web-based techniques. Our main hypothesis is that the lemma of an NE has a relatively high frequency on the World Wide Web, in contrast to the frequency of the affixed form of the NE. Although there are several papers that tell us how to use the WWW to solve simple natural language problems like [6,7], we think that this will become a rapidly growing area and more research will be carried out over the next couple of years. To the best of our knowledge, there is currently no procedure for NE lemmatisation which uses the Web as a basis for decision making.

3 Web-Based Lemmatisation

The use of online knowledge sources in Human Language Technology (HLT) and Data Mining tasks has become an active field of research over the past few years. This trend has been boosted by several special and interesting properties of the World Wide Web. First of all, it provides a practically limitless source of (unlabeled) data to exploit and, more importantly, it can bring some dynamism to applications. As online data change and rapidly expand over time, a system can remain up-to-date and extend its knowledge without the need for fine tuning or any human intervention (like retraining on up-to-date data). These features make the Web a very useful source of knowledge for HLT applications as well. On the other hand, accessing the data is feasible just via search engines (e.g. we cannot iterate through all of the occurrences of a word). There are two interesting problems here: first, appropriate queries must be sent to a search engine; second, the response of the engine offers several opportunities (result frequencies, snippets, etc.) in addition to simply "reading" the pages found.

In the case of lemmatisation we make use of the result frequencies of the search engines. In order to be able to select the appropriate lemma for each problematic NE, we applied the following strategy. In step-by-step fashion, each ending that seems to be a possible suffix is cut off the NE. As for the apparently multiple NEs separated by punctuation marks or not, every possible cut is performed; that is, in the case of *Stratford-upon-Avon*, the possibilities *Stratford + upon + Avon* (3 lemmas), *Stratford-upon + Avon*, *Stratford + upon-Avon* (2 lemmas), *Stratford-upon-Avon* (1 lemma) are generated. Then after having found all the possible lemmas we do a Google search and a Yahoo search. Our key hypothesis is that the frequency of the lemma-candidates on the WWW is high – or at least the ratio of the full form and lemma-candidate frequencies is relatively high – with an appropriate lemma and low in incorrect cases.

In order to verify our hypothesis we manually constructed four corpora of positive and negative examples for NE lemmatisation. Then, Google and Yahoo searches were performed on every possible lemma and training datasets were compiled for machine learning algorithms using the queried frequencies as features. The final decision rules were learnt on these datasets.

4 Experiments

In this section we describe how we constructed our datasets and then investigate the performance of several machine learning methods in the tasks outlined above.

4.1 The Corpora

The lists of negative and positive examples for lemmatisation were collected manually. We adopted the principal rule that we had to work on real-world examples (we did not generate fictitious examples), so our three annotators were asked to browse the Internet and collect "interesting" cases. These corpora are the unions of the lists collected by 3 linguists and were checked by the chief annotator. The samples mainly consist of person names, company names and geographical locations occurrences on web pages. Table 1 lists the size of each corpora (constructed for the three problems). The corpora are accessible and are free of charge.

The first problematic case of finding the lemma of a NE is when the NE ends in an apparent suffix (which we will call the *suffix* problem in the following). In agglutinative languages such as Hungarian, NEs can have hundreds of different inflections. In English, nouns can only bear the plural or the possessive marker *-s* or *'s*. There are NEs that end in an apparent suffix (such as *Adidas* in English), but this pseudo-suffix belongs to the lemma of the NE.

We decided to build two corpora for the suffix problem; one for Hungarian and one for English and we produced the possible suffix lists for the two languages. In Hungarian more than one suffix can be matched to several phrases. In these cases we examined every possible cut and the correct lemma (chosen by a linguist expert) became a positive example, while every other cut was treated as a negative one.

The other two lemmatisation tasks, namely the case where several NEs of the same type follow each other and they are not separated by punctuation marks and the case where one NE contains punctuation marks within are handled together in the following (and is called the *separation* task). When two NEs of the same type follow each other, they are usually separated by a punctuation mark (e.g. a comma or a hyphen). Thus, if present, the punctuation mark signals the boundary between the two NEs (e.g. ***Arsenal-Manchester*** *final*; ***Obama-Clinton*** *debate*; ***Budapest-Bécs*** *marathon*). However, the assumption that punctuation marks are constant markers of boundaries between consecutive NEs and that the absence of punctuation marks indicates a single (longer) name phrase often fails, and thus a more sophisticated procedure is necessary to locate NE phrase boundaries. Counterexamples for this naive assumption are NEs such as the *Saxon-Coburg-Gotha family*, where the hyphens occur within the NE, and sentences such as ***Gyurcsány Orbán*** *gazdaságpolitikájáról mondott véleményt.* (*'**Gyurcsány** expressed his views on **Orbán**'s economic policy.'* (two consecutive entities) as opposed to *"He expressed his views on **Orbán Gyurcsány**'s economic policy."* (one single two-token-long entity)). Without background knowledge of the participants in the present-day political sphere in Hungary, the separation of the above two NEs would a pose problem. Actually, the first rendition of the Hungarian sentence conveys the true, intended meaning; that is, the two NEs are correctly separated. As for the second version, the NEs are not separated and are treated as a two-token-long entity. In Hungarian,

Table 1. The sizes of the corpora

	Suffix Eng	Suffix Hun	Separation Eng	Separation Hun
positive examples	74	207	51	137
negative examples	84	543	34	69

however, a phrase like *Gyurcsány Orbán* could be a perfect full name, *Gyurcsány* being a family name and *Orbán* being – in this case – the first name.

As consecutive NEs without punctuation marks appear frequently in Hungarian due to the free word-order, we decided to construct a corpus of negative and positive cases for Hungarian. Still, such cases can occur just as the consequence of a spelling error in English. Hence we focused on special punctuation marks which can separate entities in several cases (***Obama-Clinton*** *debate*) but are part of the entity in others. In the separation task there are several cases where more than one cut is possible (more than two tokens in Hungarian and more than one special mark in English). In such cases we again asked a linguist expert to choose the correct cut and every incorrect cut became a negative example.

4.2 The Feature Set

To create training datasets for machine learning methods – which try to learn how to separate correct and incorrect cuts based on labeled examples – we sent queries to the Google and Yahoo search engines using their APIs[1]. The queries started and finished in quotation marks and the *site:.hu* constraint was used in the Hungarian part of the experiments. In the suffix tasks, we sent queries with and without suffixes to both engines and collected the number of hits. The original database contained four dimensional feature vectors. Two dimensions listed the number of Google hits and two components listed similar values from the Yahoo search engine.

The original training datasets for the separation tasks contained six features. Two stood for the number of Google hits for the potential first and second parts of a cut and one represented the number of Google hits for the whole observed phrase. The remaining three features conveyed the same information, but here we used the Yahoo search engine. Each of the four tasks was a binary classification problem. The class value was set to 0 for negative examples and 1 for positive ones.

Our preliminary experiments showed that using just the original form of the datasets (Non-Rate rows of Table 2) is not optimal in terms of classification accuracy. Hence we performed some basic transformations on the original data, the first component of the feature vector was divided by the second component. If the given second component was zero, then the new feature value was also zero (Rate rows in Table 2). This yielded a two dimensional dataset when we utilised both Yahoo and Google hits for the suffix classification tasks and a four dimensional for the separation tasks. Finally, we took the minimum and the maximum of the separated parts' ratios hence providing the possibility to learn rules such as *"if* ***one of the*** *separated parts' frequency ratio is higher than X"*.

[1] Google API: `http://code.google.com/apis/soapsearch/`; Yahoo API: `http://developer.yahoo.com/search/`

4.3 Machine Learning Methods

In the tests, we experimented with several machine learning algorithms. Each method was evaluated using the 10-fold-cross-validation method. Here classification accuracy was used as the evaluation metric in each experiment. We employed the WEKA [8] machine learning library to train our models. Different kinds of learning algorithms and parameter settings were examined to achieve the highest possible accuracy. The first one is a baseline method which is the naive classifier; it classifies each sample using the most frequent label observed on the training dataset.

We used the k-Nearest Neighbour [9] algorithm, which is one of the simplest machine learning algorithms. In this algorithm, an object is classified by taking the majority vote of its neighbours, with the object being assigned to the class most common amongst its k nearest neighbours. Here k is a positive integer, and it is typically not very large.

We also used the C4.5 [10] decision tree learning algorithm as it is a widely applied method in data mining. In a decision tree each interior node corresponds to an attribute; an edge to a child represents a possible value or an interval of values of that variable. A leaf represents the most probable class label given the values of the variables represented by the path from the root. A tree can be learned by splitting the source set into subsets based on an attribute value test. This process is repeated on each derived subset in a recursive manner. The recursion stops either when further splitting is not possible or when the same classification can be applied to each element of the derived subset. The M parameter of the C4.5 defines the minimum number of instances per leaf, i.e. the there are no more splits on nodes if the number of instances is fewer than M.

Logistic Regression [11] is another well studied machine learning method. It seeks to maximise the conditional probability of classes, subject to feature constraints (observations). This is performed by weighting features so as to maximise the likelihood of data.

The results obtained by using the different feature sets and the machine learning algorithms described above are presented in Table 2 for the suffix task, and in Table 3 for the separation task.

4.4 Discussion

For the English suffix task, the size of the databases is quite small, which leads to a dominance of the k-Nearest Neighbour classifier, since the lazy learners – like k-Nearest Neighbour – usually achieve good results on small datasets.

In this task the training algorithms attain their best performance on the transformed datasets using rates of query hits (this holds when Yahoo or Google searches were performed). One could say that the rate of the hits (one feature) is the best characterisation in this task. However, we can see that with the Hungarian suffix problem the original dataset characterises the problem better, and thus the transformation is really unnecessary.

The best results for the Hungarian suffix problem are achieved on the full dataset, but they are almost the same as those for untransformed Yahoo dataset. Without doubt, this is due to the special property of the Yahoo search engine which searches accent sensitively, in contrast to Google. For example, for the query *Ottó* Google finds every webpage which contains *Ottó* and *Otto* as well, while Yahoo just returns the *Ottó*-s.

Table 2. Suffix task results obtained from applying different learning methods. (The first letter stands for the language /E – English; H – Hungarian/. The second letter stands for the search engine used /B – Both; G – Google; Y – Yahoo/. NR means Non-Rate database, R means Rate database. Thus for example, H-B-NR means Hungarian problem using non-rate dataset and both search engines).

	kNN k=3	kNN k=5	C4.5 M=2	C4.5 M=5	Log Reg	Base
E-B-NR	89.24	89.24	86.71	84.18	84.81	53.16
E-B-R	93.04	89.87	91.77	92.41	73.42	53.16
E-G-NR	87.34	86.71	87.34	81.65	82.28	53.16
E-G-R	**93.67**	**93.67**	87.97	92.41	90.51	53.16
E-Y-NR	89.87	89.87	86.08	85.44	84.18	53.16
E-Y-R	91.77	91.77	87.34	91.77	88.61	53.16
H-B-NR	**94.27**	**94.27**	82.67	90.00	88.27	72.40
H-B-R	84.67	86.13	81.73	81.73	72.40	72.40
H-G-NR	85.33	85.33	82.40	82.93	83.33	72.40
H-G-R	83.60	78.40	83.60	77.60	77.60	72.40
H-Y-NR	93.73	93.73	83.87	88.13	86.13	72.40
H-Y-R	87.20	84.27	87.20	74.00	74.00	72.40

Table 3. Separation task results obtained from applying different learning methods

	kNN k=3	kNN k=5	C4.5 M=2	C4.5 M=5	Log Reg	Base
English	88.23	84.71	95.29	94.12	77.65	60.00
Hungarian	79.25	81.13	80.66	79.72	70.31	64.63

The separation problem for Hungarian proved to be a difficult task. The decision tree (which we found to be the best solution) is a one-level high tree with a split. This can be interpreted as *if one of the resulting parts' frequency ratio is high enough, then it is an appropriate cut*. It is interesting to see that among the learned rules for the English separation task, there is a constraint for the second part of a possible separation (while the learnt hypothesis for Hungarian consisted of simple *if (any) one of the resulting parts is...* rules).

5 Conclusions

In this paper we introduced corpora for the English and Hungarian Named Entity lemmatisation tasks. The corpora are freely available for further comparison studies.The NE lemmatisation task is very important for textual data indexing systems, for instance, and is of great importance for agglutinative (Finno-Ugric) languages and other languages that are rich in inflections (like Slavic languages). To handle this task, we proposed a web-based heuristic which sends queries for every possible lemma to two popular web search engines. Based on the constructed corpora we automatically derived simple decision rules on the search engine responses by applying machine learning methods.

Despite the small size of our corpora, we got fairly good empirical results and even better accuracies can probably be obtained with bigger corpora. The 91.09% average accuracy score supports our hypothesis that even the frequencies of possible lemmas provide enough information to help find the right separations. We encountered several problems when using search engines to obtain the lemma-frequencies like, the need of an accent sensitive search, difficulties in the proper handling of punctuation marks in queries and that the frequency values of results are estimated values only. We think that if offline corpora with an appropriate size were available, the frequency counts would be more precise and our heuristic could probably attain even better results.

Acknowledgments. This work was supported in part by the NKTH grant of Jedlik Ányos R&D Programme 2007 of the Hungarian government (codename TUDORKA7). The author wishes to thank the anonymous reviewers for valuable comments.

References

1. Farkas, R., Simon, E., Szarvas, Gy., Varga, D.: Gyder: Maxent metonymy resolution. In: Proceedings of the Fourth International Workshop on Semantic Evaluations (SemEval-2007), Prague, Czech Republic, pp. 161–164. Association for Computational Linguistics (2007)
2. Melĉuk, I.: Modèle de la déclinaison hongroise. In: Cours de morphologie générale (théorique et descriptive), Montréal, Les Presses de l'Université de Montréal, CNRS (edn). vol. 5, pp. 191–261 (2000)
3. Halácsy, P., Trón, V.: Benefits of resource-based stemming in Hungarian information retrieval. In: Peters, C., Clough, P., Gey, F.C., Karlgren, J., Magnini, B., Oard, D.W., de Rijke, M., Stempfhuber, M. (eds.) CLEF 2006. LNCS, vol. 4730, pp. 99–106. Springer, Heidelberg (2007)
4. Piskorski, J., Sydow, M., Kupść, A.: Lemmatization of polish person names. In: Proceedings of the Workshop on Balto-Slavonic Natural Language Processing, Prague, Czech Republic, pp. 27–34. Association for Computational Linguistics (2007)
5. Erjavec, T., Dzeroski, S.: Machine learning of morphosyntactic structure: Lemmatizing unknown Slovene words. Applied Artificial Intelligence 18, 17–41 (2004)
6. Bunescu, R.C., Pasca, M.: Using encyclopedic knowledge for named entity disambiguation. In: EACL (2006)
7. Etzioni, O., Cafarella, M., Downey, D., Popescu, A.M., Shaked, T., Soderland, S., Weld, D.S., Yates, A.: Unsupervised named-entity extraction from the web: an experimental study. Artif. Intell. 165, 91–134 (2005)
8. Witten, I.H., Frank, E.: Data Mining: Practical Machine Learning Tools and Techniques, 2nd edn. Morgan Kaufmann Series in Data Management Systems. Morgan Kaufmann, San Francisco (2005)
9. Aha, D.W., Kibler, D., Albert, M.K.: Instance-based learning algorithms. Mach. Learn. 6, 37–66 (1991)
10. Quinlan, J.R.: C4.5: Programs for Machine Learning. Morgan Kaufmann, San Francisco (1993)
11. Berger, A.L., Pietra, S.D., Pietra, V.J.D.: A maximum entropy approach to natural language processing. Computational Linguistics 22, 39–71 (1996)

Combining Multiple Resources to Build Reliable Wordnets

Darja Fišer[1] and Benoît Sagot[2]

[1] Fac. of Arts, Univ. of Ljubljana, Aškerčeva 2, 1000 Ljubljana, Slovenia
[2] Alpage, INRIA / Paris 7, 30 rue du Ch. des rentiers, 75013 Paris, France
darja.fiser@guest.arnes.si, benoit.sagot@inria.fr

Abstract. This paper compares automatically generated sets of synonyms in French and Slovene wordnets with respect to the resources used in the construction process. Polysemous words were disambiguated via a five-language word-alignment of the SEERA.NET parallel corpus, a subcorpus of the JRC Acquis. The extracted multilingual lexicon was disambiguated with the existing wordnets for these languages. On the other hand, a bilingual approach sufficed to acquire equivalents for monosemous words. Bilingual lexicons were extracted from different resources, including Wikipedia, Wiktionary and EUROVOC thesaurus. A representative sample of the generated synsets was evaluated against the gold-standards.

1 Introduction

The first wordnet was developed for English at Princeton University (PWN). Over time it has become one of the most valuable resources in applications for natural language understanding and interpretation, which initiated the development of wordnets for many other languages apart from English [1,2]. Currently, wordnets for more than 50 languages are registered with the Global WordNet Association (http://www.globalwordnet.org/). While it is true that manual construction of each wordnet is the most reliable and produces the best results as far as linguistic soundness and accuracy is concerned, such an endeavour is highly time-consuming and expensive. This is why alternative, semi- or fully automatic approaches have been proposed. By taking advantage of the existing resources, they facilitate faster and easier development of a wordnet [3,4].

Apart from the knowledge acquisition bottleneck, another major problem in the wordnet community is the availability of the developed wordnets. Currently, only a handful of them are freely available (Arabic, Hebrew, Irish and Princeton). For example, a wordnet for French has been created within the EuroWordNet (EWN) project [1], the resource has not been widely used mainly due to licensing issues. In addition, there has been no follow-up project to further extend and improve the core French WordNet since the EWN project has ended [5]. This issue was taken into account in the two recent wordnet development projects presented in this paper, the results of which will be automatically constructed (but later also manually checked) broad-coverage open-source wordnets for French (WOLF, http://wolf.gforge.inria.fr) and Slovene (SloWNet, http://nl.ijs.si/slownet).

P. Sojka et al. (Eds.): TSD 2008, LNAI 5246, pp. 61–68, 2008.
© Springer-Verlag Berlin Heidelberg 2008

The paper is organized as follows: a brief overview of the related work is given in the next section. Section 3 presents the two wordnet development projects. Section 4 presents and evaluates the created resources with a focus on a source-by-source evaluation, and the last section gives conclusions and perspectives.

2 Related Work

The relevant literature reports on several techniques used to build semantic lexicons, most of which can be divided into two approaches. Contrary to the merge approach, according to which a wordnet for a certain language is first created based on monolingual resources and then mapped to other wordnets, we have opted for the expand approach [1]. This model takes a fixed set of synsets from Princeton WordNet (PWN) and translates them into the target language, preserving the structure of the original wordnet. The cost of the expand model is that the resulting wordnets are biased by the PWN. However, due to its greater simplicity, the expand model has been adopted in a number of projects, such as the BalkaNet [2] and MultiWordNet [6], as well as EWN [1].

Research teams adopting the latter approach took advantage of a wide range of resources at their disposal, including machine readable bilingual and monolingual dictionaries, taxonomies, ontologies and others. For the construction of WOLF and SloWNet, we have leveraged three different publicly available types of resources: the JRC-Acquis parallel corpus, Wikipedia (and other related wiki resources) and other types of bilingual resources. Equivalents for monosemous literals that do not require sense disambiguation were extracted from bilingual resources. Roughly 82% of literals found in PWN are monosemous, however most of them are not in the core vocabulary. On the other hand, the parallel corpus was used to obtain semantically relevant information from translations so as to be able to handle polysemous literals. The idea that semantic insights can be derived from the translation relation has been explored by [7,8,9]. The approach has also yielded promising results in an earlier smaller-scale experiment to obtain synsets for Slovene wordnet [10].

3 Approach

This section briefly presents the approach used to construct a wordnet automatically. For a more detailed description of the approach, see [11].

In the align approach we used the SEE-ERA.NET corpus (project ICT 10503 RP), a 1.5-million-word sentence-aligned subcorpus of JRC-Acquis [12] in eight languages. Apart from French and Slovene, we used English, Romanian, Czech and Bulgarian. We used different tools to POS-tag and lemmatize the corpus before word-aligning it with Uplug [13]. This allowed us to build five multilingual lexicons that include French and four multilingual lexicons that include Slovene. They contain between 49,356 (Fr-Ro-Cz-Bg-En) to 59,020 entries (Fr-Cz-Bg-En). The next step was to assign the appropriate synset id to each entry of these lexicons. To achieve this, we gathered the set of all synset ids assigned to each literal of a given entry (apart from the French or Slovene one) in the corresponding BalkaNet wordnet [2]. Since all these wordnets share the same synset ids as PWN 2.0, the intersection of all the found synset ids is computed. The intersection of all possible senses in each language is likely to output the correct one, which can be assigned to the French or Slovene literal. Applied to the above-mentioned multilingual lexicons, this technique allowed us to build several sets of (French or Slovene) synsets (see Table 2 for quantitative data). Because tagging, lemmatization and alignment are not perfect, synsets created in this way do inherit some of these errors. However, the value of this approach lies in the fact that they cover polysemous literals from the core vocabulary, which the translation approach cannot handle (see Section 4).

For the translation approach, applied on monosemous literals from the PWN 2.0, we used the following bilingual resources:

- Wikipedia (`http://www.wikipedia.org`), a multilingual collaborative encyclopaedia. We extracted bilingual Fr-En and Sl-En lexicons thanks to inter-wiki links that relate articles on the same topic in different languages.[1]
- The French, Slovene and English Wiktionaries (`http://www.wiktionary.org`), lexical companions to Wikipedia, which contain translations into other languages.
- The Wikispecies (`http://species.wikimedia.org`), a taxonomy of living species with translations of Latin standard names into vernacular languages.
- Eurovoc descriptors (`http://europa.eu/eurovoc`) is a multilingual thesaurus used for classification of EU documents.
- For Slovene, we used a large-coverage electronic bilingual (English-Slovene) dictionary (over 130,000 entries).
- Finally, we created trivial translations by retaining all numeric literals (such as 1,000 or 3.14159...) and all Latin taxonomic terms (extracted from the TreeOfLife project – `http://www.tolweb.org`).

Because they are created by translation of monosemous literals, these synsets will on the one hand be very reliable (see Table 3), but at the same time mostly concern non-core vocabulary (see Table 1).

Synsets obtained from both approaches were merged, while preserving information on the source of each piece of information. This enabled us to perform a simple heuristic filtering according to the reliability of each source, on the diversity of sources that assign a given literal to a given synset, and on frequency information (for the sources from the align approach).

4 Results and Evaluation

4.1 Global Evaluation

We compared the merged Slovene and French wordnets to PWN, French EWN and a manually created sample of Slovene WordNet, called ManSloWNet[2]. Although we

[1] These lexicons have 307,256 entries for French and 27,667 for Slovene. The difference in size is substantial and will also lead to very different number of the generated synsets. The same is true for most other bilingual resources used in this approach.

[2] This subset of the Slovene WordNet contains all synsets from BCS1 and 2 (approx. 5,000), which were automatically translated from Serbian, its closest relative in the BalkaNet family. All the synsets were then manually corrected [14].

Table 1. WOLF and SloWNet synsets. Percentages are given compared to PWN 2.0.

Automatically generated French synsets (WOLF)

wordnet	PWN 2.0	align		transl		merged (WOLF)		French EWN	
BCS1	1,218	791	64.9%	175	14.4%	**870**	**71.4%**	1,211	99.4%
BCS2	3,471	1,309	37.7%	523	15.1%	**1,668**	**48.0%**	3,022	87.1%
BCS3	3,827	824	21.5%	1,100	28.7%	**1,801**	**47.1%**	2,304	60.2%
non-BCS	106,908	2,844	2.7%	25,566	23.9%	**28,012**	**26.2%**	15,584	14.6%
total	*115,424*	*5,768*	*5.0%*	*27,364*	*23.7%*	***32,351***	***28.0%***	*22,121*	*19.2%*

Automatically generated Slovene synsets (SloWNet)

wordnet	PWN 2.0	align		transl		merged (SloWNet)		ManSloWNet	
BCS1	1,218	618	50.7%	181	14.9%	**714**	**58.6%**	1,218	100%
BCS2	3,471	896	25.8%	606	17.4%	**1,361**	**39.2%**	3,469	99.9%
BCS3	3,827	577	15.1%	1,128	29.5%	**1,611**	**42.1%**	180	4.7%
non-BCS	106,908	1,603	1.5%	24,116	22.6%	**25,422**	**23.8%**	1	0.0%
total	*115,424*	*3,694*	*3.2%*	*26,031*	*22.6%*	***29,108***	***25.2%***	*4,868*	*4.2%*

are aware of the fact that these resources are not perfect, they were considered as gold standard for our evaluation procedure because they were by far the best resources of such kind we could obtain.

WOLF currently contains 32,351 synsets that include 38,001 unique literals. This figure is much greater than the number of synsets in French EWN (22,121 synsets could be mapped into PWN 2.0 synsets). This is directly related to the high number of monosemous PWN literals in non-core synsets (119,528 out of 145,627) that the translation approach was able to handle adequately. Moreover, French EWN has only nominal and verbal synsets, whereas WOLF includes adjectival and adverbial synsets as well. Figures for SloWNet are similar: 29,108 synsets that include 45,694 literals (to be compared with the 4,868 synsets of ManSloWNet). However, without the En-Sl dictionary that was used for Slovene, the figures would have been much lower.

In order to evaluate the coverage of the generated wordnets, we used the BalkaNet *Basic Concept Sets* [2]. Basic synsets are grouped into three BCS categories, BCS1 being the most fundamental set of senses. The results for the automatically constructed wordnets are compared to the goldstandards (see Table 1). They show that both WOLF and SloWNet have a reasonable coverage of BCS senses. They also show that our approach still does not come close to PWN, which was the backbone of our experiment. However, the generated wordnets are considerably richer than the only other wordnets that exist for French and Slovene, especially for non-BCS synsets. Moreover, although the same approach was used, and despite the use of a bilingual dictionary, SloWNet is smaller than WOLF. This is mainly because French Wikipedia is considerably larger than the Slovene one and thus yields many more monosemous synsets, which are not always found in the En-Sl bilingual dictionary.

The align approach yielded a relatively low number of synsets compared to bilingual resources, mostly because it relies on an intersection operation among several languages: if some synsets were missing in any of the existing wordnets used for comparison, there was no match among the languages and the synset could not be generated.

Interesting as well is the nature of synsets that were generated from the different sources. Basically, the align approach that handled all kinds of words resulted predominantly in core synsets from the BCS categories. On the other hand, the bilingual resources that tackled only the monosemous expressions provided us with much more specific synsets outside the core vocabulary. The align approach worked only on single words, which is why all MWEs in the resulting wordnets come from bilingual resources.

4.2 Source-by-Source Evaluation

From a qualitative point of view, we were interested in how reliable the various sources, such as the different language combinations in the align approach, wiki sources and other thesauri, were for the creation of synsets. This is why the rest of the evaluation is performed on each individual source and the reliability scores obtained will be used to generate confidence measures for the rest of the generated synsets from that source which we were unable to evaluate automatically (because they are missing in the goldstandards). We restricted this manual evaluation to nominal synsets. Verbal synsets are more difficult to handle automatically for many reasons: higher polysemy of frequent verbs, differences in linguistics systems in dealing with phrasal verbs, light verb constructions and others. These synsets, as well as adjectival and adverbial synsets, will be evaluated carefully in the future. For each source, we checked whether a given literal in the generated wordnets is assigned the appropriate synset id according to the goldstandards. We considered only those literals that are both in the goldstandard and in the evaluated resource. A random sample of 100 (literal,synset) pairs present in the acquired resource but absent in the goldstandard were inspected by hand and classified into the following categories (see Table 3):

- the literal is an appropriate expression of the concept represented by that synset id but is missing from the goldstandard (absent in GS but correct); as mentioned before, the goldstandards we used for automatic evaluation are not perfect and complete, which is why a given literal that was automatically assigned to a particular synset can be a legitimate literal missing in the goldstandard rather than an error; for example, the French literal *document* and the Slovene literal *dokument* were correctly added in the synset corresponding to PWN literal *document*: this synset was absent from French EWN altogether, whereas in ManSloWNet it only contained literal *spis*;
- the literal is not appropriate for that synset but is semantically very close to it, its hypernym or its hyponym (closely related); such cases can be considered as correct if more coarse-grained sense granularity is sufficient for a given application; for example, it might suffice to treat words, such as *ekipa* (*team*) and *skupina* (*group*) as synonyms in a particular HLT task;
- the literal is neither appropriate nor semantically related to the synset in question because it results from wrong sense disambiguation, wrong word alignment or wrong lemmatization (wrong).

The latter category contains real errors in the generated wordnets. Many of them (around 30% in Slovene data) are related to insufficient sense disambiguation at the stage of comparing wordnets in other languages. For example, the word *examination*

Table 2. Evaluation of WOLF and SloWNet w.r.t. corresponding goldstandard (GS) wordnets (French EWN and ManSloWNet). Results on the translation approach for Slovene are not shown, because they are not statistically significant (not enough data).

Source	# of (lit,synsetid) pairs	Present in GS	synset not in GS	Discrepancy w.r.t GS
WOLF				
Fr-Cz-En	1760	61.7%	7.5%	30.8%
Fr-Cz-Bg-En	1092	67.8%	4.9%	27.4%
Fr-Ro-En	2002	64.7%	8.1%	27.2%
Fr-Ro-Cz-En	1206	70.6%	5.4%	24.0%
Fr-Ro-Cz-Bg-En	796	75.5%	3.3%	21.2%
Wikipedia	368	94.0%	0.3%	5.7%
Fr Wiktionary	577	69.8%	1.0%	29.1%
En Wiktionary	365	88.5%	—	11.5%
Wikispecies	21	*90.5%*	*4.8%*	*4.8%*
EUROVOC descr.	69	67.6%	—	32.3%
SloWNet				
Sl-Cz-En	2084	53.4%	10.9%	35.6%
Sl-Cz-Bg-En	1383	59.3%	6.6%	34.1%
Sl-Ro-Cz-En	1589	57.7%	8.0%	34.3%
Sl-Ro-Cz-Bg-En	1101	61.0%	5.1%	33.9%

can mean a medical check-up. In this case, the correct Slovene translation is *preiskava*. But when the same English word is used for a school exam, it should be translated as *preverjanje znanja*, not as *preiskava*. However, the latter was aligned twice with the English word *examination* and with the Czech word *zkouška*, whose meanings include *school exam*. This leads to a non-empty intersection of synset ids in the Sl-Cz-En source, which assigns the *school exam* synset to *preiskava*. Many errors are also the consequence of wrong word alignment of the corpus. This happened a lot in cases where the order of constituents in noun phrases in one language is substantially different from the order in another language. For example, the English compound *member state*$_{\text{head}}$ is always translated in the opposite order as *država*$_{\text{head}}$ *članica* in Slovene and *état*$_{\text{head}}$ *membre* in French, and is thus likely to be misaligned. The third source of errors are lemmatization problems, much more common in Slovene than French because the Slovene tagger was trained on a smaller corpus. If a strange lemma is guessed by the lemmatization algorithm for an unknown wordform, it will most likely be filtered out by the following stages in our synset generation procedure. However, if a word is assigned a wrong but legitimate lemma, it will be treated as a possible synonym for a certain concept by our algorithm and therefore appear in the wrong synset. For example, if the word form *vode* (singular genitive form of the lemma *water*) is wrongly lemmatized as *vod* (Eng. *platoon*), it will be placed in all the *water* synsets, which is a serious error that reduces the usability of the resource. In French, some expressions with plural canonical forms, such as *affaires* (*(one's) stuff*) got lemmatized into singular (*affaire*, Eng. *affair, deal, case*), which is inappropriate for that synset.

Table 3. Manual evaluation of WOLF and SloWNet and precision of BCS synsets according to the source used for generation. Figures in italics are to be considered carefully, given the low number of (literal, synset id) pairs.

Source	Present in GS	Absent in GS but correct	**Source prec.**	Closely related	Wrong
WOLF					
Fr-Cz-En	61.7%	13.8%	**75.5%**	10.9%	13.6%
Fr-Cz-Bg-En	67.8%	12.4%	**80.1%**	9.2%	10.7%
Fr-Ro-En	64.7%	15.4%	**80.1%**	8.1%	11.8%
Fr-Ro-Cz-En	70.6%	13.3%	**84.0%**	8.4%	7.6%
Fr-Ro-Cz-Bg-En	75.5%	13.2%	**88.7%**	6.8%	4.5%
Wikipedia	94.0%	4.1%	**98.1%**	0.8%	1.1%
Fr Wiktionary	69.8%	12.2%	**82.0%**	10.7%	7.2%
En Wiktionary	88.5%	6.5%	**95.0%**	4.0%	1.1%
Wikispecies	*90.5%*	—	***90.5%***	—	*9.1%*
EUROVOC descr.	67.6%	*8.1%*	***75.7%***	*16.2%*	*8.1%*
SloWNet					
Sl-Cz-En	53.4%	7.6%	**61.0%**	4.7%	34.3%
Sl-Cz-Bg-En	59.3%	6.8%	**66.1%**	4.2%	29.7%
Sl-Ro-Cz-En	57.7%	7.5%	**65.2%**	3.8%	31.0%
Sl-Ro-Cz-Bg-En	61.0%	7.3%	**68.4%**	4.0%	27.6%

5 Conclusion

This paper has presented the two new lexico-semantic resources (wordnets) that were created automatically and are freely available for reuse and extension. The results obtained show that the approach taken is promising and should be exploited further as it yields a network of wide-coverage and quite reliable synsets[3] that can be used in many HLT applications. Some issues are still outstanding, however, such as the gaps in the hirerarchy and word sense errors.

Manual revision of the work is required for better performance of the resources in a real life setting and is being carried out. Both wordnets could be further extended by mapping polysemous Wikipedia entries to PWN with a WSD approach similar to [15]. Next, lexicosyntactic patterns could be used to extract semantically related words from either the corpus [16] or Wikipedia [17]. Moreover, Wiktionaries start handling polysemy to some extent, including by differentiating translations according to senses defined by short gloses.

References

1. Vossen, P. (ed.): EuroWordNet: a multilingual database with lexical semantic networks for European Languages. Kluwer, Dordrecht (1999)
2. Tufiş, D.: Balkanet design and development of a multilingual balkan wordnet. Romanian Journal of Information Science and Technology 7 (2000)

[3] We plan to assign to them confidence levels according to source-by-source evaluation.

3. Farreres, X., Rigau, G., Rodriguez, H.: Using WordNet for building WordNets. In: Proceedings of COLING-ACL Workshop on Usage of WordNet in Natural Language Processing Systems, Montreal, Canada (1998)
4. Barbu, E., Mititelu, V.B.: Automatic building of Wordnets. In: Proceedings of RANLP 2005, Borovets, Bulgaria (2006)
5. Jacquin, C., Desmontils, E., Monceaux, L.: French EuroWordNet, Lexical Database Improvements. In: Gelbukh, A. (ed.) CICLing 2007. LNCS, vol. 4394. Springer, Heidelberg (2007)
6. Pianta, E., Bentivogli, L., Girardi, C.: Multiwordnet: developing an aligned multilingual database. In: Proc. of the^{1st} Global WordNet Conf., Mysore, India (2002)
7. Resnik, P., Yarowsky, D.: A perspective on word sense disambiguation methods and their evaluation. In: ACL SIGLEX Workshop Tagging Text with Lexical Semantics: Why, What, and How?, Washington, DC, United States (1997)
8. Ide, N., Erjavec, T., Tufiş, D.: Sense discrimination with parallel corpora. In: Proc. of ACL 2002 Workshop on Word Sense Disambiguation (2002)
9. Diab, M.: The feasibility of bootstrapping an Arabic Wordnet leveraging parallel corpora and an English Wordnet. In: Proc. of the Arabic Language Technologies and Resources (2004)
10. Fišer, D.: Leveraging parallel corpora and existing wordnets for automatic construction of the Slovene Wordnet. In: Proc. of L&TC 2007, Poznań, Poland (2007)
11. Fišer, D., Sagot, B.: Proc. of Ontolex 2008. In: Building a free French wordnet from multilingual resources (to appear, 2008)
12. Ralf, S., Pouliquen, B., Widiger, A., Ignat, C., Erjavec, T., Tufiş, D., Varga, D.: The JRC Acquis: A multilingual aligned parallel corpus with 20+ languages. In: Proc. of LREC 2006 (2006)
13. Tiedemann, J.: Combining clues for word alignment. In: Proc. of EACL 2003, Budapest, Hungary (2003)
14. Erjavec, T., Fišer, D.: Building Slovene WordNet. In: Proc. of LREC 2006, Genoa, Italy (2006)
15. Ruiz-Casado, M., Alfonseca, E., Castells, P.: Automatic assignment of wikipedia encyclopedic entries to wordnet synsets. In: Szczepaniak, P.S., Kacprzyk, J., Niewiadomski, A. (eds.) AWIC 2005. LNCS (LNAI), vol. 3528. Springer, Heidelberg (2005)
16. Hearst, M.A.: Automatic acquisition of hyponyms from large text corpora. In: Proc. of COLING 1992, Nantes, France (1992)
17. Ruiz-Casado, M., Alfonseca, E., Castells, P.: Automatic extraction of semantic relationships for wordnet by means of pattern learning from wikipedia. In: Montoyo, A., Muńoz, R., Métais, E. (eds.) NLDB 2005. LNCS, vol. 3513. Springer, Heidelberg (2005)

Active Tags for Semantic Analysis

Ivan Habernal and Miloslav Konopík

University of West Bohemia, Department of Computer Sciences,
Univerzitní 22, CZ-306 14 Plzeň, Czech Republic
{habernal,konopik}@kiv.zcu.cz

Abstract. We propose a new method for semantic analysis. The method is based on handwritten context-free grammars enriched with semantic tags. Associating the rules of a context-free grammar with semantic tags is beneficial; however, after parsing the tags are spread across the parse tree and it is usually hard to extract the complete semantic information from it. Thus, we developed an easy-to-use and yet very powerful mechanism for tag propagation. The mechanism allows the semantic information to be easily extracted from the parse tree. The propagation mechanism is based on an idea to add propagation instruction to the semantic tags. The tags with such instructions are called *active tags* in this article. Using the proposed method we developed a useful tool for semantic parsing that we offer for free on our internet pages.

1 Introduction

Semantic analysis is a very important part of natural language processing. Its purpose is to extract and describe the meaning that is contained in a sentence. The ability to understand a sentence or an utterance is surely essential for many natural language processing tasks.

In this article we describe a method that is based upon context-free grammars (CFG) enriched with semantic tags. In this approach the grammar describes the structure of analyzed sentences and the semantic tags describe the semantic information.

The process of analyzing sentences using a grammar is called *parsing* and a result of parsing is called a *parse tree*. The parse tree describes the structure of a sentence. If we use a grammar enriched with semantic tags, the semantic information is spread across the parse tree in the semantic tags. There are several ways of extracting semantic information from the parse tree. Some of the common ways as well as the newly proposed approach will be presented in this article.

The approach described in this article relies on handwritten grammars and the semantic tags. The necessity to create the grammar rules and the tags manually could be perceived as a drawback when compared to new approaches that use stochastic algorithms for automatic learning of semantic analysis rules (e.g. [7]). However, some experiments show (e.g. [2]) that stochastic algorithms are very good at identifying wide variety of sentence types but their weakness lies in capturing complex sentence phrases such as date/time expressions or spoken numbers expressions, residence address phrases, etc. A stochastic algorithm is usually not capable to learn such complex relations. On the contrary we show that the handwritten grammars are capable of capturing these expressions

P. Sojka et al. (Eds.): TSD 2008, LNAI 5246, pp. 69–76, 2008.
© Springer-Verlag Berlin Heidelberg 2008

easily. Moreover our method is able to formalize the semantics thus it is not complicated to produce a uniform representation of the semantics for several structurally different but semantically uniform phrases.

The idea to propagate the information from bottom to top while parsing is not new and was used in many CFG-based approaches (e.g. [1]). However, in such approaches the grammar designer is forced to use a fixed syntax or ECMA scripts [4], etc. Such requirement could be useful for some tasks but generally it is limiting because the given syntax could be either too complicated or not expressive enough. The approach described in this article is designed to be general. The grammar designer is allowed to use a script or to create structured objects or any other formalism suitable for the particular application.

The rest of this article is organized as follows. At first we describe an approach of semantic analysis with semantic tags that do not use the proposed *active tags*. We show the drawback of this approach. Then we introduce and explain our approach with active tags. Afterwards we show some examples of parsing using this method. Finally we summarize the proposed method and give links where the software implementation of this method can be obtained.

2 Parsing with Semantic Tags

Formal grammars are a useful formalism for analyzing sentences. The idea to use formal grammars for semantic analysis was first introduced by Burton [3] in formalism called *semantic grammars*. This formalism freely mixes syntax and semantics in individual grammar rules. Semantic grammars hence combine aspects of syntax and semantics in a simple uniform framework. After parsing an input sentence with respect to a given semantic grammar, the semantics is stored in a parse tree. However, the parse tree contains all words from the sentence and many of them do not carry any semantic information.

The semantic tags help to solve two problems. Firstly, semantic tags help to express which words in the parse tree carry relevant semantic content and which words serve purely syntactic needs in a sentence. The semantically relevant words are associated

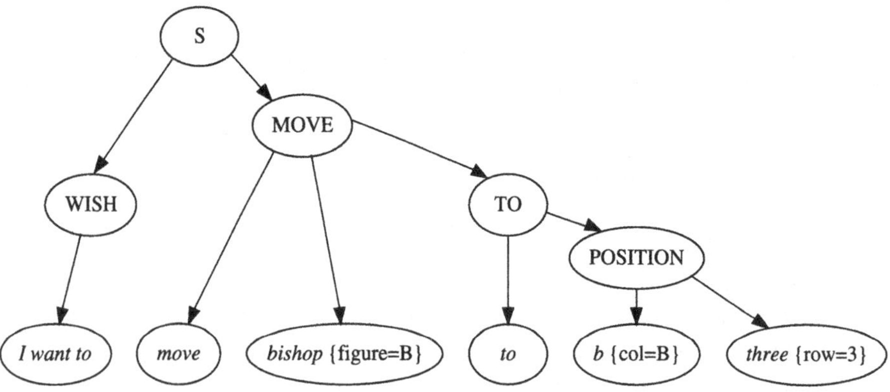

Fig. 1. Example of a parse tree with semantic tags

with the semantic tags that represent their semantics and the words that are not semantically relevant are left without the semantic tags. Secondly, they are used to formalize the semantic information contained in the words of a sentence. E.g. the word "twenty" is formalized as "20".

Figure 1 shows a parse tree for a sentence from a voice driven chess game. The semantic tags are placed in curly brackets and clearly describe the meaning contained in the sentence. The output semantic tags are `figure=B,col=B,row=3`.

The example in Figure 1 also shows a drawback of this approach. Just imagine a sentence "I want to go from b three to d three". Now the obtained set of tags will be `col=B, row=3, col=d, row=3`. In such a case it is difficult to distinguish what is the source position and what is the destination position. The next chapter not only shows how it is possible to elegantly solve this problem but it also shows that the semantic tags can deal with much more complicated problems.

3 Active Tags for Semantic Parsing

The semantic information in parse trees is spread across the tree in the semantic tags and the information need to be extracted. It is possible to go through the tree and store all the tags into a vector. Then it is easy to read the semantic information but the structural information is lost (see the chess example in previous section and the problem of the source and destination positions). Other possible approach is to leave the tags in the parse tree and extract the semantic information by a dedicated algorithm. In that way the structural information is not lost but this approach requires to write a special program code for each grammar. To avoid all these complications we developed a formalism of the so-called *active tags*. An active tag is a semantic tag enriched with a special processing instruction that controls the process of merging the pieces of semantic information in the parse tree. When the active tags are used and evaluated, the semantics is extracted from the tree in a form of one resulting tag that contains the complete semantic information.

The semantic information from the tree is joined in the following way. Each superior tag is responsible for joining the information from the tags that are placed directly below the superior tag. By a recursive evaluation of the active semantic tags, the information is propagated in a bottom-up manner in the tree.

An active tag contains at least one reference to a sub-tag. During evaluation the reference to a sub-tag is replaced with a value of the sub-tag. The reference has the following notion: #number (e.g. #2). The sub-tags are automatically numbered in the same order as stated in the grammar. Then the number in the sub-tag reference says which tag with a given number will be used to replace the reference.

The core of our method is based on replacing the tag references with actual values of the sub-tags. Despite how simple this principle may seem it is capable of creating very complicated semantic constructs.

For example see Figure 2. It shows how simple it is to solve the chess problem from Section 2. The sub-tags (`B` and `3`) of the node `POSITION` are automatically numbered and the value of the active tag in the node `POSITION` is constructed by replacing the references `#0` and `#1` with the values `B` and `3` so the string `B3` is created. Similarly

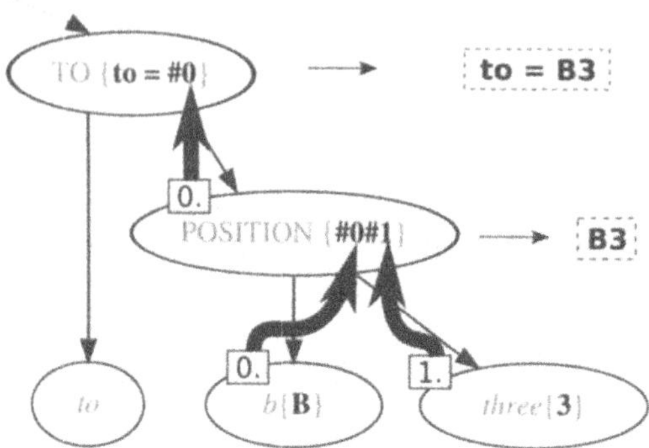

Fig. 2. A parse tree for a chess game command

the reference `#0` in the node `TO` is replaced with the first sub-tag value and the tag `from=B3` is created. And so on.

In this way one can create scripts in the Java language, Perl scripts, ECMA scripts, attribute-value pairs and arbitrary other scripts or values. The interpretation of the resulting tag is separated from the creation of the result tag.

3.1 Spoken Numbers Analysis Example

This section shows an comprehensive example that demonstrates the processing of spoken number phrases. We reduced the grammar to numbers from 0 to 999,999 because of space limitation.

The semantic grammar with active tags is defined as follows:

```
#JSGF V1.0 UTF-8;


<S> = <thousands> {#0} | <hundreds> {#0} | <one_to_hundred> {#0};

<thousands> = (<thousand> <hundreds>) {#0+#1} | <thousand> {#0} |
              (<thousand> and <one_to_hundred>) {#0+#1};
<thousand> =  (<numeral> thousand) {#0*1000} |
              (<ten_to_thousand> thousand) {(#0)*1000};
<ten_to_thousand> = (<tens> <numeral>) {#0*10+#1} | <tens> {#0*10} |
   <teen> {#0} | <hundred> {#0} | (<hundred> <one_to_hundred>) {#0+#1};

<hundreds> = <hundred> {#0} | (<hundred> and <one_to_hundred>) {#0+#1};
<hundred> = <numeral> {#0 * 100} hundred;
<one_to_hundred> = (<tens> <numeral>) {#0*10+#1} | <tens> {#0*10} |
                   <teen> {#0} | <numeral> {#0};

<tens> = twenty {2} | thirty {3} | forty {4} | fifty {5} |
         sixty {6} | seventy {7} | eighty {8} | ninety {9};
```

```
<teen> = ten {10} | eleven {11} | twelve {12} | thirteen {13} |
       fourteen {14} | fifteen {15} | sixteen {16} |
       seventeen {17} | eighteen {18} | nineteen {19};

<numeral> = one {1} | two {2} | three {3} | four {4} | five {5} |
          six {6} | seven {7} | eight {8} | nine {9};
```

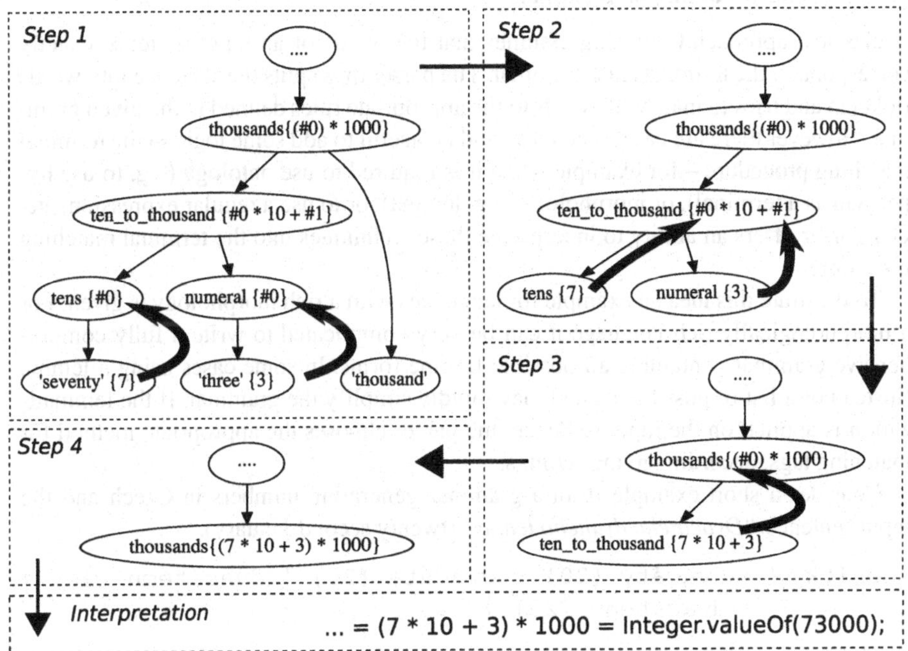

Fig. 3. An example of tag processing for spoken numbers domain

The processing of the phrase "seventy three thousand" is depicted in Figure 3. During the tags extraction phase (steps 1 to 4) the resulting tag "$(7 * 10 + 3) * 1000$" is created. Then it is simply interpreted (in our case with the BeanShell [5]) and the number 73,000 is generated.

3.2 Formal Definition of a Grammar with Active Tags

Context-Free Grammar enriched with active semantic tags is a 6-tuple $G = (N, \Sigma, \mathcal{T}, \mathcal{V}, \mathcal{R}, S)$, where

- N is a set of non-terminal symbols (or "variables"). The non-terminals are enclosed in angle brackets (e.g. $\langle$nonterminal1$\rangle$),
- Σ is a set of terminal symbols,
- $\mathcal{T}$ is a set of semantic tags. The tags are strings enclosed in curly brackets e.g. {tag1},

- $\mathcal{V}$ is a set of active semantic tags. The active tags are like nonactive tags enclosed in curly brackets but contain at least one reference denoted by # and a number e.g. {#1 tag1 #2};
- $\mathcal{R}$ is a set of production rules, each of the from $A \rightarrow X_1Y_1\ X_2Y_2\ \ldots\ X_nY_n$, where $A \in N$, $X_i \in (N \cup \Sigma)$, $Y_i \in (\mathcal{T} \cup \mathcal{V} \cup \emptyset)$,
- $S \in N$ is the start symbol.

3.3 Advanced Parsing of Terminals

A classical approach to parsing assumes that the words of an input sentence exactly correspond to the terminals in a grammar. The parser thus splits the sentence into words (tokens) and tries to match the words to the appropriate rules defined in the given grammar. However there are cases when it would be useful to add some logic to the terminal matching procedure – for example when it is required to use ontology (e.g. to use hypernym as a terminal), or morphology (e.g. lemma), or to use a regular expression, etc. Our parser offers an ability to incorporate these techniques into the terminal matching procedure.

To illustrate this idea an example for languages with a rich morphology is given. For a morphologically rich language it can be very complicated to write a fully comprehensive grammar containing all possible lexeme forms. In some cases using a lemma (instead of a list of possible forms) may rapidly simplify the grammar. If the lemmatization is applied on the input sentence, the parser chooses the appropriate method for matching tags, according to the settings.

Consider a short example from a grammar generating numbers in Czech and the input sentence *"Dvacátého druhého ledna"* (twenty second January).

```
<desitky> = dvacátý {20} | dvacátá {20} | dvacátému {20} |
            dvacátého {20} | ...
```

This fraction of the grammar shows some word forms of the word "dvacet" (dvacet means twenty in Czech). In the case of this grammar a standard parser chooses the rule generating the first word ("dvacátého") properly. The main disadvantage is that there must be an exhausting list of words covering the particular lexeme for each morphological form. Using the alternative parsing approach, we can rewrite the grammar as follows: `<desitky> = @L_dvacátý {20}`.

The `@L_` prefix means that the parser will accept the lemma. Once the input sentence is preprocessed by lemmatization, the parser obtains the following input *"Dvacátého; dvacátý druhého;druhý ledna;leden. . . "*, where the second form is a lemma.

The framework also provides matching by regular expressions which can be useful for e.g. parsing text containing symbols. For instance this approach allows parsing dates in written form (11th January 2008), where the rule generating/accepting the day of month would be `<day-of-month> = @R_[1-3]?\d\th`.

4 Practical Applications

We have developed a framework for semantic analysis with active tags in the Java programming language. The grammars are in the JSGF format [6] and the software uses our own partial implementation of JSAPI[1]. The final script obtained from the parse tree evaluation is independent on a programming language; the framework has default and native support for scripting in Java language (using BeanShell [5] – a dynamic scripting language for Java platform).

The framework has been tested in various domains. The proposed method is used as a semantic parsing algorithm in a voice-driven chess game [8]. We displayed this application in many demonstrations where we encountered many variants of how to tell a chess application what move it is supposed to do. The proposed method allowed us to quickly incorporate new variants into the semantic analysis without changing the program code of the application. Another advantage is that the algorithm provides us with consistent semantic analysis. A uniform semantic representation is created for all variants with the same meaning but possibly with completely different grammar rules.

We also develop a semantic analysis algorithm for spoken queries to an internet search engine [2]. In this application the domain is much broader and it is impossible to manually write a grammar that would cover all possible questions. Thus we use a stochastic algorithm as a main semantic analysis algorithm. However, we found that we need to perform a local parsing to identify complex phrases such as date/time expressions, spoken numbers, and others. After parsing the input with the proposed semantic parser the main stochastic parser works with these phrases as atoms without the necessity to deal with them. In this task, the proposed method proved to be a very good method of local parsing.

5 Conclusion

In this article we presented a simple and easy-to-use framework for semantic information extraction. It is based on context-free grammars and a mechanism of active tags.

The main difference between our approach and some similar approaches (e.g. [1]) is that our mechanism of propagation of semantic tags is not connected with any script. This fact simplifies the formalism and also it makes the formalism more universal because the interpretation of tag values is separated from the tag propagation mechanism. It is thus possible to use arbitrary interpretation method. We also deal with the problem of parsing morphologically rich languages and the problem of parsing inputs with symbols.

We found in our experiments that this approach is extremely helpful in building semantic analysis applications. We believe that this approach can be used in many other applications that require natural language understanding. Thus we decided to release the software implementation of the proposed semantic analysis method at our internet pages. The developed software is not burdened by any licence restriction; we provide it for free under LGPL licence.

Download

The software implementation can be downloaded from `https://liks.fav.zcu.cz/mediawiki/index.php/LINGVOParser`.

[1] Java Speech API specifies an interface for speech processing on the Java platform.

Acknowledgements

This work was supported by grant no. 2C06009 Cot-Sewing.

References

1. Hunt, A., Lucas, B., Walker, W.: Ecmascript Action Tags for JSGF (proposal) (September 1999), http://x-language.tripod.com/ECMAScriptActionTagsforJSGF.html
2. Konopík, M., Mouček, R.: Towards Semantic Analysis of Spoken Queries. In: SPECOM 2007 proceedings, Moscow (2007) ISBN 6-7452-0110-X
3. Brown, J.S., Burton, R.R.: Multiple representations of knowledge for tutorial reasoning. In: Representation and Understanding, New York, pp. 311–349 (1975)
4. ECMA. ECMAScript language specification. ISO/IEC 16262, http://www.ecma.ch/ecma1/STAND/ecma-262.htm
5. Niemeyer, P.: BeanShell 2.0 Documentation, http://www.beanshell.org/
6. Sun Microsystems, Inc. Java Speech Grammar Format specification (1998), java.sun.com/products/java-media/speech/forDevelopers/JSGF/
7. He, Y., Young, S.: Semantic processing using the Hidden Vector State model. Computer Speech and Language 19(1), 85–106 (2005)
8. Hošna, M., Konopík, M.: Voice Controlled Chess usable also for Blind People. In: SPECOM 2007 Proceedings, pp. 725–728. Moscow State Linquistic University, Moskva (2007) ISBN 6-7452-0110-X

Spoken Requests for Tourist Information

Speech Acts Annotation

Laura Hasler

University of Wolverhampton, Wulfruna St., Wolverhampton, WV1 1LY, UK
L.Hasler@wlv.ac.uk

Abstract. This paper presents an ongoing corpus annotation of speech acts in the domain of tourism, which falls within a wider project on multimodal question answering. An annotation scheme and set of guidelines are developed to mark information about parts of spoken utterances which require a response, distinguishing them from parts of utterances which do not. The corpus used for annotation consists of transcriptions of single-speaker utterances aimed at obtaining tourist information. Interannotator agreement is computed between two annotators to assess the reliability of the guidelines used to facilitate their task.

1 Introduction

When designing question answering (QA) systems which can take speech as input, it is necessary to distinguish questions or requests requiring a response from other information within utterances. Such information often serves to contextualise the request rather than requiring a response itself. Phenomena such as repetitions and repairs also need to be dealt with. One way of distinguishing the parts of utterances which require a response from the surrounding context is to annotate utterances with speech acts (SAs) information. Speech acts theories allow labels such as *request*, *question*, and *state* to be assigned to specific parts of utterances. If it is possible to reliably assign such labels to a corpus of transcribed spoken utterances, this information can be used to train and test QA systems. This paper reports on an ongoing annotation of SAs in the domain of tourism; part of QALL-ME,[1] a wider project concerned with multimodal and multilingual QA. An annotation scheme and guidelines are developed, and applied by two annotators to a corpus of transcriptions of single-speaker utterances aimed at obtaining tourist information. Section 2 briefly describes speech acts and related work. Our corpus is introduced in Section 3. Section 4 details the annotation scheme, and Sections 5 and 6 present the guidelines. Interannotator agreement is discussed in Section 7.

2 Speech Acts and Related Annotation Initiatives

The concept of *speech acts* was introduced with Austin's [1] and Searle's [8,9] investigations into the notion that actions are performed by language, claiming that rather than just making statements about the world, speakers actually *do* things with their words.

[1] `http://qallme.fbk.eu`

P. Sojka et al. (Eds.): TSD 2008, LNAI 5246, pp. 77–84, 2008.
© Springer-Verlag Berlin Heidelberg 2008

Searle developed the idea of *indirect* SAs, where one SA is performed via another, e.g., an utterance functioning as a statement on the surface can have an underlying function of requesting in a given context. More recently, related theories of *dialogue* or *conversation acts* have developed and several annotation projects, such as VERBMOBIL [14], HCRC Map Task [4], TRAINS [13], DAMSL [3], and SPAAC [11], have been undertaken.[2] These focus on a different kind of data to ours: dialogues rather than single-speaker utterances, aimed at completing particular tasks such as appointment scheduling, linking locations on a map, and booking train journeys and tickets. Their schemes cannot be carried over directly to our data because they annotate information relating to conversation management and structure, common ground establishment, updates, and backward- and forward-looking acts. However, because they annotate task-based corpora, similar to our corpus, general elements such as the labelling of requests and assertions can be borrowed from them. These need to be adapted because some tags are too specific, and some common labels such as *thank-bye* and *request-info* are not particularly suited to our data.

3 Corpus for Annotation

The SAs annotation discussed below is applied to an English language corpus collected as part of the QALL-ME project on multimodal and multilingual QA. The corpus consists of 4,501 single-speaker utterances (62,626 words) in total; 2,286 *spontaneous* utterances and 2,215 *read* utterances. *Spontaneous* utterances are produced by speakers using scenarios which prompt them to ask for certain types of information, such as the time a particular film is showing. *Read* utterances, in contrast, are predefined utterances read out by speakers. The data were produced by 113 different speakers: 88 native and 21 non-native speakers of English, and 63 female and 46 male speakers.[3] The recorded utterances were transcribed according to guidelines describing how to deal with speech phenomena, such as hesitations, repetitions and repairs. Square brackets are used to denote sound which is not part of the spoken utterance being recorded (e.g. [uhm], [laugh], [noise], [voice]), unintelligible speech([???]), and mispronunciations (e.g. hotel+[pron=rotel]). The annotation is ongoing and to date, 700 randomly selected spontaneous utterances (11,106 words) have been annotated for SAs, including an overlap to calculate interannotator agreement (Section 7).

4 Annotation Scheme

The annotation scheme marks data unsuitable for further SAs annotation using the tags <trash>, <interrupted> and <nonsense>. These tags allow the separation of "useful" units of text from those which are unsuitable for further text processing in the context of QA. *Utterances* which are suitable for SAs annotation are labelled with the <utterance> tag, and then *C-units* [2][4] within them are marked using <c-unit>. The

[2] [5] and [12] provide overviews/comparisons of various theories and projects.

[3] This information is not available for 4 speakers in the corpus.

[4] C-units are grammatical units in spoken language.

<c-unit> tag has the attributes <clausal> and <non-clausal>; <non-clausal> is further split into <pragmatic> and <specify-info>. Units within utterances which are not suitable for further processing due to their form are indicated using the <fragment> tag. *Primary speech acts* in the utterance are labelled using the <primary_speech_act> tag. The primary speech act is the surface SA, which relies on the form of the unit. Attributes of <primary_speech_act>, based on a manual analysis of our corpus, are <request>, <question>, <state>, <introduce>, <end>. If primary SAs take as an attribute <request>, <question> or <state>, the *secondary speech act* is also marked on the same span of text. This is the underlying SA function of the unit; the "real" function in a particular context. <secondary_speech_act> has the attributes <request>, <question>, <state>. The primary and secondary SAs work together to allow the <speech_act_type> tag to be assigned as a relation between them. *Speech act type* takes the attribute <direct> when primary and secondary SAs are the same, and <indirect> when they are different. The multipurpose annotation tool PALinkA [7] is used for annotation.

5 Annotation Guidelines: Utterance-Level

5.1 Unsuitable Utterances

It is not possible to annotate SAs in some cases because there is too much information missing from the beginning or end of the utterance, so that it impedes easy understanding. This is usually due to the speaker speaking before the recording started. Generally, if the main clause verb is missing, utterances are not suitable for further speech acts annotation. However, each utterance should be considered carefully as the presence of the main clause verb alone is not always a cut-and-dried basis for distinction. The tag <interrupted> is used and the utterances are not annotated further. For example:

-otel near Birmingham University that has a phone [noise]

there a beach near [voice]

There are sometimes "nonsense" utterances in our data. "Nonsense" utterances include those which do not make sense as a discrete unit because they do not contain all the information necessary for their full understanding. They also include utterances which do not make sense because the speaker does not continue their line of thought throughout their speech, either grammatically or content-wise, coherently enough to allow full understanding. The tag <nonsense> is assigned in these cases, and no further annotation is necessary. For example:

what is the what is the duration of the movie

would like to know the name of a cinema for twenty fifth of February which contains moderate sex references in a movie which I'd like to see

There are also cases where it is not possible to annotate SAs because recordings do not constitute utterances due to a lack of relevant speech (e.g. — *[???]*, *[voice]*). The tag <trash> is used, and the utterances are not annotated further.

5.2 Suitable Utterances

Where there is only potentially one word missing from the beginning of an utterance, it is usually clear what this word could be. Even if there are different interpretations available, the meaning of the utterance is still roughly the same and the SA label does not change depending on that word. The main clause verb (or part of it) should also generally be present in utterances which are suitable for further annotation of speech acts. Again, each incomplete utterance should be carefully considered as this is not always the case. The tag <utterance> is used to label suitable utterances. Examples of acceptable levels of incompleteness:

you give me the number for the pharmacy on Clark Road please

there any Leonardo DiCaprio films on on Wednesda-

Typical speech phenomena such as hesitations, repetitions and repairs appear within complete utterances which are suitable for SAs annotation. Such units are marked with the tag <fragment>, which allows them to be retained in the complete utterance but easily identified if unnecessary in future text processing. Units are only considered as fragments if they are completely abandoned, repaired or repeated within the utterance; interruptions at the beginning or end of the utterance (see Section 5.1) should not be marked in this way. For example:

{where is the} where is the Alexandra Theatre Birmingham

{[noise] to find a s-} I'm trying to find a subtitled showing of vampire diary

Our data contains speech transcribed within [square brackets] (see Section 3), which should be included in marked units with no additional annotation.

5.3 C-Units within Suitable Utterances

Before SAs are annotated, C-units need to be indicated within suitable utterances using the <c-unit> tag. A C-unit [2] is a syntactically independent clausal (with all its dependent units) or non-clausal unit found in spoken language. *Clausal* C-units are labelled with the attribute <clausal>, and *non-clausal* C-units with <non-clausal>. Only the largest units are identified; the concern is not generally with embedded C-units. The exception to this is where non-clausal C-units appear within a larger clausal C-unit. Adopting the notion of C-unit allows, for example, NPs specifying information within or after clausal units to be indicated as grammatical elements of the utterance. These elements are not fragments because they are necessary pieces of information in the utterance rather than units which are abandoned or repeated. Conjunctions linking C-units should be marked as part of the second linked unit. For consistency, square bracketed text from the transcription (see Section 3) should be treated in the same way. Discourse or pragmatic items (e.g. *OK*, *hello*, *please*, etc.) should be marked as separate C-units. The function of non-clausal C-units should be indicated by assigning either <pragmatic> or <specify-info>. Examples:

{could you give me details of The New Art Gallery in Walsall} {and does it have a car-park}

{would you be able to tell me {the bus 5 4 3} the start hours for the bus}

6 Annotation Guidelines: Speech Acts

6.1 Speech Act Text Spans

A speech act often, but not always, corresponds to a C-unit within the utterance or to the utterance itself. SAs should not include conjunctions which join C-units. Non-clausal

C-units which function to specify necessary information are not considered a speech act on their own, but as part of a larger SA in the utterance; their "specifying" status is indicated in the appropriate C-unit tag. There can be problems when it is unclear to which larger SA a non-clausal C-unit belongs. In the example below, it is not possible to reliably decide to which questioning SA the specifying non-clausal C-unit *Quality Hotel Wolverhampton* belongs. Such cases should be not be annotated for speech acts, but commented on as problematic, with any suggestions for potential annotation given.

{does it allow pets} {Quality Hotel Wolverhampton} {does it allow pets}

6.2 Speech Acts in the Corpus

Five different categories of speech acts were identified in the corpus. *Requesting* occurs when the speaker's utterance has the function of attempting to obtain a specific piece of information. *Wh-questions* and *how-questions* are therefore considered as requesting rather than questioning because they seek to obtain information rather than clarification. These SAs are labelled <request>:

who is the director of because I said so

give me the name of a nearby Thai restaurant

Questioning is distinct from requesting because at the surface level, these SAs deal with questions about the ability to give information rather than requesting that information as such, or questions which seek clarification (yes/no). Questioning in this annotation is therefore limited to interrogative structures which expect a yes/no answer. Questioning SAs are labelled <question>. For example:

can I have the phone number for a hotel in Walsall

does the Light House have wheelchair access

Stating occurs when the speaker makes a statement about the world. This can be either an indirect speech act (see Section 6.3) in its own right, or it can contextualise the request which appears elsewhere in the utterance. When this happens, the request itself often relies on the contextualising statement for full understanding. The second example below demonstrates both stating functions in the same utterance. Stating SAs are labelled <state>. For example:

{I'd like to visit the Bank} could you tell me where that is

{there's an event taking place because I said so} {I'd like to know which movie director's involved with that}

Introducing is used to signal a part of an utterance which appears before the request so that it does not start abruptly. In other dialogue annotation projects (e.g. DAMSL, SPAAC), our introducing SAs cover discourse markers which contribute to conversation management, such as *OK*, *right*, *yes*. It is interesting that such items occur in our single-speaker data and therefore they need to be assigned an appropriate label. Introducing also covers *greeting* items such as *hey*, *hello*, *hi*. These SAs are labelled <introduce>. For example:

{yes} I'd like accommodation please in Dudley with possibly a gym

{hi} can I take my pet to the Novotel in Wolverhampton at all

Ending is assigned to items which appear after the request, and functions to signal the end of an utterance, perhaps so that it does not finish abruptly. These SAs are usually labelled *thanking* or *thank-bye* elsewhere, and indeed, in our data they are represented by lexical items such as *thanks*. However, similar to introducing items, they are not considered to strictly perform this type of SA because our data do not consist of dialogues. Ending SAs are labelled <end>:

[uhm] can you please tell me if the Novotel in Wolverhampton has room facilities for disabled access {thank you}

like to see a movie on a cinema in Birmingham [uh] which duration is about 2 hours and [uhm] I prefer if it has moderate violence {thanks}

Each utterance does not always contain only one speech act. Generally, there is only one <introduce> or <end> per utterance; it is much more likely to have more than one <request> or <question>. An utterance expresses one request or question when it comprises one clausal C-unit with this SA function, and more than one when distinct clausal C-units are present. For example:

what's what's the name and address of the Thai restaurant near the Mailbox

{what is the name of a hotel with a conference room} and {what is the price}

6.3 Direct and Indirect Speech Acts

Speech act functions do not always match the literal meaning conveyed by the surface form of an utterance, hence a distinction between *direct* and *indirect* speech acts (see Section 2). Because the context of our data is simulating a search for tourist information, the majority of the utterances ultimately function to request information. This is the underlying, implicit, or secondary SA. However, the primary, or surface, SA can be and often is different. It depends on the surface structure or form of the utterance; particular forms typically correspond to particular functions (imperative: requesting/commanding, interrogative: questioning, declarative: stating/asserting). Where the primary and secondary SAs of an utterance differ, there is an *indirect* speech act, i.e., one SA is performed via another. Where they are the same, there is a *direct* speech act. For example:

give me the directions [uh] for a hotel that is near to the cinema and the theatre please (Typical function: to request/command (imperative); Primary SA: requesting; Secondary SA: requesting; SA type: direct)

I'm looking for a hotel that's in Dudley and that has self-catering available (Typical function: to state/assert (declarative); Primary SA: stating; Secondary SA: requesting; SA type: indirect)

There are cases where it is difficult to decide on the secondary SA for a span of text when the primary SA is questioning; it cannot always be reliably decided whether it is questioning or requesting. In the examples below, it could be either, depending on whether the speaker wants clarification (yes/no) or specific information (dates, times, price, etc.). It is impossible to know from the data alone. These cases are annotated as *question-question*, as it seems intuitively natural to pose a follow-up request to obtain more information as necessary. Examples:

is the puffy chair showing at Cineworld

do you have to pay to get into Reflex

7 Interannotator Agreement

There are currently 700 utterances in the corpus annotated with SAs information. To assess the reliability of the guidelines when used by different annotators, interannotator agreement was computed on a set of 300 utterances. Two annotators, both familiar with speech acts theories, used the guidelines (Sections 5 and 6) to annotate the same utterances. The Dice coefficient (see [6]) was first used to establish the level of agreement between spans of text marked as C-units, primary SAs and secondary SAs by both annotators. This is necessary because interannotator agreement can only be measured for labels attached to the same spans of text. The results of the consistency of text span marking between the two annotators are: C-units: 84.58%; primary SAs: 58.96%; secondary SAs: 64.31%.

The Kappa statistic (K) [10] was then used to calculate interannotator agreement on the attributes assigned to C-units and primary and secondary SAs within those text spans agreed upon by annotators. Although there is some discussion regarding the values which should be used to indicate reliability of annotations, it is generally held that $K>0.8$ indicates good reliability. For those C-units whose span is agreed upon by both annotators, K was calculated for the distinction between the classes <clausal> and <non-clausal>. There are 395 agreed C-unit spans and K is 0.98. For agreed non-clausal C-units when distinguishing the classes <pragmatic> and <specify-info>, K is 1. For primary speech acts with agreed span, agreement over the 5 classes <introduce>, <request>, <question>, <state>, <end> is measured. There are 245 agreed primary SA spans, and K is 0.97. For those secondary SAs with agreed span, interannotator agreement is calculated over the 3 classes <request>, <question>, <state>. There are 228 agreed secondary SA spans, and K is 0.84.

The results of the interannotator agreement calculations show that the level of agreement on text spans for primary and secondary SAs is not particularly high. This indicates that the parts of the guidelines describing SA text spans should be improved in an attempt to increase the reliability of annotation applied by different annotators. A more detailed explanation of the difference between C-unit and primary speech act spans will be given, with more examples taken from the corpus to illustrate this. However, agreement on the labelling of speech acts themselves on those spans that are agreed on is high ($K>0.8$), and therefore the guidelines used in the annotation process can be used reliably for this task. Agreement between annotators on attributes assigned to C-units is also high and therefore this annotation too can be reliably carried out using the guidelines.

8 Conclusions

This paper presented a scheme and set of guidelines developed in an ongoing annotation of speech acts in the QALL-ME corpus of transcribed spoken requests for tourist information. The scheme encodes information relating to various elements of utterances and SAs, as well as typical speech phenomena such as repetitions and repairs. The guidelines describe the aspects of the data to be annotated, and are given to facilitate annotation. Interannotator agreement was measured on 300 utterances marked by two annotators to assess the reliability of the guidelines. The results proved that they are a reliable resource to be used in the annotation of C-units and their attributes, and attributes assigned to primary and secondary SAs in our corpus. However, there is room for improvement regarding the description of marking primary and secondary SAs text spans. This will be addressed before any further annotation. For future work, the investigation of rhetorical relations between SAs in an utterance is planned.

Acknowledgements. This research is supported by the EU-funded QALL-ME project (FP6 IST-033860). The author would also like to thank Georgiana Puşcaşu for her help with computing the interannotator agreement.

References

1. Austin, J.L.: How to do things with words. In: Urmson, J.O., Sbisa, M. (eds.), 2nd edn. Clarendon Press, Oxford (1975)
2. Biber, D., Johansson, S., Leech, G., Conrad, S., Finnegan, E.: Longman Grammar of Spoken and Written English. Longman, Harlow (1999)
3. DAMSL, http://www.cs.rochester.edu/research/speech/damsl/RevisedManual/
4. HCRC Map Task, http://www.hcrc.ed.ac.uk/maptask/
5. Larsson, S.: Coding Schemas for Dialogue Moves. Technical report. Göteborg University, Department of Linguistics (1998)
6. Manning, C., Schutze, H.: Foundations of Statistical Natural Language Processing. MIT Press, Cambridge (1999)
7. Orăsan, C.: PALinkA: A highly customisable tool for discourse annotation. In: 4th SIGdial Workshop on Discourse and Dialogue, ACL 2003, pp. 39–43 (2003)
8. Searle, J.R.: Speech Acts. Cambridge University Press, Cambridge (1969)
9. Searle, J.R.: A classification of illocutionary acts. Language in Society 5, 1–23 (1976)
10. Siegel, S., Castellan, N.J.: Nonparametric Statistics for the Behavioural Sciences, 2nd edn. McGraw-Hill, New York (1988)
11. SPAAC, http://ucrel.lancs.ac.uk/projects.html#spaac
12. Stent, A.: Dialogue Systems as Conversational Partners: Applying conversation acts theory to natural language generation for task-oriented mixed-initiative spoken dialogue. Ph.D. Thesis, University of Rochester (2001)
13. TRAINS, http://www.cs.rochester.edu/research/cisd/projects/trains/
14. VERBMOBIL, http://verbmobil.dfki.de/

Affisix: Tool for Prefix Recognition*

Jaroslava Hlaváčová and Michal Hrušecký

Charles University in Prague, ÚFAL MFF, Czech Republic
hlavacova@ufal.mff.cuni.cz, michal.hrusecky@seznam.cz

Abstract. In the paper, we present a software tool Affisix for automatic recognition of prefixes. On the basis of an extensive list of words in a language, it determines the segments – candidates for prefixes. There are two methods implemented for the recognition – the entropy method and the squares method. We briefly describe the methods, propose their improvements and present the results of experiments with Czech.

1 Prefixes in Natural Language Processing

Prefixes are morphs. They are used for creating words "from the left". They have a meaning, but not always the same. It is not easy to recognize morphs automatically. It was the reason why we have adopted the following definition of prefix. It is rather technical, but expresses its linguistic sense quite well.

Definition 1. *A string of letters is a* ***prefix*** *if*

(i) it is beginning of many words,
(ii) it can be replaced with another string of letters (possibly empty) so that the result is meaningful.

There is no strict answer at a question, what is many words. We are not able to decide purely automatically, whether a string is a real linguistic prefix. We can only order the prefix candidates according to a measure. The higher the value of the measure, the greater chance that a string is a real prefix.

According to our definition, we consider not only traditional, pure prefixes, but also strings that can be viewed as components of another means of word creation – composition. However, these morphs behave as prefixes and it is reasonable to treat them so. In fact, it was the reason why we started our work – to discover these strings that are often added to the beginning of existing words to change their meaning.

Example 1. The string *euro* is prefix, as it stands at the beginning of many words (*euro-atlantický, euro-americký, euro-skeptik,...*) and it can be always either replaced by another string (*euro-krat* x *byro-, auto-, techno-, demo-krat*) or torn off (the rest of the list), yielding always a word of the language.

Prefixes are very productive in Czech, as well as in many other languages. Apart from a rather small closed set of well-known verbal prefixes, there are numbers of prefixes,

* This paper is a result of the projects supported by the grants of the Czech Academy of Sciences 1ET101120503 and 1ET101120413.

P. Sojka et al. (Eds.): TSD 2008, LNAI 5246, pp. 85–92, 2008.
© Springer-Verlag Berlin Heidelberg 2008

that modify mainly nouns, but also adjectives and adverbs (derived from adjectives). We call these prefixes "nominal". They are often used instead of adjectives – for instance *ministůl* means *malý stůl* (*minitable = small table*). These prefixes can be joined with almost every noun. They change the meaning of the nouns usually in the same way. This is difference from verbal prefixes, which can have different meanings (functions) for different base verbs. Another difference is that the set of nominal prefixes is not closed.

At the beginning of any task of NLP[1] there is a morphological analysis. It is based usually on a large morphological dictionary containing "all" words of a given language. However, in any corpus, there are words that cannot be recognized by the analysis, because they are not present in the morphological dictionary. It is natural, that computational linguists try to minimize number of unrecognized words [1]. One way can go through recognizing prefixes. If we had a complete list of prefixes, we could use it for recognition of prefixed words. Unfortunately, there is no such list of Czech prefixes. Thus, we prepared it from the data of the Czech National Corpus [2], using the Affisix tool.

The algorithm for recognition of prefixed words is then straightforward – if a beginning string of an unrecognized word is the same as a prefix, remove it and check the rest of the word. If it can be recognized by a morphological analysis, it is very probably that the whole word is a prefixed word and has the same morphological characteristics.

1.1 Notation

Let Ω denote the set of words existing in a language and let the word w be the concatenation of letters $a_1, \ldots, a_n$. We write: $w = a_1 :: \cdots :: a_n$[2].

Definition 2. ***Segmentation*** *of a word w is a division of the word into two strings p and s such that $w = p :: s$. Formally, any of the segments may be empty.*

1.2 Properties of Prefixes

Let us have a word w with the segmentation $w = p :: s$. If we find a lot of ending segments s_i such that $p :: s_i \in \Omega$ and only few beginning segments p_i such that $p_i :: s \in \Omega$ we can expect, that p is a prefix. This feature of the segmentation $p :: s$ means that p is the beginning of many words (condition (i) of the definition of prefixes) and the rest s of the word cannot have many different beginning segments – in that case, it would be a suffix or an ending.

1.3 Additional Constraints

One problem we faced was that in our results, there have appeared not only segmentations between a prefix and the rest of the word, but also between beginning of the words and a suffix or an ending. To exclude a great amount of inflectional endings, we work

[1] Natural Language Processing.

[2] We take this notation from [3].

with lemmas, not with word forms. Still, there are many common suffixes that can spoil the result with non-prefixes.

We can reduce this problem significantly by setting an additional restriction; for discovering prefixes we demand that the segmentation have to be in the first half of the word. We are aware that we can miss some prefixes (for example *counter::feit*), but the results become cleaner, as they are not spoilt by non prefixal segmentations. We can expect that in large amount of words we get the same prefix also with longer words, so in the end we would not miss it.

On the other hand, nearly any single letter satisfies our condition (i) (it can continue with many words) and short prefixes are usually already well known. So we can use additional constraint and require only the prefixes longer than one letter.

These additional constraints have a good side-efect: they reduce the number of segmentations for checking, which speeds up the processing the data.

2 Methods

The basis of our work was the article [3] where various methods were described and used for Spanish. Some of them have been already applied to recognition of Czech prefixes out of about 170,000 words [4]. The purpose of our work was to reimplement some of the methods in order to be user-friendly and reusable for other languages. Moreover, we have also added the method of difference entropy, that works better than the pure entropy.

Generally, all methods check every segmentation of every word of the input list and test if that segmentation satisfies requirements for prefixes. If so the beginning segment is considered a prefix.

2.1 Squares

One of the simplest methods used for prefix recognition is the method of squares.

Example 2. Take for example the word *disassemble* and let us discuss its segmentation after the third letter *dis::assemble*. To get a square, we need another two segments. One, which can replace the beginning segment (*dis-*), and another one for the ending segment (*-assemble*), so that the both beginning segments form words with the both ending segments. In our example, we can use *re-* as the alternate beginning segment and *-appear* as the alternate ending segment. We get the words *disassemble, reassemble, disappear* and *reappear.*

Definition 3. ***Square*** *is a quadruple of word segments* $\langle p_1, p_2, s_1, s_2\rangle$ *such that* $p_1 :: s_1 \in \Omega$, $p_2 :: s_1 \in \Omega$, $p_1 :: s_2 \in \Omega$ *and* $p_2 :: s_2 \in \Omega$.

Definition 4. *For each segmentation p::s of word w we define* ***number of squares*** *S(p) as* $S(p) = |\{\langle p, p_2, s, s_2\rangle; \langle p, p_2, s, s_2\rangle \text{ is a square}\}|$

We use the number of squares to decide whether the beginning segment is a prefix.

Possible Problems. The fact that some grammatical prefixes have common terminal letters, can cause complications. Take for example the following square ⟨*under*, *over*, *clock*, *estimate*⟩. The prefixes *under-* and *over-* have common ends. Thus, we get also other coincidental squares: ⟨*unde*, *ove*, *rclock*, *restimate*⟩ and ⟨*und*, *ov*, *erclock*, *erestimate*⟩. This can result in wrong prefixes *unde-*, *und-*, *ove-* and *ov-*.

There can be another problem with accidental squares. For example we have the square ⟨ *b-*, *m-*, *-ug*, *-all* ⟩ yielding the lemmas *bug, mug, mall, ball*, but the segments *b-*, *m-* do not play the role of prefixes in these words.

However, none of these problems seems to be relevant in case of large input data. At least in our case, we did not notice serious influence of this type of coincidences. However in case of small unrepresentative data, they can be more significant.

2.2 Entropy

This method uses the notion of entropy to decide whether a starting segment is a prefix or not. The basis of this approach consists in the fact, that a segmentation of a word just between individual morphs has considerably higher entropy than other segmentations of the word. We can use it for recognizing prefixes; the number of prefixes in a language is very small compared with number of roots that may follow. Thus, the entropy after the prefix (or generally any other morph) is much higher than the entropy in the middle of a morph. The entropy is calculated according to the formula:

$$H(p) = -\sum_{s_i \in S} p(s_i|p)\log_2 p(s_i|p)$$

where S is a set of possible end segments and $p(s_i|p)$ is the probability that s_i is an end segment, given the beginning segment p. If the entropy value $H(p)$ is high enough, there is high uncertainty about what would follow. It indicates that many words share this same starting segment, which is one of the characteristics of prefixes.

We calculate entropy values for every segmentation of a given word, starting with the segment consisting of the first letter only. Then, we add the second letter and repeat the evaluation, etc. This method is called forward entropy method. Alternatively, we can make the same procedure from the end, starting with the segment containing only the final letter of the word and continue adding the letter before the last one and so on. This method is called the backward entropy.

Sometimes, however, the higher entropy does not indicate the border between morphs correctly. This can be seen for example in Table 1, where we get the forward entropy 2.16 for the segment *neb-* which is not any prefix.

That is why we propose to use not the entropy itself, but the growth of the entropy as a better indicator.

Table 1. Comparison of forward and difference entropy methods

	n	e	b	e	z	p	e	č	í
forward entropy	1.61	3.02	2.16	2.10	0.33	0.00	0.68	1.28	
difference entropy	0.00	1.42	−0.86	−0.06	−1.78	−0.33	0.68	0.59	

2.3 Difference Entropy

Difference entropy method is just a small modification of the entropy method described above but providing considerably better results. For every segmentation, we calculate its entropy and then check the difference between the entropies of neighbouring segments.

It filters out single letters at the beginning of words. They have always high entropy because after the first letter there is always a great uncertainty how the word would continue. It does not mean, that every single letter is a prefix; it just means that every word has to begin with something. On the other hand, towards the end of the word, number of possible continuations lowers, so the entropies of segmentations are much smaller. This may mean not only that we did not find prefix but the same thing can happen in case of rare prefixes. There, even a smaller value of entropy can signify that a prefix was found.

The great entropy growth indicates that we are more than usually uncertain what would follow. This is the main difference. Not only high uncertainty, but unusually high. With this approach we get results similar to the entropy method but with threshold changing dynamically according to the neighbouring value, so it is not so dependent on the position in the word as is the (pure) entropy method.

Table 2. Comparison of forward and forward difference entropy methods

	o	d	p	ř	e	d	c	h	v	í	l	e
forward	2.55	2.92	2.16	1.43	0.00	1.55	0.00	0.00	0.00	0.00	0.00	
difference	0.00	0.38	−0.76	−0.73	−1.43	1.55	−1.55	0.00	0.00	0.00	0.00	

Example 3. Let us discuss the examples from the Czech language presented in Tables 1 and 2. We can see high values of forward entropy in the beginning of the word. With the threshold set to 2, we get segments *ne, neb, nebe*, resp. *o, od, odp*, as promising candidates for prefixes. The great uncertainty at the beginning of the words causes high entropy values, though only the some of them are real prefixes (*od* and *ne*). However, the differences between adjacent entropies for these four segments are not always so high. Thus, we should not rely on the entropy itself, but use the difference entropy instead. Then, we see great increases only after the segments *ne* and *odpřed*. In fact, the latter one is a composition of two prefixes, but we can count it as a prefix too.

The disadvantage of the difference entropy is that it filters out short prefixes (for example the prefix *od* in the Table 2). That is the reason why it is meaningful to exploit more methods for prefix recognition and combine them (see later).

3 Experiments

3.1 Data

We made several experiments with the lemmas taken from the corpus SYN2000 [2]. We removed the lemmas with the following properties:

- with upper case letters (mainly proper names, abbreviations, etc.),
- with other than Czech characters (including non-characters),

- with more than 2 consecutive equal letters (there are no such words in Czech),
- pronouns, prepositions, interjections and particles (they do not contain prefixes).

There are approximately 320 thousand different lemmas (precisely 321,168) in the list.

3.2 Results

For each segment, every method gives a value telling how good this segment is as a prefix. However these results have different scales. For example, value 3 for the entropy is very high, while 3 squares are very few. For an easier comparison, we normalize the results.

First of all, as numbers of squares have really wide range of possible results, we use their logarithms instead. Next, we take the best result from each method and divide all other results from the same method with this number. Thus, we get always values between 0 and 1.

Table 3. Examples of normalization

prefix	dfentr	||dfentr||	fentr	||fentr||	fsqrs	ln(fsqrs)	||fsqrs||	combi	||combi||
rádoby	3.07	1.00	3.07	0.98	3891	8.26	0.49	2.48	0.92
pseudo	2.95	0.96	3.00	0.96	78730	11.27	0.67	2.60	0.97

In Table 3 there are presented all the values just described[3]: *fentr* is forward entropy, *dfentr* is difference forward entropy and *fsqrs* is number of squares. We use ||.|| to mark the normalized values. The last column, *combi*, stands for a combination of the methods. This combination is sum of normalized values of the forward entropy, the difference entropy and the squares.

Table 4. The first 20 prefix candidates sorted according to combi

prefix	combi	||fentr||	||dfentr||	||fsqrs||	prefix	combi	||fentr||	||dfentr||	||fsqrs||
super	2.682	0.967	0.978	0.737	severo	2.384	0.966	0.839	0.578
pseudo	2.602	0.963	0.961	0.677	jiho	2.370	0.944	0.848	0.577
mikro	2.532	0.921	0.917	0.693	makro	2.366	0.935	0.886	0.544
sebe	2.505	0.898	0.849	0.757	elektro	2.364	0.931	0.797	0.635
rádoby	2.480	0.983	1.000	0.496	jedno	2.361	0.923	0.726	0.711
deseti	2.437	0.917	0.865	0.653	nízko	2.359	0.940	0.919	0.499
mimo	2.423	0.968	0.801	0.653	mega	2.341	0.904	0.861	0.575
hyper	2.420	0.906	0.894	0.619	vnitro	2.332	0.939	0.862	0.530
anti	2.393	0.942	0.706	0.744	spolu	2.327	0.921	0.711	0.695
roz	2.390	0.922	0.530	0.937	dvou	2.317	0.929	0.659	0.728

[3] We do not present results of backward entropy, as it was already proved that they recognize suffixes rather than prefixes [3].

3.3 Comparison

To be able to compare several different methods, we use the **precision**:

Definition 5. *Let* Γ *be a set of prefixes obtained by a method and let* Ψ *be a set of prefixes existing in a language. Then* **precision** P *is defined as*

$$P = \frac{|\Gamma \cap \Psi|}{|\Gamma|} \times 100\%$$

It tells us, how successful the method was. Table 5 shows values for the methods discussed in this paper. We took first 10/20/50/100 most promising prefixes according to each method and measured the precision. Suffix *-.m2* means the additional condition that a prefix has to be shorter than half of the word. The best results come from the combination of all the methods. However there are many ways how to combine the results. We want to explore various combinations in our future experiments. Among single methods, the best one was the difference entropy method with the additional condition on the length of the word.

Table 5. Precision of the methods for 10/20/50/100 best prefix candidates

method	10	20	50	100
combi	100%	100%	100%	96%
dfentr	100%	100%	100%	84%
dfentr.m2	100%	100%	100%	92%
fentr	100%	90%	82%	79%
fentr.m2	100%	90%	82%	79%
fsqrs	100%	85%	60%	51%
fsqrs.m2	100%	85%	60%	51%

It may be expected that we get similar lists consisting of well known prefixes from each method so precision wouldn't tell us anything interesting. However, as we can see in Table 6, the results are quite different. For example, in the top 100 prefix candidates there are only 6 identical ones between the square and the difference entropy methods.

Table 6. Cardinality of intersection of first 100 results from different methods.

	dfentr	dfentr.m2	fentr	fsqrs
dfentr	100	89	27	6
dfentr.m2	89	100	29	6
fentr	27	29	100	34
fsqrs	6	6	34	100

4 Summarization

Affisix [5] is an open-source project. All implemented methods are language independent in that sense, that they make only some general assumptions and do not depend on any characteristic feature of a particular language. However, the tests have been processed on the Czech language only. The current version also supports suffix recognition (using the same methods) but we have not experimented with them yet. Only two methods are supported so far, but we want to add other ones (see [3]). We also want to make experiments with different settings of the methods, combinations of them as well as experiments with different languages. An ultimate goal is to have a universal tool which can be used to test new hypotheses about segmentation of words, and possibly compare the word creation among different languages.

References

1. Hlaváčová, J.: Morphological Guesser of Czech Words. In: Matoušek, V. (ed.) Proc. TSD 2001, pp. 70–75. Springer, Berlin (2001)
2. Ústav Českého národního korpusu FF UK: Český národní korpus – Syn2000 (2000), http://ucnk.ff.cuni.cz
3. Urrea, A.M.: Automatic discovery of affixes by means of a corpus: A catalog of spanish affixes. Journal of Quantitative Linguistics 7, 97–114 (2000)
4. Urrea, A.M., Hlaváčová, J.: Automatic Recognition of Czech Derivational Prefixes. In: Gelbukh, A. (ed.) CICLing 2005. LNCS, vol. 3406, pp. 189–197. Springer, Heidelberg (2005)
5. Hrušecký, M.: Affisix, http://affisix.sf.net

Variants and Homographs
Eternal Problem of Dictionary Makers*

Jaroslava Hlaváčová and Markéta Lopatková

Charles University in Prague, ÚFAL MFF, Czech Republic
lopatkova@ufal.mff.cuni.cz, hlavacova@ufal.mff.cuni.cz

Abstract. We discuss two types of asymmetry between wordforms and their (morphological) characteristics, namely (morphological) variants and homographs. We introduce a concept of multiple lemma that allows for unique identification of wordform variants as well as 'morphologically-based' identification of homographic lexemes. The deeper insight into these concepts allows further refining of morphological dictionaries and subsequently better performance of any NLP tasks. We demonstrate our approach on the morphological dictionary of Czech.

1 Introduction and Basic Concepts

In many languages, there are wordforms that may be written in several ways; they have two (or more) alternative realizations. We call these alternatives variants of wordforms. By definition, the values of all morphological categories are identical for the variants. This fact complicates for instance language generation, important part of machine translation. How can a machine decide which variant is appropriate for the concrete task? This is the reason why the variants should be distinguished and evaluated within morphological dictionaries.

Homographs, wordforms that are 'accidentally' identical in the spelling but different in their meaning, can be seen as a dual problem. If their morphological paradigms differ, the necessity to treat them as separate lemmas is obvious. But even if their paradigms are the same, it is appropriate to treat them with special care because their different meanings may affect various levels of language description.

There is not a common understanding of basic concepts among (computational) linguists. Although there have been many attempts to set standards (for instance [1]), they were usually too general, especially for the purposes of inflectionally rich languages. This was the reason why we have decided to specify the basic concepts in a way that allows to cover consistently and meaningfully all special cases that may occur in languages with rich inflection. However, we are convinced that they are useful for other languages too. We support this by examples from English and Czech, as representatives of different types of languages. Our definitions are based on those in [2,3,4,5], and [6].

* This paper is a result of the projects supported by the grant of the Czech Ministry of Education MSM113200006 and the grants of the Czech Academy of Sciences 1ET101120503 and 1ET101120413.

P. Sojka et al. (Eds.): TSD 2008, LNAI 5246, pp. 93–100, 2008.
© Springer-Verlag Berlin Heidelberg 2008

Note: In the following paragraphs we use the concept of meaning. We do not try to present its exact explanation. We consider the lexical meaning an axiomatic concept. Among all the attempts to set its sufficient explanation, we find the following one as the most appropriate for our purposes: "Lexical meaning is a reflection of extralinguistic reality formed by the language system and communicative needs mediated by consciousness.", see [4].

Relations among basic concepts are visualized at the Figure 1.

Wordform is every string of letters that forms a word of a language, e.g. *flower*, *flowers*, *where*, *writes*, *written*.

Lemma is a basic wordform. Each language may have its own standards, but usually it uses infinitive form for verbs, singular nominative for nouns,... Lemma is usually used as a headword in dictionaries. Lemmas of the examples from the previous paragraph are the following strings: *flower*, *flower*, *where*, *write*, *write*.

Paradigm is a set of wordforms that can be created by means of inflection from a basic wordform (lemma). E.g. the paradigm belonging to the lemma *write* is the set {*write*, *writes*, *wrote*, *writing*, *written*}. It can be specified either by listing all the wordforms, or by a pattern (a rule according to which all inflectional forms can be derived from the basic form).

Lexical unit is an abstract unit associating the paradigm (represented by the lemma) with a single meaning. In dictionaries, the meaning is usually represented by a gloss, a syntactic pattern and a semantic characteristics (e.g. a set of semantic roles). The lexical unit can be understood as 'a given word in the given sense', see also [3].

Lexeme is a set of lexical units that share the same paradigm. We are aware that especially this term is simplified but it is sufficient for dictionaries containing all necessary information about words but at the same time, easy to use.

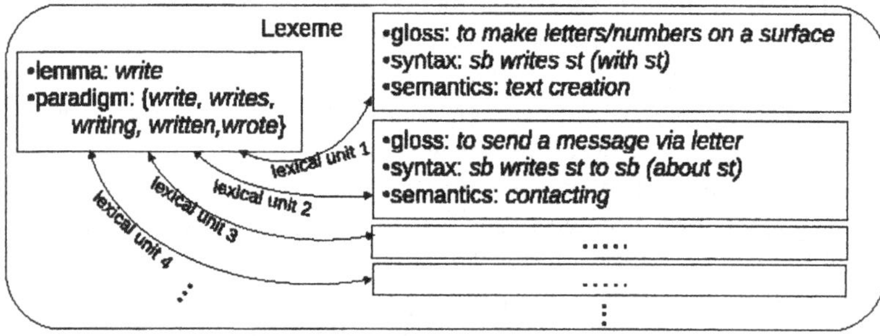

Fig. 1. Relations among basic concepts

Dictionary is a set of records, called also entries or dictionary entries, that describe individual lexemes. In other words, every dictionary entry contains a complex description of one lexeme, represented by its lemma.

Variants are those wordforms that belong to the same lexeme and values of all their morphological categories are identical. Examples: *colour/color*, *got/gotten* as past participles.

Homographs are wordforms with identical orthographic lettering, i.e. the identical strings of letters (regardless of their phonetic forms), whose meanings are (substantially) different and cannot be connected. E.g. wordform *pen* as a 'writing instrument' and wordform *pen* as an 'enclosure', *bank* as a 'bench', *bank* as a 'riverside' and *bank* as a 'financial institution'.

2 Variants

Variants often violate the so-called Golden Rule of Morphology, see [6]:

lemma + morphological tag = unique wordform

In other words: Given a lemma and a morphological tag, no more than one wordform should exist, belonging to that lemma and having that morphological tag.

This requirement is very important for unambiguous identification of wordforms in morphological and other dictionaries. If satisfied, we can use the pair <lemma, morphological tag> as the unique identifier for each wordform. However, variants often violate this rule because they have the same morphological tag and are assigned the same lemma in morphological dictionaries. That is why it is necessary to make their unique tagging clear.

2.1 Types of Variants

We define two types of variants – one affecting the whole paradigm and the second one affecting only wordforms with some combinations of morphological values. The former one is called global, the latter inflectional.

Global variants are those variants that relate to all wordforms of a paradigm, in all cases in the same way. E.g. *colour/color* and all their forms, namely *colours/colors*, and for verbs also *coloured/colored*, *colouring/coloring*. The pairs *zdvihat/zdvíhat* [to lift] or *okno/vokno* [window][1] are examples of global variants in Czech, as all wordforms of respective lemmas demonstrate the same difference, namely *i-í* at the fourth position of the first example and *o-vo* at the beginning of the second example.

Inflectional variants are those variants that relate only to some wordforms of a paradigm. E.g. variants *got/gotten* are inflectional, because they appear only in the past participle of the lemma *get*; other wordforms of the lemma *get* do not have variants. As a Czech example, we can present two forms *jazyce/jazyku* for the locative case singular of the lemma *jazyk* [language] or *turisti/turisté* for the nominative case plural of the lemma *turista* [tourist].

[1] The variants of the first example are both neutral, the latter form of the second example is colloquial.

2.2 Lemma Variants and Multiple Lemma

Inflectional variants may affect any wordform, including the basic one – lemma. That case may lead to their wrong interpretation as global variants but they should be classified as inflectional variants, in accord with the definition introduced above – the variant is not expressed by all the wordforms. Let us take the example of variants *bydlit/bydlet* [to live]. They differ in the infinitive form and in the past tense (*bydlil/bydlel*); the rest of wordforms is the same for both lemmas. Thus the variants *bydlit/bydlet* are classified as the inflectional variants.

Variants of lemmas in general, also called **lemma variants**, can be either global (when they exhibit throughout the whole paradigm), or inflectional.

Lemma variants should be treated with a particular care, as lemmas have a special position among other wordforms – they usually serve as representatives of the whole paradigms and also as labels for lexemes.

To be able to recognize and generate all wordforms of all lemma variants, we have to decide about their representative form. As the selection of a unique representative is an uneasy task (see [6]) we introduce the concept of **multiple lemma** as a set of all lemma variants. A paradigm of the multiple lemma, called **extended paradigm**, is a union of paradigms of individual lemmas constituting the multiple lemma. For example, the lemma *skútr* [scooter] has three different spellings, namely *skútr*, *skutr* and *skutr*, each having its own paradigm. The multiple lemma {*skútr*, *skutr*, *skutr*} has an extended paradigm containing all wordforms of all three members of this set.

Implementation of Multiple Lemmas. In the morphological dictionary of Czech [7], the wordforms are not listed separately, they are clustered according to their lemmas. The lemma represents the whole paradigm. However, the multiple lemma cannot represent the extended paradigm straightforwardly because a set cannot serve as unique identifier. Thus, we keep all lemma variants separately but we connect them with pointers (see Figure 2).

Fig. 2. Schema of implementation of multiple lemma

For morphological analysis,this arrangement can be used in several ways:

- En bloc: The multiple lemma becomes the representative of all wordforms belonging to all lemma variants during the compilation of the dictionary.
- Stepwise: The multiple lemmas create a separate list. The morphological analysis of input texts is processed in its traditional way, assigning a single lemma to every wordform. After the analysis, single lemmas are replaced with the multiple ones, when appropriate.
- External: The list of lemma variants is implemented externally into a software tool that processes the output of the morphological analysis.

Each of the approaches has its positive and negative aspects. We will not compare them here, as we concentrate rather on the dictionary building, not on its possible uses.

2.3 Variants as New Categories

The last thing that is necessary to solve is a compliance with the Golden Rule of Morphology. After adopting the concept of multiple lemma, it would be violated even more. Let us take the multiple lemma {*skútr*, *skutr*, *skutr*} again. The dative singular has now three possible spellings, namely {*skútru*, *skutru*, *skutru*} instead of one, that is required by the Golden Rule. In order to distinguish the variants, we introduce a new category, **Global Variant**. The value of this category differentiates a style of individual variants.[2]

As this sort of information is very hard to grasp, we prefer to express the values by means of small integers, to give at least an approximate picture of its scalable nature. Thus, the synchronic standard variants, roughly of the same stylistic characteristic, have the value of 0, 1 or 2. Following integers express the bookish or archaic variants. Integers 9, 8,... serve for the spoken or colloquial variants, see Figure 2.[3]

The inflectional variants are marked similarly – there is a new category called **Inflectional Variant** that distinguishes different variants of wordforms of a lemma. This category describes the same property of the wordform – the style, and so have the same set of values.

Each type, inflectional and global, is considered as a special category with its set of values. The necessity of two categories implies from the fact that many variants are both global and inflectional. For example, the wordforms *okny/voknama* (plural instrumental of the multiple lemma {*okno, vokno*} [window]) are global variants (because of the protective *v-* at the beginning shared by all wordforms of the lemma *vokno*) as well as inflectional variants (because of the colloquial instrumental ending *-ama*).

Values of both categories of variants become parts of morphological tags. Thus, the Golden Rule of Morphology always holds true.

3 Homographs

Homographs are wordforms with the same lettering but different meaning. As with variants, two types can be distinguished – inflectional and global ones.

As was stated in Section 2, the pair <lemma, morphological tag> should be unique for every wordform. Homographs, as two (or more) identical wordforms with different meaning, have to differ in this pair. We distinguish two possibilities – with the same lemma (type T) and with different lemmas (type L):

(T) Wordforms with the same lemma and different morphological tags, as e.g. *stopped* being both past tense and past participle. This type, which is very frequent for Czech (e.g. the wordform *hradu* [castle] being both genitive singular and accusative singular), is called syncretism.

[2] This sort of information should be rather a goal of a linguistic research than to be included in the morphological dictionary but users are accustomed to have it there.

[3] The numbering expresses the relationship among the lemma variants of the multiple lemma, it must not be used for any comparison across different multiple lemmas.

(L) Wordforms with different lemmas (as e.g. the wordform *smaž* is imperative of the two verbs – *smažit* [to fry] and *smazat* [to erase]; the wordform *ženu*, which is either the accusative singular of the noun *žena* [woman] or the first person of present tense of the verb *hnát* [to rush]).

Inflectional homographs are those homographs, where at most one homographic wordform is a lemma (all examples in three paragraphs above belong to inflectional homographs as they have no homographic lemma at all).

On the other hand, **global homographs** are those homographs that satisfy the following condition (1) and one of the conditions (2′) or (2″):

(1) They affect at least two lemmas (also called homographic lemmas), i.e. the same string of letters (lemma) represents two (or more) different lexemes.
(2′) The paradigms of the affected lemmas differ (e.g. *hnát* with verbal or nominal paradigm; *žít* with different wordforms *žil*/*žal* for the past tense (and two meanings, 'to live' or 'to mow')).
(2″) The lemmas are derived from different words (e.g. the verb *odrolovat* as 'to roll away' has the prefix *od-* and the stem *-rol-* whereas *odrolovat* as 'to crumble' has the prefix *o-* and the stem *-drol-*).

The reason for distinguishing inflectional and global homographs is obvious. Inflectional homographs do not cause any problem for implementation of dictionaries since particular homographic wordforms belong either to one lexeme (as in case (T)) or to more lexemes with different lemmas (cases (L)). Thus, particular lexemes are represented by unique headwords = lemmas. (However, inflectional homographs represent the central problem for morphological disambiguation).

For lexicographers, global homographs are more problematic. They refer to the cases when the same lemma belongs to two (or more) lexemes. They may violate the Golden Rule of Morphology (see Section 2), as in the case of homographs *žít* or *odrolovat*. Contrary to the case of variants, no morphological category can distinguish them. It is necessary to draw the line between them at the lemma side. The difference between the lemmas is marked with numeral suffixes: *žít*-1 and *žít*-2. Though homographs with different POS do not violate the Golden Rule of Morphology, it is reasonable to deal with them similarly. Thus, we have also *colour*-1 as a verb, *colour*-2 as a noun.

3.1 Global Homography Versus Polysemy

Polysemy is usually characterized as the case of a single word having two or more related meanings (based on [2]). Polysemy is treated within a single lexeme. It is a relation among particular lexical units of a lexeme, contrary to homographs that concern separate lexemes.

There is no clear cut between polysemy and homography as these concepts are based on the vague concept of meaning (see above). Unfortunately, lexicographers hesitate quite often and dictionaries are not consistent in distinguishing homographs from polysemic lexemes. For example, Czech verb *hradit* is treated differently in Czech normative dictionaries: as one polysemic lexeme with two lexical units 'to fence' and 'to reimburse', or as homographic lemma, i.e. lemma representing two different lexemes.

The requirement of the identity of lemmas on one hand and difference in (morphological) paradigms or difference in a word creation on the other hand are rather technical but solid criteria for homographs. Based on these criteria, global homography and polysemy can be distinguished consistently. Polysemy is characterized by the identity of the whole paradigms, while global homography requires identity of lemmas only. Thus we obtain the single lexeme for the verb *hradit* [to fence], [to reimburse] but two lexemes represented by the lemmas *žít*-1, *žít*-2 (see 2′). However, we have the single polysemic lexeme for the verb *odpovídat* regardless of the "distance" of its at least four lexical units 'to answer', 'to react', 'to be responsible', and 'to correspond'.

4 Duality of Variants and Homographs

The basic difference between the two concepts are illustrated on the schemas in Figure 3. For variants, the shape of the schema resembles the letter A, while for homographs it is the letter Y. The polysemy appears only at the syntactic (if applicable) or semantic levels of the schema (see the right schema). It is not surprising that these schemas resemble those introduced in [8], where they illustrate synonymy and homonymy as relations between separate layers of language description.

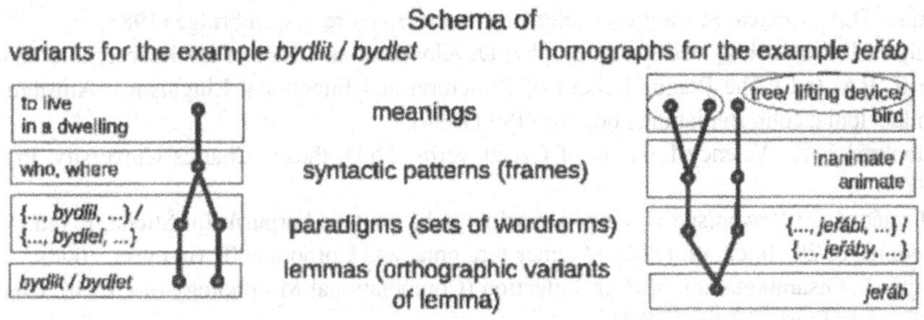

Fig. 3. Schema of variants and homographs. Parts in ellipses concern polysemy.

Let us present another example of variants and homographs to clarify the previous concepts. The word *jeřáb* can be either animate denoting a species of bird (plural nominative is *jeřábi*, with the inflectional variant *jeřábové*), or inanimate (plural nominative is *jeřáby*) having two meanings, one being a name of a tree, the second one a crane. Thus, there are two homographic lemmas *jeřáb*-1 and *jeřáb*-2. Moreover, the inanimate lemma *jeřáb*-2 is polysemic, with two meanings (but the same paradigm), see Figure 2.

Summary

We have brought a deeper insight into the problem of variants and homographs, especially those that affect lemmas.

We have also introduced a novel treatment of variants that meets the requirement of a unique wordform for each pair <lemma, morphological tag>, the so called Golden

Rule of Morphology. Variants are treated in one (extended) paradigm, specified by (multiple) lemma and morphological tag enriched with information about two types of variants, inflectional and global. A corpus user searching for all occurrences of any of these lemma variants can put into the query any lemma from the multiple lemma.

Similarly, we distinguish inflectional and global homographs. The 'morphologically-based' specification of homographs enables us to distinguish homography from polysemy consistently.

Based on a close examination of the phenomena described in this paper we have proposed an implementation of variants and homographs within a wide coverage morphological dictionary [7] used in various tasks in NLP. The proposed treatment of lemma variants enables both their subsuming under common headword, the so called multiple lemma (e.g. for querying in corpora or for IR tasks) as well as distinguishing particular wordforms (e.g. for language generation).

References

1. ISO/TC 37/SC 4: Language Resources Management – Lexical Markup Framework (LMF) (2007) Rev. 14, date 2007-06-03, `http://www.lexicalmarkupframework.org/`
2. Matthews, H.: The Concise Oxford Dictionary of Linguistics. Oxford University Press, Oxford (1997)
3. Cruse, D.A.: Lexical Semantics. Cambridge University Press, Cambridge (1986)
4. Filipec, J.: Lexicology and Lexicography: Development and State of the Research. In: Luelsdorff, P.A. (ed.) The Prague School of Structural and Functional Linguistics, Amsterdam-Philadelphia, John Benjamins, pp. 163–183 (1994)
5. Žabokrtský, Z.: Valency Lexicon of Czech Verbs. Ph.D. thesis, Charles University, Prague (2005)
6. Hlaváčová, J.: Pravopisné varianty a morfologická anotace korpusů. In: Štícha, F. (ed.) Proceedings of 2nd International Conference Grammar and Corpora 2007 (in press, 2008)
7. Hajič, J.: Disambiguation of Rich Inflection (Computational Morphology of Czech). Charles Univeristy Press, Prague (2004)
8. Panevová, J.: Formy a funkce ve stavbě české věty. Academia, Praha (1980)

An Empirical Bayesian Method for Detecting Out of Context Words

Sanaz Jabbari, Ben Allison, and Louise Guthrie

Natural Language Processing Group, Department of Computer Science
University of Sheffield, UK
{s.jabbari,b.allison,l.guthrie}@dcs.shef.ac.uk

Abstract. In this paper, we propose an empirical Bayesian method for determining whether a word is used out of context. We suggest we can treat a word's context as a multinomially distributed random variable, and this leads us to a simple and direct Bayesian hypothesis test for the problem in question. We demonstrate this method to be superior to a method based upon common practice in the literature. We also demonstrate how an empirical Bayes method, whereby we use the behaviour of other words to specify a prior distribution on model parameters, improves performance by an appreciable amount where training data is sparse.

Keywords: Word Space Models, Statistical Methods, Natural Language Processing.

1 Introduction

This paper proposes a distributional model of word use and word meaning which is derived purely from a body of text, and the application of this model to determine whether certain words are used in or out of context. In this work, a word's "context" is the ordered string of words in its proximity, and we suggest here that we can view the contexts of words as multinomially distributed random variables. We illustrate how, using this basic idea, we can formulate the problem of detecting whether or not a word is used in context as a simple Bayesian hypothesis test, with two possible and mutually exclusive hypotheses: 1) the supplied context is a context of the word in question, or 2) the supplied context is a context of some other word. We show that our proposed method is capable of outperforming a method based upon the cosine of the angle between semantic vectors of words and their contexts, and without the problem of thresholding similarity scores.

The paper also shows that the empirical Bayesian method of automatically learning appropriate prior distributions for parameters can be used within the model we propose, and we demonstrate that such a practice increases our ability to characterise word behaviour, particularly for words which do not occur with great regularity. This leads us to conclude that general patterns of word use exist to some degree or other, and can be captured by an appropriate hierarchical structure and harnessed to improve parameter estimation.

The rest of the paper is presented as follows: §2 describes the motivation behind our work and outlines some of the work in the Language Processing literature that attempt similar problem. §3 formally specifies out task, and §4 describes the methods

P. Sojka et al. (Eds.): TSD 2008, LNAI 5246, pp. 101–108, 2008.
© Springer-Verlag Berlin Heidelberg 2008

we propose and evaluate on this task. §5 describes our experimental setup, and results are presented in §6; finally, §7 ends with some concluding remarks.

2 Background

Although the application which brought us to consider this problem is a security one, in many Natural Language Processing tasks the goal is to identify the word (or a sense of a word) most consistent with some context. For instance, in word sense disambiguation, the task is to find the correct sense of a word given the context in which the word occurs; in Natural Language Generation and Machine Translation, lexical choice is one of the central issues; similarly, spelling error detection can be viewed as a task of finding words that are used out of context.

Approaches to this abstract problem typically fall into one of two possible categories. The first assumes that if a word is used in context, it is formally and logically consistent with its context: the Generative Lexicon [1] or lexical chains [2] are examples of such methods. While these approaches arguably provide much better insight into the linguistic processing capabilities of humans, as a practical approach they are handicapped by relying on large amounts of carefully hand crafted resources.

Another group of approaches uses the concept of semantic relatedness. In this paradigm, words and their contexts are treated as two separate entities defined in a common space. Judging their consistency can then be accomplished by measuring their distance in this space. These measures of distance rely either on hand-crafted semantic networks (e.g. Wordnet, Roget's thesaurus, etc.) [3,4] or on information found in corpora [5,6]. Such corpus-based methods often employ a "word-space model" into which a word and its context are mapped, and their proximity within this space represents their semantic similarity. While some models consider unconditional co-occurrence of words, others models consider co-occurrences within certain grammatical relationships [7]. Other models consider dimension-reducing transformations of the original co-occurrence matrix, such as factor analysis or Singular Value Decomposition [8].

Even assuming that the meanings of individual words are successfully represented within some word space, representing their context in the same space more of a challenge. For example, [9] defines "context vectors" to be the centroid of the vectors of content words in the context. [10] represents both words and contexts in a domain space, and uses domain vectors to represent words and contexts.

In this work, we propose a quite different approach. However, since our method is purely corpus based, we compare our work with a purely corpus based approach that uses a vector-space model to measure semantic similarity. We believe this method most accurately represents the current state of the art amongst comparable methods, and could feasibly be extended to include additional grammatical and semantic information in the same way as other vector-space methods if desired.

3 The Task: Detecting If a Word Is Out of Context

The task we attempt is exactly as follows: given some word, w and a context c_{new} in which w is purported to belong, is c_{new} consistent with w? For the purposes of this work, we treat c_{new} as a bag-of-words (i.e. words in the context are exchangeable). We further assume that a large body of text is available, which we will refer to as the background corpus, where we can observe legitimate contexts of w and all other words in which we may be interested.

w can take one of $|V|$ values, where V is the vocabulary of the problem. c_{new} is a vector of counts of dimension $|V'|$ (note that $V \supseteq V'$, since one might feasibly allow w to be any word in a dictionary/large corpus, but observe co-occurrence behaviour only with a subset of these words), with $c_{new,i}$ being the number of times the i-th word in V' occurs in c_{new}.

For the purposes of evaluation, we use recall, precision and F-measure, which is the harmonic mean of these quantities.

For the experiments we present here, we set V' as the 2,000 most frequent non-stopwords, and c_{new} is derived from the counts of these words within a 10 word window either side of the target word, so long as the window does not cross sentence boundaries.

4 The Methods

For the purposes of this paper, we advocate treating the problem as a Bayesian hypothesis test; however, the majority of work in similar areas uses vector-space measures of semantic similarity, so we compare our proposed approach to a baseline method which does exactly that. We first describe our baseline, and then concentrate on our proposed solution.

4.1 The Vector-Space Method

In light of much work on related tasks, we present a baseline method which treats the task of determining whether w is consistent with c_{new} as a judgement of the *semantic similarity* between w and c_{new}. Mapping both entities to the same space is a challenge [9] overcomes this problem by representing the semantics of a context to be the centroid of its constituent words.

For w, we observe total co-occurrence counts with all words in V' from all contexts containing w in the background corpus. The resulting vector is then normalised by the sum of its components, to leave a vector $\vec{w}$ where the sum of the components of $\vec{w}$ is unity. To represent c_{new} we perform exactly the same process for each word in V' which has non-zero count in c_{new}, and represent c_{new} by $\vec{c}_{new}$ which is the centroid of each of these vectors.

Finally, to compare the similarity of c_{new} and w we follow relatively standard practice in the area and use the cosine of the angle between $\vec{c}_{new}$ and $\vec{w}$; thus:

$$sim(w, c_{new}) = \cos(\vec{w}, \vec{c}_{new}) = \frac{\sum_i w_i \cdot c_{new,i}}{\sqrt{\sum_i w_i^2} \cdot \sqrt{\sum_i c_{new,i}^2}} \quad (1)$$

The similarity (1) has no natural threshold at which to judge whether w is consistent with c_{new}, and if we were truly advocating this method as a solution then it would be necessary to determine one, perhaps by cross validation or some other re-sampling method. However, for the purposes of comparison we post-optimise the parameter on the test data to determine its maximum possible effectiveness.

4.2 A Bayesian Hypothesis Test

The method we advocate for this problem is to treat the problem as a Bayesian hypothesis test; in contrast to a classical hypothesis test, this involves specifying the distribution of the test statistic under all hypotheses and evaluating their posterior probabilities. In this case, there are two possible hypotheses, which we will label H_1 and H_2. In plain English, they correspond to the following:

- H_1: c_{new} is a context of the word w,
- H_2: c_{new} is a context of some word other than w.

Given an observed c_{new}, we proceed by evaluating the likelihood of the two hypotheses. If $\Pr(c_{new}|H_1) > \Pr(c_{new}|H_2)$, we say that c_{new} is consistent with w; otherwise, we judge it inconsistent. These are proportional to the posterior probabilities if we assume that both hypotheses are equally probable *a priori*. Note that we simply treat c_{new} as an outcome of $\tilde{c}$, exchangeable with others in training, and thus proceed in terms of the more general variable c.

We argue that the simplest and most natural distribution of $\tilde{c}$ given any fixed w is multinomial, depending upon (vector) parameter θ_w. Thus $p(c|H_1)$ has distribution:

$$p(c|H_1) \sim Multinomial(\theta_w, \sum_i c_i) \propto \prod_i \theta_{wi}^{c_i} \tag{2}$$

Note that we omit the multinomial coefficient since it is common to both hypotheses. $p(c|H_2)$ has distribution which is simply a mixture of multinomials as above, with mixture weights being the prior probabilities of contexts from words in V. Thus:

$$p(c|H_2) = \sum_{x \in V} p(x) \cdot p(c|\theta_x) \tag{3}$$

However, the true sum of terms has prohibitive computational cost – for V derived from the 100 million word BNC, $|V|$ is approximately 600,000. Thus we look to two approximations to (3), first a more simple method and then a more accurate approximation.

Before explaining the two approximations, we address the issue of estimation of θ_w. Assume the background corpus consists of contexts of w, $\mathcal{W} = \{(c_{11} \ldots c_{1v'}), \ldots, (c_{k1} \ldots c_{kv'})\}$ such that each c_j is itself a vector of counts. To estimate θ_w we work purely in terms of the sufficient statistics $n_i = \sum_j c_{ji}$. Assuming the standard reference Dirichlet prior on θ_w and using $\hat{\theta}_w = \mathrm{E}[\theta_w|\mathcal{W}]$ leads to an estimate for the i-th component of θ_w:

$$\hat{\theta}_{wi} = \frac{n_i + 1}{n_\bullet + |V'|} \tag{4}$$

where $n_\bullet = \sum_m n_m$.

A Simple Approximation to $p(c|H_2)$. The simple approximation we use to $p(c|H_2)$ is to pool all counts for all words, and estimate a single $\theta_{\neg w}$ in the same way as (4). To be clear, let $V = \{x_1 \ldots x_v\}$ be all words in V, and let $\mathcal{X} = \{\mathcal{X}_1 \ldots \mathcal{X}_v\}$ be the set of all contexts for each of these words (note that each one of the $\mathcal{X}_j$ is a vector of vectors, as above for $\mathcal{W}$). If for each of the $\mathcal{X}_j$ we have the same $n_j = (n_{j1} \dots n_{jv'})$, which are the sufficient statistics as before, then the estimate for the i-th component of $\theta_{\neg w}$ is:

$$\hat{\theta}_{\neg wi} = \frac{\sum_j n_{ji} + 1}{\sum_j n_{j\bullet} + |V'|} \tag{5}$$

A Dirichlet Compound Multinomial Distribution. As the basis for our more accurate approximation to $p(c|H_2)$, we note that (3) is equivalent to the following:

$$p(c|H_2) = p(\theta_x) \cdot p(c|\theta_x) \tag{6}$$

That is, the mixture weights may be interpreted as a (discrete) distribution on θ_x. Whilst the true distribution on $p(\theta_x)$ is discrete in this case, a suitable continuous approximation would allow:

$$p(c|H_2) = \int p(\theta_x) \cdot p(c|\theta_x)\, d\theta_x \tag{7}$$

which is far more efficient to compute provided the distribution on θ_x is chosen to be convenient. As such, we choose the (continuous) distribution $p(\theta_x)$ to be Dirichlet. If $\theta_x \sim \mathit{Dirichlet}(\alpha)$, then (7) has solution:

$$p(c|H_2) \propto \frac{\Gamma\left(\sum_i \alpha_i\right)}{\Gamma\left(\left[\sum_i c_i\right] + \left[\sum_i \alpha_i\right]\right)} \times \prod_i \frac{\Gamma(\alpha_i + c_i)}{\Gamma(\alpha_i)} \tag{8}$$

Note that the expression is also missing the multinomial coefficient.

To estimate the α, we use the reparameterisation and estimation of the Dirichlet Compound Multinomial described in [11].

Finally, we note that the distribution $p(\theta_x)$ we have estimated above is a arguably more suitable prior on θ_w than was the reference prior in (4). Thus, we alter our estimate of θ_w from that above to be:

$$\hat{\theta}_{wi} = \frac{n_i + \alpha_i}{\sum_i (n_i + \alpha_i)} \tag{9}$$

which is the expectation of the posterior distribution of θ_w when the prior is the distribution $p(\theta_x)$ as above.

5 Experiments

Our background corpus, from which the vectors are defined and parameters are estimated, is the 100 million word British National Corpus (BNC). The number of distinct word types in this corpus is approximately 600,000.

To evaluate the models, we create a data set which simulates the word obfuscation problem. We do this by automatically substituting words of one semantic category with another, and we use semantic categories as defined by Longman's Dictionary of Contemporary English (LDOCE) [12]. We create an evaluation suite by taking sentences in which words of a particular semantic category occur, and replacing these words with words of a different semantic category – we judge these created sentences to be instances where the newly substituted word is out of context. We also retain the sentence with the original word in place, meaning that we have two copies of each of the selected sentences; one in which the target word is used in context, and one in which it is out of context.

We select four semantic categories, as defined by Longman's: Human, Plant, Movable and Non-movable. Words in each category was then divided into high, medium and low frequency bands. These correspond to words that occur at least 2,000 sentences in the BNC, between 1,000 and 2,000 sentences and between 500 and 1,000 sentences, respectively. Of course, many words occur less than 500 times, and we do intend to look at these low-frequency words; however, we assume initially that if a word is to be substituted it will be moderately frequent. The percentage of a typical document covered by words above frequency 500 in the BNC close to 100 percent, thus words below this frequency do of course occur, but none with any substantial regularity.

We substitute words from the category Human with those from Plant and vice versa, and words from the category Movable with Non-movable. We ensure that no word occurs in both Human and Plant, or Movable and Non-movable to avoid errors in evaluation, but thereafter make no attempt to disallow polysemous words so as to make the evaluation as realistic as possible. Note that this means that words may occur in more than one category, so long as it does not occur in both of the substituted categories, and particularly for the high-frequency words, words are also likely to have more than a single sense in a dictionary.

For each frequency band in each of the semantic categories, we select 500 sentences at random from the English Gigaword corpus of newswire text. This leads to a total evaluation set of 6,000 sentences, and each of these sentences is used twice as described above.

6 Results

Our results for the cosine method gives an indication of the wide variation in performance of the method as a function of the similarity threshold. The optimum thresholds are different points for different frequency bands. For the purposes of comparison with our other methods, we assume the performance of the method to be that of the threshold which maximises F-measure on the combined dataset.

Figure 1 shows performances of each of the three methods, from which it can clearly be seen that the method based upon an assumption of multinomially distributed contexts and a hypothesis test is superior to the cosine method. As a sanity check, we also note that performance degrades as a function of the frequency of the target word.

The use of the empirical Bayesian prior also has a pleasing and common-sense interpretation. For high frequency words, we see that there is little benefit in using the prior – training data is sufficiently plentiful to counteract the effects of all put the strongest prior, and such a prior is clearly not truly representative of prior knowledge which can

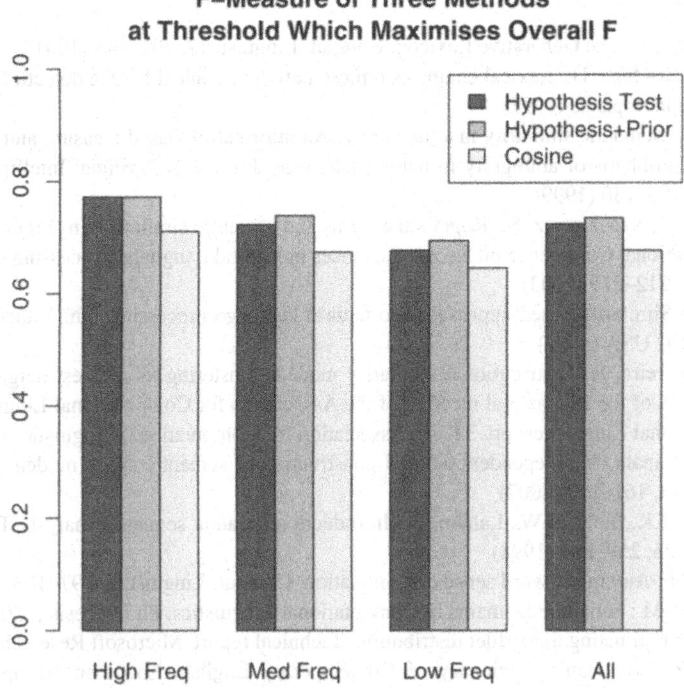

Fig. 1. Performance of the three methods, using cosine performance at the threshold which maximises F-measure on the combined dataset. *Hypothesis Test* is the method proposed in this paper with the simple approximation, while *Hypothesis Test+Prior* is the method with the more complex approximation.

be gleaned from the behaviour of other words. However, for words with lower frequency (and thus less data), the effects of specifying a more accurate prior are clearly appreciable, supplementing paucity of data for each particular word with general trends for all words.

7 Conclusion

In this paper, we have presented a method for determining whether a word is used in or out of context. The method we propose is based upon the assumption that contexts of any given word are multinomially distributed random variables. We have shown that a comparison of the likelihood of observing a new context under two simple hypotheses is an effective criterion upon which to base the decision; furthermore, the decision is not dependent upon the specification of a threshold. Finally, we note that other aspects of word behaviour, and other hypotheses, follow naturally from the model specification we employ, whereas the predominant approaches in the literature must be reinvented anew for each additional application.

References

1. Pustejovsky, J.: The Generative Lexicon. Comput. Linguist. 17, 409–441 (1991)
2. Hirst, G., St-Onge, D.: Lexical chains as representation of context for the detection and correction malapropisms (1997)
3. Resnik, P.: Semantic similarity in a taxonomy: An information-based measure and its application to problems of ambiguity in natural language. Journal of Artificial Intelligence Research 11, 95–130 (1999)
4. Jarmasz, M., Szpakowicz, S.: Roget's thesaurus and semantic similarity. In: Proceedings of the International Conference on Recent Advances in Natural Language Processing (RANLP-2003), pp. 212–219 (2003)
5. Lee, L.J.: Similarity-based approaches to natural language processing. Ph.D. thesis, Cambridge, MA, USA (1997)
6. Lee, L., Pereira, F.: Distributional similarity models: clustering vs. nearest neighbors. In: Proceedings of the 37th annual meeting of the Association for Computational Linguistics on Computational Linguistics, pp. 33–40. Association for Computational Linguistics (1999)
7. Padó, S., Lapata, M.: Dependency-based construction of semantic space models. Comput. Linguist. 33, 161–199 (2007)
8. Landauer, T.K., Foltz, P.W., Laham, D.: Introduction to latent semantic analysis. Discourse Processes 25, 259–284 (1998)
9. Schütze, H.: Automatic word sense discrimination. Comput. Linguist. 24, 97–123 (1998)
10. Gliozzo, A.M.: Semantic Domains in Computational Linguistics. Ph.D. thesis (2005)
11. Minka, T.: Estimating a dirichlet distribution. Technical report, Microsoft Research (2000)
12. Procter, P.: Longman's Dictionary of Contemporary English. Longman Group Limited (1978)

Semantic Classes in Czech Valency Lexicon
Verbs of Communication and Verbs of Exchange*

Václava Kettnerová, Markéta Lopatková, and Klára Hrstková

Charles University in Prague, MFF ÚFAL, Prague, Czech Republic
{kettnerova,lopatkova,hrstkova}@ufal.mff.cuni.cz

Abstract. We introduce a project aimed at enhancing a valency lexicon of Czech verbs with coherent semantic classes. For this purpose, we make use of FrameNet, a semantically oriented lexical resource. At the present stage, semantic frames from FrameNet have been mapped to two groups of verbs with divergent semantic and morphosyntactic properties, verbs of communication and verbs of exchange. The feasibility of this task has been proven by the achieved inter-annotator agreement – 85.9% for the verbs of communication and 78.5% for the verbs of exchange. As a result of our experiment, the verbs of communication have been classified into nine semantic classes and the verbs of exchange into ten classes, based on upper level semantic frames from FrameNet.

1 Introduction

Information on syntactic and semantic properties of verbs is crucial for a wide range of computational tasks. Lexical resources containing such information have been created with different theoretical backgrounds and with different goals. As a result, they store different information. Combining these resources is an effective way to gain more data for NLP tasks.

In this paper, we report on introducing semantic classes to VALLEX [1] while making use of FrameNet data. Our motivation has (i) a practical aspect – to provide available data for NLP tasks, such as generation, question answering, or information retrieval, and (ii) a theoretical aspect – semantic classes enable us to generalize about relations between the semantics of verbs and their syntactic behavior.

As a first step, we experimented with two groups of verbs with divergent semantic and morphosyntactic properties, verbs of communication and verbs of exchange. First, semantic frames from FrameNet were manually assigned to these verbs. Then the hierarchical network of relations in FrameNet between the semantic frames was used for sorting the Czech verbs into coherent semantic classes. Manual annotation is highly time consuming, however, it allows us to reach the desired quality.

The present paper is structured as follows: In Section 2, we introduce VALLEX and FrameNet and the motivation for our experiment. Section 3 describes our experiment with mapping the FrameNet semantic frames to the two groups of Czech verbs

* The research reported in this paper is carried under the project of the Ministry of Education, Youth and Sports No. MSM0021620838 (Objects of Research), under the grants LC536 (Center for Computational Linguistics II) and GA UK 7982/2007.

P. Sojka et al. (Eds.): TSD 2008, LNAI 5246, pp. 109–116, 2008.
© Springer-Verlag Berlin Heidelberg 2008

in VALLEX. Section 4 provides an analysis of the obtained data. Section 5 presents a method of classifying verbs in VALLEX while making use of upper level semantic frames in FrameNet. Finally, results and open questions are summarized.

2 Two Lexical Resources: VALLEX and FrameNet

In this section, we briefly characterize two lexical resources used in the project: VALLEX, which takes into account mainly syntactic criteria, and semantically oriented FrameNet.

2.1 VALLEX – Valency Lexicon of Czech Verbs

The Valency Lexicon of Czech Verbs, Version 2.5 (VALLEX 2.5)[1] provides information on the valency structure of Czech verbs in their individual senses: primarily, on the number of valency complementations, on their type (labeled by functors), and on their possible morphological forms. VALLEX 2.5 describes 2730 lexeme entries containing about 6460 lexical units (LUs), see [1]. A LU prototypically corresponds to one sense of a verb (only such senses that may be represented by different subcategorization patterns are split into separate LUs). VALLEX 2.5 applies a rather syntactic approach to valency, see [2].

Motivation for Introducing Semantic Classes to VALLEX. Semantic information that reflects the way an individual LU relates to another LU (LUs) plays a key part in NLP tasks, see esp. [3] and [4]. At present, VALLEX does not provide a sufficient insight into semantic relations among LUs.

For illustration, the verbs *prodat*^{pf} "to sell", ex. (1), and *diktovat*^{impf} "to dictate", ex. (2), share the same morphosyntactic structure and remain indistinct from each other in VALLEX, in spite of being completely different with respect to their semantics.

(1) *Petr*.ACT *prodal Pavlovi*.ADDR *motorku*.PAT
Eng. *Peter*.ACT *has sold Paul*.ADDR *the motorbike*.PAT
(2) *Ředitel*.ACT *diktoval sekretářce*.ADDR *dopis*.PAT
Eng. *The director*.ACT *has dictated a letter*.PAT *to his secretary*.ADDR

On the one hand, the information on semantic class membership of these verbs allows their differentiation. On the other hand, the semantic classes capturing the relations among LUs enable us to generalize about the morphosyntactic behavior of verbs with similar semantic properties.

2.2 FrameNet

FrameNet[2] is an on-line lexical resource for English. It documents the range of semantic and syntactic combinatory possibilities of each word in each of its senses, see [5]. As to quantitative characteristics, FrameNet contains more than 10,000 lexical units (LUs),[3] pairs consisting of a word and its meaning.

---

[1] `http://ufal.mff.cuni.cz/vallex/2.5/`

[2] `http://framenet.icsi.berkeley.edu/`

[3] For the purposes of this text, we use the same abbreviation LU both for VALLEX and FrameNet because the same concepts are concerned in principle.

The descriptive framework of FrameNet is based on *frame semantics*. Each LU evokes a particular *semantic frame* (SF) underlying its meaning. Semantic frames consist of semantic arguments, *frame elements*. FrameNet records frame-to-frame relations in the form of a hierarchical network. The relation of "Inheritance", i.e. the hyperonymy / hyponymy relation, represents the most important one – the semantics of the parent frame corresponds equally or more specifically to the semantics of its child frames.

3 Mapping Semantic Frames from FrameNet to Valency Frames in VALLEX

In this section, we report on our effort to assign the SFs from FrameNet to the VALLEX verbs of communication and verbs of exchange. As the first step, we translated each LU belonging to the selected groups from Czech to English.[4] The total number of translated Czech LUs was 341 for the verbs of communication and 129 for the verbs of exchange (C and E). These LUs correspond to 551 Czech verbs of communication and 325 verbs of exchange, if counting perfective and imperfective counterparts separately.

Two human annotators (A1 and A2) were asked to indicate whether a particular SF evoked by a translated English LU is consistent with a given Czech LU. The annotators were allowed to indicate one SF (labeled as "Unambiguous assignments of SF") or more than one SF (labeled as "Ambiguous assignments of SF") for a single Czech LU. The annotators could also conclude that no SF corresponds to a given Czech LU. For the overall statistics, see Table 1.

Table 1. Annotated data size and overall statistics on the annotations of SFs

	A1 (C / E)	A2 (C / E)
Czech LUs	341 / 129	341 / 129
SFs evoked by English LUs	610 / 171	556 / 166
Unambiguous assignments of SF	143 / 102	165 / 95
Ambiguous assignments of SF	467 / 69	391 / 71
Czech LUs without SFs	83 / 51	83 / 51

Inter-Annotator Agreement. Table 2 summarizes the inter-annotator agreement (IAA) and Cohen's κ statistics, see [6], on the total number of used SFs for the verbs of communication and the verbs of exchange. The match of answers related to SFs reaches 85.9% for the verbs of communication and 78.5% for the verbs of exchange. The κ statistics represents an evaluation metric that reflects average pairwise agreement corrected for chance agreement. Both the level 0.82 reached for the verbs of communication and the level 0.73 for the verbs of exchange represent satisfactory results [7].

4 Analysis of Semantic Frames Mapping

Statistics on Semantic Frames. 100 SFs in total (SFs regardless of the inter-annotator agreement) were mapped to 341 verbs of communication. The most frequently assigned

[4] The on-line dictionary at `http://www.lingea.cz/` was used.

Table 2. Inter-annotator agreement and κ statistics (considering the annotations of individual SFs for a given Czech LU as independent tasks)

	IAA (C / E)	κ (C / E)	IAA (C + E)	κ (C + E)
Match of SFs	85.9% / 78.5%	0.82 / 0.73	82.2%	0.77

SFs to the verbs of communication include: "Statement", "Request", "Telling", "Communication_manner", "Reporting", etc.

52 SFs in total (regardless of the inter-annotator agreement) were assigned to 129 verbs of exchange. The following SFs belong to the most frequently assigned ones: "Giving", "Getting", "Exchange", "Commerce_pay", "Theft", etc.

The number of SFs is significantly lower, if taking into account only those in which the annotators concurred – 69 for the verbs of communication and 31 for the verbs of exchange.

Ambiguous Assignment of Semantic Frames. Ambiguous annotations draw attention to the divergence in granularity of word sense disambiguation adopted by VALLEX and FrameNet. The different level of granularity represents a great setback in making one-to-one correspondences between LUs from VALLEX and those from FrameNet. In Section 5, we propose a method to overcome this difficulty.

Let us focus on the cases in which two (or more) SFs mapped to a single Czech LU are connected by the hierarchical relation of "Inheritance" – these cases reveal the finer granularity of senses applied in FrameNet. For instance, the SFs "Getting" and "Earnings_and_losses" are assigned to the single Czech LU *mít* "to get / to earn" as in *Who did you get it from?* and *He earns five thousand per month*, respectively. The SF "Earnings_and_losses" is a descendant of the SF "Getting" in the relation "Inheritance": Although the LU "to earn" from the SF "Earnings_and_losses" is semantically more specified, it inherits semantic properties from the LU "to get" evoking the SF "Getting".

FrameNet data can also be used for checking word sense disambiguation in VALLEX. The ambiguous annotations of SFs that do not arise from the finer granularity may reveal mistakes in word sense disambiguation. For instance, the SFs "Grant_permission" and "Permitting" are assigned to the Czech LU *dovolit*^{pf}, *dovolovat*^{impf} "to allow", as in *Peter has allowed me to smoke here* and *This program allows data checking*, respectively. Although the LUs appear to be semantically close, the SFs evoked by them are not in the relation of "Inheritance". Thus this Czech LU represents a candidate for being split into two distinct senses.

5 Exploiting FrameNet for Enhancing VALLEX with Semantic Classes

The hierarchical network of relations between SFs in FrameNet plays a key role in the classification of Czech LUs. The relation of "Inheritance" is of major importance as each child frame inherits semantic properties from its parent frame(s).

We made use of the upper levels of the relation of "Inheritance" for grouping LUs into coherent semantic classes: we mapped the ancestor frames from appropriate levels of the relation of "Inheritance" to Czech LUs. This method allows us to surmount the problem with the coarser granularity of verb senses in VALLEX, see Section 4.

However, the top levels of the relation of "Inheritance" cannot be exploited as they are occupied by abstract and non-lexical SFs, as e.g. "Intentionally_act" and "Reciprocality", respectively. Similarly, the top SFs describing only a very general event, as the SF "Event" (which may be understood as the core of all events), are excluded.

For instance, the SF "Giving" represents the proper level in the "Inheritance" hierarchy. Prototypically, Czech LUs to which descendant SFs of the SF "Giving" are assigned (see Figure 1) are included in the semantic class "Giving".

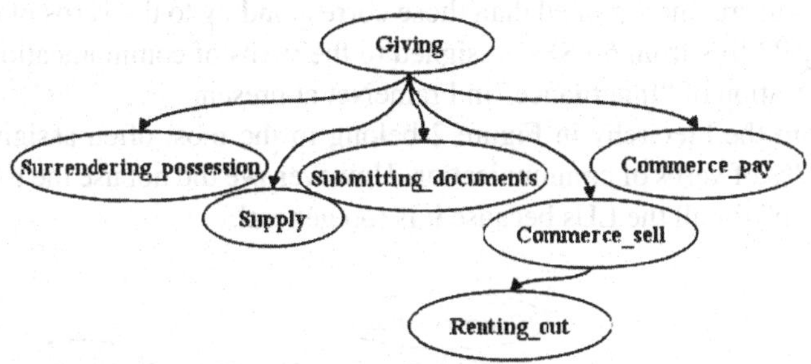

Fig. 1. The upper levels of the relation "Inheritance" of the SF "Giving"

However, if Czech LUs exhibit different morphosyntactic properties than the LUs to which the ancestor SF is assigned, we map the SF from the lower level of the relation of "Inheritance". For instance, *odměnit*^{pf} "to reward", to which the SF "Supply" is mapped, differs in its morphosyntactic properties from the LUs to which the SF "Giving" is assigned (as e.g. *odevzdat*^{pf}, "to surrender", *prodat*^{pf} "to sell", etc.). Thus "Supply" is considered as a candidate for another semantic class.

Semantic Classes for Verbs of Exchange. 31 SFs[5] assigned to the verbs of exchange correspond to 16 SFs on the upper levels of the "Inheritance" hierarchy. However, half of them cannot be exploited: six of these SFs have not yet been linked by the "Inheritance" relation and two SFs do not appear to be relevant.[6] The remaining eight SFs were accepted as semantic classes (1.–8.). We made use of two other SFs "Taking" and "Supply" from the lower levels of the "Inheritance" hierarchy, due to syntactic differences of the LUs to which these SFs are assigned (9.–10.). To conclude, we obtained the following candidates for semantic classes:

1. **"Giving"**: *odevzdat*^{pf}, "to surrender", *prodat*^{pf} "to sell", etc.,
2. **"Getting"**: *koupit*^{pf} "to buy", *přijímat*^{impf} "to receive", etc.,

[5] We took into account only the cases in which the annotators concurred.

[6] The SF "Rewards_and_punishment" is mapped only to the LU *odměnit*^{pf} "to reward". However, as this LU shares the morphosyntactic and semantic properties with the LUs to which the SF "Supply" is assigned, we include it in this semantic class. The SF "Agree_or_refuse_to_act" was assigned only to the LU *odepřít*^{pf}. However, we leave this SF aside because it does not express exchange.

3. **"Replacing"**: *nahradit*^{pf} "to replace", etc.,
4. **"Exchange"**: *měnit*^{impf} "to exchange", *vyměnit*^{pf} "to exchange", etc.,
5. **"Robbery"**: *okrást*^{pf} "to rob", *připravit*^{pf} "to rob", etc.,
6. **"Hiring"**: *najmout*^{pf} "to hire", etc.,
7. **"Transfer"**: *postoupit*^{pf}, "to hand over", *připisovat*^{impf} "to transfer", etc.,
8. **"Frugality"**: *utratit*^{pf} "to waste", etc.,
9. **"Taking"**: *krást*^{impf} "to steal", *vzít*^{pf} "to take", etc.,
10. **"Supply"**: *odměnit*^{pf} "to reward", *opatřovat*^{impf} "to provide", etc.

Semantic Classes for Verbs of Communication. The SFs corresponding to the verbs of communication are finer-grained than those corresponding to the verbs of exchange. Moreover, only 23 SFs from 69 SFs[7] assigned to the verbs of communication are connected by the relation of "Inheritance" in FrameNet at present.

The SFs from the hierarchy in Figure 2 belong to the most often assigned – they cover almost 37% of verbs of communication. However, we did not use the top level SF "Communication" for all the LUs because it is too general.

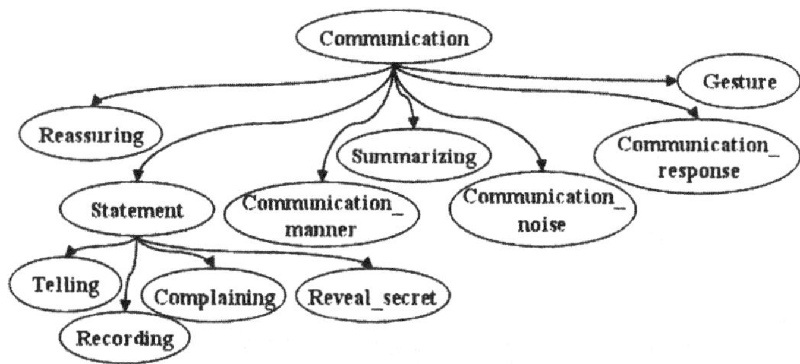

Fig. 2. The upper levels of the relation "Inheritance" of the SF "Communication"

We mapped the SF "Communication" as the semantic class to the LUs to which its descendant SFs "Communication_noise", "Communication_manner", "Gesture", and "Summarizing" were assigned. With respect to morphosyntactic properties of Czech verbs of communication, we accepted also two SFs from the lower level of this hierarchy – "Statement" and "Communication_response".[8] As a result, we obtained three semantic classes based on the above given hierarchy:

1. **'Communication'**: *šeptat*^{impf} 'to whisper', *zpívat*^{impf} 'to sing', etc.,
2. **'Statement'**: *oznámit*^{pf}, 'to announce', *tvrdit*^{impf} 'to claim', etc.,
3. **'Communication_response'**: *odvětit*^{pf} 'to reply', etc.

Three other semantic classes arose from the upper levels of the "Inheritance" hierarchy: (4) "Judgment_communication", covering also the LUs to which the SFs "Judgment_direct_address" and "Bragging" are assigned, (5) "Chatting", which includes the

[7] As in the case of the verbs of exchange, we took into account only the cases in which the annotators concurred.

[8] The SF "Reassuring" was never assigned.

LUs to which the SFs "Discussion" and "Quarreling" are assigned, and (6) "Prohibiting", which involves the LUs to which the SF "Permitting" is assigned as well.

4. **"Judgment_communication"**: *vyčítat*^{impf} "to reproach", *děkovat*^{impf} "to thank", *obvinit*^{pf} "to accuse", etc.,
5. **"Chatting"**: *diskutovat*^{biasp}, "to discuss", *hádat se*^{impf} "to quarrel", etc.,
6. **"Prohibiting"**: *zakázat*^{pf}, "to prohibit", *povolit*^{pf} "to allow", etc.

The remaining SFs frequently assigned to the verbs of communication, such as "Request"(7),[9] "Reporting" (8), and "Commitment" (9), were accepted as further candidates for separate semantic classes, despite not being linked by the "Inheritance" relation. Apparently, Czech LUs to which these SFs were assigned exhibit distinct syntactic behavior – the small differences in morphemic forms are left out of account.

7. **"Request"**: *nabádat*^{impf} "to urge", *poručit*^{pf} "to order", *prosit*^{impf} "to ask", *přemlouvat*^{impf} "to urge", *vyzvat*^{pf} "to ask", etc.,
8. **"Reporting"**: *hlásit*^{impf}, "to report", *říci*^{pf} "to tell", *udat*^{pf} "to report", etc.,
9. **"Commitment"**: *hrozit*^{impf} "to threaten", *slíbit*^{impf}, "to promise", *přísahat*^{impf} "to vow", etc.

As a result, more than 68% of verbs of communication and almost 98% of verbs of exchange in VALLEX are grouped into the semantic classes listed above. In the future, the verbs with no assigned SFs (due to the translated English LU not being covered in FrameNet) will be further examined in order to be included into the appropriate semantic class. Furthermore, we intend to add more classes to the list in agreement with the progress made in FrameNet.

6 Conclusion

We have presented an experiment in enriching a valency lexicon of Czech verbs, VALLEX, with semantic classes. We have mapped the FrameNet semantic frames to Czech verbs of communication and exchange. We have attained a satisfactory interannotator agreement, which proves the feasibility of this task.

We have exploited semantic frames from the upper levels of the relation of "Inheritance" for classifying the verbs into semantic classes. As a result, we have established nine classes for the verbs of communication, such as "Statement", "Request", "Reporting", etc., and ten classes for the verbs of exchange, such as "Giving", "Getting", "Exchange", etc. These classes cover more than 68% of verbs of communication and almost 98% of verbs of exchange.

In the future, we plan to experiment with semantic frames for other groups of verbs, especially for the verbs of motion, transport, mental action, psych verbs, etc. Moreover, we intend to make use of FrameNet frame elements as semantic labels for verb arguments.

[9] The SF "Request" was frequently mapped to the LUs along with the SF "Attempt_suasion". The LUs to which these SFs were assigned are similar in many aspects. Thus we take into account only one semantic class "Request".

References

1. Žabokrtský, Z., Lopatková, M.: Valency Information in VALLEX 2.0: Logical Structure of the Lexicon. The Prague Bulletin of Mathematical Linguistics 87, 41–60 (2007)
2. Panevová, J.: Valency Frames and the Meaning of the Sentence. In: Luelsdorff, P.L. (ed.) The Prague School of Structural and Functional Linguistics, pp. 223–243. John Benjamins, Amsterdam (1994)
3. Loper, E., Yi, S., Palmer, M.: Combining Lexical Resources: Mapping between PropBank and VerbNet. In: Proceedings of the 7th International Workshop on Computational Linguistics (2007)
4. Kingsbury, P., Palmer, M., Marcus, M.: Adding Semantic Annotation to the Penn TreeBank. In: Proceedings of the Human Language Technology Conference (2002)
5. Ruppenhofer, J., Ellsworth, M., Petruck, M.R.L., Johnson, C.R., Scheffczyk, J.: FrameNet II: Extended Theory and Practice (2006), `framenet.icsi.berkeley.edu/book/book.html`
6. Carletta, J.: Assessing agreement on classification tasks: The Kappa statistic. Computational Linguistics 22, 249–254 (1996)
7. Krippendorf, K.: Content Analysis: An Introduction to its Methodology. Sage Publications, Thousand Oaks (1980)

A Comparison of Language Models for Dialog Act Segmentation of Meeting Transcripts

Jáchym Kolář

Department of Cybernetics at Faculty of Applied Sciences,
University of West Bohemia, Univerzitní 8, CZ-306 14 Plzeň, Czech Republic
jachym@kky.zcu.cz

Abstract. This paper compares language modeling techniques for dialog act segmentation of multiparty meetings. The evaluation is twofold; we search for a convenient representation of textual information and an efficient modeling approach. The textual features capture word identities, parts-of-speech, and automatically induced classes. The models under examination include hidden event language models, maximum entropy, and BoosTexter. All presented methods are tested using both human-generated reference transcripts and automatic transcripts obtained from a state-of-the-art speech recognizer.

1 Introduction

Recent years have witnessed significant progress in the area of automatic speech recognition (ASR). Nowadays, large volumes of audio data can be transcribed automatically with reasonable accuracy. Although the application of automatic methods is extremely labor saving, raw automatic transcripts often do not have a form convenient for subsequent processing. The problem is that standard ASR systems output only a raw stream of words, leaving out important structural information such as locations of sentence or dialog act (DA) boundaries. Such locations are overt in standard text via punctuation and capitalization, but "hidden" in speech.

As proved by a number of studies, the absence of linguistic boundaries is confusing both for humans and computers. For example, Jones et al. showed that sentence breaks are critical for legibility of speech transcripts [1]. Likewise, missing sentence or DA boundaries cause significant problems to automatic downstream processes. Many natural language processing techniques (e.g., parsing, automatic summarization, information extraction and retrieval, machine translation) are typically trained on well-formatted input, such as text, and fail when dealing with unstructured streams of words. For instance, Kahn et al. reported a significant error reduction in parsing performance by using an automatic sentence boundary detection system [2].

This paper deals with automatic linguistic segmentation of multiparty meetings from the ICSI corpus [3]. The goal is to segment the meeting transcripts into meaningful utterance units. The target units herein are not defined as sentences but as DAs. Although the original manual transcripts of the ICSI corpus do contain punctuation, and thus also sentence boundaries, the punctuation is highly inconsistent. Transcribers were instructed to focus on transcribing words as quickly as possible; there was not a focus on consistency or conventions for marking punctuation. Hence, instead of using the

P. Sojka et al. (Eds.): TSD 2008, LNAI 5246, pp. 117–124, 2008.
© Springer-Verlag Berlin Heidelberg 2008

inconsistent first-pass punctuation, it was decided to employ special DA segmentation marks from the MRDA annotation project [4]. In this annotation pass, labelers carefully annotated both dialog acts and their boundaries, using using a set of segmentation conventions for the latter. For a given word sequence, the task of DA segmentation is to determine which inter-word boundaries correspond to a DA boundary. Each inter-word boundary is labeled as either a within-DA boundary or a boundary between two DAs.

There are two basic sources of information that can be used to solve the task: recognized words and prosody. Several different approaches relying on one or both of the information sources have been employed for sentence and DA segmentation [5,6,7,8,9,10]. In this paper, I focus on an effective utilization of the information contained in the recognized stream of words. Well-tuned language models (LMs) are not only important for applications where they are combined with a prosody model, but also for the applications in which we do not have access to, or cannot exploit, prosodic information.

The LM evaluation is twofold; I search both for a convenient representation of textual information and an efficient modeling approach. In terms of textual knowledge representation, I analyze contributions from word identities, parts-of-speech, and automatically induced word classes. In terms of statistical modeling, I explore three different approaches – hidden event language models, maximum entropy models, and boosting-based models. I test the methods using both reference human-generated transcripts and automatic transcripts obtained from a state-of-the-art speech recognizer. I also address the issue whether it is better to train the system on clean reference data or on data containing word recognition errors.

2 Method

2.1 Data and Experimental Setup

The ICSI meeting corpus contains approximately 72 hours of multichannel conversational English. The data were split into a training set (51 meetings, 539k words), a development set (11 meetings, 110k words), and a test set (11 meetings, 102k words). The test set contains unseen speakers, as well as speakers appearing in the training data as it is typical for the real world applications.

For model training and testing I used both human-generated reference transcripts and ASR output. Recognition results were obtained using the state-of-the-art SRI speech recognition system [11]. Word error rates for this difficult data are still quite high; the used ASR system performed at 38.2% (on the whole corpus). To generate the "reference" DA boundaries for the ASR words, the reference setup was aligned to the recognition output with the constraint that two aligned words could not occur further apart than a fixed time threshold. DA boundaries occupy 15.9% of inter-word boundaries in reference and 13.9% in automatic transcripts.

2.2 Textual Features

In this work, I do not only use simple word-based models, but also utilize textual information beyond word identities, as captured by word classes and part-of-speech tags.

I do not use chunking (or even full-parsing) features. Chunking features may slightly increase performance for well-structured speech such as broadcast news [12], but preliminary investigations showed that, because of poor chunking performance on meeting data, these features rather hurt DA segmentation accuracy on meeting speech. Hence, I did not use them in this work. The following sections describe individual groups of employed features.

Words. Word features simply capture word identities around possible DA boundaries and represent a baseline for our experiments.

Automatically Induced Classes (AIC). In language modeling, we always have to deal with data sparseness. In some tasks, we may mitigate this problem by grouping words with similar properties into word classes. The grouping reduces the number of model parameters to be estimated during training. Automatically induced classes (AIC) are derived in a data-driven way. Data-driven methods typically perform a greedy search to find the best fitting class for each word given an objective function.

The clustering algorithm I used [13] minimizes perplexity of the induced class-based n-gram with respect to the provided word bigram counts. The DA boundary token was excluded from merging, however, its statistics still affected the clustering. The algorithm works as follows. Initially, each word is placed into its own class. Then, the classes are iteratively merged until the desired number of clusters is reached. The resulting classes are mutually exclusive, i.e., each word is only mapped into a single class. In every step of the algorithm, the overall perplexity is minimized by joining the pair of classes maximizing the mean mutual information of adjacent classes

$$I(c_1, c_2) = \sum_{c_1, c_2 \in C} P(c_1, c_2) \log \frac{P(c_1, c_2)}{P(c_1)P(c_2)} \tag{1}$$

A crucial parameter of the word clustering algorithm is the target number of classes. The optimal number was empirically estimated on development data by evaluating performance of models with a different granularity. I started with a 300-class model and then was gradually decreasing the number of classes by 25 in each iteration. The optimal number of classes was estimated as 100.

I also tested a model that mixes AICs and frequent words by excluding them from class merging. This approach can be viewed as a form of back off; we back off from words to classes for rare words but keep word identities for frequent words. I have tested various numbers of left out words in combination with individual class granularities, but have never achieved better results than for the 100 classes with no excluded words.

Parts-of-Speech (POS). The AICs reflect word usage in our datasets, but do not form clusters with a clearly interpretable linguistic meaning. In contrast, part-of-speech (POS) tags describe grammatical features of words. The POS tags were obtained using the TnT tagger [14] which was tailored for conversational English. The tagger was trained using hand-labeled data from the Switchboard Treebank corpus. To achieve a better match with speech recognition output used in testing, punctuation and capitalization information was removed before using the data for tagger training [12].

Same as for AICs, I also tested mixed models. In contrast with AICs, mixing of frequent words with POS of infrequent words yielded an improvement. The reason is that while the automatic clustering algorithm takes into account bigram counts containing the DA boundary token and thus is aware of strong DA boundary indicators, POS classes are purely grammatical. By keeping the frequent words we also keep some strong boundary indicators. Optimizing the model on the development data, I ended up with 500 most frequent words being kept and not replaced by POS tags.

2.3 Statistical Language Models

Hidden Event Language Models (HELM). In speech recognition as well as in a number of other language modeling tasks, the role of the language model is to predict the next word given the word history. In contrast, the goal of language modeling in our task is to estimate the probability that an event, such as DA boundary, occurs in the given word context. Because these events are not explicitly present in the speech signal, they are called *hidden*. The hidden event LMs (HELMs) [5] describe the joint probability of words and hidden events $P(W, E)$ in an HMM. In this case, the HMM hidden variable is the type of the event (including "no-event"). The states of the model correspond to word/event pairs and the observations to words.

The model is trained by explicitly including the DA boundary as a token in the vocabulary in an n-gram LM. I used trigram LMs with modified Kneser-Ney smoothing [15]. In addition, Witten-Bell smoothing was employed for unigrams in class-based models (both AIC and POS) since the training data for these models do not contain any unigram singletons necessary for the Kneser-Ney method. During testing, the model performs the forward-backward decoding to find the DA boundaries given the word sequence. An implementation of the HELM is available in the SRILM toolkit [16].

The HELM does not allow a direct combination of multiple knowledge sources. Thus, I trained a separate model for each data stream and combined the models using a linear interpolation with weights estimated on development data.

Maximum Entropy Models. The above described HELM is a generative model. It means that during training, it does not directly maximize the posterior probabilities of the correct classes. On the other hand, Maximum Entropy (MaxEnt) [17] is a discriminative model which is trained to directly discriminate among the possible target classes. This setup avoids the mismatch between training and using the model in testing. MaxEnt framework also allows a natural combination of multiple knowledge sources within a single model, no additional model combination is necessary. MaxEnt belongs to the exponential (or log-linear) family of classifiers, i.e. the features are combined linearly, and then used as an exponent

$$P(y|x) = \frac{1}{Z(x)} \exp\left(\sum_i \alpha_i f_i(x, y)\right) \tag{2}$$

where $Z(x)$ is a normalization factor ensuring that $\sum_y p(y|x) = 1$.

An important feature of MaxEnt models is that they are prone to overfitting. To overcome this drawback, I have used smoothing with Gaussian priors that penalizes large weights. For all experiments with MaxEnt, I employed the `MegaM` toolkit.[1] For each feature group, the used features included all n-grams up to trigrams spanning across or neighboring with the inter-word boundary in question. I also added a binary feature indicating whether the word before the boundary is identical with the following word. This feature aims to capture word repetitions.

Boosting-Based Models (BoosTexter). Boosting is an aggregating machine learning method combining many weak learning algorithms to produce an accurate classifier. Each weak classifier is built based on the outputs of previous classifiers, focusing on the samples that were formerly classified incorrectly; the algorithm generates weak classification rules by calling the weak learners repeatedly in series of rounds. This approach can generally be combined with any "weak" classifier. In this work, an algorithm called BoosTexter [18] was employed.

BoosTexter was initially designed for the task of text categorization, employment of this method for tasks related to DA segmentation was firstly presented in [9,19]. The method combines weak classifiers having a basic form of one-level decision trees (stumps) using confidence-rated predictions. The test at the root of each tree can check for the presence or absence of an n-gram, or for a value of a continuous or categorical feature. Same as with MaxEnt, multiple knowledge sources can be integrated in a single model. While BoosTexter is known to be powerful when combining lexical and prosodic features within a single integral model, herein, I aim to evaluate how powerful it is when only a language model is used. In my experiments, the ICSI reimplementation of the original BoosTexter method was employed.[2] The used textual features had the same form as in the MaxEnt model.

2.4 Evaluation Metric

I measure DA segmentation performance using a "boundary error rate" (BER):

$$BER = \frac{I + M}{N} \quad [\%] \tag{3}$$

where I denotes the number of false DA boundary insertions, M the number of misses, and N the number of words in the test set.

3 Experimental Results

Table 1 presents experimental results for all three models (HMM, Maxent, and BoosTexter), all feature sets (words, AIC, POS, and POS mixed with words), and training and test conditions. The models for segmentation of human transcripts were trained on reference words. For testing on ASR data, I tried to use both true and recognized words for training, and compared performance of the models.

In reference conditions, the best models based on a single feature group were MaxEnt for mixed POS and AICs, and HELM for words. On the other hand, the models

[1] `http://hal3.name/megam`

[2] `http://code.google.com/p/icsiboost/`

Table 1. DA segmentation results for individual language models, feature sets, and experimental setups [BER %] (REF=Reference human transcripts, ASR=Automatic transcripts, AIC=Automatically Induced Classes with 100 clusters, POS=Parts-of-speech, POSmixed=Parts-of-speech for infrequent words with 500 most frequent words kept. "Chance" refers to a model which classifies all test samples as within-DA boundaries).

Model	Used Features	Train/Test Setup		
		REF/REF	**REF/ASR**	**ASR/ASR**
Chance	—	15.92%	13.85%	13.85%
HELM	Words	7.45%	9.41%	9.50%
	AIC	7.58%	9.70%	9.78%
	POS	10.62%	12.06%	11.85%
	POSmixed	7.65%	9.57%	9.59%
	Words+AIC	7.11%	9.25%	9.18%
	Words+POSmixed	7.23%	9.25%	9.31%
	Words+AIC+POSmixed	**7.02%**	**9.12%**	9.12%
MaxEnt	Words	7.50%	9.38%	9.38%
	AIC	7.42%	9.44%	9.37%
	POS	10.52%	11.79%	11.80%
	POSmixed	7.26%	9.23%	9.25%
	Words+AIC	7.19%	9.25%	9.21%
	Words+POSmixed	7.27%	9.27%	9.25%
	Words+AIC+POSmixed	**7.15%**	9.24%	**9.16%**
BoosTexter	Words	7.70%	9.52%	9.49%
	AIC	7.61%	9.50%	9.53%
	POS	10.87%	12.03%	11.13%
	POSmixed	7.68%	9.45%	9.46%
	Words+AIC	7.50%	9.42%	9.40%
	Words+POSmixed	7.66%	9.44%	9.45%
	Words+AIC+POSmixed	**7.46%**	**9.40%**	9.40%

only using POS information performed poorly. A comparison of POS and POSmixed shows that POS features are not sufficient indicators of DA boundaries and information provided by some frequent cue words is necessary to achieve satisfactory performance. In terms of a modeling approach comparison, it is interesting to observe that the generative HELM model is better in dealing with word information while the discriminative MaxEnt model better captures class information (both AIC and POSmixed). The BoosTexter model always performed worse than the other two models.

The results also indicate that an improvement is achieved when word information is combined with class information. The best result ($BER = 7.02\%$) is obtained when all three information sources are combined in the HELM model. The improvement over the baseline word-based model is statistically significant at $p < 10^{-23}$ using the Sign test. The difference between HELM and Boostexter is significant at $p < 10^{-13}$, and the difference between HELM and MaxEnt at $p < 0.02$. Of the other two models, MaxEnt outperformed BoosTexter (BER : 7.15% vs. 7.46%) which is significant at $p < 10^{-9}$.

As well as in reference conditions, MaxEnt for mixed POS was the best single model in ASR-based tests. Unlike reference conditions, MaxEnt was also the best model for capturing word information. The combination of all three knowledge sources was helpful once again, the best performing combined model was HELM ($BER = 9.12\%$) while BoosTexter was the worst. Both HELM and MaxEnt show a significant outperformance of the BoosTexter model ($p < 10^{-4}$). In contrast, the difference between HELM and MaxEnt is not significant.

A comparison of models trained on clean and erroneous data shows the following. While for HELM and BoosTexter the performance was almost the same, for the MaxEnt model, I got better results when training on automatic transcripts. However, even for the MaxEnt model, the difference in *BER* is only significant at $p < 0.08$.

4 Summary and Conclusions

I have explored the use of textual information for DA boundary detection in both human- and ASR-generated transcripts of multiparty meetings. I have analyzed contributions from word identities, parts-of-speech, and automatically induced word classes, and compared three statistical modeling approaches – HELM, MaxEnt, and BoosTexter.

Among others, the results indicate that POS information is only helpful when the most frequent words are kept and not replaced by POS tags. For both test conditions, the best results were achieved when all information sources were combined. The best performing combined model was HELM, achieving $BER = 7.02\%$ in reference and $BER = 9.12\%$ in ASR conditions. On the other hand, the boosting-based model was always the worst. While this model is powerful when combining prosodic and lexical information, it does not represent a good approach when only textual features are used and prosodic information is not accessible.

A comparison of models trained on clean and ASR data shows that for none of the models, significant improvement is achieved by training on ASR. The HELM and BoosTexter models perform approximately the same for both training setups, and the modest gain achieved by the ASR-trained MaxEnt model is not statistically significant.

Acknowledgments

This work was supported by the Ministry of Education of the Czech Republic under projects 2C06020 and ME909. In addition, I used the METACentrum computing clusters sponsored under the research program MSM6383917201. I also thank Yang Liu at UT Dallas for providing me with her models for the TnT tagger.

References

1. Jones, D., Wolf, F., Gibson, E., Williams, E., Fedorenko, E., Reynolds, D., Zissman, M.: Measuring the readability of automatic speech-to-text transcripts. In: Proc. Eurospeech, Geneva, Switzerland (2003)
2. Kahn, J.G., Ostendorf, M., Chelba, C.: Parsing conversational speech using enhanced segmentation. In: Proc. HLT-NAACL 2004, Boston, MA, USA (2004)
3. Janin, A., Baron, D., Edwards, J., Ellis, D., Gelbart, D., Morgan, N., Peskin, B., Pfau, T., Shriberg, E., Stolcke, A., Wooters, C.: The ICSI meeting corpus. In: IEEE ICASSP 2003, Hong Kong (2003)

4. Dhillon, R., Bhagat, S., Carvey, H., Shriberg, E.: Meeting recorder project: Dialog act labeling guide. Technical Report TR-04-002, ICSI, Berkeley, CA, USA (2004)
5. Stolcke, A., Shriberg, E.: Automatic linguistic segmentation of conversational speech. In: Proc. ICSLP, Philadelphia, PA, USA (1996)
6. Warnke, V., Kompe, R., Niemann, H., Nöth, E.: Integrated dialog act segmentation and classification using prosodic features and language models. In: Proc. Europeech 1997, Rhodes, Greece (1997)
7. Shriberg, E., Stolcke, A., Hakkani-Tür, D., Tür, G.: Prosody-based automatic segmentation of speech into sentences and topics. Speech Communication 32(1-2), 127–154 (2000)
8. Akita, Y., Saikou, M., Nanjo, H., Kawahara, T.: Sentence boundary detection of spontaneous Japanese using statistical language model and support vector machines. In: Proc. INTERSPEECH 2006 – ICSLP, Pittsburgh, PA, USA (2006)
9. Zimmermann, M., Hakkani-Tur, D., Fung, J., Mirghafori, N., Gottlieb, L., Shriberg, E., Liu, Y.: The ICSI+ multilingual sentence segmentation system. In: Proc. INTERSPEECH 2006 – ICSLP, pp. 117–120 (2006)
10. Kolář, J., Liu, Y., Shriberg, E.: Speaker adaptation of language models for automatic dialog act segmentation of meetings. In: Proc. INTERSPEECH 2007, Antwerp, Belgium (2007)
11. Stolcke, A., Chen, B., Franco, H., Gadde, V.R.R., Graciearena, M., Hwang, M.Y., Kirchhoff, K., Mandal, A., Morgan, N., Lei, X., Ng, T., Ostendorf, M., Sönmez, K., Venkataraman, A., Vergyri, D., Wang, W., Zheng, J., Zhu, Q.: Recent innovations in speech-to-text transcription at SRI-ICSI-UW. IEEE Trans. on Audio, Speech, and Language Processing 14(5) (2006)
12. Liu, Y.: Structural Event Detection for Rich Transcription of Speech. Ph.D. thesis, Purdue University, W. Lafayette, IN, USA (2004)
13. Brown, P., Pietra, V.D., de Souza, P., Lai, J., Mercer, R.: Class-based n-gram models of natural language. Computational Linguistics 18(4), 467–479 (1992)
14. Brants, T.: TnT – A statistical part-of-speech tagger. In: Proc. ANLP 2000, Seattle, WA, USA (2000)
15. Chen, S., Goodman, J.: An empirical study of smoothing techniques for language modeling. Technical report, Harvard University, USA (1998)
16. Stolcke, A.: SRILM – An extensible language modeling toolkit. In: Proc. ICSLP 2002, Denver, CO, USA (2002)
17. Berger, A., Della Pietra, S.A., Della Pietra, V.J.: A maximum entropy approach to natural language processing. Computational Linguistics 22(1), 39–71 (1996)
18. Schapire, R., Singer, Y.: BoosTexter: A boosting-based system for text categorization. Machine Learning 39(2–3), 135–168 (2000)
19. Kolář, J., Shriberg, E., Liu, Y.: Using prosody for automatic sentence segmentation of multi-party meetings. In: Sojka, P., Kopeček, I., Pala, K. (eds.) TSD 2006. LNCS (LNAI), vol. 4188, pp. 629–636. Springer, Heidelberg (2006)

New Methods for Pruning and Ordering of Syntax Parsing Trees

Comparison and Combination

Vojtěch Kovář, Aleš Horák, and Vladimír Kadlec

Faculty of Informatics, Masaryk University
Botanická 68a, 602 00 Brno, Czech Republic
{xkovar3,hales,xkadlec}@fi.muni.cz

Abstract. Most robust rule-based syntax parsing techniques face the problem of high number of possible syntax trees as the output. There are two possible solutions to this: either release the request for robustness and provide special rules for uncovered phenomena, or equip the parser with filtering and ordering techniques. We describe the implementation and evaluation of the latter approach. In this paper, we present new techniques of pruning and ordering the resulting syntax trees in the Czech parser synt. We describe the principles of the methods and present results of measurements of effectiveness of these methods both per method and in combination, as computed for 10,000 corpus sentences.

1 Introduction

Natural language syntactic analysis provides a basis for most of the techniques of advanced text processing ranging from intelligent search engines to semantic processing of the textual knowledge. The syntax parsing algorithms and the quality of the parsing results (derivation trees) depend heavily on the character of the language being analysed – while the quality of parsing of analytical languages (e.g. English) has already achieved nearly satisfactorily results [1,2], the analysis of free word order languages still faces many problems either in the form of huge number of rules or output trees or offers insufficient precision or coverage on corpus texts [3,4,5].

The Czech language is a representative of really free word order language. This paper concerns the Czech parser called `synt` [6] that features a developed meta-grammar of Czech and a fast parsing mechanism based on chart parsing techniques. The main part of the current development of the parser concentrates on providing techniques for working with highly ambiguous grammar. The first parts of this task are already solved – `synt` is able to work efficiently with large ambiguous grammars of tens of thousands of rules and process the input very fast. The resulting packed shared forest[1] is obtained in average within 0.07s per sentence. All possible output trees can then be ordered using *tree ranking* functionality and *n*-best trees can be printed keeping the time polynomial even when working with exponential data structures.

We have developed and implemented six methods of effective assignment of the tree ranks to the parsing structures, out of which two methods are presented in details and

[1] Similar to chart in chart parsing.

P. Sojka et al. (Eds.): TSD 2008, LNAI 5246, pp. 125–131, 2008.
© Springer-Verlag Berlin Heidelberg 2008

evaluated in this paper – *pruning chart by rule levels* and the *beautified chart* method. We describe the principle of the methods, present their impact on parsing results and on parsing time. We also show results for combination of the two methods.

2 Compared Methods

In this section, we roughly describe principles of the compared methods and outline their possible influence on the analysis result.

The actual grammar of the `synt` parser that is being edited by human experts is a form of a complex structured meta-grammar, which contains about 240 meta-rules. This grammar is then automatically expanded to two forms that are actually used for analysis and contain about 3,000 and 10,000 rules. The parser does not have any training phase used to learn the context dependencies of the input texts. All rules that guide the analysis process are developed by linguistic and computer experts.

The involved chart parsing algorithm uses a modified *Head-driven chart parser* [7], which provides very fast parsing of real-text sentences.

2.1 The Beautified Chart Method

Due to high complexity of Czech grammar constructs, the meta-grammar of Czech used in the `synt` parser generates a number of rules that have low (or "technical") impact to the linguistic interpretation of the results. These rules cause the fact that many nodes in the resulting derivation trees have mostly technical character.

The *beautified chart* method is based on removing the technical nodes from the derivation trees and therefore simplifying the trees. It was found out that more "regular" trees are often represented by only one "beautified" tree [8]. Thus, we can reduce the output derivation trees numbers by giving "beautified" trees on the parser output. This technique is not complicated for a single tree, however, to keep the analysis time in polynomial, the implementation has to work with the packed shared forest data structures rather than trees allowing us to work efficiently with very high numbers of possible output trees.

The *beautified chart method* and its implementation is detailed in [8].

2.2 Pruning Chart by Rule Levels

This method is newly implemented in the parser and has not been published before.[2] It is based on pruning the internal parser data structures according to so-called *rule levels*, see below.

A *rule level* is a function, that assigns a natural number to each grammar rule. In the following the term "rule level of a grammar rule" denotes the number resulting from an application of this function to the rule.

In the `synt` meta-grammar, a rule level in range from 0 to 99[3] is assigned to each meta-rule. The linguistic interpretation of the rule level is that the higher the level of a

[2] The method was implemented by Vladimír Kadlec in [9].

[3] Actually, the level 99 is used as a kind of "infinity" in the system. The maximum rule level number that is currently present in the meta-grammar is 9.

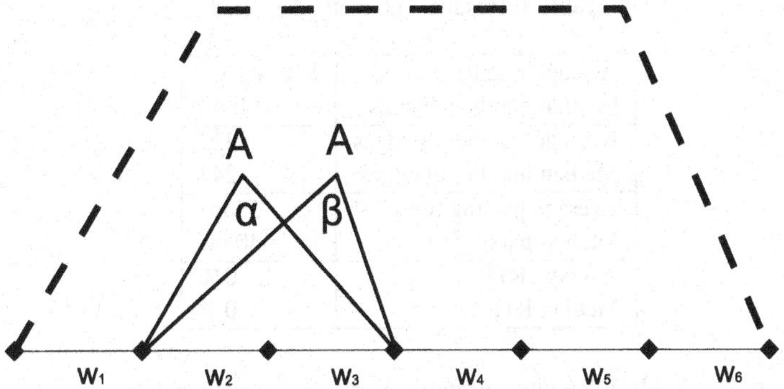

Fig. 1. Filtering by rule levels. Two sub-forests with grammar rules $l_1 : A \rightarrow \alpha$ and $l_2 : A \rightarrow \beta$ in their roots. One of them is filtered out, if these rules have different rule levels.

rule the less frequent the grammatic phenomenon described by the rule is. Rules with level 0 are also called "basic rules" and their usage is the most common. Rules with higher level are less common and sentences analysed with these rules are less fluent. However, they significantly increase robustness and coverage of the parser, so they are not to be considered as "bad" rules. The rule levels are set manually by grammar experts and are used only in specific cases (95 % meta-rules are on level 0). Even this small number of higher-level meta-rules allows to increase the coverage of the parser by about 30 %.

Before the implementation of the presented method, the rule levels were used in two ways:

1. the rule with a higher rule level received a "penalty" to the tree rank, and
2. at the beginning of the analysis the highest rule level was specified – we could thus obtain an analysis for level 0 or any higher level.

The method described in this section consists in pruning the packed shared forest structure (the result of the analysis), so that the rules with higher level number, that are not necessarily needed for the sentence acceptance, are removed from the structure. This causes that derivation trees containing less common grammatical constructions are removed from the resulting tree set and total number of trees decreases.

The idea is, that when two (or more) rules with the same non-terminal on the left hand side of the rule succeed during the parsing process for the same input, then an application of the higher level rule(s) is wrong. To be more precise, if the specific grammar rule covers the same input as another grammar rule with higher level and beginning with the same non-terminal, then the higher level rule is refused in this part of the resulting structure.

The packed shared forest structure [6] is used to represent the parsing results. We can define the filtering method in terms of the chart parsing algorithm: If there are edges $E_1 = [l_1 : A \rightarrow \bullet\alpha_\bullet, i, j]$ and $E_2 = [l_2 : A \rightarrow \bullet\beta_\bullet, i, j]$ in the chart (l_1 and l_2 are rule level numbers), then delete the edge E_1 if $l_1 > l_2$, or E_2 if $l_1 < l_2$. If $l_1 = l_2$, then both edges are kept in the chart. The Figure 1 shows an example of such rules, $w_1, w_2, \ldots, w_6$ represent words in the input sentence.

Table 1. Results – no method applied

Average number of trees	$8.49 * 10^{15}$
Median number of trees	1344
Average number of edges	322
Median number of edges	242
Average parsing time	53 ms
Median parsing time	40 ms
Average RTP	0 %
Median RTP	0 %

Table 2. Results – *pruning chart by rule levels* applied

Average number of trees	$6.38 * 10^{15}$
Median number of trees	664
Average number of edges	294
Median number of edges	223
Average parsing time	56 ms
Median parsing time	50 ms
Average RTP	37 %
Median RTP	39 %

Note that this kind of filtering is different from setting the highest rule level of the whole analysis. The presented method is local to the specific node in the parsing result – the packed shared forest.

As we said before, the rules with higher levels cover less frequent grammatic phenomena. Therefore trees removed by this method are redundant – they provide a "complicated" explanation of a phenomenon while a "straightforward" explanation is also available. This is also the main argument for the claim that we do not lose precision when applying this method.

3 Comparison and Combination of the Reducing Techniques

In this section, we describe an experiment of evaluation of the effectiveness of the methods with real corpus sentences, their comparison and evaluation of their combination.

3.1 The Experiment Setup

To evaluate the effectiveness of the methods, the analysis of testing sentences ran for four times – without the filtering techniques, with *beautified chart* enabled, with *pruning chart by rule levels* enabled and with both these methods in combination. The testing data set contained the first 10,000 sentences from the DESAM corpus [10]. The corpus contains mostly texts from Czech newspapers and all word morphological tags in the sentences were manually disambiguated. The complexity of the analyzed sentences counted by sentence words goes up to 116 words in a sentence with the average of 19 words (median 17 words) per sentence.

Table 3. Results – *beautified chart* applied

Average number of trees	$9.04 * 10^{13}$
Median number of trees	315
Average number of edges	58
Median number of edges	49
Average parsing time	78 ms
Median parsing time	50 ms
Average RTP	60 %
Median RTP	69 %

Table 4. Results – both methods applied

Average number of trees	$8.89 * 10^{13}$
Median number of trees	240
Average number of edges	55
Median number of edges	47
Average parsing time	76 ms
Median parsing time	50 ms
Average RTP	64 %
Median RTP	75 %

The following quantities were measured during the experiment:

- the total number of derivation trees,
- the total number of successful chart edges,
- the parsing time,
- the removed trees percentage (RTP).

The RTP quantity ranges from 0 to 100 and determines the effect of the particular method on one sentence – it is defined as $100 * (nf - np)/nf$ where *nf* stands for number of output trees with no reducing techniques enabled and *np* stands for number of output trees with relevant reducing technique(s) enabled.

The results of all measurements are shown in the Tables 1–4 and in the Figure 2.

3.2 Results and Interpretation

As we can see from the resulting tables, the *beautified chart* method is more successful than *pruning chart by rule levels*. However, the latter method can also significantly decrease numbers of output trees and it is also slightly less time-consuming.

We can also see that the combination of the methods gives the best results in case of number of trees reduction, so the *pruning chart by rule levels* method gives us a new advance in reducing the tree numbers. The time complexity increase is tolerable.

In global, by combining the two presented methods, we are able to reduce the output trees numbers to 20 – 25 percent of the original number without losing precision and with tolerable time-complexity increase.

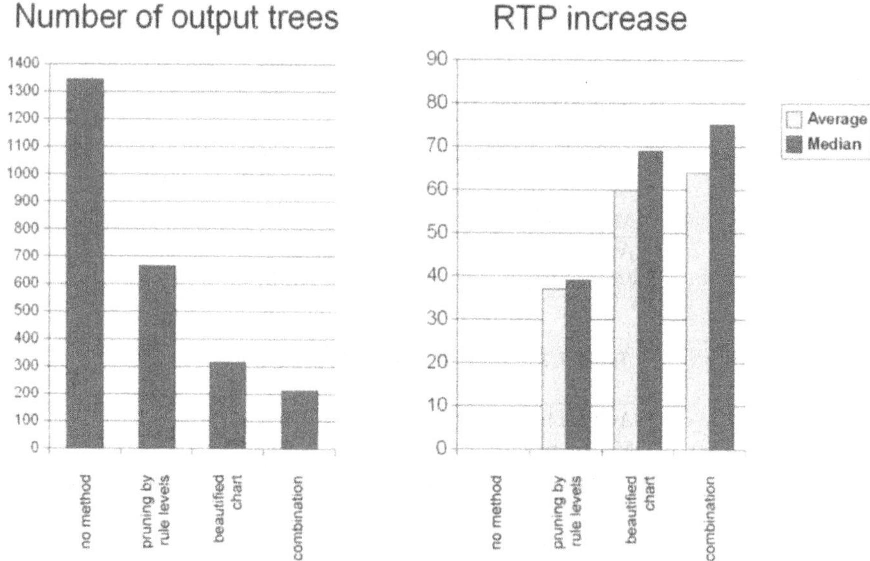

Fig. 2. An illustration of the number of trees reduction

4 Conclusions and Future Directions

In the paper, two methods that are used for enhancing the filtering and ordering of the output derivation trees in the robust rule-based Czech parser `synt` were presented. We have described the details of the methods and measured their effectiveness on large sample of corpus sentences.

We have compared these two methods separately and in combination. The results of the experiment show that the combination can prune the output derivation trees down to about 20 percent of the original numbers.

In the future development, we want to explore and evaluate other methods aimed at filtering and ordering the output syntax trees and in the best analysis selection.

Acknowledgements

This work has been partly supported by the Academy of Sciences of Czech Republic under the projects T100300414 and T100300419 and by the Ministry of Education of CR within the Center of basic research LC536 and in the National Research Programme II project 2C06009.

References

1. Baumann, S., Brinckmann, C., Hansen-Schirra, S., et al.: The muli project: Annotation and analysis of information structure in German and English. In: Proceedings of the LREC 2004 Conference, Lisboa, Portugal (2004)
2. Radford, A.: Minimalist Syntax. Cambridge University Press, Chicago (1993)

3. Horák, A., Smrž, P.: Best analysis selection in inflectional languages. In: Proceedings of the 19th international conference on Computational linguistics, Taipei, Taiwan, pp. 363–368. Association for Computational Linguistics (2002)
4. Jaeger, T., Gerassimova, V.: Bulgarian word order and the role of the direct object clitic in LFG. In: Butt, M., King, T. (eds.) Proceedings of the LFG02 Conference. CSLI Publications, Stanford (2002)
5. Hoffman, B.: The Computational Analysis of the Syntax and Interpretation of Free Word Order in Turkish. Ph.D. thesis, University of Pennsylvania, Philadelphia (1995)
6. Horák, A., Kadlec, V.: New Meta-grammar Constructs in Czech Language Parser synt. In: Proceedings of Text, Speech and Dialogue 2005, Karlovy Vary, Czech Republic. LNCS (LNAI), pp. 85–92. Springer, Heidelberg (2005)
7. Horák, A., Kadlec, V., Smrž, P.: Enhancing Best Analysis Selection and Parser Comparison. In: Sojka, P., Kopeček, I., Pala, K. (eds.) TSD 2002. LNCS (LNAI), vol. 2448, pp. 461–467. Springer, Heidelberg (2002)
8. Kovář, V., Horák, A.: Reducing the Number of Resulting Parsing Trees for the Czech Language Using the Beautified Chart Method. In: Proceedings of 3rd Language and Technology Conference, Poznań, Wydawnictwo Poznańskie, pp. 433–437 (2007)
9. Kadlec, V.: Syntactic analysis of natural languages based on context-free grammar backbone. Ph.D. thesis, Masaryk University, Czech Republic (2008)
10. Pala, K., Rychlý, P., Smrž, P.: DESAM – annotated corpus for Czech. In: Proceedings of SOFSEM 1997. LNCS, vol. 1338, pp. 523–530. Springer, Heidelberg (1997)

Web People Search with Domain Ranking

Zornitsa Kozareva[1], Rumen Moraliyski[2], and Gaël Dias[2]

[1] University of Alicante, Campus de San Vicente, Spain
zkozareva@dlsi.ua.es

[2] University of Beira Interior, Covilhã 6201-001, Portugal
rumen@hultig.di.ubi.pt, ddg@di.ubi.pt

Abstract. The world wide web is the biggest information source which people consult daily for facts and events. Studies demonstrate that 30% of the searches relate to proper names such as organizations, actors, singers, books or movie titles. However, a serious problem is posed by the high level of ambiguity where one and the same name can be shared by different individuals or even across different proper name categories. In order to provide faster and more relevant access to the requested information, current research focuses on the clustering of web pages related to the same individual. In this paper, we focus on the resolution of the web people search problem through the integration of domain information.

1 Introduction and Related Work

Most current web searches relate to proper names. However, there is a high level of ambiguity where multiple individuals share the same name and thus the harvesting and the retrieval of relevant information becomes more difficult. This issue directed current researchers towards the definition of a new task called Web People Search (WePS) [1]. The main idea of the task is to be able to group together e.g., cluster web pages that relate to different people sharing the same name.

Prior to the WePS task [2,3] focused on similar task called name discrimination. Name discrimination stands for being able to identify the different individuals behind a name, but since the number of clusters is unknown, they called the task discrimination. [3] and [4] focused on the discrimination of cities, countries, professors among other categories and conducted evaluation in Spanish, Romanian, English and Bulgarian languages. The obtained results are promising showing that the discrimination can be performed not only on multi categories but also on multilingual level. The approach of [3] used co-occurrence vectors that represent the word co-occurrences derived from the document with the ambiguous name. The hypothesis is that names appearing in different documents and having similar co-occurrence context vectors, are highly probable to refer to the same individual. [5] reported the adaptation of the vector co-occurrence approach in a web environment. [6] applied semantic information derived from latent semantic indexing and relevant domains in order to estimate the similarity among snippets containing ambiguous names.

In the WePS challenge, there were 16 participants all using different sources of information. The CU-COMSEM [7] system ranked first and it was using token-based and phrase-based information. The modeled features represent the importance of the

P. Sojka et al. (Eds.): TSD 2008, LNAI 5246, pp. 133–140, 2008.
© Springer-Verlag Berlin Heidelberg 2008

word in a document and in the whole document collection, the urls, the overlapping of noun phrases and named entities (NE). In contrast to this rich feature based system, the second best performing system IRST-BP [8] used only information from the NEs and their co-reference chaining. UBC-AS [9] system used word co-occurrence information to build a graph. The web page clustering was based on the HITS ranking. The system ranked on final position which indicates that simple word co-occurrence information is not sufficient for the WePS task.

Our main motivations in this paper are to develop a graph-based ranking criteria for synsets in a domain. The approach is flexible to the changing of the number of domain labels or the addition and deletion of synsets in *WordNet*. The second objective is to use the domain information for the resolution of WePS where each word is represented by its corresponding domains and associated weights. Web pages with similar domains are grouped together, because our hypothesis is that these pages are most likely to refer to the same individual.

The rest of the paper is organized as follows: in Section 2 we describe the *WordNet Domains* and the *eXtended WordNet* resources, Section 3 focuses on the graph construction and the synset ranking. Section 4 describes the integration of the domain information for the resolution of WePS. The approach is evaluated in Section 5 and we conclude in Section 6.

2 Domain Information

2.1 WordNet Domains

WordNet Domains[1] is a lexical resource developed by Magnini et. al, [10]. There are 167 hierarchically organized domain labels which are assigned to the *WordNet 2.0* synsets in semi-automatic way. For instance, the noun *bank* with senses 1 and 3 denoted as *bank#n#*1 and *bank#n#*3 belongs to the domain ECONOMY, *bank#n#*9 belongs to the domain ARCHITECTURE, while *bank#n#*2 belongs to the domains GEOGRAPHY and GEOLOGY. An important property of *WordNet Domains* is the ability to group together under the same domain words from different syntactic categories such as the verb *operate#v#*7 and the nouns *doctor#n#*1 and *hospital#n#*1 under the domain MEDICINE. The domains act also as semantic bridges between words which might be unrelated or far from each other in the *WordNet* hierarchy. Unfortunately, current *WordNet Domains* does not provide information about the importance of a domain to a word. Therefore, we decided to introduce a graph-based criteria which ranks a set of concepts according to their importance within a domain. Our main motivation for graph ranking is the flexibility of the technique in terms of future changes of the number of domains or the addition/deletion of synsets.

2.2 eXtended WordNet

For the creation of the graph, we used the information of the synset and the words in the gloss of the synset. According to [11], if a synset s_i has positive (negative) meaning,

[1] http://wndomains.itc.it/wordnetdomains.html

it can be interpreted as an indication that the synsets s_k defined by the terms occurring in the gloss of s_i are themselves with positive (negative) meaning. We adapted the idea for the domain ranking, assuming that the domains of the words in the gloss determine the domain of the synset and vice verse.

To create the graph, we used *eXtended WordNet*[2] because the words in the gloss are disambiguated and lemmatized. *eXtended WordNet* has been automatically generated, which means that the associations between the words and the synsets are likely to be incorrect sometimes and this can bring noise to our graph-based ranking method.

3 Capturing the Importance of a Synset in a Domain

3.1 Synset Graph Representation

We construct a graph G=$\langle N, E\rangle$, with vertex N corresponding to the *eXtended WordNet* synsets s_i and the synsets from the gloss definitions s_k. A directed edge E represents the link from s_i to s_k if the gloss of s_i contains a term in s_k. The graph is constructed from nouns, verbs, adjectives and adverbs, because they carry out the domain information. The graph has no outcoming edges for vertex whose synsets have missing glosses.

3.2 Synset Ranking with PageRank

In order to determine which synsets s_i are more important for a domain D_p, we use the well known PageRank (PR) algorithm [12]. PageRank is a random-walk model initially developed for Web page ranking. It assigns numerical weighting to the vertices in the graph with the purpose of "measuring" the relative importance within the set.

The main idea of PR is like "voting". When one vertex links to another vertex, it casts a vote. The higher the number of votes a vertex has, the higher its importance becomes. Moreover, the importance of a vertex casting a vote determines how important the vote itself is. The score of PR associated with a node N is defined as:

$$PR(N) = (1-d) * e_j + d * \sum_{N_i \in E} \frac{PR(N_i)}{outdegree(N_i)} \quad (1)$$

where $d \in [0; 1]$ is the so called damping factor and it is usually set to 0.85. The $outdegree(N_i)$ corresponds to the number of outgoing edges from vertex N_i, E is the set of vertices having edge directed to N. The values e_j sum up to unity. We set this variables to be non zero for the senses of the domain for which PR is calculated and 0 for the rest of the words senses. Thus the words from the domain have greater impact on the ranking that the others.

The algorithm is iterative, initially assuming that each node has equal weight and throughout the iterations, the vertices accumulate weight. When the algorithm converges, e.g. the weights of the vertices between two consecutive iterations change less than a certain threshold, PR lists in descending order the important synset $s_1, \ldots, s_k$ for domain D_p. We ran the algorithm for each one of the *WordNet Domains* setting e_j.

[2] xwn.hlt.utdallas.edu/

For instance, for the domain ECONOMY synset s_{10} occurs in the glosses of the synsets s_1, s_4 and s_5. As s_1, s_4, s_5 belong to the domain ECONOMY and have big weights, we assume that it is highly probable for s_{10} to belongs to the domain ECONOMY. Finally, for each domain we have a list of the synsets and a PR score which indicates how probable for the synset is to belong to the domain. The PR domain approach can be used as word sense disambiguation or sense reduction technique. For instance, the word *bank#n* has ten senses and seven domains, but PR estimated that *bank* is significantly important to the domain ECONOMY and but not so informative in the other six domains.

4 WePS with Domain Information

For each ambiguous person name, we have 100 web pages. The first step of our WePS approach consists in removing the *html* tags. Then the pages are tokenized, split into sentences and the words are POS tagged using the Montylingua toolkit [13]. We have automatically discarded sentences with length of one or two words, because they tend to refer to web links or menu titles.

4.1 WePS Domain Ranking

For each document in the ambiguous person name collection, we map the words to their corresponding domains. In order to determine the most relevant domains for a document, we introduce three different domain ranking schemes:

tf*idf: each sense s_{p,w_i} of word w_i in the documents is substituted by its *WordNet Domains*. For each domain we calculate tf*idf score which measures the relevance of the domain in the document. Evidently, a domain that is seen in all documents has low tf*idf score and is considered uninformative for the document and the whole document collection.

binary vector: each sense s_{p,w_i} of word w_i in the documents is substituted by its *WordNet Domains*. The final score of a domain in a document is the cumulative count of all words that pointed to the domain.

PR: for each sense s_{p,w_i} of word w_i in the documents, we consider only the domain with the highest PR weight. The rest of domains are discarded.

4.2 Document Similarity

In order to estimate the proximity of the web pages, we used the *cosine* and *Euclidean* distance measures. Our hypothesis is that web pages with similar domains and highly probable to refer to the same individual.

Cosine is a measure of similarity between two vectors of n dimensions by finding the angle between them. The number of dimensions in our case corresponds to the number of domains in the document collection. Given two web page vectors $WP(A)$ and $WP(B)$, the cosine similarity is the dot product $\cos(WP(A), WP(B)) = \frac{WP(A) \bullet WP(B)}{|WP(A)||WP(B)|}$. The value of 1 indicates that the two documents are very similar, while 0 indicates the documents are dissimilar.

The Euclidean distance between two web pages is the length of the path connecting them: $d = |WP(A) - WP(B)| = \sqrt{\sum_{i=1}^{n} |WP(A)_i - WP(B)_i|^2}$. In order to convert the Euclidean distance to similarity measure, we normalized it using the biggest distance and subtracted by one.

Both measures are calculated for all pairs of documents. From the obtained values we construct $N \times N$ dimensional matrix, where N corresponds to the number of document for an ambiguous name. In the WePS challenge, N is 100.

4.3 Document Clustering

The final step of our WePS algorithm consists in the clustering of the web pages according to their representation in the domain space. For the purpose we used PoBOC clustering [14] which builds a weighted graph with weights being the distances among the objects. Afterwards tight components, called poles, are found and the rest of the elements are assigned to the set of most similar poles. Thus, PoBOC decides on the number of clusters without need of a predefined threshold. PoBOC is a soft clustering algorithm and its application for the WePS tasks seems natural.

5 Experimental Evaluation

5.1 Data Description

The WePS task emerged as part of the Semeval 2007 challenge. The test data set[3] consists of 30 ambiguous person names. For each name, the organizers collected the first 100 web pages from Yahoo search engine or Wikipedia. The collected web pages have been manually annotated denoting whether a web page corresponds to given person or not. The number of clusters corresponds to the number of senses a person name has. In order to evaluate the performance of our system, we have used the official WePS evaluation script which measures the purity, inverse purity and f-score of the system.

5.2 Experimental Results

Table 1 shows the obtained WePS results for the test data set. Our systems are denoted with the initials of the different domain ranking schemes and the used document similarity measure. In the table, we show the two best and the last ranked systems in the WePS challenge as well as two rankings provided by the organizers. The ONE-IN-ONE stands for one document per cluster and has the lowest possible inverse purity for this particular data set. While the ALL-IN-ONE means that all web pages are taken to be related to the same person i.e. only one cluster, and gives the lowest possible purity measure. Observing the performance of the ONE-IN-ONE baseline, the WePS participants discussed that it is not a satisfactory result to show to the final user single pages, because even systems below the ONE-IN-ONE baseline clustered the web pages more appropriately and informatively. In the future more sensitive evaluation measures are needed.

[3] `http://nlp.uned.es/weps/task-data.html`

Table 1. Comparative performancegiven among Domain Information and WePS participants

System	Purity	Inverse Purity	F $\alpha = 0.5$
CU-COMSEM	0.72	0.88	**0.78**
IRST-BP	0.75	0.80	0.75
ONE-IN-ONE	*1.00*	*0.47*	*0.61*
AUG	0.50	0.88	0.60
PR-cos	0.87	0.47	**0.58**
tf*idf-euc	0.82	0.52	0.57
bin-cos	0.82	0.51	0.56
UBC-AS	0.30	0.91	**0.40**
ALL-IN-ONE	*0.29*	*1.00*	*0.40*

The f-score results show that the difference between our best-performing system PR-cos and the CU-COMSEM and IRST-BP participants is around 20 and 15%. Compared to UBC-AS and ALL-IN-ONE approaches, PR-cos f-score is with 18% better. In comparison to WePS approaches, our domain information approach has the advantage to provide semantic explanation for the groupings following the topic information the web pages carry out.

However, the performance of the two best WePS systems demonstrates that in the future we need to integrate knowledge from the urls, the web page titles and even whole phrases. We have also observed that the systems considering NE information reported better performance. However, in our domain approach we could not map the proper names with the domain information because such evidence is not indicated in *WordNet Domains*.

From the three different domain ranking approaches, the best performing is PR with the cosine similarity. This shows that the reduction of the number of senses using the domains is useful. In the future, we plan to study the influence of PR with more ample set of categories.

6 Conclusions

In this paper, we have presented a graph-based approach for synset ranking given a domain. The algorithm can be adapted easily to a bigger (smaller) set of categories and is flexible to future change in the number of synsets. PageRank representation shows the communication flow among the synsets and the domains.

We have demonstrated the usage of the synset-domain ranking for the resolution of the WebPS task. We have proposed three different domain ranking methods based on tf*idf, binary vector and PageRank. In order to measure the similarity between the web pages, we have employed the cosine and Euclidean distance measures. Our motivation for the usage of the domain ranking is that two documents are probable to refer to the same individual when the majority of their domains overlaps.

We have evaluated our approach with the WePS data set and we have compared the performance of our systems with WePS participants. The system ranked on 8th and 9th position from sixteen participants.

Finally, the results show that due to the varying web page contents, the domain vector space contains many zero values for shorter documents. Therefore, we want to study the amount of words necessary for the representation of the domain information in a document.

Currently, we did not take advantage of domain proximity such as `MUSIC` and `ENTERTAINMENT`, but we plan to integrate this information as well. Analysing the performances of the WePS systems proposes that named entities, urls, web page titles and noun phrases play important role.

Acknowledgements

This research has been funded by the European Union project QALLME FP6 IST-033860, Spanish Ministry of Science and Technology TEXT-MESS TIN2006-15265-C06-01 and Fundação para a Ciência e a Tecnologia (FCT) scholarship: SFRH / BD / 19909/2004.

References

1. Javier, A., Gonzalo, J., Sekine, S.: The semeval-2007 weps evaluation: Establishing a benchmark for the web people search task. In: Proceedings of the Fourth International Workshop on Semantic Evaluations (SemEval-2007), pp. 64–69 (2007)
2. Bagga, A., Baldwin, B.: Entity-based cross-document coreferencing using the vector space model. In: Proceedings of the Thirty-Sixth Annual Meeting of the Association for Computational Linguistics and Seventeenth International Conference on Computational Linguistics, pp. 79–85 (1998)
3. Pedersen, T., Purandare, A., Kulkarni, A.: Name discrimination by clustering similar contexts. In: Gelbukh, A. (ed.) CICLing 2005. LNCS, vol. 3406, pp. 226–237. Springer, Heidelberg (2005)
4. Kozareva, Z., Vázquez, S., Montoyo, A.: Multilingual name disambiguation with semantic information. In: Matoušek, V., Mautner, P. (eds.) TSD 2007. LNCS (LNAI), vol. 4629, pp. 23–30. Springer, Heidelberg (2007)
5. Pedersen, T., Kulkarni, A.: Unsupervised discrimination of person names in web contexts. In: Gelbukh, A. (ed.) CICLing 2007. LNCS, vol. 4394, pp. 299–310. Springer, Heidelberg (2007)
6. Kozareva, Z., Vázquez, S., Montoyo, A.: Discovering the underlying meanings and categories of a name through domain and semantic information. In: Proceedings of the Conference on Recent Advances in Natural Language Processing RANLP (2007)
7. Chen, Y., Martin, J.H.: Cu-comsem: Exploring rich features for unsupervised web personal name disambiguation. In: Proceedings of the Fourth International Workshop on Semantic Evaluations (SemEval-2007), pp. 125–128 (2007)
8. Popescu, O., Magnini, B.: Irst-bp: Web people search using name entities. In: Proceedings of the Fourth International Workshop on Semantic Evaluations (SemEval-2007), pp. 195–198 (2007)
9. Agirre, E., Soroa, A.: Ubc-as: A graph based unsupervised system for induction and classification. In: Proceedings of the Fourth International Workshop on Semantic Evaluations (SemEval-2007), pp. 346–349 (2007)

10. Magnini, B., Cavaglia, G.: Integrating subject field codes into wordnet. In: Proceedings of LREC-2000, Second International Conference on Language Resources and Evaluation, pp. 1413–1418 (2000)
11. Esuli, A., Sebastiani, F.: Pageranking wordnet synsets: An application to opinion mining. In: Proceedings of ACL-2007, the 45th Annual Meeting of the Association of Computational Linguistics, pp. 424–431 (2007)
12. Brin, S., Page, L.: The anatomy of a large-scale hypertextual web search engine. Computer Networks and ISDN Systems 30, 107–117 (1998)
13. Liu, H.: Montylingua: An end-to-end natural language processor with common sense (2004), http://web.media.mit.edu/~hugo/montylingua
14. Cleuziou, G., Martin, L., Vrain, C.: Poboc: an overlapping clustering algorithm. In: Application to rule-based classification and textual data, pp. 440–444 (2004)

Disyllabic Chinese Word Extraction Based on Character Thesaurus and Semantic Constraints in Word-Formation*

Sun Maosong[1], Xu Dongliang[1], Benjamin K.Y. T'sou[2], and Lu Huaming[3]

[1] The State Key Laboratory of Intelligent Technology and Systems,
Tsinghua National Laboratory for Information Science and Technology,
Dept. of Computer Sci. & Tech., Tsinghua University, Beijing 100084, China
sms@tsinghua.edu.cn

[2] Language Information Sciences Research Center, City University of Hong Kong
rlbtsou@cityu.edu.hk

[3] Beijing Information Science and Technology University, Beijing 100085, China
luhuaming@bistu.edu.cn

Abstract. This paper presents a novel approach to Chinese disyllabic word extraction based on semantic information of characters. Two thesauri of Chinese characters, manually-crafted and machine-generated, are conducted. A Chinese wordlist with 63,738 two-character words, together with the character thesauri, are explored to learn semantic constraints between characters in Chinese word-formation, resulting in two types of semantic-tag-based HMM. Experiments show that: (1) both schemes outperform their character-based counterpart; (2) the machine-generated thesaurus outperforms the hand-crafted one to some extent in word extraction, and (3) the proper combination of semantic-tag-based and character-based methods could benefit word extraction.

1 Introduction

Word extraction plays important roles in many NLP-related applications, such as text parsing, information retrieval and ontology construction. A Chinese word is composed of single or multiple Chinese characters. The task of extracting Chinese words with multi-characters from texts (in fact, a sub-task of Chinese word segmentation) is somewhat similar to that of extracting phrases (e.g., compound nouns) in English, if we regard Chinese characters as English words.

Generally, there are two kinds of approaches to word/phrase extraction, i.e., rule-based and statistic-based. The latter is currently the most widely used. In this approach, the soundness of an extracted item being a word/phrase is usually estimated by the associative strength between constituents of the item. A variety of statistical measures have been exploited to quantify the associative strength, for example, frequency and mutual information [2, 4, 7, 9, 10].

Almost all the work on Chinese word extraction, except [1], depended directly on

* This work is supported by NSFC (60321002) and 863 Project (2007AA01Z148).

P. Sojka et al. (Eds.): TSD 2008, LNAI 5246, pp. 141–151, 2008.
© Springer-Verlag Berlin Heidelberg 2008

characters in the extracted item to measure the associative strength, like [9]. These approaches ignored an important property of Chinese words: each character (one syllable in most cases) in a word usually has at least one sense, thus the sense sequence of characters involved may reflect the semantic constraint 'hidden' in the word to some extent and, consequently, the totality of sense sequences over the wordlist may constitute the complete semantic constraints underlying Chinese word-formation. We believe such constraints are helpful for validating Chinese words. The biggest advantage of considering the semantic information is it could help make some degree of inference in word extraction [1]. For example, suppose '春风' (Spring wind) and '秋风' (Autumn wind) are in the wordlist, whereas '夏风' (Summer wind) and '冬风'(Winter wind) are not (the latter two are realistic in Chinese, as in '冬风与海洋 and '夏风来袭'). All these four words bear the same sense sequence 'season+wind', so we can guess that '夏风' and '冬风' are possibly words. The idea is simple, but it is radically different from previous ones: word extraction will depend heavily on senses of characters, rather than solely on characters; the associative strength can also be determined statistically using senses of characters. A side effect of it is the data sparseness problem in word extraction may be better settled. [11] presented an novel approach with this idea. The work here will be in line with it.

To begin, we need two kinds of linguistic resources: one is a Chinese wordlist, and the other one is a thesaurus of Chinese characters. 1) A wordlist with 63,738 two-character words, THW2, is used in experiments here for learning semantic constraints underlying Chinese word-formation. The reason for choosing two-character words is that they comprise the largest proportion in a Chinese wordlist and represent the most popular word-formation of Chinese. 2) We try to construct a thesaurus of Chinese characters in two ways: the first one is 'fully hand-crafted' by linguists, and the second one is 'fully automated' using an automatic clustering algorithm.

In addition, we use R94-2K, a raw Chinese corpus with 224 million characters to train the baseline model, the character-based method, for comparison.

The test set for experiments is carefully prepared with PDA9801, a manually word-segmented and part-of-speech tagged corpus developed by Peking University, in the following steps: firstly, all distinct character bigrams excluding proper nouns are completely derived from PDA9801, yielding a list of 238,946 bigrams in total, out of which 23,725 are words and 215,221 are non-words, according to the annotation of PDA9801(This list is named TS238946 and used as the test set in [11]); secondly, bigrams in which at least one character does not occur in THW2 are deleted from TS238946. We have 237,261 bigrams left, out of which 23,690 are words and 213,571 are non-words; thirdly, we recover 8,890 words from 213,571 non-words through human judgment, resulting in 32,580 words (23,690+8,890) and 204,681 non-words (213,571-8,890). For example, the word '蜻蜓' appears twice in PDA9801: "不可/v 蜻蜓点水/i " "不/d 是/v 蜻蜓点水/i ". It will be wrongly judged as a non-word if purely depending on PDA9801 and thus needs to be recovered; and lastly, we remove all words that belong to THW2 (totally 27,634) from the list of 237,261, producing a list of 209,627, denoted TS209627, as the test set for experiments. There are 4,946 words (unknown words) and 204,681 non-words in TS209627. It is worth pointing out that the proportion of words and non-words in TS209627 is about 1:41, indicating that the task we are facing is very difficult (Also note the test set used here, TS209627, is very different from the test set used in [11], TS238946).

2 The Framework: Representing Semantic Constraints in Word-Formation by HMM

A general framework to describe Chinese word-formation from the semantic perspective is proposed in [11]. We re-state it briefly here for the ease of description.

Let *C* be the set of Chinese characters, *T* be a thesaurus over *C*, *S* be the set of semantic tags derived from *T*, *W* be the set of Chinese wordlist of two-character words, and WS be the set of pairs <word, semantic tags> over *W* in which every character in a word is assigned a unique semantic tag (though the character may possibly have multiple semantic tags in terms of *T*). Then we can construct a Hidden Markov Model (HMM) accordingly as a five-tuple, $WF_{sem} = (S, S_0, C, P_S, P_C)$, where: *S* is the set of states; $S_0 \in S$ is the set of initial states associated with the initials of semantic tag sequences of *W*; *C* is the set of output alphabet; $P_S = \{p(s_j \mid s_i)\}$ ($s_i \in S$, $s_j \in S$) is the set of transition probabilities among states; and, $P_C = \{p(c_k \mid s_i, s_j)\}$ ($s_i \in S$, $s_j \in S$, $c_k \in C$) is the set of observation probabilities. Both P_S and P_C will be trained using *WS*. This five-tuple (S, S_0, C, P_S, P_C) describes the semantic constraints in word formation underlying *W* statistically and systematically.

Given any character string c_1c_2 ($c_1 \in C$, $c_2 \in C$), the following derivations hold for $LW(c_1c_2)$, the likelihood of this string being a word, according to properties of HMM and the Bayesian theorem:

$$
\begin{aligned}
LW(c_1c_2) &= \sum_{s_1s_2} p(s_1)p(s_2 \mid s_1)p(c_1 \mid s_1, s_2)p(c_2 \mid s_1, s_2) \\
&\approx \sum_{s_1s_2} p(s_1)p(s_2 \mid s_1)p(c_1 \mid s_1)p(c_2 \mid s_2) \\
&= \sum_{s_1s_2} \frac{p(s_1, s_2)}{p(s_1)p(s_2)} p(s_1 \mid c_1)p(s_2 \mid c_2)p(c_1)p(c_2)
\end{aligned} \tag{1}
$$

where s_1s_2 ($s_1 \in S$, $s_2 \in S$) is any semantic tag sequence generated by c_1c_2 in a combinatorial way.

We ignore $p(c_1)$ and $p(c_2)$ in (1), and, for the sake of clarity, let:

$$MI^*(s_1, s_2) = \frac{p(s_1, s_2)}{p(s_1)p(s_2)} \tag{2}$$

then an equivalent of (1), denoted $LW_{MI^*}(c_1c_2)$, is consequently obtained:

$$LW_{MI^*}(c_1c_2) = \sum_{s_1s_2} MI^*(s_1, s_2)p(s_1 \mid c_1)p(s_2 \mid c_2) \tag{3}$$

Note that $MI^*(s_1,s_2)$ is exactly the inner part of $MI(s_1,s_2)$:

$$MI(s_1,s_2) = \log_2 \frac{p(s_1,s_2)}{p(s_1)p(s_2)} \tag{4}$$

We thus suppose a variation of (3) as an alternative of the likelihood:

$$LW_{MI}(c_1c_2) = \sum_{s_1s_2} MI(s_1,s_2)p(s_1 \mid c_1)p(s_2 \mid c_2) \tag{5}$$

Such a variation sounds reasonable, though the derivation from (3) to (5) does not hold mathematically.

Therefore, we have two alternatives for measuring the likelihood of c_1c_2 being a word: $LW_{MI^*}(c_1c_2)$ and $LW_{MI}(c_1c_2)$.

3 Estimation of HMM Parameters

To calculate $LW_{MI^*}(c_1c_2)$ and $LW_{MI}(c_1c_2)$ in the framework, we need to estimate parameters: $p(s_1,s_2)$, $p(s_1)$, $p(s_2)$, $p(s_1 \mid c_1)$ and $p(s_2 \mid c_2)$. We discuss this from two perspectives: using a hand-crafted thesaurus and, a machine-generated thesaurus.

3.1 Scheme 1: Parameter Estimation with a Hand-Crafted Thesaurus

A hand-crafted thesaurus of Chinese characters, THSCS (short for the Semantic Coding System for Chinese Characters), is developed [5]. It covers all 6,763 Chinese characters defined in GB-2312, a National Standard of China for Chinese character set in information exchange. In THSCS, every character is assigned its possible semantic codes (equally, semantic categories, or semantic tags) manually. The principle in designing THSCS is that its semantic hierarchy is as compatible as possible with that of TYCCL, a well-known thesaurus of Chinese words [6].

We obtain totally 1,380 semantic categories at the bottom level of THSCS (this number is not identical to its counterpart, 1,428 in TYCCL). Their distribution is not balanced. The most frequent category occurs 927 times, but a majority of categories occur only a few times: 36.4% no more than 5 times and 87% no more than 20 times. Furthermore, about 54.12% of the 6763 characters are polysemous according to THSCS. These polysemous characters are more active than those with a single category. An observation over all 32,624 two-character words in TYCCL shows that only 1.40% of them do not contain any polysemous characters.

Note that *T* and *W* in the framework are instantiated as THSCS and THW2 here.

If a manually annotated *WS* is available, the training of WF_{sem}, i.e., the training of P_S and P_C, will be very easy. Unfortunately, we do not have it yet. In fact, we only have the resources *C, T, S* and *W*. It is very tough to handcraft such a *WS* because the relevant linguistic knowledge is still incomplete, resulting in a lack of theoretical preparations. Targeting at this problem, the Baum-Welch unsupervised re-estimation algorithm [3], is explored to make some degree of approximations. The algorithm is adjusted to fit the need here, denoted SBW [11].

SBW can be expanded to a dynamic SBW, denoted SDBW, if every *w* in *W* is weighted by its string frequency in a very large corpus (here, R94-2K), that is, a frequent word should be given higher weight, and its corresponding semantic tag sequence should contribute more to word-formation, rather than weighted equally, in the process of estimating P_S and P_C ([11]).

As reported in [11]: $LW_{MI}(c_1c_2)$ outperforms $LW_{MI^*}(c_1c_2)$ under SBW, so $LW_{MI}(c_1c_2)$ is preferable; and SDBW outperforms SBW under $LW_{MI}(c_1c_2)$. Here, we simply adopt the best one, SDBW, and compare the performance $LW_{MI}(c_1c_2)$ and $LW_{MI^*}(c_1c_2)$ on TS209627 under SDBW, denoted SDBW+LWMI and SDBW+LWMI*, respectively. Fig. 1 shows that SDBW+LWMI is slightly better than SDBW+LWMI*. So the scheme 1 is fixed to SDBW+LWMI.

3.2 Scheme 2: Parameter Estimation with a Machine-Generated Thesaurus

The idea is that we cluster characters in terms of their distributions in THW2.

Suppose $C_0 \in C$ is the set of characters which occur in THW2 (l=|C_0|=5,565), for any character $c \in C_0$, let:

$b(c, c_i) = 1$ if $cc_i \in$ THW2, else $b(c, c_i) = 0$;

$b(c_i, c) = 1$ if $c_i c \in$ THW2, else $b(c_i, c) = 0$. ($c_i \in C_0$, i=1, …, l)

thus we get a distributional vector *v(c)* of *c* with dimension 2×*l*. The distance between two distributional vectors $d(v(c_j), v(c_k))$ (or, equally, the distance between two characters c_j and c_k) is then defined as $-\log(sim(v(c_j), v(c_k)))$, where the similarity of $v(c_j)$ and $v(c_k)$, $sim(v(c_j), v(c_k))$, is defined as the cosine of the angle in between.

The complete linkage clustering algorithm is used to perform automatic clustering of C_0. The algorithm is iterated until a predefined number of clusters is met. The output of the algorithm will be a partition of C_0 with *m* clusters { $u_1, \ldots\ldots, u_i, \ldots\ldots u_m$ }.

The clusters from clustering tend to behave as semantic categories, as:

(槛梃棂楣樘牖扉) (粳糯籴薏秫舂籼) (榆樟柞枞柏桉楝椴榕槭) (蛔蠕螨蠓蠹) (蚂蚱蜣螳螬) (鼋鱿鲅鲇鲈鲎鲑鲞鲟鲢鲥鲦) (岭湾峡) (裤裙衫) (胯颧肱胛胫腓骶髂髋髌) (胺醚酯炔酰) (春冬秋夏) (东南西北) (焙烘烤) (丢抛扔) (撅抿努噘) (漱涮揃盥) (琢镂镌) (凹凸) (诧骇讶悚愕迥) (峨巍嵯) (暖煦) (恍倘宛俨) (啊哎唉嗨) (俺你它她咱) (哪那这)

It is worth noting that, in our experience, less clusters mean more thorough semantic generalization – this is its advantage, however, it also dramatically increases the chance of errors in clustering and thus lead to more risk of over-generalization. We need to find a balance for this factor (please refer to Fig. 2, 3 and 4).

If the number of clusters is kept high, some characters which tend to be in the same semantic category intuitively may not be clustered together, for instance, in the case of 4,000 clusters, '鸡'(chicken), '鸭'(duck), '牛'(cow), '猪'(pig) and '兔'(rabbit) will stand alone as a separate cluster each without any relationship. To handle this problem, we make a relaxation on the partition of C_0, allowing a character to be associated with multiple clusters.

For any character *c*, let the largest similarity between *c* and centroids of $u_1,\ldots\ldots,u_i,\ldots\ldots u_m$ be $sim_0(c)$, we have: 1) *c* is said to be associated with u_i if the similarity of *c* and the centroid of u_i is larger than 0.2×$sim_0(c)$; 2) The degree of *c* being associated with u_i is proportional to the similarity of *c* and the centroid of u_i.

The normalized degree of *c* being associated with u_i will serve as $p(u_i \mid c)$, the probability of *c* belonging to u_i. In addition, the probability of u_i, $p(u_i)$, is estimated by the number of characters it contains over *l*.

In this way, the associations among '鸡', '鸭', '牛', '猪' and '兔' can be properly established.

If the clusters are viewed as semantic categories in scheme 1, (3) and (5) will be applicable to scheme 2 without any change. We thus substitute u_i with s_i herein.

We need to calculate $p(s_1,s_2)$ for (3) and (5). For a word $w=c_1c_2 \in THW2$, suppose c_i belongs to n_i clusters { $s_{i,1},\ldots,s_{i,n_i}$ } with probabilities $p(s_{i,1} \mid c_i),\ldots,p(s_{i,n_i} \mid c_i)$ obtained from the results of clustering (*i*=1,2, $n_i \geq 1$), the contribution of sequence $s_{1,j}s_{2,k}$ of *w* to the frequency counting of P_S is:

$$f(s_{1,j},s_{2,k}) = f(s_{1,j},s_{2,k}) + \frac{1}{p(s_{1,j} \mid c_1)p(s_{2,k} \mid c_2)} \quad (j=1,\ldots,n_1,\ k=1,\ldots,n_2) \tag{6}$$

Doing (6) on the whole THW2 will eventually generate $p(s_1,s_2)$.

Experiments are conducted on TS209627 to determine how many clusters and which of $LW_{MI^*}(c_1c_2)$ and $LW_{MI}(c_1c_2)$ are appropriate. The number of clusters is set at 2,000, 3,000, 4,000, 5,000 and 5,565(one cluster for each character). Fig. 2 and 3 give the results accordingly. As can be seen, the 10-point average precision of C4000+LWMI* is the highest (17.96%), and that of C4000+LWMI is the second highest (17.88%). So we fix the number of clusters to 4,000 and adopt $LW_{MI^*}(c_1c_2)$ in scheme 2, though the improvement, as shown in Fig. 4, is small compared to adopting $LW_{MI}(c_1c_2)$.This is somewhat different from what we have obtained in scheme 1: $LW_{MI}(c_1c_2)$ is better than $LW_{MI^*}(c_1c_2)$ there.

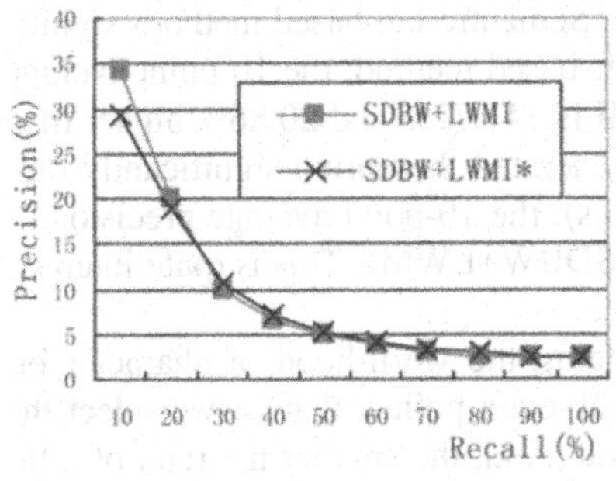

Fig. 1. Comparison of SDBW+LWMI and SDBW+LWMI*

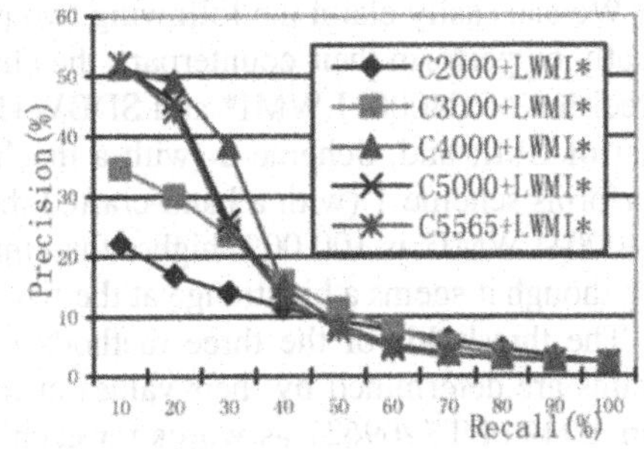

Fig. 2. Determine the number of clusters under LWMI*

Similar to SBW and SDBW, we may expand Cx000+LWMI* to dynamic Cx000+LWMI*, denoted DCx000+LWMI*, by weighting every *w* in *W* with its string frequency in R94-2K. Several experiments are conducted for comparison (Fig. 5). The number of clusters still prefers 4000, for DC4000+LWMI* outperforms both DC3000+LWMI* and DC5000+LWMI*.

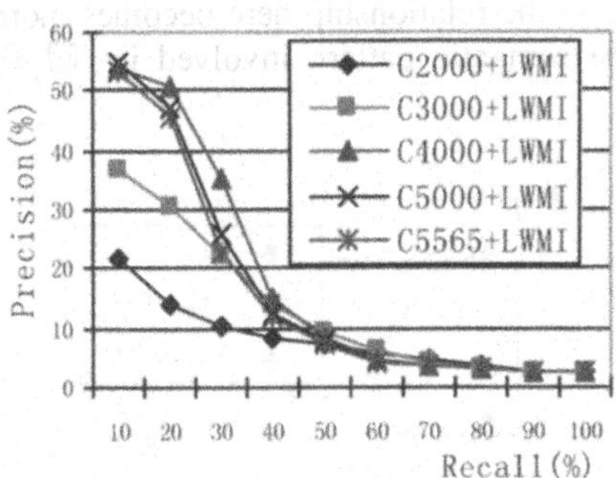

Fig. 3. Determine the number of clusters under LWMI

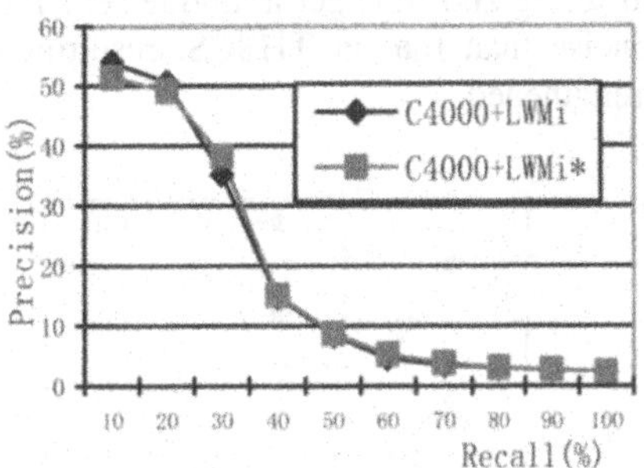

Fig. 4. Comparison of C4000+LWMI* and C4000+LWMI

However, to our surprise, DC4000+LWMI* is significantly poorer than C4000+LWMI*. This is different from what we have obtained in scheme 1: SDBW is much better than SBW there.

4 Comparison of Character-Based and Semantic-Tag-Based Methods

We compare three methods on TS209627: one is character-based mutual information (CMI), trained with R94-2K; the other two are semantic-tag-based scheme 1 (SDBW+LWMI) and scheme 2 (C4000+LWMI*). Their performances on word extraction are given in Fig. 6.

We can easily claim the following two points: Semantic-tag-based methods significantly outperform their counterpart, the character-based method: the 10-point average precision of C4000+LWMI* and SDBW+LWMI is 141.72% and 20.86% higher than that of CMI; and, Scheme 2 (with a machine-generated thesaurus) significantly outperforms scheme 1 (with a hand-crafted thesaurus): the 10-point average precision of C4000+LWMI* is 100.00% higher than that of SDBW+LWMI. This is quite interesting though it seems a bit strange at the first glance.

The thresholds of the three methods for judging the word-hood of character bigrams are determined by their values at the break-even points, that is, we select the top 4,946 of TS209627 as words for each method. And, the smaller the rank of a bigram is, the more likely it is a word.

We observe several typical examples in Table 1.

[暖春] (warm spring): '暖春' is rejected as a word by CMI, because its CMI rank is far below the break-even point. On the other hand, it is accepted by both SDBW+LWMI and C4000+LWMI*, because its ranks in both cases are above the break-even points. The reasons are: 1) '春' shares a semantic tag Ca19 (season) with '夏', '秋' and '冬' according to THSCS, and, '暖冬' in THW2 provides a 'hidden' semantic pattern to support the word-hood of '暖春' via Ca19, though the pattern is somewhat weakened due to the existence of two other 'noisy' tags of '春', Ca18(year) and Df07(desire, demand); 2) '春' is clustered with '夏', '秋' and '冬' in automatic clustering and thus get a unique semantic tag — the relationship here becomes more concise than that in THSCS, resulting in the semantic pattern involved in '暖冬' strengthened.

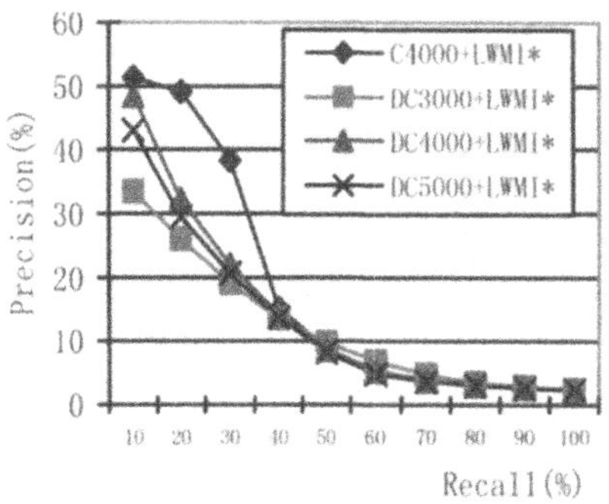

Fig. 5. C4000+LWMI* vs. Dynamic DCx000 +LWMI*

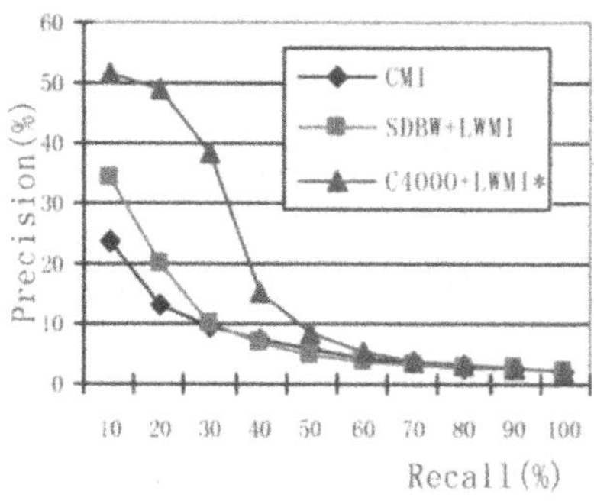

Fig. 6. Comparison of three methods

[种兔] (stud rabbit): The situation is similar to that of '暖春'. Here the word '种猪' in THW2 provides a 'hidden' semantic pattern. The distinction is that, it is the association between '兔' and '猪' rather than a cluster that serves as a bridge over '种兔' and '种猪'.

[俄军] (Russian army): Automatic clustering (C4000) fails to generate either a cluster for '俄', '苏', '美' and '日' or an association between '俄' and '苏', '美', '日', though '苏军', '美军' and '日军' do occur in THW2. The distributions of words with these characters in THW2 are too diverse to build an informative relationship.

Table 1. Typical examples of the three methods

Unknown word	CMI		SDBW+LWMI		C4000+LWMI*	
	Rank in TS209627	Judged as word by break-even point	Rank in TS209627	Judged as word by break-even point	Rank in TS209627	Judged as word by break-even point
暖春	130,249	×	4,486	√	1,422	√
种兔	9,131	×	884	√	2,741	√
俄军	14,037	×	1,074	√	43,893	×
北约	4,435	√	186,900	×	204,968	×

白俄 俄而 俄尔 俄方 俄顷 俄延 ……

复苏 流苏 屠苏 昭苏 苏北 苏剧 苏军 苏木 苏区 苏滩 苏铁 苏醒 苏绣 ……

醇美 健美 娇美 精美 俏美 完美 选美 溢美 优美 壮美 作美 姣美 媲美 美餐 美差 美钞 美称 美德 美发 美方 美籍 美金 美景 美酒 美军 美丽 美元 ……

白日 不日 初日 春日 冬日 改日 何日 吉日 假日 节日 今日 连日 烈日 落日 他日 日报 日程 日戳 日方 日记 日军 日历 日蚀 日头 日益 日元 日子 ……

By contrast, '俄', '苏', '美' and '日' share a semantic tag Di02(country) in THSCS, leading to a correct recognition of '俄军' by SDBW+LWMI.

[北约] (NATO): In fact, there does not exist a word with a hidden semantic pattern with respect to '北约' in THW2. Neither SDBW+LWMI nor C4000+LWMI* can work under this condition. However, '北约' is accepted as a word by CMI.

5 Combination of Character-Based and Semantic-Tag-Based Methods

To get a full picture of the three methods, we take a detailed look at their outputs at the break-even points. We collect the top 4,946 bigrams for each, and mix them together, forming a set with 13,460 distinct bigrams. We call this set 'the break-even set'. The break-even set is further partitioned into seven subsets as in Table 2.

As suggested by Table 2, there may exist a large room for improvement if we can combine the three methods properly. We use C4.5 [8] to do integration. TS209627 is divided into a training set (for training the decision tree) and a test set in 4:1 proportion. Experimental results on the test set are given in Table 3 ('$\oplus$' means the decision tree combination).

The combination of C4000+LWMI*$\oplus$CMI significantly outperforms all the other combinations, including the combination of all the three methods (Table 3). For comparison, we also experiment on this test set for the three methods on the individual basis (Table 4).

All combinations outperforms the three methods in the 10-point average F-measures, out of which, the performance of C4000+LWMI*$\oplus$CMI increases the most significantly: its F-measure is 261.23%, 244.40%, and 119.94% higher than the 10-point average F-measures of CMI, SDBW+LWMI and C4000+LWMI* respectively.

Table 2. Subsets of the break-even set

Set	# of words	# of non-words	Sub-total
C4000+LWMI* only	1,038	3,065	4,103
SDBW+LWMI only	454	3,533	3,987
CMI only	377	3,745	4,122
C4000+LWMI* ∩ SDBW+LWMI only	268	156	424
C4000+LWMI* ∩ CMI only	155	134	289
SDBW+LWMI ∩ CMI only	162	243	405
C4000+LWMI* ∩ SDBW+LWMI ∩ CMI	102	28	130
Sub-total	2,556	10,904	13,460

Table 3. Combination of three methods with C4.5

	Precision	Recall	F-measure
C4000+LWMI* ⊕ SDBW+LWMI ⊕ CMI	42.53	26.21	32.44
C4000+LWMI* ⊕ SDBW+LWMI	43.05	22.27	29.35
C4000+LWMI* ⊕ CMI	40.99	30.16	34.75
SDBW+LWMI ⊕ CMI	19.46	21.36	20.37

Table 4. Performance of three methods on the test set from the break-even set

Recall (%)	Precision (%)			F-measure (%)		
	CMI	SDBW+ LWMI	C4000+ LWMI*	CMI	SDBW+ LWMI	C4000+ LWMI*
10	26.92	37.55	57.65	14.58	15.79	17.04
20	13.78	21.70	51.98	16.31	20.81	28.89
30	9.64	11.06	47.59	14.60	16.16	36.80
40	6.81	6.48	19.33	11.64	11.15	26.07
50	5.45	5.04	9.13	9.84	9.17	15.44
60	4.36	3.97	5.95	8.13	7.44	10.83
70	3.37	3.11	3.97	6.42	5.95	7.51
80	2.77	2.60	3.03	5.35	5.05	5.84
90	2.41	2.46	2.57	4.69	4.78	4.99
100	2.36	2.36	2.36	4.62	4.62	4.62
Average	7.79	9.63	20.36	9.62	10.09	15.80

6 Concluding Remarks

We complete the paper with the conclusions: 1) two semantic-tag-based methods, by means of either a hand-crafted or machine-generated thesaurus, outperform their character-based counterpart; 2) a machine-generated thesaurus performs better than a hand-crafted one for word extraction; and 3) The proper combination of semantic-tag-based and character-based methods may benefit word extraction.

References

1. Chen, K.J., Chen, C.J.: A Corpus Based Study on Computational Morphology for Mandarin Chinese. In: Quantitative and Computational Studies on the Chinese Language, Hong Kong, pp. 283–305 (1998)
2. Church, K.W., Hanks, P., et al.: Using Statistics in Lexical Analysis, Lexical Acquisition: Exploiting On-line Resources to Build a Lexicon. Erlbaum, Hillsdale (1991)
3. Hajič, J.: HMM Parameters Estimation: The Baum-Welch Algorithm (2000), http://www.cs.jhu.edu/~hajic
4. Johansson, C.: Good Bigrams. In: Proc. of COLING 1996, Copenhagen, Denmark (1996)
5. Kang, S.Y., Sun, M.S., et al.: Design and Implementation of a Chinese Character Thesaurus. In: Proc. of Int'l Conf. on Chinese Computing 2001, Singapore, pp. 301–307 (2001)
6. Mei, J.J.: Tong Yi Ci Ci Lin. Shanghai Cishu Press (1983)
7. Merkel, M., Andersson, M.: Knowledge-lite Extraction of Multi-word Units with Language Filters and Entropy Thresholds. In: Proc. of RIAO 2000, Paris, France, pp. 737–746 (2000)
8. Quinlan, J.R.: C4.5: Programs for Machine Learning. Morgan Kaufmann Publishers, San Francisco (1993)
9. Sproat, R., Shih, C.L.: A Statistical Method for Finding Word Boundaries in Chinese Text. Computer Processing of Chinese and Oriental Languages 4(4), 336–349
10. Su, K.Y., Wu, M., Chang, J.S.: A Corpus-based Approach to Automatic Compound Extraction. In: Proceedings of the 32<Superscript>nd</Superscript> Annual Meeting of the Association for Computational Linguistics, Las Cruces, NM, pp. 242–247 (1994)
11. Sun, M.S., Luo, S.F., T'sou, B.K.: Word Extraction Based on Semantic Constraints in Chinese Word-Formation, Computational Linguistics and Intelligent Text Processing. In: Gelbukh, A. (ed.) CICLing 2005. LNCS, vol. 3406, pp. 202–213. Springer, Heidelberg (2005)

Statistical Word Sense Disambiguation in Contexts for Russian Nouns Denoting Physical Objects

Olga Mitrofanova[1], Olga Lashevskaya[2], and Polina Panicheva[1]

[1] Department of Mathematical Linguistics
Faculty of Philology and Arts, St. Petersburg State University
Universitetskaya emb. 11, 199034 St. Petersburg, Russia
[2] Institute of the Russian Language
Volkhonka 18/2, 119019 Moscow, Russia
alkonost-om@yandex.ru, olesar@mail.ru, ppolin@yandex.ru

Abstract. The paper presents experimental results on automatic word sense disambiguation (WSD). Contexts for polysemous and/or homonymic Russian nouns denoting physical objects serve as an empirical basis of the study. Sets of contexts were extracted from the Russian National Corpus (RNC). Machine learning software for WSD was developed within the framework of the project. WSD tool used in experiments is aimed at statistical processing and classification of noun contexts. WSD procedure was performed taking into account lexical markers of word meanings in contexts and semantic annotation of contexts. Sets of experiments allowed to define optimal conditions for WSD in Russian texts.

Keywords: WSD, Russian corpora.

1 Introduction

Ambiguity, being an inherent feature of natural language, still remains an obstacle which ought to be surmounted in a set of NLP tasks. Thus, word sense disambiguation (WSD) alongside with morphological and syntactic disambiguation plays a crucial role in successful corpora development and application. WSD implies identification of particular meanings of polysemous and/or homonymic words in contexts. Effective WSD is fairly considered to be a complicated and time-consuming procedure: even computer-aided WSD often requires manual work of linguists and proper lexicographic expertise.

The purpose of the discussed project is automatic WSD in Russian texts which includes fulfilment of certain research tasks, such as: (1) development of a WSD tool for Russian; (2) experiments on WSD in Russian texts with various parameters; (3) definition of optimal conditions for WSD in Russian.

A rich variety of reliable WSD techniques – knowledge-based, corpus-based (statistical), hybrid – have been worked out [1] and tested[1] by now.

Knowledge-based WSD is performed with the help of semantic information stored in electronic lexicographic modules (e.g., WordNet, FrameNet). Corpus-based WSD

[1] Cf.SensEval/SemEval experimental tasks and workshop materials: http://www.senseval.org, http://nlp.cs.swarthmore.edu/semeval/index.php

P. Sojka et al. (Eds.): TSD 2008, LNAI 5246, pp. 153–159, 2008.
© Springer-Verlag Berlin Heidelberg 2008

implies extraction and statistical processing of word co-occurrence data which allow to distinguish separate meanings of lexical items in contexts. Hybrid WSD brings into action both lexical resources and corpus analysis.

Major WSD techniques were enabled in experiments on semantic ambiguity resolution in Russian texts. The use of lexical databases for Russian (e.g., an electronic thesaurus RuTes [2], an RNC semantic dictionary [3]) provides rather high quality of WSD. If lexicographic information is not available, statistical WSD techniques are indispensable in processing Russian texts. As experimental data have shown, it is possible to identify word meanings in contexts taking into account POS tag distributions [4] and lexical markers [5]; hybrid WSD seems to be effective as well [6].

The scope of the project allows to carry out statistical WSD procedure in two modes – with regard to (1) lexical markers of word meanings in contexts; (2) semantic annotation of contexts – and to compare reliability of those WSD approaches. It should be noted that experiments on WSD based on semantic annotation have no precedents in Russian corpus linguistics.

2 Linguistic Data

Context samples for WSD procedure are extracted from the Russian National Corpus (RNC, `http://www.ruscorpora.ru/`) [3] which is the largest annotated corpus of Russian texts containing about 150 mln tokens. The texts included into RNC are supplied with morphological (morphosyntactic) and semantic annotation. The latter implies tagging with such categories and parameters as "concrete", "human", "animal", "space", "construction", "tool", "container", "substance", "movement", "diminutive", "causative", "verbal noun", etc. (cf. `http://www.ruscorpora.ru/en/corpora-sem.html`). Semantic tags assigned to a particular lexical item in a context account for the set of its registered meanings, so that WSD procedure is often required.

Contexts for Russian nouns denoting physical objects (such polysemous and/or homonymic words as *dom "building, private space, family, etc."*, *organ "institution, part of body, musical instrument, etc."*, *luk "onion, bow"*, *glava "head, chief, cupola, chapter, etc."*, *vid "look, form, view, kind, species, aspect"*, *kl'uč "key, clue, clef, spring, etc."*, *sovet "advice, council, etc."*, *kosa "plait, scythe, spit"*, *ploš'ad' "square, space, etc."*, etc.) serve as an empirical basis of the present study. Semantic ambiguity resolution is performed for nouns with various frequencies of particular meanings (cf. Table 1). All occurrences of considered words found in RNC were analysed with the exception of contexts for rare meanings found in less than 10 contexts (such as *dom "common space"* or *dom "dynasty"*, etc.).

Manual disambiguation is performed for a training set of contexts for a particular word, the rest of ambiguous contexts were subjected to statistical WSD.

3 WSD Procedure

A Python-based WSD software was developed to perform ambiguity resolution. To meet the need, an automatic word clustering (AWC) tool was adapted [7]. AWC tool

Table 1. Russian nouns *dom, organ, luk*: semantic annotation and frequencies of meanings (number of contexts in RNC)

Word meanings	Semantic annotation	Number of contexts in RNC
dom		3,000 (total)
dom "building"	<r:concr t:constr top:contain>	1,694
dom "private space"	<r:concr t:space>	95
dom "family"	<r:concr t:group pt:set sc:hum>	72
dom "common space"	<r:concr t:space der:shift der:metaph>	4
dom "institution"	<r:concr t:org>	292
dom "dynasty"	<r:concr pt:set sc:hum>	1
dom (merged meanings)		842
organ		834 (total)
organ "institution"	<r:concr t:org hi:class>	660
organ "part of body"	<r:concr pt:partb pc:hum pc:animal hi:class>	130
organ "musical instrument"	<r:concr t:tool:mus>	27
organ "means"	<r:concr der:shift dt:partb>	9
organ "publication"	<r:concr t:media hi:class>	8
luk		2,200 (total)
luk "onion"	<r:concr t:plant t:fruit t:food pt:aggr>	1,600
luk "bow"	<r:concr t:tool:weapon top:arc>	600

allows to form clusters of similar contexts extracted from RNC. Adjustment of AWC software for WSD purposes required implementation of machine learning and pattern recognition modules.

WSD procedure is carried out in stages. The first stage implies pre-processing of contexts. Semantically and morphologically unambiguous contexts are selected to form a training set required for machine learning, while ambiguous contexts are treated as a trial set. Machine learning is performed at the second stage. For each meaning of a word its statistical pattern is formed taking into account frequencies of (1) lexical items occurring in contexts or (2) semantic tags of context elements. Further, patterns of meanings, as well as trial contexts, are represented as vectors in a word space model. The third stage implies pattern recognition, that is selection of patterns nearest to vectors which correspond to ambiguous contexts. Distances between patterns and vectors of trial contexts are calculated. Three similarity measures were implemented, so that the user could choose between Hamming measure, Euclidean measure, and cosine measure. As a result, meanings exposed by particular patterns are automatically assigned to processed contexts.

To run a WSD program, it is necessary to choose the input files (a file with training contexts and a file containing trial contexts), select context window size, indicate whether weights should be assigned to context items, select similarity measure, and choose the output file.

4 Experimental Results

4.1 Optimal Conditions for WSD in Russian Texts

Sets of experiments on WSD in contexts for Russian nouns denoting physical objects allowed to determine optimal conditions for WSD in Russian texts.

Thorough analysis of contexts shows that appropriate choice of similarity measure (cosine measure) alongside with expansion of a training set ($S = 100 \ldots 500$ contexts) ensures over 85% (in some cases up to 95%) of correct decisions.

Cosine measure proves to be the most reliable similarity measure as it is the least sensitive to meaning frequencies. Hamming and Euclidean measures provide correspondingly 45% and 65% of correct decisions on average.

WSD experiments were performed with variable size of a training set $S = 10, 15, 55, 75, 100, 200, 500, \ldots$ (up to all contexts except for those included into a trial set). It seems that the training set should contain at least 100 unambiguous contexts, while 500 contexts provide the best results. In other cases the amount of correct decisions may be reduced because statistical patterns for meanings turn out to be rather "blurry".

WSD procedure also furnished us with additional information relevant for meaning identification, namely, lexical markers of different meanings induced from contexts (cf. Table 2).

Table 2. Lexical markers of meanings induced from contexts for the noun *organ*

Word meanings	Lexical markers
organ "institution"	učreždenije "institution", samoupravlenije "self-government", načal'nik "chief", mestnyj "local", pravoohranitel'nyj "law-enforcement", etc.
organ "part of body"	porok "defect", vrožd'onnyj "innate", etc.

4.2 WSD Based on Lexical Markers vs. WSD Based on Semantic Annotation

Experiments on WSD based on lexical markers and on semantic annotation gave rather encouraging results (cf. Table 3).

In most cases WSD based on lexical markers and on semantic annotation was equally effective, at the same time, processing of contexts taking into account semantic tags

Table 3. Results of WSD based on lexical markers and semantic annotation for the noun *luk*

	Amount of correct decisions for separate meanings		Average
	luk "onion"	*luk "bow"*	
WSD based on lexical markers	75%	90%	82.5%
WSD based on semantic annotation	75%	95%	85%

Table 4. Examples of WSD based on lexical markers and semantic annotation for the noun *luk*

luk	WSD based on lexical markers		WSD based on semantic annotation	
	Identified Meaning	Cosine measure	Identified Meaning	Cosine measure
(a) *luk "onion"* Pomn'u hleb s iz'umom, s *lukom*, s kakimi-to korenjami. ([I] remember bread with raisins, with *onion*, and with some roots.)	*luk "onion"*	0.572	*luk "onion"*	0.786
(b) *luk "onion"* Načinajut prinimat' *luk*, kapustu... ([they] begin to eat *onion*, cabbage...)	*luk "weapon"*	0.502	*luk "onion"*	0.514
(c) *luk "weapon"* Odni tugije *luki*, nad kotorymi neskol'ko čelovek spravit'sa ne mogli, "igrajuči" nat'agival'i... (Some [people] "effortlessly" bent tight *bows* with which several people couldn't cope...)	*luk "weapon"*	0.533	*luk "weapon"*	0.550
(d) *luk "weapon"* Za spinoj u nego viseli *luk* i kolčan. (There were a *bow* and a quiver behind his back.)	*luk "onion"*	0.500	*luk "weapon"*	0.517

often provides more trustworthy decisions (e.g., due to the increase of cosine measure values) and allows to evade erroneous interpretations (cf. Table 4).

Most of errors registered in WSD experiments can be explained by insufficiency of contextual information for meaning identification. WSD results for such contexts often show cosine measure values about 0.500 (cf. contexts (b) and (d), Table 4).

4.3 Analysis of Special Cases

It is hardly possible to provide unambiguous analysis of certain contexts for some polysemous nouns revealing merged meanings. For example, a noun *dom* forms pairs of meanings which are almost indistinguishable in contexts: *dom "building & personal space"*, *dom "personal space & family"*, etc. Of 3,000 contexts for a noun *dom* there

Table 5. Analysis of merged meanings for the noun *dom*: WSD based on lexical markers

dom	Manual analysis	WSD results	Cosine measure
(e)... v *dome* u Jožika topilas' peč... (... in Jožik's *home* the fire was made up...)	*dom "building & personal space"*	*dom "building"*	0.429
(f) Rodstvenniki u Livii... ludi praktičnyje... jedinstvennyj čelovek, kotoryj uvažajet jejo v etom *dome*, – eto jejo dvoreckij... (Livia's relatives ... are practically-minded people ... the only person who respects her in this *house* is her butler...)	*dom "personal space & family"*	*dom "family"*	0.452

are 842 contexts where ambiguity can't be completely resolved. In such cases WSD results compared with manual analysis allow to determine a dominating semantic feature in a pair of merged meanings (cf. contexts (e) and (f), Table 5).

In further experiments additional statistical patterns corresponding to merged meanings were introduced.

5 Conclusion and Further Work

Experiments on statistical WSD were successfully carried out for contexts of polysemous and/or homonymic Russian nouns extracted from RNC.

WSD was performed in two modes – taking into account (1) lexical markers of word meanings in contexts; (2) semantic annotation of contexts. Both approaches proved to be reliable, although in controversial cases preference should be given to WSD based on semantic annotation.

Optimal conditions for WSD in Russian texts were found out: over 85% (in some cases up to 95%) of correct decisions may be achieved due to the use of cosine measure, the size of a training set varying from 100 up to 500 contexts.

Specification of context extraction procedure may be performed. On the one hand, frequency data for analysed words may be extracted from the newly developed RNC frequency dictionary [8]. The possibility of using particular frequency measures like average reduced frequency (ARF) may be considered [9]. On the other hand, special toolkits for text processing, such as Word Sketch Engine [10], should be developed for Russian. That could be of great help in the study of selectional preferences of words in contexts [11] and, thus, in extraction of morphosyntactic and semantic information relevant for WSD tasks.

Further work implies

1. enrichment of WSD software;
2. experiments on WSD based on both lexical markers and semantic annotation of contexts;
3. experiments with changing parameters (context window size, weight assignment, expansion of trial sets of contexts, etc.);
4. verification of particular hypotheses on statistical WSD for Russian.

References

1. Agirre, E., Edmonds, Ph. (eds.): Word Sense Disambiguation: Algorithms and Applications. Text, Speech and Language Technology, vol. 33. Springer, Berlin (2007)
2. Lukaševič, N.V., Čujko, D.S.: Avtomatičeskoje razrešenije leksičeskoj mnogoznačnosti na baze tezaurusnyh znanij. In: Internet-matematika 2007, pp. 108–117. Ekaterinburg (2007)
3. Rahilina, E.V., Kobricov, B.P., Kustova, G.I., L'aševskaja, O.N., Šemanajeva Ju, O.: Mnogoznačnost' kak prikladnaja problema: leksiko-semantičeskaja razmetka v Nacional'nom korpuse russkogo jazyka. In: Kompjuternaja lingvistika i intellektual'nyje tehnologii: Trudy meždunarodnoj konferencii Dialog 2006, Moscow, pp. 445–450 (2006)

4. Azarova, I.V., Marina, A.S.: Avtomatizirovannaja klassifikacija kontekstov pri podgotovke dannyh dl'a kompjuternogo tezaurusa RussNet. In: Kompjuternaja lingvistika i intellektual'nyje tehnologii: Trudy meždunarodnoj konferencii Dialog 2006, Moscow, pp. 13–17 (2006)
5. Kobricov, B.P., L'aševskaja, O.N., Šemanajeva, O., Ju, O.: Sn'atije leksiko-semantičeskoj omonimii v novostnyh i gazteno-žurnal'nyh tekstah: poverhnostnyje fil'try i statističeskaja ocenka. In: Internet-matematika 2005: Avtomatičeskaja obrabotka web-dannyh, Moscow, pp. 38–57 (2005)
6. Toldova, S.J., Kustova, G.I., L'aševskaja, O.N.: Semantičeskije fil'try dl'a razrešenija mnogoznačnosti v nacional'nom korpuse russkogo jazyka: glagoly. In: Kompjuternaja lingvistika i intellektual'nyje tehnologii: Trudy meždunarodnoj konferencii Dialog 2008, Moscow, pp. 522–529 (2008)
7. Mitrofanova, O., Mukhin, A., Panicheva, P., Savitsky, V.: Automatic Word Clustering in Russian Texts. In: Matoušek, V., Mautner, P. (eds.) TSD 2007. LNCS (LNAI), vol. 4629, pp. 85–91. Springer, Heidelberg (2007)
8. L'aševskaja, O.N., Sharoff, S.A.: Častotnyj slovar' nacional'nogo korpusa russkogo jazyka: koncepcija i tehnologija sozdanija. In: Kompjuternaja lingvistika i intellektual'nyje tehnologii: Trudy meždunarodnoj konferencii Dialog 2008, Moscow, pp. 345–351 (2008)
9. Čermák, F., Křen, M.: Large Corpora, Lexical Frequencies and Coverage of Texts. In: Proceedings of the Corpus Linguistics Conference, Birmingham, July 14–17 (2005), `http://www.corpus.bham.ac.uk/PCLC/CermakKren05.doc`
10. Pala, K.: Word Sketches and Semantic Roles // Trudy meždunarodnoj konferencii Korpusnaja Lingvistika 2006, pp. 307–317. St. Petersburg (2006)
11. Mitrofanova, O., Belik, V., Kadina, V.: Corpus Analysis of Selectional Preferences in Russian. In: Levická, J., Garabík, R. (eds.) Computer Treatment of Slavic and East European Languages: Proceedings of the Fourth International Seminar SLOVKO 2007, Bratislava, Slovakia, October 25–27, 2007, pp. 176–182 (2007)

Where Do Parsing Errors Come From

The Case of Spoken Estonian

Kaili Müürisep[1] and Helen Nigol[2]

[1] University of Tartu, Institute of Computer Science
J. Liivi 2, 50409 Tartu, Estonia
[2] University of Tartu, Institute of Estonian and General Linguistics
Ülikooli 18, 50090 Tartu, Estonia
{kaili.muurisep,helen.nigol}@ut.ee

Abstract. This paper discusses some issues of developing a parser for spoken Estonian which is based on an already existing parser for written language, and employs the Constraint Grammar framework.

When we used a corpus of face-to-face everyday conversations as the training and testing material, the parser gained the recall 97.6% and the precision 91.8%. The parsing of institutional phone calls turned out to be a more complicated task, with the recall dropping by 3%. In this paper, we will focus on parsing nonfluent speech using a rule-based parser. We will give an overview of parsing errors and ways to overcome them.

Keywords: Parsing, Estonian language, spoken language.

1 Introduction

The work with the syntactic analyzer of Estonian started 12 years ago. The parser for written language is a stable product and has been used for the annotation of the Corpus of Estonian language[1]. Meanwhile, the Corpus of Spoken Estonian[2] [1] has been growing rapidly. Large parts of it have been morphologically disambiguated and annotated with the dialogue acts mark-up. In order to provide syntactically annotated corpus, we have adapted the parser for written language to spoken language.

In this paper, we will focus on parsing non-fluent speech using a rule based parser. An overview of the parser, its preliminary adaptation to spoken language and the description of corpora is given in the following parts of the present section.

The detection of disfluencies needs to be performed prior to parsing since ungrammatical constructions disturb the analysis of the correct parts of utterances. We outline our methodology for dealing with disfluencies in Section 2. The main modifications of rules are described in Section 3. We also give an overview of types of errors that occur when parsing spoken language, the parsing results and a thorough analysis of parsing errors.

[1] http://www.cl.ut.ee/korpused/
[2] http://www.cs.ut.ee/~koit/Dialoog/EDiC.html

P. Sojka et al. (Eds.): TSD 2008, LNAI 5246, pp. 161–168, 2008.
© Springer-Verlag Berlin Heidelberg 2008

1.1 Parser for Written Estonian

The parser for written Estonian [2] is based on the Constraint Grammar framework [3]. The CG parser consists of two modules: a morphological disambiguator and a syntactic parser. In this paper, we assume that the input (transcribed speech) is already morphologically disambiguated and the word forms have been normalized according to their orthographic forms. The parser gives a shallow surface oriented analysis to the sentence, in which every word is annotated with the tag that corresponds to its syntactic function (in addition to the morphological description). The parser for written text analyzes 88 – 90% of words unambiguously and its error rate is 2% (if the input is morphologically disambiguated and does not contain errors). Two or more tags are retained for words which are hard to analyze. The parser is a rule based parser (1200 handcrafted rules).

1.2 Adaptation of Existing Parser

We have adapted the CG parser for Estonian to the spoken language in a step-by-step manner. The outcome of the first experiment [4] with a small corpus of everyday conversations demonstrated that the task of deriving the parser for the spoken language from our previous work proved to be easier than expected. Efficient detection of clause boundaries became the key issue for successful automatic analysis, while the syntactic constraints required only minimal modification. Quite surprisingly, the performance of the parser for spoken language exceeded the performance of the original parser for written language (which can be due to simpler and shorter clauses of spoken language). The output of the parser was compared with a manually annotated corpus and the following performance was observed (the results for parsing the written language are enclosed in parentheses):

1. recall (the ratio of the number of correctly assigned syntactic tags to the number of all correct tags): 97.3% (98.5%).
2. precision (the ratio of the number of correctly assigned syntactic tags to the number of all assigned syntactic tags): 89.2% (87.5%).

Recall describes the correctness of analysis and precision illustrates the level of noise.

Increasing the corpus size did not change the results significantly. Automatic detection of disfluencies improved the recall less than 0.5% [5].

1.3 Corpora

We used morphologically disambiguated texts for the experiments described in this paper. The texts were normalized (vaguely articulated or colloquial words have the annotation of the corresponding word form in written language) and provided with some transcriptional annotation (longer pauses, falling or rising intonation).

Table 1. Training and test corpora

Type	Training	Testing
Everyday conversations	8400	6276
Institutional calls	4894	5012
Disfluency corpora	4557	-

The texts have been divided into several training sets (for generation or modification of rules) and test sets. In addition, we distinguish the corpora by genre: a half of the corpus consists of everyday face-to-face conversations with two or more participants; the other part contains a corpus of institutional phone calls to travel agencies and information desks. In order to handle disfluencies more efficiently, we have used the special corpora of disfluencies (repetitions, self-repairs, false starts). The structure of the corpora is given in Table 1.

2 Disfluency Detection

Spontaneous speech is rich of disfluencies such as partial words, filled pauses (e.g., *uh*, *um*), repetitions, false starts and self-repairs.

We have conducted an experiment [6] in which the parser was applied to the original corpora of disfluencies and to the corpora where disfluencies had been filtered out manually.

Parsing both types of disfluency corpora demonstrated that the parser gains 0.5–3% better results if the nonfluent parts of the utterances have been removed before the automatic processing (see Table 2).

Several experiments [7,8] indicate that parsing performance increases when disfluencies are removed prior to data-driven parsing. Our results prove that this statement is also valid for rule based parsing.

In order to deal with automatic filtering of disfluencies, we use an external script which employs the context of 5 words for detecting possible disfluencies and removes (comments out) these from further analysis. Removed disfluencies are restored after parsing is complete. These words are annotated with an extra tag. We prefer that utterances retain their form during the parsing. This principle is also followed by other researchers. For example, Core and Schubert [9] find that edited phrase of self-repairs should be present in the final analysis since this facilitates further semantic and pragmatic analysis of dialogues.

We use 8 different rules for detecting disfluencies which are especially efficient for handling repetitions.

2.1 Repetitions

We detect repetitions: a) where a single word is repeated in the same form (*mis mis mis sa arvad*), b) a single word is repeated but the form alters (*meil meile meeldib*),

Table 2. The impact of disfluencies in the results of the parser

Type of disfluency	Utterances	Recall	Precision
Repairs	original	94.4	84.6
	normalized	96.2	87.3
Repetitions	original	98.2	90.7
	normalized	98.6	91.8
False starts	original	97.4	90.0
	normalized	98.9	93.8

c) there is an interrupted word or a particle between repetitions (*sinna ri- sinna rüütli kaubamajja*), or d) two words are repeated in the same form (*kui sa kui sa üles jõuad*). We have to be careful with some morphological categories – repetitions of verb *be* and numerals may occur in the normal sentence, so we had to consider the part-of-speech tags.

The rule for repetitions of a single word in different forms is most problematic. Although we excluded pronouns and some combinations of cases from this rule, it can still produce errors (see 1).

(1) siis pöörate alatskivile alatskivi loss on seal
then you turn to alatskivi alatskivi castle is there
"then you turn to Alatskivi. There is a castle of Alatskivi."

Applying the rules to the corpus of repetitions showed that utterances with repetitions can be analyzed as efficiently as in a normalized corpus.

2.2 Self-repairs and False Starts

One of the self-repair indicators is the presence of truncated words (e.g. *nor-* instead of *normal*). The disfluency detector comments out these words and the repetitions around them.

False starts have been eliminated by clause boundary rules of the grammar, separating these into independent chunks. False starts are easy to recognize if they contain a verb. The parser does not try to mark up false starts with special tag, the unfinished sentence is analyzed as efficiently as possible using the existing grammatical information.

3 Modifications of Rules

The end of the dialogue turn was used as the delimiter of utterance. Although the input includes punctuation marks, they are not reliable [4], usually they describe intonation, not necessarily the end of the utterance.

Clause boundaries in written text are discovered by rules that consider conjunctions, punctuation marks and verbs. These clause boundary detection rules have been thoroughly revised since the meaning and the usage of punctuation marks have changed.

The main part of syntactic constraints remains almost unaltered. There were slight modifications in context conditions related to clause boundaries and some extra conditions regarding exceptional vocabulary.

In order to apply the parser to a new genre – information dialogues -, we added new rules for questions, temporal expressions and addresses, and removed some erroneous heuristic rules. A lot of rule development work needs still to be done for processing numeral expressions.

4 Results and Analysis of Errors

The parser gained the recall 97.6% and precision 91.8% for the benchmark corpus of everyday conversations. The results for the corpus of institutional phone calls are significantly lower (see Table 3).

Table 3. Precision and recall

	Recall	Precision
Institutional calls	96.89	86.84
Everyday conversation	97.59	91.80

About 91–95% of words get unambiguous syntactic annotation. This is a remarkable result when compared to the results of the parser for written language (at most 90%). However, unlike written language, spoken language tends to contain many particles, conjunction words, adverbs and other "easy" categories for a parser. In addition, 20% of words already have unambiguous syntactic label before the syntactic disambiguation.

4.1 Classical Types of Errors

Johannessen and Jørgensen [10] introduced the following classification for errors that occur during parsing spoken language:

1. Lexical Epanorthosis – lexical correction. It is a special case of self-repairs: one part of the string replaces another without disrupting the syntactic structure but the syntactic function has been repeated in the utterance without coordination words.

   (2) Ekskursioonid Lätti algavad kevade suve poole.
   Excursions to Latvia will start spring summer towards

   "Excursions to Latvia will start in summer"

2. Syntactic epanorthosis – syntactic correction. It is another case of self-repairs: one part of the string replaces another and introduces a new syntactic frame.

   (3) et neid ee nende majutust meil ei ole
   that them ee their accommodation we no have

   "we don't have (information about) their accomodation"

3. Ellipsis – syntactically required element is missing in the initial or medial position of utterance.

   (4) suusavarustamise aa on vä
   ski equipment aa is or
   "is there a (rental) for ski equipment?"

4. Syntactic apocope – syntactically required element is missing in the final position of utterance. Johannessen and Jørgensen [10] consider false starts as a shorter type of apocope: the speaker indicates the start of an utterance, but omits the continuation.

   (5) mis teid (huvitab)
   what you (intrested in)
   "what are you (interested in)"

5. Anacoluthon – two sentences share a common element which is a part of both sentences.

(6) seal on üks kirikki on kallaste lähedal
there is one church is kallaste near
"there is a church is near Kallaste"

6. Epizeuxis – repetitions of (sequences of) words.

Heeman and Allen [11] excluded ellipsis and anacoluthon from disfluencies. They divided repairs to three classes: a) fresh starts – the speaker abandons the utterance entirely and starts over (syntactic apocope); b) modification repairs – the speaker changes the wrong word or phrase to the other or adds a new word or phrase afterwards (*so that will total – will take seven hours to do that*) (syntactic epanorthosis); c) abridged repairs – the speaker needs time and uses filled and unfilled pauses to delay the remaining utterance. This phenomenon is not covered by Johannessen and Jørgensen. If the filled pause is not formulated as a word then abridged repairs do not cause errors. But Estonians like to use pronoun *see* (*this* in nominative case) as a filler and this hampers the analysis significantly.

4.2 Errors of the Parser

In addition to error types defined above, we distinguished three extra classes of errors typical to spoken language:

1. Anaphoric expression – the solitary noun phrase is followed by a pronoun and a correct clause or the clause with a pronoun is located before or after the solitary noun phrase:

   (7) ja isegi noh see alatskivi mõis ta ei ole nii korda tehtud
   and even noh this alatskivi manor it not be so order make
   "and even the house of Alatskivi manor has not been fixed up."

2. Agreement and subcategorization errors – the words do not agree in number or case and the speaker has not repaired the utterance.

   (8) olete te kuskilt mingid pakkumisi juba saanud
   have you somewhere some-PL-NOM offer-PL-PART already got
   "Have you got any offers from somewhere already?"

3. Vocabulary – the use of word forms in a different way. Typical examples are an extensive use of pronoun *mis* (what) and adverb *nagu* (like) and foreign language words inside Estonian phrase. One may argue that this error type is a subclass of syntactic rule errors.

The analysis of errors for benchmark corpora is given in Table 4.

The clause boundary detection errors occur in "normal" utterances – it is difficult to detect the word which starts the new clause. The errors of syntactic rules may also occur in a sentence of written language. It is specific to institutional dialogues that 25% of syntactic rule errors are caused by wrong analysis of numbers. There are a lot of dates, other time expressions, prices, and phone numbers in information dialogues. (The significant amount of ellipses are also in the numeral phrases.) 10% of syntactic rule errors for this corpus are in the dictations of e-mail addresses or web-pages. The dictation is often interrupted by feedback utterances.

Table 4. Types of errors (%, rounded up)

Error type	Information dialogue	Everyday conversation
Ellipsis	19	13
Syntactic apocope	13	20
Syntactic epanorthosis	3	3
Anacoluthon	2	1
Anaphoric constructions	2	4
Agreement and subcategorization	6	5
Vocabulary	4	5
Clause boundary detection	13	14
Syntactic rules&other	36	33

As seen in Table 4, specialities of spoken language cause about a half of errors. There is a trend that missing words and phrases cause more errors than superfluous words. The speakers are more relaxed in everyday conversations and interrupt each other quite often. Also, they can express themselves with gestures. The higher rate of apocopes in everyday conversations may be explained by these phenomena. The information dialogues contain a lot of numeral and temporal expressions which lack conjunction words quite often. This may be the reason for a bigger proportion of ellipses in information dialogues.

5 Conclusions

Transcriptions of spontaneous speech are syntactically hard to analyze, even when a human linguist is involved. It is very difficult to decide which is a correct annotation of a word in the unfinished or grammatically incorrect sentence.

Our experiments demonstrated that everyday conversations are simpler to parse, although there are more participants and the dialogues take place in face-to-face environment with supporting gestures. The parsing of institutional phone calls is a harder task due to the shortness of utterances. These may be only elliptical fragments of a sentence, often missing a verb or even reasonable syntactic structure (e.g., email addresses, phone numbers).

The work on the adaptation of morphological disambiguator to spoken language is still in progress.

References

1. Hennoste, T., Lindström, L., Rääbis, A., Toomet, P., Vellerind, R.: Tartu University Corpus of Spoken Estonian. In: Seilenthal, T., Nurk, A., Palo, T. (eds.) Congressus Nonus Internationalis Fenno-Ugristarum. Pars IV. Dissertationes sectionum: Linguistica I, Tartu, pp. 345–351 (2000)
2. Müürisep, K., Puolakainen, T., Muischnek, K., Koit, M., Roosmaa, T., Uibo, H.: A New Language for Constraint Grammar: Estonian. In: Proc. of Conference Recent Advances in Natural Language Processing, Borovets, Bulgaria, pp. 304–310 (2003)

3. Karlsson, F., Anttila, A., Heikkilä, J., Voutilainen, A.: Constraint Grammar: a Language-Independent System for Parsing Unrestricted Text. Mouton de Gruyter, Berlin (1995)
4. Müürisep, K., Uibo, H.: Shallow Parsing of Spoken Estonian Using Constraint Grammar. In: Henrichsen, P.J., Skadhauge, P.R. (eds.) Treebanking for Discourse and Speech. Proc. of NODALIDA 2005 Special Session. Copenhagen Studies in Language 32, pp. 105–118. Samfundslitteratur (2006)
5. Müürisep, K., Nigol, H.: Disfluency Detection and Parsing of Transcribed Speech of Estonian. In: Vetulani, Z. (ed.) Proc.of Human Language Technologies as a Challenge for Computer Science and Linguistics. 3rd Language & Technology Conference, Poznan, Poland, pp. 483–487. Fundacja Uniwersitetu im. A. Mickiewicza (2007)
6. Nigol, H.: Parsing Manually Detected and Normalized Disfluencies in Spoken Estonian. In: Proc. of NODALIDA 2007, Tartu (2007)
7. Charniak, E., Johnson, M.: Edit detection and parsing for transcribed speech. In: Proc. of NAACL 2001, pp. 118–126 (2001)
8. Lease, M., Johnson, M.: Early deletion of fillers in processing conversational speech. In: Proc. HLT-NAACL 2006, companion volume: short papers, pp. 73–76 (2006)
9. Core, M.G., Schubert, L.K.: A Syntactic Framework for Speech Repairs and Other Disruptions. In: Proc. of 37th Ann. Meet. of the ACL, pp. 413–420 (1999)
10. Johannessen, J.B., Jørgensen, F.: Annotating and Parsing Spoken Language. In: Henrichsen, P.J., Skadhauge, P.R. (eds.) Treebanking for Discourse and Speech. Proc. of NODALIDA 2005 Special Session. Copenhagen Studies in Language 32, pp. 83–103. Samfundslitteratu (2006)
11. Heeman, P., Allen, J.: Tagging Speech Repairs. ARPA Workshop on Human Language Technolog, pp. 187–192 (1994)

Dealing with Small, Noisy and Imbalanced Data
Machine Learning or Manual Grammars?

Adam Przepiórkowski[1,2], Michał Marcińczuk[3], and Łukasz Degórski[1]

[1] Institute of Computer Science, Polish Academy of Sciences, Warsaw
[2] Institute of Informatics, Warsaw University
[3] Institute of Applied Informatics, Wrocław University of Technology

Abstract. This paper deals with the task of definition extraction with the training corpus suffering from the problems of small size, high noise and heavy imbalance. A previous approach, based on manually constructed shallow grammars, turns out to be hard to better even by such robust classifiers as SVMs, AdaBoost and simple ensembles of classifiers. However, a linear combination of various such classifiers and manual grammars significantly improves the results of the latter.

1 Introduction

Machine learning (ML) methods gave a new stimulus to the field of Natural Language Processing and are largely responsible for its rapid development since the early 1990ies. Their success is undisputed in the areas where relatively large collections of manually annotated and balanced data of reasonably good quality are available; a prototypical such area is part-of-speech tagging.

Matters are less clear when only small amounts of noisy and heavily imbalanced training data are available; in such cases knowledge-intensive manual approaches may still turn out to be more effective. One such task is definition extraction, which may be approximated by the task of classifying sentences into those containing definitions of terms and those not containing such definitions. Previous approaches to this task usually rely on manually constructed shallow or deep grammars, perhaps with additional filtering by ML methods.

In this paper we deal with the task of extracting definitions from instructive texts in Slavic, as described in Przepiórkowski *et al.* [7]. The aim of definition extraction here is to support creators and maintainers of eLearning instructive texts in the preparation of a glossary: an automatically extracted definition is presented to the maintainer who may reject it or, perhaps after some editing, accept it for the inclusion in the glossary. It follows from this intended application that recall is more important than precision here: it is easy to manually reject false positives while it is difficult to manually find false negatives not presented by the definition extraction system.

The approach described in Przepiórkowski *et al.* [7] eschews Machine Learning and relies on manually constructed grammars of definitions. The aim of the work presented here is to examine to what extent the same task may be carried out with the use of ML classifiers. In particular, we adopt the Polish data set of Przepiórkowski *et al.* [6]

P. Sojka et al. (Eds.): TSD 2008, LNAI 5246, pp. 169–176, 2008.
© Springer-Verlag Berlin Heidelberg 2008

consisting of 10830 sentences, 546 of which are definition sentences (i.e., they are or contain definitions). Obviously, this is a relatively small data set on which to train classifiers. Moreover, the classes are heavily imbalanced, with the ratio of definitions to non-definitions $\approx$ 1:19.

To complicate matters further, it is often not clear even for humans whether a given sentence contains a definition or not: whenever a sentence describes a certain characteristic of a notion, the annotator must decide whether this characteristic is definitional or just one of many traits of the notion. Correspondingly, the inter-annotator agreement reported in Przepiórkowski *et al.* [6] is very low: when measured as Cohen's κ it is equal to 0.31 (the value of 1 would indicate perfect agreement, the value of 0 – complete randomness).

In the rest of the paper we first (§2) briefly present the manual grammar approach of Przepiórkowski *et al.* [6]. In the following two sections (§§3–4) we report on the experiments of applying ML classifiers and homogeneous ensembles of classifiers to the same data, with results uniformly worse than those of manually constructed grammars. However, the ensuing section (§5) demonstrates that a combination of ML classifiers and linguistic grammars, while still not fully satisfactory, significantly improves on the results of either approach. Sections presenting some comparisons, suggesting future work and drawing conclusions end the paper (§§6–7).

2 Manual Grammars

As described in Przepiórkowski *et al.* [6], a rather simple shallow grammar of Polish definitions, containing 48 rules (some of them consisting of rather complex regular expressions) was developed on the basis of a development corpus of 5218 sentences (containing 304 definitions[1]) and fine tuned on a thematically different corpus of 2263 sentences (with 82 definitions). The whole grammar development process took less than 2 weeks of intensive work. The resulting grammar, called GR′, and a relatively sophisticated baseline grammar B3, looking for copula and similar clues for definitions, were then tested on an unseen corpus containing 3349 sentences (with 172 definitions). The results, in terms of precision (P), recall (R), the standard F-measure (F_1), as well as two F-measures giving twice (F_2) and five times (F_5; cf. Saggion [8]) more weight to recall, are presented in Table 1(a).[2]

Using the same grammars, we evaluated them on the whole corpus (hence, also on parts of the corpus which were seen during grammar development), obtaining the results in Table 1(b). Thus, any classifier with F_2 higher than 36, when measured with the standard 10-fold cross-validation (10CV) procedure on the whole corpus, would clearly improve on these results.

[1] Note that these are definitions, not definition sentences: one sentence may contain a number of definitions and, although rarely, a definition may be split into a number of sentences.

[2] We follow Przepiórkowski *et al.* [6,7] in using F_2 as the main measure summarising the quality of the approach, but with an eye on F_5. Also, we adopt their formula for F_α as equal to $\frac{(1+\alpha)\cdot P\cdot R}{\alpha\cdot P+R}$.

Table 1. Evaluation of B3 and GR′ on (a) the testing corpus and on (b) the whole corpus

	(a) testing corpus					(b) whole corpus				
	P	R	F_1	F_2	F_5	P	R	F_1	F_2	F_5
B3	10.54	88.46	18.84	25.54	39.64	9.12	89.56	16.55	22.73	36.26
GR′	18.69	59.34	28.42	34.39	43.55	18.08	67.77	28.54	35.37	46.48

3 Single Classifiers

In the experiments reported here we assumed a relatively simple feature space: a sentence is represented by a vector of binary features, where each feature represents an n-gram, present or not in a given sentence. More specifically, after some experiments we adopted as features unigrams, bigrams and trigrams of base forms, parts of speech and grammatical cases. We chose those n-grams which were most frequent in definitions or in non-definitions. Given the 9 n-gram types (e.g., single base forms, bigrams of base forms, trigrams of cases, etc.), for each type we selected 100 most frequent n-grams of this type. As a result, each sentence is represented by a binary vector of length 781.[3]

For the experiments we used the WEKA tool [9] and its implementation of simple decision trees (ID3 and C4.5), Naïve Bayes (NB) classifiers, a simple lazy learning classifier IB1, as well as the currently more popular classifiers AdaBoost (AdaBoostM1 with Decision Stumps; AB+DS) and Support Vector Machines (nu-SVC; cf. `http://www.cs.iastate.edu/~yasser/wlsvm/`). Because of the very high prevalence of one class, we experimented with different ratios of subsampling, in each case using all definitions: 1:1 (equal number of definitions and non-definitions), 1:5 (5 non-definitions for each definition), 1:10 and 1:all (≈ 1:19, i.e., no subsampling). All experiments followed the general 10-fold cross-validation (10CV) methodology, with the corpus split randomly into 10 buckets of roughly the same size in such a way that each bucket contains roughly the same number of definitions (a balanced random split). The results are presented in Table 2.

As was expected, SVM and AdaBoost turned out to be the best classifiers for the task at hand, as measured by F_2. However, even the best classifier, based on Support Vector Machines with the 1:5 ratio of subsampling, turned out to give results significantly worse than the manual grammar GR′. Moreover, somewhat surprisingly, different ratios of subsampling turned out to be optimal for different types of classifiers: for AdaBoost the best ratio was 1:1, for C4.5, ID3, IB1 and SVM it was 1:5, while for Naïve Bayes it turned out to be 1:all (no subsampling).

4 Homogeneous Ensembles

In the next stage of experiments, homogeneous ensembles of classifiers were constructed. Experiments were conducted with the 6 types of classifiers with the best subsampling (cf. the numbers in bold in Table 2), plus additional subsampling ratios of IB1,

[3] The length is shorter than 900 because the numbers of grammatical classes, cases and bigrams of cases are smaller than 100 each.

Table 2. Performance of the classifiers for different ratio of positive to negative examples evaluated on the whole corpus with balanced random split

Classifier	Ratio	P	R	F_1	F_2	F_5	Comments
NB	1:1	9.50	60.07	16.41	21.66	31.84	
	1:5	10.53	54.58	17.65	22.79	32.16	
	1:10	10.75	51.83	17.80	22.79	31.66	
	1:all	10.94	49.82	17.94	**22.80**	31.28	
C4.5	1:1	8.25	59.89	14.50	19.41	29.31	
	1:5	14.81	30.04	19.84	**22.37**	25.65	
	1:10	19.48	16.48	17.86	17.37	16.92	
	1:all	32.35	10.07	15.36	13.07	11.38	
ID3	1:1	8.66	66.85	15.33	20.63	31.53	
	1:5	12.79	37.91	19.12	**22.91**	28.56	
	1:10	14.78	26.00	18.85	20.75	23.08	
	1:all	15.65	17.77	16.64	17.00	17.37	
IB1	1:1	9.68	50.73	16.26	21.02	29.73	
	1:5	15.94	26.19	19.82	**21.57**	23.66	
	1:10	20.00	18.86	**19.42**	19.23	19.04	
	1:all	21.85	14.28	17.28	16.15	15.16	
nu-SVC	1:1	11.79	69.05	20.14	26.37	38.16	nu=0.5
	1:5	20.75	37.55	26.73	**29.57**	33.08	nu=0.2
	1:10	27.11,	27.66	**27.38**	27.47	27.56	nu=0.1
	1:all	33.33	16.67	22.22	20.00	18.18	nu=0.05
AB+DS	1:1	11.59	68.32	19.82	**25.97**	37.63	1000 iterations
	1:5	28.13	23.44	**25.57**	24.82	24.11	1000 iterations

SVM and AdaBoost which gave promising results in other experiments, not reported here for lack of space. In case of Naïve Bayes, the best performance was obtained without subsampling, although the results were only insignificantly better than 1:5 and 1:10 subsampling, with the subsampling configurations performing better in ensembles. For this reason NB without subsampling was not considered further. The summary of the best remaining ensembles, in comparison with the two grammars, is presented in Table 3.

For each of these 9 classifiers, homogeneous ensembles (i.e., collections of classifiers of the same type) were constructed consisting of 1, 3, 5, 9 and 15 classifiers, with the final decision reached via simple voting. In most cases, with the exception of one type of IB1 and one type of AdaBoost, ensembles of 15 or 9 classifiers gave best results. Note that, again, SVM and AdaBoost gave best results and, again, while the ensemble of 9 SVMs (with 1:5 subsampling) reached F_2 close to that of GR′ (31.49 vs. 34.39/35.37), no classifier surpassed the manual approach in terms of the two F-measures favouring recall.

At this point, much more time had been spent on ML experiments than the "less than two weeks" spent by Przepiórkowski *et al.* [6] on the development of manual grammars for the same task. Of course, this does not warrant the conclusion that definition extraction should be approached linguistically rather than statistically, as many factors play a role here, including the level of expertise in grammar writing, the experience in

Table 3. Performance of the selected classifiers and the grammars evaluated on the whole corpus (10CV)

Classifier	P	R	F_1	F_2	F_5
9 × nu-SVC (1:5)	24.11	37.18	29.25	31.49	34.10
9 × AdaBoost 1000 it. (1:1)	13.24	72.34	22.39	29.08	41.48
15 × ID3 (1:5)	24.73	29.30	26.82	27.60	28.43
15 × NB (1:10)	10.75	52.01	17.81	22.81	31.71
9 × C4.5 (1:5)	24.07	24.91	24.48	24.62	24.76
15 × IB1 (1:5)	17.80	24.54	20.63	21.79	23.08
9 × nu-SVC (1:10)	30.79	26.56	28.52	27.83	27.18
1 × AdaBoost 1000 it. (1:5)	28.13	23.44	25.57	24.82	24.11
1 × IB1 (1:10)	20.00	18.86	19.42	19.23	19.04
Grammar B3	9.12	89.56	16.55	22.73	36.26
Grammar GR′	18.08	67.77	28.54	35.37	46.48

constructing classifiers, the assumed feature space, the exact character of the data, etc. Nevertheless, it seems that in case of small, noisy, imbalanced data, a manual "linguistic" approach may be a viable alternative to the dominant statistical machine learning paradigm.

5 Linear Combination of Grammars and Ensembles

If simple homogeneous ensembles of common classifiers do not give better results than manual grammar, perhaps they can be combined with the grammars to improve their results? Various such modes of combination are possible and, in a different paper, we describe some promising results of a sequential combination of the baseline grammar B3 and ML classifiers [3].

In this section we present the results of a linear combination of the 9 ensembles of classifiers introduced in the previous section, each treated as a single classifier here, with the two grammars: B3 and GR′.

In order to assign weights to particular classifiers, let us first introduce some notation. Let $D^+(x)$ mean that x is a definition, $D^-(x)$ – that x is not a definition, $D_i^+(x)$ – that x is classified as a definition by the classifier i, $D_i^-(x)$ – that x is classified as a non-definition by the classifier i, and finally, TP_i, etc. are the numbers of true positives, etc., according to the classifier i.

We can estimate the probability $p_i^+(x)$ that a given sentence x is really a definition, if the classifier i says that it is a definition, in the following way:

$$p_i^+(x) = p(D^+(x)|D_i^+(x)) = \frac{p(D^+(x) \wedge (D_i^+(x)))}{p(D_i^+(x))} \approx \frac{TP_i}{TP_i + FP_i}$$

Similarly, given that the classifier says that x is not a definition, the probability of x actually being a definition is:

$$p_i^-(x) = p(D^+(x)|D_i^-(x)) = \frac{p(D^+(x) \wedge (D_i^-(x)))}{p(D_i^-(x))} \approx \frac{FN_i}{FN_i + TN_i}$$

Let us then define $d_i(x)$ as follows:

$$d_i(x) = \begin{cases} \frac{TP_i}{TP_i+FP_i}, & \text{if } x \text{ is classified as definition} \\ \frac{FN_i}{FN_i+TN_i}, & \text{if } x \text{ is classified as non-definition} \end{cases}$$

Assuming that each of the N classifiers votes for the definitory status of x with the strength proportional to the estimated probability given above, the decision of the whole ensemble of N classifiers may be calculated as:

$$d(x) = \frac{\sum_{i=1}^{N} d_i(x)}{N}$$

If $d(x) > \delta$, the linear combination classifies x as a definition, otherwise – as a non-definition.

What is the best value of the cut-off point δ? The examination of different values close to the estimated probability that a sentence is a definition (cf. Table 4) shows that for $\delta = 0.08$, F_2 reaches almost the value of 39, significantly higher than either the F_2 for the grammar GR′ alone or the best F_2 for pure ML classifiers.[4]

It is interesting to what extent the improvement is the effect of combining various types of ML classifiers, and to what extent the presence of grammars B3 and GR′ affects the results. To this end, final experiments were performed, where three linear combinations of classifiers were trained on the part of the corpus seen when developing the grammars (cf. §2) and tested on the remaining unseen portion of the data.[5] These combinations are: the 9 ML classifiers (9ML), 9 ML classifiers and B3 (9ML+B3), and finally all 11 classifiers (9ML+B3+GR′). The best results of these combinations (i.e., for the best cut-off points) are presented in Table 5 and they clearly indicate the crucial role played by the full grammar GR′ in such heterogeneous ensembles.[6]

6 Comparisons and Future Work

We are not aware of other work of similar scope comparing and combining machine learning and linguistic approaches to definition extraction, or to other NLP tasks based on small, noisy and heavily imbalanced data, although there is a rich literature on combining inductive and manual approaches to tagging, where a similar synergy effect is

[4] In fact, in some of the other experiments, with weights assigned in less principled ways, F_2 exceeded 39. Moreover, this value is also higher than F_2 for the unanimous voting combination of B3 and GR′, where $F_2 = 37.28$, as measured on the whole corpus.

[5] This way of evaluation is unfavourable both to the grammars (they are tested on data unseen during their development) and to ML classifiers (they are trained on a smaller part of the corpus than in case of 10CV). When tested on the whole corpus, with 10CV, the best F_2 results for 9ML, 9ML+B3 and 9ML+B3+GR′ were, respectively, 35.80, 35.75 and, as already reported, 38.90. Note that the first two results, for combinations without GR′, are still lower than the results for the combination of B3 and GR′ mentioned in the previous footnote.

[6] But note that here the result of 9ML+B3+GR′ is only slightly better than that of GR′ alone as tested on the same data; cf. Table 1(a).

Table 4. Performance of the linear combination of classifiers for various values of δ as evaluated on the whole corpus (10CV)

δ	P	R	F_1	F_2	F_5
0.05	12.53	84.25	21.81	28.97	43.11
0.06	17.63	71.79	28.31	35.48	47.49
0.07	21.32	61.72	31.69	37.82	46.90
0.08	25.17	53.48	34.23	38.90	45.04
0.09	27.08	46.52	34.23	37.54	41.55
0.10	30.61	39.74	34.58	36.15	37.86
0.11	32.66	32.60	32.63	32.62	32.61
0.12	36.97	28.57	32.23	30.91	29.70
0.13	40.64	25.46	31.31	29.08	27.15

Table 5. The effect of grammars on the performance of the linear combination of classifiers, evaluated on the testing corpus

classifier	δ	P	R	F_1	F_2	F_5
9ML	0.07	18.16	40.11	25.00	28.59	33.38
9ML+B3	0.07	18.56	41.21	25.60	29.30	34.25
9ML+B3+GR′	0.06	17.19	57.14	26.43	32.20	41.19

usually observed. Previous work on definition extraction, mainly for English and other Germanic languages, usually consists of a simple sequential combination of grammatical parsing and ML filtering. Often only precision or only recall is cited, so it is difficult to directly compare our approach to these other approaches.

There are very many possible improvements to the work reported here, starting from the selection of features, through the selection of classifiers for the linear combination, to the better assignment of weights to particular (homogeneous ensembles of) classifiers. Moreover, in other work [3] we describe a sequential combination of the baseline grammar B3 and ML classifiers which achieves results comparable to GR′, but without the need for the development of a grammar more sophisticated than B3. A rather different approach worth pursing seems to be the employment of random forests [1]. Although basic random forests have already been applied in NLP with satisfactory results [10], *balanced* random forests [2], particularly well suited in heavily imbalanced classification tasks, still remain to be explored (see Kobyliński and Przepiórkowski [4]).

7 Conclusion

In the days of the – fully deserved – dominance of inductive methods, any solutions involving the manual coding of linguistic knowledge must be explicitly justified. We have shown that, in case of a task relying on very low-quality (small, noisy, imbalanced) training data, manual methods still rival statistical approaches. On the other hand, even in such difficult tasks, ML may be very useful, not as a replacement of hand-coded grammars, but as a support for them: our combination of linguistic grammars and homogeneous ensembles of various classifiers achieves results significantly higher than either of the two pure approaches. We conclude that, while firing linguists may initially increase the performance of a system, perhaps a few of them should be retained in a well-balanced heterogeneous NLP team.

References

1. Breiman, L.: Random forests. Machine Learning 45, 5–32 (2001)
2. Chen, C., Liaw, A., Breiman, L.: Using random forest to learn imbalanced data. Technical Report 666, University of California, Berkeley (2004), `http://www.stat.berkeley.edu/tech-reports/666.pdf`
3. Degórski, Ł., Marcińczuk, M., Przepiórkowski, A.: Definition extraction using a sequential combination of baseline grammars and machine learning classifiers. In: Proceedings of the Sixth International Conference on Language Resources and Evaluation, LREC 2008. ELRA, Marrakech (2008) (forthcoming)
4. Kobyliński, Ł., Przepiórkowski, A.: Definition extraction with balanced random forests. In: 6th International Conference on Natural Language Processing, GoTAL 2008, Gothenburg (2008) (forthcoming)
5. Piskorski, J., Pouliquen, B., Steinberger, R., Tanev, H. (eds.): Proceedings of the Workshop on Balto-Slavonic Natural Language Processing at ACL 2007, Prague (2007)
6. Przepiórkowski, A., Degórski, Ł., Wójtowicz, B.: On the evaluation of Polish definition extraction grammars. In: Vetulani, Z. (ed.) Proceedings of the 3rd Language & Technology Conference, Poznań, Poland, pp. 473–477 (2007a)
7. Przepiórkowski, A., Degórski, Ł., Spousta, M., Simov, K., Osenova, P., Lemnitzer, L., Kuboň, V., Wójtowicz, B.: Towards the automatic extraction of definitions in Slavic. In: [5], pp. 43–50 (2007b)
8. Saggion, H.: Identifying definitions in text collections for question answering. In: Proceedings of the Fourth International Conference on Language Resources and Evaluation, LREC 2004. ELRA, Lisbon (2004)
9. Witten, I.H., Frank, E.: Data Mining: Practical machine learning tools and techniques, 2nd edn. Morgan Kaufmann, San Francisco (2005), `http://www.cs.waikato.ac.nz/ml/weka/`
10. Xu, P., Jelinek, F.: Random forests in language modeling. In: Lin, D., Wu, D. (eds.) Proceedings of the 2004 Conference on Empirical Methods in Natural Language Processing (EMNLP 2004), pp. 325–332. ACL, Barcelona (2004)

Statistical Properties of Overlapping Ambiguities in Chinese Word Segmentation and a Strategy for Their Disambiguation*

Wei Qiao[1], Maosong Sun[1], and Wolfgang Menzel[2]

[1] State Key Laboratory of Intelligent Technology and Systems
Tsinghua National Laboratory for Information Science and Technology
Department of Computer Sci. & Tech., Tsinghua University, Beijing 100084, China
qiaow04@mails.tsinghua.edu.cn, sms@mail.tsinghua.edu.cn
[2] Department of Informatik, Hamburg University, Hamburg, Germany
menzel@informatik.uni-hamburg.de

Abstract. Overlapping ambiguity is a major ambiguity type in Chinese word segmentation. In this paper, the statistical properties of overlapping ambiguities are intensively studied based on the observations from a very large balanced general-purpose Chinese corpus. The relevant statistics are given from different perspectives. The stability of high frequent maximal overlapping ambiguities is tested based on statistical observations from both general-purpose corpus and domain-specific corpora. A disambiguation strategy for overlapping ambiguities, with a predefined solution for each of the 5,507 pseudo overlapping ambiguities, is proposed consequently, suggesting that over 42% of overlapping ambiguities in Chinese running text could be solved without making any error. Several state-of-the-art word segmenters are used to make comparisons on solving these overlapping ambiguities. Preliminary experiments show that about 2% of the 5,507 pseudo ambiguities which are mistakenly segmented by these segmenters can be properly treated by the proposed strategy.

Keywords: Overlapping ambiguity, statistical property, disambiguation strategy, domain-specific corpora.

1 Introduction

Word segmentation is the initial stage of many Chinese language processing tasks and has drawn a large body of research. Overlapping ambiguity (OA) is one of the basic types of segmentation ambiguities. A string in Chinese text is called an overlapping ambiguity string (OAS) if it satisfies following definition: suppose S is a string of Chinese characters, D is a Chinese wordlist, and S is not in D. S is an OAS if there exists a sequence of words in D denoted $w_1, w_2, \ldots, w_m (m > 2)$ that exactly cover S, and adjacent words $w_i, w_i + 1$ $(1 \leqslant i < m)$ intersect but do not cover each other. It is reported that OA constitutes 90% of segmentation ambiguities in Chinese running text [6].

* The research is supported by the National Natural Science Foundation of China under grant number 60573187 and the CINACS project.

P. Sojka et al. (Eds.): TSD 2008, LNAI 5246, pp. 177–186, 2008.
© Springer-Verlag Berlin Heidelberg 2008

Previous work on solving overlapping ambiguities can be roughly classified into two categories: rule-based and statistical approaches.

Maximum Matching (MM) can be viewed as the simplest rule-based OA disambiguation strategy. [2] indicates that MM can only achieve an accuracy of 73.1% for OA strings. A set of manually generated rules are used in [15] and reported an accuracy of 81.0%. A lexicon-based method is presented in [12], achieving an accuracy of 95.0%. A general scheme in statistical approach is to use character or word N-gram models or POS N-gram models to find the best segmentation in the candidate segmentation space of an input sentence. For example, [10] presents a character bigram method and reports an accuracy of 90.3% for OA strings. Another general scheme is to define segmentation disambiguation as a binary classification problem and use a classifier to solve it. For example, [4] uses Support Vector Machine with mutual information between Chinese character pairs as features, achieving an accuracy of 92.0% for OA strings.

It is worth noting that [11] finds that the 4,619 most frequent OA strings, which fall into an ambiguity category named "pseudo segmentation ambiguity" (see Section 2 for detail), can be disambiguated in a context-free way, and these strings can cover 53.35% of OA tokens in a news corpus. A so-called "memory-based model" is proposed to solve these OAs. The work of [5] continues in this line: totally 41,000 most frequent pseudo OA strings are identified and a solution called "lexicalized rules" is proposed. Experimental results show that it can benefit word segmentation significantly.

The research here will follow and extend that of [11] and [5]. Three basic issues remain unsolved in their work:

Basic issue 1: The corpora used in either [11] or [5] only include news data. Obviously, the findings in these works should be further validated by adopting more "appropriate" data, otherwise, they still seems to be too restricted.

Basic issue 2: Even if the above "conclusion" can really work based on the observation from more "appropriate" data, we further need to determine the stable core of pseudo OA strings, to test its coverage and check its stability in Chinese running text, both general-purpose and domain-specific.

Basic issue 3: Once the core of pseudo OA strings is determined, we need to see if there exists a disambiguation strategy that can solve them effectively.

The remainder of this paper is organized as follows: Section 2 presents the related terms. Section 3 gives distributions of OA strings based on general-purpose corpus. Section 4 observes the stability of high frequent OA strings on both general-purpose corpus and domain-specific corpora. A resulting disambiguation strategy is proposed in Section 5. Section 6 is the conclusion.

2 Related Terms

In this section, we introduce some related terms (concepts) which are defined in [9] for describing various aspects of overlapping ambiguities.

We first give five basic terms:

Length of an OAS is the number of characters it contains; Each word in an OAS is called ***Span of an OAS***; The number of spans it contains is the ***Order of an OAS***; The common part of two adjacent spans is ***Intersection of spans***; The totality of the spans of an OAS constitutes its structure and is called ***Structure of an OAS***.

Three important concepts are further introduced as follows:

Maximal overlapping ambiguity string (MOAS): Let S_1 be an OAS and occurs as a substring of a sentence S. If an OAS containing S_1 never exists in S then S_1 is called a MOAS in the context of S.

Take the sentence "他为推广普通话费尽心血"(He tried his best to popularize the Mandarin) as an example, both the string of "普通话费" and "普通话费尽心血" in this sentence are OASs whereas "普通话费" is not maximal because it is included in "普通话费尽心血".

Real segmentation ambiguity: A segmentation ambiguity is said to be real if at least its two distinct possible segmentations can be realized in running Chinese texts depending on its contexts.

For example, "其次要" is said "real", because two segmentations "其|次要(the subordination)" and "其次|要(secondly should)" can be realistic:

a.先解决其主要问题，再解决其|次要问题。(First of all we should solve the main problem, and then consider the subordination one.)

b.首先要关注整体框架，其次|要注意细节。(Firstly we should focus on the whole framework, secondly should notice the details.)

Pseudo segmentation ambiguity: A segmentation ambiguity is said to be pseudo if only one of its distinct segmentations can be realized in running text.

For example, "部长篇小说" is said "pseudo" because only the latter of its two possible segmentations, "部长(minister)| 篇(measure word)小说(novel)" and "部(measure word)|长篇小说(long novel)", can be realized in text.

The advantage of distinguishing MOAS from OAS is that the latter is comparatively isolated from its context and thus readily available for independent study. Clearly, defining an ambiguity as a maximal overlapping ambiguity provides an adequate and quite stable processing unit for further investigation of its properties, for example, whether it is real or pseudo.

3 Statistical Properties of MOAS

Targeting at the basic issue 1 and 2 mentioned in Section 1, first of all, we design and construct a huge balanced Chinese corpus, CBC. CBC is very rich in content as it contains the collection of Chinese literature since 1920's, and it is well balanced, covering rich categories such as novel, essay, news, entertainment and texts from the web. The total size of CBC is 929,963,468 characters. The Chinese wordlist we used in this paper is developed by Peking University [14], with 74,191 entries (word types), denoted CWL here. Based on CWL, we extract all of the MOASs from CBC. A total of 733,066 distinct MOAS types are obtained at last, forming a complete MOAS type set, denoted CS-MOAS. These MOAS types have 11,103,551 occurrences in CBC, covering 39,432,267 Chinese characters, which constitutes 4.24% of CBC.

We then systematically observe the statistical properties of overlapping ambiguities through their distributions in CBC.

3.1 MOAS and Zipf's Law

Figure 1(a) shows the relationship between the rank of a MOAS type and its token frequency over CS-MOAS, in a log-log scale. We can see that this relationship roughly obeys Zipf's Law, i.e., rank $\times$ $TF \approx C$, where the constant C is roughly 1.11×10^6.

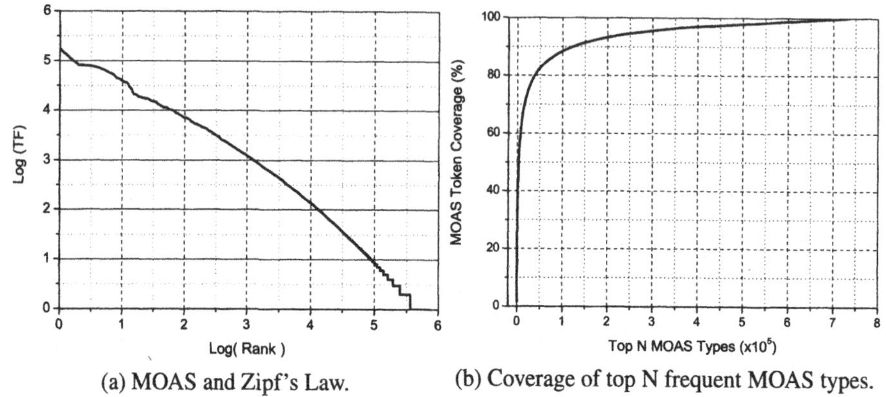

(a) MOAS and Zipf's Law. (b) Coverage of top N frequent MOAS types.

Fig. 1. MOAS and Zipf's Law

The coverage of the top N frequent MOAS types to MOAS tokens over CBC (i.e., the total occurrences of CS-MOAS in CBC) is shown in Figure 1(b), where the MOAS types are sorted by their token frequencies. It indicates that, as expected, the coverage of top 3,500, 7,000 and 40,000 frequent MOAS types is 50.78%, 60.43% and 80.39% respectively. These top frequent MOASs are thus possibly regarded as a "core" of MOAS in running text.

3.2 The Detailed MOAS Distributions

We give the detailed distributions of MOAS from following four perspectives:

Perspective 1: The distribution of MOAS length over CBC (Table 1).

As shown in Table 1, the type and token coverage of MOASs with length below 6 are as high as 98.26% and 99.72%. Obviously, these MOAS types are of higher significance in OA disambiguation.

Perspective 2: The distribution of MOAS order over CBC (Table 2).

Table 2 shows that the type and token coverage of MOASs with order 2 and 3 are 89.14% and 98.17% respectively. And that of MOASs with orders below 5 are as high as 99.63% and 99.95% respectively. They show an even more concentrated distribution compared to that of MOAS length.

Perspective 3: The distribution of intersection length over CBC (Table 3).

The distribution of intersection length is much more concentrated, with 99.66% of MOAS has intersection length of 1.

Table 1. Distribution of MOAS length over CBC

Length	#. of MOAS types	Coverage to MOAS types	Coverage to MOAS tokens
3	211,270	28.82%	55.22%
4	348,065	47.48%	36.86%
5	108,786	14.84%	5.83%
6	52,233	7.13%	1.81%
7~12	12712	1.74%	0.28%
Total	733,066	100.00%	100.00%

Table 2. Distribution of MOAS order over CBC

Order	#. of MOAS types	Coverage to MOAS types	Coverage to MOAS tokens
2	269,717	36.79%	64.49%
3	383,794	52.35%	33.68%
4	53,748	7.33%	1.27%
5	23,171	3.16%	0.51%
6~10	2,636	0.35%	0.05%
Total	733,066	100.00%	100.00%

Table 3. Distribution of MOAS intersection length over CBC

Length	#. of types	Coverage to tokens	#. of tokens	Coverage to tokens
1	1,300,736	99.66%	15,224,314	99.39%
2	4,408	0.34%	92,434	0.60%
3	30	0.00%	1,060	0.01%
Total	1,305,174	100%	15,317,808	100%

Perspective 4: The distribution of MOAS structure over CBC (Table 4).

The MOAS structure is in fact a combination of the order and length of spans of an MOAS, for example, the structure of the OAS "其次要" is (0-2,1-3)(note that the notation i-j means there exists a span which starts from location i and ends at location j of the given OAS). We totally find 97 different structure types for CS-MOAS over CBC. The top 3 major structure types are listed in Table 4.

Table 4. Distribution of MOAS structure over CBC

Structure	#. of MOAS types	Coverage to MOAS types	Coverage to MOAS tokens
(0-2,1-3,2-4)	306,627	41.83%	29.63%
(0-2,1-3)	211,270	28.82%	55.22%
(0-2,1-3,2-4,3-5)	40,482	5.52%	0.94%
others	174,687	23.83%	14.21%
Total	733,066	100%	100%

4 Stability of the Top N Frequent MOAS Types

The distributions given in Section 3 demonstrate that MOASs exhibit very strong centralization tendencies. It suggests that there may exist a "core" of MOASs with relatively small size meanwhile with quite powerful coverage capability for running texts. Regarding the top 3,500, 7,000 and 40,000 frequent MOASs as the candidates of "core", a question then comes: are these MOASs stable in Chinese running text? We observe the stability of these MOASs from following two perspectives.

4.1 Perspective 1: Stability vs. Corpus Size

Firstly, we want to study the influence of corpus size on the stability of a potential MOAS core. We randomly divide CBC into ten equal parts, each of which has a size of 194,793KB. The experiments here start from any part of CBC, with one more part added in next round.

Figure 2(a) shows that the number of MOAS types increases almost linearly with the corpus size. On the contrary, the situation is totally different in Figure 2(b), where the vertical axis stands for the number of common part between the top N frequent MOASs

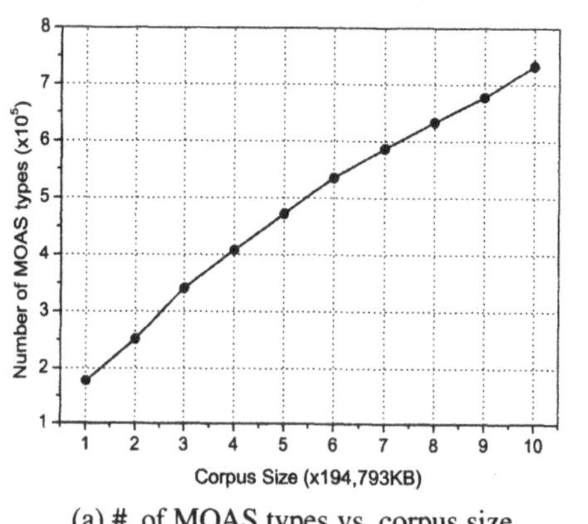

(a) #. of MOAS types vs. corpus size.

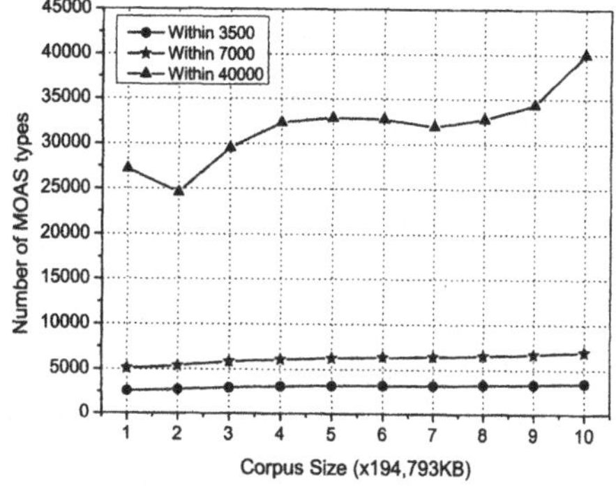

(b) #. of MOAS types falling into the top 3,500, 7,000, and 40,000 vs. corpus size.

Fig. 2. Stability vs. corpus size

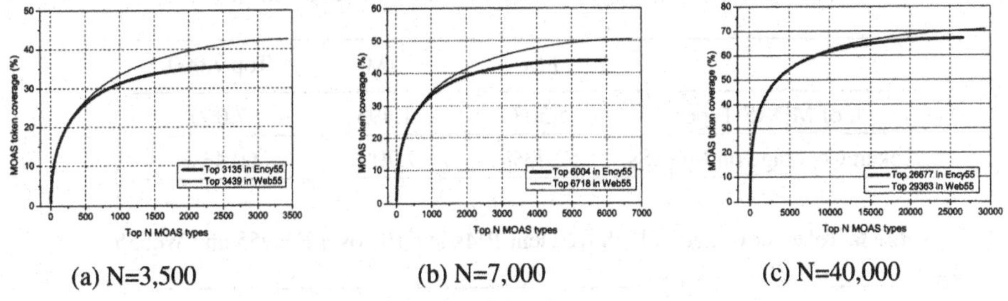

(a) N=3,500 (b) N=7,000 (c) N=40,000

Fig. 3. Coverage of top N frequent MOASs in CBC over Ency55 and Web55

in the current corpus and the top N frequent MOASs in CBC: the curves for the top 3,500 and 7,000 are almost flat, meanwhile the curve for the top 40,000 is with obvious fluctuation.

4.2 Perspective 2: Stability vs. Domain-Specific Corpora

Secondly, we try to test the stability of a potential MOAS core on domain-specific corpora. In order to do this we design and conduct the other two domain-specific corpora: Ency55 and Web55. Ency55 is the electronic version of Chinese Encyclopedia, with 90.02 million characters while Web55 is collected from the web, with 54.97 million characters. They are all organized into 55 technical domains such as Geography, Communication, Mechanics, Chemistry etc. (Both corpora are fully independent to CBC).

There are totally 168,478 and 119,663 MOAS types in Ency55 and Web55 respectively, in terms of CWL. The MOAS types in Ency55 include a total of 3,783,164 characters, covering 4.2% of the corpus, and that in Web55 includes 2,028,053 characters, covering 3.7% of the corpus.

The coverage test of the top 3,500, 7,000 and 40,000 frequent MOASs in CBC to Ency55 and Web55 is carried out, as shown in Figure 3(a), 3(b) and 3(c). As can be seen, the token coverage can still reach 35.72%, 43.84% and 67.08%, which has a 13% $\sim$ 16% drop from that of CBC (50.78%, 60.43% and 80.39%) respectively. This is another evidence for that the potential MOAS cores are quite stable in Chinese.

Due to the fact that the top 7,000 MOASs can cover 60.43% MOAS tokens in CBC, 43.84% MOAS tokens in Ency55 and 50.43% in Web55, we thus choose the top 7,000 as the core of MOASs which will serve as a basis for Section 5.

5 Disambiguation Strategy Inspired by Statistical Properties of MOAS

The top 7,000 MOASs can be further divided into two categories: pseudo MOAS (PM) and real MOAS (RM). Sometimes however it is not always easy to decide if an MOAS is pseudo or real in the strict sense. Here we make a relaxation on "pseudo": an RM which has very strong tendency to have only one segmentation way in corpus will be treated as an PM.

Table 5. Token coverage of PM and RM over MOASs in CBC

	PM	RM	Top 7,000
#. of MOAS Types	5,507	1,493	7,000
Token Coverage on MOASs	52.73%	7.70%	60.43%

Table 6. Token coverage of high frequent PMs in CBC over Ency55 and Web55

	Ency55 MOAS	Web55 MOAS
#. of Common PMs	4,342	5,079
Token Coverage on MOASs	42.21%	47.89%

Consider the RM "出国门": in almost all the cases it is realized as "出| 国门(go abroad)", while in very rare cases, there still exists a very small possibility for it to be realized as "出国| 门(the way to go abroad)", as in the sentence "他想出国门都没有(For him there is no way to go abroad)". "出国门" will thus be thought of as an PM according to the above relaxation.

In terms of the relaxation definition, 5,507 out of the top 7,000 frequent MOAS are manually judged to be PM. The token coverage of PMs and RMs on all MOASs in CBC is listed in Table 5.

Table 6 shows the token coverage of the high frequent PM in CBC to all the MOASs over Ency55 and Web55.

It is promising to see that the PMs in the top 7,000 frequent MOASs can still cover 42.21% and 47.89% MOASs in domain-specific corpora. This indicates that this set of PMs is quite stable. Thus if these PMs are solved, it can be expected about 42% of overlapping ambiguities in Chinese text can be perfectly solved.

One thing that deserves to be mentioned is, it is possible for an PM in general-purpose corpus to change into an RM in domain-specific corpora. Based on our observation on 5,507 PMs, only a few of them fall into this case.

Concerning the basic issue 3 in Section 1, since the resolution of pseudo segmentation ambiguities is independent of their contexts, a basic strategy can be formulated: for a highly frequent pseudo MOAS, its disambiguation can be performed by simply looking up a table in which its solution is pre-stored. In essence, this is an individual-based strategy, with the following merits: quite satisfactory token coverage to MOASs, full correctness for segmentation of pseudo MOASs, and low cost in time and space complexity.

Experiments have been performed to compare the performance of existing word segmenters with that of our strategy, focusing on PM resolution. Two state-of-the-art Chinese word segmenters, ICTCLAS1.0[1] and MSRSeg1.0[2], which separately scores the top one in the SIGHAN-bakeoff 2003 [8] and 2006 [1], have been chosen for comparison. For each of the 5,507 PM types, we randomly select a sample sentence which

[1] ICTCLAS 1.0.: `http://www.nlp.org.cn`

[2] MSRSeg.v1.: `http://research.microsoft.com/-S-MSRSeg`

contains the PM from Web55. The results show that proposed strategy can handle all the samples perfectly, whereas, about 2.6% of them are mistakenly segmented by ICTCLAS1.0 and 2.3% of them by MSRSeg1.0.

Here we give out some typical segmentation errors produced by ICTCLAS1.0 and MSRSeg1.0 (the underlined):

公安局 长 是 主管 这一 事故 的. (From ICTCLAS1.0)
(The police chief(公安 局长) is the person in charge of this accident).
核电站 的 特殊性 质. (From MSRSeg1.0)
(The special properties(特殊 性质) of nuclear power station).

As Conditional Random Fields (CRF) [3] is the state-of-the-art machine learning model on solving sequence labeling problem [7], the performance on PMs is also tested by using CRF. The toolkit CRF++0.50[3] is used to build our CRF-based word segmenter. The window size is set five and four tag-set is used to distinguish the position of character. The training set provided by MSRA in SIGHAN-bakeoff 2005 is used to train the CRF model. The basic feature template adopted from [13] is used. The experimental result shows that totally 2.1% of 5,507 PM types are mistakenly segmented by CRF-based word segmenter.

A typical segmentation error (the underlined) is given here:
这一 现状 先 天地 决定 了 他们 的 使命
(This situation congenitally(先天 地) makes them to take the mission).

The improvement of 2% on PMs seems trivial, but it is a net gain. The strategy could be more effective when facing running text.

6 Conclusion

In this paper the statistical properties of overlapping ambiguities are intensively studied based on observations from a very large balanced general-purpose Chinese corpus. The stability of high frequent MOAS is tested based on statistical observations on both general-purpose corpus and domain-specific corpora. A disambiguation strategy is proposed consequently. Experiments show that over 42% of overlapping ambiguities in running text can be solved without making any error. About 2% mistakes produced by state-of-the-art Chinese word segmenters on MOASs can be solved by this strategy. We are now confident to claim that the basic issues addressed in Section 1 have been settled quite satisfactorily.

References

1. Emerson, T.: The second international Chinese word segmentation bakeoff. In: Proceedings of the 4th SIGHAN Workshop, pp. 123–133 (2005)
2. Huang, C.N.: Segmentation Problems in Chinese Processing. Applied Linguistics 1, 72–78 (1997)

[3] http://crfpp.sourceforge.net/

3. Lafferty, J., McCallum, A., Pereira, F.: Conditional random fields: Probabilistic models for segmenting and labeling sequence data. In: Proceedings of 18th International Conference of ICML, pp. 282–289 (2001)
4. Li, R., Liu, S.H., Ye, S.W., Shi, Z.Z.: A method for resolving overlapping ambiguities in Chinese word segmentation based on SVM and k-NN. Journal of Chinese Information Processing 15(6), 13–18 (2001) (in Chinese)
5. Li, M., Gao, J.F., Huang, C.N., Li, J.F.: Unsupervised training for overlapping ambiguity resolution in Chinese word segmentation. In: Proceedings of SIGHAN 2003, pp. 1–7 (2003)
6. Liang, N.Y.: A Chinese automatic segmentation system for written texts – CDWS. Journal of Chinese Information Processing 1(2), 44–52 (1987) (in Chinese)
7. Peng, F.C., Feng, F.F., McCallum, A.: Chinese segmentation and new word detection using conditional random fields. In: Proceedings of COLING 2004, Geneva, Switzerland, pp. 562–568 (2004)
8. Sproat, R., Emerson, T.: The first international Chinese word segmentation bakeoff. In: Proceedings of the 2nd SIGHAN Workshop, pp. 133–143 (2003)
9. Sun, M.S., Zuo, Z.P.: Overlapping ambiguities in Chinese text. In: Overlapping ambiguities in Chinese text, pp. 323–338 (1998)
10. Sun, M.S., Huang, C.N., T'sou, B.K.Y.: 1997. Using character bigram for ambiguity resolution In Chinese word segmentation (5), 332–339 (in Chinese)
11. Sun, M.S., Zuo, Z.P., T'sou, B.K.Y.: The role of high frequent maximal crossing ambiguities in Chinese word segmentation. Journal of Chinese Information Processing 13(1), 27–37 (1999) (in Chinese)
12. Swen, B., Yu, S.W.: A graded approach for the efficient resolution of Chinese word segmentation ambiguities. In: Proceedings of 5th Natural Language Processing Pacific Rim Symposium, pp. 19–24 (1999)
13. Xue, N.W.: Chinese word segmentation as character tagging. International Journal of Computational Linguistics, 8(1), 29–48 (2003)
14. Yu, S.W., Zhu, X.F.: Grammatical Information Dictionary for Contemporary Chinese. In: Grammatical Information Dictionary for Contemporary Chinese, 2nd edition, 2nd edn. Tsinghua University Press (2003) (in Chinese)
15. Zheng, J.H., Liu, K.Y.: Research on ambiguous word segmentation technique for Chinese text. In: Language Engineering, pp. 201–206. Tsinghua University Press, Beijing (1997) (in Chinese)

The Verb Argument Browser

Bálint Sass

Pázmány Péter Catholic University, Budapest, Hungary
sass.balint@itk.ppke.hu

Abstract. We present a special corpus query tool – the Verb Argument Browser – which is suitable for investigating argument structure of verbs. It can answer the following typical research question: What are the salient words which can appear in a free position of a given verb frame? In other words: What are the most important collocates of a given verb (or verb frame) in a particular morphosyntactic position? At present, the Hungarian National Corpus is integrated, but the methodology can be extended to other languages and corpora. The application has been of significant help in building lexical resources (e.g. the Hungarian WordNet) and it can be useful in any lexicographic work or even language teaching. The tool is available online at `http://corpus.nytud.hu/vab` (username: `tsd`, password: `vab`).

1 Introduction

We "...shall know a word by the company it keeps." [1] Traditionally, corpora are investigated through concordances based on the above firthian statement. Nowadays, when corpus sizes can reach 10^9 tokens, manual processing (reading-through) of concordances is not feasible. We need corpus query systems which summarize the information from corpora and present it to the user (e.g. the linguist) in an effective and comfortable way. Since the corpus-based COBUILD dictionary was published, it has been known that such tools can give substantive support to lexicographic work.

One of the first such tools is the *Sketch Engine* [2], which is able to collect salient collocates of a word in particular predefined relations: such as salient objects of a verb, or salient adjectives preceding a noun. The Verb Argument Browser is also a corpus-summarization tool, it collects salient collocations from corpora. Compared to the Sketch Engine it implements a smaller-scale approach focusing only on verbs and arguments[1], but it has one major advantage. Namely, it can treat not just a single word but a whole verb frame (a verb together with some arguments) as one unit in collocation extraction. In other words, instead of collecting salient objects of a verb, it can collect for example salient objects of a given subject–verb pair, or even salient locatives of a given subject–verb–object triplet and so on. In such a way we can outline the salient patterns of a verb "recursively".

For the time being the Hungarian National Corpus is integrated, but the methodology can be extended to other languages, if a shallow parsed, adequately processed corpus is

[1] As there is a confusion in terminology, it should be noted that throughout this paper the term *argument* will mean complements and adjuncts together.

P. Sojka et al. (Eds.): TSD 2008, LNAI 5246, pp. 187–192, 2008.
© Springer-Verlag Berlin Heidelberg 2008

available. The simple sentence model to which such a corpus must adhere, is described in Section 2. Section 3 considers the processing steps which have been done in the case of the Hungarian National Corpus. The Verb Argument Browser tool is presented in Section 4 and the final section gives a variety of different application possibilities.

2 Sentence Model

The company of the verb are its arguments. In an abstract level, a simple sentence can be seen as one verb (in a given tense or mood or possibly modified some way) and the *set* of arguments belonging to the verb. Arguments are usually (nominal) phrases. An NP argument can be represented as the lemma of the head-word of the phrase plus the morphosyntactic function of the phrase.

Some languages take care of the order of (some) arguments. For example, in SVO languages the subject and object functions are determined by their places in the sentence. Some languages use cases and mark arguments with case markers, which usually allows fairly free argument order. For example, Hungarian has about twenty different cases. In Hungarian – just like different complements and adjuncts – the subject and the object is also marked with specific case markers, and can occur almost everywhere in the sentence. The place and order of different complements and adjuncts are far more variable also in SVO languages and they can be put as prepositional phrases (as in English) or prepositional phrases combined with cases (as in German). Order, case markers, prepositions: these are ways to determine a morphosyntactic *position* in a sentence. It is possible to automatically collect all arguments occurring in a sentence by using a (language-specific) syntactic parser.

Based on the above considerations, the model of a simple Hungarian sentence will look the following:

verb + phrase(lemma+case)–set

Let us take an example sentence: "A lány vállat vont". (The girl shoulder-OBJ pull-PAST. = The girl shrugged her shoulder.)[2] The representation according to the above model is: 'von lányØ váll-t' ('shrug girl-SUBJ shoulder-OBJ'). In this kind of description first comes the verb, then the arguments in the form: lemma, dash, case marker. There are two positions in this example sentence: a subject or nominative position (with zero case marker), and a direct object or accusative position (with case marker: '-t').

An *argument frame* consists of a verb and a list of arguments which occurs (or can occur) with the verb together. Every sentence is an instance of an argument frame. The *argument structure* of a given verb is the union of all its (important) argument frames. An example can be seen in Figure 1. The figure also illustrates that different frames often represent different meanings, and subsequently in different frames the verb should be translated differently (usually to another verb in the target language).

The frame 'von felelősség-rA -t' ('pull responsibility-ONTO OBJ' = to call sy to account) exemplifies the notion of *free* and *fixed* positions. The object position is free

[2] We will give literal word-to-word glosses and overall translation in English in parentheses, in the form you can see in this example.

'von kétség-bA -t'	('pull doubt-INTO OBJ'	= to question sg)
'von váll-t'	('pull shoulder-OBJ'	= to shrug one's shoulder)
'von -t'	('pull OBJ'	= to pull sg)
'von felelősség-rA -t'	('pull responsibility-ONTO OBJ'	= to call sy to account)

Fig. 1. Four main frames of the argument structure of verb 'von' (pull). '-t', '-bA' and '-rA' are case markers. (Note: the upper case letter signs a vowel alternation point where the exact vowel is determined by Hungarian vowel harmony).

in both languages, that means we can choose a word from a broad class of words to fill it, the meaning of the verbal construction remains the same. Conversely, the '-rA' position (the 'to' position in English) is fixed, the lemma is unambiguously defined, it is 'felelősség' (account), and we cannot change it without changing the meaning. If we treat lexically fixed NPs (as 'to account') as a part of the verb itself, we can call such constructions *multi word verbs*. They are real verbs: they usually have separate meaning, and even a full-fledged argument structure. Multi word verbs are typical units, which can be investigated using the Verb Argument Browser.

3 Preparing the Corpus

To format a corpus for the Verb Argument Browser one should build the representation of the corpus sentences according to the model desribed in Section 2.

Working with the Hungarian National Corpus we have taken the following steps. The corpus is already morphosyntactically tagged and disambiguated [3]. Natural language sentences are usually compound/complex, so the first task was to split up the sentences into units consisting of one verb frame. For completing this task we used a rule based method based on [4]. The rules were regular expressions, which mark clause boundaries on the basis of particular punctuation and conjunction patterns. Main principle was that every clause must contain one and only one verb.

Verbal forms were normalized: separated verbal prefixes were attached to the verb to form the proper lemma of the verb, verbal affixes which do not change the verb frame were removed. Noun phrases were detected by chunking with a cascaded regular grammar. The lemma and the case of the head word of the phrase were recorded. In Hungarian cases and postpositions work the same way apart from the fact that the postpositions are written as a separate word. Accordingly, postpositions were treated as case markers.

4 The Verb Argument Browser Tool

The Verb Argument Browser is useful in answering the following typical research question: What are the salient lemmas which can appear in a free position of a given verb frame? In other words: What are the most important collocates of a given verb (or verb frame) in a particular morphosyntactic position? It should be emphasized that one can give not just a verb but a whole verb frame (or a part of it), and query the lemmas occurring in an additional free position.

Mutual information is a classical measure for collocation extraction, it has a disadvantage of ranking rare items too high. To eliminate this there is an appropriate adjustment to multiply by the logarithm of frequency yielding the *salience* measure [5]. Such a measure can be used to extract collocations mentioned above: one part of the collocation is a verb (or verb frame), and the other part is a lemma in a free argument position. According to this measure a lemma in this free position is salient if it occurs more frequently with the frame than it is expected, and the lemma itself is frequent.

The user interface is shown at the top of Figure 2. The corpus can be selected in the first field. In the present version, the whole Hungarian National Corpus (187 million words) is integrated. Response times are only a few seconds even at such corpus sizes. Three smaller subcorpora are also available representing three different genres: the 11 million word corpus of the daily paper "Magyar Nemzet", the 12 million word corpus of the forums of the biggest Hungarian portal "Index" and the 11 million word corpus of selected modern Hungarian fiction. There is an option to investigate short sentences exclusively. These sentences originally consists of one verb frame, so the above splitting method (see Section 3) was not applied to them.

In the second field the verb stem can be entered. The next two lines are for the two arguments, either free (specified by a case/postposition), or fixed (specified by both case/postposition and lemma). Using the negation checkboxes it is possible to exclude some cases or some lemmas within a case. On the right side you can check the free

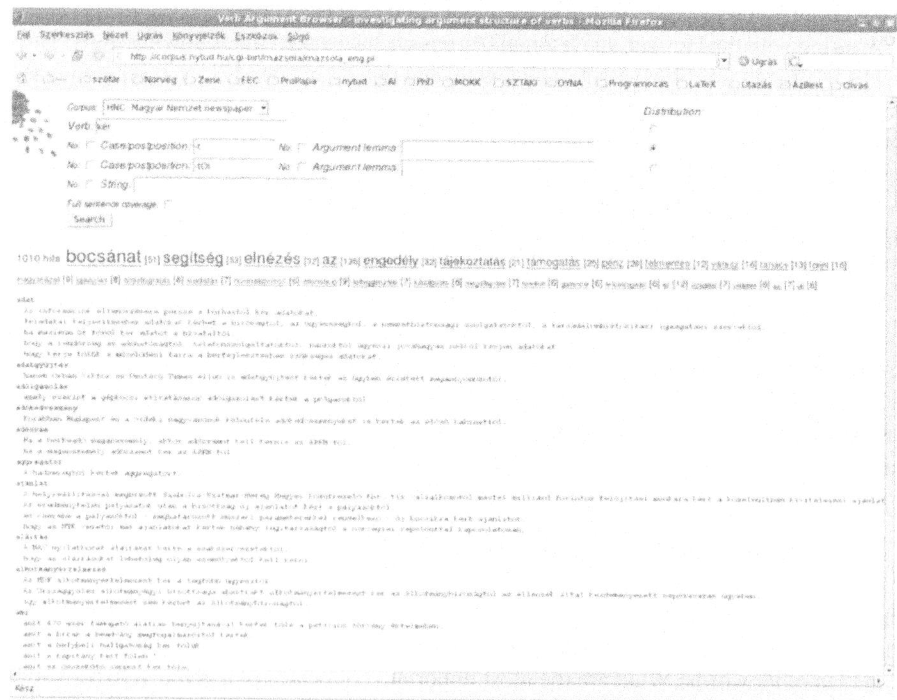

Fig. 2. Answer screen of the Verb Argument Browser. Input form is located at the top.

position, which you are interested in and want to study its salient lemmas. The query can be refined by regular expression full text search below in the *String* field.

In the example in Figure 2 we are searching for the salient lemmas of the direct object position of the frame 'kér -t -tÓl' ('ask OBJ INDIR-OBJ' = ask sy sg). We can see these lemmas in blue at the top of the answer screen (Figure 2). Salience is demonstrated by order and also by font size. Salient direct objects are: 'bocsánat' (forgiveness), 'segítség' (help), 'elnézés' (also forgiveness), 'engedély' (permission) and so on. The number in brackets shows the freqency of the lemma. If we click the lemma we can access the corpus examples listed below this "table of contents".

An illustrative example can be to compare verbs' argument structure in different text genres. If we look at the frame 'ad -t' ('give OBJ' = to give sg) in daily paper and internet forum texts, it becomes clear that these two genres have expressly distinct language. Common lemmas in the object position include 'hang' ('voice' = to give voice to sg) and 'lehetőség' ('chance' = to give sy a chance). At the same time 'hír' ('news' = to report) is diagnostic to daily paper text while 'igaz' ('justice' = to take sy's side) is diagnostic to the internet forum text.

The Verb Argument Browser is available at the following URL: `http://corpus.nytud.hu/vab` (username: `tsd`; password: `vab`). Some clickable examples can be found on the page to test the tool.

5 Applications

The Verb Argument Browser has been applied in two projects concerning the Hungarian language.

During creating the verbal part of the Hungarian WordNet [6] the tool was of significant help in two fields. On the one hand, it is important, how many different meanings a given verb has. Different frames often imply different meanings thus by analysing the argument structure of a given verb one can determine the different synsets where this verb should get into. On the other hand, the Hungarian WordNet was enriched with verbal multi-word units, which are typically a verb together with one fixed argument. Such units are directly provided by the tool, as we saw.

An online demo of a new Hungarian to English machine translation system is available at the following URL: `http://www.webforditas.hu`. During the manual development of the lexical database of this system, an essential task was to collect important lemmas in a position of a given verb frame. From the machine translation point of view important lemmas are the ones for which the verb frame has a special translation. In most cases they are exactly the same as the salient lemmas provided by our tool.

At the end we mention some other possible applications. The Verb Argument Browser can be useful in any lexicographic work, where one of the languages involved is Hungarian. Using this tool, inductive statements can be derived from language data, and an outline of a dictionary entry can emerge from these statements. It gives opportunity to take the empirical frequency of given frames into consideration: it will be possible to determine truly freqent meanings of a verb, or to order the meanings according to frequency.

Similar argument structure can entail similar meaning. To test this hypothesis it is worth generating verb classes based on argument stucture similarity and analysing semantic coherence of these classes [7,8].

According to a proposition in theoretical linguistics, which is gaining attention nowadays [9,10], the aim of grammar is not simply to separate grammatical from ungrammatical but it must explain why certain – sometimes ungrammatical – utterances occur and why certain grammatical utterances does not occur at all. The tool can provide evidence for research in this direction.

It can be of extensive use in language teaching (e.g. by showing the differences in argument structure of verbal synonyms) and also in creation of a verb frame frequency dictionary as automatic generation of verb argument profiles can be built on top of the Verb Argument Browser.

References

1. Firth, J.R.: A synopsis of linguistic theory 1930–1955. Studies in linguistic analysis, 1–32 (1957)
2. Kilgarriff, A., Rychlý, P., Smrž, P., Tugwell, D.: The Sketch Engine. In: Proceedings of EURALEX, Lorient, France, pp. 105–116 (2004)
3. Váradi, T.: The Hungarian National Corpus. In: Proceedings of the 3rd International Conference on Language Resources and Evaluation (LREC2002), Las Palmas, Spain, pp. 385–389 (2002)
4. Sass, B.: Igei vonzatkeretek az MNSZ tagmondataiban [Verb frames in the clauses of the Hungarian National Corpus]. In: Proceedings of the 4th Magyar Számítógépes Nyelvészeti Konferencia [Hungarian Conference on Computational Linguistics] (MSZNY2006), Szeged, Hungary, pp. 15–21 (2006)
5. Kilgarriff, A., Tugwell, D.: Word Sketch: Extraction and display of significant collocations for lexicography. In: Proceedings of the 39th Meeting of the Association for Computational Linguistics, workshop on COLLOCATION: Computational Extraction, Analysis and Exploitation, Toulouse, pp. 32–38 (2001)
6. Kuti, J., Varasdi, K., Gyarmati, Á., Vajda, P.: Hungarian WordNet and representation of verbal event structure. Acta Cybernetica 18, 315–328 (2007)
7. Gábor, K., Héja, E.: Clustering Hungarian verbs on the basis of complementation patterns. In: Proceedings of the ACL-SRW 2007 conference, Prague (2007)
8. Sass, B.: First attempt to automatically generate Hungarian semantic verb classes. In: Proceedings of the 4th Corpus Linguistics conference, Birmingham (2007)
9. Stefanowitsch, A.: Negative evidence and the raw frequency fallacy. Corpus Linguistics and Linguistic Theory 2, 61–77 (2006)
10. Sampson, G.R.: Grammar without grammaticality. Corpus Linguistics and Linguistic Theory 3, 1–32 (2007)

Thinking in Objects

Towards an Infrastructure for Semantic Representation of Verb-Centred Structures

Milena Slavcheva

Bulgarian Academy of Sciences
Institute for Parallel Information Processing
25A, Acad. G. Bonchev St., 1113 Sofia, Bulgaria
milena@lml.bas.bg

Abstract. This paper describes a component-driven population of a language resource consisting of semantically interpreted verb-centred structures in a cross-lingual setting. The overall infrastructure is provided by the Unified Eventity Representation (UER) – a cognitive theoretical approach to verbal semantics and a graphical formalism, based on the Unified Modeling Language (UML). The verb predicates are modeled by eventity frames which contain the eventities' components: static modeling elements representing the characteristics of participants and the relations between them; dynamic modeling elements describing the behaviour of participants and their interactions.

Keywords: UML, classifiers, verb-centred structures.

1 Introduction

The increasing demand of content processing systems requires the development of semantics-driven language resources which are highly varied, modular, possibly multilingual, linguistically and computationally plausible.

This paper describes the building of **SemInStruct** (**Sem**antically **In**terpreted Verb-Centred **Struct**ures) – a language resource of lexical semantic descriptors. The linguistic plausibility of the language resource stems from the application of underlying principles of Cognitive Linguistics [4], Construction Grammar [7] and Functional Grammar as developed by the Sankt-Petersburg linguistic circle [1,2]. Such an approach of incorporating underlying assumptions and descriptive insights from different frameworks is analogous to that of [10], where causative constructions in English are semantically represented using a cognitive lexical-paradigmatic approach. The overall infrastructure is provided by the Unified Eventity Representation (UER) [14] – a cognitive theoretical approach to verbal semantics and a graphical formalism, based on the Unified Modeling Language (UML) [13]. The UER introduces the object-oriented system design to language semantics.

The paper is structured as follows. In Section 2 the underlying modeling principles are briefly discussed. Section 3 represents the building blocks of the language resource. Section 4 discusses some advantages of the application of the object-oriented design to lexical semantics. Section 5 deals with related work and further development.

P. Sojka et al. (Eds.): TSD 2008, LNAI 5246, pp. 193–200, 2008.
© Springer-Verlag Berlin Heidelberg 2008

2 Adherence to Conceptual Modeling

SemInSruct is gradually populated by semantic descriptors of verbs, which are portioned into morphosyntactically anchored data sets. SemInSruct is a lexical resource and the semantic descriptors of the events, expressed by verbs, are built according to the principle of minimum conceptual complexity. At the same time the linking between the natural language expressions and the semantic structures is perceived according to the assumption that there is not strict distinction between lexicon and grammar: both of them form a continuum of symbolic structures (i.e., bipolar structures associating a phonological structure with a semantic structure) [10]. No rigid isomorphism is assumed between morphosyntactic and semantic categories [14].

The meanings of verbs are conceived as "concepts of events and similar entities in the mind", referred to as eventities [14]. The verb predicates are modeled by EVENTITY FRAMES, which contain the eventities' components: static modeling elements representing the characteristics of participants and the relations between them; dynamic modeling elements describing the behaviour of participants and their interactions.

The UER is based on the UML: in the classical four-layer model architecture, the UER *meta-metamodel* and the UML *meta-metamodel* coincide. The practical application of the formalism on *user data* requires the building of *model* extensions according to the criterion of cognitive adequacy: for each concept, a specific representation should be defined, which triggers modification at the *metamodel* level as well, although most of its elements are predefined in the UER.

3 Building Classifiers

At first, generic EVENTITY FRAME diagrams have been provided for classes of reflexive verb structures in Bulgarian, taking also into account their non-reflexive counterparts. The mono-lingual classification has been contrasted to classifications of reflexive verb structures in French and in Hungarian and a cross-lingual semantic classification has emerged. The cross-lingual classification has been tested on data from the three languages, Bulgarian being the source language. The semantic descriptors provided for the defined classes of verb predicates are generalized and represent parameterized EVENTITY FRAME TEMPLATES of the basic relations within the eventity types. At the same time possible user-defined extensions to the UER metalanguage have been outlined, which can be used for the specification of the semantic structures.

The next step in the development of the language resource in question has been related to the further subtyping of the verb predicates as lexicalizations of event concepts by giving values to the parameters and specifying semantic properties of the dynamic core and the static periphery of the EVENTITY FRAME. The dynamic core is a state chart depicting the state transition system of the conceptualized actions. The static periphery contains representation of the participants, their properties and relations. Figure 1 is an illustration of the modeling on the type level (please, note the type-instance distinction in the UER modeling, analogous to that in object oriented programming), representing an EVENTITY FRAME diagram of the causative TURN_EMPTY-eventity.

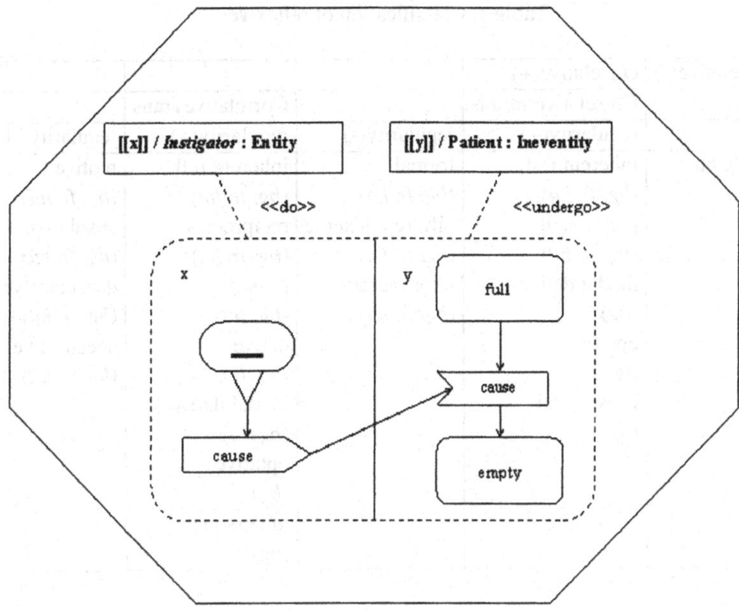

Fig. 1. Causative EVENTITY FRAME – TURN_EMPTY-eventity

Table 1 represents the anchoring classification of reflexives according to the meaning shift triggered by the combination of a verb and a reflexive marker. In Bulgarian and French non-reflexive and reflexive verb pairs are differentiated by the absence or presence of a clitic (e.g., bg. *izmâčvam* / *izmâčvam se*, fr. *torturer* / *se torturer*, *emmitoufler* / *s'emmitoufler*), while in Hungarian the verb pairs consist of words with the same root but opposing suffixes (or opposing null suffix and a reflexive suffix) positioned before the inflection (e.g., hu. *ideges-ít*(trans.) / *ideges-**ked**-ik* (refl.), *mos* (trans.) / *mos-**akod**-ik* (refl.)).

It should be noted that, in the cross-lingual comparison, the explored source pattern is the combination of a verb and a reflexive pronominal clitic in Bulgarian. Thus Table 1 contains an exhaustive classification of the semantic types in Bulgarian, linked to the morphosyntactic types in a paradigmatic lexically oriented setting. Each type label in Table 1 is provided with indications (in the line under the label) of the languages (i.e., bg, fr, hu) in which the given sense is expressed in a form containing the respective reflexive marker for each language (pronominal clitic for Bulgarian and French, suffix for Hungarian).

The conceptual structures related to the basic types of verb predicates are defined according to the following parameters: number of prominent participants and non-prominent participants involved in the eventity; semantic roles, ontological types, semantic features of the participants; association relations between the participants; relations between the participants and the action itself; the character of the action itself represented as a state-machine. The specification of properties leads to the definition of conceptual structures where features like animacy, volition, spontaneity, instigation, and sentience are relevant. Clusters of semantic features are defined which form

Table 1. Classification of reflexives

correlative(-)	correlative(+)			
	CorrelativeIntrans		CorrelativeTrans	
	regularity(+)	regularity(-)	regularity(+)	regularity(-)
bg, fr, hu	inherent refl. (*bg, fr, hu*) reciprocals (*bg, fr, hu*) modal dative (*bg*) optative (*bg*) impersonal (*bg*)	formal (*bg, fr, hu*) different lexeme (*bg, fr, hu*) idiosyncratic (*bg, fr, hu*)	inherent refl. (*bg, fr, hu*) reciprocals (*bg, fr, hu*) passive (*bg, fr*) dative (*bg, fr*) modal dative (*bg*) optative (*bg*) impersonal (*bg*)	motive (*bg, fr, hu*) absolutive (*bg, fr, hu*) deaccusative (*bg, fr, hu*) decausative (*bg, fr, hu*)

"conceptual macros". Having in mind the cognitive assumption that semantic structure (the meanings conventionally associated with words and other linguistic units) is (very closely) related to conceptual structure, we get semantic descriptors of verbs which differ from traditionally constructed lexicons. In this way lexical items are supplied "with a comprehensive knowledge representation from which a specific interpretation can be constructed" [3] in case of content processing. It becomes possible to model, for instance, conceptual types of the human-human or human-inanimate object interaction additionally to participants' relations expressed in semantic roles like Agent and Patient. Cross-linguistically, we get lexical equivalents which are cognitively motivated. In this way SemInStruct differs from the more or less traditionally compiled multilingual lexicographic resources.

For instance, let us consider some of the conceptually motivated readings of the Bulgarian verb *vključvam* "include", together with its reflexive form *vključvam se*, depending on the prominent participants, their properties and their behaviour and interaction. Possible lexicalized equivalents in French and in Hungarian are provided.

1. Causative human-human relation

```
PARTICIPANT REPRESENTATIVE: [[x]]; PARTICIPANT ROLE: Agent;
PARTICIPANT TYPE : Individual; ATTRIBUTES : <<intrinsic>> ani :
Animacy = animate; isHuman : Boolean = true; isVolitional :
Boolean = true

PARTICIPANT REPRESENTATIVE: [[y]]; PARTICIPANT ROLE: Patient;
PARTICIPANT TYPE : Individual; ATTRIBUTES : <<intrinsic>> ani :
Animacy = animate; isHuman : Boolean = true; isVolitional :
Boolean = true
```

In the dynamic core there is a cause-SIGNAL from the Agent participant, which triggers transition of the Patient participant from the state of type <<be-out>> to the state of type <<be-in>>.

Possible lexicalization in French of the semantic structure would be *englober*, *engager*, and in Hungarian *beleszámít*. The English equvalents are *include, enlist, count in.*

2. Causative human-inanimate object relation

a. Relation meaning the inclusion of an object in general

PARTICIPANT REPRESENTATIVE: [[x]]; PARTICIPANT ROLE: Agent; PARTICIPANT TYPE : Individual; ATTRIBUTES : <<intrinsic>> ani : Animacy = animate; isHuman : Boolean = true; isVolitional : Boolean = true

PARTICIPANT REPRESENTATIVE: [[y]]; PARTICIPANT ROLE: Patient; PARTICIPANT TYPE : Ineventity; ATTRIBUTES : <<intrinsic>> ani : Animacy = inanimate

In the dynamic core there is a cause-SIGNAL from the Agent, which triggers transition of the Patient participant from the state of type <<be-out>> to the state of type <<be-in>>.

Possible lexicalization in French of the above semantic structure would be *englober*, and in Hungarian *beleszámít*. The English equivalents are *include, take in.*

b. Relation with the special meaning of switching on a device

The description of the two prominent participants is the same as in 2a. The difference is in the dynamic core, where there is a cause-SIGNAL from the Agent, which triggers transition of the Patient participant from the state of type <<be-off>> to the state of type <<be-on>>.

Possible lexicalization in French of the above semantic structure would be *brancher*, and in Hungarian *bekapcsol*. In English the equvalents are *turn on, switch on; plug in.*

The next readings are associated with the reflexive form of the Bulgarian verb, that is, *vključvam se*.

3. Absolutive

PARTICIPANT REPRESENTATIVE: [[x]]; PARTICIPANT ROLE: Agent; PARTICIPANT TYPE : Individual; ATTRIBUTES : <<intrinsic>> ani : Animacy = animate; isHuman : Boolean = true; isVolitional : Boolean = true.

```
In the dynamic core there is a state transition of the single
prominent participant from the state of type <<be-out>> to the
state of type <<be-in>>.
```

Possible lexicalization in French of the semantic structure would be *s'engager*, and in Hungarian *bekapcsolódik*. In English the equvalents are *join in, take part in, participate in*.

4. Decausative

```
PARTICIPANT REPRESENTATIVE: [[y]]; PARTICIPANT ROLE: Experiencer;
PARTICIPANT TYPE : Ineventity; ATTRIBUTES : <<intrinsic>> ani :
Animacy = inanimate
```

The participant's description means that a state transition of an inanimate object was triggered by itself, the causer of the action is not pointed out.

```
In the dynamic core there is a state transition for the single
prominent participant from the state of type <<be-off>> to the
state of type <<be-on>>, since in Bulgarian the preferred
reading is related to the SWITCH ON-eventity.
```

Se brancher can be considered a possible expression in French of this semantic structure, as well as the Hungarian *bekapcsolódik*, although in this case periphrasis comes into play.

4 Packages of Model Elements

Describing the different readings of a representative number of verbs (490 Bulgarian source verbs and their equivalents in French and in Hungarian), we get objects belonging to the hierarchy of the UML metamodel of abstracting over entities. The object-oriented organization of entities supports a component-based structure of the semantic lexicon, where the linguistic knowledge can be used and reused, structured and restructured depending on the current task. There are good reasons to believe that systems adopting the object-oriented modeling approach are easier to build and maintain [9,6].

In the present work, mostly elements belonging to the *model* level are built, which can be put together in packages depending on the concepts they represent. For instance, classes of AgentParticipants, PatientParticipants, ExperiencerParticipants can be collected in packages, utilizable as suppliers of building blocks in main types of eventity frames like TransitiveFrame, IntransitiveFrame, MediumFrame. Another package, relevant to the interpretation of the verb types in Table 1, includes classes of ASSOCIATION relations between the participants: PossessionRelations, AggregationRelations, BenefitRelations. For instance, Figure 2 represents a Benefit ASSOCIATION relation.

Types of state-machines are also put together depending on the characteristic features of the actions they model. However, this is a harder and longer process, and it is better to package dynamic modeling elements on demand, that is, on a service basis, rather than as ready-made product exchange.

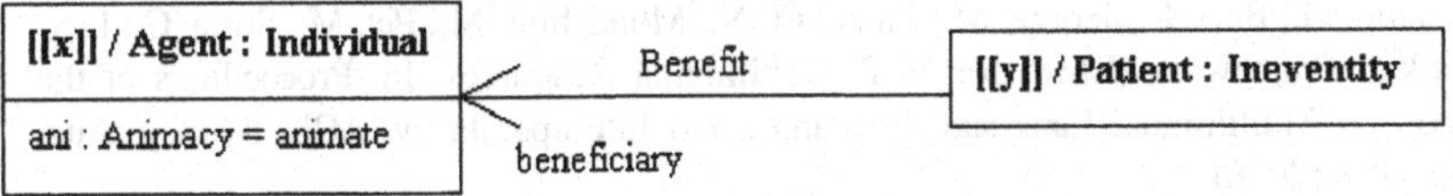

Fig. 2. Benefit ASSOCIATION relation between participants

5 Related Work and Further Development

The ideology behind the building of SemInStruct is comparable to that of CoreLex as described in [3]. CoreLex, however, is a semantic lexicon of nouns and differs in the formalization.

In a sense, SemInStruct is ontology-like. However, it differs from ontological knowledge bases like, for instance, Cyc, or industry-oriented ontologies like the KIM Ontology and KB [11], because it remains in close relation to the linguistic categories and the linguistic vocabulary. SemInStruct is not in contradiction with extensive semantic resources like Wordnet, FrameNet, PropBank and others, although it differs in the definition of the semantic structures and the type of semantic relations. So SemInStruct is somewhere in between ontologies as usually developed for industry-driven applications, and language resources definable as NLP lexicons. It should be noted that SemInStruct deals with the conceptualization related to verb predicates, while ontologies are predominantly concerned with substantive entities and named entities.

The UML is used as the basis for the Lexical Markup Framework (LMF) – a common model for the creation and use of NLP lexicons [5]. However, the LMF defines the macrostructure of the lexicons and the management of the exchange of data among different sorts of lexicons. In the work presented in this paper, the UML, as the basis of UER, is applied to modeling the internal structure of the meanings of verbal lexical items. Nevertheless, the representational format of SemInStruct makes it compatible and interoperable with a common standardized framework like LMF.

An immediate further development is related to the provision of XML representational format for SemInStruct descriptors [12,6].

In the long run SemInStruct components are utilizable in semantic annotation and in more complex Information Extraction tasks like Template Relation construction and Scenario Template production [8].

References

1. Bondarko, A.V. (ed.): Functional Grammar Theory. Category of Person. Category of Voice. Nauka Publishing House, Sankt-Petersburg (1991) (in Russian)
2. Bondarko, A.V. (ed.): Problems of Functional Grammar. Semantic Field Structures. Nauka Publishing House, Sankt-Petersburg (2005) (in Russian)
3. Buitelaar, P.: A Lexicon for Underspecified Semantic Tagging. In: Proceedings of ANLP 1997 Workshop on Tagging Text with Lexical Semantics: Why, What, and How? (1997)
4. Evans, V., Bergen, B.K., Zinken, J.: The Cognitive Linguistics Enterprise: An Overview. In: The Cognitive Linguistics Reader. Equinox Publishing Company (2006)

5. Francopoulo, G., Bel, N., George, M., Calzolari, N., Monachini, M., Pet, M., Soria, C.: Lexical Markup Framework (LMF) for NLP Multilingual Resources. In: Proceedings of the Workshop on Multilingual Language Resources and Interoperability, ACL, Sydney, Australia, pp. 1–8 (2006)
6. Francopoulo, G., Bel, N., George, M., Calzolari, N., Monachini, M., Pet, M., Soria, C.: Lexical Markup Framework: ISO Standard for Semantic Information in NLP Lexicons. In: GLDV-2007 Workshop on Lexical-Semantic and Ontological Resources, Tuebingen, Germany (2007)
7. Goldberg, A.E.: Constructions: A New Theoretical Approach to Language. In: Trends in Cognitive Science (2003)
8. Cunningham, H.: Information Extraction, Automatic, 2nd edn. Encyclopedia of Language and Linguistics. Elsevier, Amsterdam (2005)
9. Cunningham, H., Bontcheva, K.: Computational Language Systems, Architectures, 2nd edn. Encyclopedia of Language and Linguistics. Elsevier, Amsterdam (2005)
10. Lemmens, M.: Lexical Perspectives on Transitivity and Ergativity: Causative Constructions in English. John Benjamins, Amsterdam (1998)
11. Popov, B., Kiryakov, A., Kirilov, A., Manov, D., Ognyanoff, D., Goranov, M.: KIM – Semantic Annotation Platform. In: Fensel, D., Sycara, K.P., Mylopoulos, J. (eds.) ISWC 2003. LNCS, vol. 2870, pp. 834–849. Springer, Heidelberg (2003)
12. Romary, L., Salmon-Alt, S., Francopoulo, G.: Standards Going Concrete: from LMF to Morphalou. In: COLING - 2004
13. OMG Unified Modeling Language Specification, `www.omg.org/`
14. Schalley, A.C.: Cognitive Modeling and Verbal Semantics. Mouton de Gruyter, Berlin (2004)

Improving Unsupervised WSD with a Dynamic Thesaurus*

Javier Tejada-Cárcamo[1,2], Hiram Calvo[1], and Alexander Gelbukh[1]

[1] Center for Computing Research, National Polytechnic Institute, Mexico City, 07738, México
[2] Sociedad Peruana de Computación, Arequipa, Perú
jawitejada@hotmail.com, hcalvo@cic.ipn.mx, gelbukh@gelbukh.com

Abstract. The method proposed by Diana McCarthy et al. [1] obtains the predominant sense for an ambiguous word based on a weighted list of terms related to the ambiguous word. This list of terms is obtained using the distributional similarity method proposed by Lin [2] to obtain a thesaurus. In that method, every occurrence of the ambiguous word uses the same thesaurus, regardless of the context where it occurs. Every different word to be disambiguated uses the same thesaurus. In this paper we explore a different method that accounts for the context of a word when determining the most frequent sense of an ambiguous word. In our method the list of distributed similar words is built based on the syntactic context of the ambiguous word. We attain a precision of 69.86%, which is 7% higher than the supervised baseline of using the MFS of 90% SemCor against the remaining 10% of SemCor.

1 Introduction

Word Sense Disambiguation (WSD) consists on determining the sense expressed by an ambiguous word in a specific context. This task has a particular importance in querying information systems because the user may be selecting a particular set of documents based on the sense of word being used. For example, doctor has three senses listed in WordNet: (1) Person who practices medicine, (2) a person who holds Ph.D. degree from an academic institution; and (3) a title conferred on 33 saints who distinguished themselves through the orthodoxy of their theological teaching. When building multilingual querying systems, the right translation of a particular word must be chosen in order to retrieve the right set of documents.

The task of WSD can be addressed mainly in two ways: (1) supervised: applying techniques of machine-learning trained on previously hand-tagged documents and (2) unsupervised: learning directly from raw words grouping automatically clues that lend to a specific sense according to the hypothesis that different words have similar meanings if they are presented in similar contexts [2,1]. To measure the effectiveness of the state-of-the-art methods for WSD, there is a recurrent event called Senseval.

For instance, the results of Senseval-2 English all-words task are presented in Table 1. This task consists of 5,000 words of running text from three Penn Treebank and

* Work done under partial support of Mexican Government (CONACyT, SNI), IPN (PIFI, SIP). The authors wish to thank Rada Mihalcea for her useful comments and discussion.

P. Sojka et al. (Eds.): TSD 2008, LNAI 5246, pp. 201–210, 2008.
© Springer-Verlag Berlin Heidelberg 2008

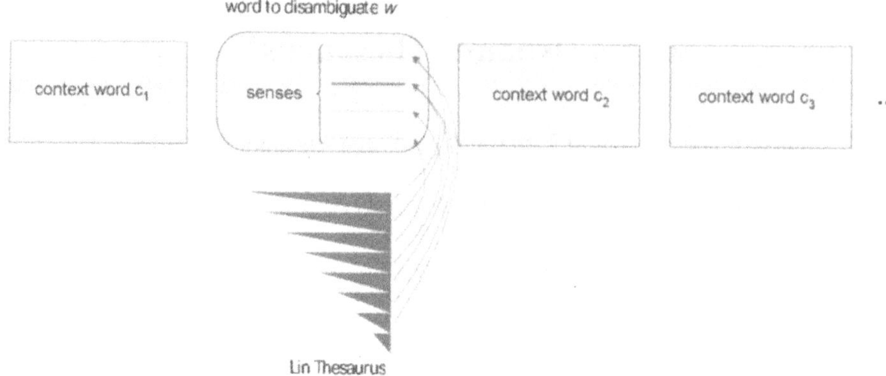

Fig. 1. Finding the predominant sense using a static thesaurus as in McCarthy *et al.*

Wall Street Journal articles. The total number of words that have to be disambiguated is 2,473. Sense tags are assigned using WordNet 1.7. The last column in the Table shows whether a particular system uses manually tagged data for learning or not. The best systems are those which learn from previously manually tagged data, however this resource is not always available for every language, and it can be a costly resource to build. Because of this, we will focus on unsupervised systems, such as UNED-AW-U2.

Table 1. Top-10 Systems of Senseval-2

Rank	Precision	Recall	Attempted	System	Uses sense tagged data?
1	0.69	0.69	100	SMUaw	Y
2	0.636	0.636	100	CNTS-Antwerp	Y
3	0.618	0.618	100	Sinequa-LIA – HMM	Y
4	0.575	0.569	98.908	UNED – AW-U2	N
5	0.556	0.55	98.908	UNED – AW-U	N
6	0.475	0.454	95.552	UCLA – gchao2	Y
7	0.474	0.453	95.552	UCLA – gchao3	Y
8	0.416	0.451	108.5	CL Research – DIMAP	N
9	0.451	0.451	100	CL Research – DIMAP (R)	N
10	0.5	0.449	89.729	UCLA – gchao	Y

Choosing always the most frequent sense for each word yields a precision and recall of 0.605. Comparing this with the results in Table 1 shows that finding the most frequent sense can be a good strategy, as the baseline of 60% would be ranked among the first 4 systems. The most frequent sense was obtained from WordNet, as the MFS is listed in the first place. The senses in WordNet are ordered according to the frequency data in the manually tagged resource SemCor [3]. Senses that have not occurred in SemCor are ordered arbitrarily. Diana McCarthy *et al.* propose in [1] an algorithm to find the prevalent sense for each word. They first build a Lin Thesaurus [2] and then each ranked term k in the thesaurus votes for a certain sense ws_i of the word. Each vote is the value

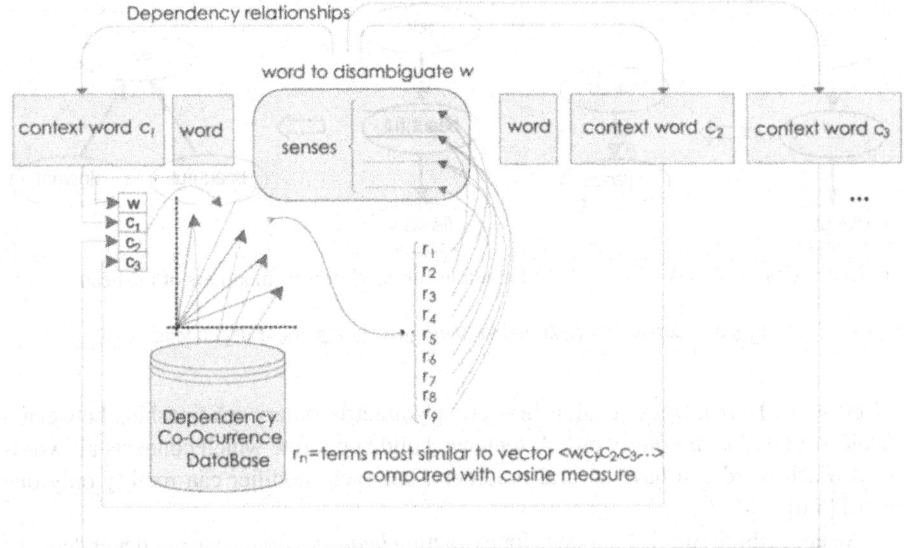

Fig. 2. Our proposal: create a dynamic thesauri based on the dependency context of the ambiguous word

of the similarity between ws_i and the sense of k (ks_j) that maximizes the similarity. Figure 1 illustrates this concept.

The method proposed by McCarthy *et al.* does not consider the context of the word to be disambiguated. The sense chosen as predominant for the word to disambiguate depends solely on the corpus used to build the thesaurus. We propose considering context words to dynamically build a thesaurus of words related to the word to be disambiguated. This thesaurus is dynamically built based on a dependency co-occurrence database (DCODB) previously collected from a corpus. Each *co-occurrent-with-context* word will vote – as in the original method – for each sense of the ambiguous word, finding the predominant sense for this word *in a particular context.* See Figure 2.

In Section 2.1 we explain how the Dependency Co-Ocurrence DataBase (DCODB) resource is built; then we explain our way of measuring the relevance of co-occurrences based on Information Theory in Section 2.2. In Sections 2.3 and 2.4, we explain relevant details of our method. In Section 3 we present an experiment and results that show that performance is as good as certain supervised methods. Finally in Section 4 we draw our conclusions.

2 Methodology

2.1 Building the Dependency Co-Occurrence Database (DCODB)

We obtain dependency relationships automatically using the MINIPAR parser. MINIPAR has been evaluated with the SUSANNE corpus, a subset of the Brown Corpus, and it was able to recognize 88% of the dependency relationships with an accuracy

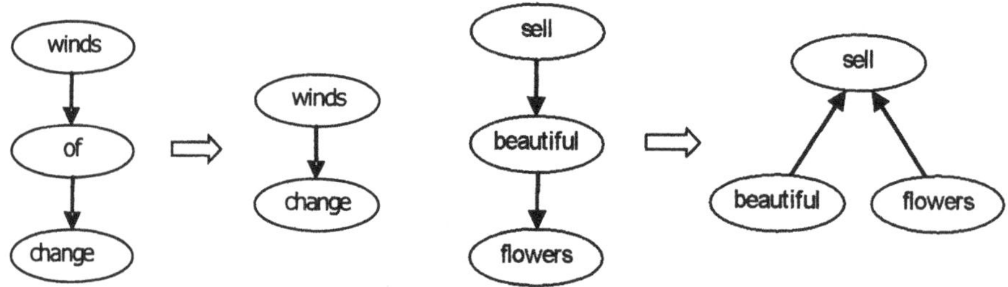

Fig. 3. Heuristics for extracting *head-governor* pairs of dependencies

of 80% [4]. Dependency relationships are asymmetric binary relationships between a *head* word and a *modifier* word. A sentence builds up a tree which connects all words in it. Each word can have several modifiers, but each modifier can modify only one word [5,6].

We apply three simple heuristics for extracting *head-governor* pairs of dependencies:

1. Ignore prepositions – see Figure 3a).
2. Include sub-modifiers as modifiers of the head – see Figure 3b).
3. Separate heads lexically identical, but with different part of speech. This helps to keep context separated.

2.2 Statistical Model

We use the wordspace model known as TF-IDF (term frecuency – inverse document frecuency). This model is usually used for classification tasks and for measuring document similarity. Each document is represented by a vector whose number of dimensions is equal to the quantity of different words that are in it. In our method, a *head* is analogous to a document, and its modifiers are the dimensions of the vector which represents it. The value for each dimension is a weight that reflects the number of co-occurrences between a modifier and a head. Thus, a vector is represented with the following equation:

$$Vector\,(head_n) = \{(mod_1, w_1), (mod_2, w_2), \ldots, (mod_n, w_n)\}$$

where:

$head_n$ is the head word,

mod_n is the name of the modifier word, and

w_n is the weight represented by the normalized number of co-ocurrences between mod_n and $head_n$.

The weight of co-ocurrence is the dot product of the normalized frequency of a head (TF) and its inverse frequency (IDF). TF shows the importance of a modifier with regard to the modified head, so that the weight of the relationship increases when the modifier appears more frequently with such *head*. TF is calculated with the following formula:

$$f_{i,j} = \frac{freq_{i,j}}{\max\left(freq_{l,j}\right)}$$

where:

$freq_{i,j}$ is the frequency of the modifier i with $head_j$, and

$\max\left(freq_{l,j}\right)$ is the highest frequency number of the modifiers of $head_j$.

IDF shows the relevance of a modifier with regard to the remaining heads in the database (DCODB), in a way that the weight of a modifier decreases if it appears more often with every other head in the DCODB; while it increases when it appears with a less number of heads. This means that highly frequent modifiers help little to discriminate when the head is represented by a vector. IDF is calculated with the equation:

$$idf_i = \log \frac{N}{n_i}$$

where:

N is the total number of heads, and

n_i is the total number of heads which co-occur with modifier i.

2.3 Disambiguating Process

Once the database has been built, we are able to begin the disambiguation process for a given word w in a context C (made up of words $C = \{c_1, c_2, \ldots c_n\}$). The first step of this process consists on obtaining a weighted list of terms related with w. The second step consist on using these terms to choose a sense of w continuing with the original algorithm proposed by McCarthy *et al.* [1]. The following sections explain these steps in detail.

Obtaining the weighted list of terms related with *w*. A word is related with another one when they are used in similar contexts. In our method this context is defined by syntactic dependencies. See Figure 2. Given an ambiguous word, w, its dependencies $c_1, c_2, c_3, \ldots$ form a vector $\vec{w} = \langle w, c_1, c_2, c_3, \ldots, w_j, \ldots \rangle$, which is compared with all vectors $\vec{r_i} = \langle r_{i,1}, r_{i,2}, r_{i,3}, \ldots, r_{i,j}, \ldots \rangle$ from DCODB using the cosine measure function:

$$cos\_measure\left(\vec{w}, \vec{r_i}\right) = \frac{\vec{w} \cdot \vec{r_i}}{\left|\vec{w}\right| \times \left|\vec{r_i}\right|} = \frac{\sum_{j=1}^{n} w_j \times r_{i,j}}{\sqrt{\sum_{j=1}^{n} (w_j)^2} \times \sqrt{\sum_{j=1}^{n} (r_{i,j})^2}}$$

The value obtained is used as a similarity weight for creating the weighted list of related terms. Note that this comparison is subject to the data sparseness problem, because the number of modifiers of an ambiguous word is between 0 and 5 – considering only one level of the syntactic tree – whereas the dimension of most vectors in the DCODB are far higher. To be able to compare both vectors, the remaining dimensions for the ambiguous word with its context, are filled with zeroes as in values for $o_1, o_2, o_3, \ldots o_n$ in Table 2. Also see Table 2 for an example of calculation of the cosine measure. Given the vector $\vec{w}$ formed by the word w and its context words (based on dependency relationships

Table 2. Example of cosine measure calculation

	←j→										
	c_1	c_2	c_3	...	c_n	o_1	o_2	o_3	...	o_m	$cos\_measure$
w	**1**	**1**	**1**	**...**	**1**	**0**	**0**	**0**		**0**	1
r_1	1	6	2	...	0	0	0	3		0	0.992165
r_2	0	4	1	...	3	4	1	1		0	0.92665
r_3	4	3	1	...	0	0	0	4		4	0.831966
...										...	
r_{13}	0	0	2	...	4	0	0	1		5	0.68319
...										...	

from the sentence where w is found), the DCODB is queried with all the r_n words to compare with each vector $\vec{r_n}$. For example, the cosine measure between $\vec{w}$ and $\vec{r_3}$ is given by:

$$cos\_measure(\vec{w}, \vec{r_3}) =$$

$$\frac{(1{\cdot}4)+(1{\cdot}3)+(1{\cdot}1)+(1{\cdot}0)+(0{\cdot}0)+(0{\cdot}0)+(0{\cdot}4)+(0{\cdot}4)}{\sqrt{1^2+1^2+1^2+1^2+0^2+0^2+0^2+0^2}+\sqrt{4^2+3^2+1^2+0^2+0^2+0^2+4^2+4^2}} = 0.831966$$

Sense Voting Algorithm. Here we describe our modifications to the sense voting algorithm proposed by McCarthy *et al.* [1]. This algorithm allows each member of the list of related terms (dynamic – in our proposal, or static – in the original form – thesaurus) to contribute for a particular sense of the ambiguous word w. The weight of the term in the list is multiplied by the semantic distance between each of the senses of a term $r_i s_j$ and the senses of the ambiguous word ws_k. The highest value of semantic distance determines the sense of w for which the term r_i votes. Once that all terms r_i have voted (or a limit has been reached), the sense of w which received more votes is selected. See Figure 4 for the pseudo-code of the algorithm.

In the following section we describe the measure of similarity used in this algorithm.

2.4 Similarity Measure

To calculate the semantic distance between two senses we use WordNet::Similarity [7]. This package is a set of libraries which implement similarity measures and semantic relationships in WordNet [8,9]. It includes similarity measures proposed by Resnik [10], Lin [2], Jiang-Conrath [11], Leacock-Chodorow [12] among others. In order to follow McCarthy *et al.* approach, we have chosen the Jiang-Conrath similarity measure as well. The Jiang-Conrath measure (jcn) uses exclusively the hyperonym and hyponym relationships in the WordNet hierarchy, and this is consistent with our tests because we are working only on the disambiguation of nouns. The Jiang-Conrath measure obtained the second best result in the experiments presented by Pedersen *et al.* in [3]. In that work they evaluate several semantic measures using the WordNet::Similarity package. The best result was obtained with the adapted Lesk measure [3], which uses information of multiple hierarchies and is less efficient.

```
Sense voting algorithm:
    foreach ambiguous word w
     with each context word ci
                build vector w =<w,c1,c2,...,cn>
        foreach vector ri in DCODB,
                calculate weight(ri)=cos_measure(w, ri)
                sort vectors r from highest to lowest cos_measure,
                        being ri the one with greatest weight()
    foreach word ri corresponding to the head of each vector ri
      foreach sense risj of word ri
                        foreach sense wsk of word w,
                        calculate a=max(a,similarity(risj,
                                wsk) weight(r))
                the sense wsk which corresponds to the last maximum
                        receives a vote of a units.
                stop if i > max_neighbors (maximum number of
                        vectors)
    return max(votes(wsk))
end
```

Fig. 4. Pseudo-code of the Sense Voting Algorithm

The Jiang-Conrath measure uses a concept of information content (IC) to measure the specificity of a concept. A concept with a high IC is very specific – for example *dessert_spoon* – while a concept with a lower IC is more general, such as *human_being*. WordNet::Similarity uses SemCor to compute the IC of WordNet concepts. The Jiang-Conrath measure is defined with the following formula:

$$dist_{jcn}(c_1, c_2) = \mathrm{IC}(c_1) + \mathrm{IC}(c_2) - 2 \times \mathrm{IC}(lcs(c_1, c_2))$$

where:

IC is the information content

lcs (*lowest common subsumer*) is the common lower node of two concepts.

3 Experiments

We tested our approach for the English language. The dependency co-occurrence database (DCODB) was built from *raw data* taken from 90% of SemCor. That is, we did not consider the sense tags. Then, we evaluated by automatically tagging the remaining 10% of SemCor, comparing with the hand-tagged senses. We have experimented varying the number of max_neighbors, that is, the number of similar vectors from the DCODB that we are considering for disambiguating each word. We have tested by using the top-10 similar, 20, 30, 40, 50, 70, 100, and 1000 most similar terms from our dynamic thesaurus. See Figure 5 for results.

The values of *precision* and *recall* are similar because the Jiang-Conrath measure always returns a similarity value when two concepts are compared. Slight deviations

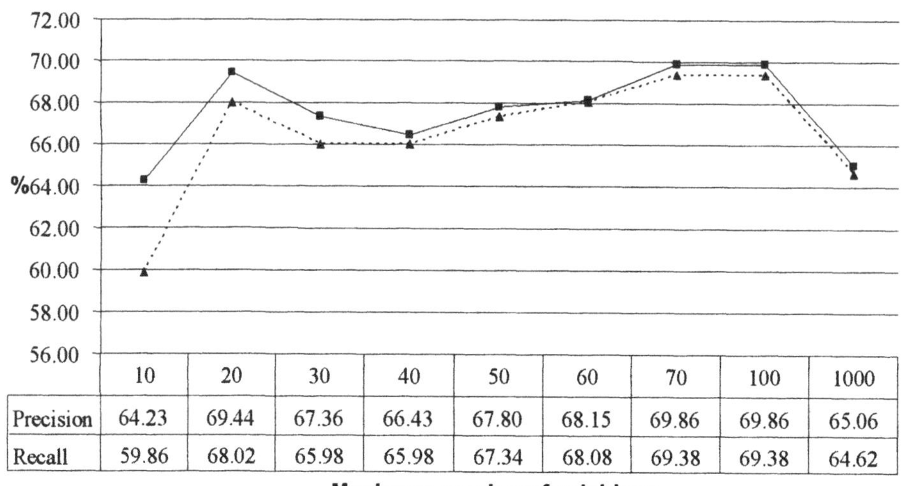

	10	20	30	40	50	60	70	100	1000
Precision	64.23	69.44	67.36	66.43	67.80	68.15	69.86	69.86	65.06
Recall	59.86	68.02	65.98	65.98	67.34	68.08	69.38	69.38	64.62

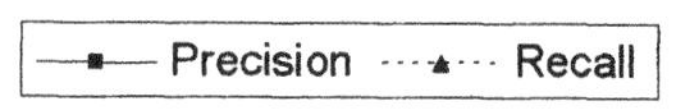

Fig. 5. Automatic tagging of 10% SemCor, varying max_neighbors

are due the inexistence of certain ambiguous words in WordNet. The best results were 69.86% by using 70 neighbors training with 90% of SemCor (used as raw text) and evaluated with the remaining 10% as gold standard. The MFS of 90% of the manually annotated SemCor corpus against the 10% as gold standard of the same corpus, yields a precision and recall of 62.84 (coverage is 100%); so our results are approximately 7% higher.

McCarthy *et al.* report 64% using a term list from a static thesaurus calculated using the Lin method. However, they present results with SENSEVAL-2 so that we cannot make a direct comparison; however, we can do a rough estimation, given that using the MFS found in SemCor yields a 62.84 while using the MFS found in SENSEVAL-2 corpus yields 60%.

4 Conclusions

The method proposed by McCarthy *et al.* [1] is used to obtain the predominant sense of an ambiguous word considering a weighted list of terms related to the ambiguous word. These terms are obtained by the Lin method for building a thesaurus [2]. This list is always the same for each ambiguous instance of a word, because it does not depends on its context. Because of this, we called this method *static thesaurus*. In our method, that list is different for each ambiguous instance of a word, depending on its syntactic (dependency) context and the corpus used for building the co-occurrence database.

The method presented disambiguates a corpus more precisely when trained with another section of the same corpus, as shown by our experiments when training with the 90% of SemCor used as raw-text, and evaluating with the remaining 10%. We compared against obtaining the most frequent sense from the same set of SemCor and evaluating with the remaining 10%. We obtained a precision of 69.86% against 62.84% of the baseline which uses supervised information, while our method is unsupervised.

As future work, we plan to carry out more experiments by comparing our automatic sense tagging with the previously hand-tagged section of the SENSEVAL-2 corpus, all-words test. This will allow us to compare directly with the systems presented at of SENSEVAL-2. We believe the results might be similar, given that using the MFS found in SemCor yields a 62.84 while using the MFS found in SENSEVAL-2 corpus yields 60%.

SemCor is a small corpus, around 1 million words, compared with the British National Corpus (approximately 100 million words). As part of our future work, we plan to train with BNC. We expect to keep the good results when evaluating with 10% SemCor, as well when evaluating with SENSEVAL-2.

On the other hand, Figure 5 is very irregular and we would not say that the optimum number of max_neighbors is always 70. In addition, terms from the weighted list (our dynamic thesaurus) are not always clearly related between them. We expect to build a resource to improve the semantic quality from such terms.

Finally, it is difficult to determine the main factor that has a greater impact in the proposed disambiguation method: the process of obtaining a weighted list of terms (the dynamic thesaurus), or the maximization algorithm. This is because the DCODB sometimes does not provide terms related with a word, and besides, the definitions for each sense of WordNet are sometimes very short. In addition, as it has been stated previously, for several tasks the senses provided by WordNet are very fine-graded, so that a semantic measure may be not accurate enough.

References

1. McCarthy, D., et al.: Finding predominant senses in untagged text. In: Proceedings of the 42nd Annual Meeting of the Association for Computational Linguistics, Barcelona, Spain (2004)
2. Lin, D.: Automatic retrieval and clustering of similar words. In: Proceedings of COLING-ACL 1998, Montreal, Canada (1998)
3. Gelbukh, A.: Using measures of semantic relatedness for word sense disambiguation. In: Patwardhan, S., Banerjee, S., Pedersen, T. (eds.) Proceedings of the Fourth International Conference on Intelligent Text Processing and Computational Linguistics, Mexico City (2003)
4. Lin, D.: Dependency-based Evaluation of MINIPAR. In: Workshop on the Evaluation of Parsing Systems, Granada, Spain (1998)
5. Hays, D.: Dependency theory: a formalism and some observations. Language 40, 511–525 (1964)
6. Mel'čuk, I.A.: Dependency syntax; theory and practice. State University of New York Press, Albany (1987)
7. Pedersen, T., Patwardhan, S., Michelizzi, J.: WordNet:Similarity – Measuring the Relatedness of Concepts. In: Proceedings of the Nineteenth National Conference on Artificial Intelligence (AAAI-2004), San Jose, CA, pp. 1024–1025 (2004)
8. Miller, G.: Introduction to WordNet: An On-line Lexical Database. Princeton Univesity (1993)
9. Miller, G.: WordNet: an On-Line Lexical Database. International Journal of Lexicography (1990)

10. Resnik, P.: Using information content to evaluate semantic similarity in a taxonomy. In: Proceedings of the 14th International Joint Conference on Artificial Intelligence, Montreal, pp. 448–453 (1995)
11. Jiang, J., Conrath, D.: Semantic similarity based on corpus statistics and lexical taxonomy. In: International Conference on Research in Computational Linguistics, Taiwan (1997)
12. Fellbaum, C.: Combining local context and WordNet similarity for word sense identification. In: Leacock, C., Chodorow, M. (eds.) WordNet: An electronic lexical database, pp. 265–283 (1998)
13. Wilks, Y., Stevenson, M.: The Grammar of Sense: Is word-sense tagging much more than part-of-speech tagging, Sheffield Department of Computer Science (1996)
14. Sahlgren, M.: The Word-Space Model Using distributional analysis to represent syntagmatic and paradigmatic relations between words in high-dimensional vector spaces. Ph.D. dissertation, Department of Linguistics, Stockholm University (2006)
15. Kaplan, A.: An experimental study of ambiguity and context. Mechanical Translation 2, 39–46 (1955)

Normalization of Temporal Information in Estonian

Margus Treumuth

University of Tartu, Estonia

Abstract. I present a model for processing temporal information in Estonian natural language. I have built a tool to convert natural-language calendar expressions to semi-formalized semantic representation. I have further used this representation to construct formal SQL constraints, which can be enforced on any relational database. The model supports compound expressions (conjunctive and disjunctive), negation and exception expressions. The system is built to work on Oracle databases and is currently running on Oracle 10g with a PHP web interface.

Keywords: Temporal reasoning, temporal reference in discourse.

1 Introduction

The goal of my current research is to design and implement the model of calendar expression recognition system for Estonian language that would serve me well as a component for more high level tasks. I have been developing dialogue systems as natural language interfaces to databases and the need for a better temporal normalization has risen from my previous work [1].

One of the most well known applications of calendar expression recognition is Google Calendar "quick add" feature. Users can "quick add" calendar events by typing standard English phrases, such as "Dinner with Michael 7pm tomorrow". Google "quick add" has not been implemented for Estonian.

There is a calendar expression recognition system freely available for Estonian (Datejs – an open-source JavaScript Date Library [2]), yet they have done the straight translation from their English system into Estonian. This approach does not work, as Estonian language is rich in morphology. A simple example of a calendar expression where Datejs would fail is "*esmaspäeval*" ("on Monday"), as this word is not in base form "*esmaspäev*" but has a suffix *-al* meaning that the word is in singular adessive.

Estonian language belongs to the Finnic group of the Finno-Ugric language family. Typologically Estonian is an agglutinating language but more fusional and analytic than the languages belonging to the northern branch of the Finnic languages. The word order is relatively free [3]. A detailed description of the grammatical system of Estonian is available in [4].

Datejs library does not support Estonian agglutinative inflections. Despite of that, Datejs is still a very good example of calendar expression recognition. Taken all this into account, you'll find quite many references to Datejs throughout this paper. Yet, several other approaches to extracting temporal information are discussed below to provide some comparison.

The system that I built is available online with a web interface and you are welcome to experiment at `http://www.dialoogid.ee/tuvastaja/`

P. Sojka et al. (Eds.): TSD 2008, LNAI 5246, pp. 211–218, 2008.
© Springer-Verlag Berlin Heidelberg 2008

2 Calendar Expressions

I used a dialogue corpus[1], a text corpus of Estonian[2] and Datejs unit tests to gather calendar expressions that were to be normalized by my system. I also added Estonian national holidays (e.g. *Jõulud* – Christmas Day) and some other less important holidays (e.g. *valentinipäev* – Valentines Day). Here are few examples, to give a basic idea of what expressions need to be handled by temporal normalization:

- *3 aastat tagasi* – 3 years ago;
- *14. veebruaril 2004 a* – February 14, 2004;
- *kell 17:00* – at 5 PM;
- *eelmisel reedel kell 8 õhtul* – last Friday at 8 PM in the evening.

Here we see how they could be normalized given that the current date is March 21, 2008:

- 3 aastat tagasi → esmaspäev, 21.03.2005;
- 14. veebruaril 2004 a → laupäev, 14.02.2004;
- kell 17:00 → reede, 21.03.2008 17:00;
- eelmisel reedel kell 8 õhtul→ reede, 14.03.2008 20:00.

3 Implementation

I have decomposed the normalization task into following sub-tasks:

1. morphological analysis and creation of n-grams,
2. matching rules to calendar expressions,
3. calculating the composition of rules based on time granularity measures,
4. validating the result against a time line model and providing output, in semi-formal notation, SQL notation, ICal format.

The database of the normalization model contains rules that are used in the process of normalizing the date expressions. I'll give an overview of the attributes of the rules and then continue by describing the sub-tasks. The normalization model makes use of the power of extended regular expressions. The rules are given as entities with the following five attributes:

REGEXP – extended regular expression for calendar expression
SEMIFORMAL – corresponding semi-formalized representation of calendar expression
SQL – corresponding SQL representation of calendar expression
ICAL – corresponding ICal representation of calendar expression
GRANULARITY – granularity of the calendar expression (MONTH; DAY; HOUR)

[1] `http://lepo.it.da.ut.ee/~koit/Dialoog/EDiC.html`
[2] `http://www.cl.ut.ee/korpused/`

The following example shows how I have defined the rule to normalize the calendar expression *"emadepäev"* – "Mother's Day". We know that the Mother's Day repeats annually on the second Sunday in May. I hereby give an example by describing a rule at first in English and then in Estonian to illustrate the differences of Estonian grammar.

Example 1. Defining a rule to normalize the calendar expression "Mother's Day".

REGEXP: Mother's Day
SEMIFORMAL: DAY= from 08.05 to 14.05 and DAY=Sunday
SQL: <see SQL constraint below this definition>
ICAL: RRULE:FREQ=YEARLY;INTERVAL=1;BYDAY=2SU;BYMONTH=5
GRANULARITY: DAY

The SQL constraint for this rule is defined as follows:

```
to_char(date, 'mm') = 5 and
to_char(date, 'dd') between 8 and 14 and
trim(to_char(date, 'day')) = 'Sunday'
```

The preceding example was translated into English for the sake of readability. The original rule in Estonian has been written as follows:

REGEXP: emadepäev
SEMIFORMAL: PÄEV=08.05 kuni 14.05 ja PÄEV=pühapäev
SQL: <same as in English>
ICAL: <same as in English>
GRANULARITY: <same as in English>

The *REGEXP* attribute contains the base form "emadepäev". There are 14 cases and 2 numbers (singular and plural) in Estonian that are formed by adding suffixes to the base form. Sometimes the base form itself also changes, e.g:

oktoober → *oktoobris* singular inessive (October → in October)
tund → *tunnil* singular adessive (hour → at this hour)

The Estonian word "emadepäev" can have the following 28 possible agglutinative inflections that also need to be matched by the rule:

-a, -ad, -ade, -adega, -adeks, -adel, -adele, -adelt, -adena, -adeni, -ades, -adesse, -adest, -adeta, -aga, -aks, -al, -ale, -alt, -ana, -ani, -as, -asid, -asse, -ast, -ata, -i, -iks, -il, -ile, -ilt, -is, -isse, -ist.

But instead of defining the rule as a long and clumsy regular expression:

emadepäev(a|ad|ade|adega|adeks|adel|adele|adelt
|adena|adeni|ades|adesse|adest|adeta|aga|aks|al|ale
|alt|ana|ani|as|asid|asse|ast|ata|i|iks|il|ile|ilt|is|isse|ist)?

We can define it simply as "emadepäev" and have the morphological analyzer do the rest.

3.1 Morphological Analysis and Creation of N-Grams

The calendar expressions in Estonian contain simple morphology and can be handled manually without fully automated morphological analysis. Yet, it is very inefficient to consider all possible agglutinations manually, I suggest not implementing semantic analysis without proper morphological analyzer [5]. No disambiguation is done in the analysis as the morphological ambiguity in Estonian temporal expressions is very low.

There are currently 59 words that were not handled correctly by the morphological analyzer, e.g *"valentinipäev"* – Valentines Day; *"jüripäev"* – St George‘s Day. This is a reasonably small amount. These words I handled with specifc regular expressions.

Let's briefly look at morphological analysis with a simple example. Given that the input would be "emadepäevaks" the output of the morphological analyzer would be:

emadepäevaks
ema+de_päev+ks //_S_ sg tr, //

The output of the analyzer shows that the word is a compound noun and that the input was in *singular translative* case. My system is able to parse this output and capture the base form without the -ks suffix.

Date expressions in Estonian language are usually made up of several words. N-grams are created to simplify the parsing process of regular expressions. The rules currently in the system contain at most three words per calendar expression. That is why I created n-grams up to level n=3 (bigrams and trigrams). As most of the rules are given in their base form, the base form n-grams where also created. The n-grams are created at run time just right after the morphological analysis.

3.2 Matching Rules to Calendar Expressions

The third step is finding a rule to match an n-gram, a word or a base form. All words, base forms and n-grams are matched against all rules. Each n-gram, word and base form is constrained to hold a reference to no more than one rule. If a rule is simultaneously referenced by trigram and by a bigram, the reference by the bigram is deleted as it is covered by the trigram. If a rule is referenced by a bigram and by a word, the reference by the word is deleted as it is covered by the bigram (see Example 2). This also means that each n-gram holds a reference to the word or base form that started the n-gram (that is the first word of n-gram). Also, each base form holds a reference to the initial word form.

Example 2. Releasing the matches that were sub-patterns of another match.

1. Consider the input *"järgmisel teisipäeval"* – "next Tuesday" – then the matches would be ***Tuesday*** and ***next Tuesday*** as there is a rule for each of these.
2. We would release the match ***Tuesday*** as it is contained in the bigram ***next Tuesday*** that was also a match.
3. Normalizing "next Tuesday" we would get *"teisipäev, 25.03.2005"* (assuming that current date is March 21, 2008).

3.3 Calculating the Composition of Rules Based on Time Granularity Measures

Conjunctive and disjunctive interpretations are also captured by my system. I have made an empirical judgment that the word "and" in Estonian natural language is mostly used to indicate that it is meant to be handled as disjunctive rather than conjunctive expression. There remains some ambiguity in this approach, yet this assumption seems to serve my purposes quite well. One of the examples of slightly ambiguous compound expression is:

"*sel esmaspäeval ja teisipäeval kell 17*"

"this Monday and Tuesday at 17:00"

Usually this is considered to be a compound disjunctive expression (and so it is handled by my system), giving the output:

(P"AEV=esmaspäev OR PÄEV=teisipäev) AND (KELL=17:00)

(DAY=Monday OR DAY=Tuesday) AND (TIME=17:00)

Yet, depending on wider context, there is a chance that the first part "this Monday" and the second part "Tuesday at 17:00" need to be handled as two separate expressions:

DAY=Monday
DAY=Tuesday AND TIME=17:00

The composition of rules based on time granularity measures is one of the distinct features of my system that has not been so widely exploited by Datejs and Google Calendar.

Above we looked at an example of conjunctive and disjunctive interpretations and saw that temporal units are composed of smaller temporal units. A larger temporal unit is a concatenation of smaller temporal units.

At this point, my system has captured all basic temporal expressions and is ready to compose them into more complex expressions that might represent duration, intervals or recurrence. This is done by taking into account the granularity measures (like hour, day, month). If there are changes in granularity level, the system determines where to add *"AND"* or *"OR"*. A formal framework for the definition and representation of time granularities was proposed by Bettini [6] especially for temporal database applications.

3.4 Validating the Result and Providing Output

Finally the resulting SQL query is validated against a background timeline to see if the conjunctive and disjunctive interpretations that were translated into SQL yield a result. This validation is done for maintenance purposes to capture all misconceptions into the system log.

The output is given by displaying the matched calendar expressions and their formal representations.

4 Experiments and Results

Here are few examples of experiments (in Estonian and in English) that did not pass tests in Datejs but were successfully normalized by my system:

- *valentinipäev* – Valentines Day (this was not even recognized in English in the Google Calendar);
- *igal laupäeval* – every Saturday (recognized as a recurring event);
- *täna ja homme* – today and tomorrow (this was recognized as a disjunctive compound expression as in *3.5* above).

Here are few examples (the test were done in Estonian, yet the examples are given in English) that did not pass tests in my system but were successfully normalized by Datejs:

- *eelmisel aprillil* – last April *(my system does not know the meaning of last – yet)*;
- *+5 aastat* – +5 years *(my system does not know the meaning of +)*.

I have used some voluntary testers and have received scores from 45% to 55% in recall, and from 75% to 85% in precision.

5 Related Work

Jyrki Niemi and Kimmo Koskenniemi [7] have discussed representing calendar expressions with finite-state transducers, which is basically the same thing that I'm doing – using extended regular expressions to capture the calendar expressions. In their work, they have presented a very good theoretical framework, yet I found no traces of any actual working implementation that was built of this framework.

The automatic time expression labeling for English and Chinese Text by Hacioglu [8] uses a statistical time expression tagger. Yet, they have trained the system on a corpus that has been tagged by a rule-based time expression tagger.

A bachelor thesis was completed by Eveli Saue [9] where a rule-based time expression tagger for Estonian was built (precision of 93% and recall 71%). The tagger does not provide normalized output or any kind of formal representation of the date expression. It just places the recognized time expressions between <TIMEX> tags. In my system I have covered most of the expressions of this tagger was able to tag. I have considered building a tagged corpus and switching to statistical approach, yet the rule-based approach seems easier to control and at the moment I don't have a corpus that would provide a considerable amount of temporal negotiation dialogues.

Saquete et al. [10] present a rule-based approach for the recognition and normalization of temporal expressions and their main point is that this way it was easy to port the system for different languages. They have used a Spanish system to create the same system for English and Italian through the automatic translation of the temporal expressions in rules. I agree with this advantage and can confirm that the porting from Estonian to English would mostly consist of automatically translating the rules. Yet, the translation from English system to Estonian system would not be an easy task as I showed above in the Datejs example.

Berglund [11] has used a common annotation scheme – TimeML (Time Mark-up Language) and has built a rule-based custom tagger to annotate text with TimeML. He has used the tagger in a system that is used to detect and order time expressions. I have decided to skip the annotation schemes (TimeML and TIMEX) in my output as an intermediate step, as it seemed more convenient for the QA task to produce output in logical expressions that are compatible with SQL.

6 Further Work

The main idea and benefit of current approach is the output of logical expressions that can be used in SQL queries. This approach of temporal normalization is well suited for a smaller domain such as natural language interfaces for relational databases. I will continue by integrating the temporal normalization tool into various natural language interfaces and scheduling applications to see how it would perform. I will extend the model by providing optional constraint relaxation.

The constraint relaxation is implemented in a dialogue system that uses the temporal extraction tool, yet the rules for constraint relaxation are not defined in the temporal extraction tool but in the dialogue system. For example, the user might mention a date to a dialogue system that would result in "not found" response. Then it would be appropriate to relax this date constraint, as in the following dialogue.

<User>: Are there any performances on Saturdays?
<System>: No, yet I found one on this Sunday...

This was an example of a constraint relaxation where the original date constraint was relaxed by adding one day. This way the users of the system can receive some alternative choices, instead of plain "not found" responses. The constraint relaxation properties can be held in the temporal extraction tool as long as they stay separate from the dialogue domain.

References

1. Treumuth, M., Alumäe, T., Meister, E.: A Natural Language Interface to a Theater Information Database. In: Proceedings of the 5th Slovenian and 1st International Language Technologies Conference 2006 (IS-LTC 2006), pp. 27–30 (2006)
2. Datejs – a JavaScript Date Library, www.datejs.com/
3. Kaalep, H.-J., Muischnek, K.: Multi-word verbs in a flective language: the case of Estonian. In: Rayson, P., Sharoff, S., Adolphs, S. (eds.) Proceedings of the EACL workshop on Multiword expressions in a multilingual context: 11th Conference of the European Chapter of the Association for Computational Linguistics, Trento, Italy, 3 April 2006, pp. 57–64. Association for Computational Linguistics (2006)
4. Erelt, M. (ed.): Estonian Language. Linguistica Uralica. Supplementary Series, vol. 1. Estonian Academy Publishers, Tallinn (2003)
5. Kaalep, H.-J.: An Estonian Morphological Analyser and the Impact of a Corpus on Its Development. Computers and the Humanities 31, 115–133 (1997)
6. Bettini, C., Jajodia, S., Wang, X.S.: Time Granularities in Databases, Data Mining, and Temporal Reasoning. Springer, Heidelberg (2000)

7. Niemi, J., Koskenniemi, K.: Representing Calendar Expressions with Finite-State Transducers that Bracket Periods of Time on a Hierachical Timeline. In: Nivre, J., Kaalep, H.-J., Muischnek, K., Koit, M. (eds.) Proceedings of the 16th Nordic Conference of Computational Linguistics NODALIDA-2007, University of Tartu, Tartu, pp. 355–362 (2007)
8. Hacioglu, K., Chen, Y., Douglas, B.: Automatic Time Expression Labeling for English and Chinese Text. In: Gelbukh, A. (ed.) CICLing 2005. LNCS, vol. 3406, pp. 348–359. Springer, Heidelberg (2005)
9. Saue E.: Automatic Extraction of Estonian Temporal Expressions (2007), math.ut.ee/~sints/bak/
10. Saquete, E., Marinez-Barco, P., Munoz, R.: Multilingual Extension of a Temporal Expression Normalizer using Annotated Corpora. In: Cross-Language Knowledge Induction Workshop (2006)
11. Berglund, A.: Extracting Temporal Information and Ordering Events for Swedish. Master's thesis report (2004)

Word Sense Disambiguation with Semantic Networks

George Tsatsaronis*, Iraklis Varlamis, and Michalis Vazirgiannis

Department of Informatics, Athens University of Economics and Business, Athens, Greece
{gbt,varlamis,mvazirg}@aueb.gr

Abstract. Word sense disambiguation (WSD) methods evolve towards exploring all of the available semantic information that word thesauri provide. In this scope, the use of semantic graphs and new measures of semantic relatedness may offer better WSD solutions. In this paper we propose a new measure of semantic relatedness between any pair of terms for the English language, using WordNet as our knowledge base. Furthermore, we introduce a new WSD method based on the proposed measure. Experimental evaluation of the proposed method in benchmark data shows that our method matches or surpasses state of the art results. Moreover, we evaluate the proposed measure of semantic relatedness in pairs of terms ranked by human subjects. Results reveal that our measure of semantic relatedness produces a ranking that is more similar to the human generated one, compared to rankings generated by other related measures of semantic relatedness proposed in the past.

Keywords: Word Sense Disambiguation, Semantic Networks, WordNet.

1 Introduction

Word Sense Disambiguation (WSD) is the task of selecting the most appropriate meaning for any given word with respect to its context. The candidate word meanings, also referred to as senses, are usually selected from a machine readable dictionary (MRD) or a word thesaurus. Several approaches have been proposed in the past and are classified depending on the resources they employ for the WSD task. Knowledge-based or dictionary-based approaches usually utilize knowledge sources like MRDs or thesauri in order to address the task. Corpus-based approaches include the use of large corpora. An alternative classification may consider the use of a training mechanism that builds a decision model (i.e. a classifier) trained on manually annotated data in order to predict the correct sense of each given word. Such approaches are considered as supervised WSD approaches. The main distinction between a supervised WSD method and an unsupervised one is in whether they use manually labelled data or not. An extensive presentation of the state of the art in WSD can be found in [1].

In this paper we propose a new knowledge-based WSD approach that does not require training. The approach considers semantic networks generated from the WordNet

* Funded by the 03ED_850 research project, implemented within the Reinforcement Programme of Human Research Manpower (PENED) and co-financed by National and Community Funds (25% from the Greek Ministry of Development- General Secretariat of Research and Technology and 75% from E.U.-European Social Fund).

P. Sojka et al. (Eds.): TSD 2008, LNAI 5246, pp. 219–226, 2008.
© Springer-Verlag Berlin Heidelberg 2008

thesaurus [2] and introduces a new measure of semantic relatedness for a pair of thesaurus' concepts[1]. Experimental evaluation of the semantic relatedness measure in 65 word pairs ranked by human subjects according to their semantic relatedness shows that our measure produces a ranking that is more similar to the human generated one, compared to other related measures of semantic relatedness proposed in the past. Furthermore, we evaluate our approach in a benchmark WSD data set, namely Senseval 2 [3], and show that it surpasses or matches previous unsupervised WSD approaches. The rest of the paper is organized as follows: Some preliminary elements, as well as related work, are discussed in Section 2. Section 3 introduces the new measure of semantic relatedness and the new WSD method. Section 4 presents the experimental evaluation and Section 5 concludes and points to future work.

2 Background and Related Work

The idea of using semantic networks to perform WSD is not new. In fact, recent research has employed the construction of rich semantic networks that utilize WordNet fully. In this section we present preliminary information concerning WordNet and semantic networks.

2.1 WordNet

WordNet is a lexical database containing English nouns, verbs, adjectives and adverbs, organized in synonym sets (synsets). Synsets can be regarded as concepts. They are connected with various edges that represent different semantic relations (see Figure 1) and sometimes cross parts of speech (POS). The proposed measure of semantic relatedness and the introduced WSD approach utilize the full range of WordNet 2.0 semantic relations. Any other thesaurus could be used as long as it provides a similar graph structure, and semantic relations like the aforementioned, that can also cross POS.

2.2 Generating Semantic Networks from WordNet

The expansion of WordNet with semantic relations that cross POS has widened the possibilities of semantic network construction from text. Early approaches [4], were based on the gloss words existing in the terms' definitions in order to build semantic networks from text. More recent approaches in semantic network construction from word thesauri [5,6] utilized the semantic relations of WordNet. These methods outperformed previous methods that use semantic networks in the *all words* WSD tasks of Senseval 2 and 3 for the English language. The evaluation in [6] revealed that the performance boost of the WSD task was mainly due to the use of the rich semantic links that WordNet offers. In this work we adopt the same semantic network construction method.

2.3 Semantic Relatedness Measures in Word Sense Disambiguation

Agirre and Rigau in [7] base their measure for sets of concepts on the individuals' density and depth and on the length of the shortest path that connects them. Resnik [8] measure

[1] *Concept* and *sense* will be used interchangeably for the remaining of the paper.

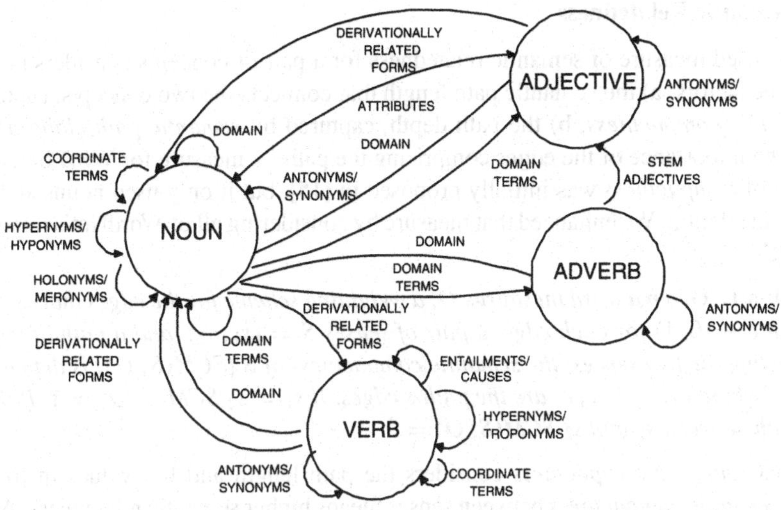

Fig. 1. Semantic relations in WordNet

for pairs of concepts is based on the information content of the deepest concept that can subsume both. Measures proposed by Jiang and Conrath [9], Hirst and St-Onge [10], Leacock and Chodorrow [11], and Lin [12], were based on similar ideas. Due to space limitations we suggest the reader to consult the analysis of Budanitsky and Hirst [13] for the majority of the aforementioned measures. All these measures are based on the noun hierarchy, whereas our measure defines the semantic relatedness between any two concepts, independently of their POS. The proposed WSD approach is based on a new measure of semantic relatedness for concept pairs, which combines in tandem the length of the semantic path connecting them, the type of the semantic edges and the depth of the nodes in the thesaurus. Experimental evaluation shows that this measure familiarizes human understanding of semantic similarity better than all other measures.

In this work we focus only to the unsupervised approaches that are based on semantic networks. Patwardhan et al. [14] modified the Lesk method to allow for the use of any measure of semantic relatedness. Mihalcea et al. [5] constructed semantic networks from WordNet and ran an adaptation of the PageRank algorithm on them, in order to address the *all words* task for the English language. Tsatsaronis et al. [6] constructed richer semantic networks and surpassed or matched the performance of the PageRank semantic networks, with a constraint spreading activation technique. We compare our work to the WSD approaches mentioned above in Section 4.2. Results show that our method surpasses or matches state of the art results for the English *all words* task in Senseval 2.

3 WSD Based on Semantic Relatedness of Terms

In this section we propose a new measure of semantic relatedness and a new WSD approach based on that measure.

3.1 Semantic Relatedness

The proposed measure of semantic relatedness for a pair of concepts considers in tandem three factors: a) the semantic path length that connects the two concepts, captured by *semantic compactness*, b) the path depth, captured by *semantic path elaboration*, and c) the importance of the edges comprising the path. A measure for WSD based on the idea of *compactness* was initially proposed in [15], but it only used nouns and the hypernym relation. We enhanced that measure by considering all of WordNet's relations and POS.

Definition 1. *Given a word thesaurus O, a weighting scheme for the edges that assigns a weight $e \in (0, 1)$ for each edge, a pair of senses $S = (s_1, s_2)$, and a path of length l connecting the two senses, the semantic compactness of S ($SCM(S, O)$) is defined as $\prod_{i=1}^{l} e_i$, where $e_1, e_2, \ldots, e_l$ are the path's edges. If $s_1 = s_2$ $SCM(S, O) = 1$. If there is no path between s_1 and s_2 $SCM(S, O) = 0$.*

Note that *semantic compactness* considers the path length and has values in [0, 1]. Higher *semantic compactness* between senses means higher semantic relatedness. Also, larger weights are assigned to stronger edge types. The intuition behind the assumption of edges' weighting is the fact that some edges provide stronger semantic connections than others. A standard way od obtaining the edges' weights can be the measurement of edges' distribution in WordNet. The frequency of occurrence of each edge type can act as its weight. The *semantic compactness* of two senses s_1 and s_2, can take different values for all the different paths that connect the two senses. Another parameter that affects term relatedness is the depth of the sense nodes comprising the path. A standard means of measuring depth in a word thesaurus is the hypernym/hyponym hierarchical relation for the noun and adjective POS and hypernym/troponym for the verb POS. A path with shallow sense nodes is more general compared to a path with deep nodes. This parameter of semantic relatedness between senses is captured by the measure of *semantic path elaboration* introduced in the following definition.

Definition 2. *Given a word thesaurus O and a pair of senses $S = (s_1, s_2)$, where $s_1, s_2 \in O$ and $s1 \neq s2$, and a path between the two senses of length l, the semantic path elaboration of the path (SPE(S,O)) is defined as $\prod_{i=1}^{l} \frac{2d_i d_{i+1}}{d_i + d_{i+1}} \cdot \frac{1}{d_{max}}$, where d_i is the depth of sense s_i according to O, and d_{max} the maximum depth of O. If $s_1 = s_2$, and $d = d_1 = d_2$ $SPE(S, O) = \frac{d}{d_{max}}$. If there is no path from s_1 to s_2, $SPE(S, O) = 0$.*

SPE is in fact the harmonic mean of the two depths normalized to the maximum thesaurus depth. The harmonic mean offers a lower upper bound than the average of depths and we think is a more realistic estimation of the path's depth. *Compactness* and *Semantic Path Elaboration* measures capture the two most important parameters of measuring semantic relatedness between terms [13], namely path length and senses depth in the used thesaurus. We combine these two measures in the definition of *Semantic Relatedness* between two senses.

Definition 3. *Given a word thesaurus O and a pair of senses $S = (s_1, s_2)$ the semantic relatedness of S (SR(S,O)) is defined as $max\{SCM(S, O) \cdot SPE(S, O)\}$.*

Note that Definition 3 can be expanded to measure the semantic relatedness for a pair of terms $T = (t_1, t_2)$, namely $SR(T, O)$. For all the pair combinations of senses that

t_1 and t_2 may be assigned, the maximum value of semantic relatedness between any two senses found is defined as the semantic relatedness of the pair of terms. In case $t_1 \equiv t_2 \equiv t$ and $t \notin O$ then semantic relatedness can be considered as 1. The semantic relatedness can only take real values in $[0, 1]$.

Algorithm 1. Word-Sense-Disambiguation(T,w,O,Θ)

Require: A set of POS-tagged terms T to be disambiguated, a word thesaurus O, a weighting scheme $w : E \rightarrow (0..1)$ for the edges of the used thesaurus and an upper threshold Θ for the maximum number of combinations examined in simulated annealing.

Ensure: A mapping of terms to senses that disambiguate them.

```
Word-Sense-Disambiguation(T,w,O,Θ)
for all terms t ∈ T do
    senses[t] =number of possible senses of t
    correct-sense[t] =random(senses[t])
end for
Minimum-MST-Weight=compute-SCH(T,correct-sense,O)
while iterations i ≤ Θ do
    Transit randomly to a neighboring assignment of senses
    Temp-MST-Weight=compute-SCH(T,correct-sense,O)
    ΔE=Minimum-MST-Weight - Temp-MST-Weight
    if ΔE > 0 then
        Transit to neighboring state with probability e^(ΔE/i)
    end if
end while
return correct-sense
```

3.2 Word Sense Disambiguation Based on Semantic Relatedness

We expand the measure of semantic relatedness between a pair of terms introduced in the previous section, to a measure of semantic coherence between a set of terms, and we use this measure to form a new knowledge-based WSD algorithm that does not require training.

Definition 4. *Given a word thesaurus O and a set of n terms* $T = (t_1, t_2, \ldots, t_n)$, *where for each* t_i, $i = 1..n$, *it holds that* $t_i \in O$, *let* $S = (s_1, s_2, \ldots, s_n)$ *be a possible assignment of senses to the terms in T. The semantic coherence of T (SCH(T,O)) is then defined as the weight of the Minimum Spanning Tree (MST) computed on the weighted undirected graph having: 1) a node for each sense in S, 2) an edge for each pair of senses* (s_i, s_j) *that are semantically related* ($SR((s_i, s_j), O) > 0$), *with an edge weight* $w_{i,j} = \frac{1}{SR((s_i,s_j),O)}$.

Based on this definition, the WSD algorithm operates as follows: From all MSTs produced for the set of terms we choose the one with the maximum semantic coherence. To alleviate the computational burden occurring by examining all possible MSTs, we use simulated annealing [16].

4 Experimental Evaluation

The experimental evaluation is bi-fold. Firstly, we compare the measure of semantic relatedness against state of the art measures, using a set of term pairs weighted by humans as a benchmark. Secondly, we evaluate the performance of the proposed WSD algorithm in the Senseval 2 benchmark collection.

4.1 Semantic Relatedness Measure Evaluation

The relatedness measures that we compare to are: Hirst and St-Onge (HS), Jiang and Conrath (JC), Leackock and Chodorow (LC), Lin (L) and Resnik (R), which are thoroughly discussed in [13]. We use the test set of 65 term pairs initially proposed by Rubenstein and Goodenough [17] and rank the term pairs using the semantic relatedness scores given by each measure and by the 51 human subjects. We measure the correlation of all rankings, including ours (SRel), using the Kendall's Tau distance measure [18]. The results are:

Table 1. Kendall's Tau distance from human rankings

	HC	**JC**	**LC**	**L**	**R**	**SRel**
Kendall's Tau	0.371	0.250	0.247	0.242	0.260	0.169

4.2 Word Sense Disambiguation Evaluation

The proposed WSD method is evaluated in Senseval 2, for the English *all words* task. The computation of the semantic relatedness between any pair of concepts requires a weighting scheme that assigns values in (0, 1). A standard weighting scheme can be the distribution of edge types in the used thesaurus. The frequency of occurrence can be the respective edge's weight. We followed that scheme, and the edge weights we produced (hypernym/hyponym edges obtained 0.57, nominalization edges 0.14, etc.) are in accordance to those stated Song et al. in [19], and the same with the ones obtained if semantic networks are constructed for each sentence in the SemCor data set. Thus, no effort for training in order to learn the edges' weights is required. We compare our approach (MST) with the standard unsupervised baseline, which randomly assigns a sense to a given word, the best reported unsupervised method in the Senseval 2 competition [3], an unsupervised approach utilizing spreading of activation on semantic

Table 2. Overall and per file accuracy on the Senseval 2 data set

	Words		MST	Baseline	SANs	Best Unsup. Senseval 2	PRSN
	Mono	Poly					
File 1 (d00)	**103**	**552**	**0.436**	**0.365**	**0.459**	unavailable	**0.439**
File 2 (d01)	**232**	**724**	**0.498**	**0.421**	**0.468**	unavailable	**0.544**
File 3 (d02)	**129**	**563**	**0.511**	**0.430**	**0.557**	unavailable	**0.542**
Overall	**464**	**1839**	**0.485**	**0.407**	**0.492**	**0.451**	**0.508**

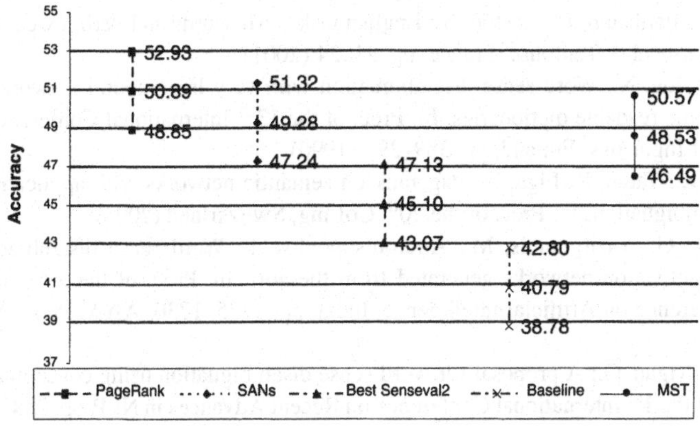

Fig. 2. Methods' accuracies with 0.95 confidence intervals

networks (SANs) [6] and an unsupervised approach executing PageRank [5] (PRSN). Reported accuracy is shown in Table 2.

Results show that the proposed WSD method surpasses the baseline and the best unsupervised method of the Senseval 2 competition. Furthermore, it matches the performance of SANs and PageRank methods, which are, to the best of our knowledge, the approaches with the best ever reported performance in unsupervised WSD overall for all POS. The difference with these two methods is in the order of magnitude of 10^{-3} and 10^{-2} respectively. The statistical significance of our results is calculated using 0.95 confidence intervals for all methods' accuracies (Figure 2).

5 Conclusions and Future Work

In this paper we introduce a new measure of semantic relatedness between senses and expand it to compute relatedness for a pair of terms. The measure combines in tandem the concepts' depth in the used thesaurus, the semantic path length that connects the two concepts and the importance of the semantic edges comprising the path. Experimental evaluation in ranking term pairs according to their relatedness, shows that our measure produces a ranking that is more similar to the human generated one, compared to rankings generated by other related measures of semantic relatedness proposed in the past. Finally, we embedded this measure into a new WSD approach that is based on measuring the weight of the minimum spanning tree connecting candidate senses. Experimental evaluation shows that our method surpasses or matches state of the art results. In the future we will investigate the impact of each of the three factors comprising our measure of semantic relatedness in the WSD task, and embed it into text retrieval and text classification models.

References

1. Agirre, E., Edmonds, P.: Word Sense Disambiguation: Algorithms and Applications. Springer, Heidelberg (2006)
2. Fellbaum, C.: WordNet, An Electronic Lexical Database. MIT Press, Cambridge (1998)

3. Palmer, M., Fellbaum, C., Cotton, S.: English tasks: All-words and verb lexical sample. In: Proc. of Senseval-2, Toulouse, France, pp. 21–24 (2001)
4. Veronis, J., Ide, N.: Word sense disambiguation with very large neural networks extracted from machine readable dictionaries. In: Proc. of the 13th International Conference on Computational Linguistics, Finland, pp. 389–394 (1990)
5. Mihalcea, R., Tarau, P., Figa, E.: Pagerank on semantic networks with application to word sense disambiguation. In: Proc. of the 20th CoLing, Switzerland (2004)
6. Tsatsaronis, G., Vazirgiannis, M., Androutsopoulos, I.: Word sense disambiguation with spreading activation networks generated from thesauri. In: Proc. of the 20th International Joint Conference on Artificial Intelligence, India, pp. 1725–1730. AAAI Press, Menlo Park (2007)
7. Agirre, E., Rigau, G.: A proposal for word sense disambiguation using conceptual distance. In: Proc. of the 1st International Conference on Recent Advances in NLP, pp. 258–264 (1995)
8. Resnik, P.: Using information content to evaluate semantic similarity. In: Proc. of the 14th International Joint Conference on Artificial Intelligence, Canada, pp. 448–453 (1995)
9. Jiang, J., Conrath, D.: Semantic similarity based on corpus statistics and lexical taxonomy. In: Proc. of ROCLING X, Taiwan, pp. 19–33 (1997)
10. Hirst, G., St-Onge, D.: Lexical chains as representations of context for the detection and correction of malapropisms. In: WordNet: An Electronic Lexical Database, Ch. 13, pp. 305–332. MIT Press, Cambridge (1998)
11. Leacock, C., Chodorow, M.: Combining local context and wordnet similarity for word sense identification. In: WordNet: An Electronic Lexical Database, Ch. 11, pp. 265–283. MIT Press, Cambridge (1998)
12. Lin, D.: An information-theoretic definition of similarity. In: Proc. of the 15th International Conference on Machine Learning, pp. 296–304 (1998)
13. Budanitsky, A., Hirst, G.: Evaluating wordnet-based measures of lexical semantic relatedness. Computational Linguistics 32(1), 13–47 (2006)
14. Patwardhan, S., Banerjee, S., Pedersen, T.: Using measures of semantic relatedness for word sense disambiguation. In: Proc. of the 4th International Conference on Inbtelligent Text Processing and Computational Linguistics, pp. 241–257. Springer, Heidelberg (2003)
15. Mavroeidis, D., Tsatsaronis, G., Vazirgiannis, M., Theobald, M., Weikum, G.: Word sense disambiguation for exploiting hierarchical thesauri in text classification. In: Proc. of the 9th PKDD, Portugal, pp. 181–192. Springer, Heidelberg (2005)
16. Cowie, J., Guthrie, J., Guthrie, L.: Lexical disambiguation using simulated annealing. In: Proc. of the 14th CoLing, France, pp. 359–365 (1992)
17. Rubenstein, H., Goodenough, J.: Contextual correlates of synonymy. Communications of the ACM 8(10), 627–633 (1965)
18. Fagin, R., Kumar, R., SivaKumar, D.: Comparing top k lists. SIAM Journal on Discrete Mathematics 17(1), 134–160 (2003)
19. Song, Y., Han, K., Rim, H.: A term weighting method based on lexical chain for automatic summarization. In: Proc. of the 5th CICLing Conference, USA, pp. 636–639 (2004)

A Pattern-Based Framework for Uncertainty Representation in Ontologies

Miroslav Vacura[1], Vojtěch Svátek[1], and Pavel Smrž[2]

[1] Faculty of Informatics and Statistics, University of Economics, Prague
W. Churchill Sq. 4, 130 67 Prague 3, Czech Republic
vacuram@vse.cz, svatek@vse.cz

[2] Faculty of Information Technology, Brno University of Technology
Božetěchova 2, 612 66 Brno, Czech Republic
smrz@fit.vutbr.cz

Abstract. We present a novel approach to representing uncertain information in ontologies based on design patterns. We provide a brief description of our approach, present its use in connection with fuzzy information and probabilistic information, and describe the possibility to model multiple types of uncertainty in a single ontology. We also shortly present an appropriate fuzzy reasoning tool and define a complex ontology architecture for well-founded handling of uncertain information.

1 Introduction

Currently available Semantic Web technology still provides inadequate foundations to handling uncertainty. One of the main achievements of the Semantic Web initiative was the standardization of a common web ontology language – OWL. While OWL datatypes provide means for including numeric uncertainty measures and necessary structural foundations ad hoc, there is no standardized way of representing uncertainty. There is a widespread opinion that for adequate representation of uncertainty in OWL some language extension is necessary, be it at the level of language syntax or of higher-level patterns.

An example of the first is Fuzzy RDF [11], which transforms the RDF model from the standard triple to the couple (value, triple), adding to the triple a "value". In contrast, Fuzzy OWL [14] is based on fuzzy description logics; a simple extension to OWL syntax has already been proposed to represent it [13]. There are also approaches based on modeling probability. BayesOWL [17] extends OWL using Bayesian networks as the underlying reasoning mechanism and probabilistic model. PR-OWL [3] is based on MEBN Theory (Multi-Entity Bayesian Network).

Probability theory is also the foundation of recently introduced probability description logics P-$\mathcal{SHIF}$(D) and P-$\mathcal{SHOIN}$(D) [9]. In this case probabilistic knowledge base consists of a PTBox (classical DL knowledge base along with probabilistic terminological knowledge) and a collection of PABoxes (encoding probabilistic assertional knowledge about a certain set of individuals). However no formal way of RDF/XML encoding is provided. This probabilistic description logics theory is basis of Pronto,[1] a

[1] http://pellet.owldl.com/pronto

P. Sojka et al. (Eds.): TSD 2008, LNAI 5246, pp. 227–234, 2008.
© Springer-Verlag Berlin Heidelberg 2008

probabilistic extension for the Pellet DL reasoner. Pronto uses OWL 1.1's annotation properties for encoding probability information in OWL RDF/XML. The drawback is that the formal semantics of annotation properties is limited.

There is also an approach based on using ontology patterns to define n-ary relations described in the "Best Practices" W3C document [5], which can be applied to uncertainty and in principle consists of creating ad hoc properties and classes for every ontology and every type of uncertain information described by the ontology. This however results in inability to use any generic uncertainty reasoner: if one needs to introduce uncertainty handling to an existing ontology, it is necessary to do re-engineering of this ontology (removing classes and properties and creating new ones), which renders the new ontology incompatible with the original one. We will show that our approach is superior to the "Best Practices" approach in most cases as it does not have such drawbacks. Our approach also provides a common OWL syntax for various kinds of uncertainty (like fuzziness or probability) without the need to extend the OWL standard.

2 Pattern-Based Representation of Fuzziness

This section will describe the method of representing fuzzy information based on Fuzzy OWL semantics [14] using our uncertainty modeling pattern. Although fuzziness isn't, exactly said, a type of uncertainty, we will show that our approach is usable to represent fuzzy information. In this example we will consider representing fuzzy information in the form of facts, i.e. A-Box entities from the description logic (DL) point of view.

The key principle of our approach to representing fuzzy information is the *separation* of crisp ontology from fuzzy information ontology. While an ontology can be built from scratch using our pattern-based approach, it is also possible to extend an existing crisp ontology (base ontology) to represent fuzzy information by creating an add-on ontology (fuzzy ontology) that only contains fuzzy information. Separation of uncertainty-related information from crisp information can be in practice realized by storing these sets of RDF triples in different RDF repositories. The performance of data querying etc. when not using uncertainty-related information would thus not be affected. This also allows for simple ways of data distribution among multiple servers with RDF repositories. Such a separation is not possible using the "Best Practices" approach [5] or using annotation properties for encoding uncertainty information.

We allow the fuzzy ontology to be in OWL Full, which may be at first sight surprising. Most ontology applications that use some kind of reasoning presuppose an OWL-DL-compliant ontology for the sake of decidability and availability of reasoning tools. In our case we only assume that the base ontology is OWL-DL-compliant. The base ontology can be used separately from the fuzzy ontology, and regular OWL DL crisp reasoning tools can be applied to it. There is no reason to apply crisp reasoning tools on the fuzzy ontology since it only includes information regarding uncertainty (although it be used for pruning the base ontology before crisp reasoning, i.e. filtering out relationships with fuzzy value under some threshold). When the user has a tool that supports fuzzy reasoning, we suppose that it either accepts our pattern-based syntax or it has some other proprietary syntax format and we need to convert the base and fuzzy

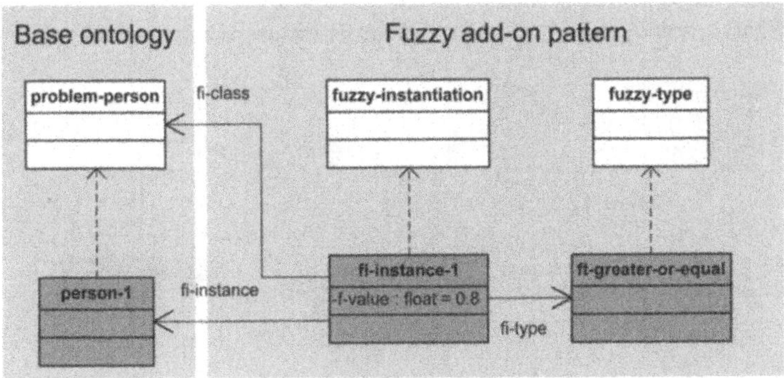

Fig. 1. Instantiation pattern

ontologies together to this format before reasoning. In either case a fuzzy ontology conforming to the OWL Full standard is not a problem.

2.1 Fuzziness in Instantiation Axioms

Instantiation axioms are assertions of form $\langle a : C \bowtie n\rangle$ – facts saying that individual a belongs to class C, where n is a level of certainty from (0, 1) and $\bowtie$ is one of $\{\leq, <, \geq, >\}$. As an example we can take a metro surveillance application that should, based on multimedia information (cameras and microphones), decide whether a person is classified as "problem person". In such a case we have an axiom saying that an instance person-1 belongs to class problem-person with some level of certainty. We introduce a few constructs that enable us to model such axioms with uncertainty by ontology patterns. For each crisp axiom of the base ontology we create a new individual belonging to class fuzzy-instantiation, which will have several properties attaching it to that crisp axiom in the base ontology and implementing uncertainty. Properties fi-instance and fi-class characterize the membership of an individual person-1 to class problem-person. Property f-type defines the type of uncertainty relation ($\bowtie$), and the datatype property f-value defines a level of certainty n. The complete pattern is depicted in Fig. 1 (individuals are grayed and classes are bright).

The following OWL code is part of the base ontology that describes an individual person-1 as belonging to class problem-person. This is part of a standard crisp OWL DL ontology and is unmodified by adding uncertainty information.

```
<rdf:Description rdf:about="#person-1">
  <rdf:type rdf:resource="#problem-person"/>
</rdf:Description>
<rdf:Description rdf:about="#problem-person">
  <rdf:type rdf:resource="http://www.w3.org/2002/
  07/owl#Class"/>
</rdf:Description>
```

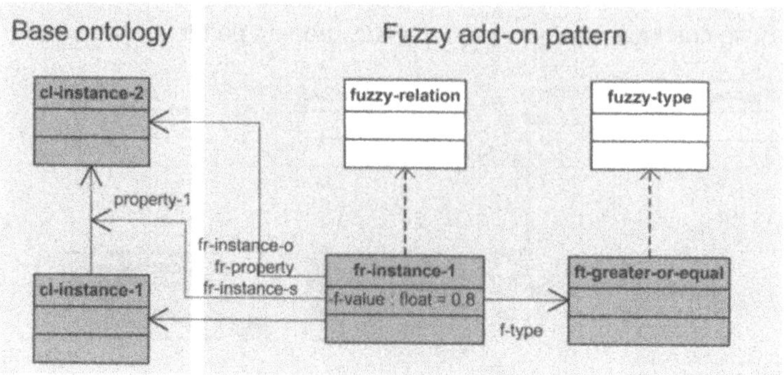

Fig. 2. Relation pattern

The next example is part of the OWL code of fuzzy add-on ontology that is separated from the base ontology and contains fuzzy information. This part shows fi-instance-1 – an individual that is used to describe fuzzy information regarding the instantiation axiom presented in the previous OWL code example. It says that the individual person-1 belongs to class problem-person with certainty level greater or equal to 0.8.

```
<rdf:Description rdf:about="#fi-instance-1">
  <f-value rdf:datatype="http://www.w3.org/2001/
   XMLSchema#float">0.8</f-value>
  <fi-class rdf:resource="#problem-person"/>
  <fi-type rdf:resource="#ft-greater-or-equal"/>
  <rdf:type rdf:resource="#fuzzy-instantiation"/>
  <fi-instance rdf:resource="#person-1"/>
</rdf:Description>
```

2.2 Fuzziness in Role Axioms

Role axioms are assertions of form $\langle (a, b) : R \bowtie n \rangle$ – facts saying that individual a and individual b are in relation R, where the level of certainty is n and $\bowtie$ is one of $\{\leq, <, \geq, >\}$. The complete pattern is depicted in Fig. 2. The OWL code of role axioms is analogous to the OWL code of instantiation axioms.

The following OWL code is part of standard role axiom definition with regards to instances cl-instance-1, cl-instance-2 and property property-1. It is again OWL-DL-compliant.

```
<class-1 rdf:ID="cl-instance-1">
        <property-1 rdf:resource="#cl-instance-2"/>
</class-1>
<class-1 rdf:ID="cl-instance-2"/>
<owl:Class rdf:ID="class-1"/>
<owl:ObjectProperty rdf:ID="property-1"/>
```

The following OWL code is part of the fuzzy add-on ontology that refers to the previous relation and adds fuzzy information to it.

```
<rdf:Description rdf:about="#fr-instance-1">
  <rdf:type rdf:resource="#fuzzy-relation"/>
  <f-type rdf:resource="#ft-greater-or-equal"/>
  <f-value rdf:datatype="http://www.w3.org/2001/
                       XMLSchema#float">0.8</f-value>
  <fr-instance-o rdf:resource="#cl-instance-2"/>
  <fr-instance-s rdf:resource="#cl-instance-1"/>
  <fr-property rdf:resource="#property-1"/>
</rdf:Description>
```

We use the property fr-instance-o to define the object individual, fr-instance-s to define the subject individual, and fr-property to define to which property we assign the fuzzy value (there can be more than one relation between the same two individuals). We also use the f-type property to define the type of relation ($\bowtie$) and again the datatype property f-value that defines the level of uncertainty n.

3 Universal Uncertainty Modeling

As we have already stated, one of major advantages of our modeling approach is that it allows us to model various kinds of uncertainty. This is not limited to modeling a different kind of uncertainty in each case but using our method one can include various kinds of uncertain information in the same ontology at the same time. Our method is also not limited to kinds of uncertainty presented in this paper but represents the core of a more general framework for handling uncertain information in ontologies, which is strictly modular and easily extensible. An illustration of handling multiple types of uncertainty is depicted in Fig. 3.

In real-world applications not only ontological modeling but also reasoning on the top of ontologies is necessary. This is why we focus on the implementation of our pattern-based representation in available reasoning engines. In the context of the EU-funded project K-Space[2], our presented fuzzy-modeling syntax is envisaged to be implemented in the FiRE reasoning engine[3].

4 Architecture Supporting Reasoning with Uncertainty

Using the components described above we can define a well-founded architecture of ontology that fully supports handling uncertainty on the basis of our uncertainty modeling framework. This architecture is based on the standard concept of well-founded ontology: the crisp ontology is aligned to the foundational ontology (in the case of CARETAKER example it is DOLCE [10]) while fuzzy and probabilistic ontologies are

[2] `http://www.k-space.eu`

[3] `http://www.image.ece.ntua.gr/~nsimou`

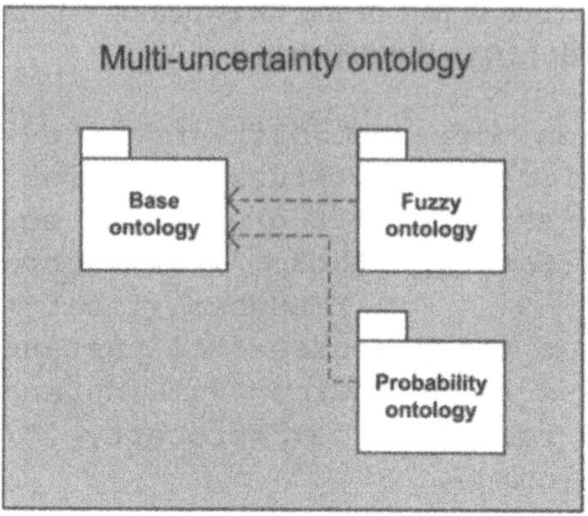

Fig. 3. Handling multiple types of uncertainty in the same ontology

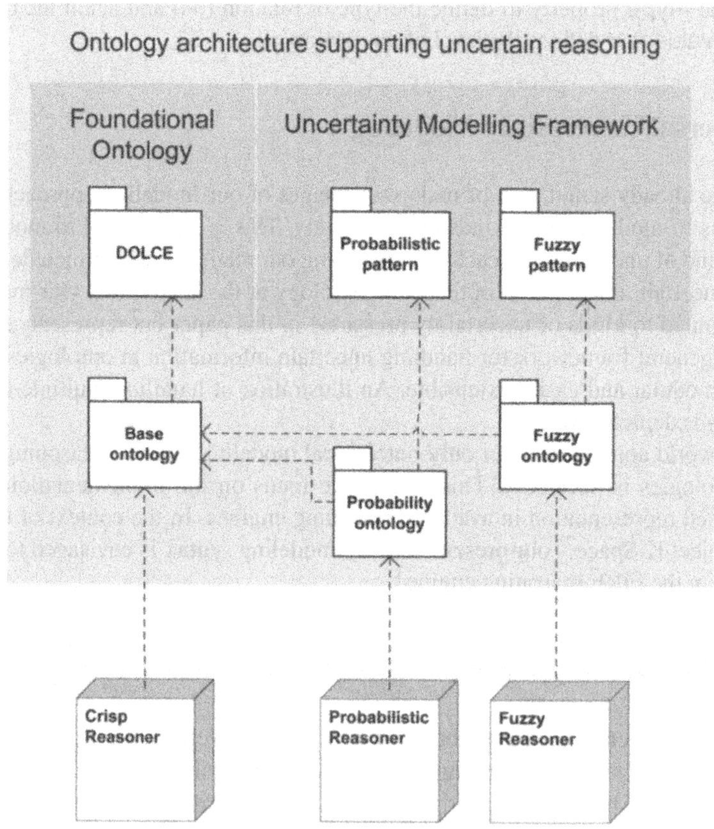

Fig. 4. Ontology architecture supporting reasoning with uncertainty

based on appropriate patterns of the uncertainty modeling framework. The architecture is modularized, so these parts of ontology are separated to independent modules. On the top of these ontologies there can be a number of different specialized reasoners. The complete scheme of the architecture is in Fig. 4.

Recent studies, undertaken e.g. within the W3C Uncertainty Reasoning for the World Wide Web Incubator Group[4], show that the requirement of combining different types of uncertainty (typically, fuzziness and probability) in the same ontological model is quite common.

5 Conclusions and Future Work

Our approach to representing uncertainty in ontologies has the following useful features: it can use existing crisp ontologies without the need to re-engineer them; it is "backward compatible" with crisp DL reasoning tools and other software; it uses strictly standard OWL language with no extensions; it can be used to represent different kinds of uncertainty (like probability and fuzziness) in the same ontology; it separates data regarding uncertainty from base data.

As uncommon feature may be seen that our fuzzy add-on is in OWL Full and not in OWL DL, but we have argued in previous sections, that thanks to strict separation of fuzzy ontology from base ontology this does not imply a problem in real-world applications.

In the future we plan to extend the coverage of our uncertainty modeling framework and to provide support and integration to other reasoning tools that can reason over uncertain information.

As one of motivations for our research we see the CARETAKER project[5], which comprises advanced approaches to recognition of multimedia data. More precisely, CARETAKER will focus on the extraction of structured knowledge from large multimedia collections recorded over networks of cameras and microphones deployed in real sites. The produced audio-visual streams, in addition to surveillance and safety issues, could represent a useful source of information if stored and automatically analyzed.

Acknowledgments. Miroslav Vacura, Vojtěch Svátek are supported by the EC under FP6, project Knowledge Space of semantic inference for automatic annotation and retrieval of multimedia content – K-Space (no.: FP6-027026), and first one also by the Czech FIGA NF VSE grant. Pavel Smrz is supported by the EU FP7 project KiWi (ref. no.: 211932).

References

1. Beckert, B., Hähnle, R., Oel, P., Sulzmann, M.: The tableau-based theorem prover 3^{TAP}, version 4.0. In: Proceedings of CADE 1996, New Brunswick, NJ, USA, pp. 303–307. Springer, Heidelberg (1996)

[4] `http://www.w3.org/2005/Incubator/urw3/`

[5] `http://www.ist-caretaker.org/`

2. Costa, P.C.G., Laskey, K.B., Laskey, K.J.: Of Klingons and Starships: Bayesian Logic for the 23rd Century. In: Uncertainty in Artificial Intelligence: Proceedings of the Twenty-first Conference. AUAI Press, Edinburgh (2005)
3. Costa, P.C.G., Laskey, K.B., Laskey, K.J.: PR-OWL: A Bayesian Framework for the Semantic Web. In: Proceedings of URSW 2005, Galway, Ireland (November 2005)
4. Giugno, R., Lukasiewicz, T.: P-SHOQ(D): A probabilistic extension of SHOQ(D) for probabilistic ontologies in the semantic web. Research Report 1843-02-06 7, INFSYS, Wien, Austria (April 2002)
5. Hayes, P., Welty, C.: Defining N-ary Relations on the Semantic Web. Technical report (2006)
6. Heinsohn, J.: Probabilistic description logics. In: Proceedings of UAI-1994, pp. 311–318 (1994)
7. Jaeger, M.: Probabilistic reasoning in terminological logics. In: Proceedings of KR-1994, pp. 305–316 (1994)
8. Koller, D., Levy, A., Pfeffer, A.: P-CLASSIC: A tractable probabilistic description logic. In: Proceedings of AAAI-1997, pp. 390–397 (1997)
9. Lukasiewicz, T.: Probabilistic description logics for the semantic web. Research report 1843-06-05, INFSYS, Technische Universitat, Wien, Austria (April 2007)
10. Masolo, C., Borgo, S., Gangemi, A., Guarino, N., Oltramari, A., Schneider, L.: WonderWeb Deliverable D17. Technical report (2002)
11. Mazzieri, M.: A fuzzy RDF semantics to represent trust metadata. In: Semantic Web Applications and Perspectives (online proceedings (2004)
12. Hayes, P., Patel-Schneider, P.F., Horrocks, I.: OWL: RDFCompatible Model-Theoretic Semantics. Technical report (2004)
13. Stoilos, G., Simou, N., Stamou, G., Kollias, S.: Uncertainty and the semantic web (2006)
14. Stoilos, G., Stamou, G., Tzouvaras, V., Pan, J.Z., Horrocks, I.: Fuzzy OWL: Uncertainty and the Semantic Web. In: Proc. of the OWL-ED 2005 (2005)
15. Stoilos, G., Stamou, G., Tzouvaras, V., Pan, J.Z., Horrocks, I.: The fuzzy description logic f-shin. In: Proceedings of URSW 2005 (2005)
16. Straccia, U.: A fuzzy description logic for the semantic web. In: Capturing Intelligence: Fuzzy Logic and the Semantic Web. Elsevier, Amsterdam (2005)
17. Zhongli, D.: A Probabilistic Extension to Ontology Language OWL. In: Proc. of the HICSS-37 (2004)
18. Zhongli, D.: A Bayesian Methodology towards Automatic Ontology Mapping. In: Proceedings of AAAI-2005 C&O Workshop (2005)

MSD Recombination for Statistical Machine Translation into Highly-Inflected Languages

Jerneja Žganec-Gros and Stanislav Gruden

Alpineon d.o.o.,
Ulica Iga Grudna 15, 1000 Ljubljana, Slovenia
jerneja@alpineon.si, staneg@alpineon.si
http://www.alpineon.si

Abstract. Freely available tools and language resources were used to build the VoiceTRAN statistical machine translation (SMT) system. Various configuration variations of the system are presented and evaluated. The VoiceTRAN SMT system outperformed the baseline conventional rule-based MT system in both English-Slovenian in-domain test setups. To further increase the generalization capability of the translation model for lower-coverage out-of-domain test sentences, an "MSD-recombination" approach was proposed. This approach not only allows a better exploitation of conventional translation models, but also performs well in the more demanding translation direction; that is, into a highly inflectional language. Using this approach in the out-of-domain setup of the English-Slovenian JRC-ACQUIS task, we have achieved significant improvements in translation quality.

1 Introduction

Machine translation (MT) systems automatically convert text strings from a source language (SL) into text strings in the target language (TL). They often allow for customization by application domain (e.g., weather), which improves the output by limiting the scope of allowable substitutions. This technique is particularly effective in domains in which formal or formulaic language is used, and therefore machine translation of government and legal documents more readily produces usable output than translation of less standardized texts or even spoken language. Some initial machine translation attempts have been reported for translation from Slovenian into English [1,2,3,4]. However, very little has been done for the opposite translation direction, from English into Slovenian [3]. We have performed experiments in both translation directions, in which especially the latter proved to be a demanding task due to the highly inflectional nature of Slovenian. This paper continues with a description of the VoiceTRAN statistical machine translation (SMT) experiment. The goal was to evaluate the performance of various SMT configuration variations against the performance of a baseline conventional rule-based MT system in order to find out, which MT system should be used in the VoiceTRAN speech-to-speech translation system. The language resources used for training the system and the system architecture are described. The "MSD-recombination" approach used for determining the word order in the target language is introduced. Finally, the efficiency of the VoiceTRAN SMT system is evaluated by standard MT metrics.

P. Sojka et al. (Eds.): TSD 2008, LNAI 5246, pp. 235–242, 2008.
© Springer-Verlag Berlin Heidelberg 2008

2 Language Resources

The following language resources were used in the experiments. The bilingual language resources always refer to the English-Slovenian language pair:

- bilingual text corpora: the VoiceTRAN application-specific corpus and two freely available corpora: JRC-ACQUIS [5] and IJS-ELAN [6];
- the monolingual FDV-IJS Slovenian corpus, collected at the University of Ljubljana, and annotated within the VoiceTRAN project;
- English-Slovenian conventional dictionaries: an in-domain dictionary of military terminology and a conventional general dictionary.

The monolingual and bilingual corpora were automatically annotated with context-disambiguated lemmas and morphosyntactic descriptions (MSDs), which included part-of-speech (POS) information [7].

3 The VoiceTRAN SMT System

The following freely available tools were used to build the VoiceTRAN SMT sys-tem. The GIZA++ toolkit [8] was used for training the VoiceTRAN translation model; the CMU-SLM toolkit [9] was used for building the language model, and the ISI ReWrite Decoder [10] was applied for translating the test sentences.

3.1 The "MSD Recombination" SMT Approach

Bilingual training data are needed to train a SMT translation model that is then able to generalize and translate new sentences. Due to the statistical nature of system training words that appear more frequently in the training corpora are more likely to develop a suitable statistical model, whereas rare words tend to be overlooked. If the training data contain too many words of the latter type, the resulting models do not perform well due to data sparsity.

This effect is even more pronounced when translating from less-inflected into more highly-inflected languages when a word in the source language can be translated by many words in the target language depending on the context.

There are several ways to tackle this problem. For example, we can build translation models based on lemmas only. This can work to a certain extent when translating in the opposite direction: from a highly-inflected language to a less-inflected one [1,4]. The lemmatized English target output often matches the correct reference translation. In contrast, when translating from English into Slovenian, translation accuracy through relying on lemmas only is rather poor.

The monolingual and bilingual corpora used in our experiments are equipped with MSD annotations (including POS information) and can be exploited to further improve the translations. Some attempts have been reported regarding the use of larger tokens in the training phase, consisting of MSDs (and POS), concatenated to their corresponding lemmas [1,4]. However, the data sparsity did not decrease significantly.

We used a different approach, which we call "MSD recombination." First, two corpora were derived from the initial training corpus.

In the first corpus, the sentences in both languages were preprocessed so that all word forms were replaced by their lemmas, using the lemmatization information provided in the source corpus. In order to derive the second corpus, all original word forms were replaced by their corresponding MSDs. These two corpora were then separately fed into the SMT training system.

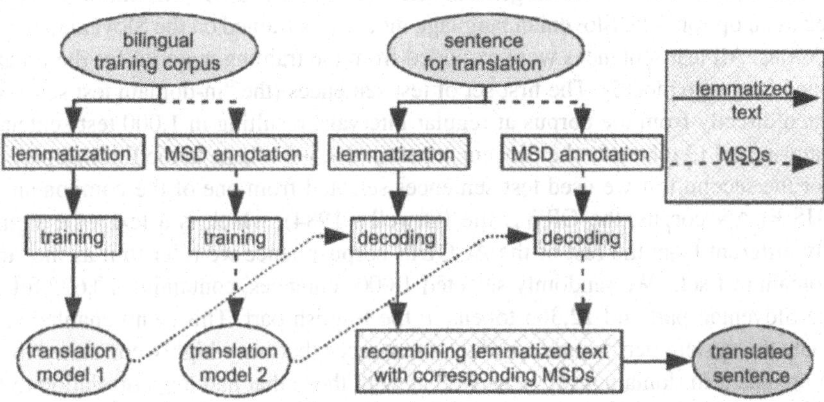

Fig. 1. Description of the "MSD recombination" translation approach. We trace how each "lemma + MSD" pair in the source language transforms into a corresponding "lemma + MSD" pair in the target language

The decoding was performed as shown in Figure 1: every test sentence was preprocessed into two sentences, in which words were replaced by lemmas in the first sentence, and by relevant MSDs in the second sentence. Then we traced how each "lemma + MSD" pair in the source language transformed into a corresponding "lemma + MSD" pair in the target language. The resulting "lemma + MSD" pair was ultimately recombined to construct the final word form in the target language.

4 Experimental Setup

We describe two SMT experiments with different test and training setups: the ACQUIS corpus experiment and the VoiceTRAN corpus experiment. In both experiments, the performance of the simple translation method, in which the sentences used for training the SMT system were taken directly from the original corpus without any prior modifications, was compared against the performance of the "MSD recombination" approach, and a baseline rule-based translation system, Presis [2], a commercial conventional bidirectional rule-based translation system for the language pair Slovenian-English. It had been adapted to the tested application domain by upgrading the lexicon with application-specific terms.

4.1 JRC-ACQUIS Experiment

The first SMT experiment was performed on the English-Slovenian part of the JRC-ACQUIS corpus [5], which is equivalent to the SVEZ-IJS corpus. All tokens contain automatically assigned context-disambiguated lemmas and MSD (including POS) annotations. Sentences longer than 25 words were discarded from the training corpus as were the test sentences. The final bilingual training corpus contained 127,475 sentences with approximately 1.04 million tokens in the Slovenian part and 1.23 million tokens in the English part. A conventional general dictionary with 140,000 translation pairs was added as an option. The Slovenian language model was trained on the Slovenian part of the corpus. All test sentences were excluded from the training material for the translation and language models. The first set of test sentences (the "in-domain test set") was selected directly from the corpus at regular intervals, resulting in 1,000 test sentences contain-ing 8,113 tokens in the Slovenian part and 9,508 tokens in the English part.

For the second test we used test sentences selected from one of the components of the IJS-ELAN corpus, the ORWL file (Orwell's 1984), which is a text type significantly different from the rest of the ACQUIS corpus; hence we refer to it as the "out-of-domain test set." We randomly selected 1,000 sentences, containing 10,622 tokens in the Slovenian part and 12,363 tokens in the English part. This setup enabled us to evaluate the system performance with test sentences that were highly correlated to the training data ("in-domain setup"), as well as with those that had low correlation to the training material ("out-of-domain setup").

4.2 VoiceTRAN Experiment

The second SMT experiment was performed on the first version of the restricted-domain VoiceTRAN parallel corpus, collected within the VoiceTRAN project and limited to government texts of the Slovenian Ministry of Defense. In comparison to the ACQUIS corpus, the VoiceTRAN sentences are more homogeneous and cover a more compact domain. Again, all tokens are annotated with automatically assigned context-disambiguated lemmas and MSD (and POS) information. The translation model was trained with 2,508 sentences, containing 23,100 tokens in the Slovenian part and 26,900 tokens in the English part. A conventional bilingual dictionary of military terminology with 18,423 entries – including multiword expressions – was added as an option.

The Slovenian language model was trained on the Slovenian part of the Voice-TRAN corpus, with the optional addition of the FDV-IJS Slovenian monolingual corpus from the same application domain, containing 302,000 sentences and 3.19 million tokens. The test sentences for the "in-domain" VoiceTRAN corpus experiment were selected from the VoiceTRAN corpus at regular intervals, resulting in 278 test sentences with 2,554 tokens in the Slovenian part and 2,951 tokens in the English part. Due to the modest size of the VoiceTRAN corpus, no "out-of-domain" tests were performed.

5 Performance Evaluation

To measure the "closeness" between the MT-generated hypothesis and human reference translations, standard objective MT metrics were used:

- GTM: General Text Matcher [11],
- NIST: a metric proposed by NIST [12], and
- METEOR: Metric for Evaluation of Translation with Explicit ORdering [13].

The scores obtained for the BLEU metric were too small to be considered reliable and they are not presented in this paper.

The SMT evaluation efforts were centered on three system variations:

1. choice of the SMT approach: "simple" vs. "MSD recombination,"
2. no dictionary vs. conventional dictionaries: a general dictionary in the JRC-ACQUIS experiment and an in-domain terminology dictionary in the VoiceTRAN experiment,
3. application domain of the test sentences: in-domain, out-of-domain.

In both tests, the performance of of the VoiceTRAN SMT system variations was compared to the efficiency of the baseline Presis rule-based system.

5.1 JRC-ACQUIS Test Results

Table 1 presents the average evaluation scores (GTM, NIST, and METEOR) of the tested SMT system configuration versus the baseline system, both for the "in-domain" as well as for the "out-of-domain" setup.

Table 1. JRC-ACQUIS evaluation scores

"In-domain" setup	GTM	NIST	METEOR
SMT: "simple"	0.36	0.91	0.29
SMT: "MSD recombination"	0.33	0.78	0.26
baseline: Presis	0.29	0.71	0.23
"Out-of-domain" setup	**GTM**	**NIST**	**METEOR**
SMT: "simple"	0.17	0.52	0.11
SMT: "MSD recombination"	0.18	0.54	0.12
baseline: Presis	0.32	0.97	0.23

In the "in-domain setup," the simple SMT approach performed slightly better than the "MSD recombination" method. Both SMT configurations outperformed the baseline rule-based MT system.

As expected, the baseline rule-based system was the best in the "out-of-domain setup," whereas here the "MSD recombination" SMT system performed better than the simple SMT system. Because the sources of the training data had been automatically tagged with lemmas and MSDs, the resulting imperfections in the training material had negative effects, especially on the "MSD recombination" translation method results. Therefore, we intend to retag the source corpora in the continuation of the project.

5.2 VoiceTRAN Test Results

Table 2 presents the average evaluation scores of the tested SMT system and training set configuration versus the rule-based baseline system for the VoiceTRAN test setup. Both tested versions of the VoiceTRAN SMT system performed better than the baseline rule-based system, which had also adapted to the VoiceTRAN application domain prior to the experiments. The addition of the in-domain terminology dictionary resulted in a substantial drop in the WER score and a rise in the other metric scores.

Table 2. VoiceTRAN evaluation scores

VoiceTRAN setup	GTM	NIST	METEOR
SMT: no dictionary	0.27	0.65	0.27
SMT: in-domain dictionary	0.40	0.86	0.35
baseline: domain-adapted Presis	0.31	0.69	0.26

Due to the modest size of the VoiceTRAN corpus, only in-domain tests were performed and only the simple translation method was evaluated. Clearly, the first version of the VoiceTRAN corpus does not contain enough training material, and there-fore an upgrade of the corpus is planned.

6 Discussion

The values obtained for measuring translation quality using the standard metrics mentioned were generally low. However, they served well enough to indicate whether we have improved the system by introducing changes into individual components of the SMT system.

In comparison to the simple translation method, the "MSD recombination" translation method did not perform well for test sentences extracted from the unprocessed corpus in the first experiment. Similar results were obtained in the VoiceTRAN experiment on VoiceTRAN corpus test sentences. In both cases, the test sentences were in-domain. The "MSD recombination" method performed better when ORWL test sentences were used, proving its potential for translation of out-of-domain sentences.

The simple translation method apparently adapted well to inflected Slovenian words, some of which were frequent enough in the training material to allow for sufficient training of the statistical model. As a consequence, when testing in-domain test sentences well correlated to the training corpus, the test set translations were translated rather well.

As expected, the "MSD recombination" translation method performed better when translating texts that were very different from the training sentence set, as was the case with the ORWL test corpus.

In both in-domain test setups, the VoiceTRAN SMT system outperformed the baseline conventional rule-based MT system. On the other hand, in the out-of-domain test setup the performance of the "MSD recombination" SMT system was also not far behind the baseline system. Therefore we may conclude the following: when lexical coverage of a SMT system has dropped sufficiently (approaching the out-of-domain scenario), the simple translation model can be replaced by the "MSD recombination" translation model, especially when translating into a highly inflected language.

Finally, we would like to mention that we found that the applied evaluation metrics are not suitable for evaluating translations from English into Slovenian, i.e. from a low-inflected source language into a highly-inflected target language. These metrics are all based on an exact comparison of entire words, which works well for English. Due to the rich inflectional paradigms in Slovenian, words that are semantically correctly translated but have the wrong ending receive a calculated score of zero. For example, a method that attributes score points for finding a correct word stem in the target language would provide a much better translation quality estimation. Nevertheless, the evaluation methods used were suitable for the purposes of our research because we were only looking for an indicator showing improvement or deterioration when using various MT systems and training set configurations.

7 Conclusion

Various configuration variations of the VoiceTRAN SMT system were presented and evaluated. In both in-domain test setups, the VoiceTRAN SMT system outperformed the baseline conventional rule-based MT system.

To increase the generalization capability of the translation model for lower-coverage out-of-domain test sentences, an "MSD-recombination" approach was pro-posed. This approach not only allows a better exploitation of conventional translation models, but also performs well in the more demanding translation direction; that is, into a highly inflected language. Using this approach in the out-of-domain setup of the English-Slovenian JRC-ACQUIS task, we have achieved significant improvements in translation quality.

Acknowledgments. The work presented in this paper was performed as part of the VoiceTRAN project supported by the Slovenian Ministry of Defense and the Slovenian Research Agency under contract No. M2-0132.

References

1. Vičič, J.: Avtomatsko prevajanje iz slovenskega v angleški jezik na osnovi statističnega strojnega prevajanja (Automatic SMT: Slovenian-English), Masters' thesis, University of Ljubljana, Slovenia (2002)
2. Romih, M., Holozan, P.: Slovenian-English Translation System. In: Proceedings of the LTC 2002, Ljubljana, Slovenia, p. 167 (2002)
3. Žganec Gros, J., Gruden, S., Mihelič, F., Erjavec, T., Vintar, Š., Holozan, P., Mihelič, A., Dobrišek, S., Žibert, J., Logar, N., Korošec, T.: The VoiceTRAN Speech Translation Demonstrator. In: Proceedings of the IS-LTC 2006, Ljubljana, Slovenia, pp. 234–239 (2006)
4. Sepesy Maučec, M., Kačič, Z.: Statistical machine translation from Slovenian to English. Journal of Computing and Information Technology 15(5), 47–59 (2007)

5. Steinberger, R., Pouliquen, B., Widiger, A., Ignat, C., Erjavec, T., Tufis, D., Varga, D.: The JRC-Acquis: A multilingual aligned parallel corpus with 20+ languages. In: Proceedings of the Fifth International Conference on Language Resources and Evaluation, LREC 2006, ELRA, Paris, pp. 2142–2147 (2006)
6. Erjavec, T.: The IJS-ELAN Slovene-English Parallel Corpus. International Journal of Corpus Linguistics 7(1), 1–20 (2002)
7. Erjavec, T.: Compilation and Exploitation of Parallel Corpora. Journal of Computing and In-formation Technology 11(2), 93–102 (2003)
8. Och, F.J., Ney, H.: A Systematic Comparison of Various Statistical Alignment Models. Computational Linguistics 29(1), 19–51 (2003), http://www.fjoch.com/GIZA++.html
9. Rosenfeld, R.: The CMU Statistical Language Modeling Toolkit, and Its Use in the 1994 ARPA CSR Evaluation. In: Proceedings of the ARPA SLT Workshop, http://www.speech.cs.cmu.edu/SLM/toolkit.html
10. Germann, U.: Greedy Decoding for Statistical Machine Translation in Almost Linear Time. In: Proceedings of the HLT-NAACL- 2003 (2003), http://www.isi.edu/licensed-sw/rewrite-decoder/
11. Turian, J.P., Shen, L., Dan Melamed, I.: Proteus Technical Report #03-005: Evaluation of Machine Translation and its Evaluation, http://nlp.cs.nyu.edu/eval/
12. Doddington, G.: Automatic Evaluation of Machine Translation Quality using N-gram Cooccurrence Statistics. In: Proceedings of the 2nd Human Language Technologies Conference, San Diego (2002)
13. Banerjee, S., Lavie, A.: METEOR: An Automatic Metric for MT Evaluation with Improved Correlation with Human Judgments. In: Proceedings of the ACL Workshop on Intrinsic and Extrinsic Evaluation Measures for Machine Translation and/or Summarization at the 43rd Annual Meeting of the Association of Computational Linguistics, Ann Arbor, Michigan (2005)

A Computational Framework to Integrate Different Semantic Resources

Qiang Zhou

Centre for Speech and Language Technologies, Research Institute of Information Tech.
CSLT, Division of Technology Innovation and Development
Tsinghua National Laboratory for Information Science and Technology
Tsinghua University, Beijing 100084, P. R. China
zq-lxd@mail.tsinghua.edu.cn

Abstract. In recent years, many large-scale semantic resources have been built in the NLP community, but how to apply them in real text semantic parsing is still a big problem. In this paper, we propose a new computational framework to deal with this problem. Its key parts are a lexical semantic ontology (LSO) representation to integrate abundant information contained in current semantic resources, and a LSO schema to automatically reorganize all this semantic knowledge in a hierarchical network. We introduce an algorithm to build the LSO schema by a three-step procedure: to build a knowledge base of lexical relationship, to accumulate all information in it to generate basic LSO nodes, and to build a LSO schema through hierarchical clustering based on different semantic relatedness measures among them. The preliminary experiments have shown promising results to indicate its computability and scaling-up characteristics. We hope it can play an important role in real world semantic computation applications.

1 Introduction

The goal of semantic computation is to identify the correct word senses and their relations in real world texts, resulting in conceptual structures that reflect various levels of semantic interpretations. Two parsing tasks are designed to deal with these problems in NLP community: word sense disambiguation (WSD) and semantic role labeling (SRL). These parsing techniques have relied on some best available language resources (LRs), such as WordNet [8], VerbNet [6], FrameNet [2] or PropBank [5], the state of the art performances are still not good enough for semantic computation in large scale real texts. An important reason is that none of these parsers can make full use of all the useful information contained in these LRs effectively.

To deal with the problem, many researches have devoted to the integration of various LRs in recent years. Shi and Mihalcea (2005) proposed a unified knowledge base to integrate FrameNet, VerbNet and WordNet [9]. Subba et. al. (2006) explored the method to integrate a robust parser with a lexicon and ontology derived from VerbNet and CoreLex [10]. Ide (2006) described a method for mapping FrameNet lexical units to WordNet senses [4]. Giuglea and Moschitti (2006) proposed a new semantic role labeling method through the integration of FrameNet, VerbNet and ProbBank [3]. All of these approaches are LR-dependent. They can not be easily expanded to other available LRs in a universal method.

P. Sojka et al. (Eds.): TSD 2008, LNAI 5246, pp. 243–250, 2008.
© Springer-Verlag Berlin Heidelberg 2008

In the paper, we propose a new solution for language resource integration and application. Its key part is the knowledge representation unit called Lexical Semantic Ontology (LSO). Each LSO can provide the complete meaning description for a word or a word group, through its sense pool and lexical relationship pool. It can be regarded as an integrated interface to link different language resources, including semantic lexicons and annotated corpus. Several LSO nodes can be further automatically constructed as a LSO schema through different semantic relatedness measures among them. Ideally, the LSO schema is a hierarchical semantic network that can re-categorize or reorganize all semantic knowledge contributed by different language resources integrated under LSO nodes.

Unlike other LR integration methods, the knowledge description framework of LSO and LSO schema are LR-independent. Each new LR can be easily integrated into the framework, through setting suitable links between their sense descriptions or lexical relationship pairs with LSO nodes and providing different semantic relatedness measures based on the LR. The LSO schema can be reconstructed to reflect the contribution of the new LR to the overall knowledge representation in the hierarchical network. Due to its computable and expandable characteristics, LSO schema can be used as a new computational lexical resource to bridge the gap between current semantic computation requirements in large-scale real texts and manually-developed semantic resources.

2 LSO and LSO Schema

Informally, a lexical semantic ontology is designed to describe different facets of the meaning of a word or a word group. Its main parts are as follows: (1) Sense pool, which comprises all possible meaning descriptions of the keywords, defined in different semantic lexicons. (2) Lexical relationship pool, which contains all possible lexical units that have special lexical relationship with the keywords in real text sentences.

A more formal definition for a LSO record can be as follows:

Definition 1: A LSO record is represented as a triple <ID, SVs, LRVs>, where:

1. ID is the LSO index. Each LSO record will be given a unique ID to represent its different types. Two types of LSO nodes are described in our current LSO framework. One is the Word-LSO node, which focuses on the meaning descriptions for a keyword. The other is the Sit-LSO node, which focuses on the meaning descriptions for a keyword group with same situation semantic codes, for example, all verbs describe a good-transferring situation. They show different description granularities.
2. SVs are vectors for the sense descriptions in different semantic lexicons. Each lexicon will be given a special SV with the following forms: $\{<s_l, w_l>\}$, where $l \in [1, n]$, n is the total number of all possible senses in the lexicon, s_l is its No. l sense description, and w_l is the frequency distributional ratio of its lexical relationship pairs (LRPs), i.e. $w_l = LRP\_Freq(S_l)/\sum_{l=1}^{n} LRP\_Freq(S_l)$. The intuition assumption under the formula is that the more language usage examples a sense have, the more it will contribute to the LSO record.
3. LRVs are different lexical relationship vectors of the key word senses. Two important lexical relationships are considered in our LSO framework: (1) syntactic

dependency relations, including subject-verb, verb-object, modifier-noun, verb-compliment, etc. (2) semantic concept relations, including semantic role and attribute-object relation, etc. They cover different lexical relations between keywords and other related words.

A LSO schema is a semantic network connected by LSO nodes with different semantic relatedness. Here, semantic relatedness is a more general concept than similarity. It includes lexical relationships such as synonym, meronym, antonym and any kind of functional relationship or frequent association. Due to the information integration advantage of LSO representation, we can flexibly select different semantic relatedness computation methods, including many resources-based measures designed for different semantic lexicons [1], and some distributional similarity based on co-occurrence lexical context [11].

Unlike many manually-summarized semantic networks, the LSO schema can be automatically constructed through the following procedures: (1) automatically connect two LSO nodes through different semantic relatedness measures; (2) automatically cluster similar LSO nodes to form a hierarchical sub-network. Starting from a mini LSO schema prototype, it can be evolved to a very large and complex network through incremental increases of new LSO nodes and introductions of new semantic relatedness computation methods. Its computable and expandable characteristics can provide us with good opportunities to explore some new knowledge-driven techniques to deal with current semantic computation problems.

A LSO schema is formally defined as follows:

Definition 2: A LSO schema can be represented as a connected network [V, E]. V is the LSO node set. E is the related edge set, where each edge with the form: <ID_i, ID_j, *RT*, *RV*> to indicate that there is a semantic relatedness, whose tag is *RT* and its related value is *RV*, between *No. i* and *No. j* LSO nodes.

3 Building LSO Schema

The big obstacle to build LSO schema is that we can't automatically extract all the needed information from current LRs directly. Although many semantic lexicons provide us with detailed meaning descriptions of a lexical entry, none of them can give us the mapping link of these meaning descriptions in other lexicons. Meanwhile, some annotated corpus can provide us with several useful lexical pairs with detailed descriptions of different syntactic or semantic relationships, or even the syntax-semantics linking information. But none of them can give us complete meaning descriptions of different lexical units in pairs. So we have to seek a new solution to deal with these problems.

We select lexical relationship pair as the breakthrough. For each lexical pair <w_i, w_j>, we extract all their usage sentences from the annotated corpus, describe their syntactic and semantic relationship in these sentences, and tag the meaning descriptions of the keywords among them with the suitable meaning definitions in all available semantic lexicons. After that, we can obtain a new lexical relationship pair (LRP) = <w_i, w_j> + <syntactic and semantic relations> + <all meaning descriptions of keywords>. All of these LRPs form a new lexical relationship knowledge base (LRKB) for LSO building.

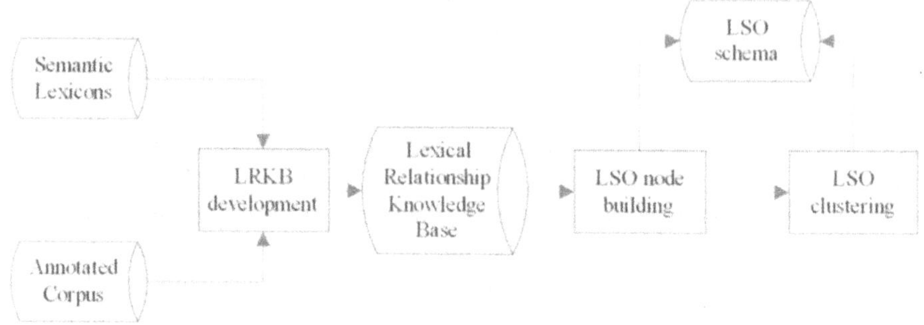

Fig. 1. Overview of the LSO building framework

The following is an annotated example for a LRKB record: "*goumai*(buy) HN1;CL1; XH1 PAT *goumai*(buy) *caidian*(television) PO TCT V N". It contains the following description information: (1) Three 'V-N' pairs annotated with 'Predicate-Object' (PO) syntactic dependency relation are extracted from Tsinghua Chinese Treebank (TCT) [14]; (2) The meaning description IDs of the key verb '*goumai*(buy)' are HN1, CL1, and XH1, corresponding to its different meaning descriptions in semantic lexicons: Hownet [15], CiLin [16] and XianHan [17] respectively; (3) The semantic role of the related noun '*caidian*(television)' is patient (PAT) in the good-transferring situation represented by the key verb *goumai*(buy).

Based on the LRKB, we can build LSO schema through the following three-step approach: (1) to generate LSO leaf nodes with detailed information of sense pool and lexical relationship pool; (2) to design efficient algorithm to compute the semantic relatedness of two LSO nodes; (3) to build LSO schemas based on the above semantic relatedness measures. Figure 1 shows the overall structure of the LSO building framework. Some detailed processing strategies and algorithms for LRKB development and LSO schema building can be found in [12] and [13]. I will only briefly introduce about a key technique among them: How to compute the semantic similarity of two LSO nodes.

The semantic similarity of two LSO nodes is the weighted arithmetic average of their sense and distributional similarity.

$$Sim(LSO_i, LSO_j) = \alpha Sense\_Sim(LSO_i, LSO_j) + (1 - \alpha) Dist\_Sim(LSO_i, LSO_j) \quad (1)$$

We can adjust different weight α to control the contribution proportion of two types of similarities. The sense similarity is the arithmetic average of the sense similarities from different semantic lexicons. Its computation formula is as follows:

$$Sense\_Sim(LSO_i, LSO_j) = \frac{1}{N} Sense\_Sim_k(LSO_i, LSO_j) \quad (2)$$

N is the total number of the semantic lexicons that can provide available sense similarities, $k \in [1, N]$. The semantic similarity of two sense description s_{ka} and s_{kb} in the *No. k* semantic lexicon can be computed as follows:

$$Sense\_Sim_k(LSO_i, LSO_j) = \sum_{a=1}^{n}\sum_{b=1}^{m} S\_Sim(s_{ka}, s_{kb}) \cdot w_{ka} \cdot w_{kb} S\_Sim(s_{ka}, s_{kb}) \quad (3)$$

The distributional similarity is computed through cosine between two lexical relationship vectors.

4 Experimental Results

We selected a subset of verb concepts that describe possession relations and transferring actions, to test the feasibility of the above LSO building framework. By analyzing all corresponding concepts and words described in current popular Chinese semantic lexicons, we manually summarized about 900 Chinese verbs and 65 situation descriptions for these possession-related V concepts. Then, we extracted 31,565 'V-N' relationship pairs with suitable syntactic relatedness in annotated corpus. Each of them were manually annotated with suitable semantic information, including the meaning description IDs of the key verbs in different semantic lexicons and the semantic role tags of the related nouns, to form a new LRKB record.

In order to build an unbiased LSO node, we must have as much as possible LRPs for a keyword. The distributional statistics of the key verbs with different LRPs in current LRKB show that about 71% key verbs have only 5.5 LRPs averagely. The data sparseness problem will bring bad effect to the description objectiveness of the generated LSO node. The other 29% key verbs can have about 105 LRPs averagely. Among them, 44 key verbs have more than 100 LRPs, and 10 key verbs have more than 500 LRPs. They can provide large enough lexical relationship knowledge for LSO relatedness computation. So we set the LRP threshold $\beta = 20$ for Verb-LSO node (i.e. a word-LSO node for a verb) building, and generated 264 Verb-LSO leaf nodes and 65 Sit-LSO leaf nodes. After hierarchical clustering on these LSO nodes, we obtained two LSO schemas: a Verb-LSO schema with 410 nodes and a Sit-LSO schema with 87 nodes. We set weight $\alpha = 0.5$ to tradeoff different contributions of sense and distributional similarity.

Figure 2 shows a snapshot of the Verb-LSO schema for 'Buy' and 'Sell' related verbs. Each oval represents a LSO node, which comprises the following indication information: its key verbs, the total number of the related nouns in its lexical relationship pool. The clustered nodes among them are showed with green background. Two types of related edges are used in the figure. The directed full lines linked from son nodes to father nodes represent hyponym relatedness. The broken lines annotated with similarity values represent the top-5 similarity measures with other LSO nodes for a special LSO node.

It can be seen that:

(1) Some fine-drawn synonym and antonym sets were automatically constructed in the Verb-LSO schema.

For the 'Buy'-related situations, we found a synonym set with the following verbs: *goumai*(buy), *caigou*(go shopping), *gouzhi*(purchase), *mai1*(buy), and *gou*(buy). For the 'Sell'-related situations, we found a synonym set with the following verbs: *mai4*(sell), *chushou*(offer for sale), *tuixiao*(market), *fanmai*(peddle), and *zhanxiao*(display and sell). They show strong intensiveness because almost all their top-5 similarity edges are linked to the LSO nodes in the same synonym set. Meanwhile, antonym relatedness between these two synonym sets was also built due to their distributional similarity in real texts.

Compared with following flat synonym sets defined in Hownet [15] and CiLin [16]: (1) Hownet: {*mai1*, *goumai*, *gouzhi*, *goujin* (purchase), *qianggou* (rush to purchase), *caigou*, *gou* ...... }; (2) CiLin: {*mai1*, *gouzhi*, *goumai*, *gou*, *zhiban*(buy), *caigou*,... }, the similarity hierarchies in the sub-schema show outstanding computational advantages. They can provide strong supports for knowledge-driven WSD and SRL of these verbs in real text sentences.

(2) Some new lexical relationship pairs were automatically generated in the clustered nodes.

Through hierarchical clustering of two similar LSO nodes, all the key verbs and related nouns covered by these nodes were combined to form a new sense pool and lexical relationship pool for the clustered LSO node. Therefore, some new lexical relationship pairs were obtained through different combinations of these key verbs and related nouns. For example, in the clustered node: {*goumai*, *caigou*, *gouzhi*, *mai1*, *gou*}, we can expect to find $5 \cdot 630 = 3150$ different lexical relationship pairs. The number is four-times bigger than the total number 782 of LRPs in all 5 single LSO nodes of {*goumai*(buy)}, {*caigou* (go shopping)}, {*gouzhi*(purchase)}, *mai1*(buy)}, and {*gou*(buy)}. They will provide us with new processing data to deal with data sparseness problem in real text semantic computation.

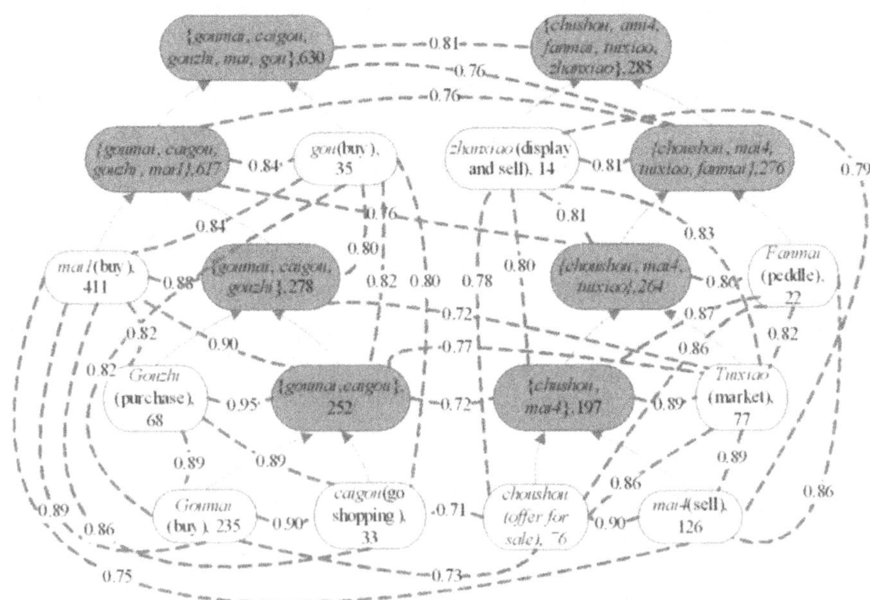

Fig. 2. A snapshot of Verb-LSO schema

5 Related Work

All the language resources have their strengths and shortcomings. So a basic idea for language resource integration is to create suitable connectivity among them so as to combine their strengths and eliminate their shortcomings.

Shi and Mihalcea (2005) proposed some automatic and semi-automatic methods to link three lexical resources: FrameNet, VerbNet and WordNet and built an improved resource in which (1) the coverage of FrameNet was extended, (2) the VerbNet lexicon was augmented with FrameNet frames and semantic roles, and (3) selectional restrictions were implemented using WordNet semantic classes. The synergistic exploration of these lexical resources showed effectiveness in building a robust semantic parser.

Giuglea and Moschitti (2006) proposed a similar method to connect VerbNet and FrameNet by mapping the FrameNet frames to the VerbNet intersective Levin class. Therefore, the advantages of the detailed semantic roles defined in FrameNet corpus and the syntactic behavior of verbs defined in VerbNet lexicon were integrated. They also used the PropBank corpus which is tightly connected to the VerbNet lexicon to increase the verb coverage and tested the effectiveness of their semantic-role labeling algorithm.

Based on the assumption that linguistic context is an indispensable factor affecting the perception of a semantic proximity between words, Li, Roth and Tu (2003) developed a contextual-sensitive lexical knowledge base called PhraseNet. The basic unit in PhraseNet is a *conset*: a word in its context, together with all relations associated with it. All these consets were automatically generated from real world corpora through an automatic parser. Meanwhile, PhraseNet can makes use of WordNet as an important knowledge source and enhance a WordNet synset with its contextual information and refine its relational structure by maintaining only those relations that respect contextual constraints. Therefore, a new knowledge base was built through integrating the syntactic relation of words extracted from real world corpora and their sense descriptions extracted from WordNet [7].

6 Conclusions

This paper proposed a unified computational framework to integrate different language resources. Its key parts are as follows: (1) A LSO knowledge representation to integrate all available knowledge in different LRs; (2) A LSO relatedness measure to tradeoff different semantic relatedness contributions from different LRs; (3) A LSO schema to reorganize semantic knowledge from different LRs in a hierarchical semantic network. Some preliminary experiments on verb concept subsets of possession relations and transferring actions have shown promising results.

There are still many key issues to be explored in the future research. First, we can add some other relatedness measures into current computational framework. They can supply new semantic related edges in current LSO schema to form a complete semantic network. Second, we can explore some knowledge-driven WSD and SRL methods based on current LSO schema. They can play following two-fold functions. On the one hand, the semantic computational experiments in real text sentences can reveal some shortcomings of current LSO schema. They can provide us with valuable information to improve current LSO description framework. On the other hand, these knowledge-driven methods can be applied to other new lexical pairs to build a semi-automatic LRKB development platform. Therefore, we can quickly and efficiently build a new large-scale LRKB for new LSO schema.

Acknowledgements. This work was supported by the National Natural Science Foundation of China (No. 60573185) and the National Hi-Tech Research and Development Program (863 plan) of China (No. 2007AA01Z173). Thanks three anonymous reviewers for their good comments and advices for the paper.

References

1. Budanitsky, A., Hirst, G.: Evaluating WordNet-based Measures of Lexical Semantic Relatedness. Computational Linguistics 32(1), 1–35 (2006)
2. Fillmore, C.J., Wooters, C., Baker, C.F.: Building a Large Lexical Databank Which Provides Deep Semantics. In: Proc. of the Pacific Asian Conference on Language, Information and Computation, Hong Kong (2001)
3. Giuglea, A.-M., Moschitti, A.: Semantic role labeling via FrameNet, VerbNet and PropBank. In: Proc. of the 21st International Conference on Computational Linguistics and 44th Annual Meeting of the ACL, Sydney, Australia, pp. 929–936 (2006)
4. Ide, N.: Making senses: bootstrapping sense-tagged lists of semantically-related words. In: Gelbukh, A. (ed.) CICLing 2006. LNCS, vol. 3878, pp. 13–27. Springer, Heidelberg (2006)
5. Kingsbury, P., Palmer, M., Marcus, M.: Adding Semantic Annotation to the Penn TreeBank. In: Proceedings of the Human Language Technology Conference, San Diego, California (2002)
6. Kipper, K., Dang, H.T., Palmer, M.: Class-Based Construction of a Verb Lexicon. In: Proc. of Seventeenth National Conference on Artificial Intelligence (AAAI-2000), Austin TX, USA, July 30–August 3 (2002)
7. Li, X., Roth, D., Tu, Y.C.: PhraseNet: Towards Context Sensitive Lexical Semantics. In: Proc. of the Seventh CoNLL Conference held at HLT-NAACL 2003, Edmonton, pp. 87–94 (2003)
8. Miller, G.A., Fellbaum, C.: Semantic Network of English. In: Levin, B., Pinker, S. (eds.) Lexical & Conceptual Semantics. Elsevier Science Publishers, Amsterdam (1991)
9. Shi, L., Mihalcea, R.: Putting Pieces Together: Comining FrameNet, VerbNet and WordNet for Robust Semantic Parsing. In: Gelbukh, A. (ed.) CICLing 2005. LNCS, vol. 3406, pp. 100–111. Springer, Heidelberg (2005)
10. Subba, R., Eugenio, B.D., Terenzi, E.: Building lexical resources for PrincPar, a large coverage parser that generates principled semantic representations. In: Proc. of LREC 2006 (2006)
11. Weeds, J.E.: Measures and applications of lexical distributional similarity. Ph.D. thesis, University of Sussex (September 2003)
12. Zhou, Q.: Develop a Syntax Semantics Linking Knowledge Base for the Chinese Language. In: Proc. of 8th Chinese Lexical Semantics Workshop, Hong Kong, May 21–23 (2007)
13. Zhou, Q.: Building Lexical Semantic Ontology for Semantic Computation. Technical Report 07-10, RIIT, Tsinghua University (2007)
14. Zhou, Q.: Build a Large-Scale Syntactically Annotated Chinese Corpus. In: Matoušek, V., Mautner, P. (eds.) TSD 2003. LNCS (LNAI), vol. 2807, pp. 106–113. Springer, Heidelberg (2003)
15. Dong, Z.D., Dong, Q.: Hownet, `www.keenage.com`
16. Jiaju, M., et al.: TongYiCi CiLin. Shanghai dictionary press, Shanghai (1983)
17. Cidian, X.H. (Contemporary Chinese Dictionary). Business Press, Beijing (1991)

Part III

Speech

"**Speech**: the expression of or the ability to express thoughts and feelings by articulate sounds: *he was born deaf and without the power of speech*."

NODE (The New Oxford Dictionary of English), Oxford, OUP, 1998, page 1788, meaning 1.

Age Determination of Children in Preschool and Primary School Age with GMM-Based Supervectors and Support Vector Machines/Regression

Tobias Bocklet, Andreas Maier, and Elmar Nöth

Institute of Pattern Recognition, University of Erlangen-Nuremberg, Germany

Abstract. This paper focuses on the automatic determination of the age of children in preschool and primary school age. For each child a *Gaussian Mixture Model* (GMM) is trained. As training method the *Maximum A Posteriori* adaptation (MAP) is used. MAP derives the speaker models from a *Universal Background Model* (UBM) and does not perform an independent parameter estimation. The means of each GMM are extracted and concatenated, which results in a so-called GMM supervector. These supervectors are then used as meta features for classification with *Support Vector Machines* (SVM) or for *Support Vector Regression* (SVR). With the classification system a precision of 83 % was achieved and a recall of 66 %. When the regression system was used to determine the age in years, a mean error of 0.8 years and a maximal error of 3 years was obtained. A regression with a monthly accuracy brought similar results.

1 Introduction

Automatic speech recognition of children's speech is much more sophisticated than speech recognition on adult's speech. This effect is often amplified by the lack of training data. Nevertheless, some approaches exist which try to compensate this drawback [1]. One remaining problem is the strong anatomic alteration of the vocal tract of children within a short period of time. An idea to face this problem is to use different acoustic models for different age classes of children [2]. The most appropriate acoustic model has to be selected before the automatic speech recognition can be performed. If the age of a child is not known a priori, the age has to be estimated out of the child's voice. This work focuses on this task.

In [3] it has been shown, that age recognition with SVM and 7 gender dependent classes outperforms different other classification ideas. The classification results of the SVM idea were in the same range as humans, and the precision even better. In this work we pick up the SVM idea and apply it to children in preschool and primary school age. As the next step we substitute the SVM in our recognition system by the SVR and are able to predict the age directly, i.e. no distinction into classes has to be done.

The outline of this paper is as follows: Section 2 describes the used corpora on which the systems are trained and tested. Section 3 gives a brief introduction into the GMM supervector idea, describes the employed cepstral features and depicts how the speaker models (GMMs) are trained. The basic ideas of SVM and SVR are summarized in Section 4 and Section 5. Section 6 shows the training and working sequence of the created systems. The results are summarized in Section 7.

P. Sojka et al. (Eds.): TSD 2008, LNAI 5246, pp. 253–260, 2008.
© Springer-Verlag Berlin Heidelberg 2008

2 Corpora

All children read the so-called PLAKSS test [4]. The test consists of 99 words. These words contain all phonemes/sounds of the German language and the most important conjunctions of them in different positions, i.e at the beginning, the end or within a word. The words are shown on a screen and are represented by pictograms. If the children are able to read they can used the shown word, otherwise they have to name the shown pictograms. Thus the PLAKSS text is very qualified for children in preschool and primary school age, which mostly are not able to read. All data was recorded with a system developed in our lab: the the PEAKS system [5].

For training and testing three different datasets have been available. One contains recordings of 38 children in preschool age. The data was recorded in a local preschool. The mean age of the children of this dataset is 5.7 ± 0.7 years.

The second dataset contains data of children in primary school age with a mean age 8.5 ± 1.4 years. The recordings took place in a local elementary school in August 2006 and April 2007. All in all 177 children were recorded in this school.

The third set of recordings was collected in an elementary school in Hannover (April 2007). 128 children have been recorded with a mean age of 8.6 ± 1.1 years.

A subset of each of these recordings was used for the experiments described in this paper. A total amount of 212 children has been chosen from all three different corpora to train the two systems. The children have been selected in order to create an almost age-balanced training set with respect to the following five classes:

- < 7 years,
- 7 years,
- 8 years,
- $9 + 10$ years,
- > 10 years.

The mean age of the training subset was 8.3 ± 2.4 years. The system was tested on a 100 speaker subset. Again the children have been selected to create an age-balanced test set. The average age in the test set was 8.4 ± 2.3.

3 GMM Supervector Idea – Metafeatures for the SVM

The creation of the GMM supervectors is shortly described in this section. The idea was first published in [6]. First feature vectors are extracted out of the speech signal. These features are then employed to train GMMs with M Gaussian densities [7]. Each GMM represents a different child. The mean values of each GMM are concatenated and used as meta features in SVMs. Each supervector is then labeled with the correct age. So each child it represented by one supervector. The principle of GMM supervector creation is displayed in Fig. 1.

3.1 Feature Extraction

As features the commonly used *Mel Frequency Cepstrum Coefficients* (MFCCs) are used. They examine the 18 Mel-bands and consider a short time window of 16 ms with

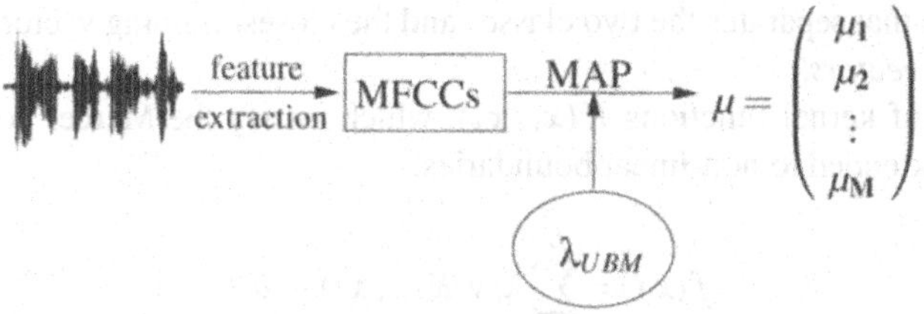

Fig. 1. Training principle of GMM supervectors. After feature extraction the speaker model (GMM) is created. The concatenation of its mean vectors creates the GMM supervector. For details see [6].

a time shift of 10 ms. The first 12 MFCCs are used and the first order derivatives are computed by a regression line over 5 consecutive frames. Th final feature vector has 24 components (log energy, MFCC(1)–(11) and the first derivatives).

3.2 Gaussian Mixture Models

The basic idea behind the GMM supervector approach is to model every speaker with a different GMM, which is a standard approach in speaker identification/verification [8]. A GMM is composed of M unimodal Gaussian densities:

$$p(\mathbf{c}|\boldsymbol{\mu}, \boldsymbol{\Sigma}) = \sum_{i=1}^{M} \omega_i \, p_i(\mathbf{c}|\boldsymbol{\mu}_i, \boldsymbol{\Sigma}_i) \tag{1}$$

$$= \sum_{i=1}^{M} \omega_i \cdot \frac{1}{(2\pi)^{D/2} |\boldsymbol{\Sigma}_i|^{(1/2)}} e^{-(1/2)(\mathbf{c}-\boldsymbol{\mu}_i)^T \boldsymbol{\Sigma}_i^{-1} (\mathbf{c}-\boldsymbol{\mu}_i)}, \tag{2}$$

where ω_i denotes the weight, $\boldsymbol{\Sigma}_i$ the covariance matrix and $\boldsymbol{\mu}_i$ the mean vector of the i-th Gaussian density.

3.3 Training

After extraction of the MFCCs a UBM is created with all the available training data, using the EM algorithm [9]. The UBM is then employed as an initial model either for an EM training or for the MAP adaptation [10]. The MAP adaptation adapts the UBM to the speaker dependent training data, so that for every child a GMM is created. MAP adaptation calculates the Gaussian mixture components by a single iteration step and combines them with the UBM parameters. The number of EM iterations for the UBM training was set to 10.

4 Classification with Support Vector Machines

The SVM [11] performs a binary classification $y \in (-1, 1)$ based on a hyperplane separation. The separator is chosen in order to maximize the distances (margin) between the hyperplane that separates the two classes and the closest training vectors, which are called *support vectors*.

By the use of kernel functions $K(\boldsymbol{x}_i, \boldsymbol{x}_j)$, which satisfy the Mercer condition, the SVM can be extended to non-linear boundaries:

$$f(\boldsymbol{x}) = \sum_{i=1}^{L} \lambda_i y_i K(\boldsymbol{x}, \boldsymbol{x}_i) + d \tag{3}$$

where y_i are the target values and $\boldsymbol{x}_i$ are the support vectors. λ_i have to be determined in the training process. L denotes the number of support vectors and d is a (learned) constant. One task of this paper is a 5-class age classification. So the binary SVM has to be extended. The simplest way is to separate each age class from all others. Therefore $N \times (N-1)/2$ classifiers are created, each of them separating two classes. The scores of these classifiers are then combined.

5 Prediction with Support Vector Regression

The general idea of regression is to use the vectors of the training set to approximate a function, which tries to predict the target value of a given vector of the test set. SVMs can also be used for regression. The method is then called SVR. A detailed description is given in [12]. Due to the fact, that no binary classification has to be performed, the so called "ε-tube" is defined. ε describes the deviation, which is allowed between the training vectors and the regression line (in positive and negative direction). Similar to the classification task, not all training vectors are needed to select the most appropriate ε-tube, but only a subset of them, i.e the support vectors. These vectors lie outside the ε-tube. Equal to the SVM labels y_i are needed for each training vector $\boldsymbol{x}_i$. In this case y_i denotes the target value, i.e. the age of a given child. The goal of SVR training is to find a function $f(\boldsymbol{x}_i) = \boldsymbol{x}_i \cdot \boldsymbol{w} + b$, that has at most ε deviation and is as flat as possible. Flatness in this case means to gather a small $\boldsymbol{w}$. This can be formulated as an optimization problem, where the norm ($||\boldsymbol{w}||$) has to be minimized. By introducing Lagrangian multipliers the optimization problem can be formulated and rewritten in a dual optimization problem. The prediction of the target value $\hat{y}$ of a given test vector $\boldsymbol{x}$ is then determined by

$$\hat{y} = \sum_{i=1}^{N} (\alpha_i - \alpha_i^*)(\boldsymbol{x}_i \cdot \boldsymbol{x}) + b. \tag{4}$$

α_i and α_i^* are Lagrangian multipliers and $\boldsymbol{x}_i$ denotes the training vectors. To extend the SVR to the non-linear case $\boldsymbol{x}_i \cdot \boldsymbol{x}$ again can be extended by kernel functions $K(\boldsymbol{x}_i, \boldsymbol{x})$.

6 Classification and Regression System

Fig. 2 shows the principle of the training and the working sequence of both systems. For the classification task a SVM is used. For the regression task the SVM is replaced by a SVR, which facilitates the possibility to predict the real age of a child in detail.

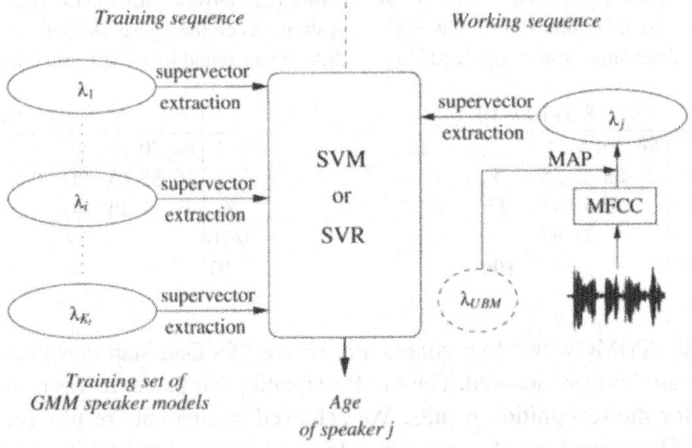

Fig. 2. Principle of the classification (SVM) and regression (SVR) system

For each speaker of the training set a GMM λ is created. The mean vectors of each GMM are concatenated and used as input vectors for the SVM/SVR training. These are denoted as GMM supervectors and can be regarded as a mapping from the utterance of a speaker (the MFCCs) to a higher-dimensional feature vector, which represents the speaker himself or characteristics of this speaker, e.g. the age. If the age (or the group membership) of a speaker j should be determined, a speaker model λ_j is derived out of the background model λ_{UBM}. Note that the background model is the same one, which is used for creating the GMMs of the training speakers. Again the GMM supervector is created for speaker j and the SVM or SVR is performed. The SVM determines the expected class membership of speaker j and the SVR directly predicts his or her age. The SVM and SVR implementations of the WEKA toolkit [13] were used.

7 Experiments and Results

This paper focuses on the automatic determination of the age of children. We performed preliminary experiments which focused on selection of the best parameters for the speaker model creation. Another set of preliminary experiments were performed to determine the best kernel for the SVM/SVR system and the combination of GMM supervectors. The results of the SVM system are described in Section 7.2 and the results of the SVR system are depicted in Section 7.3.

7.1 Preliminary Experiments

We examined the influence of the number of Gaussian densities, the training algorithm (MAP, EM) and the influence of the different forms of covariance matrices (full or diagonal) on the recognition results. In case of MAP adaptation we evaluated the use of different adaptation parameters (for details see [8]) and the kind of training, i.e. adapting all GMM parameters (ω, μ, Σ) or only the means respectively. The best results were

Table 1. Relative confusion matrix of the SVM system; The y-axis determines the actual class and the x-axis determines the recognized class

	<7	7	8	9+10	>10
<7	**60**	33		7	
7	5	**55**		35	5
8		20	**6**	47	27
9+10			3	**97**	
>10					**100**

Table 2. Relative confusion matrix of the SVR system. After the age is determined the children are assigned to the specific class.

	<7	7	8	9+10	>10
<7	**66**	20	13		
7	10	**55**	15	20	
8	6	13	**40**	40	
9+10			25	**74**	
>10				20	**80**

achieved with GMMs with MAP adaptation, where 256 Gaussian densities with full covariance matrices were adapted. The MAP adaptation parameter α was only of minor importance for the recognition results. We achieved comparable results for different values of α. The second set of experiments focused on the determination of the most appropriate kernel for SVR/SVM. We regarded linear, exponential and RBF kernels. We varied the polynomial order of the exponential kernel and the width of the radial basis function for the RBF kernel. Besides these standard kernels we also examined a GMM-based distance kernel, which is based on the KL divergence. This kernel is often used in speaker verification applications [14]. The best results were achieved with the most simple one – the linear kernel. The composition of the GMM supervector was also a task in the preliminary experiments. We focused on using other information than the mean vectors in it. So we additionally added the weights of the GMM components and the diagonal of the covariance matrices to the supervector. This results in a higher dimension, but not in a higher recognition result.

7.2 Classification Results

For the classification task the children were assigned to five different age classes (Section 2). The confusion matrix is shown in Table 1. It can be seen that all children, who belong to the class *9+10* and *>10* are classified correctly. However only 6 % children (in total 1 of 16 children) of the age of 8 are classified correctly. The recognition results of children with the age of 7 or smaller are classified correctly with about 60 %. The overall recall of the classification system was 66 % and the overall precision was 83 %. This has been calculated by an unweighted mean of the class-wise precision and recall.

7.3 Regression Results

The advantage of the regression system is, that the children do not have to be assigned to different age classes. The determination of the age is performed in two different ways: with an accuracy in years and an accuracy in months. When meassuring the accuracy in years, the maximum deviation is three years and the mean deviation was determined to be 0.8 years. We calculated the Pearson and the Spearman correlation: 0.87 and 0.83 respectively. Additionally we assigned the recognized age to the different age classes which were used for the SVM system. So we were able to build up a confusion matrix and to compare the SVM and SVR results. The confusion matrix of the SVR system is

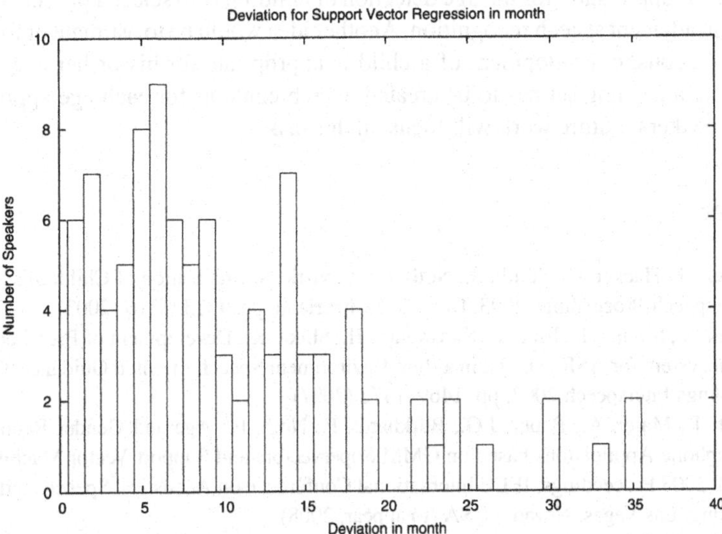

Fig. 3. Deviation between the age in months and the predicted age in months based on SVR

shown in Table 2. The overall precision of this approach was 68 % and the recall was 63 % respectively. The recall of the SVM and the SVR system are identically, but the precision of the SVM is much higher. Nevertheless, the confusion matrix of the SVR system is more balanced and way more intuitive, i.e. 40 % of the 8 year old children are classified correctly and the results for children with nine years or older are in the same range as the results for the other classes.

The second regression experiment regarded the age determination on a monthly accuracy level. Therefore the labels of the GMM supervectors have been changed. The results are shown in Table 3. The age of 65 children lies within a deviation of 12 month. The mean deviation for this experiment was 10.3 months. The maximal error was 35 months. The Pearson correlation was measured to be 0.89 and the Spearman correlation 0.83 %.

8 Outlook and Conclusion

This paper focused on the automatic determination of the age of children in preschool and primary school age. Therefore, we created GMM supervectors for every child of the training and test set and either performed a classification with SVM or a SV regression. The classification approach had problems in determining children with the age of eight years. This lack could be improved by application of the SVR and a class determination afterwards. But this system was not as precise as the SVM system.

The regression could also be performed with a monthly accuracy, which means the training labels described the age of a child in month. With this idea the age of 65 % of the children was determined with a deviation of less than 12 months.

One field of application for the age detection of children is to select a proper acoustic model for an adjacent speech recognition. Another idea would be to estimate or to assess whether the acoustic development of a child is appropriate for his or her age or not. But therefore a training set has to be created, which contains for each age appropriate reference speakers. Future work will focus on this task.

References

1. Stemmer, G., Hacker, C., Steidl, S., Nöth, E.: Acoustic Normalization of Children's Speech. In: Interspeech Proceedings 2003, Geneva, Switzerland, pp. 1313–1316 (2003)
2. Cincarek, T., Shindo, I., Toda, T., Saruwatari, H., Shika, K.: Development of Preschool Children Subsystem for ASR and QA in a Real-Environment Speech-oriented Guidance Task. In: Proceedings Interspeech 2007, pp. 1469–1472 (2007)
3. Bocklet, T., Maier, A., Bauer, J.G., Burkhardt, F., Nöth, E.: Age and Gender Recognition for Telephone Applications based on GMM Supervectors and Support Vector Machines. In: ICASSP 2008 Proceedings, IEEE International Conference on Acoustics, Speech, and Signal Processing, Las Vegas, Nevada, USA (to appear, 2008)
4. Fox, A.V.: PLAKSS - Psycholinguistische Analyse kindlicher Sprechstörungen, Swets & Zeitlinger, Frankfurt a.M (2002)
5. Maier, A., Nöth, E., Batliner, A., Nkenke, E., Schuster, M.: Fully Automatic Assessment of Speech of Children with Cleft Lip and Palate. Informatica 30(4), 477–482 (2006)
6. Campbell, W.M., Sturim, D.E., Reynolds, D.A.: Support Vector Machines Using GMM Supervectors for Speaker Verification. Signal Processing Letters 13, 308–311 (2006)
7. Bocklet, T., Maier, A., Nöth, E.: Text-independent Speaker Identification using Temporal Patterns. In: Matoušek, V., Mautner, P. (eds.) TSD 2007. LNCS (LNAI), vol. 4629, pp. 318–325. Springer, Heidelberg (2007)
8. Reynolds, D.A., Quatieri, T.F., Dunn, R.B.: Speaker Verification using Adapted Gaussian Mixture Models. Digital Signal Processing, 19–41 (2000)
9. Dempster, A., Laird, N., Rubin, D.: Maximum Likelihood from Incomplete Data via the EM Algorithm. Journal of the Royal Statistical Society, Series B (Methodological) 39(1), 1–38 (1977)
10. Gauvain, J.L., Lee, C.H.: Maximum A-Posteriori Estimation for Multivariate Gaussian Mixture Observations of Markov Chains. IEEE Transactions on Speech and Audio Processing 2, 291–298 (1994)
11. Burges, C.J.C.: A Tutorial on Support Vector Machines for Pattern Recognition. Data Mining and Knowledge Discovery 2(2), 121–167 (1998)
12. Smola, A.J., Schölkopf, B.: A tutorial on support vector regression. Statistics and Computing 14, 199–222 (2004)
13. Witten, I., Frank, E., Trigg, L., Hall, M., Holmes, G., Cunningham, S.: Weka: Practical machine learning tools and techniques with java implementations (1999)
14. Dehak, R., Dehak, N., Kenny, P., Dumouchel, P.: Linear and Non Linear Kernel GMM Support Vector Machines for Speaker Verification. In: Proceedings Interspeech 2007, Antwerp, Belgium (2007)

Language Acquisition: The Emergence of Words from Multimodal Input

Louis ten Bosch and Lou Boves

Dept Language and Speech, Radboud University Nijmegen, NL
l.tenbosch@let.ru.nl, boves@let.ru.nl
http://lands.let.ru.nl

Abstract. Young infants learn words by detecting patterns in the speech signal and by associating these patterns to stimuli provided by non-speech modalities (such as vision). In this paper, we discuss a computational model that is able to detect and build word-like representations on the basis of multimodal input data. Learning of words (and word-like entities) takes place within a communicative loop between a 'carer' and the 'learner'. Experiments carried out on three different European languages (Finnish, Swedish, and Dutch) show that a robust word representation can be learned in using approximately 50 acoustic tokens (examples) of that word. The model is inspired by the memory structure that is assumed functional for human speech processing.

Keywords: Language acquisition, word representation, learning.

1 Introduction

Language processing is one of the most complex cognitive skills of humans. Infants seem to acquire this skill effortlessly, but the large body of literature on cognition, language and memory shows that we only begin to understand the processes involved. Understanding speech is tantamount to mapping continuous speech signals to discrete concepts, which we are used to think of as a sequence of word-like elements. Infants learn to discover word-like units in speech without prior knowledge about lexical identities and despite the lack of clear word boundary cues in the signal.

In this paper we propose a computational model of word discovery that is able to learn new words and that is a plausible analogy of the way in which infants acquire their native language. Specifically, the model should explain how babies can learn new words by using already stored representations that are continuously adapted on the basis of new speech input.

Strangely enough, today automatic speech recognizers are the most elaborate computational model of speech processing. Contrary to virtually all psycholinguistic models ASR is able to handle the entire chain from speech signal to a sequence of words. However, current ASR algorithms certainly cannot claim any cognitive or ecological plausibility. At the same time, ASR systems perform substantially worse than humans [8,13]. It is widely assumed that for closing the performance gap new ASR training and matching paradigms must be explored, preferably inspired by cognitive models of human speech processing and language acquisition. In this direction, several

P. Sojka et al. (Eds.): TSD 2008, LNAI 5246, pp. 261–268, 2008.
© Springer-Verlag Berlin Heidelberg 2008

attempts have been made, e.g. learning of words [12], language acquisition [15], and incremental learning [9]. However, none of the models proposed in the literature are able to learn, adapt and generalize patterns quickly and effortlessly to recognize new variants of known words and novel words [5,16].

In this paper, we propose an embodied computational model for language acquisition that has similarities with the Cross-channel Early Lexical Learning (CELL) model [12], but differs from CELL in that it does not assume that infants represent speech in the form of a lattice of pre-defined phonemes. The current model avoids the use of pre-existing representation for decoding the information in the input signals. Instead, the representations in the model emerge from the multimodal stimuli that are presented to the model.

The structure of this paper is as follows. In the next section, we will discuss the main components of the proposed computational model and the overall architecture of the ACORNS model, while the third section deals with experiments. The final section contains a discussion and our conclusions.

2 An Embodied Model of Word Discovery

The (partially) embodied model of language acquisition and speech communication that we are developing in the FET project ACORNS [2] will contain four sub-modules, viz. sensory front-end processing, memory access and organization, information discovery and learning, and interaction in a realistic environment.

Front-end processing: In the first step, the computational model converts sensory input signals into an internal representation which is used in subsequent sub-modules for learning new patterns and for recognizing known patterns. Front-end processing may include the conversion of input signals into representation such as the MFCCs which are used in conventional ASR.

Memory organization and access: Cognitive theories of memory distinguish at least three types of memory: a sensory store in which all information is captured only for a very short time (in the order of 2 seconds), a short-term memory (also called working memory) that holds representations of sensory inputs and serves as a processing system that is able to compare new sensory inputs to previously learned patterns that are retrieved from a long-term memory. The model makes use of these types of memory and stores, retrieves and updates internal representations. At the same time this memory model supports the Memory-Prediction Theory, which holds that intelligent action is based on memorized perception-action loops.

Information discovery and integration: In the Memory-Prediction Theory it is assumed that multilayered representations are formed in which structure at a lower level map to structures at a higher level (abstraction). In the experiments reported in this paper the abstraction method is based on Non-negative Matrix Factorization (NMF) [7,4,14]. NMF is member of a family of computational approaches that represent data in a (large) matrix and use linear algebra to decompose this matrix into smaller matrices. These smaller matrices contain the information in the original matrix in a more condensed and abstract form. There are close similarities with Latent Semantic Analysis, e.g. [1].

By using NMF processes such as abstractions receive a clear interpretation in terms of linear algebraic operations. NMF is a powerful tool for discovering structure in speech data [14]. In later stages of the ACORNS project we will also experiment with other structure discovery algorithms.

Interaction and communication: In order to simulate a learning environment, the learner is endowed with the intention to learn words in order to maximize the appreciation it receives from the carer. This is done by translating the appreciation from the carer into a strive for correctly responding to the carer, which in turn is interpreted as the optimization of the interpretation of the stimulus presented by the carer. This optimization involves the Kullback-Leibler distance between the input and output of NMF.

2.1 Architecture

The ACORNS architecture (cf. Figure 1) is based on recent psycholinguistic research in speech and language processing [6].

The learner receives multimodal input consisting of an audio stream (containing infant-directed or adult-directed speech) in combination with an abstraction of the visual modality (a visual 'tag'). This tag is provided in synchrony with the speech signal. In this way, we simulate the presence of a visual sensory processing system. It is up to the learning agent to learn word-like entities from the repetitions in the audio signal and from cross-modally reoccurring systematic patterning.

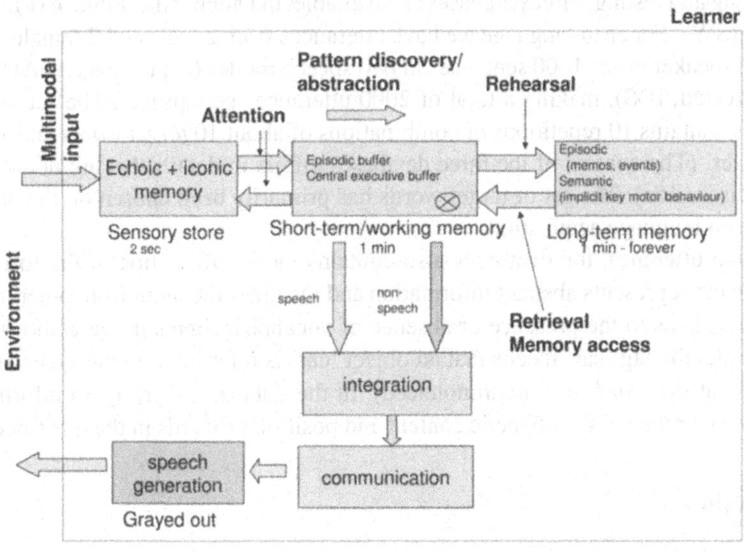

Fig. 1. Global architecture of the ACORNS system. Multimodal input is put into the sensory store. The sensory store, short-term/working memory and long term memory have different decay times. Two feedback loops are foreseen: one internal, governing the intrinsic processes and one external, in which the carer provides feedback to the model.

In its current implementation, the model makes use of a simplified version of attention and rehearsal mechanisms. The attention and rehearsal mechanisms operate on representations stored in memory, and transform stored representations into possibly more abstract representations. We interprete attention as a process that reduces the part of the input stream that must be analyzed and is therefore indispensable to keep the computation load manageable, to reduce the storage into short-term (working) memory, and to reduce the ambiguity to be resolved during the search.

3 Experiments

A series of learning experiments has been conducted, inspired by phenomena observed in literature on language acquisition by young infants. In the first experiment, we have investigated the effect of a new speaker on the adaptation of already trained word representations. The arguments in favor of episodic representations used in psycholinguistics [3] suggest that different representation may be formed for different speakers. That means that representations that conflate episodes pertaining to several speakers corresponding to the same semantic object may only form on higher levels in the hierarchy. Another experiment deals with the effect of a new language (L2) on word representations that are trained on L1.

3.1 Material

For training and testing, three databases are available, in Dutch (NL), Finnish (FIN), and Swedish (SW). For each language we have utterances from 2 male and 2 female speakers. Each speaker utters 1000 sentences in two speech modes (adult-directed, ADS, and infant-directed, IDS), making a total of 2000 utterances per speaker. The set of 1000 sentences contains 10 repetitions of combinations of about 10 *target words* and 10 carrier phrases. (The content of the three databases differs in details that are not relevant for this discussion). The set of target words has primarily been chosen on the basis of literature on language acquisition.

For each utterance, the databases also contains meta-information in the form of a 'tag'. The tag represents abstract information and idealizes the input from other modalities. It translates to the presence or absence of vocabulary items in the audio stream. For example, the tag 'car' means that an object 'car' is referred to in the speech signal (and not that the *word* 'car' is pronounced). In the database, there is no information available about the words, phonetic content and position of words in the utterances.

3.2 Results

The result of the experiments are shown in figures in which the horizontal axis represents the number of utterances (tokens) presented during training. The vertical axis represents the accuracy of the learners replies. The accuracy is defined as the number of correct responses (defined by comparing the learners reply with the ground truth in the multimodal stimulus by the carer), divided by the total number of replies.

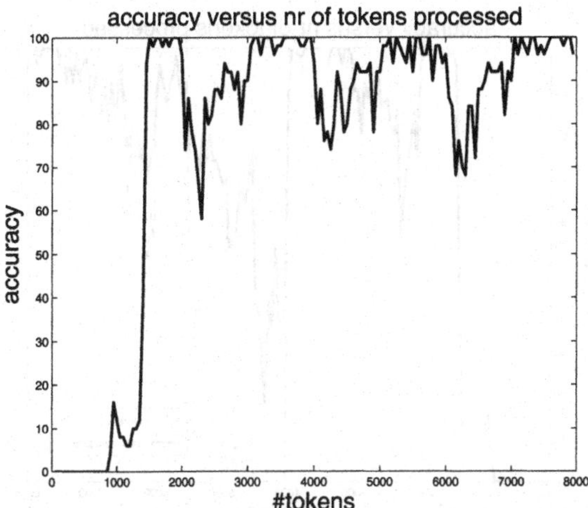

Fig. 2. (Dutch, speaker-blocked). Multimodal input data are presented blocked, speaker-by-speaker. Each time a new speaker starts (around number of tokens = 0, 2000, 4000, 6000), a drop in performance can be seen. Within about 1000 tokens (that is, approximately 100 tokens per word) the performance is back on its previous level. The decrease in performance is mainly due to different voice and speech characteristics which require an adaptation by the learning model.

4 Discussion and Conclusion

The computational model presented in this paper shows that learning relations between speech fragments and higher-order concepts can be accomplished with a general purpose pattern discovery technique. The performance of the learner depends on a number of factors – such as the ordering of the data (stimuli), the blocking per speaker, speaker changes, and multi-lingual training.

The learner is able to learn a limited set of concepts and classify a new stimulus in terms of one of these concepts. The learner needs a number of tokens before it can make a reliable representation. During the learning, it is able to gradually improve the quality of its internal representations, by minimizing the Kullback-Leiber distance between the observed data and the internal representations.

A second characteristic of the learner's behavior is the adjustment to a new speaker. As soon as a new speaker starts interacting with the learner, the internal representations are adapted to accommodate the speaker characteristics. Moreover, the learner reuses already stored representations whenever possible. This is particularly clear in the multilingual experiment based on the semantic tags (Figure 4).

The speech database contains infant- and adult directed speech. This distinction is now not used, but other experiments have shown that the learner is able to distinguish these styles.

The computational model illustrates the relevance of various issues that are known to play a role in (models of) human speech processing. One of these issues is how words get activated (and to what extent), the second with the way how competition

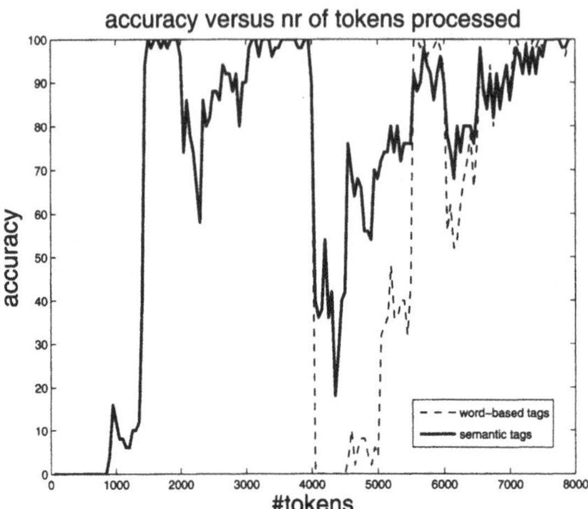

Fig. 3. (multilingual, speaker-blocked). Results of a multilingual experiment. First, two Dutch speakers are presented, then two Swedish speakers. The speakers are NL female, NL male, SWE female and SWE male. In the end the model is able to recognize Dutch and Swedish target words. The solid line represents the case were tags are language-dependent (word-based); the dashed line represents the ecologically more plausible case in which the tags are language-independent (semantic).

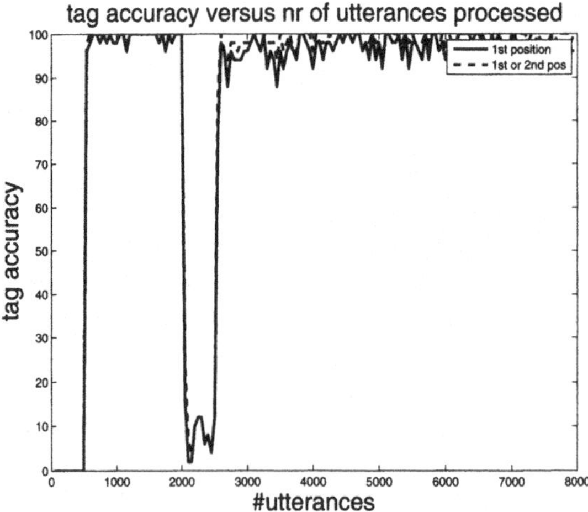

Fig. 4. The learner is first exposed to input from one Dutch speaker (the primary carer). After 2000 utterances from the primary carer, three other persons (2 males, 1 female) start interacting with the learner. The utterances from the new speakers are presented in random order. At first, the learner has difficulties adapting to new speakers, but it catches up after some 500 examples. The solid line indicates the accuracy of the 1-best reply of the learner; the dashed line shows the accuracy of the 2-best replies.

may act during the word search. In the current model, the activation of lexical items is separated from the actual competition. This is similar to Shortlist, one of the widely used computational models for human word processing [11]. Shortlist is a two-stage model in which activation of words by incoming speech input is separated from competition between the activated words. Other than Shortlist, however, the current model plays out the entire lexicon, while in Shortlist the network in which competition plays a role is constructed from only those words supported by the input. In the current model, competition is not explicitly implemented. Instead, it emerges from the parallel search among multiple candidates. This is in line with earlier findings e.g. obtained with another model of human word processing TRACE [10]. TRACE showed that competition is not a necessary consequence of parallel processing.

One of the research lines that will be pursued in the near future deals with the mechanisms that underly the emergence of words as a function of utterance-based training. In a genuine communicative setting, the learner must be able to learn not only from the presented multimodal stimuli, but also from the feedback that she receives from the carer. This will open the possibility of investigating the effect of corrective feedback on the learning process in more detail than is possible now. This will both enhance the ecological and cognitive plausibility of the computational model.

The second research line that will be exploited is also directly related to the cognitive plausibility. This research line deals with the use of *semantically related* tags that are presented to the learner in combination with the speech signal. In the current interaction model, the tags represent high-level references to objects that the learner receives and processes with 100 percent certainty. We aim at a model of a learner that receives multimodal input (speech and semantic tags) in such a way that the construction and adaptation of new representations is entirely controlled by the learner's internal learning mechanisms.

Acknowledgments. This research was funded in part by the European Commission, under contract number FP6-034362, in the ACORNS project (www.acorns-project.org).

References

1. Bellegarda, J.R.: Exploiting Latent Semantic Information for Statistical Language Modeling. Proc. IEEE 88, 1279–1296 (2000)
2. Boves, L., ten Bosch, L., Moore, R.: ACORNS _ towards computational modeling of communication and recognition skills. In: Proceedings IEEE-ICCI 2007 (2007)
3. Goldinger, S.D.: Echoes of echoes? An episodic theory of lexical access. Psychological Review 105, 251–279 (1998)
4. Hoyer, P.O.: Non-negative matrix factorization with sparseness constraints. Journal of Machine Learning Research 5, 1457–1469 (2004)
5. Johnson, S.: Emergence. Scribner, New York (2002)
6. Jones, D.M., Hughes, R.W., Macken, W.J.: Perceptual organization masquerading as phonological storage: Further support for a perceptual-gestural view of short-term memory. J. Memory and Language 54, 265–281 (2006)
7. Lee, D.D., Seung, H.S.: Algorithms for non-negative matrix factorization. Advances in Neural Information Processing Systems 13 (2001)

8. Lippmann, R.: Speech Recognition by Human and Machines. Speech Communication 22, 1–14 (1997)
9. Maloof, M.A., Michalski, R.S.: Incremental learning with partial instance memory. Artificial intelligence 154, 95–126 (2004)
10. McClelland, J.L., Elman, J.L.: The TRACE model of speech perception. Cognitive Psychology 18, 1–86 (1986)
11. Norris, D.: Shortlist: A connectionist model of continuous speech recognition. Cognition 52, 189–234 (1994)
12. Roy, D.K., Pentland, A.P.: Learning words from sights and sounds: a computational model. Cognitive Science 26, 113–146 (2002)
13. Sroka, J.J., Braida, L.D.: Human and machine consonant recognition. Speech Communication 44, 401–423 (2005)
14. Stouten, V., Demuynck, K., Van hamme, H.: Automatically Learning the Units of Speech by Non-negative Matrix Factorisation. In: Interspeech 2007, Antwerp, Belgium (2007)
15. Werker, J.F., Curtis, S.: PRIMIR: a developmental framework for of infant speech processing. Language Learning and Development 1, 197–234 (2005)
16. Werker, J.F., Yeung, H.H.: Infant speech perception bootstraps word learning. TRENDS in Cognitive Science 9, 519–527 (2005)

A Tagged Corpus-Based Study for Repeats and Self-repairs Detection in French Transcribed Speech

Rémi Bove

LIF-CNRS, Université de Provence, 13621 Aix-en-Provence, France
Remi.Bove@univ-provence.fr

Abstract. We present in this paper the results of a tagged corpus-based study conducted on two kinds of disfluencies (repeats and self-repairs) from a corpus of spontaneous spoken French. This work first investigates the linguistic features of both phenomena, and then shows how – from a corpus output tagged with Tree-Tagger – to take into account repeats and self-repairs using word N-grams model and rule-based pattern matching. Some results on a test corpus are finally presented.

Keywords: Disfluencies detection, tagging, spoken language syntax, French transcribed speech, rule-based pattern matching, word N-grams model.

1 Introduction

Spontaneous speech differs from written text. One difference is the presence of disfluencies (such as repeats, self-repairs, word-fragments, filled pauses, etc.) which occur frequently. Indeed, oral speech, as opposed to written text, is produced "online" and therefore has the specificity of retaining the traces of its elaboration. It is by fits and starts and later syntactic and/or lexical readjustements that oral spontaneous speech is elaborated. It is never produced in a smooth fashion which could be compared to edited writing, that is to say a revised, corrected and perfect form.

In the past years, researches on spoken French have thrived, in fields such as syntactic studies [2,12], psycholinguistic [3,9,10], computational aspects of human-computer dialogue [1], etc. Thus, a certain number of regular features have been identified in disfluencies as we will show in point 2.

We here investigate repeats like:

"*il y a* **des** *quand même* **des** *clients qui sont* **un peu un peu** *stressés*"
(*there are some nevertheless some customers who are a little a little stressed*)

and self-repairs such as:

"*à cette époque* **j'avais j'étais** *en maîtrise*".
(*at that time I had I were in Master*)

Our work is based on a sample of about 8500 words of the CRFP[1]. The sample contains utterances from ten speakers (different ages, sexes and origins) and three spoken situations (private, public, and professionnal) with major prosodic movements annotated. The orthographical transcription was entirely hand-coded by experts with an

[1] Corpus de Référence du Français Parlé, for more information, cf. [5].

P. Sojka et al. (Eds.): TSD 2008, LNAI 5246, pp. 269–276, 2008.
© Springer-Verlag Berlin Heidelberg 2008

extremely strict multiple listening/validation (and corrections) cycle. Our processings were then performed on a tagged version of this corpus. We arbitrarily chose to use TreeTagger ([13]) because it was easy to install and use in our experiment, efficient, and especially reconfigurable.

We here focus on the detection of repeats and self-repairs. We first present linguistic features of the phenomena investigated (2), then we explain the detection strategies (3). The last points are dedicated to the evaluation results (4) and conclusion (5). The strategies presented are particularly intended to later carry out and facilitate shallow parsing of disfluencies or other kinds of downstream machine speech processings.

2 Linguistic Features of Repeats and Self-repairs

Spontaneous speech provides many new challenges for spoken language systems (particularly in speech recognition). One of the most critical is the prevalence of repeats and self-repairs.

2.1 Repeats

Repeats are disfluencies where the speaker repeats successively one word or group of words, one or many times. The structure of repeats comprises two main elements: a "repeatable" (the elements which will be repeated) and one (or more) "repeated". They can be defined as shown in Figure 1.

On the morphosyntactic level, repeats mostly involve function words (9 repeats out of 10) [8] which, most of the time, are monosyllables as 29.5% are determiners, 26% pronouns and 13% prepositions in our corpus. These function words partake of the structuring of language and shape content words into syntactic units. Like Blanche-Benveniste [2], we have been able to check that repeats are subjected to syntactic constraints: they mainly appear at the beginning of phrases and their structure remains stable, that is to say the simple syntactic frame – without any lexical content – appears first, and the lexical filling comes second.
Examples:

"*c'est toujours euh en* **en en** *vigueur ça hein*" (#repeatable = 1; #repeated = 2)
(*it is always er in in in force that eh*)
"*on parle de* **on parle de** *l' ouvrée*" (#repeatable = 3; #repeated = 1)
(*one speaks about one speaks about the wrought*)

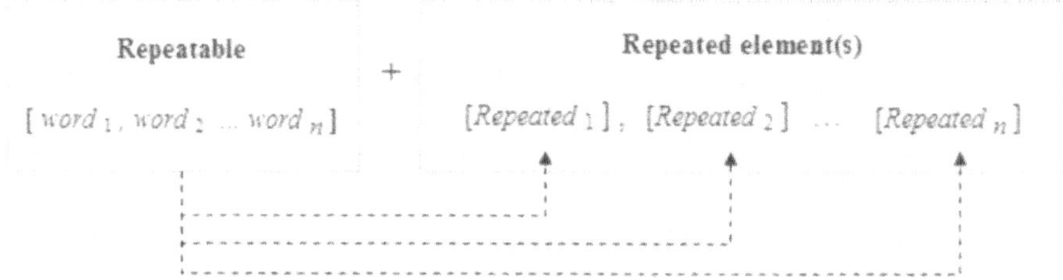

Fig. 1. Repeat structure

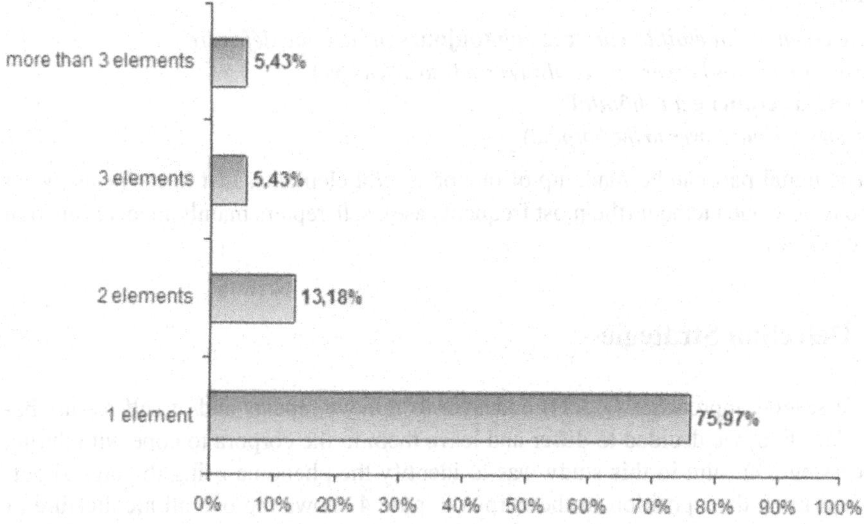

Fig. 2. Distribution of repeated elements

While examining our corpus, we noted that there are seldom a lot of elements in the repeatable. The following diagram shows the distribution of the number of repeated elements in our corpus.

In most cases (89%), there is only one or two elements included in the repeatable. It was that clue which encouraged us to use a word N-grams model (indeed, word bigram model enabled us to detect most of the repeats).

2.2 Self-repairs

Self-repairs are disfluencies where some of the words that the speaker utters need to be modified in order to correctly understand the speaker's meaning. From a structural point of view, there is a modification operated between the initial part (*X*) of self-repairs and the corrected part (*Y*). The correction can relate to gender, number, morpho-syntactic category, etc.

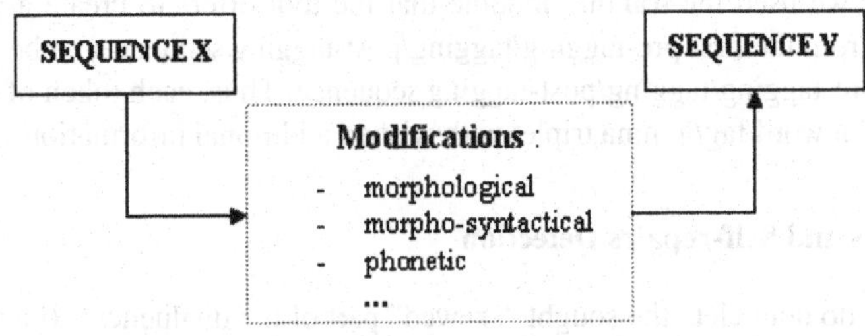

Fig. 3. Self-repair structure

Examples:

"*cette chienne qui était là euh* **très** *euh* **toujours** *prête à me défendre*"
(*this dog which was here er very er always ready to defend me*)
"**on va on retourne** *à l' hôpital*"
(*one goes one turns over to the hospital*)

The initial part can be made up of one or several elements. Just like repeats, when there is only one element (the most frequent case), self-repairs mainly involve function words (79%).

3 Detection Strategies

While several approaches ([7,11]) endeavor to remove repeats and/or self-repairs before detection, we decided to differ and leave them in the corpora to cope with during processing. Our aim in this study was to identify the phenoma using the tagged output and mark their position in the corpus. Figure 4 shows the overall architecture of our processing. Next, we briefly present the tagger we used as a working basis before giving some examples of the strategies used for detection.

3.1 Tagging

Like many probabilistic taggers, TreeTagger uses a decision tree to automatically determine the appropriate size of the context and to estimate the part-of-speech transition probabilities. It annotates the text with part-of-speech and lemma information. Our works thus are based on TreeTagger output. However, we carried out modifications of the standard TreeTagger version in order to adapt scripts to pre- and post- tagging phases.

Pre-tagging phase: This step enables (among others) both to tokenize the input corpus and to format prosodic information (in representing it by a non-word token) so that it is not tagged (TreeTagger ignores SGML tags).

Post-tagging phase: Many scripts then rewrite the output of the tagging so that it can be manipulated as wished for later processing. This rewriting step basically consists in changing or modifying the tagset which is under-specified with the standard version of TreeTagger (addition of gender, number, lemma for proper names, acronyms, etc.).

Moreover, we used the training module that the tool offers to create a new set of parameters (from the first pre-tagging/tagging/post-tagging sequence) to be embedded in a second pre-tagging/tagging/post-tagging sequence. Thus, each token of the output is made up of a word/tag/lemma triplet with wished additional information.

3.2 Repeats and Self-repairs Detection

Although we do not delete the sought "skewed" part of the disfluencies (i.e the repeatable for repeats or the initial part for self-repairs), we allow for the same kind of clues as [7] (for example) to identify the phenomena. Indeed, the two schemes presented

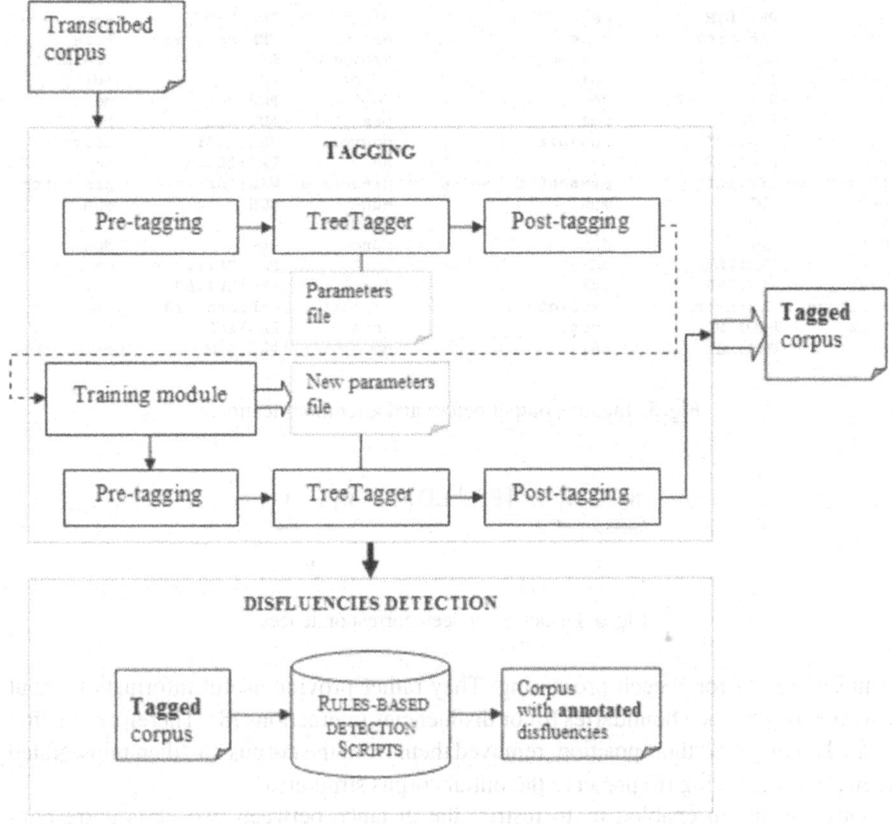

Fig. 4. Architecture of tagging and detection processing

in 2.1 and 2.2 correspond to a "canonical" structure of repeats and self-repairs. However, looking at the disfluencies patterns in our developpement corpus, we noted (and in many other studies on disfluencies too) that there are often elements included between the two parts of the disfluencies. For example, there can be editing words [ED] (*ben*, *donc*, etc.) or filled pauses [FP] (*euh*, *hum*), which are called in many conceptions "interregnum" [11].
Examples:

"*après* **bon ben** *après les choses se sont mises en place*" (edited word)
(*after well well after the things have been set up*)

"*d'être* **euh** *de pas pouvoir répondre euh aux questions ce qui est quand même gênant*" (filled pause)
(*to be er not to be able to answer er the questions which is nevertheless awkward*)

In this way, these elements move away the couples "repeatable/repeated" or "initial part/corrected part" of self-repairs. That causes missed matching and deteriorates the detection performance. Note that filled pauses and editing words do not in fact function

c'	PRO:DEM	ce	c'	PRO:DEM	ce
est	VER:pres	être	est	**VER:conj:s3**	être
vraiment	ADV	vraiment	vraiment	ADV	vraiment
idiot	ADJ	idiot	idiot	ADJ	idiot
de	PRP	de	de	**PRP:de**	de
pas	ADV	pas	pas	**ADV:pas**	pas
pouvoir	VER:infi	pouvoir	pouvoir	VER:infi	pouvoir
se	PRO:PER	se	se	**PRO:CLI:3**	se
présenter	VER:infi	présenter	présenter	VER:infi	présenter
euh	INT	euh	euh	**EUH**	euh
+	SYM	+	**<PAUSE>**		
donc	ADV	donc	donc	ADV	donc
on	PRO:PER	on	on	**PRO:CLI:s3**	on
leur	PRO:PER	lui	leur	**PRO:CLI:p3**	lui
explique	VER:pres	expliquer	explique	**VER:conj:s3**	expliquer
tout	PRO:IND	tout	tout	PRO:IND	tout
ça	PRO:DEM	cela	ça	PRO:DEM	cela

Fig. 5. Tagger's output before and after modifications

$$\underbrace{w_i \ldots w_j}_{x} \;\ldots\; [\mathrm{FP} \,\|\, \mathrm{ED}] \;\ldots\; \underbrace{w_i \ldots w_j}_{y}$$

Fig. 6. Distance between correspondences

as noise useless for speech processing. They rather provide useful information about detection of sentence boundaries [6] or disfluencies interactions [8]. Therefore, we first located them, noted their position, removed them from the corpus and then reintegrated them after processing (to preserve the initial corpus structure).

This precaution enabled us to restrict the distance between two parts correspondences. Figure 6 shows the distance between two corresponding disfluency parts. The intervals x and y are sequences of the words that respectively occur between the first part of disfluency and the second part. Thus the relevant word correspondences are those that are adjacent, and which improve the detection performance. The detection of repeats consists in searching identical word N-grams. So, when a correspondence is found, the number of words contained in the repeatable needs to be limited. From our test corpus, we found that 6 words, excluding filled pauses and editing terms, are

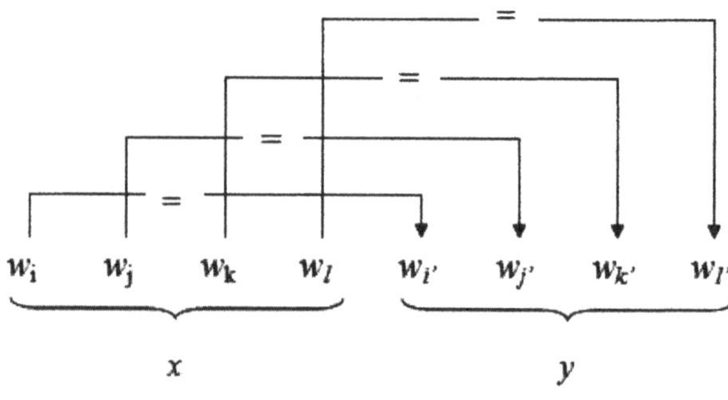

Fig. 7. Word N-grams for repeats

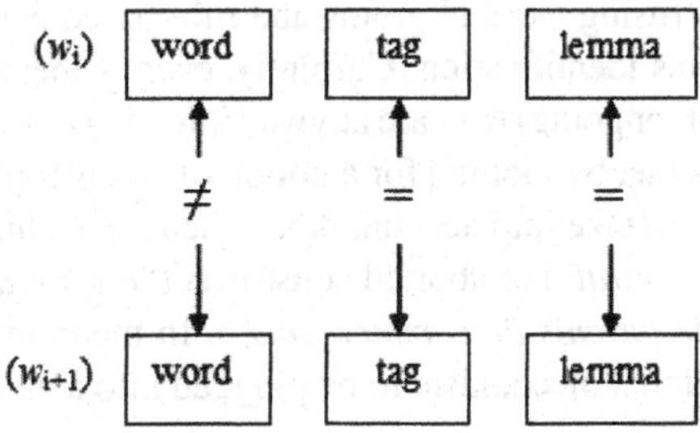

Fig. 8. Example of self-repair pattern

sufficient. In a repeat, the repeatable matches (and then the set of the repeated) can have at most 6 words (i.e $x \leqslant 6$ and $y \leqslant 6$).

Concerning self-repairs, we build the pattern on the fly using set of constraints rules to envisage relevant cases. For example, we developed a simple rule where one of the possible self-repair patterns of w_i and w_{i+1} consisted in identical tags and lemma but with different words.

4 Evaluation

We evaluated our processing on a second sample of the CRFP corpus (more than 2,100 words) different from the developpement corpus. The following table shows the classic performance measures for both phenomena.

	Precision	***Recall***	***F1***
Repeats	91%	97%	94%
Self-repairs	83%	57%	67%

Fig. 9. Results of disfluencies matching

Obviously, some detection errors remains. There are cases of ambiguous repeats as the emphases repeats (example: "*j'ai été en contact* **très très** *jeune*"). The main difficulty with self-repairs concerns their structural aspect. Indeed, this phenomenon is very similar with cases when the speaker simply lists elements, and the second one (and the following ones) is not a repair of the first one (example: "**des plantes des fleurs** *euh extrêmement euh euh rares*"), i.e, from a grammatical point of view, the cases of juxtaposition or coordination. Addition of a semantic control after detection could be a solution in order to rule out the non-relevant cases.

5 Conclusion

In this paper we have described the results of a tagged corpus-based study led on repeats and self-repairs from a corpus of spoken French. This study enabled us to show that different strategies (using word N-grams and rule-based pattern matching) can be drawn up for disfluencies identification. Currently, even if these rules are not entirely optimal (our work is still ongoing) they are anyway intended to keep the number of false positives as low as possible, by looking for a notion of "well-formness" for repeats and self-repairs. We foresee to take into account other phenomena like word-fragments ("*il y a un* **com-** *un conseil central*") or aborted constructs ("*c'est vrai que je travaille dans* **le** # *enfin je suis née de parents de commerçants*"). In medium terms, these stategies then constitute an additional mechanism to be plugged into a given framework.

References

1. Adda-Decker, M., Habert, B., Barras, C., Adda, G., Boula De Mareuil, P., Paroubek, P.: A disfluency study for cleaning spontaneous speech automatic transcripts and improving speech language models. In: DIsfluencies in Spontaneous Speech conference, pp. 67–70. Göteborg University, Sweden (2003)
2. Blanche-Benveniste, C.: Approches de la langue parlée en français. Collection L'essentiel Français, Editions OPHRYS, Paris (2000)
3. Clark, H.H., Wasow, T.: Repeating words in spontaneous speech. Cognitive Psychology 37, 201–242 (1998)
4. Core, M., Schubert, L.: A syntactic framework for speech repairs and other disruptions. In: 37th Annual Meeting of the Association for Computational Linguistics, College Park, pp. 413–420 (1999)
5. Delic, E.: Présentation du Corpus de Référence du Français Parlé. Recherches Sur le Français Parlé 18, 11–42 (2004)
6. Engel, D., Charniak, E., Jonhson, M.: Parsing and disfluency placement. In: ACL conference on Empirical Methods in Language Processing, vol. 10, pp. 49–54 (2002)
7. Heeman, P.A., Allen, J.: Detecting and correcting speech repairs. In: 32nd Annual Meeting of the Association for Computational Linguistics, pp. 295–302 (1994)
8. Henry, S., Campione, E., Véronis, J.: Répétitions et pauses (silencieuses et remplies) en français spontané. In: 15th Journées d'Etude sur la Parole, pp. 261–264 (2004)
9. Levelt, W.J.M.: Monitoring and self-repair in speech. Cognition 14, 41–104 (1983)
10. Lickley, R.: Detecting disfluency in spontaneous speech. Ph.D. thesis, University of Edinburgh. Scotland (1994)
11. Liu, Y., Shriberg, E., Stolcke, A.: Automatic disfluency identification in conversational speech using multiple knowledge sources. In: EUROSPEECH 2003, Geneva, Switzerland, pp. 957–960 (2003)
12. Martinie, B.: Etude syntaxique des énoncés réparés en français parlé. Thèse d'état, Université Paris X-Nanterre, France (1999)
13. Schmid, H.: Probabilistic Part-of-Speech Tagging Using Decision Trees. Revised version, original work. In: International Conference on New Methods in Language Processing, pp. 44–49 (1994)

Study on Speaker Adaptation Methods in the Broadcast News Transcription Task

Petr Červa, Jindřich Žďánský, Jan Silovský, and Jan Nouza

SpeechLab, Technical University of Liberec
Studentská 2, 461 17 Liberec 1, Czech Republic
{petr.cerva,jindrich.zdansky,jan.silovsky,jan.nouza}@tul.cz
http://itakura.kes.vslib.cz/kes/indexe

Abstract. This paper deals with the use of speaker adaptation methods in the broadcast news transcription task, which is very difficult from speaker adaptation point of view. It is because in typical broadcast news, speakers change frequently and their identity is not known in the time when the given program is being transcribed. Due to this fact, it is often necessary to use some unconventional speaker adaptation methods here which can operate without the knowledge of speaker's identity and/or in an unsupervised mode too. In this paper, we compare and propose several such methods that can operate both in on-line and off-line modes in addition and we show their performance in a real broadcast news transcription system.

Keywords: Supervised, unsupervised, speaker adaptation, model combination, broadcast news transcription.

1 Introduction

In recent years, the broadcast news transcription (BNT) task has attracted a lot of attention and various systems have been developed for different languages. Although theirs transcription accuracy makes them practically usable yet, it is still not fully satisfactory and one of the most challenging remaining problems is the acoustic variability across various speakers appearing in each program and the fact that the identity of all speakers in each transcribed stream is unknown when the stream is being transcribed. This fact complicates not only the process of acoustic model building but also the use of conventional supervised speaker adaptation methods, which are widely applied in other speech recognition tasks (like text dictation) to improve the accuracy of the acoustic model and reduce the WER of the system for one particular speaker.

Although there exist at least one way described below how to use supervised speaker adaptation methods in the BNT task, it is not very practical but rather complicated: at first, it is necessary to create speaker adapted acoustic models for a set of so called "key" or "training" speakers who occur often in the given type of program. The next problem is how to determine when to use the model of each speaker. Of course, speaker change detection methods followed by speaker recognition and verification can be adopted for this purpose, but unfortunately, the failure of the speaker recognition and verification can bring two types of errors both leading to a significant degradation of the accuracy

P. Sojka et al. (Eds.): TSD 2008, LNAI 5246, pp. 277–284, 2008.
© Springer-Verlag Berlin Heidelberg 2008

during speech recognition. The first type of an error is a false rejection, when the candidate speaker from speaker recognition should not be but is rejected and a speaker independent (SI) model is used instead of his/her adapted model. The latter opposite case is known as a false acceptance. Here, the candidate speaker should not be but he/she is accepted and the system performs speech recognition with an adapted model of a person which is not speaking at the given moment. Because of all previously mentioned reasons, it is necessary to adopt such adaptation methods for the BNT task in practice, which are able to create the adapted model for each speaking person without the knowledge of its identity. A typical method that can be used for this purpose is a speaker selection training – SST [1]. This method is based on a set of SD models, which are created off-line for a group of training speakers. Its main idea is to find a cohort of training speakers, who are close in the acoustic space to the target speaker, and to build the adapted model from models belonging to cohort speakers.

In practice, SST can be performed in several ways in dependency on the given type of task. All the SD models can be used for calculating of the likelihood of the testing utterance to find the cohort of the nearest speakers [1]. Alternatively, speaker identification based on Gaussian mixture models (GMMs) can be utilized for this purpose too [2]. After the cohorts are found, there exist also several possibilities how to create the adapted model for speech recognition. When at least some transcribed training data for cohort speakers are available, they can be transformed to better map the test speaker's acoustic space. The adapted model is then created by reestimation [1] on these transformed data. Another possibility in this case is to determine the weighting coefficients for combination by MAP or ML estimation [3]. The model combination can be also performed on-line and in unsupervised way when sufficient HMM statistics are stored during the phase of SD models training [2].

In this paper, we adopt the idea of SST to prose effective adaptation methods that could be suitable for both BNT tasks.

2 Proposed Speaker Adaptation Methods

In principle, there exist two main possibilities how to perform adaptation and transcription of broadcast news from speed of response point of view. If it is necessary to have the text transcription available as soon as possible (for simultaneous lettering for example) then the system must operate in the on-line mode. On the contrary, if there is just demand on accuracy of the transcribed text, the system can run in the off-line mode and the acoustic stream can be processed in several recognition passes.

Although the adaptation approaches that can be adopted in both these cases are different, we suggest always that the input for adaptation is formed by one utterance or single speaker speech segment that was automatically cut off from the given audio stream and that the output is the acoustic model (a set of phoneme HMMs) which should fit best to the given target (test) speaker.

2.1 Adaptation in the On-Line Mode

In the on-line mode, the adaptation is complicated by the fact that there is not available any information about the identity of the speaker. It is also not possible to create the phonetic transcription of each segment to perform common unsupervised adaptation (like MLLR for example), because the process of speech recognition takes too long.

On-Line Model Combination Approach. Here, we use a set of speaker dependent (SD) models, speaker identification and a fast linear combination of models that are selected during speaker identification. The mean vectors of models of nearest speakers are weighted by factors derived from the ordered list of candidates provided by the speaker identification procedure. Unlike the approach described in [3], ours does not need any HMM statistics to be stored and any phonetic transcription to be determined in a first stage. Our framework allows speech recognition of each segment to be performed with minimal overload in a single pass.

The SD models are prepared in the phase of system's training and for a set of persons (key speakers) occurring frequently in the given type of broadcast news. The amount of data available for these persons varies from several tens of seconds to tens of minutes, so it is necessary to use adaptation instead of maximum likelihood training for SD model building. In our case, the adaptation is based on MLLR [4] and MAP [5] and it is performed only for mean vectors (due to the lack of adaptation data). In this baseline approach, we use general SI models as prior sources for this adaptation.

The initial part of the adaptation process is speaker identification. For the given speech segment, a likelihood score is calculated for each of the key speakers that are represented by their GMMs (also created off-line during training). Then we utilize the scores from speaker identification for forming the cohorts of nearest speakers. We chose N speakers with the highest scores that were identified in the previous step. In this phase, there is not available any information about the phonetic transcription of the utterance, so any classical estimation (like ML or MAP) can not be applied to calculate the weights for linear combination. The combination can only be based on knowledge of prior information. As this information, we use the similarity in the acoustic space between cohort speakers and test speaker. The linear combination is performed just for mean vectors; the other parameters are the same for all cohort speakers and correspond to values of the SI model. The combined mean vector $\boldsymbol{\mu}_m^{SA}$ of the $m-th$ Gaussian component can be expressed as

$$\boldsymbol{\mu}_m^{SA} = \mathbf{S}_m \boldsymbol{\lambda}, \tag{1}$$

where $\mathbf{S}_m = [\boldsymbol{\mu}_m^1, \boldsymbol{\mu}_m^2, \ldots, \boldsymbol{\mu}_m^N]$ is the matrix of N cohort speakers, $\boldsymbol{\mu}_m^n$ is the m-th mean vector belonging to the n-th cohort speaker and $\boldsymbol{\lambda} = [\lambda_1, \lambda_2, \ldots, \lambda_N]$ is the estimated vector of weights.

In our approach, first the speakers in the cohort are sorted in ascendant order according to their likelihood scores obtained in speaker identification. Then only one global weight is calculated for the n-th speaker as

$$\lambda_n = \frac{n}{\sum_{j=1}^{N} j} \tag{2}$$

The Equation 2 ensures that the mean vectors of the nearest speaker will have N-times higher weight than those of the most distant speaker and also that $\sum_{n=1}^{N} \lambda_n = 1$ and $\lambda_n > 0 \; \forall n$.

Gender Dependent Model Selection. The easiest and natural way how to perform adaptation on unknown speaker is to try to identify the gender of the speaker and then to use a male or a female model for speech recognition. The gender of the speaker can be identified using GMMs of both genders: the higher value of likelihood determines if the unknown test speaker is a male or a female.

On-Line Model Combination Approach with Gender Identification. The adaptation approaches described below can be combined easily by incorporating the information about the gender to the model combination process. At first, the models of key speakers are again created using MAP and MLLR adaptation of mean vectors, but this time, gender dependent models are used as prior sources. These GD models are trained off-line in several iterations of the standard EM algorithm and they may have different numbers of Gaussian components (mixtures) due to this reason. This number depends for each acoustic model on the available amount of gender specific training data. The gender of the speaker is identified during the phase of speaker identification as in the previous section. The cohort of N nearest speakers is then formed just by speakers with the same gender as it was identified. This constraint on the same gender is not only natural but also practical because the two genders differ in the number of Gaussian components and only models belonging to the same gender have the same structure and theirs corresponding Gaussian components can combined. The linear combination of models is performed in the same way as described in the first approach: only the mean vectors are combined, the other parameters are copied from the prior GD model. This fact also represents the main advantage of this method against the first approach, because the variances from the given GD model perform significantly better than the ones from the SI model.

2.2 Adaptation in the Off-Line Mode

The advantage of the text transcription in the off-line mode from adaptation point of view is that there is enough time to perform one or more recognition passes to create and utilize the phonetic transcription of each speech segment being transcribed. It is also possible to use on-line adaptation methods described in the previous chapter, before the first speech recognition pass is performed. After the first speech recognition pass, the created phonetic transcription can be used for common maximization techniques to calculate more proper weights for linear combination of models of cohort speakers or it is possible to use it for a conventional adaptation method like MLLR. The final adapted model is then used for the second speech recognition pass to create the text transcription of the segment.

Off-Line Model Combination Approach. In this case, the recognized phonetic transcription is used to estimate more accurate weights for combination of means. Variances and mixture weights are again copied from the corresponding prior model (SI or rather GD), because the adaptation is unsupervised and the transcription is not accurate enough to update them too. This time, not only one global weigh is estimated for each speaker, but all Gaussian components of all acoustic models are split into a binary regression tree. During adaptation, the tree is searched down from the root toward the leaves while calculating the vector of weights only for those nodes where sufficient amount of the adaptation data is available. In this case, we use ML estimation to find $\boldsymbol{\lambda}$. Given the sequence of adaptation observations $\mathbf{X} = \{\mathbf{x}_1, \mathbf{x}_2, \ldots, \mathbf{x}_T\}$, where T is the number of frames, and assuming that all observations are independent, the goal is to find $\boldsymbol{\lambda}$ according to

$$\underset{\lambda}{\operatorname{argmax}} \sum_{m=1}^{M} \sum_{t=1}^{T} log\ p(\mathbf{x}_t | \boldsymbol{\lambda}). \tag{3}$$

where M is the number of all Gaussian components associated with observation $\mathbf{x}_t$. The complex solution of Equation 3 can be found in [3] for example.

3 Experimental Evaluation

In the preliminary experiments not presented in this paper, the proposed adaptation methods were tested on several hours of manually segmented broadcast news recorded from radio as well as from TV. We used these recordings as a development set to find the optimal values of N for both on-line and off-line adaptation modes. The results of these experiments showed us that in the on-line mode, the best way is to combine just models of several tens of speakers while in the off-line mode, the number of speakers should be as high as possible because in this case, the weights for model combination are calculated optimally using ML criterion.

3.1 Experimental Setup

In this paper, we present results that were obtained during experimental evaluation of a broadcast news transcription system that was developed for Czech language. Here, the input broadcast news audio stream is not segment into individual segments, but it is processed as a whole and just speech and non-speech segments and speaker change points are detected. Moreover, gender and speaker identification is performed too. The advantage of this approach is that the recognition errors caused by the segmentation unit are eliminated.

The system operates with a lexicon containing 314k items and with a language model based on smoothed bigrams estimated mainly on Czech newspaper texts. It uses models of 41 Czech phonemes and 7 types of noise. These are represented by three-state context independent CDHMM with up to 100 Gaussian components per state. The feature vector is composed of 39 MFCC parameters (13 static coefficients and their first and second derivatives). For speaker recognition, we used GMM with 256 mixtures and just 12th order MFCC (excluding the c0 coefficient). The speech database used for acoustic model training contained 49 hours of speech recordings. From about 1,000 speakers occurring in it, 190 women and 310 men were selected as key speakers. A different amount of adaptation data (from 18 seconds to 25 minutes) was available for them. The evaluation was done on a test database containing 90 minutes of recordings of various Czech TV stations (like NOVA, Prima and CT24). These were obtained within the COST278 project [7] and contained 13759 words.

3.2 Results in the On-Line Adaptation Mode

The results of the first performed experiment are presented in Table 1. At first, we performed just speech/non speech, gender and speaker change detection. The error rate of speech and gender detection module was 0.8 % and 1.2 % respectively. The used speaker change detection approach was based on the BINSEG method [6]. It did not recognize 13 % of all change points that should be recognized and 17 % of all detected change points were wrong.

Next, we also used speaker identification followed by verification and we tried to find out, if the person in the given speech segment is one of those from our set of key speakers. If the verification was not successful, we used only the GD model. The EER (Equal Error Rate) of the speaker verification module was 12.5 %. At last, we focused on model combination. We started with SI cohorts and combination based on sufficient HMM statistics collected during training [2] (MC1 method). Then we used SI cohorts again but the combination was based on the approach proposed in the chapter 2.1 (MC2 method). Finally, we used the same method for model combination but this time with GD cohorts (MC3). The value of N was 25 in all cases.

Table 1. Comparison of various adaptation methods in the on-line transcription mode. The WER of SI models was 23.4 %.

used models or method	GD	SD	MC1	MC2	MC3
WER [%]	21.3	21.2	22.1	21.4	20.8

The results show that the highest reduction of WER was reached by the use of GD models. Speaker verification did not bring much better results. The most effective way to reduce WER is the model combination approach, but only if it is based on gender dependent cohorts and weighting according to the proposed scheme. In this case, the WER was reduced relatively by 11 % against SI model.

3.3 Results in the Off-Line Adaptation Mode

In the second set of experiments (Table 2), we used gender dependent model combination to create transcription of each speech segment in the first pass. This transcription was then used a) to calculate more accurate weights for model combination and b) for adaptation based on MLLR. In the latter case, the model created in the on-line adaptation mode was used as a prior source for transformation. After both types of adaptation, the final adapted model was used for the second speech recognition pass.

The results of this experiment show us that the model combination method performs in this task much better than MLLR. While the WER against the first pass increased when MLLR was used, the model combination approach was able to reduce the WER significantly (relatively by 20 % for N = 190). The reached reduction of WER rose with increasing value of N, because for high values of N, a lot of cohort speakers were selected from the database while their weights were calculated optimally by ML estimation. The model combination approach is also faster than MLLR due to its lower computation complexity.

Table 2. Values of WER [%] after speaker adaptation in the off-line mode

value of N	5	25	50	75	10	150	190
MLLR method	22.2	21.5	21.4	21.6	21.5	21.6	21.7
ML based model combination	21.3	20.4	19.7	19.5	19.4	19.0	18.7

The Influence of the Lexicon Size that is Used in the First Recognition Pass. The main disadvantage of two pass approach is that the total needed computation time (approximately 3x real time) is much higher than the time needed for on-line transcription with SI models. It is because in the two pass scheme, speech recognition is performed two times and a little time is necessary for speaker recognition and ML based combination of mean vectors too. As the most performance affecting part is speech recognition, the high computation complexity of the proposed approach could be eliminated significantly by the use of a smaller lexicon in the first recognition pass. Due to this reason the last presented set of experiments (Table 3) consists in a measuring how much the recognition accuracy of the two-pass approach depends on the size of the lexicon that is used to create the phonetic transcription in the first pass.

Table 3. WER after the first and second speech recognition pass for different sizes of the lexicon that was used in the first pass

lexicon size in the first pass	314 k	200 k	100 k	50 k	10 k
WER [%] after the first pass	23.4	27.3	29.0	32.9	55.3
WER [%] after the second pass	18.7	18.8	19.0	19.1	19.0

In this set of experiments, we used gender dependent model combination for on-line adaptation before the first speech recognition pass. Then the created transcription was used for ML based combination of means as in the previous chapter. Finally, speech recognition in the second pass was performed with our largest available 314 k lexicon.

The results of this experiment showed us that the size of the lexicon used in the first pass has only a minor impact on the final WER of the second pass and that it is really possible to reduce the size of the lexicon in the first pass in order to reach computational savings. Even in the case, when the lexicon containing just 10k of words was used in the first pass so that the WER of this pass was really high (55 %), the final WER was near the same as when the 314 k lexicon was used in both passes.

4 Conclusion

In this paper, we proposed and experimentally verified several speaker adaptation methods that can be suitable for speech recognition tasks where speakers change frequently. This happens in particular in broadcast news, parliament debates, talk-shows, etc. To show the real performance of all these methods, we did not perform the experiments on manually segmented data, but within a real broadcast news transcription system, which uses an automatic speaker change point detection algorithm.

We tried to perform the adaptation in the on-line as well as in the off-line mode. In both of them we reported significantly improved results although the identities of target speakers were not known and the adaptation process was unsupervised.

In the on-line mode, the highest reduction of WER against SI models was reached by the use of a fast model combination approach based on gender and speaker identification. The WER was reduced from 23.4 % to 20.8 % (relatively by 11 %).

In the latter case, the two-pass adaptation framework reduced the WER on the level of 19 % (relatively by 20 %). We also showed that the total computation time needed for this method can be reduced significantly (on the level of 2× real time) when only a small vocabulary is used to create the phonetic transcription in the first speech recognition pass.

Acknowledgments. This work was supported partly by Czech Academy of Science (grant no. 1QS108040569) and partly by Czech Science Foundation (grant no. 102/07/P430).

References

1. Padmanabhan, M., Bahl, L., Nahamoo, D., Picheny, M.: Speaker Clustering and Transformation for Speaker Adaptation in Speech Recognition Systems. IEEE Transactions on Speech and Audio Processing 6(1), 71–77 (1998)
2. Yoshizawa, S., Baba, A., Matsunami, K., et al.: Evaluation on Unsupervised Speaker Adaptation Based on Sufficient HMM Statistics of Selected Speakers. In: Proc. of Eurospeech 2001, vol. 2, pp. 1219–1222 (2001)
3. Huang, C., Chen, T., Chang, E.: Adaptive model Combination for dynamic speaker selection training. In: Proc. ICSLP 2002, vol. 1, pp. 65–68 (2002)
4. Leggetter, C.J., Woodland, P.C.: Flexible Speaker Adaptation Using Maximum Likelihood Linear Regression. In: Proc. ARPA Spoken Language Technology Workshop, pp. 104–109 (1995)
5. Gauvain, J.L., Lee, C.H.: Maximum A Posteriori Estimation for Multivariate Gaussian Mixture Observations of Markov Chains. IEEE Trans. SAP 2, 291–298 (1994)
6. Ždánský, J.: BINSEG: An Efficient Speaker-based Segmentation Technique. In: Proc. of Interspeech 2006, Pittsburgh, USA (2006)
7. Vandecatseye, A., et al.: The COST278 pan-European Broadcast News Database. In: Proc. of LREC 2004, Lisbon (2004)

Spoken Document Retrieval Based on Approximated Sequence Alignment

Pere R. Comas and Jordi Turmo

TALP Research Center, Technical University of Catalonia (UPC)
pcomas@lsi.upc.edu, turmo@lsi.upc.edu

Abstract. This paper presents a new approach to spoken document information retrieval for spontaneous speech corpora. The classical approach to this problem is the use of an automatic speech recognizer (ASR) combined with standard information retrieval techniques. However, ASRs tend to produce transcripts of spontaneous speech with significant word error rate, which is a drawback for standard retrieval techniques. To overcome such a limitation, our method is based on an approximated sequence alignment algorithm to search "sounds like" sequences. Our approach does not depend on extra information from the ASR and outperforms up to 7 points the precision of state-of-the-art techniques in our experiments.

1 Introduction

Since affordable technology allows the storage of large masses of audio media, more and more spoken document sources become available to public access. This great body of spoken audio recordings is mainly unaccessible without accurate techniques of retrieval. Spoken document retrieval (SDR) is the task of retrieving passages from collections of spoken documents according to a user's request or query.

Classically, the approach to SDR problem is the integration of an automatic speech recognizer (ASR) with information retrieval (IR) technologies. The ASR produces a transcript of the spoken documents and these new text documents are processed with standard IR algorithms adapted to this task.

There is a vast literature on SDR for non spontaneous speech. For example, TREC conference had a spoken document retrieval task using a corpus composed of 550 hours of Broadcast News. TREC 2000 edition concluded that spoken news retrieval systems achieved almost the same performance as traditional IR systems [4]. Spontaneous speech contains disfluencies that can barely be found in broadcast news, such as repetition of words, the use of onomatopoeias, mumbling, long hesitations and simultaneous speaking. Little research has been done for spontaneous speech audio, like telephone conversations, lectures and meetings.

In this paper, we present a novel method for spontaneous speech retrieval. The rest of this paper is organized as follows. Section 2 reviews SDR literature. Section 3 describes our approach and Sections 4 and 5 presents the experiments and compares the results achieved by out approach and state-of-the-art approaches. Finally, Section 6 concludes.

P. Sojka et al. (Eds.): TSD 2008, LNAI 5246, pp. 285–292, 2008.
© Springer-Verlag Berlin Heidelberg 2008

2 State of the Art

Traditional text retrieval techniques assume the correctness of the words in the documents. Automatic Speech Recognition introduces errors that challenge traditional IR algorithms. Nevertheless, results show that a reasonable approach to SDR consists in taking the one-best output of ASR (i.e., the most probable sequence of words that generates the input audio) and performing IR on this transcript. It works reasonably well when recognition is mostly correct and documents are long enough to contain correctly recognized query terms. This is the case of TREC 2000 evaluation on Broadcast News corpora [4].

The Spoken Document Retrieval track in CLEF evaluation campaign uses a corpus of spontaneous speech for cross-lingual speech retrieval (CL-SR) [11,13]. CL-SR corpus is the Malach corpus, which is composed of nearly 600 hours of spontaneous speech from interviews with Holocaust survivors. This is a more general scenario than former TREC tracks.

Approaches to SDR can be classified in two categories according to their use of ASR-specific data. Some methods only use the one-best output as is, therefore it is independent of the specific ASR characteristics. Other methods take advantage of additional information supplied by the ASR. Some ASRs may output additional information (it depends on its implementation) such as confidence scores, n-best output, full lattices. The use of this information or other ASR-error models makes dependant of a concrete ASR.

ASR Independent Retrieval
Most of participants in TREC and CL-SR evaluations use ASR independent methods since no additional ASR information is available.

Top ranked participants in CL-SR, see [2,7,5,19], used a wide range of traditional text based IR techniques. Good results were achieved with term-based ranking schemes such Okapi BM25 [14], Divergence From Randomness [3] and Vector Space Models [15]. Most of the work done by the participants was focused on investigating the effects of meta-data, hand-assigned topics, query expansion, thesauri, side collections and translation issues. Some participants used n-gram based search instead of term search. For n-gram search, text collection and topics are transformed into a phonetic transcription, then consecutive phones are grouped into overlapping n-gram sequences, and finally they are indexed. The search consists in finding n-grams of query terms in the collection. Some experiments show how phonetic forms helps to overcome recognition errors. Some results using phonetic n-grams are reported in [6] showing only slightly improvements.

ASR Dependant Retrieval
Experimental results show that the traditional approach consisting of ASR and IR is not much effective if the task requires the retrieval of short speech segments in a domain with higher word error rate. In this cases, other approaches to SDR have been proposed. Most try to improve retrieval performance using additional information specific to the ASR. For example, Srinivasan and Petkovic [18] use an explicit model of the ASR error typology to address the OOV problem. First, they use two ASRs to generate a word transcript and a phonetic transcript of the input audio. Then they build a phone

confusion matrix that models the probability of ASR mistaking any phone for a different one. Finally, the retrieval step uses a Bayesian model to estimate the probability that the phonetic transcript of a speech segment is relevant to the query term.

Another common approach is the use of ASR lattices to make the system more robust to recognition errors. The lattice is an internal ASR data structure which contains all possible outputs given the audio input. For example, experiments in [17] report an improvement of 3.4% in F_1 measure in Switchboard corpus using a combination of word-lattices and phone-lattices as search space. The use of word-lattices alone cannot overcome the problem of OOV words.

3 Our Approach

In this paper, we present a novel method for spontaneous speech retrieval. This method is ASR independent. Our hypothesis to deal with SDR is that, given the presence of word recognition errors in the automatic transcripts, occurrences of query terms in spontaneous speech documents can be better located using approximated alignment between the phone sequences that represent the keywords and the words in the transcripts.

Following this hypothesis, we have implemented PHAST (PHonetic Alignment Search Tool), an IR-engine over large phone sequences. For the sake of efficiency, PHAST is based on the same principles used in BLAST [1], which has been successfully applied to identify patterns in biological sequences: searching small contiguous subsequences (hooks) of the pattern in a biological sequence and extending the matching to cover the whole pattern. Algorithm 1 shows a general view of PHAST.

Algorithm 1: PHAST algorithm
Parameter: $\mathcal{D}$, document collection
Parameter: $\mathcal{KW}$, keywords

```
for all d ∈ D, w ∈ KW do
  while h = detection_φ(w, d) do
    s = extension_φ(w, h, d)
    if relevant(s, h) then
      update tf(w, d)
    end if
  end while
end for
Rank collection D
```

It is a two-step process: first, keyword term frequency is computed using phonetic similarity, and second, a standard document ranking process takes place. This process is language independent, given the phone sequences. The input data is a collection of documents transcribed into phone sequences $\mathcal{D}$, and a set of keywords phonetically transcribed $\mathcal{KW}$. In the next sections, the ranking process and the functions $detection_{\phi}()$, $extension_{\phi}()$ and $relevant()$ are described.

Most of the state-of-the-art ranking functions can be used to build the document ranking from the tf scores computed by PHAST. The only condition is that these functions can deal with non-integer values as term frequency. We have tested several different ranking functions as shown in Section 4.

Global alignment		Semi-local alignment
---juniks--sʌ- -----------n		---juniks--sʌn -----------
ɪzəjuniksɛtsʌmwəʊrksteɪʃən		ɪzəjuniksɛtsʌmwəʊrksteɪʃən

Fig. 1. How global and semi-local affects the alignment of the phonetic transcription of keyword "UNIX-sun" and the sentence "is a unique set some workstation"

Function $detection_\phi(w, d)$**:** This function detects hooks h within document d considering keyword w and using the searching function ϕ. Similarly to Altschul et al. [1], function ϕ has been implemented as follows. Given a set of phonetically transcribed keywords, a deterministic finite automaton DFA$_k$ is automatically built for each keyword k in order to recognize all its possible substrings of n phones. For instance, given $n = 3$ and the keyword "alignment", which is phonetically transcribed as [əlaɪnmɪnt][1], there are seven phone substrings of length three (3-grams): əla, laɪ, aɪn, ɪnm, nmɪ, mɪn and ɪnt. One DFA is automatically built to recognize all seven 3-grams at once. Using these DFAs, the collection is scanned once to search for all the hooks.

Function $extension_\varphi(w, h, d)$**:** After a hook h is found, PHAST uses φ to extend it in document d and to compute its score value s. Function φ has been implemented with a phonetic similarity measure due to the success achieved in other research domains [8]. Concretely, we have used a flexible and mathematically sound approach to phonetic similarity proposed by Kondrak [9]. This approach computes the similarity $\Delta(a, b)$ between two phone sequences a and b using the edit distance implemented with a dynamic programming algorithm. This implementation includes two new operations of compression and expansion that allow the matching of two contiguous phones of one string to a single phone from the other. (e.g., [c] sounds like the pair [tʃ] rather than [t] or [ʃ] alone). It also allows a semi-local alignment to prevent excessive scattering, its effect is depicted in Figure 1.

The cost of the edit distance operations considers a measure of inter-phoneme similarity which is based on phone features. The features we have used are based on those used in [10] and enhanced with extra sounds from Spanish.

Score value s is finally computed by normalizing the similarity $\Delta(a, b)$ by the length of the matching. n is the length of the longest string, either a or b:

$$s = \frac{\Delta(a, b)}{\frac{\Delta(a,a)}{n} \cdot length(a, b)}$$

Function $relevant(s, h)$**:** This judges how the occurrence of w at h with score s is relevant enough for term frequency. Given matching score s and a fixed threshold t, tf is updated only if $s > t$. Initial experiments have shown that, on one hand, the best results are achieved when low scoring matchings are filtered out, and on the other hand, the best results are achieved with $tf \leftarrow tf + s$ rather than $tf \leftarrow tf + 1$. This helps to filter false positives, specially for very common syllables.

[1] We have used the international phonetic alphabet (IPA) notation for phonetic transcriptions.

4 Experimental Setting

We have performed an indirect evaluation of SDR considering IR in the framework of Question Answering (QA). QA is the task of finding exact answers to user questions formulated in natural language in a document collection. Document retrieval is a main step in QA, it discards documents with small probability of containing the answer. We have evaluated document and passage retrieval.

Empirical studies [12] show how better results in QA are achieved using a dynamic query adjusting method for IR. First the question is processed to obtain a list of keywords ranked according a linguistically motivated priority. Then some of the most salient keywords are sent to the IR engine as a boolean query. A word distance threshold t is also set in order to produce passages of high keyword density. All documents containing those passage are returned as an unordered set. If this set is too large or small, keywords and t may be altered iteratively. This ranking algorithm is used as a baseline for our experiments.

For a proper evaluation of SDR for QA we need a corpus of spontaneous speech documents with both manual and automatic transcripts. Manual transcript is an upper bound of the system performance and allows to calculate the drop off due to word error rate. CL-SR corpus is very interesting for this task, but unfortunately it lacks of manual transcripts and its use is restricted to CLEF evaluation campaign.

We have conducted experiments using a set of 76 keyword sets extracted from natural language questions with a corpus of 224 transcripts (more than 50,000 words) of automatically transcripted speeches from the European and Spanish parliaments.[2] Automatic transcripts have an average word error rate of 26.6%. We expect that the correct answer to the question is contained in one or more of the documents returned in the TOPn. In this setting we are not judging the relevance of the documents to a certain topic but the number of queries returning the answer over the total number of queries.

We call DQ_{ref} to the baseline ranking algorithm over reference corpus, DQ_{auto} is the same over the automatic transcribed corpus. The difference between both shows the performance fall-out due to ASR action. Baseline systems return an unordered set of documents, DQ_{ref} returned an average of 3.78 documents per query and DQ_{auto} an average of 5.71. Therefore we have chosen P3 and P5 as our main evaluation measures. P1 is also provided.

We have set up four systems for term detection: Words (WRD), 3-grams of characters (3GCH), 3-grams of phones (3GPH) and PHAST.

These systems have been used for automatic transcripts combined with DQ and three standard document ranking functions: Okapi BM25 (BM25), vector space models (VSM), and divergence from randomness (DFR).

We have conducted a 5-fold crossvalidation. For each fold the full question set has been randomly split in two subsets: a development set of 25 questions and a test set of 51 questions. For each fold the best parameter setting has been selected and applied to the test set. The best parameters for each ranking function have been the following. BM25: values in $(0, 1]$ for a and $[0, 0.1]$ for b. DFR: best model has been $I(n)LH1/H2$ [3] in almost any experiment. VSM: the *nsn* scheme [16] was the best in almost any

[2] Transcripts where provided by TALP Research Center within the framework of TC-STAR project `http://www.tc-star.org`

experiment. For PHAST there are also two tunable parameters. From an empirical basis, we have fixed $r = 0.80$ and $n = 4$ for both passage and document retrieval experiments. The results are reported in Section 5.

5 Results

5.1 Document Retrieval

Table 1 shows the results of the holdout validation. The baseline system DQ has been used with reference manual transcripts (DQ_{ref}) and with automatic transcripts (DQ_{auto}). Also traditional word-based retrieval has been tested over the reference and automatic transcripts as $WORD_{ref}$ and $WORD_{auto}$ respectively. The n-gram based retrieval has been used over the automatic transcripts ($3GCH_{auto}$ and $3GPH_{auto}$). PHAST obtains better results than any other system working on automatic transcripts.

We have used *precision at x* as evaluation measure. It is defined as the number of queries returning a gold document within the top x results of the ranking. As we have noted in Section 4, the baseline system does not return a ranked list of documents but an unordered set of documents judged relevant. This is why only one result has been reported in Table 1 for DQ. DQ_{ref} returned an average of 3.78 documents per query and DQ_{auto} returned an average of 5.71 documents per query. Therefore, we have chosen precision at 3 (P3) and precision at 5 (P5) as our main evaluation measures. We also provide P1 for the sake of completeness. In this setting, precision and recall measures are equivalent since we are interested in how many times the IR engine is able to return a gold document in the top 3 or 5 results.

For each system we include the average holdout validation P1, P3 and P5 for the three weighting schemes and five systems. The results are discussed in terms of P5 for an easier comparison with DQ. Similar conclusions may be achieved with P3.

Precision loss between DQ_{ref} and DQ_{auto} is 26.3 points. This is due solely to the effect of ASR transcription. For $WORD_{ref}$, the best result is 67.45%, 16.5 points behind DQ_{ref}. With automatic transcripts $WORD_{auto}$ loses 21.3% with respect to $WORD_{ref}$, this loss is comparable to the 26.3% for DQ. The best result of $WORD_{ref}$ (at P5) is still worse than DQ_{ref}, these results support what stated in Section 4: better results in QA-oriented retrieval would be achieved with DQ rather than traditional ranking techniques.

The family of n-gram systems outperforms $WORD_{auto}$ and DQ_{auto} by almost 10 points, but they are still 2 points behind $WORD_{ref}$ and 19 behind DQ_{ref}. In terms of

Table 1. Results of document retrieval. Results are in percentage.

	Okapi BM25			Vector Space Model			Divergence from Rand.		
System	P1	P3	P5	P1	P3	P5	P1	P3	P5
DQ_{ref}					**84.21**				
DQ_{auto}					**57.89**				
$WORD_{ref}$	43.92	57.25	65.10	36.86	52.15	60.39	45.88	59.60	**67.45**
$WORD_{auto}$	38.03	51.37	54.50	31.37	49.02	54.90	36.46	52.94	**56.07**
$3GCH_{auto}$	16.47	52.94	**65.10**	8.84	34.50	50.19	10.98	46.67	59.29
$3GPH_{auto}$	23.53	47.45	**58.82**	8.62	30.58	44.31	13.72	41.96	56.07
$PHAST_{auto}$	48.62	71.37	**75.29**	31.37	56.47	65.47	46.67	67.06	72.15

P1 and P3, n-gram scores are behind WORD$_{auto}$ ones. PHAST outperforms DQ$_{auto}$ in 18.7 points and it is behind DQ$_{ref}$ by 10.5. In P3, PHAST has still the best performance overall, 15.5 points behind DQ$_{ref}$. PHAST also outperforms 3GCH$_{auto}$ by 10 points, 3GPH$_{auto}$ by 17 and WORD$_{ref}$ by 7.8.

PHAST is better than to WORD, 3GCH and 3GPH approaches in two aspects. When the ASR missrecognizes one of the keywords (e.g., a proper name) it is impossible for WORD to find this term, and this information is lost. Thus, PHAST outperforms WORD in term matching capabilities allowing an approximate matching of terms. This implies a raising in coverage. The n-gram approach also improves coverage and allows approximate matching but it has no control over n-grams distribution in the text, so it lacks of a high precision (3GPH and 3GCH only outperforms WORD at P5). PHAST provides more precise and meaningful term detection.

5.2 Passage Retrieval

Table 2 shows the results of our experiments. DQ$_{ref}$ and DQ$_{auto}$ are the baseline algorithm over manual reference transcripts and automatic transcripts respectively. DQ$_{PHAST}$ is the same baseline using PHAST algorithm for term detection.

Recall is the number of queries with correct answer in the returned passages. Precision is the number of queries with correct answer if any passage is returned.

There is a 40 point loss between automatic and manual transcripts in precision and recall. In average, DQ$_{ref}$ has returned 3.78 passages per query while DQ$_{auto}$ has returned 5.71. In automatic transcripts DQ$_{auto}$ obtains worse results even returning more passages than in reference transcripts. This is due to the fact that DQ$_{auto}$ drops more keywords (uses an average of 2.2 per query) to build the passages than DQ$_{ref}$ (uses an average of 2.9). Since a substantial number of content words are ill-transcribed, it is easier to find a passage containing n keywords than containing $n + 1$. In fact, DQ$_{auto}$ only uses just one keyword in 24 queries, while DQ$_{ref}$ does it in 10 queries.

This results show how term detection is decisive for passage building. The difference between DQ$_{auto}$ and DQ$_{ref}$ in passage retrieval is 40% while it is "only" 29% in document retrieval. Passage retrieval adds a new constraint to the task of document retrieval: the keywords must be close together to be retrieved. Therefore, any transcript error changing a keyword in the transcript may prevent the formation of a passage. Because of its lack of redundancy, passage retrieval is less robust than document retrieval.

DQ$_{PHAST}$ returns an average of 3.80 passages, almost the same than DQ$_{ref}$, using 2.69 keywords. It surpasses DQ$_{auto}$ by 18% in precision and 17% in recall, taking an intermediate place between DQ$_{auto}$ and DQ$_{ref}$. The differences among DQ$_{PHAST}$, DQ$_{auto}$ and DQ$_{ref}$ are similar in passage and document retrieval.

Table 2. Results of passage retrieval. Precision, recall and average number of passages returned per query.

System	Precision	Recall	Passages
DQ$_{ref}$	86.56%	76.31%	3.78
DQ$_{auto}$	46.77%	38.15%	5.71
DQ$_{PHAST}$	64.61%	55.26%	3.80

6 Conclusions

In this paper we have presented a novel approach to spoken document retrieval. We can overcome part of automatic speech recognition errors using a sound measure of phonetic similarity and a fast search algorithm based on phonetic sequence alignment. This algorithm can be used in combination with traditional document ranking models. The results show similar improvement in passage retrieval and in document retrieval tasks. Our approach significantly outperforms other standard state-of-the-art systems by 18 and 7 points for passage retrieval and document retrieval respectively.

References

1. Altschul, S., Gish, W., Miller, W., Meyers, E.W., Lipman, D.J.: Basic local alignment search tool. Journal of Molecular Biology 215, 403–410 (1990)
2. Alzghool, M., Inkpen, D.: University of Ottawa's participation in the CL-SR task at CLEF 2006. In: CLEF 2006 (2006)
3. Amati, G., Van Rijsbergen, C.J.: Probabilistic models of information retrieval based on measuring the divergence from randomness. In: TOIS 2002 (2002)
4. Garofolo, J., Auzanne, G., Voorhees, E.: The TREC spoken document retrieval track: A success story. In: TREC 2000 (2000)
5. Inkpen, D., Alzghool, M., Islam, A.: Using various indexing schemes and multiple translations in the CL-SR task at CLEF 2005. In: CLEF 2005 (2005)
6. Inkpen, D., Alzghool, M., Jones, G., Oard, D.W.: Investigating cross-language speech retrieval for a spontaneous conversational speech collection. In: HLT-NAACL (2006)
7. Jones, G.J.F., Zhang, K., Lam-Adesina, A.M.: Dublin city university at CLEF 2006: Cross-language speech retrieval (CL-SR) experiments. In: CLEF 2006 (2006)
8. Kessler, B.: Phonetic comparison algorithms. Transactions of the Philological Society 103, 243–260 (2005)
9. Kondrak, G.: A new algorithm for the alignment of phonetic sequences. In: NAACL 2000 (2000)
10. Kondrak, G.: Algorithms for Language Reconstruction. Ph.D. thesis, University of Toronto (2002)
11. Oard, D.W., Wang, J., Jones, G.J.F., White, R.W., Pecina, P., Soergel, D., Huang, X., Shafran, I.: Overview of the CLEF-2006 cross-language speech retrieval track. In: CLEF 2006 (2006)
12. Paşca, M.: High-performance, open-domain question answering from large text collections. Ph.D. thesis, Southern Methodist University, Dallas, TX (2001)
13. Pecina, P., Hoffmannovâ, P., Jones, G.J.F., Zhang, Y., Oard, D.: Overview of the CLEF-2007 cross-canguage speech retrieval track. In: CLEF 2007 (2007)
14. Robertson, S.E., Walker, S., Spärck-Jones, K., Hancock-Beaulieu, M., Gatford, M.: Okapi at TREC-3. In: TREC-3 (1995)
15. Salton, G. (ed.): Automatic text processing. Addison-Wesley, Reading (1988)
16. Salton, G., Buckley, C.: Term weighting approaches in automatic text retrieval. Technical report, Cornell University (1987)
17. Saraclar, M., Sproat, R.: Lattice-based search for spoken utterance retrieval. In: HLT-NAACL 2004 (2004)
18. Srinivasan, S., Petkovic, D.: Phonetic confusion matrix based spoken document retrieval. In: SIGIR 2000 (2000)
19. Wang, J., Oard, D.W.: CLEF-2005 CL-SR at Maryland: Document and query expansion using side collections and thesauri. In: CLEF 2005 (2005)

Pitch Accents, Boundary Tones and Contours: Automatic Learning of Czech Intonation

Tomáš Duběda[1] and Jan Raab[2]

[1] Institute of Phonetics, Charles University in Prague
[2] Institute of Formal and Applied Linguistics, Charles University in Prague
dubeda@ff.cuni.cz, raab@ufal.mff.cuni.cz

Abstract. The present paper examines three methods of intonational stylization in the Czech language: a sequence of pitch accents, a sequence of boundary tones, and a sequence of contours. The efficiency of these methods was compared by means of a neural network which predicted the f_0 curve from each of the three types of input, with subsequent perceptual assessment. The results show that Czech intonation can be learned with about the same success rate in all three situations. This speaks in favour of a rehabilitation of contours as a traditional means of describing Czech intonation, as well as the use of boundary tones as another possible local approach.

Keywords: Pitch accents, boundary tones, contours, Czech, prosody, automatic learning, neural networks.

1 Rationale

Systematic descriptions of Czech intonation available are based exclusively on contours (e. g. [2,10,8]), while no systematic tonal approach has been proposed thus far. Aside from this, the contour-based models are heavily biased towards the nuclear parts of intonation phrases (i.e. mostly final parts, with important phonological and paralinguistic functions), considering prenuclear material (i.e. parts preceding the nucleus) implicitly as uninteresting. The only extensive account of prenuclear intonation, oriented towards speech synthesis, can be found in [9].

From a theoretical viewpoint, the current lack of tone-based models might be more than an effect of historical inertia (most traditional approaches of intonation have been contour-based, while discrete targets are a relatively recent innovation): in [7], for instance, it is argued that Czech intonation is resistant to modeling by means of local events, and that the relevant building stone is the stress unit with its holistic contour. In addition to the pitch accent/stress unit dichotomy, a third theoretical possibility would be that intonation is based on boundary tones at either edge of the stress unit (cf. J. Vaissière's theory of "English as a stress language" and "French as a boundary language", [11]). It is precisely these three hypotheses that we test in the present study.

Czech is a non-tonal language with word stress located always on the first syllable of stressable words (cf. examples below). Content words may form stress units with following grammatical words. In some cases, stress units may start with one or more unstressed syllables due to preceding clitics. Czech accents are felt as weak by

P. Sojka et al. (Eds.): TSD 2008, LNAI 5246, pp. 293–301, 2008.
© Springer-Verlag Berlin Heidelberg 2008

many foreign listeners, which may be due to low tones which frequently accompany the stressed syllable, as well as to the troublesome interference of segmental vowel length.

2 Goal, Material and Methodology

2.1 Goal

The question underlying this research is whether intonation is better anchored as a sequence of pitch accents, as a sequence of boundary tones, or as a sequence of contours. The tool for this comparison is a neural network which should learn how to predict the f_0 curve with the input provided. The output will then be assessed numerically and perceptually. A similar – though not identical – approach to intonational stylization can be found e. g. in [3].

We shall thus compare the target approach with the contour approach, and, within the target approach, compare the importance of pitch accents and boundary tones. We shall study non-nuclear intonation only. In this manner, there are three types of input to the neural network, responding to three theoretical assumptions:

1. Prenuclear intonation in Czech can be modeled as a sequence of pitch accents (PAs). We use bitonal, phonetic pitch accents which typically code intonational changes within a 3-syllable window centred around the stressed syllable, each tone expressing the change from one syllable to the other:

S	H		H	H		S	L	H	H	
ˈpru:	mɲɛr	nɛ:	ˈvi:	nɔ	sɪ	ˈt͡ʃɛs	ki:x	ˈvɪ	ɲɪt͡s	...

Průměrné výnosy českých vinic...
'Average yields of Czech vineyards...'

There are three relative tones: Higher, Same and Lower. The first tone of the intonation phrase is always (formally) S. Thick frames mark stress units.

2. Prenuclear intonation in Czech can be modeled as a sequence of boundary tones (BTs). We use two phonetic boundary tones in each stress unit: one at the beginning and one at the end. In this way, we typically code a 3-syllable window centered around the last syllable of each stress unit:

S		L	H		S	S	L	H	H	
ˈpru:	mɲɛr	nɛ:	ˈvi:	nɔ	sɪ	ˈt͡ʃɛs	ki:x	ˈvɪ	ɲɪt͡s	...

3. Prenuclear intonation in Czech can be modeled as a sequence of contours (CONTs). We use one contour in each stress unit, only reflecting unit-internal intonational behaviour:

HL			H			L		H		
ˈpru:	mɲɛr	nɛ:	ˈvi:	nɔ	sɪ	ˈt͡ʃɛs	ki:x	ˈvɪ	ɲɪt͡s	...

The inventory of attested contours is: H (rising), HL (rising-falling), L (falling) and S (flat).

2.2 Material

The material was a set of manually-annotated recordings of a text read by 5 speakers of standard Czech (3 men and 2 women). The length of the text was ± 900 syllables, out of which ± 700 were prenuclear, i.e. relevant for our study. The total number of syllables processed was ± 3 500.

The tonal annotation, as exemplified above, was carried out by the first author on a perceptual basis with simultaneous inspection of the manually-checked f_0 curve. The labeling was perception-based and syllable-synchronized (each syllable to be annotated received exactly one tone). Three relative tones were used (Higher, Lower and Same). The terms "pitch accents" and "boundary tones" as used in this paper are not part of a phonological model, and should better read "phonetic PAs/BTs" or "pre-theoretical PAs/BTs".

2.3 Network Input

The task of the neural network is to predict intonation with each of the three inputs (PAs, BTs and CONTs). The prediction runs syllable by syllable, from left to right. At each stage, the network knows the real f_0 value of the preceding syllable.

We are dealing with three types of input, of which PAs and BTs are perfectly comparable, whereas CONTs differ in several respects, mainly because the contour spreads over several syllables. Instead of taking the stress unit for the basic unit of input, we decided to maintain the syllable-by-syllable approach, but to enrich it in the case of CONTs by rough positional information, so that the network can see the relative position of the syllable within the stress unit. This is in fact not an artificial solution because in the case of PAs and BTs, the network also gets some kind of positional information (tonal annotation is present on specific syllables only), and it also holds in terms of the quantity of information provided, which is a prerequisite for transparent comparison (cf. the bits approximation below).

Along with tonal and, in the case of CONTs, positional data, the input also contains scaling information, consisting of the average f_0 over the intonation phrase, and the f_0 measured on the first syllable of the nucleus. This latter value was necessary to guarantee a good matching between the end of the prenuclear part, which was the object of prediction, and the beginning of the intonation nucleus, which was not modeled but simply copied from the original sentence. The proportion between the two scaling values also serves as a rough coding of the overall declination.

The input parameters for each syllable are summarized in Table 1.

The predicted parameter is the f_0 of the processed syllable, as measured in the centre of the syllable nucleus.

In terms of bits of information (cf. the novel view of prosody as a channel for information coding in [6]), the input for each syllable can be approximated as displayed in Table 2 (scaling information and history, being identical for all three situations, are omitted here):

Table 1. Input parameters for an individual syllable in the three training/testing situations. In tonal targets, "0" means "no target" (a syllable without a tone).

	PAs	BTs	CONTs
Tonal information	- tonal target (H; L; S; 0) of the processed syllable		- contour of the stress unit to which the processed syllable belongs (H; L; HL; S)
Positional information			- relative position of the syllable within the stress unit (syllable number/ unit length)
Scaling information	- mean f_0 of the intonation phrase - f_0 of the first nuclear syllable		
History	- f_0 of the preceding syllable (does not apply to phrase-initial syllables)		

Table 2. Bit approximation of input parameters

Type of input	Information	Total
PAs and BTs	1 tone out of 4, i.e. 2 bits	2 bits
CONTs	1 contour out of 4, i. e. 2 bits, but identical for the whole stress unit (average length 3.1 syllables), thus 2 / 3.1 = 0.65 bits 1 syllable position out of average 3.1, i. e. $\log_2 3.1$ = 1.63 bits	2.28 bits

This rough estimation makes it evident that adding positional information compensates for the tonal underspecification of CONTs, making the comparison of the three situations more realistic. Despite the fact that the network can react differently to positional information which is implicit, as in PAs and BTs, and to that which is explicit, as in CONTs, it was hard to imagine any other satisfactory solution. As it is with the setting described, CONTs have a slightly richer input than PAs and BTs (a difference of 0.28 bits for the differing part).

Additionally, we used two more situations which should serve as a reference delimiting the quality bottom and ceiling of our prediction: the NULL version (no tonal information, only positional information, scaling and history), and the ORIG version (original sentence).

2.4 Network Architecture

For the training, we used the Matlab software. We tried many different configurations of the network. We chose such a configuration, which gave satisfactory results provided it was simple enough (therefore it should generalize). Here we present detailed description of our final configuration.

The network was trained in 300 iterations. We ran every training 5 times and averaged the output in order to reduce individual diversions of a single run.

For PAs and BTs, the network was composed of 5 input neurons, and for CONTs, of 6 input neurons. The first three neurons represented three possible tonal descriptions of the current syllable (0 or 1), the next two represented scaling values (see Table 1). An extra input neuron with a relative position in the stress unit was added to CONTs. We used one hidden layer composed of 50 neurons. There was one neuron at the output. All pitch values at the input were converted into semitones to make the pitch changes independent of the pitch range.

Since we had a relatively small set of data, we used so-called "cross-validation" for the training and application of the network. We divided the data into thirds. We trained the network on 2/3 of the data and applied it to the remaining 1/3. This process was repeated 3 times with different thirds for application. Thus we obtained predicted values for the whole data.

2.5 Perceptual Evaluation

The 10 selected sentences were re-synthesized with the f_0 values obtained as the average of 5 runs of the network, for each of the four inputs (PAs, BTs, CONTs and NULL). The technique used was PSOLA under Praat [1]. In this way, we obtained 40 predicted items (10 sentences x 4 inputs). Values for nuclear stress units were not predicted, but copied from the original. This set was completed by the 10 unmodified sentences (ORIG).

The perceptual assessment of these sentences was carried out by means of an interactive application where each question contained two versions of the same sentence. There were two main possibilities as to how to formulate the question: either in terms of similarity with the original, or in terms of naturalness, independent of the original. We decided to adopt the second approach because it is less demanding for the listeners: instead of choosing a point on a scale, they only have to compare the naturalness of two same-worded sentences. Also, the question of similarity with the original is partly answered by the acoustic measures (see below).

Since using all two-term combinations of 5 sets of 10 sentences would have led to 10 * 10 = 100 questions, which would have made the test extremely lengthy, we only retained all two-term combinations of PAs, BTs and CONTs (30 questions), and added 10 combinations of NULL with a balanced selection of PAs, BTs and CONTs, and another 10 combinations of ORIG with a balanced selection of PAs, BTs and CONTs. This augmented the number of evaluated pairs of sentences to 50. The order of sentences, speakers and versions was randomized.

The instruction was: "In each of the pairs of sentences, decide whether sentence A sounds more natural, equally natural, or less natural than sentence B. Only intonation should be taken into account." This three-term choice, which includes the "same" answer, prevents the listeners from being categorical when they hear no or almost no difference, and makes the data richer because the "same" answers can be filtered when a more categorical approach is needed.

There were 31 respondents, all BA-level students of linguistics or modern languages with no hearing impairments, speaking Czech as their only mother tongue.

3 Results

In an informal inspection of the output, it turned out that the described network meets well with our expectations: in most cases, it reacts correctly to categorical tonal input

(e. g., for an H tone in input, it predicts a rise), it generates declination, and it ensures a rather good fit with the nuclear contour. The predicted prenuclear intonation seems clumsy at certain points, but generally, it leads to a clear demarcation of words. Some sentences are nearly perfect. Generally speaking, the predicted intonation is somewhat flatter compared to the original, which is a product of the network learning as well as of the averaging over five runs. As expected, the NULL version is very monotonous. The nuclear parts, copied from the original, contribute rather strongly to the overall quality impression.

3.1 Numeric Evaluation of the Output

We evaluated the difference between the original f_0 shape and the predicted one in terms of correlation coefficients. Only prenuclear parts were taken into account. The results are displayed in Table 3.

Table 3. Performance of the prediction in acoustic terms

Prediction type	f_0 correlation with the original
PAs	0.48
BTs	0.50
CONTs	0.53
NULL	0.37

The overall correlation seems to be rather poor. PAs, BTs and CONTs cluster around 0.50, with only slight differences between them. NULL has the worst score, as expected. Inter-speaker variability is relatively small, except for BTs, where the highest correlation is 0.64 and the lowest 0.32. For ORIG, the correlation would be, of course, 1.00.

3.2 Perceptual Evaluation of the Output

The perceptual assessment of PAs, BTs and CONTs altogether against NULL (prediction with no tonal information) and ORIG (original sentences) is schematized in Figure 1. The vertical division line in each rectangle expresses the preference ratio: the further it is from a variant, the better this variant was assessed, e.g. there was a 54% preference of PAs/BTs/CONTs over NULL. The error lines correspond to inter-speaker standard deviation.

There was a 70% preference for ORIG over PAs/BTs/CONTs, which indicates that the listeners were able to separate original sentences from the predicted ones. On the other hand, the NULL versions were assessed only slightly worse than PAs/BTs/CONTs.

Fig. 1. Performance of the prediction in perceptual terms – PAs/BTs/CONTs vs. NULL and ORIG

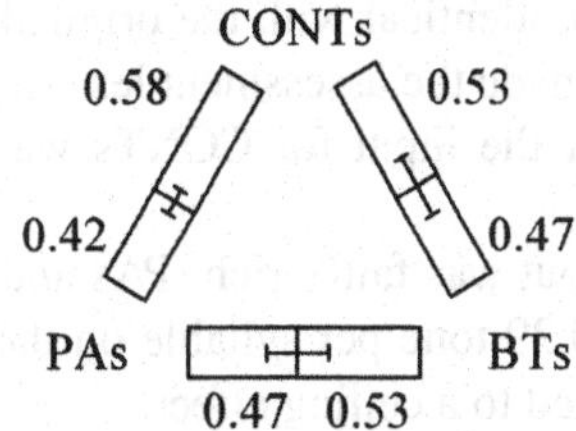

Fig. 2. Performance of the prediction in perceptual terms – PAs vs. BTs vs. CONTs

This shows, among other things, that the positional information alone is sufficient to predict acceptable intonation. However, if we compare this fact with the acoustic results contained in Table 2, we can conclude that the listeners were rather tolerant towards versions with low-variability intonation, as predicted without tonal information.

The perceptual assessment of pairs of versions (PAs, BTs and CONTs between them) is displayed in Figure 2.

The scheme shows that the only significant difference was found between CONTs and PAs, where CONTs were assessed better. The other pairs of versions show only insignificant preferences, which are however in line with the results contained in Table 2. To test the robustness of Figure 2, we calculated the same data a) without the "same" answers (i.e. we only counted answers "A is better" or "B is better"); b) without outlying listeners (exhibiting low correlation with average results); c) without listeners who were bad at telling the difference between ORIG and the other versions. In all three cases, the obtained values confirmed the setting displayed in Figures 1 and 2.

When examining speakers individually, one obtains the following orders of preferences:

- CONTs were assessed better than PAs in all speakers;
- CONTs were assessed better than BTs in 3 speakers out of 5;
- BTs were assessed better than PAs in 2 speakers, the same in 1 speaker, and worse in 2 speakers.

4 Conclusion

Before giving a general account of the results, we should point out possible sources of chaos and biases (some of which have been mentioned above):

- the most off-centre perceptual assessment is 0.70, expressing the preference of ORIG over PAs/BTs/CONTs; this indicates that the prediction led generally to rather acceptable results, which outranked the original sentences in 30% of the cases; the rest of the differences lies in the somewhat narrow interval 0.42–0.58;
- the listeners may have used different strategies to assess the output: among other things, it seems that they preferred flat intonation with tiny variations over rich excursions where the risk of sounding false is greater;
- the network processing may have been too gross to account for differences which are rather small;

- nuclear parts of intonation, identical with the original, increased considerably the quality of the output, and made the assessment less sharp;
- one should not forget that the input for CONTs was slightly overspecified (see Section 2.3);
- generally speaking, the input was fairly rich (PAs and BTs: 0.64 tone per syllable on the average; CONTs: 0.39 tone per syllable on the average + positional information), which may have led to a ceiling effect.

Bearing all these facts in mind, one can summarize the results as follows:

- Surprisingly, within our methodology, Czech intonation can be learned with about the same success rate by means of PAs, BTs and CONTs. CONTs seem to be slightly more successful than the other two types of input, but this fact may be a product of a slightly richer input.
- With regard to the hypothesis that prenuclear intonation can be universally best described by means of PAs, our results speak in favour of a rehabilitation of CONTs as a traditional means of describing Czech intonation, as well as of the use of boundary tones as another possible local approach.
- However, we were not able to demonstrate within the given setting that either of the three approaches was really worse than the other two.
- Since our results may be affected by the method used, namely by the network function, we should test alternative methods, especially HMM, to validate or refine these preliminary conclusions.
- Our conclusions are relevant to the level of automatic learning. Implications regarding the processes of intonation coding and decoding in humans can only be indirect.

Acknowledgments. This research was carried out under the GAČR 405/07/0126 and the Information Society 1ET101120503 grants.

References

1. Boersma, P.: Praat, a system for doing phonetics by computer. Glot International 5(9/10), 341–345 (2001)
2. Daneš, F.: Intonace a věta ve spisovné češtině [Intonation and the sentence in standard Czech]. Praha: ČSAV (1957)
3. Demenko, G., Wagner, A.: The stylization of intonation contours. Speech Prosody 2006. Dresden, TUD (2006)
4. Duběda, T.: Structural and quantitative properties of stress units in Czech and French. In: Braun, A., Masthoff, H.R. (eds.) Festschrift for Jens-Peter Köster, pp. 338–350. Steiner, Stuttgart (2002)
5. Hirst, D., Di Cristo, A. (eds.): Intonation Systems. A Survey of Twenty Languages. Cambridge University Press, Cambridge (1998)
6. Kochansky, G.: Prosody beyond fundamental frequency. In: Sudhoff, S., et al. (eds.) Methods in Empirical Prosody Research, pp. 89–122. Walter de Gruyter, Berlin (2006)
7. Palková, Z.: Einige Beziehungen zwischen prosodischen Merkmalen im Tschechischen. XIVth Congress of Linguists, Berlin, vol. I, pp. 507–510 (1987)

8. Palková, Z.: Fonetika a fonologie češtiny [Phonetics and phonology of Czech]. Praha, Karolinum (1997)
9. Palková, Z.: The set of phonetic rules as a basis for the prosodic component of an automatic TTS synthesis in Czech. In: Palková, Z., Janíková, J. (eds.) Phonetica Pragensia X, pp. 33–46. Praha, Karolinum (2004)
10. Romportl, M.: Studies in Phonetics. Academia, Praha (1973)
11. Vaissière, J.: Language-Independent Prosodic Features. In: Cutler, A. (ed.) Prosody. Models and Measurements, pp. 53–66 (1983)

On the Use of MLP Features for Broadcast News Transcription*

Petr Fousek, Lori Lamel, and Jean-Luc Gauvain

Spoken Language Processing Group, LIMSI-CNRS, France
{fousek,lamel,gauvain}@limsi.fr

Abstract. Multi-Layer Perceptron (MLP) features have recently been attracting growing interest for automatic speech recognition due to their complementarity with cepstral features. In this paper the use of MLP features is evaluated in a large vocabulary continuous speech recognition task, exploring different types of MLP features and their combination. Cepstral features and three types of Bottle-Neck MLP features were first evaluated without and with unsupervised model adaptation using models with the same number of parameters. When used with MLLR adaption on a broadcast news Arabic transcription task, Bottle-Neck MLP features perform as well as or even slightly better than a standard 39 PLP based front-end. This paper also explores different combination schemes (feature concatenations, cross adaptation, and hypothesis combination). Extending the feature vector by combining various feature sets led to a 9% relative word error rate reduction relative to the PLP baseline. Significant gains are also reported with both ROVER hypothesis combination and cross-model adaptation. Feature concatenation appears to be the most efficient combination method, providing the best gain with the lowest decoding cost.

1 Introduction

Over the last decade there has been growing interest in developing automatic speech-to-text transcription systems that can process broadcast data in a variety of languages. The availability of large text and audio corpora on the Internet has greatly facilitated the development of such systems, which nowadays can work quite well on unseen data that is similar to what has been used for training. However, there is still a lot of room for improvement for all system components including the acoustic front end, the acoustic, pronunciation and language models. One promising research direction is the use of MLP features in a large speech recognition task, in this case, the transcription of Arabic broadcast news from the DARPA GALE task.

Features for speech-to-text obtained from neural networks have recently been included as a component of a state-of-the-art LVCSR systems [1]. They are known to contain complementary information to cepstral features, which is why most often both features are used together.

* This work was in parts supported under the GALE program of the Defense Advanced Research Projects Agency, Contract No. HR0011-06-C-0022 an in parts by OSEO under the Quaero program.

P. Sojka et al. (Eds.): TSD 2008, LNAI 5246, pp. 303–310, 2008.
© Springer-Verlag Berlin Heidelberg 2008

Conventional neural network systems such as TANDEM [2] and TRAP [3] use three-layer MLPs trained to estimate phone posterior probabilities at every frame, which are then used as features for a GMM/HMM system. They are sometimes referred to as *probabilistic features*. The size of the MLP output features is reduced by a principal components analysis (PCA) transform. However, this might not necessarily be the optimal choice, especially when the dimensionality reduction is severe. The recently proposed *bottle-neck features* override this issue by employing four or five-layer MLPs and using outputs of a small hidden layer as features [4,5]. Not only does it allow for an arbitrary vector size, it also suggests using more MLP training targets for better discriminability.

Probabilistic features have never been shown to consistently outperform cepstral features in LVCSR. However, they can markedly improve the performance when used in conjunction with them. A number of multi-stream combination techniques have been successfully used for this purpose, four of which are studied in this work. These are MLP combination, feature concatenation, model adaptation and ROVER voting.

In this work, the bottle-neck architecture was used to deliver three types of MLP features, which differ in their input speech representations. Acoustic models are estimated using the three feature sets, and their performance is compared to a baseline system using PLP features. Different methods to combine the MLP and PLP features are explored, as well as combination of system outputs, with the goal of learning the most effective combination methods.

2 Arabic BN Task Description

The speech recognizer is a development version of the Arabic speech-to-text system component used in the AGILE participation in the GALE'07 evaluation. The transcription system has two main components, an audio partitioner and a word recognizer [6]. The audio partitioner is based on an audio stream mixture model, and serves to divide the continuous stream of acoustic data into homogeneous segments, associating cluster, gender and labels with each non-overlapping segment. The recognizer makes use of continuous density HMMs for acoustic modeling and n-gram statistics for language modeling. Each context-dependent phone model is a tied-state left-to-right CD-HMM with Gaussian mixture observation densities where the tied states are obtained.

Word recognition is performed in one or two passes, where each decoding pass generates a word lattice with cross-word, position-dependent, gender-independent acoustic models, followed by consensus decoding with 4-gram and pronunciation probabilities [6,7]. Unsupervised acoustic model adaptation is performed for each segment cluster using the CMLLR (Constrained Maximum Likelihood Linear Regression) and MLLR [8] techniques prior to second decoding pass.

A subset of the available Arabic broadcast news data was used to train acoustic models for the development system. This subset is comprised of 389 hours of manually transcribed data distributed by the Linguistic data consortium. These data were used to train the baseline gender-independent acoustic models, without maximum-likelihood linear transform (MLLT) or speaker-adaptive training (SAT). The models cover 30k contexts with 11.5k tied states, and have 32 Gaussians per state.

The language models were trained on corpora comprised of about 10 million words of audio transcriptions and 1 billion words of texts from a wide variety of sources. The recognition word list contains 200k non-vocalized, normalized entries. The language models result from the interpolation of models trained on subsets of the available data. The summed interpolation weights of the audio transcriptions is about 0.5. The pronunciation lexicon is represented with 72 symbols, including 30 simple consonants, 30 geminate consonants, 3 long and 3 short vowels, 3 vowels+tanwin, plus 3 pseudo phones for non-linguistic events (breath, filler, silence).

The test data is comprised of about 3 hours of broadcast news data referred to in the GALE community as the bnat06 development set. The out-of-vocabulary rate with this word list is about 2%, and the devset perplexity with a 4-gram language model is about 660.

3 MLP Features

Neural network feature extraction consists of two steps. The first step is *raw feature extraction* which constitutes the input layer to the MLP. Typically this vector covers a wide temporal context (100–500 ms) and therefore is highly dimensional. Second, the raw features are processed by the MLP followed by a PCA transform to yield the *HMM features*.

Two different sets of raw features are used, 9 frames of PLPs (9xPLP) and time-warped linear predictive TRAP (wLP-TRAP) [9]. The first set of raw features is based on the PLP features used in the baseline system which are mean and variance normalized per speaker. At each 10 ms frame, the MLP input is obtained by concatenating 9 successive frames of 13 PLP features (including energy) plus their first and second order derivatives (Δ and Δ^2), centered at the current frame. The feature vector has $9 \times 39 = 351$ values and covers a 150 ms window.

The second set of features is obtained by warping the temporal axis in the LP-TRAP feature calculation. Linear prediction is used to model the Hilbert envelopes of 500 ms long energy trajectories in auditory-like frequency sub-bands [10]. The input to the MLP are 25 LPC coefficients in 19 frequency bands, yielding $19 \times 25 = 475$ values which cover a 500 ms window. The naming conventions adopted for the various features sets are given in Table 1 along with how the raw features relate to the HMM features.

The bottle-neck architecture is based on a four layer MLP with an input layer, two hidden layers and an output layer. The second layer is large and it provides the necessary modeling power. The third layer is small, its size is equal to the required number of features. The output layer computes the estimates of the target class posteriors. Instead of using these posteriors as features, a PCA transform is applied to the outputs of the small hidden layer neurons (prior to a non-linear sigmoid function). A layer size of 39 was used in order to be able to more easily compare the performance of the MLP features to the PLP features.

Probabilistic MLPs are typically trained with phone targets. Since the size of the bottle-neck layer is independent of the number of output targets, it is quite easy to increase this number to improve the discrimination capacity of the MLP. Since there are often more differences between the states of the same phone than between different

Table 1. Naming conventions for MLP features and how the raw input features relate to the HMM features

ID	*Raw features (#)*	*HMM features (#)*
PLP	–	PLP+Δ+Δ^2 (39)
MLP_{9xPLP}	9x(PLP+Δ+Δ^2) (351)	MLP (39)
MLP_{wLP}	wLP-TRAP (475)	MLP (39)
MLP_{comb}	9x(PLP+Δ+Δ^2) + wLP-TRAP (826)	MLP (39)

states in the same position of different phones, it could be effective to replace the phone targets to by phone state targets. The phone state segmentations were obtained via a forced alignment using three-state triphone HMMs, with 69 phones and 3 non-linguistic units. The number of MLP targets was therefore increased from 72 to 210.

Since the MLP training is time-consuming, the MLP size and the amount of training data needs to be properly balanced. It is known that more data and/or more parameters always help, but at certain point the gain is not worth the effort. Table 2 gives the word error rate as a function of the amount of MLP training data. MLPs of a constant size (1.4 million parameters) were trained on various amounts of data using the 9xPLP raw features by gradually adding data from more speakers. HMMs were trained on the full 389 hour data set for all conditions.

The top part of the table gives WERs for phone targets. It can be seen that the improvement obtained by using additional data rapidly saturates with only a negligible gain when increasing the data by a factor of 10 (from 17 to 170 hours). The lower part of the table corresponds to using state targets. The change from phone targets to state targets brought a 2.4% relative reduction in WER (from 25.3% to 24.7%) with a MLP trained on a 17-hour data subset. The phone trained MLP correctly classified about 55% of unseen frames, whereas the state trained MLP was correct on about 50% of the frames. Given that the number of classes has tripled, it indicates that the state targets are indeed a good choice. In contrast to the phone targets, training state targets benefits from the additional data, with relative error rate reductions of 2-3% (24.2 to 23.4). The reference WER with the baseline PLP system, trained on the same data with the same model configuration, is 25.1%.

Since at the time of preparing this paper, training the MLP with wLP-TRAP features on the full corpus was not finished, the subsequent experiments were carried out using the MLP trained on 63 hours of speech recorded during the period from 2000 to 2002. Though not shown in the paper, partial experiments with the MLP trained on the full corpus show consistent improvements in performance over the values reported in the following sections.

4 System Combination

Experiments with system combinations were carried on using four types of features as listed in Table 1. The fourth feature set is obtained by combining the 9xPLP and the wLP-TRAP inputs to the MLP. All the four basic features were first evaluated without and with unsupervised acoustic model adaptation, as shown in the first four entries

Table 2. Word error rates (%) for phone and state based MLP as a function of the amount of training data (using 9xPLP raw features). All the HMMs are trained on the full 389 hours.

MLP targets	*MLP train data*	*WER (%)*
phones	1.5 hrs	27.3
	17 hrs	25.3
	170 hrs	25.0
states	17 hrs	24.7
	63 hrs	24.2
	301 hrs	23.4
PLP baseline		25.1

Table 3. Performance of PLP and MLP features, MLP combined features and feature concatenation

#	*Features*	*WER (%)*	
		1-pass	*2-pass*
1	PLP	25.1	22.5
2	MLP_{9xPLP}	24.2	22.7
3	MLP_{wLP}	25.8	23.1
4	MLP_{comb}	23.8	21.9
5	PLP + MLP_{9xPLP}	22.7	21.2
6	PLP + MLP_{wLP}	21.7	20.4
7	MLP_{9xPLP} + MLP_{wLP}	22.2	21.0

in Table 3. The baseline performance of the standard PLP features with adaptation is 22.5%. Without adaptation, the MLP_{9xPLP} features are seen to perform a little better (about 4% relative) than PLP, but with adaptation both MLP_{9xPLP} and MLP_{wLP} are slightly worse than PLP. This leads us to conclude that MLLR adaptation is less effective for MLP features than for PLP features. The MLP_{comb} (the fourth entry in the table) is seen to perform better than PLP both with and without adaptation and suggests that combining raw features at the input to the MLP classifier is effective.

Next three means of fusing the information coming from the cepstral and the MLP features were evaluated. The simplest approach is to concatenate together the features at the input to the HMM system (this doubles the size of the feature vector, $2 \times 39 = 78$ features) and to train an acoustic model. Three possible pairwise feature concatenations were evaluated and the results are given in the lower part of Table 3. These concatenated features all substantially outperform the PLP baseline, by up to 9% relative, showing that feature concatenation is a very effective approach. Given the significantly better performance of the PLP + MLP_{wLP} features over the PLP + MLP_{9xPLP} and MLP_{9xPLP} + MLP_{wLP} features, the three-way concatenation was not tested as it was judged to be not worth the increased computational complexity needed to deal with the resulting feature vector size (3×39).

Two other more computationally expensive approaches were studied, cross model adaptation and ROVER [11]. Table 4 gives some combination results using cross

Table 4. Comparing cross-adaptation and ROVER for combining multiple systems

	WER (%)	
Combined systems	*1-pass*	*2-pass*
$3 \rightarrow 1$	25.8	21.5
$1 \rightarrow 3$	25.1	22.0
$7 \rightarrow 1$	22.2	20.7
$1 \rightarrow 7$	25.1	21.2
$1 \oplus 2 \oplus 3$	22.3	20.6
$1 \oplus 3$	23.3	21.0
$5 \oplus 6$	21.2	19.9
$1 \oplus 6 \oplus 7$	21.0	19.7

adaptation (top) and ROVER (bottom). The first entry is the result of adapting the PLP models with the hypotheses of the MLP_{wLP} system. The second entry corresponds to the reverse adaptation order, i.e. the MLP_{wLP} are adapted using the hypotheses of the PLP system. The next two entries use cross adaptation on top of feature concatenation. In the first 3 cases, cross adaptation reduces the WER (note that the 2nd pass error rates must be compared with those in Table 3). Larger gains are obtained when the PLP models are used in the second pass, supporting the earlier observation that MLLR adaptation is more effective for PLP features than for MLP features. This may be because the MLP already removes the variability due to the speaker or because other, perhaps non-linear, transformations are needed to adapt MLP features. The WERs in the bottom part of the table result from ROVER combination of the first or second pass hypotheses of the listed systems. ROVER combination of the three basic features performed better than the best pair-wise cross-adaptation amongst them ($3 \rightarrow 1$) however, neither combination outperformed the simple feature concatenation WER of 20.4% (entry 6 in Table 3). ROVER also helps when applied jointly with other combination methods (see the last two rows in Table 4), beating the baseline PLP system by up to 12% relative. This best ROVER result however requires 6 decoding passes!

It is interesting to observe that the PLP features are generally best combined with MLP_{wLP}, even though the $\text{MLP}_{\text{9xPLP}}$ gives better score than MLP_{wLP}. This may be due on one side to the fact that the $\text{MLP}_{\text{9xPLP}}$ features are derived from the PLPs, and on the other side that there is a larger difference in time spans between the standard PLP and the wLP-TRAP features.

Table 5. Best results after MLLR adaptation for different types of system combination of PLP and MLP_{wLP} features

Features	*WER (%)*	*Comment*
PLP + MLP_{wLP}	20.4	best feature concatenation
$1 \oplus 3$	21.0	best ROVER (1-3)
$\text{MLP}_{\text{wLP}} \rightarrow$ PLP	21.5	best cross-adaptation
MLP_{comb}	21.9	MLP combination

Table 5 summarizes the best results after adaptation obtained for each combination method with PLP and MLP_{wLP}. The systems are sorted by WER in ascending order. It appears that feature concatenation is a very efficient combination method, as it not only results in the lowest WER for 2 front-ends but it also has the lowest cost.

5 Summary

Three novel MLP feature sets derived using the bottle-neck MLP architecture have been evaluated in the context of an LVCSR system. One feature set is based on nine frames of PLP features and their derivatives, with a temporal span of 150 ms. The other feature set is an improved version of LP-TRAP and has a longer temporal span of 500 ms. Different schemes have been used to combine these two MLP feature sets with PLP features to determine the most effective approach.

Experiments were carried out on the Gale Arabic broadcast news task. When used with MLLR adaption, the MLP features perform as well or even slightly better than a standard PLP based front-end. Doubling the feature vector by combining the two feature sets led to a 9% relative WER reduction relative to the PLP baseline. Combining the same feature sets via cross-model adaptation or ROVER also gave improvement but to a lesser degree.

Feature concatenation appears to be the most efficient combination method, providing the best gain at the lowest decoding cost. In general, it seems best to combine features based on different time spans as they provide high complementarity.

It should be noted that as shown in the paper, the MLP system accuracy can be further improved by training the MLP on more data.

References

1. Zhu, Q., Stolcke, A., Chen, B.Y., Morgan, N.: Using MLP features in SRI's conversational speech recognition system. In: INTERSPEECH 2005, pp. 2141–2144 (2005)
2. Hermansky, H., Ellis, D., Sharma, S.: TANDEM connectionist feature extraction for conventional HMM systems. In: ICASSP 2000, Istanbul, Turkey (2000)
3. Hermansky, H., Sharma, S.: TRAPs - classifiers of TempoRAl Patterns. In: ICSLP 1998 (November 1998)
4. Grézl, F., Karafiát, M., Kontár, S., Černocký, J.: Probabilistic and bottle-neck features for LVCSR of meetings. In: ICASSP 2007, April 2007, pp. 757–760. IEEE Signal Processing Society, Hononulu (2007)
5. Grézl, F., Fousek, P.: Optimizing bottle-neck features for LVCSR. In: ICASSP 2008, Las Vegas, ND (2008)
6. Gauvain, J., Lamel, L., Adda, G.: The LIMSI Broadcast News Transcription System. Speech Communication 37(1-2), 89–108 (2002)
7. Lamel, L., Messaoudi, A., J.L.G.: Improved Acoustic Modeling for Transcribing Arabic Broadcast Data. In: Interspeech 2007, Antwerp, Belgium (2007)
8. Leggetter, C., Woodland, P.: Maximum likelihood linear regression for speaker adaptation of continuous density hidden Markov models. Computer Speech and Language 9(2), 171–185 (1995)

9. Fousek, P.: Extraction of Features for Automatic Recognition of Speech Based on Spectral Dynamics. PhD thesis, Czech Technical University in Prague, Faculty of Electrical Engineering, Prague (March 2007)
10. Athineos, M., Hermansky, H., Ellis, D.P.: LP-TRAP: Linear predictive temporal patterns. In: ICSLP 2004 (2004)
11. Fiscus, J.: A Post-Processing System to Yield Reduced Word Error Rates: Recogniser Output Voting Error Reduction (ROVER) (1997)

Acoustic Modeling for Speech Recognition in Telephone Based Dialog System Using Limited Audio Resources

Rok Gajšek, Janez Žibert, and France Mihelič

Faculty of Electrical Engineering, University of Ljubljana,
Tržaška 25, SI-1000 Ljubljana, Slovenia
rok.gajsek@fe.uni-lj.si, janez.zibert@fe.uni-lj.si,
france.mihelic@fe.uni-lj.com
http://luks.fe.uni-lj.si/

Abstract. In the article we evaluate different techniques of acoustic modeling for speech recognition in the case of limited audio resources. The objective was to build different sets of acoustic models, the first was trained on a small set of telephone speech recordings and the other was trained on a bigger database with broadband speech recordings and later adapted to a different audio environment. Different adaptation methods (MLLR, MAP) were examined in combination with different parameterization features (MFCC, PLP, RPLP). We show that using adaptation methods, which are mainly used for speaker adaptation purposes, can increase the robustness of speech recognition in cases of mismatched training and working acoustic environment conditions.

Keywords: Environment adaptation, acoustic modeling, robust speech recognition.

1 Introduction

In building of an automatic dialog system for retrieving information [11] a development of reliable speech recognizer is a key element as all other modules depend heavily on its accuracy. Performance of the speech recognizer in applications where speech input is acquired through telephone is even more crucial since it is known that error rates are higher compared to the systems that are designed to work in a studio-like environment. A problem that we were faced with was the lack of recorded telephone speech in Slovenian language that could be used for training acoustic modules. Acquiring a large amount of telephone recordings would be time consuming, therefore we decided to use a smaller database and examine different adaptation methods which were reported to give good results for environmental adaptation [8] and compare these to new models trained on small amount of telephone speech.

Since perceptual linear prediction coefficients (PLP) [6] are known to perform better than mel-frequency cepstral coefficients (MFCC) [1] when dealing with noisy audio, they were our first choice. In [7], some modifications to original PLP coefficients were presented, that should increase the accuracy of recognizing noisy speech data. Hence, they were also examined in our approaches.

The evaluation of different acoustic models were performed with a simple phoneme-speech recognizer, which was able to recognize Slovene phonemes. Since we were

P. Sojka et al. (Eds.): TSD 2008, LNAI 5246, pp. 311–316, 2008.
© Springer-Verlag Berlin Heidelberg 2008

mostly interested in the effects of adaptation and feature parameterizations, the recognizer was built using only acoustic models incorporated in a 0-gram language model, where each phoneme was equally represented.

2 Speech Databases

2.1 Voicetran Database

The original acoustic model was built by using Voicetran database [9], which consists of 3 databases Gopolis, K211d and VNTV in total duration of 12.6 hours. The majority of recordings (Gopolis, K211d) were acquired in a controlled acoustic environment by using high quality microphones. The second, smaller part (VNTV), consists of weather news reports taken from broadcast news recordings. All material has been recorded at sample rate of 16 kHz. The recordings were split to training (90%) and testing part (10%).

2.2 Woz Database

During a Wizard-of-oz experiment [11], which was primarily used for building a dialogue manager module [5], 1.5 hours of telephone speech were recorded. The database consists of recordings of 76 people (38 male, 38 female with different accents) interacting with the system in order to obtain desired weather information. For recording purposes a Dialogic DIVA Server BRI–2M PCI card was used with sampling rate of 8kHZ and 8 bit nonlinear A-law encoding, which is standard in ISDN communications. The calls were made through different types of lines (GSM, ISDN and analog line) and they were taken from various acoustic environments with different background conditions. As previously with Voicetran database the recordings were split into training (90%) and test (10%) part.

2.3 Evaluation Database

For the evaluation purposes another 20 minutes of speech was recorded by 5 different people using the same equipment as for the Woz recordings. Each of them read 40 sentences, 15 of which were weather reports, 15 were enquires about the weather and 10 sentences were selected from the transcription of the conversation from previous Woz experiment. At the end each speaker was asked to describe by its own words the weather conditions. These last utterances were treated as spontaneous speech. The recordings were split into sentences and evaluated for each speaker. This material is referred to as the Eval database throughout the paper.

Both sets of telephone recordings have a sample rate of 8 kHz, which forced us to limit the filter bank to 4 kHz during feature extraction for all three databases.

3 Acoustic Modeling

Our goal was to compare, on one hand the acoustic model trained on a small set of target recordings in telephone environments and the adaptation of a acoustic model that was trained on a larger broadband database. Therefore the first step was to train a simple monophone HMM acoustic model using Voicetran database. In the next step the Woz database was used for building another acoustic model. As expected, a model trained on Woz recordings performed better on telephone recordings, but its performance was not adequate due to limited amount of training data. Therefore, we decided to additionally explore standard adaptation techniques to increase the recognition results.

For building and testing of the acoustic models the HTK toolkit [10] was used in all experiments. In addition changes to the PLP parametrization method (RPLP) were implemented to the HTK feature extraction module as presented in [7].

3.1 Parametrization

Different parametrization methods were put to the test to establish which gives better results when working with microphone audio. In all cases, 12 coefficients and short-term energy were calculated with their Δ and $\Delta\Delta$ components. MFCC coefficients are a standard for speech parametrization. They are known to give better results when working with a high quality recordings, but when the recordings are of lower quality (e.g. background noise) perceptual linear prediction coefficient should be a better choice. In [7] some modifications to the standard PLP method were presented, where most noticeable difference is a new filter bank with a large number of filters. In our work we used 128 filters, whereas for MFCC and PLP a standard 24 filters were used. In initial comparison of PLP and revised PLP (RPLP) the latter was found to give better results. Comparison of phoneme recognition results by using MFCC and RPLP features is presented in Table 1. As can be seen the RPLP features produced a smaller phone error rate (PER), when they were implemented in the acoustic model from the Voicetran data on the test part of both Voicetran and Woz recordings. As expected, the recognition rates are not as good when the Voicetran model without any adaptation was used on the Woz recordings, but the results are still in favor of RPLP features.

Table 1. Phone error rates for different acoustic models and parameterizations

Type	Voicetran model	
	Voicetran	Woz
MFCC	25.84%	64.62%
RPLP	**24.09%**	**63.93%**

3.2 Adaptation Methods

Two standard adaptation methods – Maximum Likelihood Linear Regression (MLLR) [2] and Maximum a posteriori adaptation (MAP) [4] – were tested. For MLLR, the constrained version (also known as CMLLR) [3] was used, meaning that the same transformation is used for both variance and mean (Equations 1, 2):

$$\hat{\mu} = A^{'}\mu - b^{'} \tag{1}$$

$$\hat{\Sigma} = A^{'}\Sigma A^{'T} \tag{2}$$

where $A^{'}$ and $b^{'}$ present the transformation, μ is the old mean vector and Σ is the old covariance matrix. $\hat{\mu}$ and $\hat{\Sigma}$ are the new mean vector and covariance matrix.

A method of transforming only the means (MLLRMEAN) [10] was also evaluated.

MLLR adaptation method is known for producing better results in comparison to MAP adaptation when small amounts of data is available. When the amount of data increases the MAP adaptation prevails [10]. Since we used 1.5 hours of audio for adaptation, we expected better results with MAP.

4 Performance Evaluation

4.1 Monophone Models

In achieving our goal to build a robust telephone speech recognizer the first step was to assess the performance of an already build Voicetran recognizer. Since this acoustic model was built by using only microphone recordings and no adaptation was performed, the phoneme recognition results on telephone recordings were not satisfactory. Better results were achieved when a new model was built by using Woz database. The comparison of the phone error rates is shown in Table 2.

Table 2. Phone error rates for Woz and Voicetran acoustic models on Woz and Eval databases

Database	Woz	Voicetran
Woz	**39.59%**	63.93%
Eval	**50.69%**	52.15%

Next step was to evaluate different adaptation methods. Constrained MLLR methods, where only means were adapted (MLLRMEAN), and MAP adaptations were used to adapt the Voicetran model by using the Woz database recordings. Only RPLP version of the Voicetran acoustic model was used since it has been shown from the previous experiments (Table 1), that this parametrization gives superior results.

Table 3. Phone error rates for different types of adaptation

Database	Voicetran adaptation		
	CMLLR	MLLRMEAN	MAP
Woz	44.39%	41.25%	**38.7%**
Eval	47.68%	46.69%	**44.78%**

When comparing Tables 2 and 3 smaller phone error rates are obtained in the case of evaluation database. Another important issue is that adapted Voicetran model gives better results (44.78%) in comparison to the model, trained one a small data set of third-party target microphone recordings (50.69%). From Table 3 it can also be seen that MAP produces better acoustic adaptation to the new environment. Another interesting observation is that transforming only means with MLLR adaptation gives better results than regular constrained MLLR, where also variances are adapted.

4.2 Triphone Models

The results examined with the adaptation of a monophone acoustic model are still not adequate for a reliable recognizer performance. Thus, we also tested context-depended triphone models. In this case, we applied the same experimental settings as in previous case. Comparison of the performances of different speech recognizers is shown in Table 4. Since the Woz database was used for the adaptation and for the evaluation, the error rates in that case are lower. The results on the Eval database give more realistic view on the recognition accuracy. As was in the case of the monophone model the adaptation method yield better results.

Table 4. Phone error rates for triphone acoustic models

Database	Woz	Voicetran	Voicetran-MAP
Woz	27.17%	47.4%	28.3%
Eval	43.9%	46.14%	**37.65%**

4.3 Word Recognition

In our experiments so far, only recognition of phonemes was evaluated. The reason was that we wanted to evaluate changes only in acoustic models without any influence of language models. Throughout our experiments we found that a suitable combination of parametrization and adaptation techniques were RPLP features with MAP adaptation. Based on these representations of speech data a speech recognition system was built and tested. We were able to get the word error rate (WER) to 32.03% when testing on Woz database, but it should be stated that weighting the actual language model had a noticeable effect on the recognition performance (higher weight on the language model gave better results). The language model – used at the moment – was a simple bigram model constructed from Woz database transcriptions without any smoothing procedure with relatively small application dependent (enquiries about weather) vocabulary of about 2073 words and low perplexity of 21.

5 Conclusion

In the paper we presented different techniques of acoustic modeling in the case of limited audio resources. Our goal was to built a robust speech recognition engine, which will be included in a telephone based automatic dialogue system for weather information retrieval. The obstacle we had to overcome was a limited amount of telephone recordings. The solution was to built a speech recognition system from broadband speech data and adapt its acoustic models to a different – telephone – environment. We achieved this by applying a standard adaptation technique MAP in combination with a modified version of PLP coefficients (RPLP). This procedure gave a 14% relative increase of phoneme recognition accuracy when using monophone acoustic model and 18% when working with triphone model.

We have also shown that adaptation methods MLLR and MAP mainly used in speaker adaptation, provide a reliable methods for compensation of changes in the acoustic environment.

Future work will be focused on developing an adequate language model, which is expected, that in combination with the developed acoustical models will increase the overall speech recognition accuracy in our telephone based dialog system.

References

1. Davis, S.B., Mermelstein, P.: Comparison of parametric representations for monosyllabic word recognition in continuously spoken sentences. IEEE Trans. Acoust., Speech, Signal Processing ASSP-28(4), 357–365 (1980)
2. Gales, M., Pye, D., Woodland, P.: Variance compensation within the MLLR framework for robust speech recognition and speaker adaptation. In: Proc. ICSLP 1996, Philadelphia, USA, vol. 3, pp. 1832–1835 (1996)
3. Digalakis, V.V., Rtischev, D., Neumeyer, L.G.: Speaker adaptation using constrained estimation of Gaussian mixtures. IEEE Transactions SAP 3, 357–366 (1995)
4. Gauvain, J.L., Lee, C.H.: Maximum a-posteriori estimation for multivariate Gaussian mixture observations of Markov chains. IEEE Transactions SAP 2, 291–298 (1994)
5. Hajdinjak, M., Mihelič, F.: The wizard of Oz system for weather information retrieval. In: Matoušek, V., Mautner, P. (eds.) TSD 2003. LNCS (LNAI), vol. 2807, pp. 400–405. Springer, Heidelberg (2003)
6. Hermansky, H., Brian, H., Wakita, H.: Perceptually based linear predictive analysis of speech. In: ICASSP 1985, pp. 509–512 (1985)
7. Höning, F., Stemmer, G., Hacker, C., Brugnara, F.: Revising perceptual linear prediction (PLP). In: Proceedings of INTERSPEECH 2005, pp. 2997–3000 (2005)
8. Maier, A., Haderlein, T., Nöth, E.: Environmental Adaptation with a Small Data Set of the Target Domain. In: Sojka, P., Kopeček, I., Pala, K. (eds.) TSD 2006. LNCS (LNAI), vol. 4188, pp. 431–437. Springer, Heidelberg (2006)
9. Mihelič, F., et al.: Spoken language resources ad LUKS of the University of Ljubljana. International Journal of Speech Technology 6(3), 221–232 (2003)
10. Young, S., et al.: The HTK Book (for HTK version 3.4). Cambridge University Engeneering Department (2006)
11. Žibert, J., Martinčić-Ipšić, S., Hajdinjak, M., Ipšić, I., Mihelič, F.: Development of a bilingual spoken dialog system for weather information retrieval. In: Proceedings of EUROSPEECH 2003, pp. 1917–1920 (2003)

Performance Evaluation for Voice Conversion Systems

Todor Ganchev, Alexandros Lazaridis, Iosif Mporas, and Nikos Fakotakis

Wire Communications Laboratory, Dept. of Electrical and Computer Engineering,
University of Patras, 26500, Rion-Patras, Greece
tganchev@ieee.org, {alaza,imporas,fakotakis}@upatras.gr

Abstract. In the present work, we introduce a new performance evaluation measure for assessing the capacity of voice conversion systems to modify the speech of one speaker (source) so that it sounds as if it was uttered by another speaker (target). This measure relies on a GMM-UBM-based likelihood estimator that estimates the degree of proximity between an utterance of the converted voice and the predefined models of the source and target voices. The proposed approach allows the formulation of an objective criterion, which is applicable for both evaluation of the virtue of a single system and for direct comparison (benchmarking) among different voice conversion systems. To illustrate the functionality and the practical usefulness of the proposed measure, we contrast it with four well-known objective evaluation criteria.

Keywords: Performance evaluation, voice conversion, speaker identification.

1 Introduction

The automatic modification of the speech signal originating from one speaker (source) so that it sounds as pronounced by another speaker (target) is referred to as *voice conversion* (VC). The VC process utilizes a small amount of training data from the source and target speakers to create a set of conversion rules. Subsequently, these rules are used to transform other speech, originating from the source speaker, to sound like produced by target speaker [1]. Over the past years various VC methods have been introduced. The separate conversion of the spectral envelope (vocal tract) and the spectral detail (excitation-residual signal) is the most commonly used approach. Research in the field of speaker recognition has shown that, in order to recognize a speaker, the spectral envelope alone contains enough information. In addition, the use of spectral detail in VC systems provides a much more natural sounding speech.

Different VC approaches offer different trade-offs. Various subjective and objective evaluation tests have been developed for measuring the VC performance. By using these evaluation measures, either the sound quality of the converted speech or the capacity of a system to convert the identity of the source speaker to the target one, can be assessed. These tests are implemented as subjective listening evaluations (time consuming – expensive) or as objective evaluations (often not accounting for the perceptual quality, not intuitive).

Two subjective tests are most often used to evaluate the performance of VC systems. The first one is the force-choice ABX test [2], which assesses the VC capacity of a

P. Sojka et al. (Eds.): TSD 2008, LNAI 5246, pp. 317–324, 2008.
© Springer-Verlag Berlin Heidelberg 2008

system. In this test listeners are presented with stimuli A, B and X and are asked to judge if X sounds closer to A or to B. The second one is the mean opinion score (MOS) test [3], by which the sound quality of the converted speech is evaluated in a 5-point scale (5: excellent, 4: good, 3: fair, 2: poor, 1: bad).

In addition, there are various objective performance metrics, which estimate the differences among the source, target and converted speech signals. Two of the most frequently used measures are the signal-to-noise ratio (SNR) and the relative spectral distortion (SD). The SNR estimates the ratio between the energy of the transformed speech and the energy of the difference between the transformed and the target speech. The latter one measures the distance between the converted speech and the target speech in contrast to the distance between the source speech and the target speech. This objective error has been used with different interpretations, such as: log spectral distance, magnitude spectral distance or spectral distance between feature vectors (e.g. Mel-Frequency Cepstral Coefficients).

In the present work, we introduce a new objective measure for the assessment of the capacity of VC systems. Based on GMM-UBM estimator, the proposed approach defines a normalized performance evaluation metric. Using the proposed evaluation measure, assessment of the performance of a single VC system or a direct comparison (benchmarking) among multiple systems can be performed.

Our work bears some resemblance with [4], in which a speaker identification (SID) approach is used so as to evaluate the quality of the converted voice. However, in the present development the speaker models are built and used in a different manner, which results in a more appropriate and intuitive performance estimation measure.

2 Related Work

A common evaluation measure in the area of speech processing is the signal-to-noise ratio (SNR). The segmental SNR is defined as the ratio between the energy of the target signal and the energy of the difference between the converted and target signals for the mth frame. Measured on a dB-scale, the SNR is defined as:

$$SNR\,(t,c) = 10\log_{10}\frac{\sum_{n=1}^{N}\left|H_t^m(e^{j2\pi n/N})\right|^2}{\sum_{n=1}^{N}\left(\left|H_c^m(e^{j2\pi n/N})\right| - \left|H_t^m(e^{j2\pi n/N})\right|\right)^2} \tag{1}$$

where N is the size of the Fourier transform $H(e^{j2\pi n/N})$, t is the original target speech signal and c is the converted speech signal. Higher values for the SNR indicate a better VC system.

One of the most commonly used objective error measures is the spectral distortion (SD) between two speech signals. The spectral distortion error $E_{SD}(A,B)$ between the spectra of two signals A and B is defined as:

$$E_{SD}(A,B) = \frac{1}{M}\sum_{m=1}^{M}\sqrt{\frac{1}{N}\sum_{n=1}^{N}\left(20\log_{10}H_A^m(e^{j2\pi n/N}) - 20\log_{10}H_B^m(e^{j2\pi n/N})\right)^2}. \tag{2}$$

Here M is the number of frames and N is the size of the Fourier transform $H(e^{j2\pi n/N})$ computed for each segment m of the signal.

In order to fulfil the need of having a normalized error across different speaker combinations, Kain [5] established an error metric referred to as *performance index*, P_{SD}, which is defined as:

$$P_{SD} = 1 - \frac{E_{SD}(t(n), c(n))}{E_{SD}(t(n), s(n))} \quad (3)$$

Here $E_{SD}(t, c)$ is the spectral distortion error (2) between the target and the converted speech, and $E_{SD}(t, s)$ is the spectral distortion error between the target and the source speech.

In [4], the author employed a SID system to assess the outcome of VC. Specifically, the author assumed that if a SID system can select the target over the source speaker it means that the VC is successful. According to [4], for a given input utterance X, the log-likelihood ratio of the target speaker to the source speaker is estimated as:

$$\theta_{st} = \log \frac{p(X|\lambda_t)}{p(X|\lambda_s)} = \log p(X|\lambda_t) - \log p(X|\lambda_s) \quad (4)$$

Here λ_t and λ_s are the target and the source speaker models, respectively. A good VC system is characterized by positive values of θ_{st}, and a negative value of θ_{st} indicates a poor VC system.

3 The Proposed Performance Measure

Following the operation logic of the subjective force choice ABX test [2], we propose a related objective evaluation of the VC capacity that we refer to as objective ABX test. Specifically, by estimating the degree of match between the predefined speaker models and samples of the converted (X), source (A) and target (B) voices, we are able to assess the degree of success in the transformation from voice A to B. This scheme allows evaluating the capacity of a single VC system, or comparing the VC performance among different systems (i.e. benchmarking of VC systems.)

The overall idea of the proposed objective ABX test is presented in Figure 1. Specifically, the left part of the figure illustrates the VC process and the right one the SID process. During the training phase, the VC system builds models for the source and target speakers which serve for deriving the VC rules. During the operation of the VC system, these rules are utilized for the transformation of the source speech, which results to the converted speech X. For a good VC system it is expected that the characteristics of speech utterance X will match those of the target speaker B.

The SID process relies on predefined speaker models A, B and a reference model that are built during the training phase. The reference model, which represents the world of all speakers, is trained from large speech corpora (hundreds or thousands of speakers) and should be general enough not to interfere with the individual models of the target speakers. The speaker-specific individual models can be trained from the same training datasets, once used for training the models of the VC system or from another representative dataset. During operation, the SID system processes the input speech and computes the degree of resemblance to the predefined speaker models. The outcome of

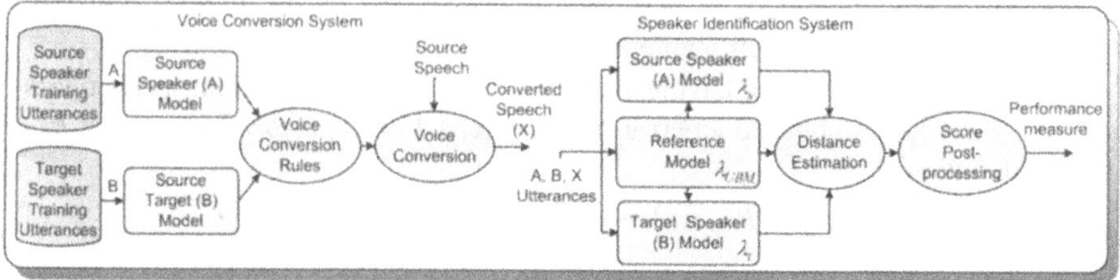

Fig. 1. Block diagram of the proposed objective ABX test

this process is a likelihood-based measure (score), and eventually a hard decision about the class belonging of the input.

Various implementations of the proposed evaluation scheme for the assessment of the capacity of VC systems can be obtained, depending on the manner the speaker models are built and used. In the following, we outline the implementation based on GMM-UBM likelihood estimator [6]. In this particular case, initially a world model, known as Universal Background Model (UBM), is trained from a large amount of speech data, typically involving population of hundreds of speakers. At the next step, speaker-specific models λ_s and λ_t are built for the source and target speakers, A and B, respectively. This is performed through a data-dependent maximum a posteriori (MAP) adaptation of the UBM by utilizing the training data of speakers A and B. Once the models are trained, the SID system is ready for operation.

The objective ABX test is performed as follows: a stimulus X, which is the result of the conversion of voice A to B, is used as the input of the SID system. Firstly, X is evaluated against the UBM, and the top-C scoring mixture components are found. Subsequently, the degree of similarity between the input X and the predefined models of the source and target voices, A and B, respectively, is computed only for the top-C components.

Next, the log-likelihood ratios $\Lambda_s(X)$ and $\Lambda_t(X)$ computed for the source and target models A and B respectively, normalized with respect to the UBM, are utilized to compute the ratio S_X:

$$S_X = \frac{\Lambda_t(X)}{\Lambda_s(X)} = \frac{\log p\,(X|\lambda_t) - \log p\,(X|\lambda_{UBM})}{\log p\,(X|\lambda_s) - \log p\,(X|\lambda_{UBM})}. \tag{5}$$

S_X is indicative of the quality of VC, and therefore can be used as a criterion for evaluating the VC capacity of a system, or for a direct comparison among different systems.

In the case of a direct comparison among different VC systems, only the input utterances X can be processed (please refer to Figure 1). However, here we consider processing all A, B, X utterances, which allows the normalization of the score S_X and assessment of the VC capacity of a single system. The scores S_A and S_B are computed by estimating (5) for utterances A and B, respectively.

The rationale behind the proposed objective ABX test is in the estimation of the ratio between two relative measures, the log-likelihood ratios $\Lambda_t(X)$ and $\Lambda_s(X)$, computed for a given converted utterance X. These ratios represent the normalized, with respect to λ_{UBM}, distances from the target and the source models λ_t and λ_s, respectively. Thus, for a given input X, we can think of S_X as the ratio between the distances to the target and source models.

Evaluating (5) for a large number of utterances X allows improving the robustness of the performance estimation. Computing the mean value and the standard deviation of the resulting S_X scores provides the ground for fair judgment of the VC capacity for a given method or for comparison among methods. Normalizing the distance between the means of the S_X distributions for the target and converted utterances with respect to the distance between the centres of the distributions for the source and target utterances leads to a more intuitive performance measure, which we have named D_{NORM}:

$$D_{NORM} = \frac{\mu_t - \mu_X}{\mu_t - \mu_s}. \tag{6}$$

Here μ_s, μ_t and μ_X are the mean values of the distributions for the scores S_A, S_B and S_X, respectively. $D_{NORM} \in [0, 1]$, where value $D_{NORM} = 0$ means a very good VC system (the distribution of S_X scores for the converted voice has the same mean value as the one for the target voice). In opposite, value $D_{NORM} = 1$ indicates useless VC system – the transformed voice has the same mean value of the score distribution as the source voice.

4 Experimental Setup

The proposed evaluation methodology is illustrated by experimenting with two voice conversion algorithms, namely, the (i) vocal tract length normalization (VTLN) [7] and (ii) spectral envelope GMM mapping [8] in combination with residual prediction model [9]. These methods were chosen as they are illustrative for the VC capacity of two different VC approaches. The first method is based on VTLN approach performed in the frequency domain (FD-VTLN), which aims at warping the frequency axis of the phase and magnitude spectrum. The second one is based on a probabilistic approach using GMM, describing and mapping the source and target feature distributions. In the remaining of this work, we will refer to these algorithms as to *voice conversion system 1* (VC system 1) and *voice conversion system 2* (VC system 2), respectively.

In the present experimental setup, we consider male-to-male VC. Specifically, utilizing the CMU Arctic database [10], we performed conversion from voice *bdl* to voice *rms*. The VC models were trained with 25 sentences from each speaker. Once the two VC systems were trained, we processed another 100 files for each of the two VC algorithms. These datasets served as the test sentences, X, during the evaluation process.

The speech parameterization for the SID process is based on a frame size of 40 milliseconds. The feature vector consists of 40 MFCC coefficients [11], log-energy, and the $\log(f - f_{\min})$. Here f is the fundamental frequency estimated through the autocorrelation-based algorithm [12] and the constant f_{min} =55 Hz is the smallest value of the fundamental frequency that the pitch estimation algorithm can detect for the specified frame size. The speech parameters were estimated 100 times per second.

For the SID systems, we have built the λ_{UBM} model from all utterances of the first 100 male speakers from the training set of the TIMIT [13] database. The speaker-specific models λ_s and λ_t were obtained through MAP adaptation on the UBM. In both cases the MAP adaptation was performed by using the same 25 sentences that were used earlier for training the VC systems. The size of the training dataset was found to affect the quality of the models, but no influence from the actual selection of training utterances was observed.

The λ_{UBM} and the speaker-specific models λ_s and λ_t have size of 256 components. After the evaluation of the input utterance against the entire UBM, only the indexes of the top-5 components (C=5) were fed to the speaker-specific models.

5 Experiments and Results

In order to contrast the operation of the proposed objective ABX test with other objective evaluation measures, we firstly compare it to the SID method used in [4]. Afterwards, we perform a comparison to other widely-used objective quality assessment measures, presented in Section 2. In all experimentations we employ the two VC systems mentioned in Section 4, namely VC systems 1 and 2, and use the same sets of one hundred test utterances A, B and X.

In Figure 2 we present performance comparison between the two VC systems of interest in terms of the θ_{st} scores [4] and the proposed S_X measure. Specifically, in Figure 2 *a*) and *b*) we present the distributions of the θ_{st} scores for the converted speech

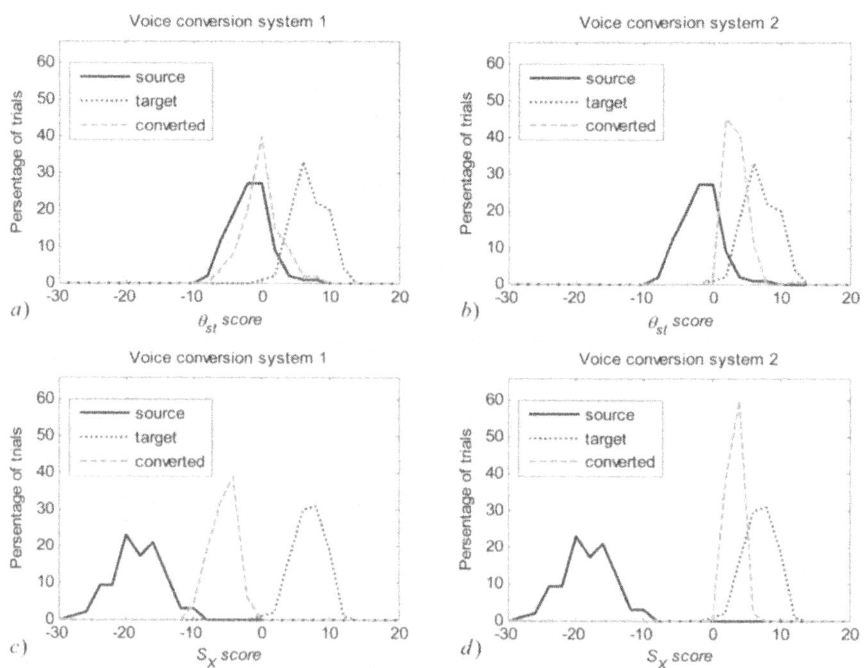

Fig. 2. Comparative results for the scores [4] and the proposed scores. Plots a) and b) present the scores for VC systems 1 and 2; and plots c) and d) present the scores for VC systems 1 and 2, respectively.

Table 1. Performance results for VC systems 1 and 2 for different objective measures

Performance measure	VC system 1	VC system 2
Segmental SNR (eq.1)	1.609	4.376
Performance index P_{SD} (eq.3)	0.032	0.135
θ_{st} scores (eq.4)	−0.024	3.313
S_X scores (eq.5)	−5.548	3.198
Distance D_{NORM} (eq.6)	0.492	0.144

X (dashed green line) for the two VC systems, 1 and 2, respectively. Although a direct comparison of the VC capacity of the two systems can be performed solely on the basis of the distributions obtained for the X utterances, for the purpose of better illustration we also present the distributions for the source and target utterances, A and B, respectively, when scored against the model built for the target voice B. In Figure 2 *a*) and *b*) these are plotted with solid blue line for the source A and dotted red line for the target B, respectively. As plots 2 *a*) and 2 *b*) present, the VC system 2 has distribution of the θ_{st} scores closer to that of the target, and therefore it demonstrates higher VC capacity. However, since the distributions of scores for the source and target voices overlap, the resolution of this setup is not advantageous, and the θ_{st} scores are not very intuitive.

In Figure 2 *c*) and 2 *d*) we present the distributions of the S_X scores obtained for all X utterances for the VC systems 1 and 2, respectively. Again although the VC capacity of these systems can be judged directly by comparing the distributions of the S_X scores (dashed green line), we present also the distributions obtained for the source and target utterances, A and B, respectively. As plots 2 *c*) and 2 *d*) present the distributions of scores for the source and target utterances do not overlap. This is due to the advantage that the reference model λ_{UBM} provides. Specifically, instead of computing the ratio between the likelihoods obtained for models λ_s and λ_t (as it is done in [4]), in the proposed method we compute the ratio between the normalized, with respect to λ_{UBM} log-likelihood, ratios. This permits better resolution of the experimental setup, and therefore, better resolution of the measure of the VC capacity.

As plots 2 c) and 2 d) present the VC system 2 offers significantly better performance, when compared to VC system 1. In terms of SID performance this is expressed as follows: The output utterances X of VC system 1 were recognized as the target voice B in 80 % of the cases, while the corresponding performance for the VC system 2 was 100 % recognition rate.

For the purpose of comprehensiveness we also provide results for the objective measures described in Section 2. Specifically, in Table 1 we present the average segmental SNR, performance index P_{SD}, θ_{st} and S_X scores, as well as the normalized distance D_{NORM}. All values are averaged for the 100 test utterances.

As Table 1 presents, the proposed D_{NORM} offers a very convenient measure of the VC performance, even for evaluating the capacity of a single system. For instance, for a good system (i.e. VC system 2) the distance D_{NORM} between the means of the distributions of the S_X scores for the target and converted voices is relatively small ($D_{NORM} = 0.144$). In opposite, for a poor VC system this distance is significant. As we can see in Figure 2 *c*), the distribution of the S_X scores for VC system 1 is midway

($D_{NORM} = 0.492$) the distance between the distributions of the source and target. Thus, VC system 1 preserves much of the characteristics of the source voice.

6 Conclusion

We proposed a new methodology for objective evaluation of the capacity of VC systems referred to as objective ABX test. This method resulted in an intuitive performance measure D_{NORM} that represents the normalized distance between the distributions of the S_X scores for the target and converted voices. The D_{NORM} measure is convenient for evaluating the capacity of a single voice conversion system, or for direct comparison among different systems.

Acknowledgments. This work was supported by the PlayMancer project (FP7-ICT-215839-2007), which is funded by the European Commission.

References

1. Abe, M., Nakamura, S., Shikano, K., Kuwabara, H.: Voice conversion through vector quantization. In: Proc. ICASSP 1988, USA, pp. 655–658 (1988)
2. Kreiman, J., Papcun, G.: Comparing, discrimination and recognition of unfamiliar voices. Speech Communication 10(3), 265–275 (1991)
3. Methods for subjective determination of transmission quality, Tech. Rep. ITU-T Recommendation P.800, ITU, Switzerland (1996)
4. Arslan, L.M.: Speaker transformation algorithm using segmental codebooks (STASC). Speech Communication 28(3), 211–226 (1999)
5. Kain, A.: High resolution voice transformation. Ph.D. dissertation, OGI, Portland, USA (2001)
6. Reynolds, D.A., Quatieri, T.F., Dunn, R.B.: Speaker verification using adapted Gaussian mixture models. Digital Signal Processing 10(1-3), 19–41 (2000)
7. Sündermann, D., Ney, H., Höge, H.: VTLN-based cross-language voice conversion. In: Proc. ASRU 2003, USA, pp. 676–681 (2003)
8. Stylianou, Y., Cappé, O., Moulines, E.: Continuous probabilistic transform for voice conversion. IEEE Trans. Speech and Audio Processing 6(2), 131–142 (1998)
9. Sündermann, D., Bonafonte, A., Ney, H., Höge, H.: A study on residual prediction techniques for voice conversion. In: Proc. ICASSP 2005, USA, vol. 1, pp. 13–16 (2005)
10. Kominek, J., Black, A.: The CMU ARCTIC speech databases for speech synthesis research. Technical Report CMU-LTI-03-177, Carnegie Mellon University, Pittsburgh, PA (2003)
11. Slaney, M.: Auditory toolbox. Version 2. Technical Report #1998-010, Interval Research Corporation (1998)
12. Rabiner, L.R., Cheng, M.J., Rosenberg, A.E., McGonegal, C.A.: A comparative performance study of several pitch detection algorithms. IEEE Trans. Acoust. Speech & Signal Proc. 24(5), 399–418 (1976)
13. Garofolo, J.: Getting started with the DARPA-TIMIT CD-ROM: An acoustic phonetic continuous speech database. National Institute of Standards and Technology (NIST), USA (1998)

Influence of Reading Errors on the Text-Based Automatic Evaluation of Pathologic Voices

Tino Haderlein[1,2], Elmar Nöth[1], Andreas Maier[1,2], Maria Schuster[2], and Frank Rosanowski[2]

[1] Universität Erlangen-Nürnberg, Lehrstuhl für Mustererkennung (Informatik 5) Martensstraße 3, 91058 Erlangen, Germany
Tino.Haderlein@informatik.uni-erlangen.de
http://www5.informatik.uni-erlangen.de
[2] Universität Erlangen-Nürnberg, Abteilung für Phoniatrie und Pädaudiologie Bohlenplatz 21, 91054 Erlangen, Germany

Abstract. In speech therapy and rehabilitation, a patient's voice has to be evaluated by the therapist. Established methods for objective, automatic evaluation analyze only recordings of sustained vowels. However, an isolated vowel does not reflect a real communication situation. In this paper, a speech recognition system and a prosody module are used to analyze a text that was read out by the patients. The correlation between the perceptive evaluation of speech intelligibility by five medical experts and measures like word accuracy (WA), word recognition rate (WR), and prosodic features was examined. The focus was on the influence of reading errors on this correlation.

The test speakers were 85 persons suffering from cancer in the larynx. 65 of them had undergone partial laryngectomy, i.e. partial removal of the larynx. The correlation between the human intelligibility ratings on a five-point scale and the machine was $r = -0.61$ for WA, $r \approx 0.55$ for WR, and $r \approx 0.60$ for prosodic features based on word duration and energy. The reading errors did not have a significant influence on the results. Hence, no special preprocessing of the audio files is necessary.

1 Introduction

Although less than 1% of all cancers affect the larynx, it is necessary to provide proper rehabilitation therapies since speech is the main means of communication. In the USA, 10,000 new cases of laryngeal cancer are diagnosed each year [1]. In severe cases total laryngectomy has to be performed, i.e. the removal of the entire larynx. In early and intermediate stages, usually partial laryngectomy is sufficient, and at least one of the vocal folds or the vestibular folds can be preserved (see Figure 1). Dependent on the location and size of the tumor, the voice may sound normal before and after surgery. However, hoarse voices are very common.

In speech therapy and rehabilitation, a patient's voice has to be evaluated by the therapist. Automatically computed, objective measures are a very helpful support for this task. Established methods for objective evaluation, however, analyze only recordings of sustained vowels in order to find irregularities in the voice (see e.g. [2,3]). However, this

P. Sojka et al. (Eds.): TSD 2008, LNAI 5246, pp. 325–332, 2008.
© Springer-Verlag Berlin Heidelberg 2008

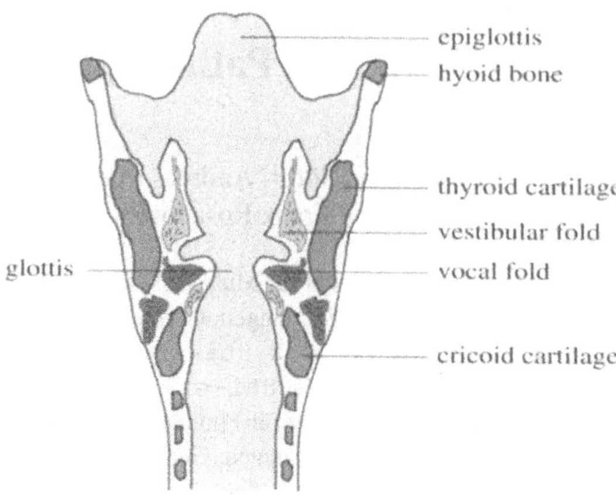

Fig. 1. Anatomy of an intact larynx

does not reflect a real communication situation because no speech but only the voice is examined. Criteria like intelligibility cannot be evaluated in this way. For this study, the test persons read a given standard text which was then analyzed by methods of automatic speech recognition and prosodic analysis. A standard text was used especially in view of the prosodic evaluation because the comparability of results among the patients is reduced when the utterances differ with respect to duration, number of words, percentage of different phone classes, etc.

For speech after total laryngectomy, where the patients use a substitute voice produced in the upper esophagus, we showed in previous work that an automatic speech recognition system can be used to rate intelligibility [4]. The word accuracy of the speech recognizer was identified as suitable measure for this task. It showed a correlation of more than $|r| = 0.8$ to the human evaluation. However, these results relied on the assumption that the recognition errors were only caused by the acoustic properties of the voices. Another source of error are reading errors. When the recognized word sequence is compared to the text reference, a patient with a high-quality voice might get bad automatic evaluation results due to misread words. This problem could be solved by replacing the text reference by a transliteration of the respective speech sample, but this method is not applicable in clinical practice.

In this paper, we examined how severe the influence of reading errors is on the results of automatic evaluation and the correlation to human evaluation results. In Section 2, the speech data used as the test set will be introduced. Section 3 will give some information about the speech recognition system. An overview on the prosodic analysis will be presented in Section 4. Section 5 contains the results, and Section 6 will give a short outlook on future work.

2 Test Data

The test files were recorded from 85 patients (75 men, 10 women) suffering from cancer in different regions of the larynx. 65 of them had already undergone partial

Table 1. File statistics for the speech corpora with and without reading errors

	duration					words	vocabulary
	total	avg.	st. dev.	min.	max.		
with errors	89 min	63 s	18 s	43 s	144 s	9519	71+187
without errors	82 min	58 s	15 s	40 s	125 s	9151	71+83

laryngectomy, 20 speakers were still awaiting surgery. The former group was recorded on the average 2.4 months after surgery. The average age of all speakers was 60.7 years with a standard deviation of 9.7 years. The youngest and the oldest person were 34 and 83 years old, respectively.

Each person read the text "Der Nordwind und die Sonne", a phonetically balanced text with 108 words (71 disjunctive) which is used in German speaking countries in speech therapy. The English version is known as "The North Wind and the Sun" [5]. The speech data were sampled with 16 kHz and an amplitude resolution of 16 bit.

In order to obtain a reference for the automatic evaluation, five experienced phoniatricians and speech scientists evaluated each speaker's intelligibility according to a 5-point scale with the labels "very high", "high", "moderate", "low", and "none". Each rater's decision for each patient was converted to an integer number between 1 and 5.

Due to reading errors, repetitions and remarks like "I don't have my glasses with me.", the vocabulary in the recordings did not only contain the 71 words of the text reference but also 187 additional words and word fragments. 27 of the files were error-free. In all other samples, at least one error occurred (see Figure 2). In order to determine the influence of these phenomena on the evaluation results, a second version of the data set was created by removing the additional words where possible. In total, 368 (3.9%) of the 9519 words were eliminated from the original speech samples.

Since the text flow was supposed to be preserved, misreading of single words without corrections, i.e. word substitutions, were not removed. This means that for instance the correction "*M*or- Nordwind" was reduced to "Nordwind" while the word "Nordw*u*nd" without correction was left unchanged. Also breaks in words, like "gel- -ten" were not changed when the full word was not repeated. This explains why the corrected files, further denoted as "without errors", still contain 83 out-of-text words and word fragments (see Table 1).

3 The Speech Recognition System

The speech recognition system used for the experiments was developed at the Chair of Pattern Recognition in Erlangen [6]. It can handle spontaneous speech with mid-sized vocabularies up to 10,000 words. The system is based on semi-continuous Hidden Markov Models (HMM). It can model phones in a context as large as statistically useful and thus forms the so-called polyphones, a generalization of the well-known bi- or triphones. The HMMs for each polyphone have three to four states; the codebook had 500 classes with full covariance matrices. The short-time analysis applies a Hamming window with a length of 16 ms, the frame rate is 10 ms. The filterbank for the Mel-spectrum consists of 25 triangle filters. For each frame, a 24-dimensional feature vector is

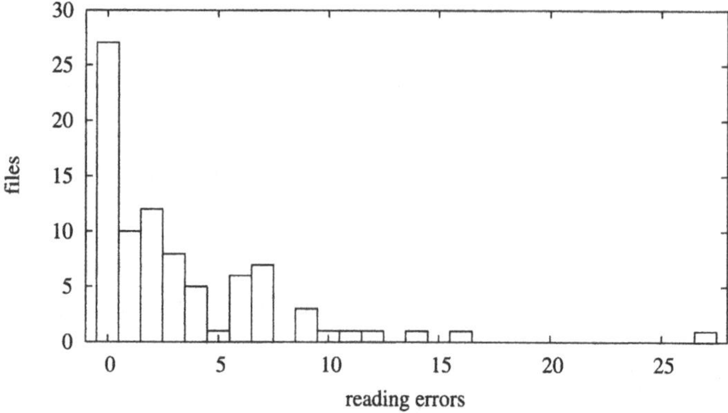

Fig. 2. Absolute number of reading errors in the 85 text samples

computed. It contains short-time energy, 11 Mel-frequency cepstral coefficients, and the first-order derivatives of these 12 static features. The derivatives are approximated by the slope of a linear regression line over 5 consecutive frames (56 ms). A unigram language model was used so that the results are mainly dependent on the acoustic models.

The baseline system for the experiments in this paper was trained with German dialogues from the VERBMOBIL project [7]. The topic in these recordings is appointment scheduling. The data were recorded with a close-talking microphone at a sampling frequency of 16 kHz and quantized with 16 bit. About 80% of the 578 training speakers (304 male, 274 female) were between 20 and 29 years old, less than 10% were over 40. 11,714 utterances (257,810 words) of the VERBMOBIL-German data (12,030 utterances, 263,633 words, 27.7 hours of speech) were used for the training and 48 (1042 words) for the validation set, i.e. the corpus partitions were the same as in [6].

The recognition vocabulary of the recognizer was changed to the 71 words of the standard text. The uttered word fragments and out-of-text words were not added to the vocabulary because in a clinical application it will also not be possible to add the current reader's errors to the vocabulary in real-time.

4 Prosodic Features

In order to find automatically computable counterparts for subjective rating criteria, we also use a "prosody module" to compute features based upon frequency, duration and speech energy (intensity) measures. This is state-of-the-art in automatic speech analysis on normal voices [8,9,10].

The prosody module takes the output of our word recognition module in addition to the speech signal as input. In this case the time-alignment of the recognizer and the information about the underlying phoneme classes (like *long vowel*) can be used by the prosody module. For each speech unit which is of interest (here: words), a fixed reference point has to be chosen for the computation of the prosodic features. We decided in favor of the end of a word because the word is a well-defined unit in word recognition, it can be provided by any standard word recognizer, and because this point can be more easily defined than, for example, the middle of the syllable nucleus in word accent position. For each reference point, we extract 95 prosodic features over intervals which contain one single word, a word-pause-word interval or the pause between two words. A full description of the features used is beyond the scope of this paper; details and further references are given in [11].

Besides the 95 local features per word, 15 global features were computed per utterance from jitter, shimmer and the number of voiced/unvoiced (V/UV) decisions. They cover each of mean and standard deviation for jitter and shimmer, the number, length and maximum length each for voiced and unvoiced sections, the ratio of the numbers of voiced and unvoiced sections, the ratio of length of voiced sections to the length of the signal and the same for unvoiced sections. The last global feature is the standard deviation of the fundamental frequency F_0.

We examined the prosodic features of our speech data because for substitute voices after total laryngectomy we had found that several duration-based features showed correlations of up to $|r| = 0.72$ between human and automatic evaluation of intelligibility [12, p. 117]. The agreement was measured as the correlation between the mean value of the respective feature in a recording and the average expert's intelligibility rating for that file.

5 Results

The absolute recognition rates for the pathologic speakers when using a unigram language model were at about 50% for word accuracy (WA) and word recognition rate (WR; see Table 2). This was expected since the speech recognizer was trained with normal speech because the amount of pathologic speech data was too small for training. On the other hand, the recognizer simulates a "naïve" listener that has never heard pathologic speech before. This represents the situation that speech patients are confronted with in their daily life.

The average WA and WR rose only non-significantly when the reading errors were removed from the audio files. However, in some cases the recognition results got slightly worse when reading errors – mainly at the beginning of the files – were removed (see Figure 3). The benefit of these sections for channel adaptation were obviously higher than the harm caused by the out-of-text utterances.

The agreement among the human raters when judging the speakers' intelligibility was $r = 0.81$. This value was computed as the mean of all correlations obtained when one of the raters was compared to the average of the remaining four. The correlation between the average human and the automatic evaluation for all 85 speakers was about $r = -0.6$ (see Table 2). The coefficient is negative because high recognition rates came from "good" voices with a low score number and vice versa. There is no significant difference in the correlation for the speech data with and without reading errors.

The human-machine correlation was also computed for the subgroup of speakers whose WA and WR were better after error correction and for the subgroup of the remaining speakers. 50 speakers showed improved WA; the correlation was $r = -0.67$ both for the files with and without errors. The other 35 patients reached $r = -0.52$

Table 2. Recognition results for the speech corpora with and without reading errors (85 speakers) and correlation between automatic measure and human intelligibility rating *(rightmost column)*

	measure	avg.	st. dev.	min.	max.	correl.
with errors	WA	48.0	17.2	3.4	81.3	−0.61
without errors	WA	49.3	17.0	10.1	81.3	−0.61
with errors	WR	53.2	15.3	9.1	82.2	−0.56
without errors	WR	54.1	15.4	9.1	82.2	−0.55

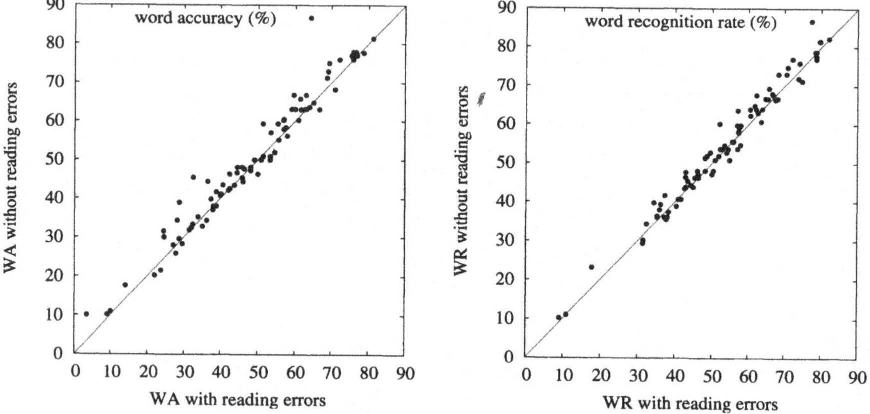

Fig. 3. Word accuracy *(left)* and word recognition rate *(right)* before and after removing reading errors; all dots above the diagonal mean better results afterward

Table 3. Correlation r between selected prosodic features and human intelligibility ratings; presented are criteria with a correlation of $|r| \geq 0.5$

feature	correlation	
	with errors	without errors
ratio of duration of unvoiced segments and file length	+0.51	+0.53
duration of silent pause after current word	+0.54	+0.53
normalized energy of word-pause-word interval	+0.62	+0.59
normalized duration of word-pause-word interval	+0.63	+0.60

before and $r = -0.51$ after elimination of the reading errors. 49 speakers showed improved WR in the repaired files; the correlation dropped slightly from $r = -0.54$ to $r = -0.53$. The other 36 patients reached $r = -0.59$ in the original files and $r = -0.57$ in the files without errors.

It was expected that for recordings with a lot of reading errors the word recognition rate would achieve higher correlation to the human rating. This was based on the assumption that human raters are not affected in their judging of intelligibility when a speaker utters words that are not in the text reference. However, the correlation for WR was in all experiments of this study smaller than for the word accuracy.

For the agreement between human evaluation and prosodic features, the findings of [12] were confirmed. The same prosodic features as for substitute voices showed the highest correlation also for the 85 speakers of this study (see Table 3). The duration and pause-based features are highly correlated to the human intelligibility criterion since non-fluent pathologic speakers often show low voice quality and hence low intelligibility. The high correlation to the word energy can be explained by irregular noise in low quality voices which are again less intelligible.

6 Conclusion and Outlook

In speech therapy and rehabilitation, a patient's voice has to be evaluated by the therapist. Speech recognition systems can be used to objectively analyze the intelligibility of pathologic speech. In this paper, the influence of reading errors and out-of-text utterances on human-machine correlation was examined. For this purpose, the effects named above were removed from the original speech samples where possible, and the correlation between speech expert and automatic speech recognizer was compared for the files with and without errors. The correlation showed no significant difference. Hence, the reading errors in text recordings do not have to be eliminated before automatic evaluation.

For the improvement of the correlation to human evaluation, adaptation of the speech recognizer to pathologic speech should be considered. However, for substitute voices after total laryngectomy, the adaptation enhances the recognition results but not the agreement to the human reference [13]. Another approach considering the recognition rates is to include the words of frequently occurring out-of-text phrases, like "I forgot my glasses.", into the recognition vocabulary of the recognizer. This is part of future work.

Acknowledgments

This work was partially funded by the German Cancer Aid (Deutsche Krebshilfe) under grant 107873. The responsibility for the contents of this study lies with the authors.

References

1. American Cancer Society: Cancer facts and figures 2000, Atlanta, GA (2000)
2. Makeieff, M., Barbotte, E., Giovanni, A., Guerrier, B.: Acoustic and aerodynamic measurement of speech production after supracricoid partial laryngectomy. Laryngoscope 115(3), 546–551 (2005)
3. Fröhlich, M., Michaelis, D., Strube, H.W., Kruse, E.: Acoustic voice analysis by means of the hoarseness diagram. J. Speech Lang. Hear. Res. 43(3), 706–720 (2000)
4. Schuster, M., Haderlein, T., Nöth, E., Lohscheller, J., Eysholdt, U., Rosanowski, F.: Intelligibility of laryngectomees' substitute speech: automatic speech recognition and subjective rating. Eur. Arch. Otorhinolaryngol. 263(2), 188–193 (2006)
5. International Phonetic Association (IPA): Handbook of the International Phonetic Association. Cambridge University Press (1999)

6. Stemmer, G.: Modeling Variability in Speech Recognition. Studien zur Mustererkennung, vol. 19. Logos Verlag, Berlin (2005)
7. Wahlster, W. (ed.): Verbmobil: Foundations of Speech-to-Speech Translation. Springer, Berlin (2000)
8. Nöth, E., Batliner, A., Kießling, A., Kompe, R., Niemann, H.: Verbmobil: The Use of Prosody in the Linguistic Components of a Speech Understanding System. IEEE Trans. on Speech and Audio Processing 8(5), 519–532 (2000)
9. Chen, K., Hasegawa-Johnson, M., Cohen, A., Borys, S., Kim, S.-S., Cole, J., Choi, J.-Y.: Prosody dependent speech recognition on radio news corpus of American English. IEEE Trans. Audio, Speech, and Language Processing 14, 232–245 (2006)
10. Shriberg, E., Stolcke, A.: Direct Modeling of Prosody: An Overview of Applications in Automatic Speech Processing. In: Proc. International Conference on Speech Prosody, Nara, Japan, pp. 575–582 (2004)
11. Batliner, A., Buckow, A., Niemann, H., Nöth, E., Warnke, V.: The Prosody Module [7], pp. 106–121
12. Haderlein, T.: Automatic Evaluation of Tracheoesophageal Substitute Voices. Studien zur Mustererkennung, vol. 25. Logos Verlag, Berlin (2007)
13. Haderlein, T., Steidl, S., Nöth, E., Rosanowski, F., Schuster, M.: Automatic Recognition and Evaluation of Tracheoesophageal Speech. In: Sojka, P., Kopeček, I., Pala, K. (eds.) TSD 2004. LNCS (LNAI), vol. 3206, pp. 331–338. Springer, Heidelberg (2004)

Czech Pitch Contour Modeling Using Linear Prediction

Petr Horák

Department of Digital Signal Processing and Speech Synthesis
Institute of Photonics and Electronics, Academy of Sciences of the Czech Republic
Chaberská 57, CZ-182 51 Praha 8, Czech Republic
horak@ufe.cz

Abstract. Present Czech TTS systems can produce synthetic speech with high intelligibility but low naturalness. The difference between natural and synthetic speech is still too high. Naturalness of the synthetic speech is given by the signal modeling and by the prosody modeling. This paper deals with the improving of the synthetic prosody modeling especially with the improving of the intonation modeling. A mathematical model of the pitch contour modeling can significantly limit the complexity of intonational rules creation and increase the naturalness of resulting synthetic speech. The linear prediction inonational model has been implemented into TTS system Epos for practical use. This built-in inonational model uses excitation by rules and provides in conjunction with a new triphone time domain inventories more naturalness synthetic speech than previous direct intonational rules.

1 Introduction

Even though prosody is attributed to intensity, intonation and duration of speech, it seems that in most languages intonation is the main contribute to prosodic information. A model describing pitch contours would then be of extreme importance in text-to-speech synthesis.

The idea of LPC prosody modeling is based on describing of the complex sentence F0 contour by a few LP coefficients and simple linear prediction error signal. The LP coefficients should contain speaker dependencies while prediction error signal information about relevant prosody changes [1]. To describe prosody contours on syllabic level with sufficient accuracy we need channel with bandwidth approximately 5 Hz. Such channel can be created using 10 Hz F0 contour sampling frequency and 4 LP coefficients [2].

We use the pitchdetector with 125 Hz output F0 contour sampling frequency. To decrease we use decimation of original F0 contour by 1:10. The F0 contour from the pitchdetector must be first interpolated (removing of the zeros and pitchdetector errors) and filtered by the antialiasing filter. Filtered F0 contour signal is analyzed by linear predictor to obtain 4 LP coefficients and linear prediction error (residual) signal. The prediction error signal can be simplified by using rectangular approximation and transformation of the approximated error signal to the differential error signal.

P. Sojka et al. (Eds.): TSD 2008, LNAI 5246, pp. 333–339, 2008.
© Springer-Verlag Berlin Heidelberg 2008

2 Pitch Contour Analysis

The linear prediction intonation model is based on LP analysis (closely described in [4]) and synthesis of the pitch contour. The pitch contour can be estimated only at voiced parts of speech signal hence the pitch contour is non-continuous signal. The continuous signal is required for further LP analysis. Therefore the values of pitch are estimated at unvoiced parts of speech by linear interpolation. The example of the speech signal with pitch contour and its continuous interpolation is illustrated on Figure 1.

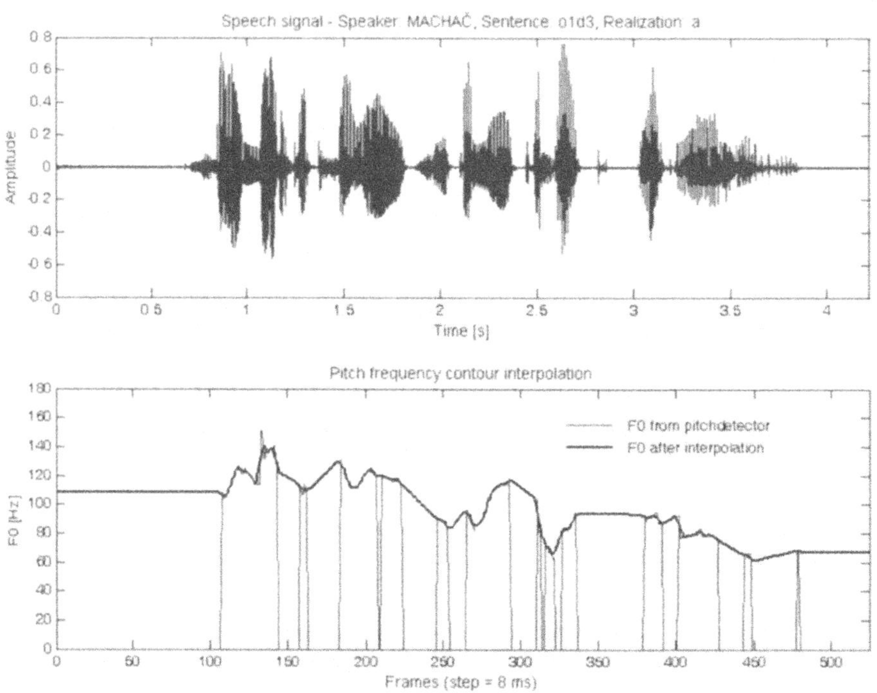

Fig. 1. Interpolation of pitch contour

The pitch contour is obtained by pitch-detector with period 8 ms (sampling frequency 125 Hz). The band about 5 Hz is efficient for human perception of intonation [2]. Therefore decimation with ratio 1:10 is applied to the pitch contour. The DC component is subsequently removed from the pitch signal. The filtered pitch contour (low-pass filter 0–5 Hz) is described on Figure 2.

The pitch contour after decimation is illustrated on Figure 3a. The autocorrelation method is used for estimation of LP coefficients together with the Le Roux-Guegen algorithm [5] and transformation of LP coefficients [4]. The residual signal is subsequently estimated using LP coefficients and inverse filter. The residual signal we can see on Figure 3b.

The F0 contour can be resynthesized from approximated error signal (obtained by summation of the differential approximated error signal) and 4 LPC coefficients. The

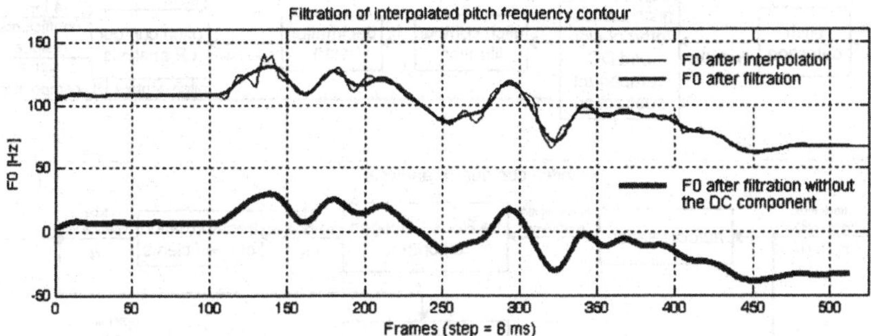

Fig. 2. Filtration of interpolated pitch contour (without DC component)

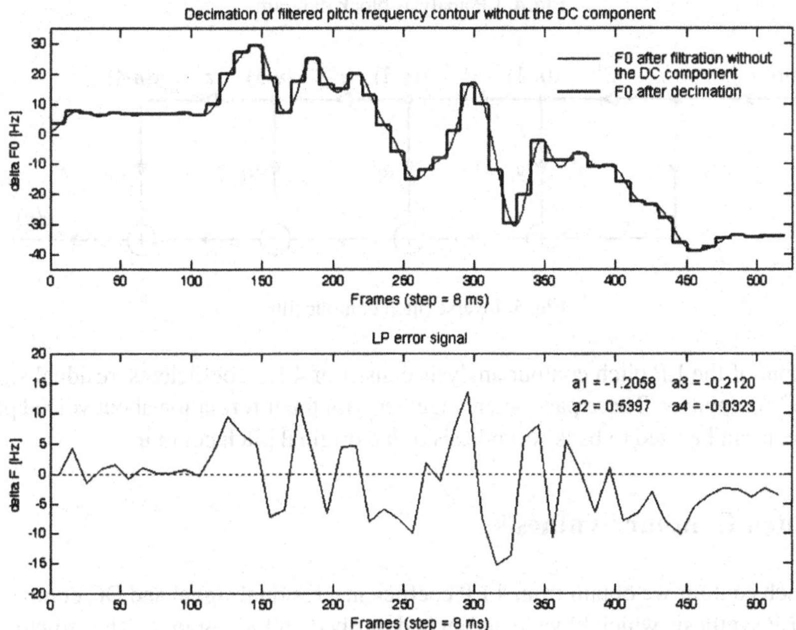

Fig. 3. Decimation of filtered pitch contour and LP analysis

original F0 contour is obtained by oversampling 10:1 and followed filtering of resynthesized F0 contour.

The block diagram of pitch contour analysis is described on Figure 4a and detailed illustration of LP analysis is on Figure 4b. The inverse filter block diagram used to the residual signal estimation is illustrated on Figure 5.

The linear predictive F0 contour synthesis and resynthesis was tested by including to the speech vocoder based on the same principle (LPC analysis and resynthesis of the speech signal). Listening tests proved that from simplified error signal we could reconstruct applicable prosody contour.

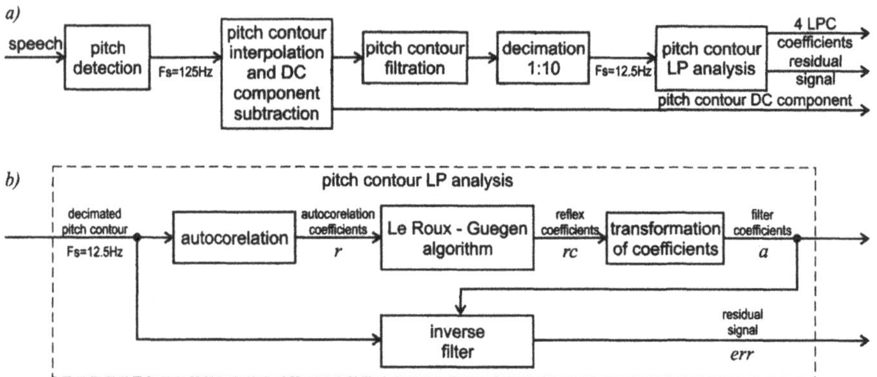

Fig. 4. LP analysis block diagram

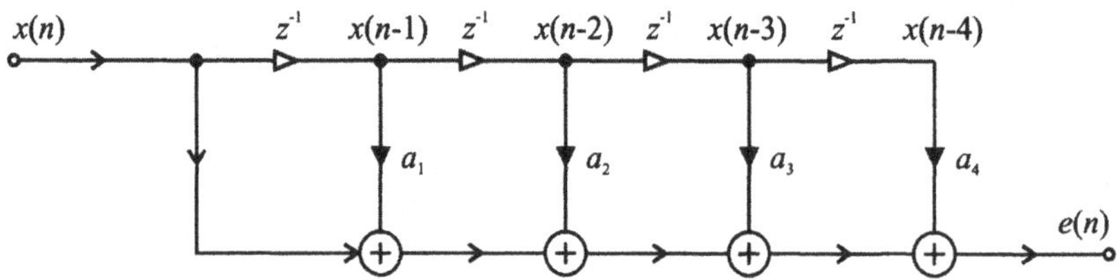

Fig. 5. Inverse pitch contour filter

Output of the LP pitch contour analysis consist of 4 LP coefficients, residual signal and DC component. These parameters together with the information about voiced parts of speech can be used to back resynthesis of the original pitch contour.

3 Pitch Contour Synthesis

The pitch contour we obtain from 4 LP coefficients, residual signal and DC component using LP synthesis which block diagram is described on the Figure 6. The structure of the reconstruction (synthesis) filter illustrated on Figure 7.

The continuous pitch contour on the output of reconstruction filter is back resampled with the ratio 1:10 (to the 125 Hz sampling frequency) and filtered by the reconstruction

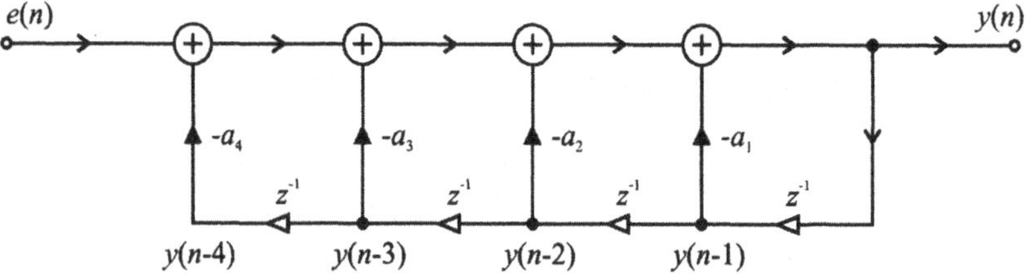

Fig. 6. Reconstruction (synthesis) pitch contour filter

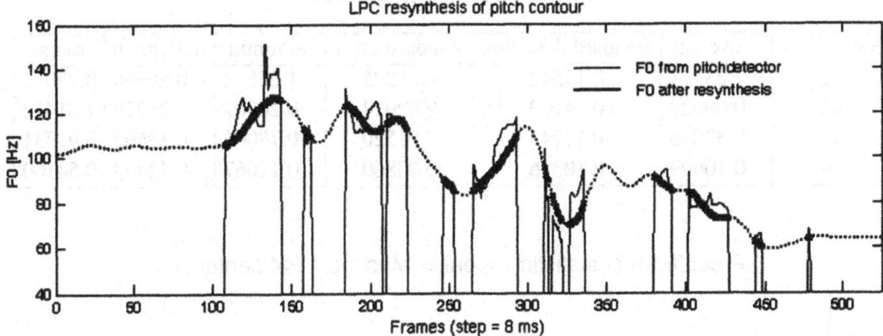

Fig. 7. LP resynthesis of pitch contour

filter (we us the same as in analysis). The DC component is consequently added and the information about voiced/unvoiced parts of speech is applied.

4 TTS Implementation

The linear prediction intonation model has been implemented into TTS system Epos [3]. The block diagram of Epos with implemented LP intonation model is illustrated on Figure 8. The excitation of LP intonation model is generated by rules based on the text analysis. The demonstration of the pitch contour resynthesis using the linear prediction is on the Figure 7.

The approximated differential linear prediction error signal contains relevant information of fundamental frequency contour changes. This information can be used, in relation to text analysis, as help at proposing and confirmation of new prosody rules.

The residual signal used for excitation of the synthesis filter is generated by rules and approximated with the rectangle signal. The LP coefficients are estimated from training speech database from native speaker Machač. The estimated LP coefficients for speaker Machač is described in Table 1 and on Figure 9.

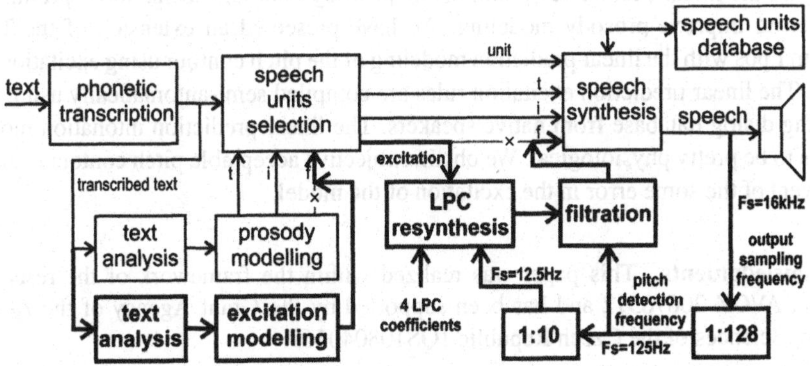

Fig. 8. TTS system Epos with implemented LP pitch contour generation

Table 1. LP coefficients evaluation

coeffficient	average	standard deviation	standard error	minimum	maximum	range
a1	−1.23761	0.15542	0.01295	−1.65578	−0.86500	0.79078
a2	0.60009	0.24162	0.02014	0.20307	1.21729	1.01422
a3	−0.32046	0.18243	0.01520	−0.78066	0.19665	0.97731
a4	0.10699	0.10326	0.00860	−0.11667	0.42493	0.54160

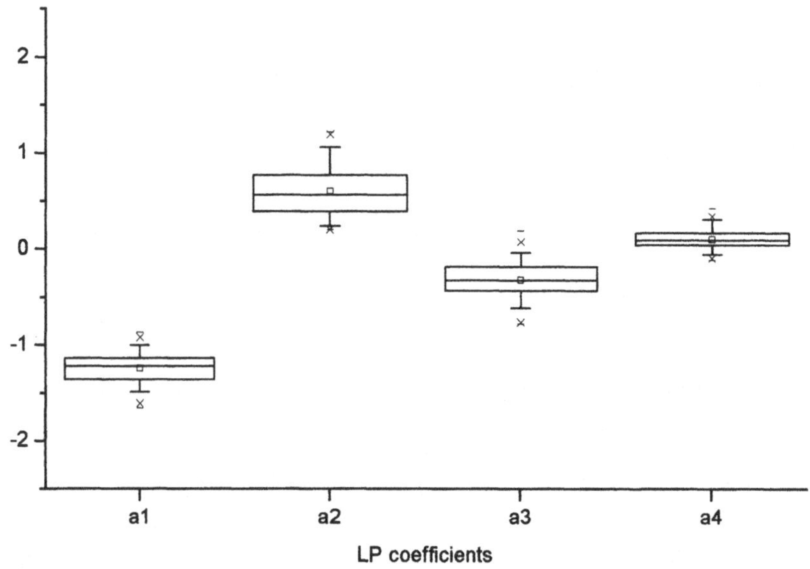

Fig. 9. LP coefficients evaluation

5 Conclusion

Using of the linear predictive coding of the prosody contours seems to be promising method to improve prosody modeling. We have presented an extension of the TTS system Epos with the linear prediction modeling of the pitch contour using excitation by rules. The linear prediction excitation rules are compiled semi-automatically using the training dialog database from native speakers. The linear prediction intonation model seems to be pretty physiological. We obtain subjective acceptable pitch contour even in the event of the some error in the excitation of the model.

Acknowledgments. This paper was realized within the framework of the research project AVOZ 20670512 and has been supported by the Grant Agency of the Academy of Sciences of the Czech Republic 1QS108040569.

References

1. Mersdorf, J., Rinscheid, A., Bruggen, M., Schmidt, K.U.: Kodierung von Intonationsverläufen Mittels Linearer Prediktion. In: Studientexte zur Sprachkommunikation, Achte Konferenz Elektronische Sprachsignalverarbeitung, Cottbus, Germany, August 25–27, 1997, pp. 169–176 (1997)
2. Horák, P.: Modeling of the Suprasegmental Characteristics of the Spoken Czech Using Linear Prediction. Ph.D. Thesis, Czech Technical University, Prague, 116 pages (2002), http://epos.ufe.cz/publications/2002/ph_dis.pdf
3. Hanika, J., Horák, P.: Epos A New Approach to the Speech Synthesis. In: Proceedings of the First Workshop on Text, Speech and Dialogue TSD 98, Brno, Czech Republic, September 23–26, 1998, pp. 51–54 (1998)
4. Markel, J.D., Gray, A.H.: Linear Prediction of Speech. Springer, New York (1976)
5. Leroux, J., Guegen, C.: A Fixed Point Computation of Partial Correlation Coefficients. In: IEEE Transactions on ASSP, June 1977, pp. 257–259 (1977)

Energy Normalization in Automatic Speech Recognition

Nikša Jakovljević[1], Marko Janev[1], Darko Pekar[2], and Dragiša Mišković[1]

[1] University of Novi Sad, Faculty of Engineering, Trg Dositeja Obradovića 6
21000 Novi Sad, Serbia
jakovnik@uns.ns.ac.yu
[2] AlfaNum ltd., Trg Dositeja Obradovića 6
21000 Novi Sad, Serbia

Abstract. In this paper a novel method for energy normalization is presented. The objective of this method is to remove unwanted energy variations caused by different microphone gains, various loudness levels across speakers, as well as changes of single speaker loudness level over time. The solution presented here is based on principles used in automatic gain control. The use of this method results in relative improvement of the performances of an automatic speech recognition system by 26%.

1 Introduction

Acoustic variability is one of the main contributors to degradation of performances of automatic speech recognition (ASR) systems. A systematic review of what has been done in this area can be found in [1]. One of the standard features used in ASR systems is the energy of the speech signal. The energy can be helpful for phoneme discrimination (e.g. vowels vs consonants; voiced vs unvoiced).

Unwanted energy variations are caused by many factors, the dominant one being background noise. Background noise drastically changes the sound level of silent segments, and on the other hand slightly changes the sound level of loud segments. Several solutions for this problem are proposed in [3,4,5,6].

Unwanted energy variations can also be caused by: (1) different microphone gains, (2) different microphone placement, (3) variations in loudness levels across different speakers as well as (4) changes in loudness level of a single speaker over time.

A silent environment, where this ASR system will be used, results in bigger influence of other causes of energy variations. Automatic gain control which is used on sound cards does not have satisfactory effects, and for that reason a certain variation of gain control for energy normalization should be applied in the ASR front-end block.

In this paper a novel energy normalization procedure based on automatic gain control principles is presented. The paper is organized as follows. In the following section a short description of existing methods will be presented along with reasons why they could not be successfully applied in this case. The proposed algorithm will be presented in Section 3. Experimental results and a brief description of the used ASR system will be given in Section 4. Finally, in Section 5 conclusions and possible future work will be presented.

P. Sojka et al. (Eds.): TSD 2008, LNAI 5246, pp. 341–347, 2008.
© Springer-Verlag Berlin Heidelberg 2008

2 Energy Normalization Methods in ASR Systems

In this section a brief review of existing energy normalization methods along with comments on their ability to achieve the newly established goal is presented. Energy normalizaton needs to eliminate energy variations caused by variations in loudness levels across different speakers and changes in loudness level of a single speaker over time. An additional constraint is the requirement for real-time processing.

One of the first energy normalization methods is cepstral mean normalization (CMN). Since the first cepstral coefficient is energy, CMN can be viewed as an energy normalization method, which is the usual approach in this framework. As it is well known, the basic goal of CMN is the elimination of the influence of the channel. The success of CMN depends on the duration of the averaging interval. The longer interval means that the average CMN value is less dependent on the spoken phone sequence, thus the effects of CMN are better. On the other hand, if the interval is too short, CMN may result in degradation of the ASR performance. The latter is the reason why CMN cannot be used in real-time applications. Similar conclusions hold for cepstral variance normalization (CVN). Both CMN and CVN can reduce energy dispersion caused by variations in loudness level across speakers, but cannot compensate energy variations within a single utterance. They both presume that loudness level does not change over the course of the utterance. Details about CMN and CVN can be found in [2].

Log-energy dynamic range normalization (ERN) reduces energy dispersion caused by different levels of background noise. This method is based on the fact that the same noise results in small changes in log energy on high-energy segments (e.g. stressed vowel) but in drastic changes on low-energy segments (e.g. silence, occlusions of stops and affricates). It is assumed that all utterances of all speakers have the same maximum energy as well as dynamic range. Under this assumption, the target minimum log energy value is set based on the estimated maximum energy in the utterance and the assumed dynamic range. If the minimum log energy in a single utterance is less than the calculated target minimum log energy, the energy should be rescaled to a specified target range, otherwise nothing should be done.

More details about this procedure can be found in [3,4]. It is clear that ERN is not the solution for the problem of energy normalization as defined in this paper.

3 Method Description

The input parameter for the energy normalization block is frame energy. Frames are 30 ms long and shifted by 10 ms, extracted by application of a Hamming window function. In this way central samples of the frame carry greater weight than those near the boundaries. In order to carry out energy normalization of a single utterance, it is necessary to track peak energy of successive speech segments. The standard way to achieve this is using IIR systems defined by:

$$E_p(n) = \gamma E_p(n-1) + (1-\gamma)E(n) \tag{1}$$

where $E_p(n)$ is the peak energy at the n-th frame, $E(n)$ is the energy at the n-th frame and γ is the memory coefficient. From Equation 1 it follows that the values of the memory coefficient lie in the interval (0, 1], and that any other value has no physical meaning. Since it should be desirable to "catch" peak energy fast and to change it slowly once it is "caught", the value of the memory coefficient should be specified separately for segments with rising and falling energy. In the case of rising energy, the value of the memory coefficient should be small. On the other hand, during the segments with falling energy, the value of the memory coefficient should be relatively close to 1. In this way the resulting peak energy would change very slowly, but quickly enough to avoid omitting the next maximum if it is smaller than the previous one. In this paper the values of the memory coefficient of $\gamma_r = 0.30$ for rising and $\gamma_f = 0.99$ for falling energy were adopted. The normalization process consists of dividing the current energy value $E(n)$ with the current peak tracker value $E_p(n)$ i.e.:

$$E_n(n) = E(n)/E_p(n) \tag{2}$$

where $E_n(n)$ is the normalized value of the energy at the n-th frame. The meaning of the other variables is the same as in Equation 1. Since peak energy decreases constantly over silent segments, the normalization strategy described above results in the increase of the noise level at silent segments, as shown in Figure 1. One way to overcome this problem is the implementation of separate energy normalization procedures for silent segments and for segments with speech. This approach requires the introduction of a function for automatic detection of speech activity. A simple solution presented in [7] is applied in this paper. Two additional track curves are used, one for fast energy tracking ($E_f(n)$) and one for slow energy tracking ($E_s(n)$). If $E_f(n) > E_s(n)$, the n-th frame is a part of a speech segment, otherwise it is a part of a non-speech segment. The values of $E_f(n)$ and $E_s(n)$ are calculated by:

$$E_i(n) = \gamma E_i(n-1) + (1-\gamma)E(n) \tag{3}$$

where the index i can be either f or s. The meaning of the other parameters is the same as in Equation 1. As well as in the case of peak energy tracking, the value of the memory coefficient γ depends on whether the energy is rising or falling. Consequently there are 2 values of the memory coefficient for each of the trackers. The memory coefficients for the slow tracker in the case of rising and falling energy are marked by γ_{sr} and γ_{sf}, while those for the fast tracker are marked by γ_{fr} and γ_{ff}, with the following relations holding:

$$\begin{aligned} \gamma_{sr} &\leq \gamma_{sf} < 1 \\ \gamma_{fr} &\leq \gamma_{ff} < 1 \\ \gamma_{fr} &\leq \gamma_{sr} \\ \gamma_{ff} &\leq \gamma_{sf} \end{aligned} \tag{4}$$

Acceptable performances are achieved with the following values of memory coefficients: $\gamma_{sr} = 0.85$, $\gamma_{sf} = 0.95$, $\gamma_{fr} = 0.80$ and $\gamma_{ff} = 0.90$.

The algorithm for automatic detection of speech activity mentioned above requires that the maximum noise energy be set. A silent environment, where this ASR system will be used, provides low noise level, and thus the value of the maximum noise level can be set easily. In this specific case its value is 10^{-4}.

The rule for calculation of normalized energy is thus modified. One level of peak energy is used for silent segments (marked by $E_{ps}(n)$) and the other level of peak energy

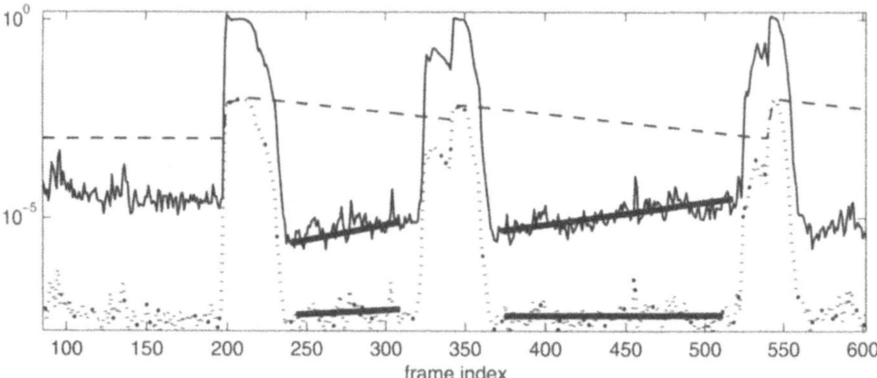

Fig. 1. Frame energy (*dotted line*), normalized energy (*solid line*) and peak energy (*dashed line*) of the speech signal. The decrease of the peak energy (normalization level) on the silent segments (*numbered from 230 to 320 and 370 to 520*) results in an increase of the normalized noise energy.

for speech segments ($E_{pv}(n)$). The value of the $E_{pv}(n)$ is the same as the value of the $E_p(n)$ defined by Equation 1. The value of the $E_{ps}(n)$ is the value of the E_{pv} in the last speech frame before the current silent segment. This new rule for energy normalization can be summarized as:

$$E_n(n) = \begin{cases} E(n)/E_{pv}(n) \text{ for speech segments} \\ E(n)/E_{ps}(n) \text{ for silent segments} \end{cases} \quad (5)$$

Another possibility is to normalize energy at silent segments with respect to the current maximum energy value within the whole utterance. Although the noise level is almost the same in the whole utterance and therefore normalization with respect to the same level should have much more sense, the uncertainty of speech activity detection at some segments with consonant-vowel transitions makes this alternative solution impossible. Disregard of this fact leads to errors such as the one shown in Figure 2.

An additional requirement that a speech segment last at least n_{min} frames is introduced in order to ensure that the value of the $E_{ps}(n)$ be the actual peak value of the last speech frame. This constraint implies that $E_{ps}(n)$ is changed only if the last n_{min} successive frames are detected as speech. In this specific case, the value of n_{min} is 3. Since the maximum signal energy rarely occurs at the beginning of the word, it is necessary to calculate the values of peak energy at several following frames as well, therefore a tolerable delay D is defined. The estimation of the maximum energy is better if the number of following frames included in the calculation is greater. On the other hand, the application demands for real-time processing set the upper bound of delay. The value of the delay D of 10 frames i.e. 100 ms was adopted.

To avoid the possibility that after a longer silence interval, energy normalization is performed with initial values related to silent frames, the minimum value of the peak energy E_0 is set. If the value of the peak energy $E_{pv}(n)$ becomes less than E_0, $E_{pv}(n)$ is set to E_0. Too great a value of E_0 at speech segments with smaller energy can result in inappropriate normalization. For this application the value of 10^{-3} was adopted.

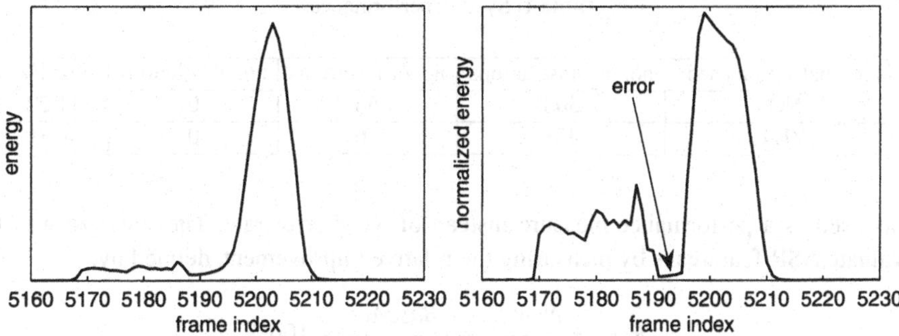

Fig. 2. Energy (*left*) and resulting normalized energy (*right*) on the same speech segment. A small part of the speech segment is detected as silence (*frames from 5191 to 5194*). That misclassification results in unacceptable variation of the normalized energy for the speech segment caused by different normalization levels for the speech and silent segments.

4 A Description of the ASR and Achieved Performances

Two systems are created to estimate performances of the proposed method for energy normalization. These two systems differ only in energy coefficients. Feature vectors in both systems, baseline system and the system which uses the proposed method for energy normalization, contain 34 coefficients, where 32 of them are common (16 Mel-frequency spectral envelope coefficients and their derivatives) for both systems. In the baseline system log energy and the derivative of energy are used instead of log of normalized energy and the derivative of normalized energy. More details about Mel-frequency spectral envelope coefficients are given in [8].

The particular software application where the ASR system will be used requires a simple grammar. The ASR input is a sequence of consonant vowel (CV) words. Between words there are silent intervals. Such a grammar structure leads to a reduced number of contexts (the left context for consonants and the right context for vowels are silence) as well as a reduction of feature dispersion caused by coarticulation effects. Although the number of possible words is small, the basic modelling unit is a context dependent phone.

The ASR system is based on hidden Markov models (HMM) and Gaussian mixture models (GMM). Instead of a diagonal covariance matrix, each Gaussian distribution is determined by a full covariance matrix. It was sufficient that an HMM state be modelled by a single Gaussian distribution, because of the reduction of the number of different contexts.

The training corpus contains 7 hours of speech. The audio files are encoded in the PCM format (22050 Hz, 16 bits per sample). As already noted, the noise level is negligible.

The results are presented in Table 1. The test corpus contains 2513 CV words i.e. 5026 phonemes. With regard to specific purposes of the ASR, phone error rate (PER)

Table 1. System performances

Is normalization used?	no. of substitutions	no. of insertions	no. of deletions	PER [%]
NO	508	56	0	11.22
YES	401	16	0	8.30

was used as a performance measure instead of word error rate. The standard way to evaluate ASR features is by measuring the relative improvement, defined by:

$$R.I. = \frac{NewAcc - BaseAcc}{100 - BaseAcc} \times 100\% \tag{6}$$

where $NewAcc$ is the accuracy of a system with new features and $BaseAcc$ is the accuracy of a baseline system. Relative improvement in terms of PER is defined by:

$$R.I. = \frac{BasePER - NewPER}{BasePER} \times 100\% \tag{7}$$

For the results presented in Table 1 the relative improvement is 26%. This improvement is caused by the reduction of energy dispersion in phone models.

5 Conclusion and Future Work

In this paper a new method for on-line energy normalization is presented. The method is based on principles used in automatic gain control. Its objective is to reduce the dispersion of energy in an HMM state, caused by variable microphone gains, loudness levels across speakers and loudness level of a single speaker over time. A relative improvement of about 26% was achieved on the test set. The test set contains high quality audio files (22050 Hz, 16 bits per sample) with low noise level.

The future research should include an applicability study on the use of this method for telephone quality speech signals with a higher noise level. A possible solution could be a combination of this method with some of the methods presented in Section 2. These steps demand the existence of an ASR system for English and an Aurora 2 test set intended for such research.

References

1. Benzeghiba, M., De Mori, R., Deroo, O., Dupont, S., Jouvet, D., Fissore, L., Laface, P., Mertins, A., Ris, C., Rose, R., Tyagi, V., Wellekens, C.: Impact of Variabilities on Speech Recognition. In: Proc. SPECOM 2006 (2006)
2. Togneri, R., Toh, A.M., Nordholm, S.: Evaluation and Modification of Cepstral Moment Normalization for Speech Recognition in Additibe Babble Ensemble. In: Proc. SST 2006, pp. 94–99 (2006)
3. Lee, Y., Ko, H.: Efective Energy Feature Compensation Using Modified Log-energy Dynamic Range Normalization for Robust Speech Recognition. IECIE Trans. Commun. Anal. E90-B(6), 1508–1511 (2007)

4. Zhu, W., O'Shaughnessy, D.: Log-energy Dynamic Range Normalization for Robust Speech Recognition. In: Proc. ICASSP 2005, vol. 1, pp. 245–248 (2005)
5. Zhu, W., O'Shaughnessy, D.: Using Noise Reduction Andspectral Emphasis Techniques to Improve ASR Performance Innoisy Conditions. In: Proc. ASRU 2003 (2003)
6. Zhu, W., O'Shaughnessy, D.: Incorporating Frequency Masking Filtering in a Standard MFCC Feature Extraction Algorithm. In: Proc. ICSP 2004, vol. 1, pp. 617–620 (2004)
7. Hänsler, E., Schmidt, G.: Acoustic Echo and Noise Control. New Jersey. Wiley, Chichester (2004)
8. Jakovljević, N., Mišković, D., Sečujski, M., Pekar, D.: Vocal Tract Normalization Based on Formant Positions. In: Proc. IS LTC 2006 (2006)

HMM-Based Speech Synthesis for the Greek Language

Sotiris Karabetsos, Pirros Tsiakoulis, Aimilios Chalamandaris, and Spyros Raptis

Institute for Language and Speech Processing (ILSP) / R.C. "Athena",
Voice and Sound Technology Department
Artemidos 6 & Epidavrou, Maroussi, GR 15125, Athens, Greece
{sotoskar,ptsiak,achalam,spy}@ilsp.gr
http://www.ilsp.gr

Abstract. The success and the dominance of Hidden Markov Models (HMM) in the field of speech recognition, tends to extend also in the area of speech synthesis, since HMM provide a generalized statistical framework for efficient parametric speech modeling and generation. In this work, we describe the adaption, the implementation and the evaluation of the HMM speech synthesis framework for the case of the Greek language. Specifically, we detail on both the development of the training speech databases and the implementation issues relative to the particular characteristics of the Greek language. Experimental evaluation depicts that the developed text-to-speech system is capable of producing adequately natural speech in terms of intelligibility and intonation.

Keywords: HMM, Speech Synthesis, Text to Speech, Greek Language, Statistical Parametric Speech Synthesis, Hidden Markov Model.

1 Introduction

Nowadays, the most common approach for achieving high quality near-natural speech synthesis is the corpus-based unit selection technique. In principle, this method presumes no adoption of any specific speech model and simply relies on runtime selection and concatenation of speech units from a large speech database using explicit matching criteria [1]. However, despite the dominance of corpus-based unit selection text to speech systems, there is an increased interest for speech synthesis based on an efficient model-based parametric framework. The capability of producing synthetic speech of high naturalness and intelligibility through a proper manipulation of the parameters of the model, offers significant advantages not only in controlling the synthesis procedure but also in easily adapting the technology in different languages, contexts and applications [2].

In parametric speech synthesis, the most common model is the, so-called, source-filter model of speech production [3]. Early developed speech synthesis techniques (e.g. formant synthesis) adopted this model in a rule-based framework so as to achieve general purpose speech synthesis. Recently, the same model has been efficiently adopted in a data-driven approach utilizing the statistical framework of the Hidden Markov Models (HMM) leading to the HMM-based speech synthesis [4,5]. Although the HMM-based speech synthesis framework is still not superior to the corpus-based approach, recent

P. Sojka et al. (Eds.): TSD 2008, LNAI 5246, pp. 349–356, 2008.
© Springer-Verlag Berlin Heidelberg 2008

results have established it as one of the most successful parametric speech synthesis techniques [6,7].

Consequently, the HMM-based speech synthesis framework has been successfully applied in different languages where the emphasis is mostly put on investigating the particular language requirements in order to efficiently model the contextual information. Besides the Japanese language, examples of text to speech systems in other languages using HMM-based speech synthesis include, English [8], German [9], Chinese [10], Slovenian [11], Portuguese [12], Spanish [13], Korean [14], Arabic [15], to name a few. In most of these cases, the resulting text to speech system was found to perform well in terms of achieved quality. In this work, we describe the adaption, the implementation and the evaluation of the HMM speech synthesis framework for the case of the Greek language. Specifically, we report on both the development of the training speech databases and the implementation issues relative to the particular characteristics of the Greek language. The developed text to speech system is assessed by comparative subjective tests using both diphone and corpus-based unit selection speech synthesis systems. To our knowledge, this is the first report on the exploitation of the HMM speech synthesis framework for the Greek language.

The rest of the paper is organized as follows: in Section 2 the HMM framework for speech synthesis is briefly discussed while Section 3 describes its adaption for the case of the Greek language. Section 4, provides the experimental assessment and Section 5 concludes the work.

2 The HMM Speech Synthesis Framework

The schematic representation of the HMM-based speech synthesis framework is illustrated in Figure 1. As mentioned, the data-driven methodology is followed since the model is trained on a pre-recorded and annotated speech database. Thus, the framework consists of both a training part and a synthesis part.

The responsibility of the training part is to perform the statistical modeling of speech, based on characteristics extracted from the speech database. In particular, the extracted characteristics relate not only to prosodic (source) and vocal tract (spectral representation) information, but also extend on contextual and durational modeling [16,17]. More specifically, the HMM-based speech synthesis framework performs simultaneous modeling of pitch and spectrum taking into account the dynamics of both quantities as well. Spectral representation utilizes Mel-based cepstral coefficients while prosody is represented as logF0. Multi Space probability Distribution (MSD) modeling is performed to alleviate the problem of non continuous pitch values in unvoiced regions. Moreover, context clustering is performed using decision trees so as to fully exploit the contextual information in lexical and syntactic level [17]. Duration models are also constructed according to the method described in [16].

At the synthesis part, the input text is analyzed and converted to a context-dependent phoneme sequence. Next, the synthesis utterance is represented by the concatenation of the individual context-dependent HMMs. The speech signal is synthesized (reconstructed) from spectral coefficients and pitch values using the MLSA filter. Speech generation is guided in a way that the output probability for the HMMs is maximized [5,8].

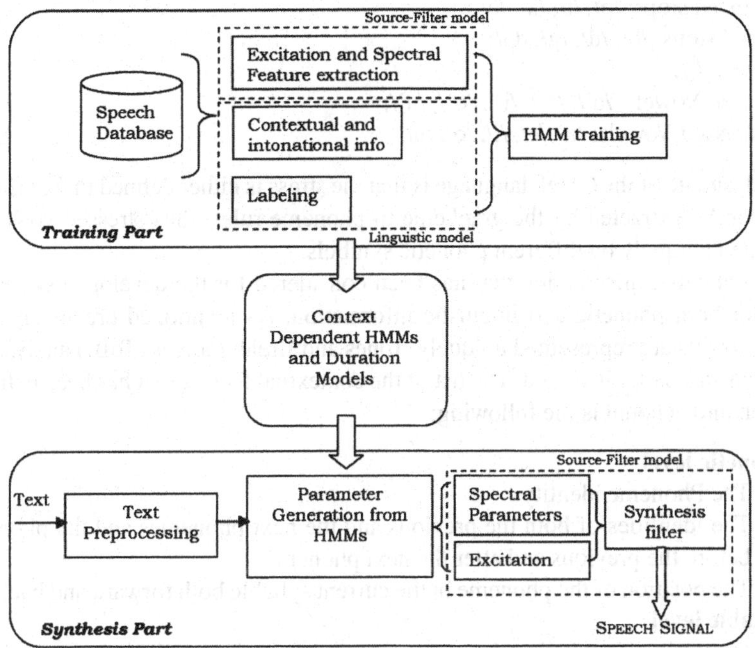

Fig. 1. The training and synthesis parts of the HMM speech synthesis framework

The inclusion of both spectral and prosodic dynamics (delta and acceleration coefficients) ensures a smooth and pragmatic speech generation. Since the source-filter model of speech production and reconstruction is employed, various types of excitation signals may be utilized which may result "buzzy" ([8]) and high quality speech [18].

3 HMM Speech Synthesis for the Greek Language

The HMM-based speech synthesis system for the Greek language follows the framework depicted in Figure 1. Its adaption considers the particular characteristics of the Greek language and mainly focuses on the analysis and the contextual modeling since contextual information is language dependent. However, the HMM framework provides a general setup for sufficient context modeling that can be easily adopted for different languages.

For the phonemic representation of Greek phonemes, a set of 37 phonemes was adopted that are divided in 26 consonants and allophones, 5 unstressed vowels, 5 stressed vowels and one for the silence (sentence beginning, sentence ending and pause). The set of the 37 phonemes define nine classes as follows:

- Unvoiced fricatives: /f/, /x/, /X/, /T/, /s/;
- Voiced fricatives: /v/, /J/, /j/, /D/, /z/;
- Liquids: /l/, /r/, /L/;
- Nasals: /m/, /M/, /n/, /N/, /h/;

- Unvoiced stops: /p/, /t/, /k/, /c/;
- Voiced stops: /b/, /d/, /g/, /G/;
- Silence: /-/;
- Stressed Vowels: /a'/, /e'/, /i'/, /o'/, /u'/;
- Unstressed Vowels: /a/, /e/, /i/, /o/, /u/.

An advantage of the Greek language is that the stress is either defined in free running text or can be extracted by the grapheme to phoneme rules, thus stressed vowels are represented uniquely by different phonetic symbols.

The contextual information that has been considered for the developed system accounts for both phonetic and linguistic information. As mentioned previously, since stressed vowels are represented uniquely, Tones and Break Indices (ToBI) analysis and annotation was partly involved. The list of the contextual factors that has been extracted and taken into account is the following:

- **Phonetic level**
  - The Phoneme identity
  - The identities of both the previous and the next phonemes and the phonemes before the previous and after the next phoneme
  - The position of the phoneme in the current syllable both forward and backward
- **Syllable level**
  - Determination whether the current, next and previous syllable is stressed
  - Number of phonemes in current, next and previous syllable
  - Position of the syllable both in phrase and word
  - Number of vowels in the syllable
  - Number of stressed syllables in the current phrase before and after the current syllable, and the number of syllables between the current and the previous and between the current and the next stressed syllables.
- **Word level**
  - The number of syllables in the current, next and previous word
  - The position of the word in the phrase (both forward and backward) and the sentence
- **Phrase level**
  - The number of syllables in the current, next and previous phrase
  - The position of the phrase in the sentence (both forward and backward)
- **Sentence level**
  - The number of phrases in the sentence
  - The number of words in the sentence
  - The number of syllables in the sentence

For the training of the HMMs, two speech databases were considered: one using a female and the other using a male speaker. Since both of the produced systems performed equivalently in terms of achieved quality, only the results for the female speaker are shown in the next section. The training speech database consisted of 1200 phonetically balanced reading-style utterances. The sampling frequency was 16KHz with 16 bits resolution. The segmentation and the labelling of the databases was performed automatically (by using HMMs) and manually refined. Notice that the same speech databases were utilized for the construction of the relevant corpus-based unit selection text to speech system. For the HMM synthesis, parameter extraction was carried out using a 25msec Blackman-windowed frame length at a 5msec frame rate. The spectral representation considers mel-based cepstral coefficients together with their delta and delta-delta values including the zeroth coefficient. Pitch is represented by using logF0 along with delta and delta-delta values as well. The HMM topology utilized was a 5-state left-to-right with no skip. Based on the database and the contextual factors considered, the HMM speech synthesis system resulted a total number of full context models of 116432, with 1293 questions for the construction of the state clustering decision trees.

4 Experimental Evaluation

The assessment of the produced HMM-based speech synthesis system was based on conducting small scale listening tests. The system was compared against two speech synthesis systems namely, a diphone-based and a corpus-based unit selection system. The diphone-based system has only one instance per diphone. Evaluation is based on the Mean Opinion Score (MOS) concerning the naturalness and the intelligibility (clearness) on a scale of one to five where one stands for "bad" and five stands for "excellent". A total of 15 no-domain specific sentences were synthesized by the three systems. Each sentence was 5 to 10 words long and was not included in the training database. A group of 10 listeners, comprised by both speech and non-speech experts were asked to express their opinion for each sentence in a MOS scale. For every sentence, the three versions (produced by each system) were presented to each listener in random order each time. Every group could be heard by the listeners more than once. The results of the test are summarized in Table 1, where the MOS is shown for both naturalness and intelligibility The results depict that the HMM-based system performs slightly better than the diphone-based one, in terms of overall quality. On the other hand, the corpus-based unit selection system outperforms both systems in overall quality. However, the HMM-based system achieves a good score in the achieved intelligibility and a fair score in naturalness. Nevertheless, the MOS result on naturalness, depends mostly on the deterioration due to vocoder-like speech generation and secondly on the resulting intonation. Thus, the adoption of a more sophisticated manipulation for the source-filter model (e.g. as in [18]) and the enhancement of prosodic information, could lead to a high quality HMM-based speech synthesis system. An example of HMM-based synthesis is shown in Figure 2, where both the spectrogram and the pitch contour are illustrated for the Greek utterance "kata ti Jno'mi mu eksi'su

Table 1. MOS-based comparative evaluation of the HMM-based speech synthesis system. The system is compared against both a diphone-based and corpus-based general purpose text to speech systems.

SYSTEM	NATURALNESS	INTELLIGIBILITY
HMM-based	3.9	4.2
Diphone-based	3.7	3.8
Corpus-based	4.5	4.6

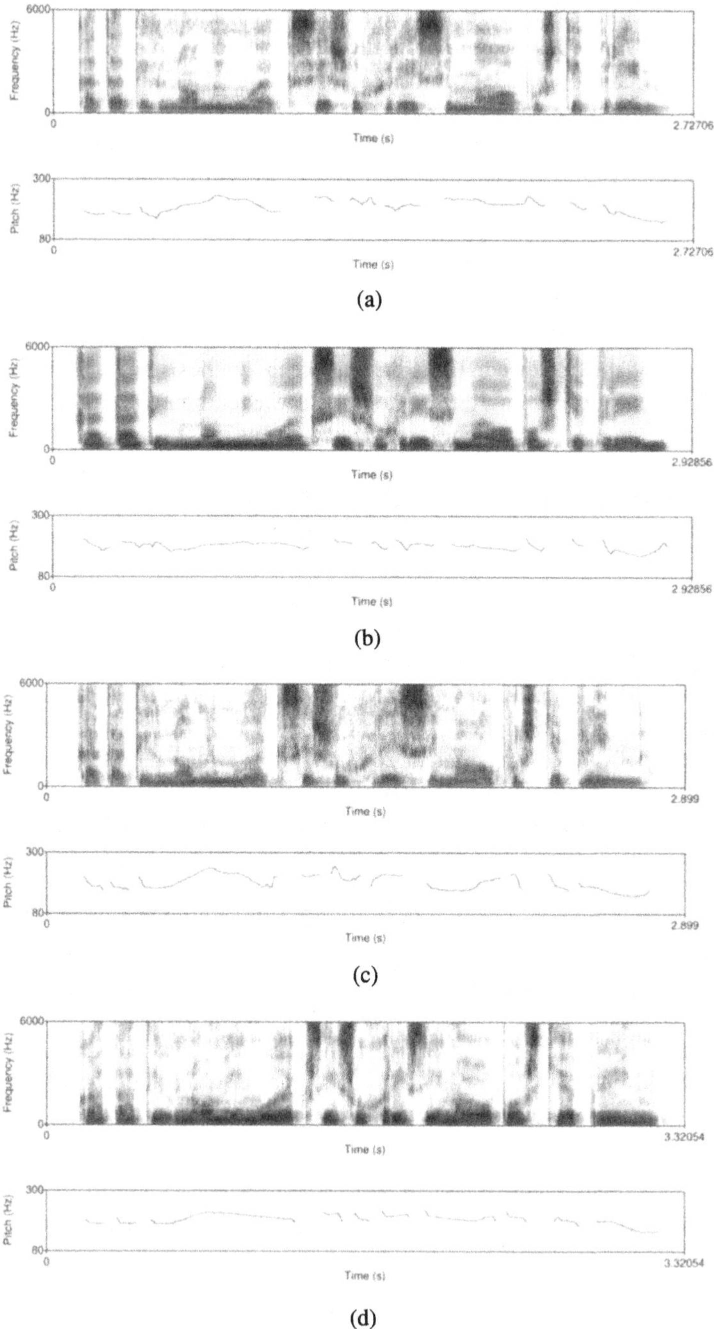

Fig. 2. Example of spectrogram and pitch contour of a synthesized utterance: a) natural speech signal, b) HMM-based, c) corpus-based and d) diphone-based

xri'sima o'pos ce pri'n". The figure also shows the original signal along with the synthesized version using the diphone-based and the corpus-based system. It is seen that HMM-based synthesis produces smooth speech signal but tends to follow a monotone pitch contour similarly to the diphone-based case. The opposite situation is true for the corpus-based approach. This result was expected since, in the training part, prosodic information was taken into account for stressed vowels only. Some examples of synthetic utterances produced by the three text-to-speech systems are available at the following URL address: `http://speech.ilsp.gr/TSD2008/samples.htm`

5 Conclusions and Further Work

HMM-based speech synthesis provides an efficient model-based parametric method for speech synthesis that is based on a statistical framework. In this work, the HMM-based speech synthesis framework was adapted for the development of text to speech system for the Greek language. The performance of the system has been evaluated through comparative listening tests, against a diphone-based and a corpus-based unit selection system. The MOS results have shown that the HMM system performs better than the diphone-based system but worse then the corpus-based one. This is mainly due to the reconstruction process for speech generation which leads to a vocoder-like style of speech. However, these results are encouraging since this is the first HMM speech synthesis system built for the Greek language and many improvements are possible. Further work entails the adoption of a more sophisticated model for speech reconstruction and the inclusion of more prosodic properties in order to increase the naturalness of the produced speech.

Acknowledgments. This work has been partially supported through Community and National funding under the "Information Society" programme of the Greek General Secretariat of Research and Technology, and through the Operational Programme Interreg III Á / PHARE CBC Greece Bulgaria 2000–2006. We would also like to thank all the persons involved in the listening tests and contributed to this work.

References

1. Hunt, A., Black, A.: Unit Selection in a Concatenative Speech Synthesis System Using a Large Speech Database. In: ICASSP 1996, Atlanta, pp. 373–376 (1996)
2. Black, A., Zen, H., Tokuda, K.: Statistical parametric speech synthesis. In: International Conference on Acoustics, Speech and Signal Processing (ICASSP 2007), Hawaii, pp. 1229–1232 (2007)
3. Quatieri, T., F.: Discrete Time Speech Signal Processing, Principles and Practice. Prentice Hall, Upper Saddle River (2002)
4. Tokuda, K., Masuko, T., Yamada, T.: An Algorithm for Speech Parameter Generation from Continuous mixture HMMs with Dynamic Features. In: Proc. of Eurospeech (1995)
5. Tokuda, K., Yoshimura, K., Masuko, T., Kobayashi, T., Kitamura, T.: Speech Parameter Generation Algorithms for HMM-based Speech Synthesis. In: International Conference on Acoustics, Speech and Signal Processing (ICASSP 2000), pp. 1315–1318 (June 2000)

6. Zen, H., Toda, T.: An overview of Nitech HMM-based speech synthesis system for Blizzard Challenge 2005. In: Proc. of Interspeech 2005, Lisbon, pp. 93–96 (2005)
7. Zen, H., Toda, T., Tokuda, K.: The Nitech-NAIST HMM-based speech synthesis system for the Blizzard Challenge 2006. In: Proc. Blizzard Challenge 2006 (2006)
8. Tokuda, K., Zen, H., Black, A.: An HMM-based speech synthesis system applied to English. In: Proc. of IEEE Speech Synthesis Workshop 2002 (IEEE SSW 2002) (September 2002)
9. Krstulovic, S., Hunecke, A., Schroeder, M.: An HMM-Based Speech Synthesis System applied to German and its Adaptation to a Limited Set of Expressive Football Announcements. In: Proc. of Interspeech 2007, Antwerp (2007)
10. Qian, Y., Soong, F., Chen, Y., Chu, M.: An HMM-based Mandarin Chinese text-to-speech system. In: Huo, Q., Ma, B., Chng, E.-S., Li, H. (eds.) ISCSLP 2006. LNCS (LNAI), vol. 4274, pp. 223–232. Springer, Heidelberg (2006)
11. Vesnicer, B., Mihelic, F.: Evaluation of the Slovenian HMM-based speech synthesis system. In: Sojka, P., Kopeček, I., Pala, K. (eds.) TSD 2004. LNCS (LNAI), vol. 3206, pp. 513–520. Springer, Heidelberg (2004)
12. Maia, R., Zen, H., Tokuda, K., Kitamura, T., Resende, F., G, Jr.: Towards the development of a Brazilian Portuguese text-to-speech system based on HMM. In: Proc. of Eurospeech 2003, pp.2465–2468, Geneva (2003)
13. Gonzalvo, X., Iriondo, I., Socor, J., Alas, F., Monzo, C.: HMM-based Spanish speech synthesis using CBR as F0 estimator. In: ISCA Tutorial and Research Workshop on Non Linear Speech Processing - NOLISP 2007 (2007)
14. Kim, S.-J., Kim, J.-J., Hahn, M.-S.: HMM-based Korean speech synthesis system for hand-held devices. IEEE Trans. Consumer Electronics 52(4), 1384–1390 (2006)
15. Abdel-Hamid, O., Abdou, S., Rashwan, M.: Improving Arabic HMM based speech synthesis quality. In: Proc. of Interspeech 2006, Pittsburg, pp. 1332–1335 (2006)
16. Yoshimura, T., Tokuda, K., Masuko, T., Kobayashi, T., Kitamura, T.: Simultaneous modeling of spectrum, pitch and duration in HMM-based speech synthesis. In: Proc. of Eurospeech 1999, pp. 2347–2350 (September 1999)
17. Yamagishi, J., Tamura, M., Masuko, T., Tokuda, K., Kobayashi, T.: A context clustering technique for average voice models. IEICE Trans. Inf. & Syst. E86-D(3), 534–542 (2003)
18. Yamagishi, J., Zen, H., Toda, T., Tokuda, K.: Speaker-Independent HMM-based Speech Synthesis System – HTS-2007 System for the Blizzard Challenge 2007. In: Proc. of Blizzard Challenge 2007 workshop, Bonn, pp. 1–6 (2007)

Advances in Acoustic Modeling for the Recognition of Czech

Jiří Kopecký, Ondřej Glembek, and Martin Karafiát

Speech@FIT, Faculty of Information Technology,
Brno University of Technology, Božetěchova 2, 612 66 Brno, Czech Republic
{kopecky,glembek,karafiat}@fit.vutbr.cz

Abstract. This paper presents recent advances in Automatic Speech Recognition for the Czech Language. Improvements were achieved both in acoustic and language modeling. We mainly aim on the acoustic part of the issue. The results are presented in two contexts, the lecture recognition and SpeeCon+Temic test set. The paper shows the impact of using advanced modeling techniques such as HLDA, VTLN and CMLLR. On the lecture test set, we show that training acoustic models using word networks together with the pronunciation dictionary gives about 4–5% absolute performance improvement as opposed to using direct phonetic transcriptions. An effect of incorporating the "schwa" phoneme in the training phase shows a slight improvement.

Keywords: Automatic Speech Recognition, LVCSR system, acoustic modeling, HLDA, VTLN, CMLLR, lectures recognition.

1 Introduction

In the framework of e-learning, more and more lectures and seminars are recorded, streamed to the Internet and stored to archives. To add value to the recordings, users are allowed to search in the records of the lectures and browse them efficiently. Large vocabulary continuous speech recognition (LVCSR) is used to produce recognition lattices to cope with standard word and phrase indexing and search.

Although a lot of work has been done in the Czech domain in the past years ([4,9], advanced techniques of acoustic modeling, such as HLDA, VTLN, CMLLR, discriminative training, etc., have been studied more thoroughly for English tasks. Our work aims at incorporating these techniques for the Czech spontaneous speech, especially lectures recognition.

In Section 2, description of advanced techniques used for acoustic modeling is presented. Description of all data is given in Section 3. Section 4 contains some information about the used recognizer through our experiments. Section 5 shows achieved results on different test sets. The paper concludes with a summary and states future work in Section 6.

2 Acoustic Modeling Techniques

The investigated techniques apply standard speech recognition based on context-dependent Hidden Markov models (CD-HMM) [13]. The following techniques were

P. Sojka et al. (Eds.): TSD 2008, LNAI 5246, pp. 357–363, 2008.
© Springer-Verlag Berlin Heidelberg 2008

used in our experiments. Their setup was based on previous experiments run on the English tasks [11].

2.1 HLDA

Heteroscedastic linear discriminant analysis (HLDA), which was first proposed by N. Kumar [5,6], can be viewed as a generalization of Linear Discriminant Analysis (LDA). LDA is a data driven technique looking for linear transformation allowing for dimensionality reduction of features. Like LDA, HLDA assumes that classes obey multivariate Gaussian distribution, however, the assumption of the same covariance matrix shared by all classes is relaxed. HLDA assumes that n-dimensional original feature space can be split into two statistically independent subspaces: While in p useful dimensions (containing discriminatory information), classes are well separated, in $(n - p)$ nuisance dimensions, the distributions of classes are overlapped. In our case, the classes are the Gaussian mixture components.

2.2 CMLLR

Maximum likelihood linear regression (MLLR) is an adaptation technique based on estimating linear transformations for groups of model parameters by maximizing the likelihood of the adaptation data [2]. Unlike MLLR, which allows different transforms for the means and variances, *constrained MLLR* (CMLLR) aims at estimating a single transformation for both the means and variances. This constraint allows to apply CMLLR online by transforming the features [3].

2.3 VTLN

Vocal tract length normalization (VTLN) is a speaker based normalization based on warping of frequency axis by speaker dependent warping factor [7]. The normalization of vocal tract among the speakers has a positive effect on reduction of inter-speaker variability. The warping factor is typically found empirically by a searching procedure which compares likelihoods at different warping factors. The features are repeatedly coded using all warping factors in searching range, typically 0.8–1.2, and the one with best likelihood is chosen.

3 Data

3.1 Training Data

Czech SpeeCon[1] is a speech database collected in the frame of EC-sponsored project "Speech Driven Interfaces for Consumer Applications". The database consists of 550 sessions, each comprising one adult speaker. The sessions were recorded in four different environments: office, home, public place, car. Speakers taking part in recordings

[1] `http://www.speechdat.org/speecon/index.html`

were selected with respect to achieve specified coverage regarding gender, age, and speaker dialects.

The content of the corpus is divided into four sections: free spontaneous items (an open number of spontaneous topics out of a set of 30 topics), elicited spontaneous items, read speech (phonetically rich sentences and words, numbers, digits, times, dates, etc.), and core words. Out of this set, we chose free spontaneous items, and a subset of read speech comprising phonetically rich sentences and words.

The database was annotated orthographically including correcting the phonetic form of utterances. To ensure maximum quality, all transcriptions were automatically checked for syntax, spelling, etc. These checks were based on comparison with already checked lexicon. Selected annotations were hand checked, especially for usage of annotation marks.

Temic is a Czech speech data collection comprising 710 speakers collected for the TEMIC Speech Dialog Systems GmbH in Ulm[2] at Czech Technical University in Prague in co-operation with Brno University of Technology and University of West Bohemia in Plzen. Speaker coverage and content of the items are similar to SpeeCon. The audio data were all recorded in car under different conditions and in different situations (e.g., engine on, engine off, door slam, wipers on, etc.). The annotation systems used in these databases were unified without loss of significant information.

Utterances matching the following criteria were pruned out: non-balanced and short utterances (e.g. city names, numbers), broken utterances (containing misspelled items, uncertain internet words, etc.). We ended up with 59 hours of data, 56 hours of which were left for training.

3.2 Test Data

We created two different test sets through the work on Czech recognition system:

- SpeTem test set – contains about 3 hours of speech and is derived from the same corpus as the training data.
- Lecture test set – the target domain of our work is decoding of lectures. Hence we have chosen two lectures recorded and transcribed on our faculty as the second test set: The first lecture from the "Information Systems Project Management" (IRP) course in total time 1.6 hours of speech and the second lecture from "Multimedia" (MUL) course containing about 1 hour.

3.3 Language Model Data

We used a general bigram language model (LM) for all decoding of our experiments and acoustic model comparison. Furthermore, SpeTem test set was expanded by a trigram LM. Both LM's were trained on the Czech National Corpus [1]. The subset chosen for training contains nearly 500M word forms. This is an extremely heterogeneous corpus that consists of texts pertaining to different topics and thus can serve as the basis of a general language model. The corpus contains 2.8M different word forms. At the stage

[2] `http://www.temic-sds.com/english`

of vocabulary construction, we included in the vocabulary only those words that appear in the corpus at least 30 times. That resulted in the smaller vocabulary of 350K words. Even such a vocabulary is presently considered as extremely large for LVCSR tasks. However, we did not want to reduce it any further because inflectional nature of the Czech language calls for larger vocabularies (as compared to English) in recognition of continuous speech [12]. Good-Turing discounting with Katz backoff was used for language model smoothing, singleton N-grams were discarded.

3.4 Data Processing

The phonetic alphabet uses 43 different phonetic elements which are covered in phonetically rich material. It covers 29 consonants, 11 vowels, 3 diphones. The monophone set further includes one special model for silence and also all speaker or background noises and finally the last model representing short pauses between words. Encoding of the phonetic forms uses modified SAMPA[3].

Originally, the handling of "schwa" was rudimentary and for example in spelled items, there were only plosive models (such as "t", "d", etc.) followed directly by silence; we excluded the phoneme "schwa" and mapped it to the silence model. Because of our training databases contain precious phonetic transcriptions, we could use them directly for the training. Acoustic models trained on this base will be called as *version .v0*. For all experiments, the baseline system was trained using HTK tools [13],

The following work led us to complete our phoneme set with the "schwa" phoneme. Presently, we are also using our own training toolkit STK [4] developed at Speech@FIT group, which allows training from phoneme networks. We created them from word transcriptions and pronunciation dictionary included in training databases and used them in the training process instead of straight phonetic string. This approach allows more freedom by choosing the correct pronunciation variant of each word. This multi-pronunciation occurs mainly in foreign and non-literary words in our training set. The influence of this newer acoustic models (marked as *version .v1*) is investigated in Section 5.

It was not clear what exactly brings the improvement achieved by acoustic models in version .v1 – the new phoneme schwa or training from networks? Therefor we decided to train another acoustic models (*version .v2*) where the schwa was mapped on silence model again but the training process was done from phoneme networks. This work is still in the beginning, therefore Table 2 is not complete yet. However it has been shown, that training using network, brings most of the improvements.

4 Recognizer

Speech features are 13 PLP coefficients augmented with their first and second derivatives (39 coefficients in total) with cepstral mean and variance normalization applied

[3] `http://www.phon.ucl.ac.uk/home/sampa/home.htm`

[4] Lukáš Burget, Petr Schwarz, Ondřej Glembek, Martin Karafiát, Honza Černocký: STK toolkit, `http://speech.fit.vutbr.cz/cs/software/hmm-toolkit-stk-speech-fit`

Table 1. Comparison of different advanced techniques in acoustic modeling on SpeTem test set

Acoustic models	2gram decoding	3gram expansion
xwrd.sn2	22.33	20.92
xwrd.sn2.hlda	21.46	19.35
xwrd.sn2.cmllr	20.38	18.17
xwrd.sn2.vtln0	20.47	18.48
xwrd.sn2.vtln1	20.24	18.20
xwrd.sn2.vtln2	20.19	18.34
xwrd.sn2.vtln3	20.31	18.43
xwrd.sn2.vtln4	20.31	18.26
xwrd.sn2.vtln5	20.41	18.41
xwrd.sn2.vtln1.hlda	19.87	17.45
xwrd.sn2.vtln1.cmllr	**19.03**	**16.93**
xwrd.sn2.vtln1.cmllr.hlda	19.28	17.30

Table 2. Results achieved on the Lecture test set

Acoustic models	IRP			MUL		
	.v0	.v1	.v2	.v0	.v1	.v2
xwrd.sn2	52.78	48.12	48.59	61.79	56.33	57.57
xwrd.sn2.hlda	51.48					
xwrd.sn2.cmllr	51.38					
xwrd.sn2.vtln0	45.69	**42.47**		**54.12**	54.19	
xwrd.sn2.vtln1	45.44					
xwrd.sn2.vtln2	44.93					

per conversation side. Acoustic models are based on left-to-right 3 state cross-word triphone HMMs with states tied according to phonetic decision tree clustering. Number of tied states was tuned around 4,000 in the phase of state clustering. After one phase of retraining, clustering was performed once more.

5 Experimental Results

All results are presented in terms of word error rate (WER).

As can be seen in Table 1, each technique gives some improvement – about 1% (from HLDA), almost 2% (thanks to CMLLR and VTLN) absolutely. Vtln0 represent the acoustic vtln-models trained on the output from non-vtln models; vtln1-5 was trained on previous iteration of the vtln models. We also tried to combine these techniques. For this purpose we used vtln1 models and CMLLR, HLDA transformations. The results are presented in the second part of Table 1. Surprisingly, the combination of both transformations with VTLN performs worse than its combination separately. However, we are still able to improve our results by 4% by using these advanced modeling techniques.

Table 2 show some results on lecture test set in WER [%] by using only 2gram decoding network. Presently, we are working on this issue, so the table is not complete so far. Three versions of acoustic models are compared:

- .v0 acoustic models without schwa trained straight from phoneme strings,
- .v1 acoustic models with schwa trained from phoneme network,
- .v2 acoustic models without schwa trained from phoneme network.

What we can see is the significant improvement by using VTLN adaptation. The change of training method from phoneme string to phoneme network helps too. The influence of new "schwa" model is not so noticeable but even now we can see a little improvement of acoustic models even though "schwa" is omitted in the decoding part.

6 Conclusions

We have used some advanced acoustic modeling techniques, which were successfully tested in the English LVCSR. Not only their effect was visible on the SpeTem test, but mainly on the target lecture test set, where the baseline system gave poor results.

The effect of network training is eminent in lecture decoding, therefore we need to complete our experiments. Another improvement is expected from discriminative training of our acoustic models [8] and usage of posterior features [10]. We also expect improvement by integrating the "schwa" phoneme into decoding networks.

Acknowledgments

This work was partly supported by Ministry of Trade and Commerce of Czech Republic under project FT-TA3/006 and by Ministry of Interior of Czech Republic under project VD20072010B16. The hardware used in this work was partially provided by CESNET under project No. 201/2006.

References

1. Český národní korpus (Czech National Corpus). Technical report, Ústav Českého národního korpusu FF UK, Praha, Česká republika (2005)
2. Dempster, A.P., Laird, N.M., Rubin, D.B.: Maximum Likelihood from Incomplete Data via the EM Algorithm. Journal of the Royal Statistical Society 39(1), 1–38 (1977)
3. M. Gales. Maximum Likelihood Linear Transformations for HMM-based Speech Recognition. Technical Report CUED/FINFENG/TR291, Cambridge University (1997), `http://citeseer.ist.psu.edu/article/gales98maximum.html`
4. Nouza, J., Žďánsky, J., Červa, P., Kolorenč, J.: Continual On-line Monitoring of Czech Spoken Broadcast Programs. In: International Conference on Spoken Language Processing (Interspeech 2006 – ICSLP 2006), Pittsburgh, USA, pp. 1650–1653 (2006)
5. Kumar, N.: Investigation of Silicon-Auditory Models and Generalization of Linear Discriminant Analysis for Improved Speech Recognition. Ph.D. thesis, John Hopkins University, Baltimore (1997)

6. Kumar, N., Andreou, A.G.: Heteroscedastic discriminant analysis and reduced rank HMMs for improved speech recognition. Speech Communcation 26, 283–297 (1998)
7. Lee, L., Rose, R.: Speaker Normalization Using Efficient Frequency Warping Procedures. In: Proc. ICASSP 1996, Atlanta, GA, USA, May 1996, pp. 339–341 (1996)
8. Povey, D.: Discriminative Training for Large Vocabulary Speech Recognition. Ph.D. thesis, Cambridge University Engineering Dept (2003)
9. Psutka, J.: Komunikace s počítačem mluvenou řečí. Academia, Praha (1995)
10. Grézl, F., Karafiát, M., Kontár, S., Černocký, J.: Probabilistic and bottle-neck features for LVCSR of meetings. In: Proc. IEEE International Conference on Acoustics, Speech and Signal Processing (ICASSP 2007), Hononulu, USA, pp. 757–760 (2007)
11. Thomas, H., Vincent, W., Lukáš, B., Martin, K., John, D., Jithendra, V., Giulia, G., Mike, L.: The AMI System for the Transcription of Speech in Meetings. In: Proc. IEEE International Conference on Acoustics, Speech and Signal Processing (ICASSP 2007), Hononulu, USA, pp. 357–360 (2007)
12. Whittaker, E.W.D.: Statistical Language Modelling for Automatic Speech Recognition of Russian and English. Ph.D. thesis, Cambridge University (2000)
13. Young, S., Jansen, J., Odell, J., Ollason, D., Woodland, P.: The HTK book. Entropics Cambridge Research Lab, Cambridge, UK (2002)

Talking Head as Life Blog

Ladislav Kunc[1], Jan Kleindienst[2], and Pavel Slavík[1]

[1] Department of Computer Science and Engineering, FEE, CTU in Prague
Karlovo náměstí 13, Praha 2, 121 35, Czech Republic
[2] IBM Česká Republika
V Parku 2294/4, Praha 4, 148 00, Czech Republic
{kuncl1,slavik}@fel.cvut.cz, jankle@cz.ibm.com

Abstract. The paper describes an experimental presentation system that can automatically generate dynamic ECA-based presentations from structured data including text context, images, music and sounds, videos, etc. Thus the Embodied Conversational Agent acts as a moderator in the chosen presentation context, typically personal diaries. Since an ECA represents a rich channel for conveying both verbal and non-verbal messages, we are researching ECAs as facilitators that transpose "dry" data such as diaries and blogs into more lively and dynamic presentations based on ontologies. We constructed our framework on an existing toolkit ECAF that supports runtime generation of ECA agents. We describe the extensions of the toolkit and give an overview of the current system architecture. We describe the particular Grandma TV scenario, where a family uses the ECA automatic presentation engine to deliver weekly family news to distant grandparents. Recently conducted usability studies suggest the pros and cons of the presented approach.

Keywords: Multimodal, ECA, TV, presentations, interactions, blog.

1 Introduction

Blogging is an increasingly popular way of capturing and sharing information. We are designing a presentation system that can turn individual blog entries (or any relevant structured data) into a multimodal presentation where the talking head acts as a moderator. Our tool should be able to prepare such presentations with minimal manual effort. We based our work on the already existing ECAF toolkit [5] that supports runtime generation of ECA agents; however, we needed to make some extensions to the tool.

Embodied conversational agents (ECA) represent a rich channel for conveying both verbal and non-verbal messages in human-computer interaction. ECAs humanize the computer-user interface and that makes them well acceptable for common users. By expressing emotions, ECAs enrich the user experience even further [1]. Hence we could utilize the human-like capability of ECAs to turn "dry" data such as diaries and blogs into more lively and dynamic presentations. This can be achieved by transforming the individual blog items such as text messages, images, music, video into a multimodal stream running on the background and the ECA acting on the foreground to connect the individual pieces through narration.

P. Sojka et al. (Eds.): TSD 2008, LNAI 5246, pp. 365–372, 2008.
© Springer-Verlag Berlin Heidelberg 2008

1.1 Background

The usage of digital actors in multimodal presentations is conditional on the existence of the right techniques to build virtual agents. Some fields of ECA research adopt a 2D representation of an agent, but the animation of a 3D virtual avatar is much more challenging.

The main part of the human body people recognise is the face. The ability to show facial expressions is very important for the virtual agent to be believable. The scale of agent expressions is affected by the technique used for facial animation. There are many techniques of facial animation in the 3D space. The first attempts on face simulation were carried out by Parke and were based on the interpolation of vertices between extreme positions [6]. The interpolation of 3D meshes is not efficient because every expression needs its own mesh. Thus parametric models have appeared; such as the widely known Waters extended parametric model based on physics pseudo-muscle simulation [9].

An ECA as a multimodal output interface has several communication channels to control, e.g. eyes, head position, expressions, etc. Therefore a proper control language for generating ECAs is needed. Example languages that satisfy this need are SMILE-AGENT [4] or the ECAF XML language [5].

There is a lack of research literature on the problem of using ECAs as presentation narrators. The concept and importance of ECAs for new multimedia and TV services is described in the work [8]. There is also a possibility of setting up ECA agents as conversational characters in multimodal game environments [3]. The ability of dialog guidance between user and virtual agent is stressed in the Rea-agent interface [2].

Later in the text we describe a complete multimodal framework for our concept of an agent acting as narrator in a multimodal presentation and technologies that can be used to create such a presentation automatically.

2 Key Features Implemented

Our goal is to take the data of user activities as input and use it to automatically build ECA-guided presentations. There are two primary technology blocks we used in building the prototype. The first one is related to the visual ECA system; the second one deals with the presentation templates for a specific domain.

2.1 Visual ECA

The base of the visual ECA system is the ECAF talking head toolkit [5]. This toolkit covers nine human communication channels, e. g. human voice, head pointing, head position, face expressions, etc. To make the presentations more appealing, we needed to extend the existing ECAF toolkit by implementing some new functions. These are the main extensions we have implemented:

Background Music. We added commands into the ECAF command set to support playback of background music. This combines with the already existing ECAF support for playing sounds on the foreground, so that we can now e.g. mix pre-recorded prompts with background music.

Background Video. We intend to support videos as first-class parts of a presentation. Hence we extended ECAF to handle video playback on the background as an additional channel. The video is played with the help of Microsoft DirectShow API [3], which supports plenty of video types such as MPEG or AVI.

Dynamic Image Presentation. To introduce a notion of motion in displaying static images, we move images along a predefined trajectory. This creates the effect of image animations (Figure 1). First the static image is zoomed in at the beginning and then moved on the screen to bring the user closer to the middle of the image. The ECAF command supports parameters for controlling the speed of animation.

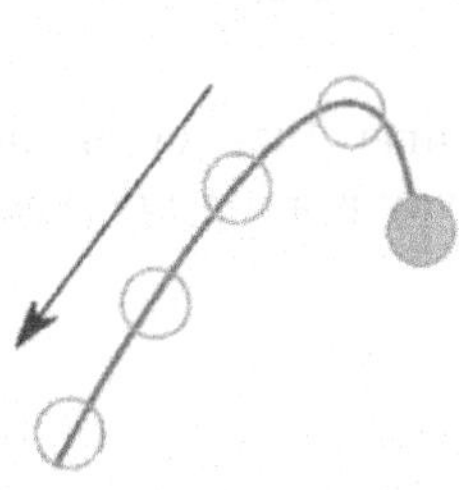

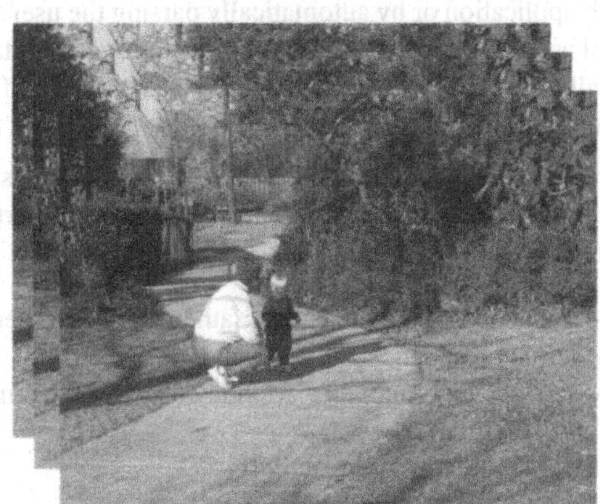

Fig. 1. Dynamic background photos changing trajectory

Movable Pointer. There are parts of the presentation where we need to draw the user's attention to a specific position, e.g. by pointing to particular coordinates on the background image. We extended ECAF by a new channel which represents a movable pointer. The position and angle of the pointer can be controlled independently of the head movements.

2.2 Scenario Templates and Ontology Handlers

Scenario templates and ontologies are the main new functions added to our system. Together they define a "storyline" along which data are presented and contribute to the naturalness of the presented data. This subsystem is currently work-in-progress and only a simple template implementation exists at the moment.

Scenario Templates. Scenario templates are prescriptions how specific data should be delivered to the user as a multimodal presentation. They define content that can be presented verbally and non-verbally. Utterances are examples of spoken content.

Imagine these utterances as building blocks of the whole presentation. For example, in the case of a Grandma TV template, each member of the family has a parameterized textual template that describes his/her activities. There are also template parts devoted to a family group, e.g. parents versus children activities. The text is also enriched by expression, position or modal ECAF tags, partially automatically based on hints present in the text. The tags control the expressivity and appearance of the talking head.

Ontology Handlers. We use ontologies to generate gluing sentences between the parts of the presentation where one template follows another.

3 Data Feeds

Structured data for producing a presentation could be obtained through a specialized web application or by automatically parsing the user's personal calendar application, as well as the other sources. We also plan to use some standard Internet feeds, for example Really Simple Syndication 2.0 (RSS) [10]. These feeds could be translated into our internal feed structure.

Activity Feeds. Input data are user's specific events sorted by time. These events are called activities in our data representation. One particular activity is a tuple of these parameters:

- person type and name – e.g. father Frank, mother Jenny;
- activity frequency – weekly, monthly;
- part of week, month – Monday, Tuesday, beginning and end of month, etc. (or precise time);
- textual description of activity;
- sound attachment;
- photos;
- videos;
- happy/sad modifier.

These activities and other internal structures are represented in an XML language. Bellow is an example of an activity entry to describe e.g. a business trip of a dad:

Example of an Activity Entry

```
<activity>
  <type>weekly</type>
  <when>Wednesday</when>
  <what>business trip</what>
  <where>Boston</where>
  <sound>C:\sounds\trip_msg.wav</sound>
  <pictures>
    <picture>C:\images\my.jpg</picture>
    <picture>C:\images\boston.jpg</picture>
  </pictures>
  <headmodifiers>
```

```
    <mood>happy</mood>
  </headmodifiers>
</activity>
```

The description of the business trip is an example of a "single activity". Talking Head Life Blog also supports activities performed by multiple people as the "group activities". The internal structure of these activities is almost the same as for single activities.

Input Feed. These activities determine the output of the presentation system and represent the feed of the application as you can see in Figure 2 (a).

4 Talking Head Life Blog Architecture

In this section we present Talking Head Life Blog architecture and design. The Talking Head Life Blog is designed on a mock-up Talking Head scenario in the ECAF language [5].

The key parts of the Talking Head Life Blog architecture design are depicted in Figure 2 (b). At the heart of the application is the design block of ontology handlers (briefly introduced in Section 2.2) that process input data.

Our system recognises input activities in the form of texts, sounds, pictures and videos. These activities are acquired from the user through a specialized web application or possibly from the user's calendar data. There are two essential parameters that configure the system – KnowledgeBase and Scenario Templates.

The KnowledgeBase is a database of sentences that are used to glue together individual presentation blocks. We use the KnowledgeBase data to increase the naturalness of speech content, e.g. through using certain colloquial phrases, etc. At the current stage of development of the system, there are two groups of such filling sentences:

Openings (eg. "Hi grandma, I have regular weekly family news for you".) – Randomly selected sentences for commencing the presentation.

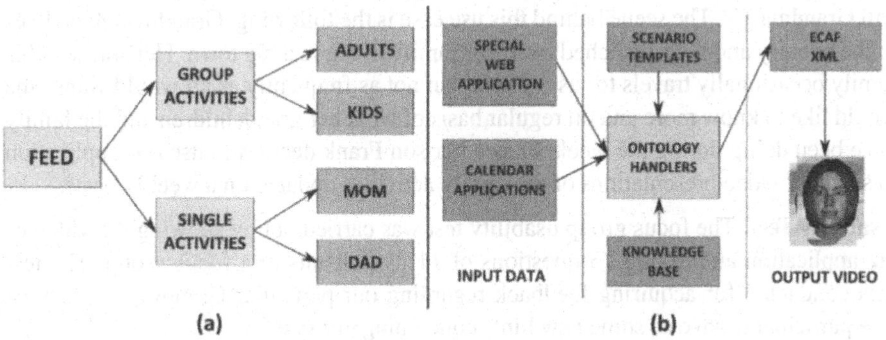

Fig. 2. (a) Organization of an application input feed XML; (b) Talking Head Life Blog architecture scheme

Comments (e.g. “Does it sound like a fun?”, “Sounds great, huh?”) – These sentences are randomly chosen to “color” the presentation of images or video.

Once initiated, the entire process is controlled by the Scenario Templates. When the Talking Head Life Blog processes input activities, it assigns one utterance of the scenario template to a single activity. The text in utterances includes `<fill>` tags. The tags specify where to put input activity parameters. One example is the father business trip template presented above. The `<fill>` tags will also guide the rendering of attached multimedia content.

Example of one utterance

```
<utterance who="father">
  <fill data="name" />, your son, went to
  <fill data="what"> to <fill data="where">.
  <fill data="sound" />
  <fill data="pictures" />
</utterance>
```

Generally, the Talking Head Life Blog application uses an appropriate scenario template and a master presentation scaffolding and in the process tries to fit the input activities into this template. The ontology handlers glue these utterances into the final presentation form using the sentences from the KnowledgeBase. The output of this process is a generated ECAF XML language script. Subsequently this script could be interpreted by the ECAF talking head toolkit. This ECAF toolkit outputs the presentation as a real time video application and/or as a Windows Media video file. Finally the generated multimodal video can be displayed on any terminal with video playback capability such as a PC, phone, car computer, etc.

5 Case Study

We demonstrate the use of the presentation engine on a social inclusion service that we call Grandma TV. The scene behind this use case is the following: Grandma Olivia lives in the country and feels detached from her family living in a big town. Her son with his family occasionally travels to visit Olivia, but not as frequently as all would want. She would like to know more and on regular basis of what her grandchildren and the family have been doing during the week. Hence her son Frank decides to use our application to send her video presentations of the family activities updated on a weekly basis.

Usability Test. The focus group usability test was carried out by showing the video of the application and asking 13 questions of 11 participants in a focus group. The test was conducted for acquiring feedback regarding our particular Grandma case study. The participants gave us some new hints concerning our system.

First, the feedback indicates that the idea of having the presentation moderated by the ECA is generally interesting and valuable. Second, the feedback also indicates the participants were intrigued by the range of output modalities presented by the ECA. Simultaneously they agreed that the head behaviour addresses adequately the content of

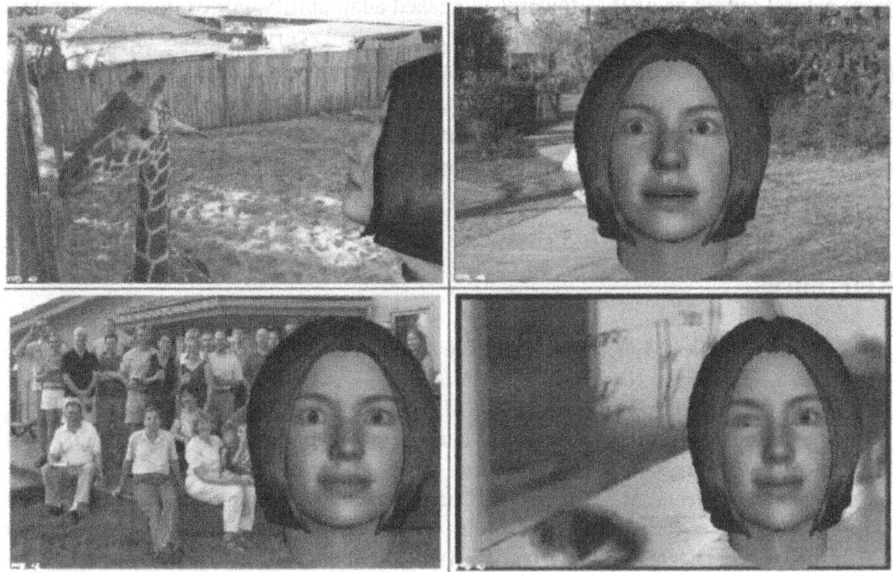

Fig. 3. Grandma case study guided presentation

the presentation. The majority of participants state that the new implemented method of dynamic picture presentation enriches the presentation of photos very lively. Finally, the duration (approx. 3 minutes) of the presentation was largely approved to be sufficient.

The usability test also helped identify the shortcomings of the current design. The talking head sometimes hides parts of the presented information. This point should be solved by proper sizing or making the head slightly transparent at problematic places. Another issues refer to the preliminary state of the whole system. There should be more head gestures, expressions and the rendering of the eyes should be improved. The current text-to-speech synthesis will also benefit from an upgrade.

6 Conclusion and Future Work

In this paper we have introduced a work aiming at the automatic generation of dynamic ECA-based presentations from structured user data. The work is based on an existing ECAF toolkit that we have extended with a set of visually appealing gestures to support TV-like presentation, where the talking head acts as a narrator.

We have described one particular use case, the Grandma scenario, where a family uses the automatic presentation engine to deliver weekly family news to distant grandparents. We also conducted preliminary usability tests. They have proved that the idea is generally valid, but there is a need of improvement in the area of making face expressions more believable and speech synthesis more realistic.

The scenario templates, which we have introduced in the text, are key for making the presentations more natural as well as more tailored for the particular user's preferences. Hence our research effort will aim towards employing ontologies to generate more natural output as well as towards increased adaptability. For example, a grandma does not want to be told that her grandson plays football every Thursday in each weekly presentation. In addition, future versions should incorporate some amount of interactivity, so that the user could e.g. select only parts of the presentation that are interesting for him/her. The usability tests have shown that there are such users. One of the best solutions may be to divide a presentation into chapters and simply let the user choose the desired ones.

References

1. Bates, J.: The Role of Emotion in Believable Agents. In: Communications of ACM 37-7, pp. 122–125. ACM Press, New York (1994)
2. Cassell, J., Bickmore, T.W., Billinghurst, M., Campbell, L., Chang, K., Vilhjalmsson, V.V., Yan, H.: Embodiment in Conversational Interfaces: Rea. In: Proceedings of CHI 2009, pp. 520–527. ACM Press, New York (1999)
3. Corradini, A., Mehta, M., Bernsen, N., Charfuelan, M.: Animating an interactive conversational character for an educational game system. In: 10th Int. Conf. on Intelligent User Interfaces, San Diego, California, pp. 183–190 (2005)
4. Koray, B., Not, E., Zancanaro, M., Pianesi, F.: Xface Open Source Project and SMIL-Agent Scripting Language for Creating and Animating Embodied Conversational Agents. In: 15th Int. Conf. on Multimedia 2007, pp. 1013–1016. ACM Press, New York (2007)
5. Kunc, L., Kleindienst, J.: ECAF: Authoring Language for Embodied Conversational Agents. In: 10th Int. Conf. on Text, Speech and Dialogue, pp. 206–213. Springer, Heidelberg (2007)
6. Parke, F.I.: Computer Generated Animation of Faces. In: ACM annual conference 1972. ACM Press, New York (1972)
7. Pesce, M.D.: Programming Microsoft® DirectShow® for Digital Video and Television. Microsoft Press (2003)
8. Thalmann, N.M., Thalmann, D.: Digital actors for interactive television. IEEE Special Issue on Digital Television, Part 2 83(7), 1022–1031 (1995)
9. Waters, K.: A Muscle Model for Animating Three-Dimensional Facial Expression. In: Computer Graphics, SIGGRAPH 1987, vol. 21, pp. 17–24. ACM Press, New York (1987)
10. Really Simple Syndication 2.0 specification, `http://www.rssboard.org/rss-specification`

The Module of Morphological and Syntactic Analysis SMART

Anastasia Leontyeva and Ildar Kagirov

St. Petersburg Institute for Informatics and Automation of RAS
SPIIRAS, 39, 14th line, St. Petersburg, Russia
{karpov,ronzhin,an_leo}@iias.spb.su
http://www.spiiras.nw.ru/speech

Abstract. The present paper presents the program of morphological and syntactic analyses of the Russian language. The principles of morphological analysis are based on some ideas by A.A. Zaliznyak. The entire glossary is to be kept as sets of lexical stems and grammatical markers, such as flexions and suffixes. A great deal of stems is represented by a few variants, owing to in-stem phonological alternations. Such approach helps to reduce the glossary and retrench the time of searching a particular word-form. Syntactic analysis is based on the rules of Russian. During the processing of some texts, base syntactic groups were singled out. The module could be used as a morphology analyzer or as a syntactic analyzer. This module provides information about the initial wordform, its paradigm, its transcription and compose a vocabulary for Russian continuous speech recognizer. The module is also used for syntactic analysis of Russian sentence.

Keywords: Morphological analysis, syntactic analysis, word processing, speech decoder.

1 Introduction

Nowadays creation of automatic speech recognition systems is one of the top themes in applied linguistics. However, there is still a lot of work to do. Besides, it is obvious that speech recognition systems are to be based on different principles, depending on the language concerned.

Most of models were elaborated for languages with syntax-oriented grammar (such as English). As a result, they use "morphs → words" recognition principle. Indeed, this approach is very good, assuming that the language has a rudimentary inflection system at least. But it is inapplicable to languages of "rich" morphology, such as Russian, which possesses a large glossary, overblown by means of inflexion [1,2]. For example, the English word "table" has two wordforms, while the Russian word "stol" has ten wordforms.

To reduce the vocabulary size, complexity of language model (LM) and decoding time the various sub-word unites instead of whole words were investigated [3]. The difference of the existing approaches consists of type of the using sub-word units and methods for unit composing and word decomposition in the process of recognition and training correspondingly [4]. Two main approaches could be marked here: (1) rule-based segmentation methods (2) unsupervised data-driven word splitting [5]. Due to

P. Sojka et al. (Eds.): TSD 2008, LNAI 5246, pp. 373–380, 2008.
© Springer-Verlag Berlin Heidelberg 2008

inflections, ambiguity and other phenomena, it is not trivial to automatically split the words into meaningful parts. The former approach can perform this splitting, but requires vast handcraft and careful programming of grammar rules [6], while the latter automatically find morpheme-like units (statistical morphs) using the unsupervised machine learning method. It was found out that the unsupervised data-driven word splitting algorithm Morfessor presented in [5] was very effective in recognition of Finnish, Estonian and Turkish.

Not only language modeling makes difficulties for applying the morph-based decoder of the inflective languages, but also lack of speech and text training data. By this reason grammatical morph-based models are more perspective because they can generate all possible wordform sequences which may not contain in a limited training text data. The present article briefly sketches one of the grammar based approach to word splitting and main principles of the automatic morphological analysis.

2 Morphological Analysis: An Approach to the Inflectional Languages

For the purpose of reducing the time of input text interpretation a program module of morphological analysis is to be added. This is a good way to avoid dictionary overflowing: no oblique word-forms are kept in the vocabulary, because they are described by the Russian inflection rules. Provided with these, the module can build up any oblique form or deal with any word-form from grammatical point of view.

The morphological module finds the lexical stem and the flexion of all the inflective words (as units of morpheme language level). All the lexical stems are reserved for the vocabulary, while the flexions (i.e. grammatical markers) form up paradigms according to grammatical rules. Thus a range of word-forms with different flexions are treated as a single unit of the vocabulary. In fact, the function of the morphological program is to get rid of all the language regularities (i.e. grammar) in the input, preserving all the irregular (i.e. lexis) only. Besides, it helps to find grammatical meanings, embedded in word-forms.

The basic units, the program operates with, are morphs. Morph is a syntagmatic unit; therefore morphs are to be sorted on positional criteria. Any "morphological" language has strict rules of morph distribution besides, such as combinatory description, position within the limits of a word-form etc. So, each listed morph is to be matched with some grammatical rules. These principles allow sorting out types of morphs as follows:

1. Derivational affixes (including interfixes). It is necessary to take into account the part of speech of the analyzed word-form, when looking for an derivational affix.
2. Grammatical markers (i.e. flexions and grammatical suffixes). Taking consideration of the part of speech is necessary too; besides, this procedure requires information of grammatical categories of the whole word-form.
3. Roots. The most important about the roots is that each word-form has them.

3 Morphology Rules: Paradigms

There are strict rules in Russian, in accordance with them flexions and grammatical suffixes are added to stems. In general, the rules match sets of grammatical markers with classes of words (in traditional terms, types of declension or conjugation). These rules were described by A.A. Zaliznyak in the theoretical introduction to "Grammatical Vocabulary of the Russian Language", with classes of words and affixes matched with special marks added to each lexeme. For instance, "v'in'jetka fem 3*a" (Eng. "vignette") means that this lexeme is sorted into the 3-th type of feminine substantive declension, with vowel-alternation. The next step is to look for fem 3*a substantive flexion paradigm. The same marks are used in the morphological program.

The whole set of grammatical affixes and their grammatical semantics can be depicted as a binary matrix P with size N×K, where N – number of grammar marker $\{x_1, x_2, \ldots, x_N\}$, K – number of grammar characteristics $\{g_1, g_2, \ldots, g_K\}$. The values of the matrix elements are calculated by the following rule: $p_{ij} = 1$ only if morphological differential characteristic g_i and a treated grammar marker x_j are matching, otherwise $p_{ij} = 0$.

Different forms of one lexeme can have different stems, besides different flexions and grammatical suffixes. There are two types of such stem-variation: suppletive forms, alternations in stem. All the cases of the first type are unique and appertain to the vocabulary, while in-stem alternations are described by alternation rules. A database of stems was created, which presents all the variants of each stem. Each individual variant has "valency" to restricted set of flexions. Stems used by the morphological program were generated according to in-stem alternation rules from the "Grammar Dictionary" by A.A. Zaliznyak.

There are several types of in-stem alternation. Words with alternation would have one or two additional stems according to alternation type. Taking into account alternations the size of vocabulary increased lightly. But using stems instead of words significantly reduces vocabulary size and time of decoding. The received results are presented below.

4 The Structure of the Module of Morphological Analysis

The main idea of morphological module is a base wordform divides into the lexical stem and the flexion according to rules of Russian language. The module has a flexible structure (Figure 2). It can be used as a single program for automatic text processing or as a part of Russian speech recognizer. The types of input and output data are determinated according to mode of operation. The input data can consist of a wordform, a list of wordforms or stem index. Task manager defines procedures required for corresponding task. If the input data consist of wordforms, then the search module divides a base wordform into the lexical stem and the flexion, and finds index of stem in databases. If the input data consist of stem indexes, search module finds the stem in databases according to its index. Then it defines part of speech and uses one of processing modules: the adjective processing module, the verb processing module, the noun processing module. The first module processes adjective and other parts of speech, which decline in a similar way (e.g. participles). The second module processes verbs and its derivative forms (e.g. adverbial participle). The third module processes nouns, abbreviations and invariable words (e.g. adverb). The processing modules use own databases and grammatical rules. The output forming module defines procedures which forms the output data. The transcribing module produces wordform transcription using information about stress and phonetic rules (Ronzhin, Karpov, 2007). The paradigm forming module synthesizes a paradigm for the base wordform. The grammatical markers defining module analyses markers of the stem and defines grammatical characteristics. The final module forms list of stem indexes. Thus, if the input data consist of wordforms, the output data may contain all possible paradigms of the wordform, transcription or information about grammatical characteristics. If stem indexes input the morphological module, output data contain wordform with proper grammatical characteristics. Such mode of operation is used when the morphological module is a part of Russian speech recognition system.

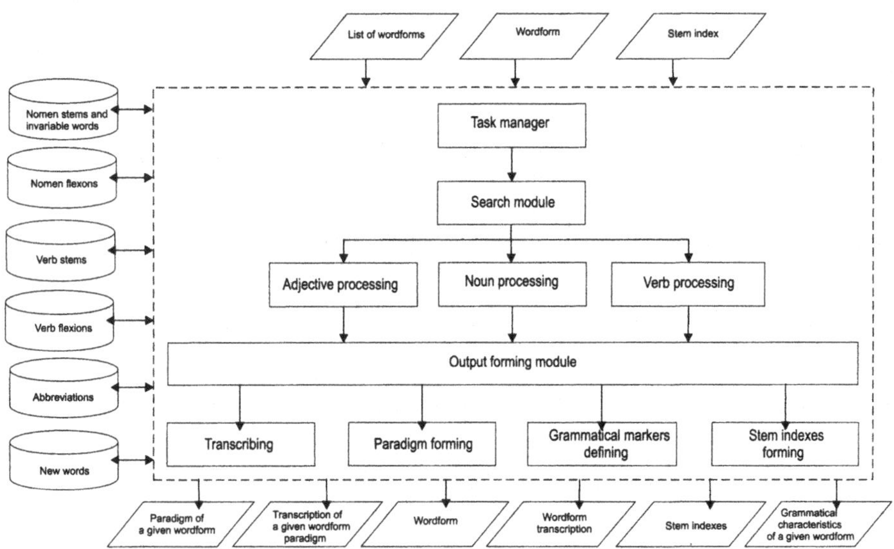

Fig. 1. The structure of the morphological module

Using the module of morphological analysis for composing a vocabulary for Russian continuous speech recognizer, it was received results shown as Table 1. The number of stems and flexions increased after transcribing due to soft consonants and stressed vowels. For example, (Zaliznjak, 1980), used as a base vocabulary, contains 97473 different stems and flexions (proper nouns were excluded from consideration). After taking into account alternation and transcription the number of stems increased almost twice as much. Table 1 shows how the number of stems increases in each processing stage.

Advantages of stem-based vocabulary in comparison with word-based vocabulary are shown as Table 2.

Table 1. Vocabulary processing results for Russian continuous speech recognizer

	Variable				Invariable	Total
	Nouns		Verbs			
	Number of stems	Number of flexions	Number of stems	Number of flexions		
Number of base wordforms	66224	60	27799	199	3191	97473
Number of wordforms, taking into account alter-nation	86337	60	35263	199	3191	125050
Wordform tran-scription	133123	118	51219	629	3191	188280

Table 2. Vocabulary processing results for Russian continuous speech recognizer

	Variable		Invariable	Total
	Nouns	Verbs		
Number of stems	86337	35263	3191	124791
Number of flexions	60	199	-	259
Number of word-forms	1608093	2356234	3191	3967518
Number of differ-ent wordforms	776797	1316058	3191	2096046

Using the word-based approach for Russian language (and other inflective languages) the size of vocabulary is over 2 millions units. The stem-based approach allows greatly reduce number of vocabulary units, which is about 120 thousands.

5 The Structure of the Module of Syntactic Analysis

The morphological module is included in structure of the module of syntactic analysis. The syntactic module is used for sentence analysis. Subject and verb groups are singled out during the analysis, and relationships between sentence constituents are defined. At the first stage of development the syntactic module is used for the analysis simple sentences. In Russian language it could be defined following syntactic groups (Table 3).

The following table of symbols is used: Con – conjunction, Adj – adjective, ASF – adjective short form, Aux – auxiliary verb, N – noun, V – verb, Adv – adverb, Inf – infinitive, P – preposition, NG – Noun group, VG1 – Verb group1, VG12 – Verb group2, PG – Preposition group. Arrows show relationships between sentence constituents.

During the analysis the module uses syntactic database. In input text the module finds a sentence (a part of the text between two parameters of the end of the sentence – a point, an exclamation mark, a question mark). Then the syntactic groups (shown in

Table 3. Syntactic groups which could be marked out in Russian sentences

Noun group	Verb group1	Verb group2	Adjective group	Preposition group
{N}	{V}	{(VG1)→(NG)}	{ASF}	{P→(NG)}
{N→N}	{V→Adv}	{(NG)←(VG1)}	{Adj}	
{Adj→N}	{Adv←V}	{(VG1)→(PG)}	{ASF→Adv}	
{Adj←N→N}	{Adv←V→Adv}	{(PG)←(VG1)}	{Adv←ASF}	
{N - Con - N}	{Aux→ASF}	{(NG)←(VG1)→ (NG)}	{ASF→(NG)}	
		{(NG)←(VG1)→ (PG)}	{ASF→ {Inf→(NG)}}	
		{(VG1)→(NG) (PG)}	{(PG)←ASF}	
		{(VG1)→{Inf→ (NG)}}	{(PG)←(AG)}	
		{(VG1)→(AG)}		

Table 3) are singled out. Allocation of groups is carried out by the analysis of morphological parameters of wordforms and their position rather each other. During the morphological analysis all possible stems of the wordform are defined. If several stems are corresponding to the wordform, the module generates all possible hypotheses. It is necessary to note that in Russian language a word could have the same wordforms in more than one case. The type of syntactic group also depends on the wordform case. The head of subject group is a noun in nominative case or a personal pronoun. The head of predicate group is a verb or adjective short form. The structure of the syntactic module is shown as a Figure 2.

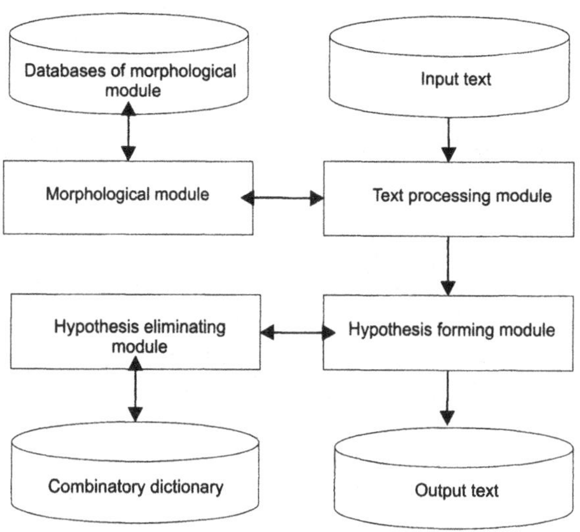

Fig. 2. The structure of the syntactic module

Input text consists of the set of simple sentences. Each sentence word-by-word enters the module of text processing. All possible stems of wordform are defined, and coincidence of nominative, genitive is accusative cases are checked. Then the hypothesis forming module generates necessary quantity of hypothesis. When all words in the sentence have been considered, hypothesizes enter the module of its elimination. This module deletes wrong and recurring hypothesizes and outputs others in a text-file. It is necessary to note that it could be several correct variants of sentence parse in Russian language. The correct analysis depends on some parameters including the meaning of the sentence. During the hypothesis elimination, grammatical agreement of sentence parts is checked. As a result of if output text-file, consisting of parsing sentences, forms. An example of syntactic analysis you could see below.

Input sentence: I go for a walk every day.

Output sentence: [I]ng [go [for a walk] [every day]ng]vg.

It has been developed the first version of the syntactic analyzer. Syntactic groups were singled out using the initial database, consisting of about 500 simple context-free sentences. The module was tested on the database, consisting of other context-free 500 sentences. The obtained results were following: 32% of sentences were processed incorrectly. The main cause of the incorrect analysis is that during the analysis of a random text, the more complicated syntactic groups were found out (26% of sentences). Another reason of errors is incomplete vocabulary of recognition system (6% of sentences). So our present task is addition both the vocabulary and the list of syntactic groups to develop more flexible syntax analyzer.

6 Conclusion

The present article is a short description of the syntax and morphology analyzer for Russian texts. The best way to describe an inflectional language is to present it as an inflectional system with flexions, stems and paradigms. This approach can be applied to the Russian language due to the brilliant opus by A.A. Zaliznyak. The developed program includes two modules: the module of morphological analysis and the module of syntactic analysis. The first module could successfully used for Russian speech recognition. It allows to reduce the size of decoder's vocabulary. The syntax module is in course of development. The first stage is completed: the module could analyze simple sentences successfully. The module allows us to correct wrong flexions of words, which may occur during the speech recognition, due to analysis relationships between sentence constituents.

References

1. Gelbukh, A., Sidorov, G.: Approach to construction of automatic morphological analysis systems for inflective languages with little effort. In: Computational Linguistics and Intelligent Text Processing (CICLing 2003). LNCS, vol. 2588, pp. 215–220. Springer, Heidelberg (2003)
2. Sokirko, A.V.: Morphological modules on the web. In: Proc. Dialog 2004, Moscow (2004), www.aot.ru

3. Carki, K., Geutner, P., Schultz, T.: Turkish LVCSR: Towards Better Speech Recognition for Agglutinative Languages. In: Proc. ICASSP 2000, Istanbul, Turkey (2000)
4. Kneissler, J., Klakow, D.: Speech recognition for huge vocabularies by using optimized subword units. In: Proc. Eurospeech 2001, Aalborg, Denmark, pp. 69–72 (2001)
5. Kurimo, M., Creutz, M., Varjokallio, M., Arisoy, E., Saraclar, M.: Unsupervised segmentation of words into morphemes – Morpho Challenge 2005, Application to Automatic Speech Recognition. In: Proc. Interspeech 2006, Pittsburgh, USA (2006)
6. Zaliznjak, A.A.: Grammatical dictionary for Russian language. Rus. jaz, Moscow (1980)
7. Dressler, W.U., Gagarina, N.: Basic questions in establishing the verb classes of contemporary Russian – Essays in Poetics, Literary History and Linguistics. In: Fleishman, L., Ivanov, F.V.V., et al. (eds.), OGI, Moscow, pp. 754–760 (1999)
8. Ronzhin, A.L., Karpov, A.A.: Russian Voice Interface. Pattern Recognition and Image Analysis 17(2), 321–336 (2007)

An Extension to the Sammon Mapping for the Robust Visualization of Speaker Dependencies

Andreas Maier, Julian Exner, Stefan Steidl, Anton Batliner, Tino Haderlein, and Elmar Nöth

Universität Erlangen-Nürnberg, Lehrstuhl für Mustererkennung (Informatik 5)
Martensstraße 3, 91058 Erlangen, Germany
andreas.maier@informatik.uni-erlangen.de

Abstract. We present a novel method for the visualization of speakers which is microphone independent. To solve the problem of lacking microphone independency we present two methods to reduce the influence of the recording conditions on the visualization. The first one is a registration of maps created from identical speakers recorded under different conditions, i.e., different microphones and distances in two steps: Dimension reduction followed by the linear registration of the maps. The second method is an extension of the Sammon mapping method, which performs a non-linear registration during the dimension reduction procedure. The proposed method surpasses the two step registration approach with a mapping error ranging from 17 % to 24 % and a grouping error which is close to zero.

1 Introduction

The facets of voices and speech are very complex. They comprise various stationary and dynamic properties like frequency, energy, and even more complex structures such as prosody. In order to comprehend these characteristics the high dimensionality of the speech properties has to be reduced. Therefore, the visualization in two or three dimensions was shown to be very effective in many fields of application:

- The visualization often allows to gain further insight in the structure of the data. For example the speaking style like loudness and rate-of-speech of different speakers can be analyzed [1].
- The visualization can also be used to select a subset of representative training speakers which cover all of its areas to reduce the number of training speakers. In a first step few data of many speakers are collected of which only the representative ones are included for a second recording session in order to collect more data. In this manner the recognition performance stays in the same range as if the second session would have been done with all speakers [2].
- The visualization can reveal the relations between patients with voice disorders in different graduations [3]. This gives the medical personal a better understanding of the different disorders. Projection of a new speaker allows to compare him to the other speakers.

P. Sojka et al. (Eds.): TSD 2008, LNAI 5246, pp. 381–388, 2008.
© Springer-Verlag Berlin Heidelberg 2008

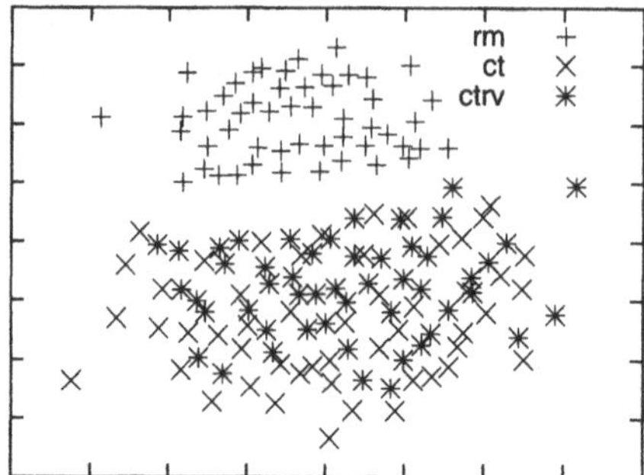

Fig. 1. 51 speakers recorded simultaneously with three different microphones: remote (rm), artificial reverberation (ctrv), and close talk (ct). ct and ctrv form one cluster while the rm forms another cluster. All three clusters contain the same speakers.

A problem for the visualization of speech data is the fact that the recording conditions have a great impact. The recording conditions consist mainly of the used microphone, the distance between the microphone and the speaker, and the acoustical properties of the recording location. If a speaker uttering a sentence was recorded simultaneously by multiple microphones of different quality at different distances, the points representing the same speaker are spread across the result of the visualization. Fig. 1 gives an extreme example using the Sammon mapping: The speakers form two clusters although the speakers were recorded simultaneously. This is caused by the acoustic difference between the two microphones which were chosen for the recording. The two corresponding representations of the same speaker are far away from each other in this visualization. The dominating factor is the microphone. In general, all visualizations of data collected in different acoustic conditions show similar effects in different graduations depending on the discrepancy between the acoustics.

If applied in a medical environment, for example with our fully automatic internet speech evaluation software [4], recordings are often performed at multiple locations simultaneously, e.g. in multi-site studies. Therefore, a method is desirable which removes or reduces these recording differences. The mismatch of the recording conditions can be reduced if a set of known calibration data is replayed with a standardized loudspeaker at a new location. In this manner, the effect of the new microphone and the recording conditions can be "learned" and removed from the visualization. In this paper we chose for simultaneously recorded data as we wanted to exclude the disturbances which might be created by the playback with a loudspeaker.

In order to create a visualization of the data the dimension has to be reduced to a two- or three-dimensional space. As a representation of a speaker we chose for the parameters of a speaker-adapted speech recognizer. Furthermore, the map should present same speakers at the same or at least a very close position, i.e., minimize the recording influences. To minimize the interferences of the recording conditions, two approaches are presented: The first one employs the standard Sammon mapping for the dimension reduction and a linear transformation of the data points in the low dimensional domain in order to project corresponding ones as close to each other as possible, i.e., a registration of the maps. The second one extends the Sammon mapping by a grouping term which causes the same speakers to be projected as close to each other as possible i.e. it uses the prior knowledge about the group membership, and punishes the distance of points belonging to the same group already during the dimension reduction.

All methods were evaluated using the Aibo database. It consists of children speech recorded with a head-set microphone and the microphone of a video camera. A third recording condition was simulated using artificial reverberation. The Aibo data show very strong differences between the recordings conditions and are therefore ideal for the demonstration of our method.

2 Material

The database used in this work contains emotional speech of children. In a Wizard-of-Oz experiment children in the age of 12 to 14 years were faced with the task to control a Sony AIBO^{TM} robot by voice [5]. In total 51 pupils (21 male and 30 female) of two different schools were recorded in the German language. The whole scenery was recorded by a video camera in order to document the experiment and a head-mounted microphone (UT 14/20 SHURE UHF). The close talking version is referred to as *ct*. From the sound track of the video tape a second version of the AIBO corpus was extracted: a distant-talking version (*rm*) was obtained. In this manner no second manual transliteration was necessary because the transcription of the distant-talking and the close-talking version is the same. The distance between the speaker's position and the video camera was approximately 2.5 m. 8.5 hours of spontaneous speech data were recorded.

Artificial reverberation is used to create disturbances which resemble those caused by reverberation in a real acoustic environment. It is applied to the signal directly before the feature extraction. The idea is to convolve the speech signal with impulse responses characteristic for reverberation in typical application scenarios e.g. a living room. Thus a reverberated signal can be computed. These impulse responses can be measured in the proposed target environment or generated artificially. In current research the artificial reverberation was found to improve the robustness of speech recognizers to acoustic mismatches [6]. We applied the same twelve impulse responses as in the previously mentioned work.

In this manner the recordings from the close talk microphone were artificially reverberated to simulate another recording condition (*ctrv*). This way, three speech recognizers are adapted for each of the children from a speaker-independent one and are used for the creation of the visualizations.

3 Methods

3.1 Reduction of Dimensionality

The Sammon transformation (ST) is a nonlinear method for mapping high dimensional data to a plane or a 3-D space [7]. As already mentioned, the ST uses the distances between the high dimensional data to find a lower dimensional representation – called map in the following – that preserves the topology of the original data, i.e. keeps the distance ratios between the low dimensional representation – called star in the following – as close as possible to the original distances. Doing so, the ST is cluster preserving. To ensure this, the function e_S is used as a measurement of the error of the resulting map (2-D case):

$$e_S = s \sum_{p=1}^{N-1} \sum_{q=p+1}^{N} \frac{(\delta_{pq} - \theta_{pq})^2}{\delta_{pq}} \quad \text{with} \tag{1}$$

$$\theta_{pq} = \sqrt{(p_x - q_x)^2 + (p_y - q_y)^2} \tag{2}$$

δ_{pq} is the high dimensional distance between the high dimensional features p and q stored in a distance matrix $\boldsymbol{D}$, θ_{pq} is the Euclidian distance between the corresponding stars p and q in the map. For the computation of the high dimensional distance between two speech recognizers we use the Mahalanobis distance [8] as in [1]. s is a scaling factor derived from the high dimensional distances:

$$s = \frac{1}{\sum_{p=1}^{N-1} \sum_{q=p+1}^{N} \delta_{pq}} \tag{3}$$

The transformation is started with randomly initialized positions for the stars. Then the position of each star is optimized, using a conjugate gradient descent library [9]. In [2] this method is referred to as "COSMOS" (COprehensive Space Map of Objective Signal).

3.2 Reduction of the Influence of the Recording Conditions in the Visualization

The first approach to reduce the influence of the recording conditions, is the use of a linear registration. The idea is to use utterances from several speakers and record them under different conditions. Then the features generated from the recordings are transformed into a 2-D (or 3-D) map. The map is split according to the recording conditions ($h_1 \ldots h_H$), and afterwards a linear registration is applied, aiming to reduce the distance between the stars belonging to one speaker. The objective function for the registration is

$$e_{\text{REG}}(h_i, h_j) = \frac{1}{N_m} \sum_{i=1}^{N_m} \theta_{p^{h_i} p^{h_j}} \tag{4}$$

for the two maps recorded with microphone h_i and h_j, each consisting of $N_m = \frac{N}{H}$ stars. $\theta_{a_i b_i}$ is the Euclidian distance between the star p^{h_i} of the map from h_i and star p^{h_j} of microphone h_j.

$$n_i' = A n_i + t \tag{5}$$

with the transformation matrix A and the translation vector t.

The error is minimized using gradient descent. For the projection of a new star into a map the dimensionality has to be reduced first according to the Sammon mapping. Then, the registration can be performed according to Equation 5.

A non-linear registration approach can be included into the optimization process of the Sammon mapping: To minimize the distance between stars belonging to the same speaker additional information about the group affiliation is used, i.e., stars representing the same speaker form a group. Therefore, a grouping error is introduced to extend the objective function. The recording is of the same structure as for the linear approach. A group weight g_{ij} indicates whether the stars, respectively the high dimensional features, belong to the same group. Thus, $g_{ij} = 1$, if the feature vector j corresponds to speaker i, else $g_{ij} = 0$. Remember that one speaker is recorded in our application by multiple microphones, so there are more recordings for one speaker.

The original error function of the Sammon mapping is altered such that it reduces the distance between stars that belong to the same group. So a new error function e_Q is formed:

$$e_Q = s \sum_{p=1}^{N-1} \sum_{q=p+1}^{N} \left[Q g_{pq} \theta_{pq} + (1-Q)(1-g_{pq}) \frac{(\delta_{pq} - \theta_{pq})^2}{\delta_{pq}} \right] \tag{6}$$

g_{pq} is the group indicator and Q is the weight factor which balances the standard Sammon error to the additional error term. Again, gradient descent is applied to optimize the error criterion.

In allusion to the name of the method for speaker visualization as presented in [1] we refer to our method as QMOS.

3.3 Quality Metrics for the Visualization

The measurement of the quality of a visualization is a very difficult task. In our case we decided to use two measurements for the evaluation:

- Sammon Error e_S: The remaining error computed by the Sammon error function according to Equation 1. This error is used to describe the loss of the mapping from the high dimensional space to the low dimensional space. In the literature this term was shown to be a crucial factor to describe the quality of a representation [1,2,3].
- Grouping Error e_{Grp}: The average distance between stars belonging to the same group (on a map with normalized coordinates in an interval between 0 and 1).

$$e_{\text{Grp}} = \frac{1}{N} \sum_{i=1}^{N-1} \sum_{j=i+1}^{N} \theta_{ij} g_{ij} \tag{7}$$

Both errors are relative to the maximal error and can therefore also be interpreted as percentages, i.e. an error of 0.14 corresponds to 14 % of the maximal error.

4 Results

For all experiments the speech data, together with a transliteration of the spoken text, is used to adapt a speech recognizer for each speaker, using MLLR adaption of the Gaussian mixture density for the output probabilities. The mixture densities are used to compute distances or directly as features.

Table 1. Metrics for maps created from all data with the different visualization methods: Both versions of COSMOS have a very high grouping error. QMOS surpasses both methods in grouping and Sammon error while having a lower grouping error.

method	e_S	e_{Grp}
COSMOS	**0.09**	0.40
COSMOS + reg.	0.21	0.21

method	Q	Error e_S	e_{Grp}
QMOS	0.00	**0.09**	0.40
QMOS	0.60	0.09	0.36
QMOS	0.75	0.16	0.07
QMOS	0.87	**0.18**	**0.01**
QMOS	0.96	0.24	**0.00**

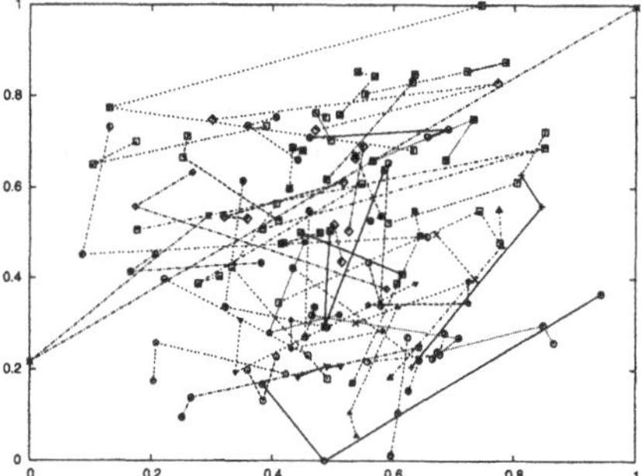

Fig. 2. Visualization computed with COSMOS and a linear registration: Points representing the same speaker are connected with lines. The visualization method does not yield a good visualization. The speakers are almost randomly distributed in each map.

Evaluation was performed with and without linear registration. Table 1 shows the results. The method with the best Sammon error is of course the Sammon mapping. The visualizations of the registered maps can be seen in Fig. 2. The linear method fails to project the same speakers close to each other. The visualization cannot be interpreted properly.

Since the QMOS method is dependent on the weighting factor Q it has to be determined experimentally. Table 1 shows the dependency between the group and the Sammon error. The trade-off between grouping accuracy and reduction of the Sammon error has to be determined. The effect of the weight on the visualization is shown in Fig. 3. The optimal value of the group error is at $Q = 0.87$ with a grouping error of only 0.01. At that position the trade-off between grouping and Sammon error is also very good. Note that there are several configurations of Q which yield a very low sum of grouping and Sammon error i.e. one can choose from several optimal values Q depending on the problem one wishes to visualize.

5 Discussion

We evaluated the visualization methods for speakers in different acoustic conditions. The example we chose is difficult since the differences of the acoustic conditions in

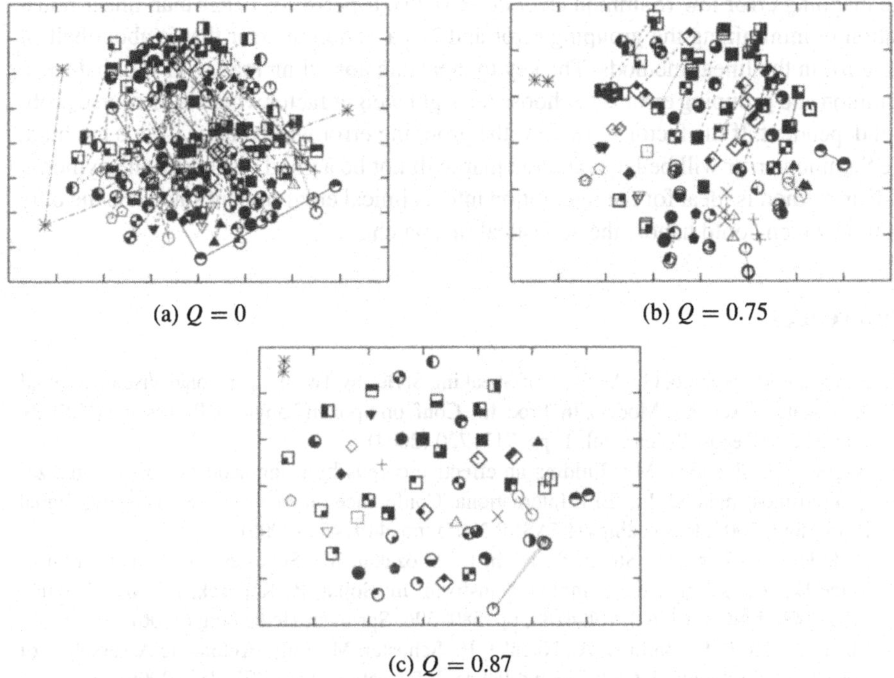

(a) $Q = 0$

(b) $Q = 0.75$

(c) $Q = 0.87$

Fig. 3. Extended ST on Aibo data with three different weight factors: The points belonging to one speaker are connected with lines. (a) shows the same map as Fig. 1.

the AIBO database are large. Unfortunately, a signal-to-noise ratio cannot be computed between all versions of the AIBO corpus since the data is not always frame-matched. As reported in [10] the baseline recognition rates for matched conditions differ a lot (77.2 % WA on ct, 63.1 % WA on ctrv, and 46.9 % WA on rm). They are even worse if training and test set are not from the same version (12.0 % WA with the ct recognizer on the rm test set). Using these data we created a suitable mapping task since we wanted to create a method for visualization which can also handle extreme cases.

The linear method for the registration could only yield maps in which the corresponding speakers are in a corresponding region, i.e., the speakers are not projected into the opposite site of the map. Investigation of other linear methods such as PCA, LDA, and ICA showed no better results. A configuration with a lower group or Sammon error than the proposed method could not be obtained.

6 Conclusion

We successfully created a new method for the robust visualization of speaker dependencies: Using our novel method it is possible to create a single map although the data was collected with different microphones. The method can even handle very strong differences in the acoustic conditions.

It was shown that the QMOS method is a good method to reduce the influence of recording conditions on a visualization (grouping error was almost zero) while keeping the mapping error low (Sammon error $e_S < 0.25$). It performs better than linear registration in minimizing the grouping error and has a Sammon error that is about half of the error in the linear methods. The key to create an appealing map with well balanced Sammon and grouping error is to choose the right weight factor which is of course problem dependent. If the factor is too low, the grouping error will be large, if it is too high, the Sammon error will be large and the map will not be a good visualization anymore.

Our method is ideal for the integration into a clinical environment since it is the only method which could handle the acoustical mismatches.

References

1. Shozakai, M., Nagino, G.: Analysis of Speaking Styles by Two-Dimensional Visualization of Aggregate of Acoustic Models. In: Proc. Int. Conf. on Spoken Language Processing (ICSLP), Jeju Island (Rep.of Korea), vol. 1, pp. 717–720 (2004)
2. Nagino, G., Shozakai, M.: Building an effective corpus by using acoustic space visualization (cosmos) method. In: IEEE International Conference on Acoustics, Speech, and Signal Processing, 2005. Proceedings (ICASSP 2005), pp. 449–452 (2005)
3. Haderlein, T., Zorn, D., Steidl, S., Nöth, E., Shozakai, M., Schuster, M.: Visualization of Voice Disorders Using the Sammon Transform. In: Sojka, P., Kopeček, I., Pala, K. (eds.) TSD 2006. LNCS (LNAI), vol. 4188, pp. 589–596. Springer, Heidelberg (2006)
4. Maier, A., Nöth, E., Batliner, A., Nkenke, E., Schuster, M.: Fully Automatic Assessment of Speech of Children with Cleft Lip and Palate. Informatica 30(4), 477–482 (2006)
5. Batliner, A., Hacker, C., Steidl, S., Nöth, E., D'Arcy, S., Russell, M., Wong, M.: You stupid tin box - children interacting with the AIBO robot: A cross-linguistic emotional speech corpus. In: Proceedings of the 4th International Conferen e of Language Resources and Evaluation LREC 2004, ELRA edn., pp. 171–174 (2004)
6. Maier, A., Haderlein, T., Nöth, E.: Environmental Adaptation with a Small Data Set of the Target Domain. In: Sojka, P., Kopeček, I., Pala, K. (eds.) TSD 2006. LNCS (LNAI), vol. 4188, pp. 431–437. Springer, Heidelberg (2006)
7. Sammon, J.: A nonlinear mapping for data structure analysis. In: IEEE Transactions on Computers C-18, pp. 401–409 (1969)
8. Mahalanobis, P.C.: On the generalised distance in statistics. In: Proceedings of the National Institute of Science of India 12, pp. 49–55 (1936)
9. Naylor, W., Chapman, B.: WNLIB Homepage (2008) (last visited 17/01/2008), `www.willnaylor.com/wnlib.html`
10. Maier, A., Hacker, C., Steidl, S., Nöth, E., Niemann, H.: Robust parallel speech recognition in multiple energy bands. In: Kropatsch, G., Sablatnig, R., Hanbury, A. (eds.) DAGM 2005. LNCS, vol. 3663, pp. 133–140. Springer, Heidelberg (2005)

Analysis of Hypernasal Speech in Children with Cleft Lip and Palate

Andreas Maier[1,2], Alexander Reuß[1], Christian Hacker[1], Maria Schuster[2], and Elmar Nöth[1]

[1] Universität Erlangen-Nürnberg, Lehrstuhl für Mustererkennung (Informatik 5)
Martensstraße 3, 91058 Erlangen, Germany
[2] Universität Erlangen-Nürnberg, Abteilung für Phoniatrie und Pädaudiologie
Bohlenplatz 21, 91054 Erlangen, Germany
andreas.maier@informatik.uni-erlangen.de

Abstract. In children with cleft lip and palate speech disorders appear often. One major disorder amongst them is hypernasality. This is the first study which shows that it is possible to automatically detect hypernasality in connected speech without any invasive means. Therefore, we investigated MFCCs and pronunciation features. The pronunciation features are computed from phoneme confusion probabilities. Furthermore, we examine frame level features based on the Teager Energy operator. The classification of hypernasal speech is performed with up to 66.6 % (CL) and 86.9 % (RR) on word level. On frame level rates of 62.3 % (CL) and 90.3 % (RR) are reached.

1 Introduction

In clinical practice it is desirable to objectively quantify the severity of speech disorders by non-invasive means. The state-of-the-art techniques to measure hypernasality today are quite invasive since the patients' nasal and oral airflows have to be measured. This is usually done with devices like a Nasometer [1] which is placed between the mouth and the nose in order to separate both airflows. This procedure is complicated – especially with children.

Non-invasive methods exist [2,3], however, their application demands a lot of manual preprocessing since these methods can only be applied to sustained vowels or consonant-vowel combinations. In the literature the segmentation is usually done manually which costs a lot of time and effort. In order to close this diagnostic gap we want to investigate, if a fully automatic evaluation system can be applied for such a task. Therefore, an automatic speech recognition system is used to segment the audio data into words. To improve the automatic segmentation the transliteration of the speech data was supplied to the speech recognition system. In a next step, of course, we will replace the manual transliteration with an automatic speech recognition system. To train and evaluate an automatic classifier, a speech therapist labeled all words either as "hypernasal" or "normal".

This research is being integrated into our automatic speech evaluation platform [4] environment, which is a web application to analyze and evaluate various speech disorders. At the moment our system can already judge the speaker's intelligibility. To

P. Sojka et al. (Eds.): TSD 2008, LNAI 5246, pp. 389–396, 2008.
© Springer-Verlag Berlin Heidelberg 2008

achieve this, it uses a speech recognizer and calculates either word accuracy (WA) or word recognition rate (WR) as a measure for intelligibility [5].

The aim of this work is to take our system one step further. It is our intention to enable the ability to detect specific speech disorders. In the following sections, we describe facial clefts and analyze different features for the detection of hypernasality.

2 Cleft Lip and Palate

Cleft lip and palate (CLP) is a frequent congenital alteration of the face with a prevalence of about one in 750 to 900 births [6]. Today, the visible effects of CLP can be surgically lessened. If a grown up patient has been treated well, it is hardly noticeable, that he had a facial cleft as a child. Apart from surgical interventions, the patient also receives speech therapy. This is necessary, because the alteration can have a major impact on the patient's vocal tract and can lead to various speech disorders. The most common is hypernasality which is caused by enhanced nasal air emissions. The first formant is less distinct while antiresonances and nasal formants appear [7]. As a consequence, vowels are perceived with a characteristic nasal sound. However, speakers also have problems with other phonemes: Fricatives can not be pronounced correctly and plosives are weakened [8].

During the speech therapy of a patient with hypernasality due to cleft lip and palate, an automatic system to detect hypernasal speech would be very useful because it can make the treatment easier by providing a way to keep track of the patient's progress.

3 Classification System

All experiments use a Gaussian mixture model (GMM) classifier according to the following decision rule:

$$k = \operatorname*{argmax}_{\kappa} \; P(\Omega_\kappa) \cdot P(\boldsymbol{c} \mid \Omega_\kappa) \tag{1}$$

Here, κ denotes a class, $\boldsymbol{c}$ is our feature vector and Ω_κ is the event that the current observation belongs to class κ. In our case, there are only two classes: $\kappa \in \{\text{nasal}, \text{normal}\}$.

The probability $P(\boldsymbol{c} \mid \Omega_\kappa)$ is approximated by a mixture of M Gaussian densities $\mathcal{N}(\boldsymbol{c}, \boldsymbol{\mu}_m, \boldsymbol{\Sigma}_m)$ with mean vectors $\boldsymbol{\mu}_m$ and covariance matrices $\boldsymbol{\Sigma}_m$:

$$P(\boldsymbol{c} \mid \Omega_\kappa) \approx \sum_{m=1}^{M} a_m \mathcal{N}(\boldsymbol{c}, \boldsymbol{\mu}_m, \boldsymbol{\Sigma}_m)$$

$$\text{with} \sum_{m=1}^{M} a_m = 1 \tag{2}$$

Our classifier is trained by calculating a_m, $\boldsymbol{\mu}_m$ and $\boldsymbol{\Sigma}_m$ by means of the expectation maximization algorithm [9].

3.1 Word Dependent A-Priori Probabilities

We estimate the prior probabilities $P(\Omega_{\text{nasal}})$ by calculating the frequency of hypernasality for every word w from the training set. If w is never marked as hypernasal in our training set, this would lead to a zero probability which means a record of w can never be classified as hypernasal. Therefore, we interpolate by equally distributing the probability mass of the words that are marked as hypernasal less than two times:

$$P_w(\Omega_{\text{nasal}}) \approx \begin{cases} \dfrac{\#\text{nasal}_w}{\#\text{total}_w} & \text{if } \#\text{nasal}_w > 1 \\[2ex] \dfrac{\sum\limits_{z \in S} \#\text{nasal}_z}{\sum\limits_{z \in S} \#\text{total}_z} & \text{else} \end{cases}$$

S denotes the subset of words that were marked less than two times. $\#\text{nasal}_w$ is the number of times the word w was marked and $\#\text{total}_w$ is the number of times w occurs.

3.2 Pronunciation Features

Pronunciation features, as described in [10], were designed to rate a speaker's pronunciation. They are used for measuring the progress when learning a foreign language. In this work, we study these features' applicability to the detection of hypernasal speech. More precisely, we only analyze a subset of these features that is based on phoneme confusion probabilities on word level. To calculate these phoneme confusion features we compare the result of the forced alignment of every word to the result of a phoneme recognizer. The phoneme recognizer uses semi continuous hidden Markov models and a 4-gram language model. It is based on MFCCs calculated every 10 ms with a frame size of 16 ms. From these informations phoneme confusion matrices $\boldsymbol{C}$ are built. They contain for every pair of phonemes a, b the probability that a was detected by the recognizer when there should be b according to the forced alignment, i.e.,

$$C_{a,b} = P(a \mid b) \tag{3}$$

From the training set, we calculate two confusion matrices: one for the hypernasal data and one for the normal data. We need to recalculate these matrices in every iteration of the LOO evaluation (cf. Section 4) because, in order to obtain representative results, the current test speaker may not be involved in the generation of our phoneme confusion model. The quotient Q is calculated for every frame:

$$Q = \frac{P_{\text{nasal}}(a \mid b)}{P_{\text{normal}}(a \mid b)} \tag{4}$$

From these frame-wise results, we calculate the following features for the word level:

- PC1: mean of Q,
- PC2: maximum of Q,
- PC3: minimum of Q,
- PC4: scatter of Q,
- PC5: median of Q.

3.3 Cepstral Features

Furthermore, we investigate cepstral features, more accurately Mel frequency cepstral coefficients (MFCCs). These features are calculated frame wise and consist of the short time signal energy, 11 MFCCs and 12 dynamic features in a context of ± 2 frames i.e. 56 ms. Using these features, we train a frame based classifier to analyze hypernasal vs. normal speech. However, the expert annotations have all been performed on the word level. Thus, the expert labels have to be mapped onto the frame level. In a first approach, we simply label every frame with the respective word label; then a frame based classifier is trained. As it is known for every frame, which word it was taken from, we can still use the word based a-priori probabilities (cf. Section 3.1) as we did before.

To compare the classification results with previous investigations with word based pronunciation features, the evaluation is again performed on the word level. This means, that in the test phase a decision for the class of a word is derived from all the classification results for its frames. There are several ways of making a word level decision from the frame level (e.g. mean or median) but the best results were achieved using the maximum frame wise classifier score as the classification hypotheses for the whole word.

As mentioned above, during training all frames are labeled the same as the word they belong to. However, if a word is labeled "hypernasal" that does not mean that *every* part of this word is hypernasal. So we got normal frames labeled as hypernasal in our training procedure. We tackle that issue using a bootstrapping algorithm (similar to [11]). We train our frame wise classifier just like we did before. Then we classify the training data and relabel the frames of the hypernasal words with the hypothesis of the frame based classifier. This process is iterated a fixed number of times. We choose two iterations per word as preliminary experiments showed that more iterations do not yield further improvements.

3.4 Teager Energy Operator

The next feature we evaluate is the Teager Energy operator (TEO) [2]. It is defined as:

$$\psi[x(n)] = x^2(n) - x(n+1)x(n-1) \tag{5}$$

$x(n)$ denotes the time domain audio signal. The TEO's output is called the Teager Energy profile (TEP).

As already described in [2], the TEP can be used to detect hypernasal speech because it is sensitive to composite signals. When normal speech is lowpass filtered in a way that the maximum frequency f_{lowpass} is somewhere between the first and the second formant, the resulting signal mainly consists of the first formant. However, doing the same with hypernasal speech results in a composite signal due to the strong anti formants. If we now compare the lowpass filtered TEP to the TEP of the same signal that was bandpass filtered around the first formant we should see more difference in case of a hypernasal signal. We measure that difference with the correlation coefficient of these TEPs. In the following, the bandpass filter covers the frequency range ±100 Hz around the first formant estimated using PRAAT [12].

We got the following experimental setup: We use the correlation between both TEPs as a feature and calculate it for every frame. Then, we apply a phoneme recognizer to determine the phoneme, the frame belongs to. The classifier is trained per vowel with the features from all frames of the training words that were associated with that vowel. Afterwards, these classifiers are tested with the frames from the words of the test speaker that were assigned to the respective vowel.

4 Evaluation

As our data set is rather small we use leave-one-speaker-out (LOO) evaluation to rate our classifiers. There are much more normal words than hypernasal words, so recognition rate (RR) is not very meaningful. Therefore, we calculate the mean recognition rate per class (CL) as well. It is the average of the recalls for the classes "hypernasal" and "normal".

$$\mathrm{CL} = 0.5(\mathrm{REC}_{\mathrm{nasal}} + \mathrm{REC}_{\mathrm{normal}}) \tag{6}$$

where $\mathrm{REC}_{\mathrm{nasal}}$ is the recall of the class "hypernasal" and $\mathrm{REC}_{\mathrm{normal}}$ the recall of the class "normal".

5 Data

The data we use consists of recordings of 3 girls and 10 boys (5 to 11 years old) with cleft lip and palate recorded during the PLAKSS test [13]. Pictures are shown to the children which illustrate the words the children should speak. There are 99 different words, that contain all phonemes of the German language at 3 different positions in the word: at the beginning, at the end, and in the center. The single words were extracted from the recordings using forced alignment and were labeled as "hypernasal" or "normal" by an experienced speech therapist.

Since some children skipped some words and others had a quite low intelligibility some words could not be properly segmented by the forced alignment. These words had to be excluded from the data. In total we got 771 words. 683 of them are labeled "normal" and 88 are labeled "hypernasal". As some phonemes are more likely to be mispronounced due to hypernasality than others, the probability $P_w(\Omega_{\mathrm{nasal}})$ for being marked hypernasal is different for every word.

In order to keep the data as realistic as possible slight errors in the forced alignment were kept in the database (some samples were cut off at the beginning or at the end of the word). The audio files were stored with a sampling rate of 16 kHz and quantized with 16 bit per sample.

6 Results

LOO evaluation of the single pronunciation features with our Gaussian classifier (we choose $M = 1$, more densities have shown to decrease the rates in this case) leads to the results shown in Table 1 (a). The idea behind choosing these features is that hypernasal speakers have problems pronouncing specific phonemes (plosives, some fricatives and some vowels). Therefore, the phonemes a recognizer does not identify properly should, to some extend, be similar for the nasal speakers. The class-wise recognition rate of

Table 1. Results on word level obtained with different features. MFCCs and pronunciation features yield feasible results.

FEATURE	RR	CL
PC1	**86.9 %**	57.7 %
PC2	86.4 %	52.4 %
PC3	82.7 %	**64.1 %**
PC4	86.8 %	56.3 %
PC5	86.1 %	57.7 %

(a) pronunciation features

DENSITIES	RR	CL
2	70.6 %	**65.1 %**
5	74.7 %	64.4 %
10	**75.5 %**	64.9 %
15	**75.5 %**	63.9 %

(b) MFCCs (no bootstrap)

DENSITIES	RR	CL
2	82.1 %	**66.6 %**
5	82.3 %	64.8 %
10	82.2 %	64.7 %
15	**83.1 %**	65.2 %

(c) MFCCs (bootstrap)

Table 2. Results for frame wise Teager Energy features for different vowels and best cutoff frequency

VOWEL	RR	CL	CUTOFF f_{lowpass}
/9/	**90.3 %**	62.5 %	1900 Hz
/a/	80.0 %	59.2 %	1000 Hz
/i:/	84.8 %	60.1 %	1900 Hz
/o:/	87.0 %	**63.2 %**	1000 Hz
/O/	88.2 %	55.6 %	1900 Hz
/u:/	90.0 %	55.6 %	1000 Hz
/U/	89.4 %	56.1 %	1900 Hz

up to 64.1 % verifies our assumption. As our training sets are rather small the confusion probabilities $P(a|b)$ can not be calculated very exactly. Therefore, we expect even better results in future experiments with more training data.

Testing the word level classification system based on frame-wise MFCCs as described in Section 3.3 leads to the results shown in Table 1 (b) and Table 1 (c). It can be seen that bootstrapping slightly improves the class wise recognition rate while considerably improving the total recognition rate.

The results of the frame-wise classification of the Teager Energy correlation feature as described in Section 3.4 were also promising. The formant frequencies for the bandpass were extracted with "Praat" automatically and a bandpass with a bandwidth of 200 Hz around the first formant was performed for the one TEP. For the other TEP we run 4 series of tests with lowpass cutoff frequencies $f_{\text{lowpass}} = 1000$ Hz, 1300 Hz, 1600 Hz, and 1900 Hz. Then, the correlation between both TEPs was determined and fed to the classification procedure. Table 2 shows the best results of for each of the vowels which appeared in the test data (vowels in SAMPA notation).

The results show, that the TEO can be used to classify hypernasal speech of children with cleft lip and palate. However, the rates are not as good as [2] might let expect. We see two reasons for this: first, we have no phoneme level annotation (a similar problem to what we discussed before regarding the MFCCs), second this concept does not work as good with children as their formants are harder to find (more detection errors) than those of adults. Due to the difficulties in the determination of the cutoff frequencies and that the TEP is only suitable for vowels we did not study their performance further on word level, yet.

7 Discussion and Outlook

In the results section encouraging results for the classification of hypernasality in children's speech from automatically segmented audio data were presented: The class-wise recognition rate CL reaches up to 66.6 % and the recognition rates RR is in one case even 89.4 (53.4 % CL). We explain this effect with the fact that nasality detection in children's speech is more difficult than in adults' speech. Misdetection of normal speech as hypernasal speech, however, happened rarely in the best classifiers.

We still see some room for improvement in our future work. As our data set is relatively small, classification results could be greatly enhanced by using more data (further recordings were already performed). This will help estimating the prior probabilities for the GMM classifier and the confusion matrices for the pronunciation features. The combination of multiple features will also improve the performance of the classification.

Another possibility to improve the results is the usage of a phoneme level annotation. This will be a sensible step, because only some phones of a hypernasal word show nasal characteristics. The other phones of the realization might still be perceived as normal. The recognition rates of the MFCCs and the TEO features could benefit from it. Moreover, we want to investigate whether the TEP features can be enhanced in order to be applicable on word or speaker level, since their classification rates look quite promising. We expect to be able to use the techniques presented here soon in clinical practice.

8 Summary

In this study we could show that the classification of hypernasality in children's speech on automatically segmented data is possible. We described the evaluation of several features regarding their suitability to classify hypernasal speech of children with cleft lip and palate. On word level, class-wise recognition rates of up to 66.6 % and global recognition rates of 86.9 % were achieved. First, we extracted pronunciation features based on phoneme confusion statistics. With these, we reached a CL of up to 64.1 % and a RR of 86.9 %. MFCC features were best in CL. We extracted them frame-wise and derived a word level decision from that. With a bootstrapping approach, we improved the annotation which led to rates of up to 66.6 % CL and 83.1 % RR. Finally, we studied the TEO on frame level. It was tested using separate classifiers for the frames belonging to different vowels which were identified using a simple phoneme recognizer. The results showed, that the TEO's performance varied for different phonemes and that it does not work as well in our scenario as in preceeding works with adult speakers and manually segmented consonant-vowel and consonant-vowel-consonant clusters.

References

1. Kay Elemetrics Corporation, New York, USA, Instruction manual of the nasometer Model 6200-3, IBM PC Version (1994)
2. Cairns, D., Hansen, J.H.L., Riski, J.: A Noninvasive Technique for Detecting Hypernasal Speech Using a Nonlinear Operator. IEEE Transactions on Biomedical Engineering 43(1), 35–45 (1996)

3. Zečević, A.: Ein sprachgestütztes Trainingssystem zur Evaluierung der Nasalitiät, Ph.D. thesis, Universität Mannheim, Germany (2002)
4. Maier, A., Nöth, E., Batliner, A., Nkenke, E., Schuster, M.: Fully Automatic Assessment of Speech of Children with Cleft Lip and Palate. Informatica 30(4), 477–482 (2006)
5. Nöth, E., Maier, A., Haderlein, T., Riedhammer, K., Rosanowski, F., Schuster, M.: Automatic Evaluation of Pathologic Speech – from Research to Routine Clinical Use. In: Matoušek, Mautner (eds.) [14], pp. 294–301
6. Tolarova, M., Cervenka, J.: Classification and birth prevalence of orofacial clefts. Am. J. Med. Genet. 75(2), 126–137 (1998)
7. Fant, G.: Nasal Sounds and Nasalization. In: Acoustic Theory of Speech Production, The Hague, The Netherlands, Mouton (1960)
8. Braun, O.: Sprachstörungen bei Kindern und Jugendlichen, Kohlhammer, Stuttgart, Germany (2002)
9. Bocklet, T., Maier, A., Nöth, E.: Text-independent Speaker Identification using Temporal Patterns. In: Matoušek,, Mautner (eds.) [14], pp. 318–325.
10. Hacker, C., Cincarek, T., Gruhn, R., Steidl, S., Nöth, E., Niemann, H.: Pronunciation feature extraction. In: Pattern Recognition, 27th DAGM Symposium, Vienna, Austria, August 30–September 2005. LNCS, pp. 141–148. Springer, Berlin (2005)
11. Abney, S.P.: Bootstrapping. In: Proceedings of the 40th Annual Meeting of the Association for Computational Linguistics (ACL), San Francisco, USA, pp. 360–367 (2002)
12. Boersma, P.: Praat, a system for doing phonetics by computer. Glot International 5(9/10), 341,345 (2001)
13. Fox, A.V.: PLAKSS – Psycholinguistische Analyse kindlicher Sprechstörungen. In: Swets, Zeitlinger, Frankfurt, A.M. (eds.) Swets & Zeitlinger, Frankfurt a, Harcourt Test Services GmbH Frankfurt a.M, Germany (2002)
14. Matoušek, V., Mautner, P. (eds.): 10th International Conf. on Text, Speech and Dialogue (TSD). LNCS (LNAI), vol. 4629. Springer, Berlin (2007)

Syllable Based Language Model for Large Vocabulary Continuous Speech Recognition of Polish

Piotr Majewski

University of Łódź, Faculty of Mathematics and Computer Science
ul. Banacha 22, 90-238 Łódź, Poland
pmajewski@math.uni.lodz.pl

Abstract. Most of state-of-the-art large vocabulary continuous speech recognition systems use word-based n-gram language models. Such models are not optimal solution for inflectional or agglutinative languages. The Polish language is highly inflectional one and requires a very large corpora to create a sufficient language model with the small out-of-vocabulary ratio. We propose a syllable-based language model, which is better suited to highly inflectional language like Polish. In case of lack of resources (i.e. small corpora) syllable-based model outperforms word-based models in terms of number of out-of-vocabulary units (syllables in our model). Such model is an approximation of the morpheme-based model for Polish. In our paper, we show results of evaluation of syllable based model and its usefulness in speech recognition tasks.

Keywords: Polish, large vocabulary continuous speech recognition, language modeling, sub-word units, syllable-based units.

1 Language Modeling

Statistical language modeling is based on predicting the next word (or sub-word unit) in context of previous words. Such modeling is surprisingly efficient in the fields of automatic speech recognition. However, there is a limit of using very broad context of previous words and very large corpora. The performance of n-gram language model, which is based on statistics of sequences of different word forms, is not satisfied in case of inflectional languages (like Polish and other Slavic languages) or agglutinative languages. The problem lies in multiple forms of a single lexeme in case of inflectional languages and a great freedom in a word creation in case of agglutinative languages. To adjust language models for such languages one should use a huge corpus containing all of the possible inflectional or derivational forms of words. Otherwise, there may be a lot of unknown words (which did not occur in a training set) during model evaluation process. The same condition of training set is required in languages, which are characterized by free words order, e.g. Polish. In this case, the training set should contain the largest possible words order variants set. But even in large corpora many n-grams or word forms have statistically non-significant frequency.

P. Sojka et al. (Eds.): TSD 2008, LNAI 5246, pp. 397–401, 2008.
© Springer-Verlag Berlin Heidelberg 2008

2 Sub-Word Based Models

The answer for part of the earlier mentioned shortcomings is a model based on sub-word units. For Slavic languages, morpheme-based models has been proposed [2,6]. They are based on decomposition of words into morphemes: stems and ending, thus linguistics-based algorithms are needed for enabling such decomposition. The number of possible morphemes is always smaller than the number of possible words, thus their average frequency in the corpus is higher than average frequency of word in the same corpus. It allows for reduction of out-of-vocabulary ratio and to limit the search space in recognition tasks using such language models.

Syllables are another promising sub-word units, especially for syllable-based languages, like Chinese [10]. In [4], a mixed, syllable- and word-based model was used for recognition of German. Syllable could be considered the natural sub-word unit in Polish too, although linguists do not consider syllable a natural functional unit in Polish, but the conventional way of segmentation of speech only [7]. Syllable-based n-gram model has shorter context than word-based model with identical order n. Such context can include only one or two words. Short segments like syllables could potentially be identical with morphemes (especially endings, which are usually shorter), giving us morphology-based statistical language model rather than syntax- or semantic-based n-gram word model.

In this paper we investigate language models based on the following scheme of a syllable in Polish [7]:

$$[C]^*V[C]^* \ . \tag{1}$$

As we can see a syllable consists of onset (0 or more consonants), nucleus (vowel) and coda (0 or more consonants). It is trivial to point out the nuclei in a word, but selection of borders between syllables could have some variants. We chose the simple set of rules based on [11]: single consonant between vowels belongs to following syllable; group of consonants should be divided such that following syllable should get maximum possible number of consonants, but they should be allowed in a word onset in Polish.

3 Tools and Data

The tool used in experiments is open source SRI Language Modeling toolkit [9]. The corpus used for construction of the n-gram based language model is the Enhanced Corpus of Frequency Dictionary of Contemporary Polish [5]. It consists of sample texts (essays, news, scientific texts, novels and dramas) published between 1963 and 1967. Every word in this corpus was annotated with corresponding lemma and grammatical codes. Corpus data contain about 540,000 words. After analysis of the corpus we noted it contains 85,373 different word tokens, as many as 36,239 (57.6%) were noticed only once. Analysis of grammatical codes included in the corpus showed it contains 38,883 lexemes, but many of them 18,128 (46.6%) are represented only once. This means that 20,755 lexemes is represented at least twice in a set of 49,134 inflected forms. Of course, such a small corpus does not allow us to measure probability of higher order n-gram possible in the language well. To adapt the model to various test sets the smoothing techniques should be used to model n-grams not present in corpus.

The word-based corpus has been further processed using simple word splitting algorithm described earlier, resulting in syllable-based corpus. After processing there were 8,489 different syllables in the corpus. 2,038 of them were present only once.

4 Experiments

4.1 Syllable-Based Model

To evaluate both word-based and syllable-based models the corpus was divided into training set (about 92% of samples) and test set (about 8% of samples). Training set was used to construct 2-gram, 3-gram, 4-gram and 5-gram syllable-based Katz backoff model with Good-Turing smoothing. The same training set was used to construct corresponding word-based n-gram models. The results obtained for both models are shown in Table 1.

Table 1. Perplexity for syllable-based and word-based models

N-gram order	Syllable-based model	Word-based model
2-gram	104.1240	2104.12
3-gram	64.1258	2097.86
4-gram	64.1569	1960.37
5-gram	66.7397	1960.95

Results show that such small corpus allows us to use only low order n-grams, because higher order n-grams are statistically non-significant.

Metric used in this evaluation is the perplexity, defined as:

$$\mathrm{Perp}_{N-\mathrm{gram}}\left(w_1, w_2, \ldots, w_W\right) = \left[\prod_{i=1}^{W} P\left(w_i | w_{i-1}, \ldots, w_{i-N+1}\right)\right]^{-\frac{1}{W}} . \quad (2)$$

Perplexity shows geometric mean of reciprocal of probability of a token in the test set giving the model, so these results are not comparable to each other, because of different sets of tokens and different number of token classes. The word-based test set consists of 44,669 tokens and syllable based test set consists of 100,150 tokens. It means statistical word consists of 2.242 syllables. To compare syllable-based and word-based perplexities we should "scale" the syllable-based perplexity by raising it to the power 2.242. Such scaled perplexities are shown in Table 2.

These results show very small dependency of syllables in the process of word composition. Such model is very robust and allocates high probability to nonsense words.

The resulted perplexity compared to perplexity in some English language models (170 for bigram, 109 for trigram [3]) is quite high, but it is more comparable to perplexity in some Finnish trigram language models (4,300 for word-based model and 15 for syllable-based model [8]). The corpus used in this evaluation is quite small, so using a bigger corpus could help obtain better results.

Table 2. Perplexity for word-based and scaled syllable-based models

Language model used	Word-based	Scaled syllable-based
2-gram	2104.12	33,369.45
3-gram	2097.86	11,255.57
4-gram	1960.37	11,267.81
5-gram	1960.95	12,310.32

The main advantage of syllable-based model is out-of-vocabulary unit ratio. 9.26% words from test set were not present in training set and only 0.17% syllables from test set were not present in training set.

4.2 Morpheme Approximation Model

As we proved, syllable-based model can not model dependencies between words well. Thus, we evaluated a "lexeme-based" model, in which all words were replaced with their lemmas. Of course, such a model can not be used directly in real speech recognition tasks. It could only be useful as a second stage in the multiple-pass recognition system. An evaluation of lexeme-based model could be used to compare resulted perplexity with other models. Order of lexemes shows us true dependency between words in a phrase, based on allowed syntax of Polish and semantic connections between words.

Table 3. Perplexity for word-based, lexeme-based and stem approximation models

Language model used	Perplexity
Word-based 2-gram	2104.12
Lexeme-based 2-gram	1256.53
Stem approximation syllable-based 2-gram	1291.40
Word-based 3-gram	2097.86
Lexeme-based 3-gram	1229.38
Stem approximation syllable-based 3-gram	1266.94

Table 3 shows significant improvement of perplexity of lexeme-based model, compared to word-based model.

We modified our syllable-based model to approximate it to the morpheme-based model. We constructed the n-gram model based on "approximate stems". For one-syllable words we assumed that such words consist only of a stem and their ending has zero length. For multi-syllable words we assumed that the last syllable is equivalent to ending and preceding syllables form the stem. We constructed a model based only on "approximate stems" (the model assigns the stem a probability based on previous stems), and we ignored the prediction of ending, to be able to compare results with previous models (Table 3). Results showed that model based on the stem approximation is comparable with the lexeme-based model.

4.3 Conclusions and Future Work

In this article we demonstrate that syllables are useful sub-word units in language modeling of Polish. Syllable-based model is a very promising choice for modeling language in many cases such as small available corpora or highly inflectional language. The main advantage of this model is a very small number of out-of-vocabulary units – syllables. Obtained perplexity, especially scaled to word average length is quite high so it is necessary to develop more sophisticated model than n-gram Katz back-off with Good-Turing smoothing. There are phenomena in inflectional languages such as agreement, which could not be modeled well by such simple n-gram model. The Factored Language Model [1] could help to model such phenomena better. The advanced syllable-based model could be very convenient candidate in the field of large vocabulary continuous speech recognition.

It would be interesting to compare our word-to-syllable splitting algorithm with an algorithm resulting in morpheme-based segmentation or algorithm of data-driven unsupervised automatic segmentation [8].

References

1. Bilmes, J., Kirchhoff, K.: Factored Language Models and Generalized Parallel Backoff. In: Human Language Technology Conference, Edmonton (2003)
2. Byrne, W., Hajič, J., Ircing, P., Krbec, P., Psutka, J.: Morpheme based language models for speech recognition of Czech. In: International Conference on Text Speech and Dialogue, Brno (2000)
3. Jurafsky, D., Martin, J.H.: Speech and Language Processing: An Introduction to Natural Language Processing, Computational Linguistics, and Speech Recognition. Prentice-Hall, Englewood Cliffs (2000)
4. Larson, M., Eickeler, S.: Using Syllable-based Indexing Features and Language Models to improve German Spoken Document Retrieval. In: Proceedings of the 8th European Conference on Speech Communication and Technology (EUROSPEECH), Geneva, pp. 1217–1220 (2003)
5. Ogrodniczuk, M.(ed): Enhanced Corpus of Frequency Dictionary of Contemporary Polish, `http://www.mimuw.edu.pl/polszczyzna/pl196x/index_en.htm`
6. Rotovnik, T., Sepesy, M.M., Kačič, Z.: Large vocabulary continuous speech recognition of an inflected language using stems and endings. Speech Communication 49, 437–452 (2007)
7. Sawicka, I.: Fonologia. In: Gramatyka współczesnego języka polskiego, t. Fonetyka i fonologia, Instytut Języka Polskiego PAN, Kraków (1995)
8. Siivola, V., Hirsimäki, T., Creutz, M., Kurimo, M.: Unlimited Vocabulary Speech Recognition Based on Morphs Discovered in an Unsupervised Manner. In: Proceedings of the 8th European Conference on Speech Communication and Technology (EUROSPEECH), Geneva, pp. 2293–2296 (2003)
9. Stolcke, A.: SRILM – an extensible language modeling toolkit. In: Proc. Intl. Conf. on Spoken Language Processing, Denver (2002)
10. Xu, B., Ma, B., Zhang, S., Qu, F., Huang, T.: Speaker-independent Dictation of Chinese Speech with 32K Vocabulary. In: Proc. Intl. Conf. on Spoken Language Processing, Philadelphia, vol. 4, pp. 2320–2323 (1996)
11. Wierzchowska, B.: Fonetyka i fonologia języka polskiego. Ossolineum, Wrocław (1980)

Accent and Channel Adaptation for Use in a Telephone-Based Spoken Dialog System

Kinfe Tadesse Mengistu and Andreas Wendemuth

Cognitive Systems Group, FEIT-IESK, Otto-von-Guericke University
Universitätsplatz 2, 39106 Magdeburg, Germany
{Kinfe.Tadesse,Andreas.Wendemuth}@ovgu.de

Abstract. An utterance conveys not only the intended message but also information about the speaker's gender, accent, age group, etc. In a spoken dialog system, these information can be used to improve speech recognition for a target group of users that share common vocal characteristics. In this paper, we describe various approaches to adapt acoustic models trained on native English data to the vocal characteristics of German-accented English speakers. We show that significant performance boost can be achieved by using speaker adaptation techniques such as Maximum Likelihood Linear Regression (MLLR), Maximum a Posteriori (MAP) adaptation, and a combination of the two for the purpose of accent adaptation. We also show that promising performance gain can be obtained through cross-language accent adaptation, where native German speech from a different application domain is used as enrollment data. Moreover, we show the use of MLLR for telephone channel adaptation.

Keywords: Accent adaptation, accented speech, channel adaptation, cross-language accent adaptation, MAP, MLLR, native speech.

1 Introduction

A spoken utterance naturally conveys speaker-specific information such as accent, gender, age group etc. along with the intended message. These by-product information can be utilized to improve speech recognition in spoken dialog systems where there is considerable mismatch between training and actual usage environments. For instance, a speaker-independent (SI) model built using speech samples from a large group of native speakers of English would perform very poorly when tested with non-native speakers with typical accent. In automatic speech recognition systems, a mismatch in accent between the speakers used in testing and training can lead to over 30% increase in word error rate (WER) [1]. It has also been reported in [2] that the word error rate is about 3–4 times higher on strongly Japanese-accented or Spanish-accented English speakers than on native speakers on the same task.

The performance of a speech recognition engine is also affected by the quality of speech data, which in turn is affected by speaking environment, background noise and the communication channel. For instance, the performance of a SI model trained on microphone-recorded data degrades considerably when used in a telephone-based spoken dialog system.

P. Sojka et al. (Eds.): TSD 2008, LNAI 5246, pp. 403–410, 2008.
© Springer-Verlag Berlin Heidelberg 2008

In this paper, we investigate various approaches to improve the performance of the speech recognition engine in our bilingual (German and English) telephone-based spoken dialog system. The experiments described are mainly the use of speaker adaptation techniques for the purpose of accent and channel adaptation.

The rest of the paper is organized as follows. In Section 2, a brief overview of previous studies in the domain of accent adaptation is given. Section 3 presents the baseline SI models for both English and German. A brief description of the speaker adaptation techniques used in this paper is given in Section 4. In Section 5 the data used in the experiments are described and in Section 6 we describe the experiments performed and the results obtained. Finally, concluding remarks are presented in Section 7.

2 Previous Work in the Field

Accent adaptation in a spoken dialog system is an important topic of research as people may need to communicate with a dialog system in a language which is not their native. A number of studies have been done in this topic and a brief overview of related researches in the field is presented below.

An obvious approach to obtain better performance for accented speech is to train an accent-specific model using speech samples from speakers with the target accent. It has been shown in [3] that training on a relatively small amount of German-accented English from the Verbmobil conversational meeting-scheduling task resulted in significant WER reduction than using a large amount of native English training material. In [3] and [4], training on a merger of in-domain native and accented data has been shown to improve performance on accented speech. It has also been shown in [4] that applying a few additional forward-backward iterations with accented data on a well-trained seed model improves performance.

The standard speaker adaptation techniques such as Maximum Likelihood Linear Regression (MLLR) and Maximum a Posteriori (MAP) adaptation are other popular approaches used to deal with accented speakers. MLLR adaptation has been successfully used in [3] on German-accented English and [4] on Japanese-accented English. However, both used a single global transform for all models. Another study in [5] used 65 separate transforms but on individual test speakers and not on a group of target users.

On the German-accented English task described in [3], it is shown that MAP performs better at decreasing WER than MLLR when more adaptation data is available. However, it has not been shown whether combining MAP and MLLR could yield further performance gain.

Using the native language of accented speakers as enrollment data to adapt the parameters of a SI acoustic model trained on native English has also been explored in [3,4], and [6]. However, in [4] using native data for adaptation did not improve recognition and in some cases it resulted in large performance degradation. In [3], the use of bilingual model; i.e., a model trained on a merger of data from large amount of native English and native German that share a common phoneme set was investigated and only slight improvement in recognition was observed. In [6] accent adaptation without accented data was investigated to adapt an English model to Cantonese-accented English and promising improvement in phoneme accuracy has been reported.

As can be observed, previous applications of MLLR to a group of German-accented speakers, use only a single global transform to adapt all models. It has also not been shown whether combining MAP and MLLR could be more useful. Therefore, we show that MLLR with multiple transforms gives improved results for accent adaptation. We also show that using MLLR transformed model as an informative prior for MAP adaptation boosts performance when enough enrollment data is available. Moreover, cross-language accent adaptation is further investigated where native German speech from a different domain is used to adapt a SI native US-English model and promising results are obtained. We also try to capture variability due to both gender and accent by adapting separate native gender-dependent models to the target accent. Finally, we show the performance gain obtained by using MLLR adaptation technique to adapt acoustic models trained on microphone-recorded data to the charactersitcs of the telephone channel.

3 Speaker Adaptation Techniques

The accent-adapted models are generated using supervised, Maximum Likelihood Linear Regression (MLLR), and Maximum a Posteriori (MAP) adaptation techniques implemented in HTK [7].

3.1 Maximum Likelihood Linear Regression (MLLR)

Maximum Likelihood Linear Regression (MLLR) [8] is a transformation-based method that uses adaptation data to estimate linear transformations of model parameters to maximize the likelihood of the adaptation data. If the adaptation data is limited, a single global transform can be used to adapt all Gaussians; otherwise, when more data is available and more rigorous transformation is required, it is possible to cluster acoustically similar model parameters together into regression classes. Then a separate transform can be estimated for each class of Gaussians. One of the virtues of MLLR is that it makes adaptation of distributions for which there were no observations in the adaptation data possible.

The transformation of the transition probabilities and the mixture component weight will have little effect on the final performance [9]. However, transformation of the diagonal covariance matrix can give significant performance improvement.

3.2 Maximum a Posteriori (MAP) Adaptation

Maximum a Posteriori (MAP) adaptation is a model-based approach that maximizes the a posteriori probability using prior knowledge about the model parameter distribution. Given good informative priors and large amount of adaptation data, MAP can perform better than MLLR. MAP re-estimates model parameters and more data is required for better re-estimation of model parameters. Using MLLR transformed means as the priors for MAP adaptation, can yield further performance gain.

4 Baseline Models

The English baseline speaker-independent acoustic model is built using 15 hours of native US-English telephone speech spoken by 100 speakers (72 female and 28 male) from the DARPA 2001 Communicator Evaluation Corpus [10].

The speech data used for building the baseline German acoustic model is a German domain-dependent speech database ©ERBA; "Erlanger Bahn Anfragen", from Bavarian Archive for Speech Signals (BAS).[1] The data consists of 15 hours of speech spoken by 101 speakers (40 female and 61 male) recorded with close-talking microphone in an office environment in the domain of train information inquiries.

For both systems, we used Mel-Frequency Cepstral Coefficients (MFCCs) as acoustic features. The MFCCs are computed by performing pre-emphasis on the acoustic waveform, dividing the incoming waveform into overlapping blocks of 20 ms, multiplying each block by a Hamming Window, followed by removing the DC offset from each windowed excerpt of the waveform. Then the Fast Fourier Transform (FFT) of the windowed signal is calculated and the power spectrum is fed into a series of 24 Mel-Frequency filterbank channels. Discrete Cosine Transform (DCT) is applied to the logarithm of the filterbank outputs. Finally, the first (Δ) and second ($\Delta\Delta$) time differences between parameter values over successive frames are computed to better model temporal variation of the speech spectrum. As a result, a feature vector is generated every 10ms containing 39 dimensions comprising 13 cepstral components including the 0th order coefficient and the corresponding delta (Δ) and delta-delta ($\Delta\Delta$) coefficients. Cepstral Mean Subtraction (CMS) is applied to the German features as the training utterances are long enough[2] to benefit from CMS. It has been reported in [11] that for CMS to be useful utterances must be longer than 2–4 seconds. As a significant amount of utterances in the English corpus are very short, we do not apply CMS to the English features.

As the vocabularies used in the application domains are fairly limited (about 1,200 distinct words for English and about 950 unique words for German), a back-off bigram language model is built on the transcriptions of the training utterances.

In the experiments described here, we used well-trained 3-state context-independent monophone models with 32 Gaussians per state. It has been reported in [12] that triphones trained on native speech are not appropriate for use with non-native speakers. In fact, our preliminary experiments also revealed that monophones outperform triphones in recognizing non-native speech.

5 Data Preparation

We recorded German-accented English speech from 15 male and 15 female native German speakers over the telephone. The enrollment data consists of 600 utterances from 10 male and 10 female speakers.[3] The test-set also consists of 600 utterances

[1] `http://www.phonetik.uni-muenchen.de/Bas/`

[2] Average duration of an utterance is ERBA database is 5.3 sec.

[3] 30 utterances from each speaker.

recorded from 5 male and 5 female speakers.[4] The prompts for the enrollment and test data are drawn from the transcriptions of the native English training and test data, respectively.

We also recorded 600 German utterances from 10 male and 10 female native German speakers over the telephone using prompts from the transcriptions of the test-set of the German speech corpus. These data are used for two purposes – namely, for cross-language accent adaptation and testing the performance of the simulated SI model on actual telephone speech (Section 6.2). In addition, for the purpose of adapting simulated telephone quality speech to the telephone channel, we recorded 300 utterances from 5 male and 5 female native German speakers over the telephone. The prompts are drawn from the transcriptions of the German training set.

6 Experiments and Results

6.1 Accent Adaptation

In order to deal with the variability due to gender as well, we use gender-dependent models in addition to the speaker-independent model.

Table 1 summarizes the performance of the SI model, and the gender-dependent baseline models on the German-accented English test data. The word error rate of the SI monophone model on a same-sized test-set from 5 male and 5 female native speakers with the same 600 utterances is 27.29%.

Table 1. Performance of the baseline models on accented speech

Acoustic Model	% WER
Speaker-Independent (SI)	48.0
Gender-Dependent (Female)	49.29
Gender-Dependent (Male)	42.94

As can be observed in Table 1, the performance of the baseline models on accented speech is very poor. This is clearly attributed to the mismatch between the training and test data. Therefore, the following series of experiments are performed in order to improve the performance of the models on accented speech.

To start with, we combined the German-accented English adaptation data with the native English training set and re-trained a SI acoustic model on the merger. The resulting model gives 36.21% WER yielding 11.79% absolute (24.56% relative) WER reduction.

Then, we investigated the speaker adaptation techniques – namely, MLLR, MAP and MLLR followed by MAP, where we use German-accented English as adaptation data. For MLLR, optimal performance was obtained with 42 Regression classes where both means and diagonal covariances are transformed. Table 2 summarizes the results obtained.

[4] 60 utterances from each speaker.

Table 2. Performance after MLLR, MAP and MLLR+MAP adaptation

Acoustic Model	MLLR (%WER)	MAP (%WER)	MLLR+MAP (%WER)
Speaker-Independent (SI)	30.41	35.11	27.54
Gender-Dependent (Female)	29.64	40.65	29.64
Gender-Dependent (Male)	21.35	36.38	21.01

As can be seen in Table 2 MLLR alone resulted in 17.59% absolute (36.64% relative) WER reduction with the SI model. MAP alone gives a relatively smaller improvement as the adaptation data is not large enough. However, applying MAP on the MLLR transformed models resulted in 20.46% absolute (42.62% relative), 21.93% absolute (51.07% relative), and 19.65% absolute (39.86% relative) WER reduction with SI, male and female models, respectively. However, MAP after MLLR did not improve performance significantly to the gender-dependent models, this is most likely due to data insufficiency as the adaptation data is split into two to adapt the two gender-dependent models.

We believe that the useful information for accent adaptation can also be captured from one's native speech. Therefore, the use of native German speech data to adapt native English acoustic model to German accent is investigated. In this paper, we refer to this technique as cross-language accent adaptation.

In order to use cross-language accent adaptation, we first constructed an approximate mapping between the phoneme sets of German and English. We then built an auxiliary pronunciation dictionary that defines the pronunciation of the German words in the adaptation set with English phonemes. To produce monophone transcriptions, we force-aligned the adaptation data using the English SI model and the auxiliary dictionary. Then we performed MLLR adaptation with 2 global transforms[5] where both means and covariances are transformed. Table 3 summarizes the results obtained.

Table 3. Performance after cross-language accent adaptation using MLLR, MAP and MLLR+MAP

Acoustic Model	MLLR (%WER)	MAP (%WER)	MLLR+MAP (%WER)
Speaker-Independent (SI)	39.01	43.88	35.83
Gender-Dependent (Female)	40.0	45.50	36.75
Gender-Dependent (Male)	30.56	40.61	29.49

As can be seen in Table 3, when MAP is applied on MLLR transformed means and covariances we obtain, 12.17% absolute (25.35% relative), 13.45% absolute (31.32% relative), and 12.54% absolute (25.44% relative) WER reduction for the SI, male and female models, respectively. In fact, the improvement we could get using cross-language approach is relatively less than what we achieved using with-in language adaptation as the phoneme mapping is not accurate as some German phonemes do not have a counterpart in English and vice versa. Even then, the results are promising and may be helpful in cases where it is hard to collect accented speech in a given application domain.

[5] Only 2 transforms; i.e., one for silence and another for speech models, give optimal performance.

6.2 Channel Adaptation

In the absence of enough amount of telephone-recorded data in a given application domain, one can use simulated telephone quality speech to train SI models for use in a telephone-based spoken dialog system.

Simulating telephone quality speech involves introducing the obvious effects of the telephone channel into the microphone-recorded data. These include downsampling the audio data to 8 kHz and applying a low-pass filter with a cutoff frequency of 3,400 Hz and a high-pass filter of 300 Hz to approximate the band-limiting effects of the telephone channel. Furthermore, in order to approximate the loss due to the logarithmic encoding in the telephone channel, the 16-bit quantized signals are converted to a-law companded signal and back to linearly quantized 16-bit signal. A SI model is trained on the simulated training data. Table 4 shows the performance of the model.

Table 4. Performance on "simulated" and "actual" telephone speech

Test Data	% WER
Simulated telephone quality speech	15.86
Actual telephone speech	23.80

As can be seen, the performance degrades significantly when the model is tested with actual telephone speech. This could be due to some "uncaptured" effects of the telephone channel. To compensate for this we applied MLLR adaptation using the adaptation data described in Section 5 and obtained 4.39% absolute (18.44% relative) WER reduction with 42 regression classes where both mean and diagonal covariance parameters are transformed.

7 Conclusions

In this paper, we described the use of speaker adaptation techniques to compensate for the degradation due to accent and channel mismatch. It has been shown that both within-language and cross-language accent adaptation techniques are effective methods to deal with accented speech. Cross-language accent adaptation can be particulary useful when accented data are unavailable. It has also been shown that multiple transforms, and transformations of both mean and diagonal covariance are more productive. Furthermore, applying MAP on top of MLLR transformed models give the best result in cases where we have enough adaptation data. The performance gain due to MLLR adaptation to compensate for channel mismatch has also been described.

References

1. Huang, C., Chang, E., Chen, T.: Accent Issues in Large Vocabulary Continuous Speech Recognition. Microsoft Research China, Technical Report, MSR-TR-2001-69 (2001)
2. Tomokiyo, L.M.: Recognizing Non-native Speech: Characterizing and Adapting to Non-native Usage in Speech Recognition. Ph.D. thesis, Carnige Mellon University (2001)

3. Wang, Z., Schultz, T., Waibel, A.: Comparison of Acoustic Model Adaptation Techniques on Non-native Speech. In: IEEE International Conference on Acoustics, Speech, and Signal Processing (ICASSP), pp. 540–543 (2003)
4. Tomokiyo, L.M., Waibel, A.: Adaptation Methods for Non-native Speech. In: Proceedings of the Workshop on Multilinguality in Spoken Language Processing, Aalborg (2001)
5. Huang, C., Chang, E., Zhou, J., Lee, K.: Accent Modeling Based on Pronunciation Dictionary Adaptation for Large Vocabulary Mandarin Speech Recognition. In: Proceedings of International Conference on Spoken Language Processing (ICSLP), pp. 818–821 (2000)
6. Liu, W.K., Fung, P.: MLLR-Based Accent Model Adaptation without Accented Data. In: Proceedings of International Conference on Spoken Language Processing (ICSLP), pp. 738–741 (2000)
7. Young, S., Evermann, G., Gales, M., Hain, T., Kershaw, D., Liu, X.A., Moore, G., Odell, J., Ollason, D., Povey, D., Valtchev, V., Woodland, P.: The HTK Book. Revised for HTK Version 3.4. Cambridge University Engineering Department, Cambridge (2006)
8. Leggetter, C., Woodland, C.P.: Flexible Speaker Adaptation Using Maximum Likelihood Linear Regression. In: Proceedings of Eurospeech 1995, pp. 1155–1158 (1995)
9. Leggetter, C., Woodland, C.P.: Maximum Likelihood Linear Regression for Speaker Adaptation of Continuous Density Hidden Markov Models. Computer Speech and Language 9, 171–185 (1995)
10. Walker, M., Aberdeen, J., Sanders, G.: 2001 Communicator Evaluation. Linguistic Data Consortium, Philadelphia (2003)
11. Alsteris, L.D., Paliwal, K.K.: Evaluation of the Modified Group Delay Feature for Isolated Word Recognition. In: Proceedings of International Symposium on Signal Processing and Its Applications (ISSPA), pp. 715–718 (2005)
12. He, X., Zhao, Y.: Model Complexity Optimization for Non-native English Speakers. In: Proceedings of Eurospeech 2001, vol. 2, pp. 1461–1464 (2001)

Efficient Unit-Selection in Text-to-Speech Synthesis

Aleš Mihelič and Jerneja Žganec Gros

Alpineon d.o.o.,
Ulica Iga Grudna 15, 1000 Ljubljana, Slovenia
{ales,jerneja}@alpineon.si
http://www.alpineon.si

Abstract. This paper presents a method for selecting speech units for polyphone concatenative speech synthesis, in which the simplification of procedures for search paths in a graph accelerated the speed of the unit-selection procedure with minimum effects on the speech quality. The speech units selected are still optimal; only the costs of merging the units on which the selection is based are less accurately determined. Due to its low processing power and memory footprint requirements, the method is applicable in embedded speech synthesizers.

1 Introduction

In unit-selection speech synthesis, limitations in computational processing power and memory footprint used in embedded systems affect the planning of the unit-selection process. The selection of speech units is the part of concatenative or corpus-based speech synthesis that can exert the most influence on the speed of the entire speech synthesis process.

It is necessary to find a favorable compromise between the size of the speech corpus and the computational complexity of the unit-selection procedure. If the unit-selection procedure is very simplified and thus also very fast, a selection of units in a larger speech corpus can be performed in the same amount of time. Oversimplification of the procedure can, however, result in the selection of inappropriate speech units and therefore reduce the speech quality despite using a larger corpus.

The paper is structured in the following way. In Section 2, unit-selection is introduced as a graph-search problem. An overview of unit-selection methods is presented. The unit-selection procedure with which we succeeded in accelerating the speed of the procedure without significantly affecting the speech quality is presented in Section 3. This is achieved by simplifying the calculation of the concatenation cost and thus creating conditions enabling a specific structure of the algorithm for finding the optimal path in the graph. The evaluation of the speed and the speech quality of the proposed unit-selection procedures is presented in Section 4.

2 Unit-Selection

The problem of finding the optimal sequence of recorded units for quality speech signal synthesis can be presented as finding an optimal path in a graph. This kind of presentation clearly demonstrates the problem of selecting speech units and, at the same time,

P. Sojka et al. (Eds.): TSD 2008, LNAI 5246, pp. 411–418, 2008.
© Springer-Verlag Berlin Heidelberg 2008

enables the use of recognized procedures for solving this problem. Each vertex of the graph represents a basic speech unit from the speech corpus. The basic speech segments may be allophones, diphones, triphones, or any other basic speech unit. The graph is divided into individual levels. The first level contains the initial vertices; that is, all basic speech units in the speech corpus that correspond to the first basic speech unit in the input character sequence that needs to be synthesized.

The edge between the vertices determines the possibility of merging the basic speech units represented by the connected vertices. In this kind of graph, finding the optimal speech unit sequence can be defined as finding the optimal path between any initial vertex in the graph and any final vertex in the graph, whereby the edges between the graph's vertices determine the possible paths.

2.1 The Concatenation Cost

A speech signal is formed by merging or concatenating speech units from the prerecorded speech corpus. During the process of merging, audible speech signal discontinuities can occur. We try to evaluate the influence of signal discontinuity on the speech quality through the cost of concatenation.

There are several possible approaches to evaluating the influence of concatenation on the speech quality. The simplest method is to define the cost as "0" for concatenating speech units that directly follow one another in the speech corpus, and to define the cost as "1" for all other speech unit combinations. The use of the cost "0" in units that directly follow one another in the speech corpus is logical because they are already linked together and therefore merging is not necessary. With the use of the cost "1" in units that do not follow one another in the speech corpus, all the concatenations were equally evaluated, regardless of the characteristics of the units being merged. With this kind of concatenation cost, the procedure for finding the optimal speech unit sequence would select the sequence with the smallest number of mergers, regardless of the type of speech units.

A better evaluation of the influence of concatenation on speech quality is achieved if the cost of speech unit merging depends on the allophones that are concatenated. Similar to the previous approach, the cost "0" is defined for the merging of speech units that directly follow one another in the speech corpus.

The most accurate evaluation of the influence of concatenation on speech quality is achieved by taking into account the phonetic features of both units merged when calculating the concatenation cost. A great deficiency of this method of determining the cost of merging is its numeric complexity. As concatenation costs are determined individually for every pair of basic speech units from the speech corpus, they are impossible to calculate in advance.

To solve this problem, we propose a compromise solution that is considerably faster, and nonetheless partly takes into account the phonetic features of concatenated speech units, is determining the concatenation cost in advance for the individual groups of basic speech units from the speech corpus. In this approach, all the basic speech units in a speech corpus are classified into groups on the basis of their phonetic features such that the speech units within an individual group phonetically resemble one another to the best extent possible. This is achieved by using clustering techniques. The concatenation costs are calculated in advance for all group combinations and saved.

3 Speech Unit Selection Method with a Simplified Cost of Merging

This section proposes a simplified speech unit selection method that is very fast and thus appropriate for implementation in embedded systems. The basic simplification in this method is that the cost of merging two speech segments depends only on the phonemes that are being joined by merging. If merging is carried out at the center of the phonemes, such as in diphonic synthesis, the cost of merging for each phoneme is defined in the center of the phoneme.

If merging is carried out at the phoneme boundaries, the cost of merging must be defined for all the sequences of two phonemes that can occur in speech. These costs of merging can be defined in advance and are not calculated during synthesis. In addition to these costs of merging, it is presumed that the cost of merging equals "0" if the segments that are being merged directly follow one another in the speech corpus, regardless of the phonemes joined at the concatenation point.

The graph used in speech unit selection is created as described in the previous paragraph. It comprises N levels, whereby each level corresponds to exactly one basic speech segment in the input sequence that is to be synthesized. At level k of the graph, which corresponds to the speech segment S_k, q_k vertices are located; at level $k+1$, which corresponds to the speech segment S_{k+1}, q_{k+1} vertices are located; and so forth.

Every vertex E_k^i $(1 \le i \le q_k)$ at level k of the graph represents a specific recording of the speech segment S_k in the speech corpus. For every vertex E_k^i, the cost of fit of the prosodic features of the corpus speech segment represented by the vertex is also calculated, as well as the required prosodies for the speech segment S_k in the input sequence. This cost is labeled $C^P(E_k^i)$. The vertices are connected by linking every vertex E_k^i $(1 \le i \le q_k)$ at level k with all the vertices E_{k-1}^j $(1 \le j \le q_{k-1})$ at level $k-1$. The cost of the connection between vertices E_k^i and E_{k-1}^j equals the cost of merging the speech corpus segments represented by the vertices. This is labeled $C_L(E_{k-1}^j, E_k^i)$.

In finding the optimal path in the graph it must be established which path between any initial vertex E_1^i $(1 \le i \le q_1)$ and any final vertex E_N^i $(1 \le i \le q_N)$ of the graph has the lowest cost.

The cost of the entire path is calculated by adding the costs of merging or the costs of edges between the vertices traversed (C_L), and the costs of fit of the prosodic features or the costs of the vertices visited (C_P). Thus, at every level k $(1 \le k \le N)$ of the graph only one of the vertices E_k^i $(1 \le i \le q_k)$ must be selected, or only one of the speech segments in the speech corpus that will be used in speech synthesis. This vertex is labeled $E_k^{x(k)}$. The cost of the optimal path in the graph can be expressed as:

$$C = \min_{x(1),x(2),\ldots,x(N)} \left(C_P(E_1^{x(1)}) + \sum_{k=2}^{N} C_P(E_k^{x(k)}) + C_L(E_k^{x(k)}, E_{k-1}^{x(k-1)}) \right). \quad (1)$$

The cost of the optimal path as the function of selecting a vertex $x(k)$ at the individual level of the graph is a decomposable function. If the cost of the optimal path between the graph's initial vertices and the vertex E_k^i at level k of the graph is labeled $C_O(E_k^i)$, and if the cost of the optimal path between the graph's initial vertices and any k-level vertex is labeled C_k, the following applies:

$$C_k = \min_{x(k)} \left(C_O(E_k^{x(k)}) \right) \tag{2}$$

and

$$C_O(E_k^i) = C_P(E_k^i) + \min_{x(k-1)} \left(C_L(E_k^i, E_{k-1}^{x(k-1)}) + C_O(E_{k-1}^{x(k-1)}) \right). \tag{3}$$

It can be seen that the function of the cost can be defined recursively or that the cost of the path to vertex E_k^i at level k of the graph depends only on the cost of the prosodic fit for vertex E_k^i and the costs of optimal paths to the vertices of the previous level $(C_O(E_{k-1}^i))$, to which the costs of merging are added.

In optimizing such a function, dynamic programming can be used to find the optimal path in the graph. This method simplifies the search for the optimal path by dividing it into searches for partial optimal paths for every level of the graph.

In practice, the procedure is designed such that four parameters are defined for every vertex of the graph. The first, parameter $I(E_k^i)$, is an index of the basic speech unit in the speech corpus represented by the vertex. This parameter is already defined for the vertex at the start of the procedure, when the graph is being created. The second parameter equals the cost of fit of prosodic features $C_P(E_k^i)$, which is also calculated when creating the graph. The third parameter equals the lowest cumulative cost or the lowest cost of the path between any initial vertex and the current $C_O(E_k^i)$ vertex. This cost is calculated during the optimal path calculation procedure. The fourth parameter is an index of the vertex $P(E_k^i)$ from the previous level of the graph located on the optimal path between the initial vertices and the current vertex. This parameter is also calculated during the graph search procedure.The procedure begins by defining the cost of fit of the prosodic features of the same vertices for the lowest cumulative cost of initial vertices:

$$C_O(E_1^i) = C_P(E_1^i), \quad (1 \le i \le q_1). \tag{4}$$

In the initial vertices, the indicator of the vertex from the previous level of the graph is set to "0" because initial vertices have no precursor. Then the lowest cost of the path to individual vertices at the second level of the graph is defined:

$$C_O(E_2^i) = C_P(E_2^i) + \min_{j=1}^{q_1} \left(C_L(E_2^i, E_1^j) + C_O(E_1^j) \right), \quad (1 \le i \le q_k). \tag{5}$$

In addition, the (j) index of the vertex at the previous level of the graph located on this path with the lowest cost is recorded. This procedure is repeated sequentially for all remaining levels of the graph:

$$C_O(E_k^i) = C_P(E_k^i) + \min_{j=1}^{q_{k-1}} \left(C_L(E_k^i, E_{k-1}^j) + C_O(E_{k-1}^j) \right), (1 \le i \le q, 2 \le k \le N). \tag{6}$$

The cost of the optimal path is the lowest among the costs of optimal paths to individual final vertices of the graph:

$$C = \min_{j=1}^{q_N} \left(C_O(E_N^j) \right). \tag{7}$$

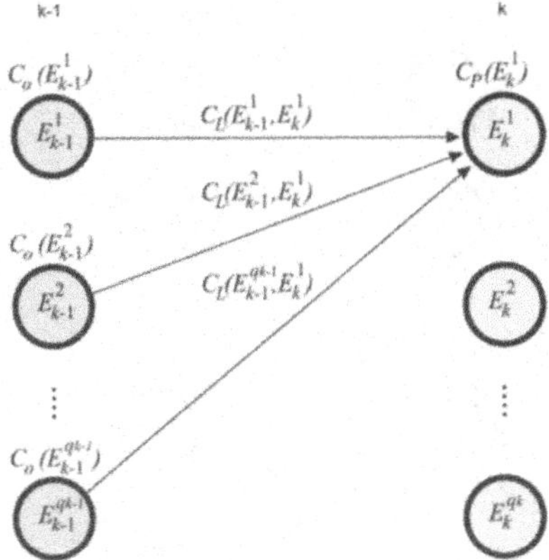

Fig. 1. The cost of the optimal path to vertex E_k^i depends on the costs of optimal paths to the vertices at the graph's previous level $C_O(E_{k-1}^i)$, the costs of merging $C_L(E_{k-1}^j, E_k^i)$, and the cost of fit of the prosodic features $C_P(E_k^i)$; k represents the level of the graph

The optimal final vertex is the final vertex with the lowest cumulative cost. After the procedure is concluded, the sequence of vertices located on the optimal path is compiled by tracing in reverse the indices of vertices at the previous levels of the graph $P(E_k^i)$ that were saved during the procedure.

With the simplification of the cost of merging introduced in this procedure, the concatenation costs can be determined in advance, so that the cost of merging $C_L(E_{k-1}^j, E_k^i)$ depends only on the type (phonetic group) of speech segments S_k and S_{k-1}. This also means that all the costs of the edges between the vertices of the graphs E_{k-1}^j and E_k^i are the same for any j and i. This does not apply only if the speech segments represented by vertices E_{k-1}^j and E_k^i, directly follow one another in the speech corpus. In this case, the cost of merging equals 0.

This means that the calculation of the lowest cost of the path can be further simplified. The recursive equation for calculating the lowest cost of the path to vertex E_k^i is shown in Equation (2).

Taking into account these simplifications and that $C_L(S_{k-1}, S_k)$ is always a positive number, Equation (3) can be rewritten as:

$$C_O(E_k^i) = C_P(E_k^i) + \begin{cases} \min\left(C_O(E_{k-1}^J), \min_{j=1}^{q_{k-1}} C_L(S_{k-1}, S_k) + C_O(E_{k-1}^j)\right), & \\ \qquad\qquad\qquad\qquad \text{if } \exists\, J; I(E_k^i) - I(E_{k-1}^J) = 1 \\ \min_{j=1}^{q_{k-1}} \left(C_L(S_{k-1}, S_k) + C_O(E_{k-1}^j)\right), & \text{otherwise} \end{cases} \tag{8}$$

Because the calculation of the minimum in the equation above does not depend on i, this calculation can be performed only once for all the vertices incident to the same level S_k of the graph. Equation (8) can now be expressed as:

$$C_O(E_k^i) = C_P(E_k^i) + \begin{cases} \min\left(C_O(E_{k-1}^J), C_0'(S_k)\right), & \text{if } \exists\, J; I(E_k^i) - I(E_{k-1}^J) = 1 \\ C_0'(S_k), & \text{otherwise} \end{cases} \tag{9}$$

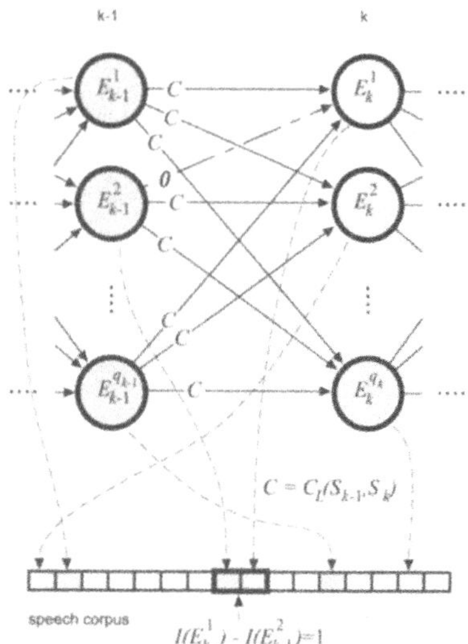

Fig. 2. The cost of merging two speech segments directly following one another in the speech corpus equals 0; *k* represents the level of the graph. The concatenation costs of all other segments depend only on the type (group) of speech segments that are being merged, and are therefore the same for all the connections between two levels of the graph.

It can be established that, by using the unit-selection procedure with the simplified concatenation cost described above, only one calculation of the minimum is required for every level of the graph, and only one sum and one comparison for every vertex of the graph. The time required to calculate the optimal path increases almost linearly with the increase in the size of the speech corpus.

4 Evaluation

Two versions of embedded concatenative speech synthesis for Slovenian using two different methods for selecting speech units were compared according to computational speed. The underlying speech corpus, which was used in both experiments, is described in [9].

The first version used a unit-selection method with a simplified cost of merging described to select speech units described in Section 3, while the second used the widely-used simplified search method for speech unit selection, which does not search through all the possible paths in the graph, but limits itself only to the most promising ones. Using the second method, the quality of synthesized speech was slightly lower than the quality of the synthesized speech in the first procedure.

The search time in both speech unit selection methods increases linearly with the length of the utterance that is to be synthesized, and also linearly with the size of the speech corpus, which is an improvement compared to traditional procedures for find-ing paths in the graph used in speech unit selection of concatenative or corpus speech synthesis. As anticipated, both methods operated increasingly more slowly when in-creasing the sentence size for which they were seeking the segments required for synthesis. The simplified search method for speech unit selection is faster than the

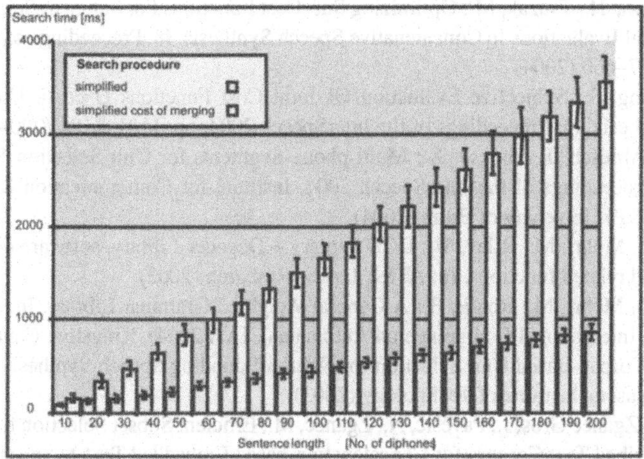

Fig. 3. Comparison of computational speed for two unit-selection search methods. As anticipated, the simplified search method for speech unit selection is faster than the speech unit selection method with a simplified cost of merging described in Section 3 because it does not search through all the possible paths in the graph, but limits itself only to the most promising ones. The search speed increases with the length of the sentence for which the procedure must find a suitable speech unit sequence in the speech corpus.

speech unit selection method with a simplified cost of merging described in Section 3 because it is less complex. As shown in Figure 3, it can find segments for synthesizing shorter utterances twice as fast, and segments for longer sentences four times as fast.

5 Conclusion

Limitations in computational processing power and memory footprint used in embedded systems affect the planning of the unit-selection process. This article presents a new method for selecting speech units in polyphone concatenative speech synthesis, in which simplifications of procedures for finding the path in the graph increase the speed of the speech unit-selection procedure with minimum effects on the speech quality. The units selected are still optimal; only the costs of merging the units on which the selection is based are less accurately determined. Further evaluations in terms of measuring computational speed and assessment of the resulting speech quality by comparing the proposed method to other speech unit selection methods are planned in future.

Due to its low computational speed and memory footprint requirements, the method is suitable for use in embedded speech synthesizers.

References

1. Vesnicer, B., Mihelič, F.: Evaluation of the Slovenian HMM-Based Speech Synthesis System. LNCS, pp. 513–520. Springer, Heidelberg (2006)
2. Campbell, W.N.: Processing a speech corpus for CHATR synthesis. In: Proceedings of the ICSP, Seul, Korea, pp. 183–186 (1997)

3. Toda, T., Kawa, H., Tsuzak, M.: Optimizing Sub-Cost Functions For Segment Selection Based On Perceptual Evaluations In Concatenative Speech Synthesis. In: Proceedings of the ICASSP 2004, pp. 657–660 (2004)
4. Vepa, J., King, S.: Subjective Evaluation Of Joint Cost Functions Used In Unit Selection Speech Synthesis. In: Proceedings of the InterSpeech 2004, pp. 1181–1184 (2004)
5. Breuer, S., Abresch, J., Phoxsy, X.: Multi-phone Segments for Unit Selection Speech Synthesis. In: Proceedings of the InterSpeech 2004, Institute for Communication Research and Phonetics (IKP) University of Bonn (2004)
6. Allauzen, C., Mohri, M., Riley, M.: DCD Library – Decoder Library, software collection for decoding and related functions. In: AT&T Labs – Research (2003)
7. Allauzen, C., Mohri, M., Roark, B.: A General Weighted Grammar Library. In: Proceedings of the Ninth International Conference on Automata (CIAA 2004), Kingston, Canada (2004)
8. Yi, J.R.W.: Corpus-Based Unit Selection for Natural-Sounding Speech Synthesis. Ph.D. Thesis, Massachusetts Institute of Technology (2003)
9. Mihelič, A., Žganec Gros, J., Pavešič, N., Žganec, M.: Efficient Subset Selection from Phonetically Transcribed Text Corpora for Concatenation-based Embedded Text-to-speech Synthesis. Inf. MIDEM 36(1), 19–24 (2006)

Prosodic Events Recognition in Evaluation of Speech-Synthesis System Performance

France Mihelič[1], Boštjan Vesnicer[1], Janez Žibert[1], and Elmar Nöth[2]

[1] University of Ljubljana, Faculty od Electrical Engineering,
Tržaška 25, SI-1000 Ljubljana, Slovenia
{mihelicf,bostjan.vesnicer,janez.zibert}@fe.uni-lj.si
[2] University Erlangen-Nürnberg, IMMD5
Martensstrasse 3, 91058 Erlangen, Germany
noeth@informatik.uni-erlangen.de

Abstract. We present an objective-evaluation method of the prosody modeling in an HMM-based Slovene speech-synthesis system. Method is based on the results of the automatic recognition of syntactic-prosodic boundary positions and accented words in the synthetic speech. We have shown that the recognition results represent a close match with the prosodic notations, labeled by the human expert on the natural-speech counterpart that was used to train the speech-synthesis system. The recognition rate of the prosodic events is proposed as an objective evaluation measure for the quality of the prosodic modeling in the speech-synthesis system. The results of the proposed evaluation method are also in accordance with previous subjective-listening assesment evaluations, where high scores for the naturalness for such a type of speech synthesis were observed.

Keywords: Speech synthesis, system evaluation, prosody.

1 Introduction

Modern speech synthesizers are able to achieve high intelligibility; however, they still suffer from rather unnatural speech. Recently, to increase the naturalness of speech-synthesis systems, there has been a noticeable shift from diphone-based toward corpus-based unit-selection speech synthesis [3]. In such systems the emphasis is more on engineering techniques (searching, optimization, statistical modeling) than on Linguistic-rule development [8]. Many of these algorithms are borrowed from the automatic-speech-recognition (ASR) community. of speech databases (e.g. [7]). However, to achieve a reasonable performance with such corpus-based systems large computational resources are required, and these are often not available when dealing with embedded systems.

In accordance with current trends an HMM-based approach to speech synthesis for Slovene was implemented. In contrast to other corpus-based speech-synthesis systems, a system using acoustic models trained on a large speech database using HMM techniques provides a relatively small memory footprint, comparable to – or even smaller than - the footprint of embedded diphone-based systems [5], and demands no special computational load.

P. Sojka et al. (Eds.): TSD 2008, LNAI 5246, pp. 419–426, 2008.
© Springer-Verlag Berlin Heidelberg 2008

Subjective listening-evaluation tests of the HMM-based system showed a surprisingly high level of prosodic quality for such synthesized speech, although no special prosody modeling was used in the system [13]. In our present research we tried to apply the established methods already used in ASR for prosodic-events recognition to evaluate the prosodic content in the synthesized speech.

The rest of the paper is organized as follows. Section 2 overviews the main idea behind the HMM-based speech-synthesis system. In Section 3 we present the prosodic labeling and databases used for the training and the evaluations. Section 4 gives a short description of the prosodic features and classification procedure for the detection of prosodic events in a speech-synthesis system. The results of the experiments are discussed in Section 5. Finally, concluding remarks and plans for future work are presented in the last section.

2 HMM-Based Speech Synthesis

The selected HMM-based approach for speech synthesis differs from other approaches because it uses the HMM framework not only for the segmentation and labeling of the database but also as a model for speech production. The method was originally proposed in [10] and later extended by Yoshimura et al. [14].

For a reliable estimation of the parameters of the statistical model a speech parametrization is required. Since we want to be able to synthesize high-quality speech, the parameters should contain enough information for the reconstruction of speech, which should be perceptually similar to the original. For this purpose, the *source-filter* theory of speech production [9] is applicable. In order to follow this theory, the parameters of the *vocal tract* transfer function and the *excitation* (f_0) need to be estimated. We used MFCC and log f_0 parameters along with their Δ and $\Delta\Delta$ counterparts.

The procedures for estimating the parameters of HMMs are well known from the field of ASR, and can also be applied in our case. A difference here is that additional parameters of the excitation should be modeled. For this purpose, new types of HMMs are introduced. They are called multi-space-distribution (MSD) HMMs, and are presented in [12].

For duration modeling we estimate the duration densities only when the training process is already finished [14]. Evaluation tests using this kind of duration prediction give comparable results [13] to the previously developed two-stage duration model using the diphone-concatenation technique [4].

A detailed description of the speech parametrization and the HMM structure that we used is given in [13].

3 Prosody Labeling

We have decided to investigate the possibility of generalizing the described speech-synthesis approach to enable some simple prosody modeling. Therefore, the same speech corpora that were used for the training of the HMM acoustic models used in the speech synthesis were manually annotated with syntactic-prosodic boundaries and word accents in the utterances. These corpora consist of 578 sentences (37 minutes of speech) uttered by the speaker 02m from the Slovenian Weather Forecast Speech Database (VNTV) [7]. Recently, when SiBN speech corpora were recorded and annotated [16] we acquired additional records of weather forecasts from the same speaker consisting of an additional 253 sentences (17 min of speech). These data were also prosodically annotated in the same manner. The syntactic-prosodic boundaries along the lines of [1] were annotated for the transliterations of speech utterances, and word accents were labeled via acoustic perceptual sessions. We used the same three-class annotation prosodic annotation for the prosody boundaries and the word accents as in our previously reported work [6] (Table 1 and Table 2).

Table 1. Syntactic-prosodic boundary labels

M3:	Clause boundaries
M2:	Constituent boundaries likely to be marked prosodically
M0:	Every other word boundary

Table 2. Word accent labels

PA:	The most prominent (primary) accent within a prosodic clause
SA:	All other accented words are marked as carrying secondary accent
UA:	Unaccented words

An example of the labeled text utterance is given below[1]:

V prihodnjih SA *dneh bo vroče* PA M3. *Ob koncu* SA *tedna* PA M2 *pa se bo vročini* SA *pridružila še* M2 *soparnost* SA M3.
English translation: *In the following days it will be hot. During the weekend the hot weather will be accompanied by sultry conditions.*

6363 syntactic-prosodic boundary markers and word-accent markers for VNTV database and 2940 of them for SiBN were determined. Frequencies for particular class markers are given in Table 3.

4 Prosodic Features Selection and Classification

Classification was performed using prosodic feature sets derived from duration segmental characteristics on the word level, speaking rate, energy and pitch. The duration and energy features were additionally normalized by applying normalization procedures based on statistical parameters that were estimated from the training data [2] (pp. 38–59).

The energy and pitch features were based on the short-term energy and f0 contour, respectively. Additionally, we used the mean and the median as features [2] (pp. 59–62).

[1] M0 and UA labels are not indicated in the example.

Table 3. Distribution of syntactic-prosodic boundary and word-accent markers in absolute figures for VNTV and SiBN databases

Syntactic-prosodic boundaries			Word-accent markers		
	VNTV	SiBN		VNTV	SiBN
M3	906	406	PA	906	407
M2	1343	538	SA	1433	750
M0	4114	1996	UA	4024	1783

We derived the same features set (95 features) that were proposed in the experiments on German speech [2] (page 103).

The prosody events were detected by support vector machines (SVMs). In our previous studies a classification with neural networks was also performed [6], but we gained a significant improvement in terms of recognition scores when SVMs were applied. The LIBSVM software [17] was used for the training and recognition tests. We used the RBF kernel with $C = 2^3$ and $\gamma = 2^{-5}$. All the data were linearly scaled into the $[-1, +1]$ range during the pre-processing stages.

5 Experimental Results

Our first experiments on the recognition of prosody events were made on the same data set (02m VNTV part) that was used for the training of the HMM-based Slovene speech-synthesis system. We used the data text transcription and the Viterbi forced alignment method to label the speech on the phoneme- and word-duration level. Afterwards, 95 dimensional feature vectors were computed for each word of the uttered speech.

5.1 Cross Validation on Natural Speech

The first evaluation was performed on natural speech. The cross-evaluation results gave an 81% average overall recognition rate and a 75% class-wise recognition rate for the detection of syntactic-prosodic boundaries, and a 69% overall recognition rate and a 66% class-wise recognition rate for word-accent detection. The results proved the consistency of the speech-data labeling procedure and the appropriateness of using SVM for classification process and served also to determine the applicable pair of RBF-kernel parameters reported in Section 4.

5.2 Recognition of the Prosodic Events in the Synthesized Speech Counterpart

In our next experiment a new database consisting of the synthesized speech generated from text transcriptions of speech-data recordings used for the training of the speech-synthesis system was acquired. In this way we obtained the synthesized-speech database counterpart of the original natural-speech database (02m VNTV part). We were also able to use the same prosodic markers as in the previous experiment. To determine the prosodic feature vectors for synthesized speech this database was also labeled on the phoneme- and word-duration levels. Surprisingly, the cross-evaluation check on this database gave entirely comparable results to those obtained from the natural-speech counterpart. And even when we were using the natural-speech database for training and its synthesized counterpart for testing, we still got relatively high recognition scores, with an 83.8% overall recognition rate, a 70.7% class-wise recognition rate for prosodic boundaries, and a 77.5% overall recognition rate and a 69.9% class-wise recognition rate for word accents. The confusion matrices for each class are given in Table 4[2] and Table 5.

Table 4. Confusion matrices for syntactic-prosodic boundaries recognition for synthesized speech. The recognition system was trained on the natural speech counterpart.

actual/predicted	M0	M2	M3
M0	89.0%	10.4%	0.6%
M2	19.8%	77.0%	3.2%
M3	5.9%	38.1%	46.0%

Table 5. Confusion matrices for word-accent recognition for synthesized speech. The recognition system was trained on the natural-speech counterpart.

actual/predicted	UA	SA	PA
UA	85.9%	10.4%	3.7%
SA	23.8%	66.6%	9.6%
PA	22.1%	20.7%	57.2%

The relatively high recognition results could be due to the method used to acquire the HMM acoustical models for sub-word units of the speech-synthesis system. In our case, triphone units that represented an acoustic model of each phoneme of Slovene with a specific left and right phoneme context were modeled [13]. Since our training database was very domain specific (nearly consecutive season weather forecasts) consisting of only 770 different words, we could expect that the training material for the specific triphone unit was obtained in a large part from uttered word(s) in a specific prosodic context. Based on this assumption, synthesized words in the same context – we have used text transcriptions of training speech material – possess similar basic prosodic attributes in terms of duration, pitch and energy, and could therefore enable an appropriate prosodic impression.

Since in this case we got a very close match of the prosodic features between natural and synthetic speech, the question arises as to what degradation in terms of automatic prosodic-events recognition could we expect if some different text in the same domain of weather forecasts were to be synthesized?

5.3 Recognition of the Prosodic Events in Synthesized Speech from Test Set

As mentioned in Section 3, we have recently acquired new speech material from the same speaker that was used for the training of our speech-synthesis system. This speech

[2] In syntactic-prosodic boundaries recognition we did not count the recognition of the M3 marker at the end of each utterance, since the recognition of this marker is trivial.

material includes 267 different words (35%) that were not uttered in the training database, and almost all the uttered sentences were different from those in the training set. On the basis of its text transcription we were again able to produce its synthesized counterpart. Afterwards, prosodic feature vectors for this new synthetic speech data were computed.

In our next recognition experiments the database that consisted of the synthesized speech from the VNTV database was used to train the SVM classifier, and this new data was used for the recognition tests. In this case we still got a relatively close match between the annotation of the syntactic-prosodic boundaries and the recognition results, which is shown in Table 6[3]. The overall recognition rate was 75% and the class-wise recognition rate was 66%.

With the recognition of word accents we encountered strong confusion[4] between the primary accent (PA) and other accented words (SA) in comparison with the human labeler's annotations, and therefore a stronger degradation in the recognition scores, where we got a 62% overall recognition rate and 53% class-wise recognition rate. Nevertheless, the automatic distinction between the accented words and those without accent still remains relatively high (Table 7), showing an adequate prosodic content of the test part of the synthesized speech.

Table 6. Confusion matrices for syntactic-prosodic boundaries recognition for the test part of the synthesized speech. The recognition system was trained on the VNTV synthesized speech.

actual/predicted	M0	M2	M3
M0	80.5%	15.2%	4.3%
M2	29.7%	61.0%	9.3%
M3	20.3%	23.5%	56.2%

Table 7. Confusion matrices for word accents recognition for the test part of synthesized speech. Recognition system was trained on the VNTV synthesized speech.

actual/predicted	UA	PA and SA
UA	70.1%	29.9%
PA and SA	19.1%	80.9%

6 Discussion

On the basis of the presented results, the prosodic quality of domain-constrained synthesized speech produced by an HMM-based speech-synthesis system is relatively high, even though no additional prosodic modeling was included in the system. This could be an important issue when dealing with embedded systems where an expansion of the system model usually leads to an additional demand for a larger memory footprint and more computational power.

[3] As in Table 4 we did not count the recognition of the M3 marker at the end of each utterance.

[4] This confusion could also partly depend on the inconsistency of the human labeler, since there is a 4-year gap between the labeling of the training and test sets.

A further improvement in the prosody modeling could be achieved if the prosodic markers in the training set were to be used to build different prosody-specific acoustic models o f sub-word units. Note, however, that in this case the text-to-speech system should be able to extract prosodic markers from the text that is to be synthesized. Some previous studies indicate [1] that automatic prosodic annotation based on appropriate statistical n-gram modeling gives reasonably good results and could be used for this task.

After analyzing the recognition results, the automatic recognition of prosodic events was also shown to be an effective tool for the objective evaluation of the performance of speech-synthesis systems on the prosody-modeling level and could therefore also be used for side-by-side comparisons of the different systems. The application of such a type of evaluation in comparison with subjective evaluation tests – where a set of representative listeners should be gathered for each evaluation test, questionnaires prepared, filled in and analyzed – is much less time consuming and could be easily reproduced.

7 Conclusion

We have described experimental results from an automatic recognition of prosodic events in the synthesized speech produced by an HMM-based speech-synthesis system. The results indicate that such kinds of tests could be used as an objective measure for the evaluation of prosody modeling in speech-synthesis systems. They also confirm a relatively good impression of naturalness with HMM-based speech synthesis, which was also noticed during previously performed subjective listening tests. The results of this study also suggest that it could be worthwhile to make use of the presented (or similar) prosodic annotations of training speech corpora for constructing prosody-specific sub-word acoustic models. However, the effectiveness of such an approach should be further investigated, which we plan to do in our future experiments.

References

1. Batliner, A., Kompe, R., Kießling, A., Mast, M., Niemann, H., Nöth, E.: M = Syntax + Prosody: A syntactic-prosodic labelling scheme for large spontaneous speech databases. Speech Communication 25, 193–222 (1998)
2. Buckow, J.: Multilingual Prosody in Automatic Speech Understanding. Logos Verlag Berlin (2004)
3. Campbell, N., Black, A.: Prosody and the Selection of Source Units for Concatenative Synthesis. In: van Santen, J., Sproat, R., Olive, J., Hirschberg, J. (eds.) Progress in Speech Synthesis, pp. 279–282. Springer, Heidelberg (1996)
4. Gros, J.: A two-level duration model for the Slovenian speech. Electrotechnical Review 66(2), 92–97 (1999)
5. Mihelič, A., Gros, Ž., Pavešić, N., Žganec, M.: Efficient subset selection from phonetically transcribed text corpora for concatenation-based embedded text-to-speech synthesis. Informacije MIDEM 36(1), 19–24 (2006)
6. Mihelič, F., Gros, J., Nöth, E., ibert, J., Pavešić, N.: Spoken Language Resources at LUKS of the University of Ljubljana. Journal of Speech Technology 6, 221–232 (2003)

7. Mihelič, F., Gros, J., Dobrišek, S., Žibert, J., Pavešić, N.: Spoken Language Resources at LUKS of the University of Ljubljana. International Journal of Speech Technology 6, 221–232 (2003)
8. Ostendorf, M., Bulyko, I.: The Impact of Speech Recognition on Speech Synthesis. In: Proc. of the IEEE Workshop on Speech Synthesis (2002)
9. Rabiner, L., Huang, B.-H.: Fundamentals of Speech Recognition. Prentice Hall, Englewood Cliffs (1993)
10. Tokuda, K., Kobayashi, T., Imai, S.: Speech parameter generation from HMM using dynamic features. In: Proc. of ICASSP, vol. 1, pp. 660–663 (1995)
11. Tokuda, K., Yoshimura, T., Masuko, T., Kobayashi, T., Kitamura, T.: Speech Parameter Generation Algorithms for HMM-based Speech Synthesis. In: Proc. ICASSP, vol. 3, pp. 1315–1318 (2000)
12. Tokuda, K., Masuko, T., Miyazaki, N., Kobayashi, T.: Multi-Space Probability Distribution HMM. IEICE Transactions on Information and Systems E85-D(3), 455–464 (2002)
13. Vesnicer, B., Mihelič, F.: Evaluation of Slovenian HMM-Based Speech Synthesis System. In: Sojka, P., Kopeček, I., Pala, K. (eds.) TSD 2004. LNCS (LNAI), vol. 3206. Springer, Heidelberg (2004)
14. Yoshimura, T., Tokuda, K., Masuko, T., Kobayashi, T., Kitamura, T.: Duration Modeling for HMM-based Speech Synthesis. In: Proc. ICSLP, vol. 2, pp. 29–32 (1998)
15. Zemljak, M., Kačič, Z., Dobrišek, S., Gros, J., Weiss, P.: Computer-based Symbols for Slovene Speech. Journal for Linguistics and Literary Studies 2, 159–294 (2002)
16. Žibert, J., Mihelič, F.: Development of Slovenian broadcast news speech database. In: Proceedings of Fourth International Conference on Language Resources and Evaluation, Lisbon, Portugal, pp. 2095–2098 (2004)
17. Chang, C.-C., Lin, C.-J.: LIBSVM: a library for support vector machines, `www.csie.ntu.edu.tw/~cjlin/libsvm`

UU Database: A Spoken Dialogue Corpus for Studies on Paralinguistic Information in Expressive Conversation

Hiroki Mori[1], Tomoyuki Satake[1], Makoto Nakamura[2], and Hideki Kasuya[3]

[1] Graduate School of Engineering, Utsunomiya University, 321-8585 Japan
[2] Faculty of International Studies, Utsunomiya University, 321-8505 Japan
[3] International University of Health and Welfare, 324-8501 Japan

Abstract. The Utsunomiya University (UU) Spoken Dialogue Database for Paralinguistic Information Studies, now available to the public, is introduced. The UU database is intended mainly for use in understanding the usage, structure and effect of paralinguistic information in expressive Japanese conversational speech. This paper describes the outline, design, building, and key properties of the UU database, to show how the corpus meets the demands of speech scientists and developers who are interested in the nature of expressive dialogue speech.

1 Introduction

Paralinguistic information plays a key role especially in spoken dialogue. A number of speech databases have been developed for spoken dialogue study. Although such databases are useful for investigating some aspects of spoken dialogue, they are not specialized for paralinguistic information research. Moreover, the existing spoken dialogue databases do not explicitly provide annotations for emotional states of speakers, which form a part of paralinguistic information.

This paper illustrates the outline, design, building, and key properties of the Utsunomiya University (UU) Spoken Dialogue Database for Paralinguistic Information Studies. The building of the UU database involved task design, recording, and annotation. Because there is no common framework for annotating paralinguistic information that spontaneous speech may deliver, a psychologically-motivated annotation method based on abstract dimensions is examined. Then, the procedure of experiments for annotating emotional states is described. Finally, statistics of the results of the annotation experiments are shown to confirm the effectiveness of the database design.

2 Task Design

To collect speech samples of really natural and spontaneous dialogue, extensive recording of daily conversation (the "Pirelli-calendar" approach [1]) might be a more suitable way than laboratory recording of task-oriented dialogue. Nevertheless, assuming a task is often a preferred way in recording dialogues because of its efficiency. Careful choice of the task for recording dialogues is therefore important to enable a database to be used for the investigation of rich emotional expression or pragmatic communication.

In this research, the objectives of task design were the following:

P. Sojka et al. (Eds.): TSD 2008, LNAI 5246, pp. 427–434, 2008.
© Springer-Verlag Berlin Heidelberg 2008

- To stimulate expressively-rich and vivid conversation;
- To involve opinion exchanges, negotiations and persuasions, and to stimulate the active participation of both speakers;
- To enable the subjects to participate with genuine interest, to improve motivation.

Conventional tasks for dialogue research, such as the map task [2,3], meet some of these criteria. In the design of the map task, however, neither emotion nor expressivity was of main concern. Similarly, other existing tasks are considered to be insufficiently interesting to the subjects.

We therefore devised a "4-frame cartoon sorting task." In this task, four cards each containing one frame extracted from a cartoon are shuffled, and each participant has two cards out of the four, and is asked to estimate the original order without looking at the remaining cards.

The task meets the above objectives because it involves such processes as information elicitation and explanation, as well as discussion to reach an agreement. The variety of expressiveness is also confirmed by the statistical investigation of paralinguistic information annotation described in Section 6.

The usefulness of the 4-frame cartoon sorting task was revealed through the building process of the UU database. Main advantages found include:

- It is relatively easy to prepare materials for the task.
- The difficulty of the task can be controlled to some extent by the complexity of the original cartoon.
- In the task, the participants need to explain not only what the characters in the cartoon are saying but also describe the situation. This makes the utterances more spontaneous.
- The goal is easy to understand because the story of a 4-frame cartoon should be relatively clear.
- The task amazingly motivates the participants because cartoons are familiar and generally popular. (This assumption was supported by the fact that all the participants wanted to know the correct order of the original cartoons even after their sessions.)

3 Dialogue Recording

The current release of the UU database includes natural and spontaneous dialogues of college students consisting of seven pairs(12 females, 2 males). The participants and pairing were selected carefully to ensure that both people in each pair were of the same grade and able to get along well with each other.

The cartoons for the 4-frame cartoon sorting task were prepared by gathering those in which merely reading the script was not sufficient to understand the story and some explanation of the scene was necessary. Each pair participated in three to seven independent sessions, using different cartoon materials for different sessions.

The sessions were performed without eye contact, allowing speech only. The recording was carried out in a soundproof room. The participants wore headsets (Sennheiser HMD-25), through which the partner's voice could be heard. The objective of the experiment (to build a spoken dialogue database) was not informed to the participants in advance. The speech data was digitally recorded by DAT. The sampling frequency was 44.1 kHz. In total, dialogue speech was recorded for 27 sessions, lasting about 130 minutes. The recordings were then transcribed with some markups (e.g. backchannel responses, fillers, discourse markers, etc.) The whole speech signal was segmented into 4840 utterances, where an utterance is defined as a speech continuum bounded by either silence (> 400 ms) or slash unit boundaries [4,5].

4 Annotation of Emotional States

The term "emotion" is used in multiple senses. Cowie [6] carefully distinguished "full-blown emotion" (or "emotion" in the narrow sense) from "underlying emotion," which refers to an aspect of mental states that may influence a person's thoughts and actions but the actions are more or less under control. Not surprisingly, it is rather rare to find full-blown emotion in task-oriented dialogues like those in this research. Participants, on the other hand, show various (vocal and facial) expressions that possibly reflect their underlying emotion. Furthermore, one might even simulate emotion that has some communicative role, or to conform to a socially expected or preferred manner of behavior. In this paper, we use the term "emotional states [7]" to refer to any kind of emotion and emotion-like states described above.

Emotional states form a part of paralinguistic information, which can be interpreted as meaningful information delivered along with spoken language. Paralinguistic information is assumed to be classified further according to whether it is categorical or non-categorical. For example, illocutionary acts which are marked by specific prosodic patterns (e.g. affirmation vs. interrogation) are apparently categorical. In contrast, it is natural, or at least safer, to consider emotional states to be non-categorical because the range of our interest is not limited to full-blown emotions. We attempted to annotate the non-categorical part of paralinguistic information (which includes emotional states) of the recorded speech in a manner coincident with continuity, rather than category labels [7].

We used the following six scales for evaluating emotional states:

(1) pleasant-unpleasant,
(2) aroused-sleepy,
(3) dominant-submissive,
(4) credible-doubtful,
(5) interested-indifferent,
(6) positive-negative.

These items were chosen in order to capture several aspects of emotional states, each of which is based on a psychological background described below. Note that the choice of scales does not mean they are sufficient for evaluating any kinds of emotional states; rather, they are necessary for examining the effectiveness of our database design from the viewpoint of richness in expressivity.

Personal emotional state. Items (1) and (2) are for measuring the emotional state of a speaker. The levels of pleasantness and arousal are the most common measures of emotional states and are repeatedly adopted in emotion studies (e.g., Russell & Bullock [8]; Russell, Weiss & Mendelsohn [9]) since Schlosberg (1952) [10].

Interpersonal relationship. Items (3) and (4) concern interpersonal relationships. Dominance (or control) is one of the most studied interpersonal relationships [11,12,13], and it was found that we can correctly decode the dominance of a person via nonverbal behavior including paralinguistic cues [14]. Many studies examined the effect of credibility (or confidence) and deception of the speaker [15,16], and found that each of these is one of the most important factors affecting the outcomes of a persuasive communication. Because both speakers can be less confident in information gap tasks, doubtful utterances are rather common even though they do not intend to cheat each other.

Attitude. Items (5) and (6) are related to the attitude of the speaker to the partner. Attitude is the belief that guides or directs an individual's perception and behavior, which is a critical factor in considering human interaction as well as interpersonal relationships. Positivity includes positive/negative attitudes such as liking, intimacy and affiliation [17,18,19]. Previous studies of interpersonal behaviors revealed that interest [22] and the positive attitude are one of the keys affiliating the participants of a conversation [20,21].

Regardless of the causes and effects of the aspects mentioned above, all the annotations of emotional states provided by the UU database are designed to be effect-oriented. In other words, the annotators are to judge "what impression the speaker's way of speaking will have on the listener," rather than "what kind of emotion or message causes the speaker's way of speaking."

5 Statistical Nature of Paralinguistic Information Annotation

One of the major difficulties in annotating paralinguistic information for the UU database was that there was no rigid theoretical foundation that justifies its evaluation and description. Therefore, the authors first tried to explore the nature, at least the statistical nature, of the ratings of emotional states by a large number of annotators.

5.1 Experiment 1: Consistency

The stimuli in this experiment included twelve utterances picked from the corpus, four of which have a high degree of prominence, four with intermediate, and four with low. The entire set of stimuli was composed of $12 \times 8 = 96$ utterances, so each utterance appeared eight times. The order of presentation was randomized. The subjects consisted of 13 paid persons in their 20s. They were asked to evaluate the perceived emotional states of the speakers for each utterance on a 7-point scale. In evaluating the pleasant-unpleasant scale, for example, 1 corresponds to extremely unpleasant, 4 to neutral, and 7 to extremely pleasant.

Figure 1 shows the standard deviation (SD) of the emotional state ratings for each individual utterance.

The horizontal axis indicates the annotator IDs. The lines in boxes show the lower quartile, median, and upper quartile of the SD values for the twelve utterances. The figure shows that most of the SD values were as low as 0.5. For example, if a subject

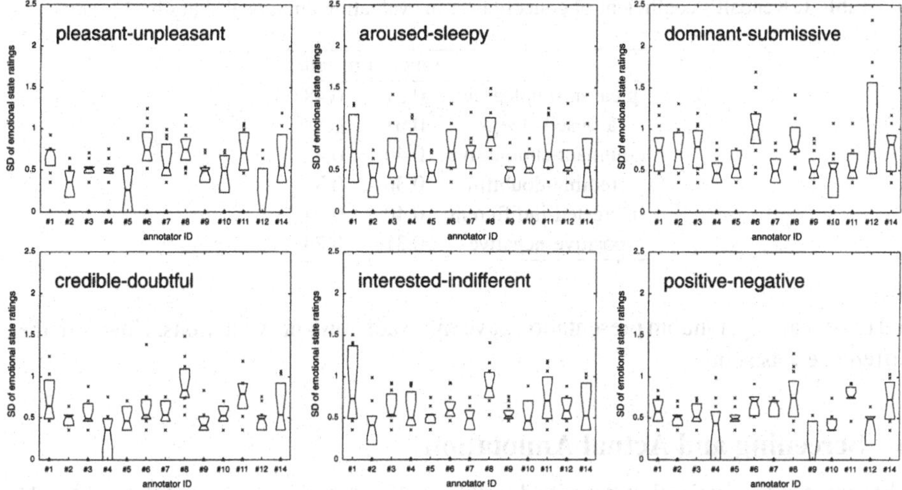

Fig. 1. Distribution of the standard deviation of the emotional state ratings for identical utterances

gave ratings of 4, 4, 4, 4, 4, 5, 5, 5 for an utterance, its SD was 0.518. This implies that most subjects could provide rather consistent ratings for the same stimulus. Some annotators (e.g. #8) were inconsistent for most evaluation items, while others (e.g. #2, #9) were relatively consistent.

Among the six scales, ratings for dominant-submissive showed high deviation as a whole. Especially, annotator #12 rated the dominant-submissive scale in a very inconsistent manner for several utterances. The ratings for the utterances were concentrated at 2 (very submissive) and 6 (very dominant). This could be attributed to the utterances' ambiguity in terms of dominance. For example, one of the utterances expressed strong agreement with her partner's proposal. This could be dominant because the agreement dominated the direction of dialogue; at the same time, it could be submissive because she followed her partner's idea.

5.2 Experiment 2: Inter-rater Agreement

The stimuli in this experiment were 215 utterances of two sessions by two pairs. The subjects consisted of 14 paid persons including all the subjects for Experiment 1. Two orders of presentation were used. At first, subjects were presented the stimuli in a random order. Following the emotional state evaluation for 215 utterances, they were presented the same utterance set in the correct order of the original sessions. In the random condition, the subjects did not know the context of each utterance; in the in-order condition, they did.

Table 1 lists Kendall's coefficient of concordance. The results show that the inter-rater agreement in evaluating emotional states was fair, especially for the first three evaluation items.

Mori et al. [23] showed that the order of presentation had an effect on ratings for most subjects. The results shown in Table 1 imply that the effect of presentation order

Table 1. Kendall's coefficient of concordance in evaluating emotional states by 14 raters

	random	in-order
pleasant-unpleasant	0.51	0.48
aroused-sleepy	0.61	0.58
dominant-submissive	0.46	0.45
credible-doubtful	0.36	0.33
interested-indifferent	0.40	0.34
positive-negative	0.31	0.29

did exist, namely, random presentation gave more concordant evaluations. However, the difference was subtle.

6 Screening and Actual Annotation

The statistical investigation described in the previous section suggested that we could obtain fairly reliable paralinguistic information annotation if we could employ good annotators. To select annotators for rating paralinguistic information for the UU database, we first recruited six female adults, and engaged them in a screening test. They were asked to evaluate an utterance set that was composed by unifying the sets described in Sections 5.1 and 5.2.

Figure 3 shows a short dialogue fragment from the UU database. The actual database is a set of XML structured documents.

The screening criteria were as follows.

Criterion 1: Consistency. The variance of ratings for the same stimulus should be as low as that of a "good annotator" described in Section 5.1.

Criterion 2: Correlation with the average. The tendency of ratings should not differ much from those of the great majority described in Section 5.

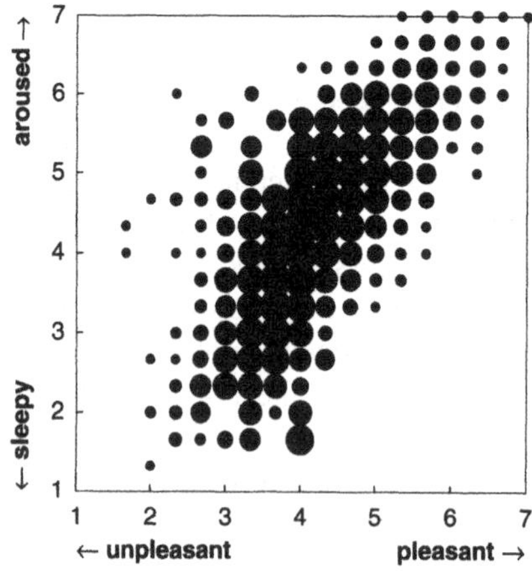

Fig. 2. Distribution of rated emotional states for the corpus. The circle area is proportional to the number of occurrences.

Utterance	pleasantness	arousal	dominance	credibility	interest	positivity
R: {breath}Huh/	2.7	5.3	3.7	3.3	5.3	3.3
R: Hang on/	2.3	3.0	3.7	2.3	5.0	2.7
R: I don't get it, I don't/	2.3	4.7	3.3	3.0	5.3	3.0
L: Er your A what is it/	5.0	5.7	5.3	4.7	5.7	4.3
R: A is Snoopy, holding a ball, and the boy is, y'know, (D sashite) reaching out his hand/	5.7	6.3	5.7	6.0	6.0	6.0
L: {B uh}{laugh}	5.7	5.0	3.0	5.3	5.3	5.3
L: {B uh}	5.7	4.7	3.0	5.3	5.0	5.3
L: {B uh}	3.3	2.7	2.3	4.3	5.0	4.7
L: Ah-hah, (D ku), it's the other way round, isn't it/	4.7	6.3	5.3	4.3	6.0	4.3
L: The little one first fetched and passed it to Snoopy, then Snoopy fetched it/	5.0	6.0	5.7	4.7	5.7	5.0
R: {B uh}	3.7	4.3	3.0	4.3	4.3	4.7
R: Oh maybe that's it/	6.0	6.7	5.3	6.7	6.7	6.3
L: Should be that one/	5.7	6.3	5.3	6.0	6.3	6.3
L: Maybe/	5.7	5.7	4.7	5.7	5.3	6.0
R: That's it that's it that's it that's it/	6.3	6.3	4.7	6.7	6.7	6.7

Fig. 3. An excerpt of a dialogue and its corresponding paralinguistic information annotation (averaged values). {B}: backchannel response, (D): disfluent portion.

Criterion 3: Distinction of evaluation items. The annotator should independently interpret each evaluation item and its paralinguistic aspect.

As a result, two of the six annotators were screened out because they did not meet Criteria 1 and 3. Finally, three qualified annotators were selected to evaluate the rest of the corpus. The whole annotation job took approximately four months. Complete (3 of 3) and partial (2 of 3) agreement were 22.09% (chance: 2.04%) and 83.92% (chance: 38.78%), respectively. Kendall's coefficient of concordance ranged from 0.56 to 0.81.

Figure 2 shows the distribution of averaged ratings for pleasant-unpleasant and aroused-sleepy dimensions. The ratings are distributed over a broad range, which means the database covers a wide variety of expressive speech.

7 Conclusion

The full distribution of the UU database comes with 2-channel waveform and XML files which provide orthographic and phonetic transcription, temporal information, and paralinguistic information ratings for all utterances.

The UU database is the world's first speech corpus specifically designed for studies on paralinguistic information in expressive dialogue speech. The UU database is distributed by the NII Speech Resources Consortium, and is also freely downloadable from the UU database website, `http://uudb.speech-lab.org/`

References

1. Campbell, N.: The JST/CREST ESP project – a mid-term progress report. In: 1st JST/CREST Intl. Workshop Expressive Speech Processing, pp. 61–70 (2003)
2. Anderson, A.H., Bader, M., Bard, E.G., Doherty, G., Garrod, S., Isard, S., Kowtko, J., McAllister, J., Miller, J., Sotillo, C., Thompson, H., Weinert, R.: The HCRC Map Task Corpus. Lang. Speech 34, 351–366 (1991)
3. Horiuchi, Y., Nakano, Y., Koiso, H., Ishizaki, M., Suzuki, H., Okada, M., Naka, M., Tutiya, S., Ichikawa, A.: The Design and Statistical Characterization of the Japanese Map Task Dialogue Corpus. J. JSAI 14, 261–272 (1999)
4. JSAI SIG of Corpus-Based Research for Discourse and Dialogue: Japanese Slash Unit (Utterance Unit) Labeling Manual Ver. 2.0.1 (2002)
5. Mori, H.: Basic Unit in Annotating Paralinguistic Information for Conversational Speech. JSAI SIG Notes SIG-SLUD-A 603, 9–14 (2007)
6. Cowie, R., Douglas-Cowie, E., Tsapatsoulis, N., Votsis, G., Kollias, S., Fellentz, W., Taylor, J.G.: Emotion Recognition in Human-Computer Interaction. IEEE Signal Processing Magazine 18, 32–80 (2001)
7. Cowie, R., Cornelius, R.R.: Describing the Emotional States that are Expressed in Speech. Speech Communication 40, 5–32 (2003)
8. Russell, J.A., Bullock, M.: Multidimensional Scaling of Emotional Facial Expressions: Similarity from Preschoolers to Adults. J. Pers. Soc. Psychol. 48, 1290–1298 (1985)
9. Russell, J.A., Weiss, A., Mendelsohn, G.A.: Affect Grid: A Single-Item Scale of Pleasure and Arousal. J. Pers. Soc. Psychol. 57, 493–502 (1989)
10. Schlosberg, H.: The Description of Facial Expressions in terms of Two Dimensions. J. Exp. Psychol. 44, 229–237 (1952)
11. Burgoon, J.K., Buller, D.B., Woodall, W.G.: Nonverbal Communication: The Unspoken Dialogue. Harper and Row, New York (1989)
12. Palmer, M.T.: Controlling Conversation: Turns, Topics and Interpersonal Control. Commun. Monogr. 56, 1–18 (1990)
13. Stang, D.J.: Effect of Interaction Rate on Ratings of Leadership and Liking. J. Pers. Soc. Psychol. 27, 405–408 (1973)
14. Rosenthal, R., Hall, J.A., DiMatteo, M.R., Rogers, P.L., Archer, D.: Sensitivity to Nonverbal Communication: The PONS Test. Johns Hopkins University Press, Baltimore (1979)
15. Keeley, M., Hart, A.: Nonverbal Behaviors in Dyadic Interaction. In: Duck, S.W. (ed.) Dynamics of Relationships. Understanding Relationship Processes, vol. 4, pp. 135–162. Sage, Newbury Park (1994)
16. Miller, G.R., Stiff, J.B.: Deceptive Communication. Sage, Newbury Park (1993)
17. Argyle, M., Dean, J.: Eye Contact, Distance, and Affiliation. Sociometry 28, 289–304 (1965)
18. Patterson, M.L.: A Functional Approach to Nonverbal Exchange. In: Feldman, R.S., Rime, B. (eds.) Fundamentals of Nonverbal Behavior, pp. 458–495. Cambridge University Press, New York (1991)
19. Warner, R.M., Malloy, D., Schneider, K., Knoth, R., Wilder, B.: Rhythmic Organization of Social Interaction and Observer Ratings of Positive Affect and Involvement. J. Nonverbal Behav. 11, 57–74 (1987)
20. Argyle, M., Alkema, F., Gilmour, R.: The Communication of Friendly and Hostile Attitudes by Verbal and Non-Verbal Signals. Eur. J. Soc. Psychol. 1, 385–402 (1972)
21. Mehrabian, A., Ksionsky, S.: A Theory of Affiliation. Heath & Co., Lexington (1974)
22. Duck, S.W.: Social Context of Relationships. Understanding Relationship Processes, vol. 3. Sage, Newbury Park (1993)
23. Mori, H., Aizawa, H., Kasuya, H.: Consistency and Agreement of Paralinguistic Information Annotation for Conversational Speech. J. Acoust. Soc. Jpn. 61, 690–697 (2005)

Perceptually Motivated Sub-band Decomposition for FDLP Audio Coding*

Petr Motlíček[1,2], Sriram Ganapathy[1,3], Hynek Hermansky[1,2,3], Harinath Garudadri[4], and Marios Athineos[5]

[1] IDIAP Research Institute, Martigny, Switzerland
{motlicek,hynek,ganapathy}@idiap.ch
[2] Faculty of Information Technology, Brno University of Technology, Czech Republic
[3] École Polytechnique Fédérale de Lausanne (EPFL), Switzerland
[4] Qualcomm Inc., San Diego, California, USA
hgarudad@qualcomm.com
[5] International Computer Science Institute, Berkeley, California, USA
msa24@columbia.edu

Abstract. This paper describes employment of non-uniform QMF decomposition to increase the efficiency of a generic wide-band audio coding system based on Frequency Domain Linear Prediction (FDLP). The base line FDLP codec, operating at high bit-rates (~136 kbps), exploits a uniform QMF decomposition into 64 sub-bands followed by sub-band processing based on FDLP. Here, we propose a non-uniform QMF decomposition into 32 frequency sub-bands obtained by merging 64 uniform QMF bands. The merging operation is performed in such a way that bandwidths of the resulting critically sampled sub-bands emulate the characteristics of the critical band filters in the human auditory system. Such frequency decomposition, when employed in the FDLP audio codec, results in a bit-rate reduction of 40% over the base line. We also describe the complete audio codec, which provides high-fidelity audio compression at ~66 kbps. In subjective listening tests, the FDLP codec outperforms MPEG-1 Layer 3 (MP3) and achieves similar qualities as MPEG-4 HE-AAC codec.

1 Introduction

A novel speech coding system, proposed recently [1], exploits the predictability of the temporal evolution of spectral envelopes of speech signal using Frequency Domain Linear Prediction (FDLP) [2,3]. Unlike [2], this technique applies FDLP to approximate relatively long segments of the Hilbert envelopes in individual frequency sub-bands. This speech compression technique was later extended to high quality audio coding [4], where an input audio signal is decomposed into N frequency sub-bands. Temporal envelopes of these sub-bands are then approximated using FDLP applied over relatively long time segments (e.g. 1, 000 ms).

* This work was partially supported by grants from ICSI Berkeley, USA; the Swiss National Center of Competence in Research (NCCR) on "Inter active Multi-modal Information Management (IM)2"; managed by the IDIAP Research Institute on behalf of the Swiss Federal Authorities, and by the European Commission 6th Framework DIRAC Integrated Project.

P. Sojka et al. (Eds.): TSD 2008, LNAI 5246, pp. 435–442, 2008.
© Springer-Verlag Berlin Heidelberg 2008

Since the FDLP model does not represent the sub-band signal perfectly, the remaining residual signal (carrier) is further processed and its frequency representatives are selectively quantized and transmitted. Efficient encoding of the sub-band residuals plays an important role in the performance of the FDLP codec and this is largely dependent on frequency decomposition employed.

Recently, an uniform 64 band Quadrature Mirror Filter (QMF) decomposition (analogous to MPEG-1 architecture [5]) was employed in the FDLP codec. This version of the codec achieves good quality of reconstructed signal compared to the older version using Gaussian band decomposition [4]. The performance advantage was mainly due to the increased frequency resolution in lower sub-bands and critical sub-sampling resulting in the minimal number of FDLP residual parameters to be transmitted. However, a higher number of sub-bands resulted in higher number of AR model parameters to be transmitted. Hence, the final bit-rates for this version of the codec were significantly higher (~136 kbps) and therefore, the FDLP codec was not competitive with the state-of-the-art audio compression systems.

In this paper, we propose a non-uniform QMF decomposition to be exploited in the FDLP codec. The idea of non-uniform QMF decomposition has been known for nearly two decades (e.g. [6,7]). In [8], Masking Pattern Adapted Subband Coding (MASCAM) system was proposed exploiting a tree-structured non-uniform QMF bank simulating the critical frequency decomposition of the auditory filter bank. This coder achieves high quality reconstructions for signals up to 15 kHz frequency at bit-rates between $80 - -100$ kbps per channel. Similar to [8], we propose sub-band decomposition which mimic the human auditory critical band filters. However, proposed QMF bank differs in many aspects, such as in the prototype filter, bandwidths, length of processed sequences, etc. The main contrast between the proposed technique and conventional filter bank occurs in the fact that the proposed QMF bank can be designed for smaller transition widths. As the QMF operates on long segments of the input signal, the additional delay arising due to sharper frequency bands can be accommodated. Such a flexibility is usually not present in codecs operating on shorter signal segments.

Proposed version of non-uniform QMF replaces original uniform decomposition in FDLP audio codec. Overall, it provides a good compromise between fine spectral resolution for low frequency sub-bands and lesser number of FDLP parameters to be encoded. Other benefits of employing non-uniform sub-band decomposition in the FDLP codec are:

- As psychoacoustic models operate in non-uniform (critical) sub-bands, they can be advantageously used to reduce the final bit-rates.
- In general, audio signals have lower energy in higher sub-bands (above 12 kHz). Therefore, the temporal evolution of the spectral envelopes in higher sub-bands require only small order AR model (employed in FDLP). A non-uniform decomposition provides one solution to have same order AR model for all sub-bands and yet, reduces the AR model parameters to be transmitted.
- The FDLP residual energies are more uniform across the sub-bands and hence, similar post-processing techniques can be applied in all sub-bands.

The proposed technique becomes the key part of the high-fidelity audio compression system based on FDLP for medium bit-rates. Objective quality tests highlight the importance of the proposed technique, compared to the version of the codec exploiting uniform sub-band decomposition. Finally, subjective evaluations of the complete codec at ~66 kbps show its relative performance compared to the state-of-the-art MPEG audio compression systems (MP3, MPEG4 HE-AAC) at similar bit-rates.

This paper is organized as follows. Section 2 describes proposed non-uniform frequency decomposition. Section 3 discusses the general structure of the codec and mentions achieved performances using objective evaluations. Section 4 describes subjective listening tests performed with the complete version of the codec operating at ~ 66 kbps is given. Finally, Section 5 concludes the paper.

2 Non-uniform Frequency Decomposition

In the proposed sub-band decomposition, the 64 uniform QMF bands are merged to obtain 32 non-uniform bands. Since the QMF decomposition in the base line system is implemented in a tree-like structure (6-stage binary tree [4]), the merging is equivalent to tying some branches at any particular stage to form a non-uniform band. This tying operation tries to follow critical band decomposition in the human auditory system. This means that more bands at higher frequencies are merged together while maintaining perfect reconstruction. The graphical scheme of the non-uniform QMF analysis bank, resulting from merging 64 bands into 32 bands, is shown in Figure 1.

Application of non-uniform QMF decomposition is supported by the Flatness Measure (FM) of the prediction error power E_i (energy of the residual signal of AR model in each sub-band i) computed across N QMF sub-bands. FM is defined as:

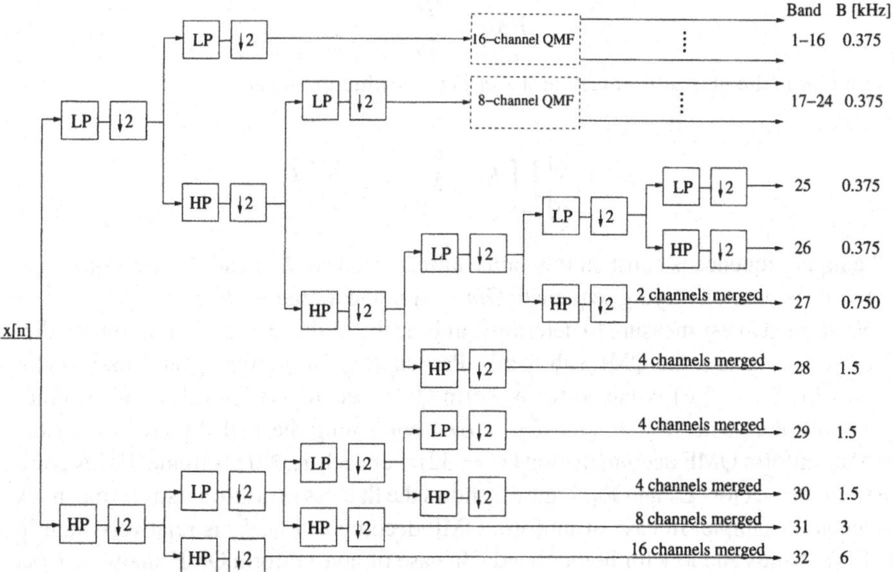

Fig. 1. The 32 channel non-uniform QMF derived using 6-stage network. Input signal $x[n]$ is sampled at 48 kHz. LP and HP denote Low-Pass and High-Pass band, respectively. ↓ 2 denotes down-sampling by 2. B denotes frequency bandwidth.

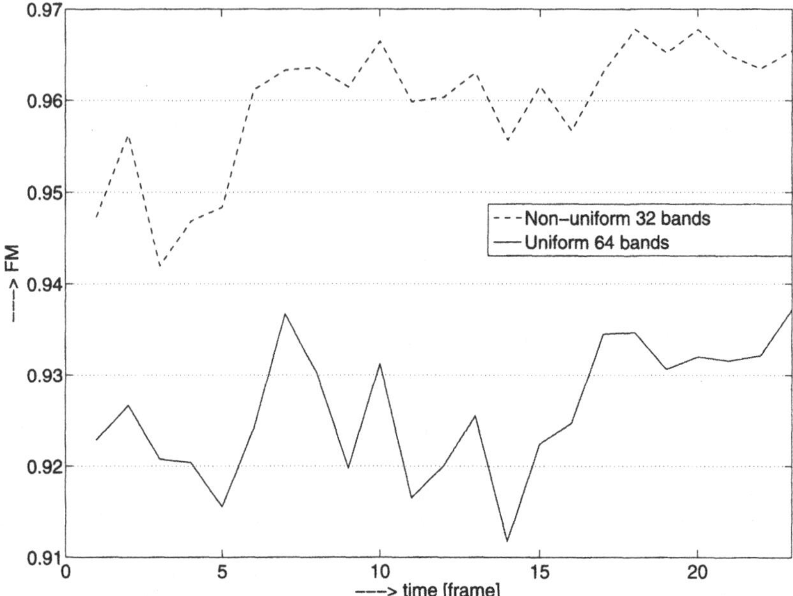

Fig. 2. Comparison of the flatness measure of the prediction error **E** for 64 band uniform QMF and 32 band non-uniform QMF for an audio recording

$$FM = \frac{Gm}{Am}, \tag{1}$$

where Gm is the geometric mean and Am is the arithmetic mean:

$$Gm = \sqrt[\frac{1}{N}]{\prod_{i=1}^{N} E_i}, \quad Am = \frac{1}{N}\sum_{i=1}^{N} E_i. \tag{2}$$

If the input sequence is constant (contains uniform values), Gm and Am are equal, and $FM = 1$. In case of varying sequence, $Gm < Am$ and therefore, $FM < 1$.

We apply flatness measure to determine uniformity of the distributions of the prediction errors E_i across the QMF sub-bands. Particularly, for a given input frame, vector $\mathbf{E_u} = (E_1, E_2 \ldots E_N)$ obtained for uniform QMF decomposition ($N = 64$) is compared with the vector $\mathbf{E_n} = (E_1, E_2 \ldots E_N)$ containing the FDLP prediction errors for non-uniform QMF decomposition ($N = 32$). For each $1,000$ ms frame, FM is computed for the vectors $\mathbf{E_u}$ and $\mathbf{E_n}$. Figure 2 shows the flatness measure versus frame index for an audio sample. In case of uniform QMF decomposition, E_i is relatively high in the lower bands and low for higher bands. In case of non-uniform QMF analysis, E_i at higher bands is comparable to those in the lower frequency bands. Such FM curves can be seen for majority of audio samples.

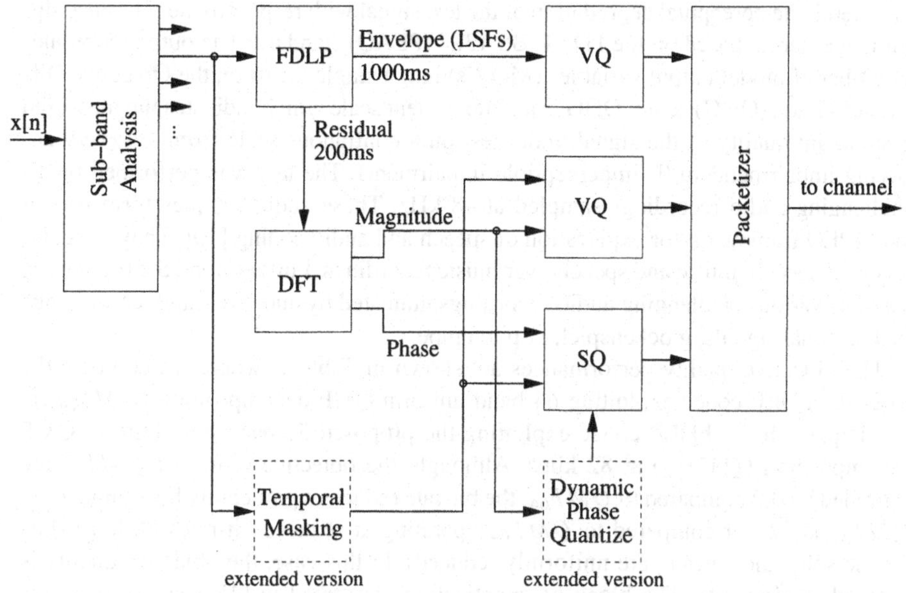

Fig. 3. Scheme of the FDLP encoder

A higher flatness measure of prediction error means that the degree of approximation provided by FDLP envelope is similar in all sub-bands. Therefore, non-uniform QMF decomposition allows uniform post-processing of FDLP residuals.

3 Structure of the FDLP Codec

FDLP codec is based on processing long (hundreds of ms) temporal segments. As described in [4], the full-band input signal is decomposed into frequency sub-bands. In each sub-band, FDLP is applied and Line Spectral Frequencies (LSFs) approximating the sub-band temporal envelopes are quantized using Vector Quantization (VQ). The residuals (sub-band carriers) are processed in Discrete Fourier Transform (DFT) domain. Its magnitude spectral parameters are quantized using VQ, as well. Phase spectral components of sub-band residuals are Scalar Quantized (SQ). Graphical scheme of the FDLP encoder is given in Figure 3.

In the decoder, quantized spectral components of the sub-band carriers are reconstructed and transformed into time-domain using inverse DFT. The reconstructed FDLP envelopes (from LSF parameters) are used to modulate the corresponding sub-band carriers. Finally, sub-band synthesis is applied to reconstruct the full-band signal.

3.1 Objective Evaluation of the Proposed Algorithm

The qualitative performance of the proposed non-uniform frequency decomposition is evaluated using Perceptual Evaluation of Audio Quality (PEAQ) distortion measure [9].

In general, the perceptual degradation of the test signal with respect to the reference signal is measured, based on the ITU-R BS.1387 (PEAQ) standard. The output combines a number of model output variables (MOV's) into a single measure, the Objective Difference Grade (ODG) score. ODG is an impairment scale which indicates the measured basic audio quality of the signal under test on a continuous scale from −4 (very annoying impairment) to 0 (imperceptible impairment). The test was performed on 18 challenging audio recordings sampled at 48 kHz. These audio samples form part of the MPEG framework for exploration of speech and audio coding [10]. They are comprised of speech, music and speech over music recordings. Furthermore, the framework contains various challenging audio recordings indicated by audio research community, such as tonal signals, glockenspiel, or pitch pipe.

The objective quality performances are shown in Table 1, where we compare the base line FDLP codec exploiting 64 band uniform QMF decomposition (QMF_{64}) at 136 kbps with the FDLP codec exploiting the proposed 32 band non-uniform QMF decomposition (QMF_{32}) at 82 kbps. Although, the objective scores of QMF_{32} are degraded by 0.2 compared to QMF_{64}, the bit-rate reduces significantly by around 40%. QMF_{32} is further compared to QMF_{64} operating at reduced bit-rates 88 kbps (bits for the sub-band carriers are uniformly reduced). In this case, the objective quality is reduced significantly. The block of quantization, described in [4], was not modified during these experiments.

4 Subjective Evaluations

For subjective listening tests, the FDLP codec (described in Section 3) exploiting non-uniform QMF decomposition (described in Section 2) is further extended with a perceptual model, a block performing dynamic phase quantization and a block of noise substitution.

The qualitative performance of the complete codec utilizing proposed non-uniform QMF decomposition is evaluated using MUSHRA (MUlti-Stimulus test with Hidden Reference and Anchor) listening tests [12] performed on 8 audio samples from MPEG audio exploration database [10]. The subjects' task is to evaluate the quality of the audio sample processed by each of the conditions involved in the test as well as the uncompressed condition and two or more degraded Anchor conditions (typically low-pass filtered at 3.5 kHz and 7 kHz).

Table 1. Mean objective quality test results provided by PEAQ (ODG scores) over 18 audio recordings: the base line FDLP codec at 136 kbps, the base line codec operating at reduced bit-rates 88 kbps, and the codec exploiting proposed 32 band non-uniform QMF decomposition at 82 kbps.

bit-rate [kbps]	136	88	82
system	QMF_{64} Uniform	QMF_{64} Uniform	QMF_{32} Non-Uniform
ODG Scores	−1.04	-2.02	−1.23

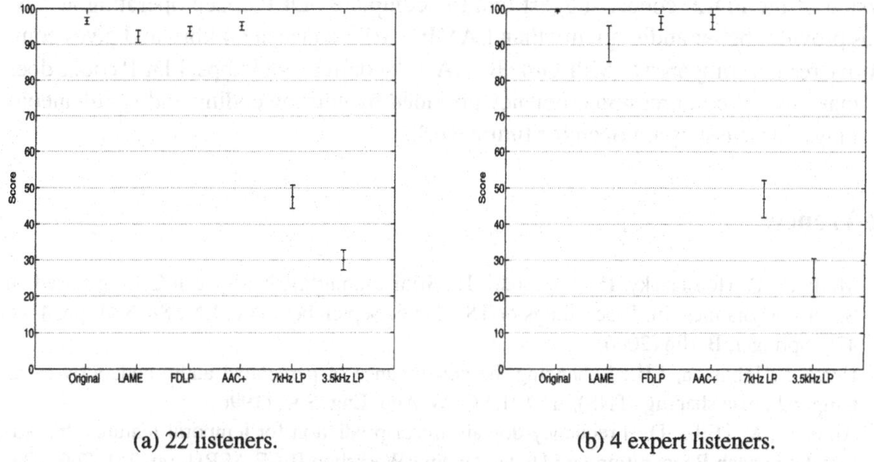

(a) 22 listeners.

(b) 4 expert listeners.

Fig. 4. MUSHRA results for 8 audio samples using three coded versions (FDLP, HE-AAC and LAME MP3), hidden reference (original) and two anchors (7 kHz and 3.5 kHz low-pass filtered)

We compare the subjective quality of the following codecs:

- Complete version of the FDLP codec at ~66 kbps.
- LAME – MP3 (MPEG 1, layer 3) at 64 kbps [13]. Lame codec based on MPEG-1 architecture is currently considered the best MP3 encoder at mid-high bit-rates and at variable bit-rates.
- MPEG-4 HE-AAC, v1 at ~64 kbps [14]. The HE-AAC coder is the combination of Spectral Band Replication (SBR) [15] and Advanced Audio Coding (AAC) [16] and was standardized as High-Efficiency AAC (HE-AAC) in Extension 1 of MPEG-4 Audio [17].

The cumulative MUSHRA scores (mean values with 95% confidence) are shown in Figures 4(a) and (b). MUSHRA tests were performed independently in two different labs (with the same setup). Figure 4(a) shows mean scores for the results from both labs (combined scores for 18 non-expert listeners and 4 expert listeners), while Figure 4(b) shows mean scores for 4 expert listeners in one lab.

These figures show that the FDLP codec performs better than LAME-MP3 and closely achieves subjective results of MPEG-4 HE-AAC standard.

5 Conclusions and Discussions

A technique for employing non-uniform frequency decomposition in the FDLP wide-band audio codec is presented here. The resulting QMF sub-bands closely follow the human auditory critical band decomposition. According to objective quality results, the new technique provides bit-rate reduction of about 40% over the base line, which is mainly due to transmitting few spectral components from the higher bands without affecting the quality significantly. Subjective evaluations, performed on the extended version of the FDLP codec, suggest that the complete FDLP codec operating at ~66 kbps provides better audio quality than LAME – MP3 codec at 64 kbps and gives competitive results compared to MPEG-4 HE-AAC standard at ~64 kbps. FDLP codec does not make use of compression efficiency provided by entropy coding and simultaneous masking. These issues are open for future work.

References

1. Motlicek, P., Hermansky, H., Garudadri, H., Srinivasamurthy, N.: Speech Coding Based on Spectral Dynamics. In: Proceedings of TSD 2006, September 2006. LNCS/LNAI, pp. 471–478. Springer, Berlin (2006)
2. Herre, J., Johnston, J.H.: Enhancing the performance of perceptual audio coders by using temporal noise shaping (TNS). In: 101st Conv. Aud. Eng. Soc. (1996)
3. Athineos, M., Ellis, D.: Frequency-domain linear prediction for temporal features. In: Automatic Speech Recognition and Understanding Workshop IEEE ASRU, pp. 261–266 (December 2003)
4. Motlicek, P., Ganapathy, S., Hermansky, H., Garudadri, H.: Scalable Wide-band Audio Codec based on Frequency Domain Linear Prediction, Tech. Rep., IDIAP, RR 07-16, version 2 (September 2007)
5. Pan, D.: A tutorial on mpeg audio compression. IEEE Multimedia Journal 02(2), 60–74 (summer, 1995)
6. Charbonnier, A., Rault, J.-B.: Design of nearly perfect non-uniform QMF filter banks. In: Proc. of ICASSP, New York, NY, USA (April 1988)
7. Vaidyanathan, P.: Multirate Systems And Filter Banks. Prentice Hall Signal Processing Series, Englewood Cliffs, New Jersey 07632 (1993)
8. Theile, G., Stoll, G., Link, M.: Low-bit rate coding of high quality audio signals, In: Proc. 82nd Conv. Aud. Eng. Soc. (March 1987) (preprint 2432)
9. Thiede, T., Treurniet, W.C., Bitto, R., Schmidmer, C., Sporer, T., Beerends, J.G., Colomes, C., Keyhl, M., Stoll, G., Brandenburg, K., Feiten, B.: PEAQ – The ITU Standard for Objective Measurement of Perceived Audio Quality. J. Audio Eng. Soc. 48, 3–29 (2000)
10. ISO/IEC JTC1/SC29/WG11: Framework for Exploration of Speech and Audio Coding, MPEG2007/N9254, Lausanne, Switzerland (July 2007)
11. Ganapathy, S., Motlicek, P., Hermansky, H., Garudadri, H.: Temporal Masking for Bit-rate Reduction in Audio Codec Based on Frequency Domain Linear Prediction., Tech. Rep., IDIAP, RR 07-48 (October 2007)
12. ITU-R Recommendation BS.1534: Method for the subjective assessment of intermediate audio quality (June 2001)
13. LAME MP3 codec, `http://lame.sourceforge.net`
14. 3GPP TS 26.401: Enhanced aacPlus General Audio Codec, General Description
15. Dietz, M., Liljeryd, L., Kjorling, K., Kunz, O.: Spectral Band Replication, a novel approach in audio coding. In: AES 112th Convention, Munich, DE (May 2002); Preprint 5553
16. Bosi, M., Brandenburg, K., Quackenbush, S., Fielder, L., Akagiri, K., Fuchs, H., Dietz, M., Herre, J., Davidson, G., Oikawa, Y.: ISO/IEC MPEG-2 Advanced Audio Coding. J. Audio Eng. Soc. 45(10), 789–814 (1997)
17. ISO/IEC, Coding of audio-visual objects Part 3: Audio, AMENDMENT 1: Bandwidth Extension, ISO/IEC Int. Std. 14496-3:2001/Amd.12003 (2003)

Forced-Alignment and Edit-Distance Scoring for Vocabulary Tutoring Applications

Serguei Pakhomov[1], Jayson Richardson[2], Matt Finholt-Daniel[3], and Gregory Sales[3]

[1] University of Minnesota, 308 Harvard Street SE, Minneapolis, MN, 55455
pakh0002@umn.edu

[2] University of North Carolina, 601 South College Road, Wilmington, NC
jrichardson@uncw.edu

[3] Seward Incorporated, 2200 East Franklin Ave., Minneapolis, Minnesota 55404
{mfinholt,gsales}@sewardinc.com

Abstract. We demonstrate an application of Automatic Speech Recognition (ASR) technology to the assessment of young children's basic English vocabulary. We use a test set of 2935 speech samples manually rated by 3 reviewers to compare several approaches to measuring and classifying the accuracy of the children's pronunciation of words, including acoustic confidence scoring obtained by forced alignment and edit distance between the expected and actual ASR output. We show that phoneme-level language modeling can be used to obtain good classification results even with a relatively small amount of acoustic training data. The area under the ROC curve of the ASR-based classifier that uses a bi-phone language model interpolated with a general English bi-phone model is 0.80 (95% CI 0.78–0.82). The point where both sensitivity and specificity are at their maximum is where sensitivity is 0.74 and the specificity is 0.80 with 0.77 harmonic mean, which is comparable to human performance (ICC=0.75; absolute agreement = 81%).

Keywords: Automatic speech recognition, vocabulary tutor, sub-word language modeling.

1 Introduction

Accurate measurement of young children's reading ability and knowledge of the English vocabulary is an essential but challenging component of automated reading assessment and tutoring systems. Available pronunciation tutoring systems may be categorized into those that simply provide the pronunciation models that the learners can imitate and those that perform an analysis of the learners' speech. Projects in the latter category include FLUENCY [1], LISTEN [2,3,5,4], EduSpeak [6], and a multi-center collaboration between the University of Colorado, Boulder, the University of California, San Diego and the Oregon Graduate Institute [7,8], among others. These systems are based on Automatic Speech Recognition (ASR) in a text-independent paradigm where the goal is to provide feedback on any utterance the learner may produce. In text-dependent applications the system is trained to provide feedback on a specific set of utterances. An advantage of the text-dependent approach is that it enables more accurate

P. Sojka et al. (Eds.): TSD 2008, LNAI 5246, pp. 443–450, 2008.
© Springer-Verlag Berlin Heidelberg 2008

pronunciation scoring and finer control over corrective feedback to facilitate "bottom-up" perception training. Such training is important to help learners develop their perceptual abilities that are, in turn, critical to the development of accurate pronunciation [9].

While reading tutoring applications typically provide word-level scoring and feedback, a recent study by Hagen, et al. [10] suggests that fine-grained analysis at the level of sub-word units is important for reading tutoring systems. Accurate sub-word unit analysis provides the necessary information for highly specific feedback to the learner thus making the tutoring more effective. In this paper, we present an evaluation of several acoustic and language modeling approaches aimed at maximizing the accuracy of an isolated word, text-dependent vocabulary tutoring system that relies on sub-word level analysis.

2 Methods

2.1 Vocabulary Tutoring System

Our vocabulary tutoring system relies on processing the output of an ASR engine in a text-dependent paradigm where the children are expected to read isolated words from a predefined vocabulary of the top 4000 most frequent English words [11]. For this project we focused on 100 words (10 clusters of 10 words) distributed throughout the 4000. While the words are read in isolation, they are presented to the learners in context at several points during the tutorial consisting of a series of lessons designed to elucidate the usage of the essential reading vocabulary. Each lesson begins with an assessment, where 10 words are presented to the learner who is asked to read them into a standard close-talking microphone. Subsequently, the words are processed (details to follow) and the output, classified as "acceptable" or "unacceptable", is shown to the learner via a red or a green dot on a console in the user interface. "Unacceptable" words are then defined and presented in appropriate contexts through audio and text and associated with representative images. The learner may practice each word with the immediate feedback from the speech classification component of the system. In the last part of the lesson, the learner undergoes another assessment on the same 10 words initially presented at the start of the lesson. The direct feedback of the classification results to the learner requires that the ASR-based speech classification is as accurate as possible, as even a small number of errors may result in high motivational or cognitive costs [4]. The system architecture consists of a central server that runs the computation intensive components and a web-based client with the graphical user interface.

2.2 ASR Component

The ASR component is based on the Hidden Markov Model Toolkit (HTK 3.4) [12]. We trained three speaker independent acoustic models using two datasets: the CMU Kids corpus [13] and a corpus of speech recordings collected locally (LOCAL) specifically for this application.

Acoustic Modeling: The Carnegie Mellon University (CMU Kids) corpus consists of 5,180 sentences read aloud by children of ages 6–11 [1]. This corpus has been previously used for the LISTEN project [3,2] and consists of two sets of speakers – those that were deemed to be either "good" or "poor" readers. Since our system is intended to model the standard American English speech, we used the former subset of the CMU Kids corpus consisting of 3,333 utterances by 44 speakers (duration = 97 min.) The LOCAL corpus consists of recordings of 100 words spoken in isolation by 10 children – 3 boys and 7 girls, grades 2–6 (duration = 31 min.). The speech was recorded at 16 kHz at 16 bit depth with a high quality microphone (Shure SM7B). We also used a combination of these two corpora (CMU+LOCAL) representing a hybrid that combines continuous speech with isolated speech in addition to increasing the corpus size.

Mel-frequency Cepstral Coefficient (MFCC) type acoustic models for each of the three corpora were trained, each consisting of a set of tied-state continuous density triphone Hidden Markov Models (HMMs). Vocal Tract Length Normalization (VTLN) with warping frequency factor set to 1.0, 19 filterbank channels and 10 cepstral features were used with cepstral mean normalization turned on. The audio was sampled at 10 millisecond intervals with a Hamming window of 25 milliseconds.

Language Modeling: We trained three types of phoneme level language models. The first type (FIXED) used each of the word's dictionary pronunciation to construct a network with fixed transitions between individual phonemes. Here is an example of a network for the word "train": (sil t r ey n sil). Using this type of network forces the recognizer to find the best match between the input audio and the HMMs that correspond to the nodes in the network (a.k.a. forced alignment). The second type (FREE) consisted of a phoneme loop network where any phoneme may occur in any sequence or not at all with mandatory silences at the start and end of the utterance: (sil {(aa |... |zh)} sil). This type of modeling is independent of the dictionary pronunciations of individual words and uses the same phoneme list as in acoustic modeling. The third type (BIPHONE) consisted of a word-independent bi-phone back-off model trained using the dictionary pronunciations of the top most frequent 4000 words and linearly interpolated with a word-dependent bi-phone model. The models were interpolated using a weight 0.2 resulting in the final model representing 80% of the word-dependent transition probabilities and 20% of word independent probabilities. The BIPHONE model is intended to bias the recognizer towards the word the learner is expected to speak but also to provide the flexibility in recognizing erroneous speech.

Post-processing (scoring): The classification of the ASR output into "acceptable" and "unacceptable" using the FIXED model was performed based on the acoustic confidence scores normalized by the duration of the phoneme. This methodology was previously used for detection of pronunciation errors in language learners' utterances [14]. The output of the recognizer using the other two types of language models was measured by computing the edit distance [15] between the sequence of phonemes in the ASR output and those in the dictionary pronunciation. To account for word length variablity, the raw edit distance scores were normalized by the number of phonemes in

[1] There was only one 11 year old in the corpus, thus it is more representative of the younger age range.

the reference dictionary pronunciation. Silences were excluded and diphthongs were treated as two separate symbols. For example, if the recognizer's output for the word "railroad" was [r ey l d h ow w d] while the dictionary pronunciation was [r ey l r ow d], the edit distance score was 3 resulting in the normalized score of 0.375. Classification decisions were made using simple thresholding.

Reference standard: The reference standard for evaluating the system was created in "field" conditions where 16 children who did not participate in training of the acoustic models were asked to use the prototype of the system in their normal environment. A total of 2935 utterances representing the selection of 100 words and different stages during the lessons were collected. Each utterance was manually classified by 3 independent judges familiar with the objectives of the tutoring system but not the details of the ASR component. The majority vote was used to compile the final reference standard.

Evaluation: We used the Intra-Class Correlation Coefficient (ICC) statistic [16] to measure inter-rater agreement and determine the reliability of the reference standard. We used receiver-operator characteristics (ROC) analysis to evaluate our tutoring system following Mostow (Mostow, 2006), who argues that the use of ROC analysis is clearer than traditional word-error rates in evaluating automated reading tutoring systems. In addition, ROC analysis is well suited for binary classification problems where accuracy on both positive (sensitivity) and negative (specificity) categories is important. Sensitivity is the ratio of correctly classified positive examples to the total number of positive examples, while specificity is the ratio of the number of correctly classified negative examples to the total number of negative examples.

3 Results

3.1 Reference Standard

We used the majority vote to divide the 2935 audio samples these into 1619 (55%) "acceptable" and 1316 (45%) "unacceptable". The ICC for the judgments of all three raters was 0.75 (95% CI 0.74–0.77). Of the 2935 judgments by the three raters, in 1273 cases all three agreed on the "acceptable" rating and in 1114 cases all three agreed on the "unacceptable" rating. Thus the overall absolute agreement was 81%.

3.2 Language Modeling

The results of comparisons between the three different language modeling approaches are illustrated in Figure 1(a). The CMU+LOCAL acoustic model was used to generate the curves. The area under the ROC is 0.74 (95% CI 0.72–0.71) for the FIXED model, 0.62 (95% CI 0.60–0.64) for the FREE model and 0.80 (95% CI 0.78–0.82) for the BIPHONE model. Using the FREE and FIXED approaches yielded curves with approximately the same area; however, the shapes of the curves are very different where the curve for the FREE model is further away from the diagonal for high specificity but low sensitivity values. The FIXED model produced the opposite effect with the curve being further away from the diagonal for low specificity but high sensitivity values. The

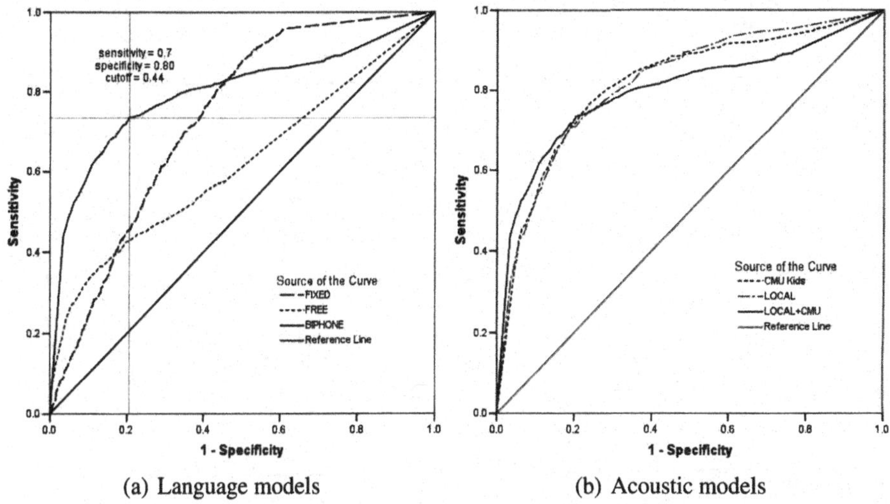

(a) Language models (b) Acoustic models

Fig. 1. ROC curves representing the test results for language and acoustic models

curve for the BIPHONE model has the greatest area and shows a balance between sensitivity and specificity. The peak performance (highest sensitivity AND specificity) is observed at the cutoff of 0.44 resulting in the sensitivity of 0.74 and specificity of 0.8.

Figure 1(b) shows the ROC curves that represent the three different acoustic models: CMU Kids, LOCAL and CMU+LOCAL. The area under the ROC curve is 0.82 (95% CI 0.80–0.83) for the CMU Kids model, 0.82 (95% CI 0.80–0.84) for the LOCAL model and 0.80 (95% CI 0.78–0.82) for the CMU+LOCAL combination. The areas under ROC curve are very similar but with slightly different shapes where the CMU+LOCAL model tends to have higher sensitivity when the specificity is also higher (>0.8) than the other two taken separately.

3.3 Individual Learner Performance

We stratified the results obtained with the BIPHONE language model, CMU+LOCAL acoustic model and the threshold of 0.44. This resulted in sensitivity and specificity point estimates for each individual learner presented in Table 1. The majority of these children were rated as High or Medium ability readers by their teachers and were not English as Second Language (ESL) learners. There is a significant association between the teacher ratings (dichotomized by combining Medium and High ratings) and the ESL status (chi-square = 7.27; p = 0.007). Thus, the teachers rated non-ESL learners as better readers. There is also a significant association between the ESL status and the classification accuracy dichotomized using both sensitivity and specificity >0.70 as the threshold (chi-square = 6.35; p = 0.012). Thus, non-ESL learner were more likely to have better classification accuracy. No association was found between sex, grade level and classification accuracy.

Student	Teacher Rating*	Grade	ESL (Y/N)	Sex	Reference Standard Ratings† PASS N corr.	N total	FAIL N corr.	N total	Sens	Spec	F-score§
S1	H	3	N	f	156	184	28	43	0.85	0.65	0.74
S2	M	4	Y	f	110	132	64	129	0.83	0.50	0.62
S3‡	L	2	Y	f	70	85	140	180	0.82	0.78	0.80
S4	M	4	N	f	118	153	89	118	0.77	0.75	0.76
S5	M	3	Y	m	143	178	63	96	0.80	0.66	0.72
S6	L	1	Y	m	8	14	107	113	0.57	0.95	0.71
S7	H	1	N	f	20	23	51	56	0.87	0.91	0.89
S8	M	2	N	f	36	42	27	30	0.86	0.90	0.88
S9	M	3	N	f	122	153	28	41	0.80	0.68	0.74
S10	M	3	N	f	151	188	37	45	0.80	0.82	0.81
S11	L	1	Y	m	4	6	45	48	0.67	0.94	0.78
S12	M	2	Y	m	23	62	46	57	0.37	0.81	0.51
S13	M	2	N	m	112	152	44	58	0.74	0.76	0.75
S14	L	4	Y	m	64	109	138	152	0.59	0.91	0.71
S15	L	4	Y	f	51	136	63	72	0.38	0.88	0.53
S16	M	1	N	m	2	2	69	70	1.00	0.99	0.99
Overall					**1190**	**1619**	**1039**	**1308**	**0.74**	**0.80**	**0.77**

* Teacher rating legend (High = H, Low = L, Medium = M)

† These columns represent the raw number of samples in the test set rated as "acceptable" (PASS, N Total) or "unacceptable" (FAIL, N Total) by the majority vote. "N corr." represents the number of samples the ASR based classifier identified correctly for each of these two categories

§ F-score is the harmonic mean between sensitivity and specificity F = 2*(Sens*Spec)/(Sens+Spec)

‡ Shaded rows represent students for whom the classification is both sensitive and specific (both > 0.70)

Fig. 2. Individual learner demographic and reading performance characteristics compared to the vocabulary tutor classification accuracy results

4 Discussion

In this paper we showed that on a relatively simple task such as isolated word classification for a vocabulary tutor it is possible to achieve high classification accuracy even with limited acoustic training data. The acoustic models were trained on less than 2 hours of speech; however, a general bi-phone probabilistic language model trained on 4000 English words most frequent in children's reading curriculum and interpolated with bi-phone models trained on individual words was able to compensate successfully for the small size of the acoustic training data. This finding is important because obtaining high quality acoustic data is a time consuming and costly process. Our results indicate that it is possible to "bootstrap" a system with a small corpus of acoustic data and in the meantime continue collecting additional data.

The reference standard is highly reliable. The ICC of 0.75 is in the "substantial" agreement range (0.61–0.8) [16]. The reference standard is also of substantial size (n = 2935) which makes the test results highly significant (p <0.0001). Nevertheless, the test set does have some disagreements thus creating a ceiling for accuracy of 81.0% (95% CI 79.8–82.7) where all three raters were unable to agree.

The performance of the ASR component using the CMU+LOCAL acoustic model and the BIPHONE language model was close to human. Out of the 2935 words spoken in the pilot, three human raters agreed 81% of the time. Out of the same set of words the ASR component had only 4% relative less agreement with the human raters than human raters agreed with each other. The comparison between acoustic models indicated that all three models had comparable performance in terms of their areas under the ROC. The fact that the CMU+LOCAL model exhibits higher sensitivity at higher specificity values, suggests that adding a small amount of task specific (i.e. isolated word recognition) acoustic data to a continuous speech corpus results is a model that is more sensitive but less specific.

The association between the learners' ESL status and the ASR-based classification accuracy indicates that the ESL learners may not be represented by the acoustic model as well as native English speakers. Table 1 shows that both the ESL and non-ESL learners were classified with similar specificity (mean = 0.80 for both groups) but that the ESL learners were classified with lower sensitivity (mean = 0.63 for ESL vs. 0.84 for non-ESL). Our application is intended to represent the standard American English pronunciation and thus the lower sensitivity on ESL speech is likely to constitute a positive outcome but requires a more detailed investigation.

Limitations: The current study did not measure the improvement in learners' performance after using the system and will be addressed in future work. While we used nearly 3000 test samples for this study, they represent only 16 learners from a select set of schools. As the vocabulary tutoring system is used more widely, we will be able to expand the test base to a larger number of learners and a more diverse population thus making the results more generalizable. Another potential limitation is the difference in the level of background noise between the audio used for training the acoustic models collected in a quiet environment and the "real world" samples. It is possible that modeling background noise may further improve the results. The current system did not take into account prosodic information such as duration, stress or fundamental frequency, which may be important predictors of whether the child is familiar with the prompted word. Word duration has been previously associated with reading ability [4]. In our test set, the "unacceptable" samples tended to be significantly longer (mean = 2.76 seconds) than the "acceptable" ones (mean = 2.27 seconds; $p < 0.0001$), justifying further investigation of prosodic characteristics.

References

1. Eskenazi, M.: Using Automatic Speech Processing for Foreign Language Pronunciation Tutoring: Some Issues and a Prototype. Language Learning and Technology 2, 62–67 (1999)
2. Mostow, J., Roth, S., Hauptmann, A., Kane, M.: A Prototype Reading Coach that Listens. In: Proceedings of the twelfth national conference on Artificial intelligence, vol. 1, pp. 785–792 (1994)
3. Mostow, J., Aist, G., Burkhead, P., Corbett, A., Cuneo, A., Eitelman, S., Huang, C., Junker, B., Sklar, M.B., Tobin, B.: Evaluation of an Automated Reading Tutor that Listens: Comparison to Human Tutoring and Classroom Instruction. Journal of Educational Computing Research 29, 61–117 (2003)
4. Mostow, J.: Is ASR accurate enough for automated reading tutors, and how can we tell? In: Proceedings of the Ninth International Conference on Spoken Language Processing (Interspeech 2006 – ICSLP), Special Session on Speech and Language in Education, pp. 837–840 (2006)
5. Heiner, C., Beck, J.E., Mostow, J.: Automated Vocabulary Instruction in a Reading Tutor. In: Intelligent Tutoring Systems, pp. 741–743. Springer, Berlin (2006)

6. Franco, H., Neumeyer, L., Digalakis, V., Weintraub, M.: Automatic Scoring of Pronunciation Quality. Speech Communication 30, 83–93 (1999)
7. Cole, R., van Vuuren, S., Pellom, B., Hacioglu, K., Ma, J., Movellan, J., Schwartz, S., Wade-Stein, D., Ward, W., Yan, J.: Perceptive Animated Interfaces: First Steps Toward a New Paradigm for Human Computer Interaction. Proceedings of the IEEE 91, 1391–1405 (2003)
8. Cole, R., Wise, B., van Vuuren, S.: How Marni teaches children to read. Education Technology 47, 14–18 (2006)
9. Akahane-Yamada, R., Tohkura, Y., Bradlow, A.R., Pisoni, D.B.: Does training in speech perception modify speech production? In: Proceedings of International Conference of Speech and Language Processing, pp. 606–609 (1996)
10. Hagen, A., Pellom, B., Cole, R.: Highly accurate children's speech recognition for interactive reading tutors using subword units. Speech Communication 49, 861–873 (2007)
11. Carroll, J.B., Davies, P., Richman, B.: Word frequency book. Houghton Mifflin Co., Boston (1971)
12. Young, S., Kershaw, D., Odell, J., Ollason, D., Valtchev, V.: The HTK Book Version 3.4. Cambridge University, Cambridge (2006)
13. Eskenazi, M., Mostow, J.: The CMU KIDS Speech Corpus. In: Corpus of children's read speech digitized and transcribed on two CD-ROMs, with assistance from Multicom Research and David Graff, Linguistic Data Consortium, University of Pennsylvania, Pittsburgh, PA (1997)
14. Eskenazi, M.: Detection of foreign speakers' pronunciation errors for second language training – preliminary results. In: Proceedings of the Fourth International Conference on Spoken Language (ICSLP 1996), vol. 3, pp. 1465–1468 (1996)
15. Levenshtein, V.: Binary codes capable of correcting deletions, insertions, and reversals. Soviet Physics Doklady 10, 707–710 (1966)
16. Shrout, P.E., Fleiss, J.L.: Intraclass Correlations: Uses in Assessing Rater Reliability. Psychological Bulletin 86, 420–428 (1979)

Exploiting Contextual Information for Speech/Non-Speech Detection

Sree Hari Krishnan Parthasarathi, Petr Motlíček, and Hynek Hermansky

IDIAP Research Institute, Martigny
Swiss Federal Institute of Technology at Lausanne (EPFL), Switzerland
{hari.parthasarathi,petr.motlicek,hynek}@idiap.ch

Abstract. In this paper, we investigate the effect of temporal context for speech/ non-speech detection (SND). It is shown that even a simple feature such as full-band energy, when employed with a large-enough context, shows promise for further investigation. Experimental evaluations on the test data set, with a state-of-the-art multi-layer perceptron based SND system and a simple energy threshold based SND method, using the F-measure, show an absolute performance gain of 4.4% and 5.4% respectively. The optimal contextual length was found to be 1000 ms. Further numerical optimizations yield an improvement (3.37% absolute), resulting in an absolute gain of 7.77% and 8.77% over the MLP based and energy based methods respectively. ROC based performance evaluation also reveals promising performance for the proposed method, particularly in low SNR conditions.

Keywords: Speech/non-speech detection, modulation spectrum, temporal context.

1 Introduction

The primary objective of our work is to design a simple speech/non-speech detection (SND) algorithm that can be implemented on low power devices. Historically, short-term energy has been one of the most important features for SND [1]. In this paper, we study the effect of long temporal context on signal energy for SND using a data-driven approach. Two recent studies of SND, [6] and [5], exploit temporal context using modulation spectrum on multiple spectral bands.

In our approach, the weights of the context around the frame-to-be-classified, are obtained using Linear Discriminant Analysis (LDA). This method gives us an interpretation in terms of a filter in the modulation spectral domain. However, it is well-known that when the features are correlated and when the dimension is large, the LDA covariance matrices are not estimated well. Many solutions to this problem of regularizing the covariance matrix exist [3]. The details of our regularization method are provided in Section 2.

The rest of the paper is organized as follows. Section 2 discusses the proposed method, with and without regularization, in detail. Description of the experimental evaluation and the data set is provided in Section 3. Finally, we draw some conclusions in Section 4.

P. Sojka et al. (Eds.): TSD 2008, LNAI 5246, pp. 451–459, 2008.
© Springer-Verlag Berlin Heidelberg 2008

2 Obtaining the Weights of the Temporal Context: Proposed Method

2.1 Features

The first step is to obtain feature vectors for the two classes. This is done as follows: for each speech signal in the data set, the logarithmic full-band energy is computed using a rectangular analysis window of length and shift 25 ms and 10 ms, respectively. Feature vectors are extracted by considering overlapping windows (i.e., shift of 10 ms) on this temporal trajectory. This method of extracting features introduces a context around the frame under consideration.

To better understand the choice of this feature vector, we briefly discuss the characteristics of the speech and non-speech data: the mean speech and non-speech vectors (obtained from the training data set, Section 3) are shown in Figure 1. These vectors are 1010 ms long (101 frames at 10 ms frame rate). It can be seen that these vectors are quite distinct for speech and non-speech. Further, these vectors are easily interpretable. Since speech frames have higher energy than non-speech on an average, the mean speech vector shows a pronounced peak at the center. The converse is true for the mean non-speech vector.

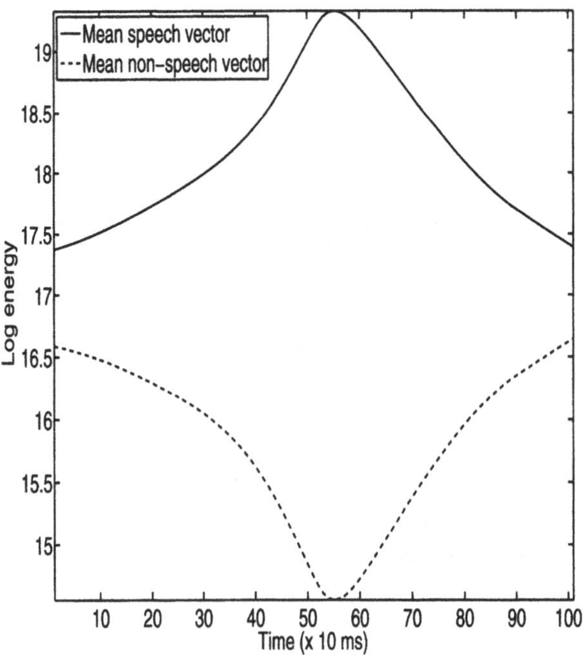

Fig. 1. Mean speech and non-speech vectors

2.2 Training LDA to Obtain the Weights of the Context

In this section, we obtain the weights of the context around the the frame that is to be classified, using LDA [3]. The label of the class at the center of the feature vector determines the training targets. LDA is used to obtain the weight vector for classification. This is the weight vector obtained without regularization.

However, since the features are highly correlated and are in a high dimension, the estimate of the within-class covariance matrices becomes poor. One of the solutions to this problem is to first perform dimensionality reduction using Principal Component Analysis (PCA) and then employ LDA [3]. On the otherhand, it is well known that PCA is asymptotically equivalent to Discrete Cosine Transform (DCT) for Markov-1 signals if the correlation coefficient is close to 1[1]. Therefore, we perform dimensionality reduction by projecting the features obtained using the method described in Section 2.1 onto the first few DCT bases. LDA is then used to estimate the contextual weights in this space. The weights in the subspace are projected back on the entire DCT bases to estimate the weight vector in the original space. This is the weight vector obtained with regularization.

For comparison, Figure 2(a) shows the weight vectors (flipped left to right, for interpretation as an impulse response) obtained by LDA and regularized LDA. It can be observed that regularizing the LDA using DCT yields a smooth weight vector. Further, it is also shown experimentally that the overall performance increases when regularized LDA is used (Section 3).

We now analyze the shape of the impulse response: the valley at the center can be understood from the fact that the mean speech and non-speech vectors suggest that the dimensions most important for classification are the 20 frames around the center. This can be interpreted that a context of 300 to 400 ms around the center is important for classification. Also, note that reducing the feature vector dimension to one, reduces the method to energy thresholding.

Determination of SND Boundaries. During testing, the feature vectors ($\mathbf{x}_i$) of the speech signal are computed as described in Section 2.1. The vectors are projected on to $\mathbf{w}$. These projected values are then compared with the threshold (θ), to determine the class.

3 Experiments and Evaluation

We investigate three questions in this section: (a) How does the proposed method compare with the state-of-the-art? Section 3.2 and 3.3 address this question. (b) Does utilizing context yield better performance? To answer this, the significance of the contextual window is studied in Section 3.4. (c) Does regularization improve classification performance? Towards this, regularized LDA is compared with LDA without regularization (Section 3.5) for the optimal length obtained in Section 3.4.

[1] Our experiments also showed virtually identical weight vectors obtained by LDA after either PCA or DCT.

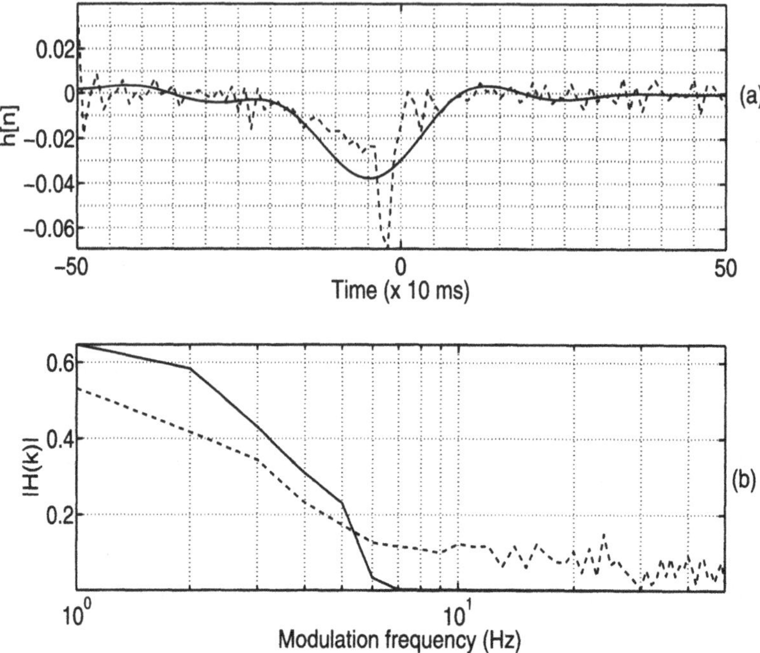

Fig. 2. (a) Dotted and solid lines indicate impulse responses obtained by LDA and regularized LDA respectively. (b) Dotted and solid lines indicate magnitude responses obtained by LDA and regularized LDA respectively.

3.1 Experimental Setup

Experiments are conducted on a subset of the NIST meeting room corpus [4]. Data obtained from close-talking microphones are used for the experiments. The sampling rate and the quantization of the data are 16 kHz and 16 bits respectively. The training and testing sets consist approximately of 1 and 3 hours of data respectively. The overall ratio of non-speech to speech segments is 46% : 54%. The labels for the training and testing data are obtained by forced-alignment of ASR phoneme models [2]. All phonemes except "sil" are considered "speech", while the "sil" regions are labeled as "non-speech". Since the data used in this study is from close-talking microphones, the signals are relatively clean. To study the effect of noise on the SND systems, babble noise from NOISEX-92 database [7] is added at various SNR levels.

To evaluate the importance of temporal context for SND, we compare the proposed method (long temporal context) with two short-term methods (a) a state-of-the-art multi-layer perceptron (MLP) based method [2] and (b) a simple short-term energy threshold based method. The MLP based method uses 12 MF-PLP coefficients along with their first and second derivatives. To these, the following auxiliary features are added: normalized energy from all channels, signal kurtosis, mean cross-correlation and maximum normalized cross-correlation. The MLP is trained on 98 hours

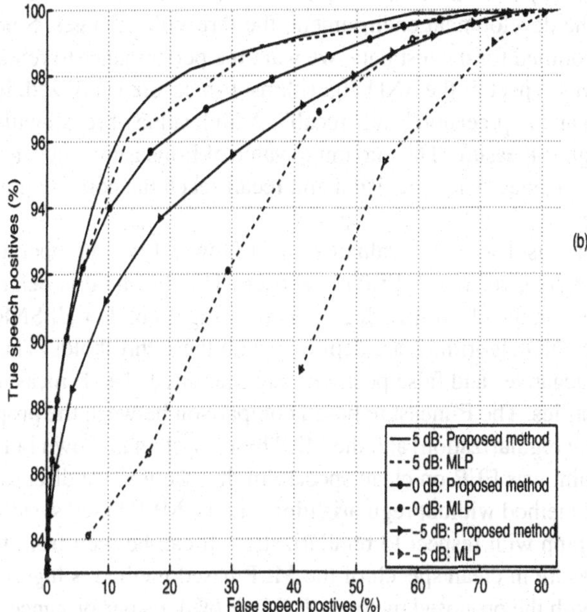

Fig. 3. Comparison of proposed and MLP based system in 5, 0 and −5 dB SNR

of training data, with a hyperbolic tangent hidden activation function and soft-max output activation function. The energy based method computes short-term log energy and uses a threshold to make the SND decisions.

3.2 Comparison with MLP Based System Using ROC Curves

In the first set of experiments, a context of 1000 ms is used for the proposed method (without regularization). A comparison of the proposed method with the MLP based system using the Receiver Operating Characteristics (ROC) curve method is done. To plot the ROC curve, "true speech positives" and "false speech positives" are computed by varying the thresholds of the methods. The noisy data is obtained by adding babble noise from NOISEX-92 database at four different SNR levels: 10 dB, 5 dB, 0 dB and −5 dB. The results are shown in Figure 3. It illustrates that the proposed system performs better than the MLP based system in significantly noisy conditions. We attribute the performance of the proposed method to the usage of long-term contextual information.

Experiments also revealed that the MLP based method performs better than the proposed method when the environment is relatively less noisy. This is indeed not surprising because the MLP based method is trained on many hours of meeting room data and consequently performs well when the testing conditions match the training conditions.

3.3 Comparison with MLP Based System Using F-Measure

The ROC method of evaluating algorithms does not measure the sensitivity of the methods to thresholds. As as illustration, for any SND method, the threshold is set for a particular operating point on the development data. When the testing environment is different from the development environment, the threshold changes. Since the threshold cannot be modified for the test data, we want the performance to remain the same.

To evaluate this aspect of the SND algorithms, F-measure is utilized. It is defined as the harmonic mean of "precision" and "recall". A high value of recall with a high precision, yields a high F-measure. The maximum value of F-measure that can be obtained is 1. This value is obtained when precision and recall reach the corresponding maximum values of 1 each.

The F-measure is used for evaluation as follows: first, an operating point on the ROC (say, equal error rate – EER) is chosen. The threshold is determined for all the SND methods on the clean speech at this operating point. For all SNR levels in the test data set, the SND algorithms are deployed with these thresholds. The metrics, true positives, false negatives and false positives, are measured. The F-measure is obtained from these quantities. The F-measure based comparison between the proposed method (with and without regularization) and the MLP based system is shown in Figure 4. Here the operating point was EER on clean speech. In this section, we discuss the comparison of proposed method without regularization and the MLP based system. Section 3.5 discusses this graph with respect to regularization. It can be seen from the figure that while the F-measure in clean speech of the MLP based method is high ($EER = 2\%$) in comparison with the proposed method ($EER = 6\%$), its performance at lower SNR levels drops below the proposed method. It indicates that the proposed method is less sensitive to thresholds than the MLP based method.

Computation of the mean F-measure over different SNRs (Clean, 10 dB, 5 dB, 0 dB and -5 dB), shows that the mean F-measure of the proposed method(0.914) at all SNR levels is higher than that of the MLP based method(0.87) by about 4.4% absolute performance. Further, we observe that the F-measure of the MLP based method drops by 24% (absolute) from clean speech to -5 dB SNR. In comparison, the proposed method drops only by 12% for the same change in environment, again indicating the robustness of the thresholds.

3.4 Determination of Optimal Length

In the second set of experiments, the length of the context is varied to identify the optimal length for the proposed method. Studying the effect of different contextual lengths inherently includes a comparison with the short-term energy based method as well.

The optimal length is obtained by varying the lengths of the contextual information at a particular SNR. Again the performance is measured using F-measure. The lengths of the feature vector (number of dimensions = number of full-band energy frames) studied in terms of number of frames of context were: 1 (energy threshold, with no context), 3, 5, 13, 27, 51, 101, 201 and 401.

Figure 5 shows the F-measure plots for various contextual lengths at five different SNR levels. From this plot, it can be observed that, for clean speech, using longer temporal context does not improve the performance. On the other hand, as the noise level increases, the temporal context becomes important. Further, it can be observed that the optimal context is around 101 frames (1000 ms). Also, it can be seen that the

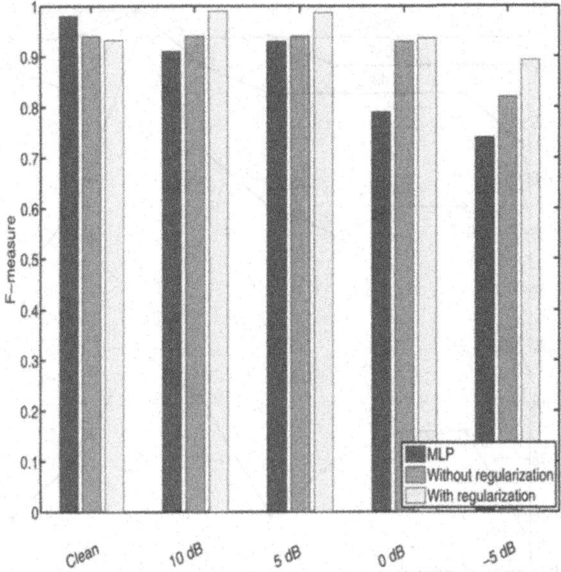

Fig. 4. Comparison of MLP based system and the proposed method (with and without regularization) using F-measure

simple energy based method is the most sensitive algorithm to changes in noise level and that when the context is increased to 1000 ms, the performance increases by 5.4% absolute. Indeed, this result is not surprising.

Also, the performance of the mean MLP method is better than that of the mean simple energy based method. Further, in clean environment, the MLP method outperforms the simple energy based method. On the other hand, at −5 dB SNR, MLP method performs worse than the energy based method, as the training and the testing conditions are badly mismatched.

3.5 Experiments with Regularized LDA

Experiments with the optimal weighted context (101 frames) derived using regularized LDA is discussed. The results of experiments, with and without regularization, is shown in Figure 4. This figure shows that, except in the case of clean speech, regularization improves the performance. Computation of the mean F-measure over all the SNR conditions for the two cases are: (a) 94.77% with regularization (b) 91.4% without regularization. It shows that regularized LDA improves the performance by 3.37% absolute over LDA without regularization. The overall improvement obtained by regularized LDA over MLP based system is 7.77% (absolute). Further, the regularized LDA yields an absolute improvement of 8.77% over the energy based method.

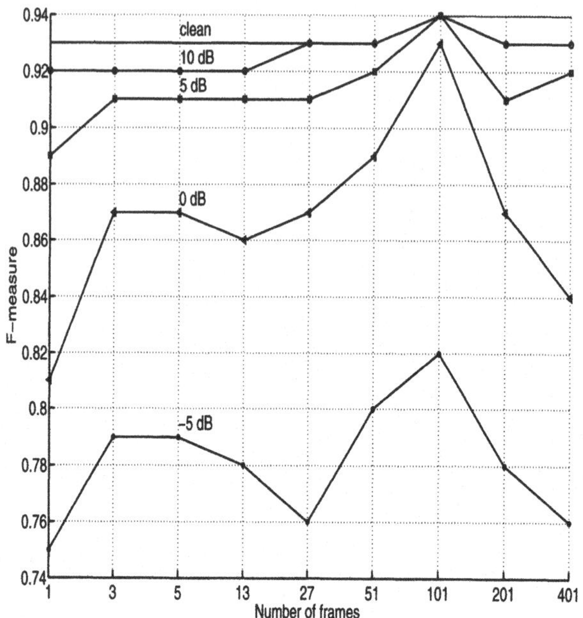

Fig. 5. Determination of optimal lengths at 5 different SNR levels: clean, 10 dB, 5 dB, 0 dB, −5 dB

4 Conclusion

We have presented a method for SND that employs full-band energy with long contextual information. This method utilizes LDA to obtain the weights of the context. The proposed method is compared with a state-of-the-art MLP based SND and an energy based system. In terms of F-measure, it shows an absolute performance gain of 4.4% and 5.4% respectively over these methods. Further,to improve the estimation of the within class covariance matrices in LDA, regularization using DCT is performed. This modification yields an absolute improvement of 7.77% and 8.77% over the MLP based and energy based method respectively.

It shows that even a simple feature such as full-band energy, when utilized with a large-enough context, is promising. In future work, we wish to investigate the importance of contextual information for sub-band energies.

Acknowledgements

This work was supported by the MIFAVO (NSF Project Micropower integrated face and voice detection, grant number: 200021-112354/1) and Detection and Identification of Rare Audio-Visual Cues (contract numbers of DIRAC is: FP6-0027787) projects; managed by the IDIAP Research Institute on behalf of Swiss Federal Authorities.

References

1. Atal, B.S., Rabiner, L.R.: A pattern recognition approach to voiced-unvoiced-silence classification with applications to speech recognition. IEEE Trans. on Acoust., Speech and Signal Process (1976)
2. Dines, J., Vepa, J., Hain, T.: The segmentation of multi-channel meeting recordings for automatic speech recognition. In: Int. Conf. on Spoken Language Processing (Interspeech ICSLP), Pittsburgh, USA, pp. 1213–1216 (2006)
3. Fukunaga, K.: Introduction to Statistical Pattern Recognition. Academic Press, London (1990)
4. Garofolo, J.S., Laprun, C.D., Michel, M., Stanford, V.M., Tabassi, E.: The NIST meeting room pilot corpus (2004)
5. Maganti, H.K., Motlicek, P., Perez, D.G.: Unsupervised speech/non-speech detection for automatic speech recognition in meeting rooms. In: IEEE Int. Conf. on Acoustics, Speech, and Signal Processing (ICASSP) (2007)
6. Mesgarani, N., Slaney, M., Shamma, S.A.: Discrimination of speech from nonspeech based on multiscale spectro-temporal Modulations. IEEE Transactions on Audio, Speech and Language Processing 14, 920–930 (2006)
7. Varga, A.P., Steeneken, H.J.M., Tomlinson, M., Jones, D.: The noisex-92 study on the effect of additive noise on automatic speech recognition. Tech. Report DRA Speech Research Unit (1992)

Identification of Speakers from Their Hum

Hemant A. Patil, Robin Jain, and Prakhar Jain

Dhirubhai Ambani Institute of Information and Communication Technology,
Gandhinagar, Gujarat, India
{hemant_patil,robin_jain,prakhar_jain}@daiict.ac.in
http://www.daiict.ac.in

Abstract. Automatic Speaker Recognition (ASR)is an economic method of biometrics because of the availability of low cost and powerful processors. An ASR system will be efficient if the proper *speaker-specific* features are extracted. Most of the state-of-the-art ASR systems use the natural speech signal (either read speech or spontaneous speech) from the subjects. In this paper, an attempt is made to identify speakers from their *hum*. The experiments are shown for Linear Prediction Coefficients (LPC), Linear Prediction Cepstral Coefficients (LPCC), and Mel Frequency Cepstral Coefficients (MFCC) as input feature vectors to the polynomial classifier of 2nd order approximation. Results are found to be better for MFCC than LP-based features.

Keywords: Speaker recognition, linear prediction, mel cepstrum, hum, polynomial classifier.

1 Introduction

Recognizing a person's voice with the help machines has been an active area of research for several decades now, but still remains active, as the problem is difficult. Even though its accuracy is lower than the fingerprints or retinal scans, speech remains the simplest method of input for biometric and hence keeps the researchers interest alive [10]. One of the first studies in ASR has been reported by Kersta [8]. In this paper, we propose a speaker recognition system for identification of speakers based on their hum. A hum is sound made by singing a wordless tone with the mouth completely closed, forcing the sound to emerge from the nose. To hum is to produce such sound, most often with a *melody*. A hum has a particular timbre (or sound quality), usually monotone or slightly varying tones. Recently, there has been a growing interest in designing Query by Humming (QBH) system for music retrieval applications [1,7]. Figure 1 and Figure 2 shows the hum (and their corresponding pitch contour, formant contour and spectrogram) produced for a Hindi song, *viz.*, 'Chalte Chalte Mere Ye Geet Yaad Rakhna', by two male speakers of age 21 years. The recording was done for both speakers with the same experimental setup discussed in next section. It is evident from the plots that the pattern of hum signal, pitch contour, formant contour and spectrogram for each speaker are *distinct*. This motivates us to investigate whether we can use hum produced by the speakers for voice biometrics problem.

P. Sojka et al. (Eds.): TSD 2008, LNAI 5246, pp. 461–468, 2008.
© Springer-Verlag Berlin Heidelberg 2008

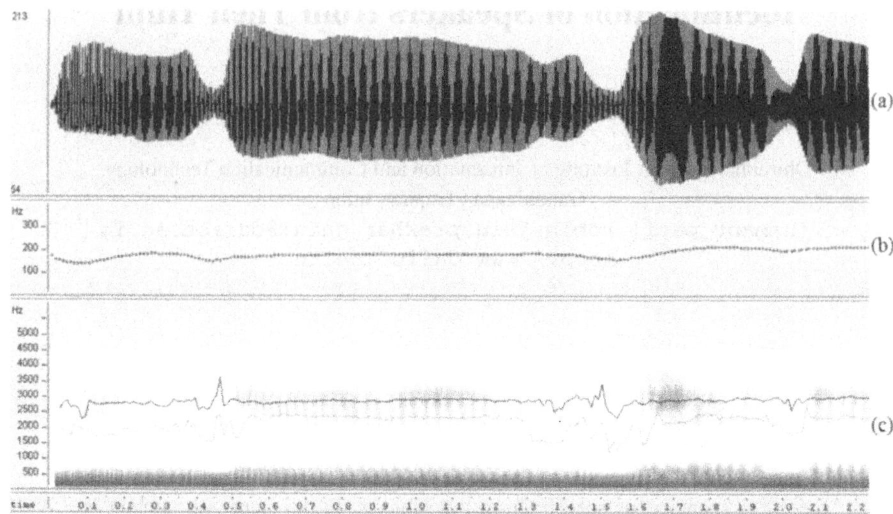

Fig. 1. (a) Hum for a song, *viz.*, 'Chalte Chalte...' by a male speaker A, (b) Pitch contour of hum shown in (a), (c) Spectrogram and formant contour of hum shown in (a)

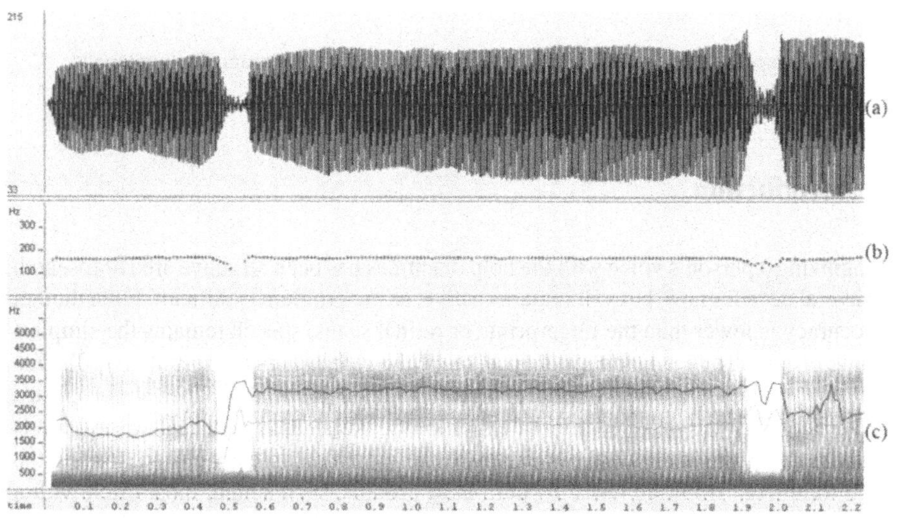

Fig. 2. (a) Hum for a song, *viz.*, 'Chalte Chalte...' by a male speaker B, (b) Pitch contour of hum shown in (a), (c) Spectrogram and formant contour of hum shown in (a)

2 Experimental Setup

The database is prepared from 10 subjects (6 male and 4 female) in the radio room of Dhirubhai Ambani Institute of Information and Communication Technology. Subjects were asked to hum for 15 most popular songs of legendry late Kishore Kumar (a famous singer in Hindi cinema) out of which 4 hums from each subject were kept for testing and remaining hums were kept for machine training. Table 1 shows the details of our corpus.

Table 1. Humming Database Description for ASR system

Option	Detail
No. of speakers	10 (6 male and 4 female)
No. of sessions	1
Data type	Hum for a song
Sampling rate	22,050 Hz
Sampling format	1-channel, 8-bit resolution
Type of hum	Humming for 15 popular songs of legendry late Kishore Kumar in Hindi cinema.
Application	Speaker Recognition from humming corpus
Training hum	Hums for 11 songs
Testing hum	Hums for 4 songs
No. of repetitions	None
Training segments	20s, 40s, 60s.
Testing segments	1s, 2s, 3s, 4s, 5s.
Microphone	SCHURE microphone
Acoustic environment	Radio Room of DA-IICT.

3 Speech Features and Classifier Used

In this paper, Linear Prediction Coefficients (LPC), Linear Prediction Cepstral Coefficients (LPCC) [2,3] and Mel frequency Cepstral Coefficients (MFCC) [5] are used as 12-dimensional *spectral* feature vectors and polynomial classifier [4] is used to build speaker models for all the ASR experiments. In next sub-sections, their computational details are discussed.

3.1 LP-Based Features

Speech sample *s(n)* is expressed by the difference equation

$$s(n) = \sum_{k=1}^{p} a_k s(n-k) + Gu(n) \tag{1}$$

where $\{a_k\}_{k=1}^{p}$ are called as LPC and G is gain term in LP model. The optimal values of $\{a_k\}_{k=1}^{p}$ are obtained by using the least square error formulation of the linear prediction. This involves solutions of the normal equations given by

$$\sum_{k-1}^{p} a_k R(n-k) = -R(n); \qquad n = 1, \ldots, p \tag{2}$$

where R(n) = $\sum_m s(m)s(m-n)$ is the autocorrelation function, which is a 2nd order statistics as proposed in Yegnanarayana et al. [11]. *LPC are known to be dependent on text material used for the recording* [2] *and hence in some sense the syllables, phonemes in speech and in turn are supposed to carry speaker-specific information.* Given that all the poles $z = z_i$ are inside the unit circle and the gain is 1, the causal LP cepstral coefficients (LPCC) of *H(z)*(vocal tract system function) is given by [9]:

$$LPCC(n) = \frac{1}{n}\sum_{i=1}^{p} |r_i|^n cos(\theta_i n),\ n > 0,\ with\ \ z_i = r_i\ exp(j\theta_i) \tag{3}$$

where z_i's are the poles of LP transfer function. *Since phoneme spectra in a particular language are produced with significant contribution of vocal tract spectral envelope, LPCC are also speaker-specific. Moreover, they can track speaker-specific information better than LPC.*

3.2 MFCC Features

MFCC is developed to mimic human perception process for hearing. For computing MFCC, we warp the speech spectrum into Mel frequency scale. This Mel frequency warping is done by multiplying the magnitude of speech spectrum for a preprocessed frame by magnitude of triangular filters in Mel filterbank followed by log-compression of sub-band energies and finally DCT. Davis and Mermelstein proposed one such filterbank to simulate this in 1980 for speech recognition application [10]. The frequency spacing of the filters used in Mel filterbank is kept linear up till 1 kHz and logarithmic after 1 kHz. The frequency spacing is designed to simulate the *subjective spectrum* from *physical spectrum* to emphasize the human perception process. *Thus, MFCC can be a potential feature to identify phonetically distinct languages (because for phonetically similar languages there will be confusion in MFCC due to its dependence of human perception process for hearing).*

3.3 Polynomial Classifier

Due to Weierstrass-Stone approximation theorem, polynomial classifiers are universal approximators to the optimal Bayes classifier. They are processed by the polynomial discriminant function [4,6]. The basic structure of the classifier is shown in Figure 3. Every speaker *i* has speaker-specific vector w_i, to be identified during training and the output of a discriminant function is averaged over time resulting in a score for every w_i [4].

The score is then given by

$$s_i = \frac{1}{N}\sum_{i=1}^{N} w^T p(x_i) \tag{4}$$

where $x_i = i^{th}$ input test feature vector, w = speaker model and p(x) = vector of polynomial basis terms of the input test feature vector.

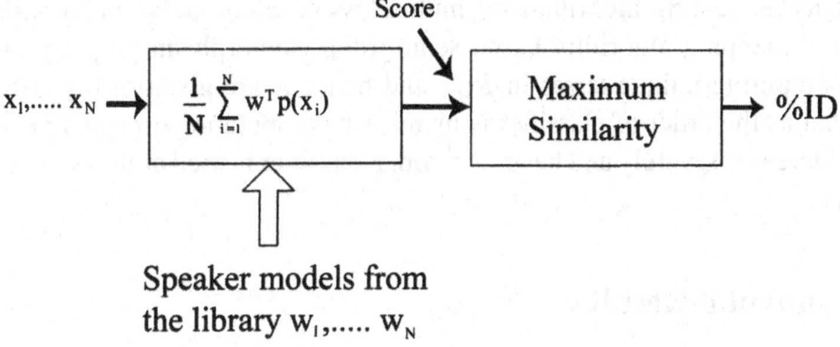

Fig. 3. The polynomial classifier structure for speaker identification

Training polynomial classifier is accomplished by obtaining the optimum speaker model for each speaker using discriminatively trained classifier with mean-squared error (MSE) criterion, i.e., for a speaker's feature vector, an output of one is desired, whereas for an impostor data an output of zero is desired.

For the two-class problem, let w_{spk} be the optimum speaker model, ω the class label, and y(ω) the ideal output, i.e., y(spk)=1 and y(imp)=0. The resulting problem using MSE is

$$w_{spk} = arg\ min_w\ \ E\left[\left(w^T p(x) - y(\omega)\right)^2\right] \tag{5}$$

where E[.] means expectation over X and ω. This can be approximated using the training feature set as

$$w_{spk} = arg\ min_w\ \ E\left[\sum_{i=1}^{N_{spk}} |w^T p(x_i) - 1|^2 + \sum_{i=1}^{N_{imp}} |w^T p(y_i)|^2\right] \tag{6}$$

where $x_1, \ldots, x_{Nspk}$ are speaker's training data and $y_1, \ldots, y_{Nimp}$ is the impostor data. This training algorithm can be expressed in matrix form. Let $M_{spk} = \left[p(x_1)\ p(x_2)\ \ldots p(x_{Nspk})\right]^T$ and a similar matrix for M_{imp}. Also let, $M = \left[M_{spk}\ \ M_{imp}\right]^T$ and thus the training problem in Equation 6 is reduced to the well-known linear approximation problem in normed space (Hilbert space) as

$$w_{spk} = arg\ min_w\ \ \|Mw - o\|_2 \tag{7}$$

where o consists of N_{spk} ones followed by N_{imp} zeros. We define $R_{spk} = M_{spk}^T\ M_{spk}$ and define R_{imp} similarly. Also define $R = R_{spk} + R_{imp}$. Thus the problem can be solved using the method of normal equations,

$$\left(R_{spk} + R_{imp}\right) w_{spk} = M_{spk}^T 1\ \ => \ \ w_{spk} = R^{-1}\left(M_{spk}^T 1\right) \tag{8}$$

The details of training algorithm for multi-class problem, polynomial basis determination and mapping algorithm based semi-group isomorphism property of monomials for computing unique terms in R_{spk} and hence R are given in [4]. It has been observed that as the order of classifier is increased, computation time and memory requirement increases severely and hence 2nd order classifier is used as the basis for all the experiments.

4 Experimental Results

Results are shown for ASR experiments as average success rates (average computed over single testing segments of 1s, 2s, 3s, 4s, and 5s, i.e., average of 5 success rates) for different training (TR) durations of 20s, 40s and 60s. In this work, the success rates are defined as

$$SR = \frac{100N_c}{N_t} \tag{9}$$

where N_c is the number of correctly identified speakers/testing segments and N_t is the total number of speakers used for machine learning. Feature analysis was performed using 12th order LPC on a 23.2 ms frame with an overlap of 50%. Each frame was pre-emphasized with the filter $1-0.97z^{-1}$, followed by Hamming window and then the mean value is subtracted from each speech frame. Similar pre-processing steps were performed for MFCC. The results are shown in Table 2 – Table 4 for different testing and training durations with 2nd order polynomial approximation whereas average success rates (over testing hum duration) for different feature sets (FS) is shown in Table 5.

Some of the observations from the results are as follows:

- On the whole, success rates are found to be more sensitive to training hum duration than testing hum duration. Results show that 60s of humming duration is not at all sufficient for good ASR performance.
- Average success rates for MFCC are higher than LPC and LPCC. This may be due to the fact that hums are mostly nasalized sounds, which are difficult to model by all-pole model of the vocal tract whereas due to human perception process for hearing we perceive hum for a song and since MFCC is designed to mimic human

Table 2. Success Rates (%) for LPC with 2nd Order Approximation

TEST(sec)	20s	40s	60s
1	60	80	90
2	70	80	80
3	80	90	90
4	90	100	100
5	90	90	90
Av.	78	88	90

Table 3. Success Rates (%) for LPCC with 2nd Order Approximation

TEST(sec)	20s	40s	60s
1	80	70	70
2	70	80	90
3	80	100	90
4	70	80	80
5	70	80	80
Av.	74	82	82

Table 4. Success Rates (%) for MFCC with 2nd Order Approximation

TEST(sec)	20s	40s	60s
1	90	90	90
2	90	100	100
3	100	100	100
4	90	90	90
5	100	100	100
Av.	94	96	96

Table 5. Avg. Success Rates (over testing hum durations) with 2nd Order Approximation

	TR		
FS	20s	40s	60s
LPC	78	88	90
LPCC	74	82	82
MFCC	**94**	**96**	**96**

perception process for hearing, it performs better for this problem. This is probably due to the fact that the spectral details in short-time spectrum are not captured very well in LP spectrum and hence possibly LP-based are not able to capture perceptual information (e.g. in some sense melody in hum produced by a speaker). On the other hand, for computing MFCC, short-time signal spectrum is warped to Mel scale to mimic human perception process for hearing and hence possibly MFCC performs better than LP-based features.

5 Conclusion

In this paper, a novel approach to speaker recognition is proposed for identification of speaker from their *hum*. The problem addressed is original and may have interesting technological applications. The relative performance of different system features, *viz.*, LPC, LPCC and MFCC is compared and LP spectral analysis is used to justify the results. The work on larger population size is under progress.

Acknowledgement

The authors of this paper would like to thank the authorities of DA-IICT Gandhinagar for giving their support to carry out this research work. They would like also to thank Prof. B. Yegnanarayana of IIIT Hyderabad and Prof. S. R. M. Prasanna of IIT Guwahati for their valuable suggestions during this research work.

References

1. Adams, N.H., Bartsch, M.A., Wakefield, G.H.: Note Segmentation and Quantization for Music Information Retrieval. IEEE Trans. Audio., Speech and Language Processing 14(1), 131–141 (2006)
2. Atal, B.S., Hanuaer, S.L.: Speech analysis and synthesis by linear prediction of the speech wave. J. Acoust. Soc. Amer. 50, 637–655 (1971)
3. Atal, B.S.: Effectiveness of linear prediction of the speech wave for automatic speaker identification and verification. J. Acoust. Soc. Amer. 55, 1304–1312 (1974)
4. Campbell, W.M., Assaleh, K.T., Broun, C.C.: Speaker recognition with polynomial classifiers. IEEE Trans. on Speech and Audio Processing 10, 205–212 (2002)

5. Davis, S.B., Mermelstein, P.: Comparison of parametric representations for monosyllabic word recognition in continuously spoken sentences. IEEE Trans. Acoust., Speech and Signal Processing 28, 357–366 (1980)
6. Duda, R.O., Hart, P.E., Stork, D.G.: Pattern Classification and Scene Analysis, 2nd edn. Wiley-Interscience, Chichester (2001)
7. Jang, J.-S.R., Lee, H.-R.: A General Framework of Progressive Filtering and Its Application to Query by Singing/Humming. IEEE Trans. Audio., Speech and Language Processing 16(2), 350–358 (2008)
8. Kersta, L.G.: Voiceprint Identification. Nature 196, 1253–1257 (1962)
9. Oppenheim, A.V., Schafer, R.W.: Discrete-Time Signal Processing. Prentice-Hall, Englewood Cliffs (1989)
10. Patil, H.A.: Speaker Recognition in Indian Languages: A Feature Based Approach. Ph.D. thesis, Department of Electrical Engineering, IIT Kharagpur, India (2005)
11. Yegnanarayana, B., Prasanna, S.R.M., Zachariah, J.M., Gupta, C.S.: Combining evidence from source, suprasegmental and spectral features for a fixed-text speaker verification system. IEEE Trans. Speech Audio Processing 13(4), 575–582 (2005)

Reverse Correlation for Analyzing MLP Posterior Features in ASR

Joel Pinto, Garimella S.V.S. Sivaram, and Hynek Hermansky

IDIAP Research Institute, Martigny
École Polytechnique Fédérale de Lausanne, Switzerland
{joel.pinto,sgarimel,hynek}@idiap.ch

Abstract. In this work, we investigate the reverse correlation technique for analyzing posterior feature extraction using an multilayered perceptron trained on multi-resolution RASTA (MRASTA) features. The filter bank in MRASTA feature extraction is motivated by human auditory modeling. The MLP is trained based on an error criterion and is purely data driven. In this work, we analyze the functionality of the combined system using reverse correlation analysis.

1 Introduction

Posterior based features figure prominently in the current state-of-the-art large vocabulary continuous speech recognition systems [1,2]. Here, a multilayered perceptron is discriminatively trained on conventional features (MFCC, PLP, etc.) to estimate the posterior probability of phonemes for every frame (typically 10 ms). The posterior probabilities are used as features in subsequent modeling and hence the name posterior features. The posterior features can be used either stand alone [3] or in conjunction with other traditional features [4].

While posterior based features have shown to improve the ASR performance, understanding of its working is limited as neural networks are considered black-boxes and the trained weights do not reflect any properties of speech/features. After the MLP is trained, its properties are typically not further analyzed. It would be useful to develop techniques that would allow to evaluate the trained MLP other than applying it in the target ASR system. This paper aims to contribute to the development of such objective evaluation techniques.

The trained MLP is treated as a nonlinear "black box" in a manner similar to the treatment of the nonlinear perceptual systems in biology. Namely, the reverse correlation technique [10], often applied for obtaining the linear time-invariant (LTI) approximation of the unknown system under consideration [10]. In this work, the MLP is trained using MRASTA [5] features. As shown in Figure 1, we treat the MRASTA filters followed by MLP as the unknown system taking critical band energies as input and estimating posterior probabilities at the output. We consider MRASTA features because (a) average stimuli derived from reverse correlation analysis can be compared to the expected time-frequency pattern and interpreted in terms of formant energies, and (b) have successfully been applied in various state-of-the-art ASR systems [4] and hence the usefulness of the analysis.

P. Sojka et al. (Eds.): TSD 2008, LNAI 5246, pp. 469–476, 2008.
© Springer-Verlag Berlin Heidelberg 2008

To draw analogy to the reverse correlation studies in physiology [10], we can loosely compare the MRASTA-MLP system to the human auditory system. The variable frequency response in MRASTA feature extraction attempts to emulate the property that each particular higher level neuron in the auditory cortex is the most sensitive to a particular modulation frequency of the signal [7,8,9]. Since we do not know exactly how the human brain is integrating this information to perceive speech sounds, we conveniently assume that the MLP learns the transformation. However, human auditory system is far superior compared to the simple MRASTA-MLP system. For example, humans do not perceive random time frequency pattern (away from the speech classes) as speech sounds whereas, MLP could assign a high posterior probability depending on its distance from decision boundary. This model deficiency clearly shows up in the reverse correlation experiments using white noise stimulus (Section 3.3). One way to overcome this deficiency is to use generative models for speech (or phonemes) such as GMM, as it restricts the boundary of a speech classes.

The rest of the paper is organized as follows. In Section 2, we briefly describe the MRASTA-MLP system that we analyze in this paper. In Section 3, we review the reverse correlation technique and use the same to analyze the basic system for various stimuli, namely speech and white noise. Section 4 describes the deficiency of the MRASTA-MLP system in white noise analysis and discusses the generative GMM model.

2 MRASTA-MLP System

The block diagram of a posterior feature extraction using MRASTA features is shown in Figure 1.

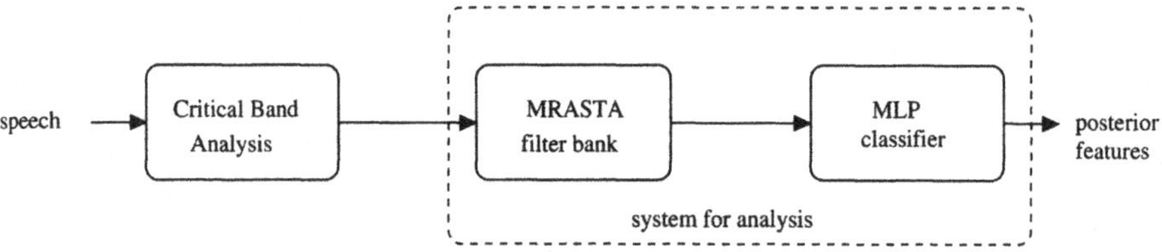

Fig. 1. Block diagram of computing posterior features using MRASTA feature extraction

2.1 Critical Band Analysis

Speech is first frame blocked into 25 ms windows with a frame shift of 10ms. Spectral analysis is performed on the windowed speech signal and energies in the critical bands are computed. The center frequency and bandwidth of the critical bands are based on the perceptual modeling of speech. The trajectory of the log-energy in each of the 19 critical bands is then filtered independently using a bank of MRASTA filters.

2.2 MRASTA Filters

MRASTA filters [5] are zero-mean, 101-tap finite impulse response filters whose shape is that of either the first or second derivative of a Gaussian function. The variance of the Gaussian function controls the resolution of each filter. Our implementation of an MRASTA filter-bank includes 8 first derivatives and 8 second derivatives of Gaussian functions with standard deviations between 8ms and 130 ms. Furthermore, the frequency derivatives are appended to the base features.

2.3 MLP Classifiers

We consider a three layered MLP classifier, where the features presented at the input layer are projected to a higher dimensional hidden layer. The nodes in the output layer represent the phoneme classes. The hidden nodes have a static non-linearity function such as sigmoid, tanh etc. The output layer has a softmax nonlinearity, which enforces the constraint that the outputs sum to unity. Cross entropy error criterion is used to train the MLP. It has been shown that MLPs with sufficient capacity estimate the Bayesian *a posteriori* probability provided that, the network is trained on sufficient training data and classes are taken with the correct *a priori* probabilities [6].

3 Reverse Correlation

Reverse correlation can be used to identify linear time-invariant (LTI) systems. If an LTI system is presented with white noise as input and yields spikes at the output, its impulse response function can be recovered by a simple spike-triggered average of the noise stimulus preceding the spikes. Section 3.1 describes the theory of reverse correlation for a linear system. In 3.2, we investigate its possible extension to analyzing a MLP using speech signal as input. In section 3.3, we apply reverse correlation by presenting white noise as input to the system.

3.1 Reverse Correlation on LTI System

Suppose that an unknown linear system with impulse response $h(t)$ and frequency response $H(\omega)$ is to be identified. Suppose that when the system is presented with white noise, spikes are produced at times times $t_1, t_2 \cdots t_N$. Denoting $x(t)$ and $y(t)$ as the input and output to the system, the power spectrum of the system can be written as

$$H(\omega) = \frac{S_{xy}(\omega)}{S_{xx}(\omega)}, \tag{1}$$

where, $S_{xy}(\omega)$ is the cross power spectral density and $S_{xx}(\omega) = \sigma^2$ is the power spectral density of the white noise input. Hence, the impulse response of the unknown system can be written as

$$\begin{aligned} h(t) &= \frac{1}{\sigma^2} r_{xy}(t) = \frac{1}{\sigma^2} \int_{-\infty}^{\infty} x(\tau - t) y(\tau) d\tau \\ &= \frac{1}{\sigma^2} \int_{-\infty}^{\infty} x(\tau - t) \sum_{k=1}^{N} \delta(\tau - t_k) d\tau \end{aligned}$$

$$= \frac{1}{\sigma^2} \sum_{k=1}^{N} x(t_k - t)$$

This is the reverse-correlation formula which states that the impulse response $h(t)$ of an LTI system can be obtained as the average of the stimulus preceding the spikes.

Reverse correlation analysis is valid only for a linear system that produces spikes when presented with white noise input. Since the MRASTA-MLP system is a nonlinear system with memory, its impulse response is not defined. Nevertheless, this method can be used to estimate an average pattern in the time-frequency (critical band energy) plane that represents patterns likely to trigger the output neuron for a phoneme. In this direction, we perform reverse correlation studies using actual speech signal and white noise as input. This is explained in the following sections.

3.2 Reverse Correlation on MLP (Speech Input)

We present speech signal from the test set and average all time-frequency patterns that give a posterior probability greater than certain threshold (e.g. 0.9) for a particular phoneme. Reverse correlation analysis on the TIMIT database shows that the average time-frequency pattern thus obtained is consistent with the expected time-frequency pattern derived using the ground truth label information as shown in Figure 2. While the average pattern obtained by reverse correlation analysis is consistent with the expected pattern, this is in the average sense (first order approximation) and this does not indicate that the trained system is perfect. Moreover, such a result is not surprising as the neural network is trained to do so.

Reverse correlation analysis using speech as input will reveal the behavior of the system for time-frequency patterns that closely match those that are seen during training. This analysis will not reveal the true functionality of the system as the stimulus space is restricted to be speech like. Reverse correlation analysis with white noise as critical band energies would reveal the behavior of the system in the average sense. White noise analysis is also motivated by the following two factors. Firstly, in the reverse correlation analysis explained in Section 3.1, impulse response of a linear system can be estimated

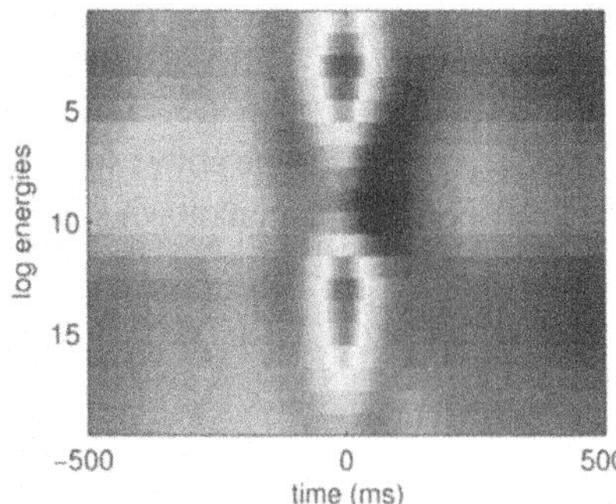

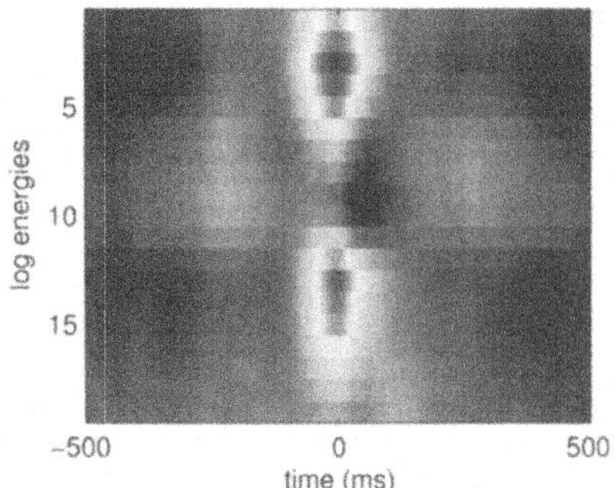

Fig. 2. The true average time-frequency pattern (left) and the average pattern estimated by reverse correlation analysis for the phoneme /iy/

as the average of the noise stimulus preceding the spikes. Secondly, in physiology experiments, spectro-temporal receptive field (STRF) of a neuron can be estimated for white noise stimulus by using reverse correlation technique [10].

3.3 Reverse Correlation on MLP (White Noise Input)

We present uniform noise as critical band energies to the MRASTA-MLP system and perform reverse correlation analysis. The minimum and maximum value of the uniform noise for each critical band is estimated from the training data. In this way, we bound the stimulus space. Noise is presented as critical band energies and not as the actual speech signal. This is because we are interested in identifying response of the system that estimates posterior probabilities from time frequency plane as this can be compared to the formant structure observed in a spectrogram.

Experiments were conducted on the TIMIT database. The average stimuli pattern obtained by reverse correlation is noisy and a plot similar to Figure 2 will not be informative. Hence, we plot the trajectories of the individual critical bands obtained from reverse correlation as shown in Figure 3. It can be observed from the figure that the trajectories obtained from reverse correlation have similar shape to the expected trajectory for all phonemes. This enables us to devise strategies to compare different systems (e.g. trained on different amounts of data, different capacity, various languages, etc.) without having to actually run ASR experiments.

The average pattern is still very noisy when compared to the one derived using speech as input. This can be attributed to the inherent nature of modeling in the MLP as explained in the following section. On the other hand, human auditory system is robust to white noise and will not associate noise patterns to any phoneme.

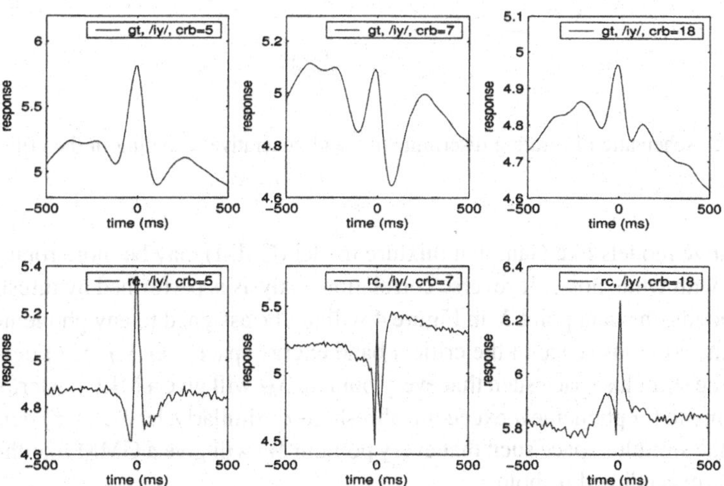

Fig. 3. Critical band trajectories for phoneme /iy/, estimated based on ground truth (gt) (top) and reverse correlation (rc) (bottom) for critical bands 5, 7, and 18

4 Generative vs. Discriminative Modeling

An MLP is trained using an error criterion which minimizes the classification error on the training set. This is achieved by adjusting the decision boundaries to maximally separate the data points corresponding to the classes. This leaves huge voids within the stimulus space, where a posterior probability of close to unity is assigned to data points even falling away from its distribution. Figure 4 is the block schematic diagram illustrating discriminative and generative modeling in the critical band space. Here, the data point X falls outside the data points of phonemes $P1$ and $P2$. However, the MLP will assign it to class $P2$ with probability close to unity. This is reason why reverse correlation analysis with white noise fails to give a time-frequency pattern close the one computed using ground truth in Figure 2. On the contrary, human auditory system is robust to white noise and will not associate noise patterns to any phoneme.

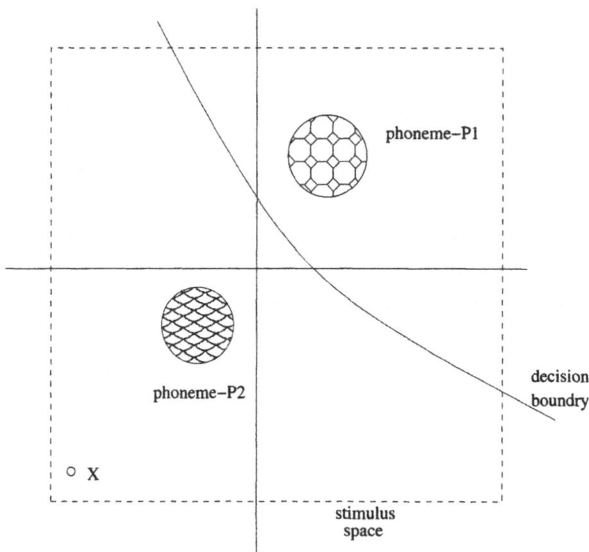

Fig. 4. Block schematic illustrating discriminative and generative modeling in the critical band space

Generative models like Gaussian mixture model (GMM) may be more robust when presented with white noise. If reverse correlation analysis is performed by thresholding the likelihoods, the data point X in Figure 4 will not be assigned to any phoneme class. Let S be the stimulus space in the critical band energy space. Let $S_M(q, \tau)$ denote the subset of the stimulus space such that every point in S_M will give a MLP posterior probability estimate for phoneme q exceeding threshold τ. Similarly, let $S_G(q, \tau)$ denote the subset of the stimulus space such that every point in S_G will give a GMM likelihood for phoneme q exceeding threshold τ.

$$S_M(q, \tau) = \{x \subset S \mid P(q|x) > \tau\} \quad (2)$$
$$S_G(q, \tau) = \{x \subset S \mid p(x|q) > \tau\} \quad (3)$$

In the case of generative GMM model, by selecting sufficiently high threshold τ, the volume of S_G can be shrunk so that reverse correlation analysis will give an average pattern close to the one obtained with speech input. On the other hand, in the case of discriminative MLP, even though a high τ (close to unity) is fixed, the volume of S_M will be still large as points far of from decision boundary will give an high posterior probability. Reverse correlation studies on GMM model is practically impossible as the volume of S_G will be significantly smaller than stimulus space S especially as the dimension of the feature vector increases. If infinite noise samples are generated, then we can expect an average pattern close to that obtained with speech input.

5 Conclusions

In this work, we present preliminary experiments on the use of reverse correlation for analyzing the system consisting of MRASTA filter banks followed by an MLP. Reverse correlation was performed using two stimuli sources namely, speech and white noise. In the case of speech stimuli, as expected the average time frequency pattern obtained by reverse correlation is close to the expected pattern derived from ground truth. Even in the case of white noise stimuli, the reverse correlation gives time-frequency patterns which are similar to the expected patterns. Reverse correlation with white noise input assumes significance as this could lead to various strategies to analyzing different MLPs (trained on different data sizes, different capacities, different languages, etc.) without actually having to run ASR experiments. In this work, we chose MRASTA feature extraction. In general, reverse correlation analysis can be applied to any feature extraction technique.

Acknowledgements

This work was supported in parts by the Swiss National Science Foundation under the Indo-Swiss joint research program KEYSPOT, the European Union under the DIRAC integrated project, contract No. FP6-IST-027787 as well as DARPA under the GALE program, contract No. HR0011-06-C-0023. Any findings and conclusions expressed in this material are those of the authors and do not necessarily reflect the views of funding agencies.

References

1. Zhu, Q., Stolcke, A., Chen, B., Morgan, N.: Using MLP Features in SRI's Conversational Speech Recognition System. In: Proc. of Interspeech, pp. 2141–2144 (2005)
2. Zhu, Q., Chen, B., Morgan, N., Stolcke, A.: On Using MLP Features in LVCSR. In: Proc. of Interspeech, pp. 921–924 (2004)
3. Hermansky, H., Ellis, D.P.W., Sharma, S.: Tandem connectionist feature extraction for conventional HMM systems. In: Proc. of ICASSP (2000)
4. Valente, F., et al.: Hierarchical Neural Networks Feature Extraction for LVCSR system. In: Proc. of Interspeech (2007)

5. Hermansky, H., Fousek, P.: Multi-resolution RASTA filtering for TANDEM-based ASR. In: Proc. of Interspeech, pp. 361–364 (2005)
6. Richard, M.D., Lippmann, R.P.: Neural Network Classifiers Estimate Bayesian a posteriori Probabilities. Neural Computation 3, 461–483 (1991)
7. Depireux, D.A., Simon, J.Z., Klein, D.J., Shamma, S.A.: Spectro-temporal response field characterization with dynamic ripples in ferret primary auditory cortex. Journal of Neurophysiology 85, 1220–1234 (2001)
8. Theunissen, F.E., Sen, K., Doupe, A.J.: Spectral-Temporal Receptive Fields of Nonlinear Auditory Neurons Obtained Using Natural Sounds. Journal of Neurophysiology 20, 2315–2331 (2000)
9. Kleinschmidt, M., Gelbart, D.: Improving Word Accuracy with Gabor Feature Extraction. In: Proc. of ICSLP, Colorado, USA (2002)
10. Klein, D.J., Depireux, D.A., Simon, J.Z., Shamma, S.A.: Robust Spectrotemporal Reverse Correlation for the Auditory System: Optimizing Stimulus Design. Journal of Computational Neuroscience 9, 85–111 (2000)

Acquisition of Telephone Data from Radio Broadcasts with Applications to Language Recognition*

Oldřich Plchot, Valiantsina Hubeika, Lukáš Burget, Petr Schwarz, and Pavel Matějka

Speech@FIT, Brno University of Technology, Czech Republic,
{iplchot,burget,schwarzp,matejkap}@fit.vutbr.cz,
xhubei00@stud.fit.vutbr.cz

Abstract. This paper presents a procedure of acquiring linguistic data from the broadcast media and its use in language recognition. The goal of this work is to answer the question whether the automatically obtained data from broadcasts can replace or augment to the continuous telephone speech. The main challenges are channel compensation issues and great portion of unspontaneous speech in broadcasts. The experimental results are obtained on NIST LRE 2007 evaluation system, using both NIST provided training data and data, obtained from broadcasts.

Keywords: Language Identification (LID), Broadcast data, Phone call detection, Channel compensation.

1 Introduction

We introduce a process of automatic acquisition of speech data from the various media sources for the language identification task. The last editions of NIST Language Recognition (LRE) evaluations have shown that both acoustic and phonotactic approaches have reached a certain maturity level in both modeling of target languages and dealing with the influences of different channels. However we are still facing the common problem: the lack of training data. There is no good or large enough database of training data for many languages including even languages like Thai, which is spoken by 65 million speakers. Also, there is an increasing demand to recognize languages from smaller and less populous regions (many of them relevant for security of defense domain). For some of these languages no standard speech resources exists.

This work aims at solving this problem using the data acquired from public sources, such as satellite and Internet TVs and radios, which contain conversational speech or telephone calls. This approach should provide us with large amount of data that we

* This work was partly supported by European projects AMIDA (FP6-033812), Caretaker (FP6-027231) and MOBIO (FP7-214324), by Grant Agency of Czech Republic under project No. 102/08/0707 and by Czech Ministry of Education under project No. MSM0021630528. The hardware used in this work was partially provided by CESNET under project No. 162/2005. Lukáš Burget was supported by Grant Agency of Czech Republic under project No. GP102/06/383.

P. Sojka et al. (Eds.): TSD 2008, LNAI 5246, pp. 477–483, 2008.
© Springer-Verlag Berlin Heidelberg 2008

Table 1. Overview of different channels. DVB stands for Digital Video Broadcasting – Terrestrial, Cable and Satellite. By parallel recording we mean the possibility of acquiring more broadcasts simultaneously using one recording device (i.e. one DVB-S receiver).

	Inet. radio	**DVB-T**	**DVB-C**	**DVB-S**	**Analog**
Languages	approx. 100	1 – 3	approx. 5	20 – 30	3 – 5
Quality	variable	good	good	good	bad
Parallel recording	yes	yes	yes	yes	no

expect will lead to improved performance for languages included in our present systems and to capability of processing languages that we were unable to recognize due to absence of the data.

First, the obtained data has to be preprocessed in order to acquire clean speech segments or individual phone calls. The task is to examine both sets of obtained data (wideband speech segments and phone calls) by training and evaluating the systems for languages with a lack of standard training data and on the basis of the results conclude, whether these data can be used to improve existing LRE systems.

The main challenge is channel compensation, as the obtained data are acoustically very different from the conversational telephone speech (CTS) commonly used in LRE. Broadcast data contain a great deal of unspontaneous speech as well. Further task is to explore how unspontaneous speech affects current LRE systems (which are supposed to be trained on spontaneous data). The notion of channel compensation will therefore have to be extended to cope with these factors.

We have done experiments with Thai language so far and we are planning to extend this work to the broad range of other languages.

2 Data Acquisition

There is unlimited source of speech data available from the broadcast media. We can acquire data from several sources, each of which has different channel parameters, quality and number of available languages. The list of available sources in the Czech Republic are shown in Table 1 [1].

All of the listed sources except Internet radios are geographically dependent regarding location. The quality of different Internet sources vary a lot and it is important to carefully choose them. We have used an archive[1] of Voice of America Internet radio to obtain the Thai data, as the Thai language is not available from DVB-S in the Czech Republic.

This particular data of VoA were obtained in MP3 format, bitrate is 24 Kbit/s, sampling rate 22,050 Hz, 16 bit encoding, mono. Original media data we obtained include a great portion of music and speech with the music in background. We have to deal with this problem and select only clean speech segments. Also we should deal with the problem of a low speaker variability in the obtained data, for instance as it is common in news programmes, which are moderated by the same speaker constantly. However we have not investigated this potential problem so far.

[1] FTP server 8475.ftp.storage.akadns.net directory /mp3/voa/eap/thai

3 Experiment

3.1 Selection of Scenario

Two main scenarios face the issue of the lack of training data. It can be either the situation when we have not enough training data from standard sources or when we do not have any data for particular language at all. In NIST LRE 2007 evaluation we can simulate the first scenario with Thai where we have only 1.5 hour of training data (see Table 2), and we can easily simulate the second scenario by removing these data.

We downloaded about 250 hours of radio data. We decided to select only phone calls, because they match the target CTS data and we can successfully detect them. The phone calls usually contain no music and we suppose they represent conversational speech. We detected 10 hours of telephone calls in the obtained data which is about 4% of the original recordings. Since there were some English calls, the phonotactic LID (from LRE2005 BUT system [2]) was used to detect Thai versus English. Consequently we listened to all samples with low recognition confidence to verify they are not English.

3.2 Detecting Phone Calls

Our phone call detector is based on the fact that a telephone channel acts like a bandpass filter, which passes energy between approximately 400 Hz and 3.4 KHz. On the other

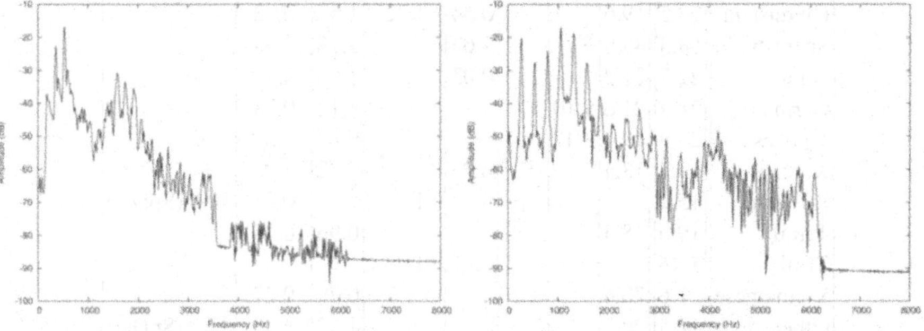

Fig. 1. Power Spectral Density of telephone call in the broadcast (left figure) and wideband speech (right figure)

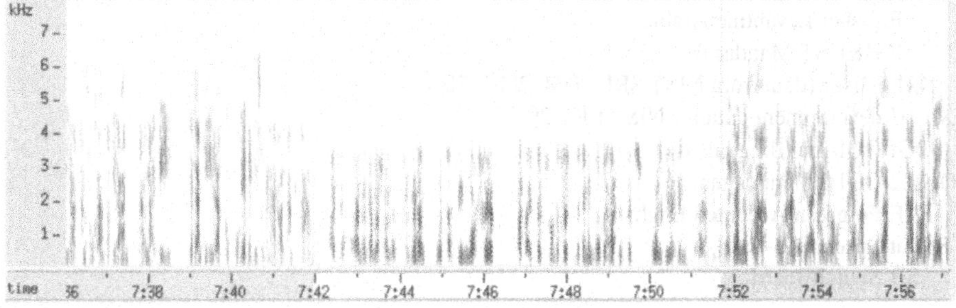

Fig. 2. Phone Call in a Radio Broadcast

hand, regular wideband speech contains significant energy up to around 5 KHz. Common media sources like satellite radio or Internet radios are usually sampled at 22 kHz so it supports this bandwidth, which means that if we place a phone call into the regular radio transmission, we will see a significant change in the spectrum (see Figure 2).

For the detection, we first resample the signal to commonly used 16 kHz. The signal is divided into frames of 512 samples with no overlap and Fourier spectrum is computed for each frame. To detect boundary between wideband and telephone speech, we concentrate on the frequency range between 2350 and 4600 Hz. The power spectral density (PSD) in this range was used (see Figure 1). At first the PSD was normalized to zero mean and unit variance. Then values in the first half and values in the second half of the PSD were summed. A ratio between these two sums was compared with a threshold and the decision was made. If the sum from higher frequencies is bigger than the sum from lower frequencies, there is more energy and it is a wide band speech.

Table 2. Training data in hours for each language and source

	sum	CF	CH	F	SRE	LDC07	OGI	OGI22	Other
Arabic	212	19.5	10.4	175	5.93	1.45		0.33	
Bengali	4.27				2.86	1.42			
Chinese	93.2	41.7	1.64	17.2	44.9	4.2	0.87	0.85	
English	264	39.8	4.68	162	34.9		6.77	0.52	15.6 (FAE)
Hindustani	23.5	19.6			0.64	1.32	1.53	0.42	
Spanish	54.3	43.8	6.71		2.63		1.18	0.38	
Farsi	22.7	21.2			0.03		1.00	0.42	
German	28.2	21.6	5.10				1.12	0.38	
Japanese	23.9	19.1	3.47				0.87	0.35	
Korean	19.7	18.4			0.09		0.72	0.5	
Russian	15.1				3.38	1.33		0.43	10.0 (SpDat)
Tamil	19.6	18.4					0.96	0.26	
Thai	**1.45**				0.15	1.23			
Vietnamese	21.6	20.6					0.79	0.27	
Other	62.5	20.7					1.10	3.29	37.4 (SpDat)

CF CallFriend
CH CallHome
F Fisher English Part 1.and 2.
F Fisher Levantine Arabic
F HKUST Mandarin
SRE Mixer (data from NIST SRE 2004, 2005, 2006)
LDC07 development data for NIST LRE 2007
OGI OGI-multilingual
OGI22 OGI 22 languages
FAE Foreigen Accented English
SpDat SpeechDat-East
SB SwitchBoard

3.3 System Description

We use the best single performing system from our NIST LRE 2007 submission to perform the following experiments. The inspiration for this system came from our GMM system for speaker recognition [3] which follows conventional Universal Background Model-Gaussian Mixture Modeling (UBM-GMM) paradigm [4] and employs number of techniques that have previously proved to improve GMM system performance [5]. This system was chosen because it can easily compensate for the channel distortion.

Table 2 lists the corpora (distributed by LDC and ELRA) used to train our systems.

Our system uses the popular shifted-delta-cepstra (SDC) [6] feature extraction, where 7 MFCC coefficients (including coefficient C0) are concatenated with SDC 7-1-3-7, which totals in 56 coefficients per frame.

Vocal-tract length normalization (VTLN) [7] performs simple speaker adaptation. VTLN warping factors are estimated using single GMM (512 Gaussians), ML-trained on the whole CallFriend database (using all the languages). The model was trained in standard speaker adaptive training (SAT) fashion in four iterations of alternately re-estimating the model parameters and the warping factors for the training data.

Each language model is obtained by traditional *relevance MAP* adaptation [8] of UBM using enrollment conversation. Only means are adapted.

In verification phase, standard Top-N Expected Log Likelihood Ratio (ELLR) scoring [8] is used to obtain verification score, where N is set to 10. However, for each trial, both language model and UBM are adapted to channel of test conversation using simple eigenchannel adaptation [3] prior to computing the log likelihood ratio score.

3.4 Channel Compensation

As the data used in the system were obtained from different databases and therefore recorded over different channels, eigenchannel adaptation [9] was applied in order to compensate the channel distortion. In language detection task, channel variability may comprehend not only variability in the telephone channel or type of microphone but also session or speaker variability.

4 Results and Discussion

The results are evaluated using standard metrics: Detection Error Tradeoff (DET) curve (see Figure 3), Decision Cost Function (DCF) and Equal Error Rate (EER) [10]. Several experiments were run using the original telephone data provided by NIST and the telephone data acquired from radio. All experiments were done on *10 second segments*. Two Thai models were trained on both sets respectively. Other 13 language-dependent GMMs were shared by both systems.The results are presented in Table 3 for the Thai language only (Thai DCF row) and for the complete systems containing all 14languages. The "NIST" column stands for a system with Thai trained on 1.45 hours of CTS data available from LDC, the "Radio"denotes systems trained on 10 hours telephone data obtained from broadcasts.

Table 3. Results. Channel comp. stands for the system with eigenchannel compensation. Channel comp. +4 stands for the system where to the former eigenchannels 4 additional eigenchannels computed only on Thai data (NIST and Radio) were added.

	No channel comp. NIST	No channel comp. Radio	Channel comp. NIST	Channel comp. Radio	Channel comp. +4 Radio
DCF all lang.	12.83	13.66	7.30	7.56	7.47
EER all lang.	13.02	13.80	7.41	7.65	7.59
Thai DCF	7.81	11.61	3.93	6.05	5.97

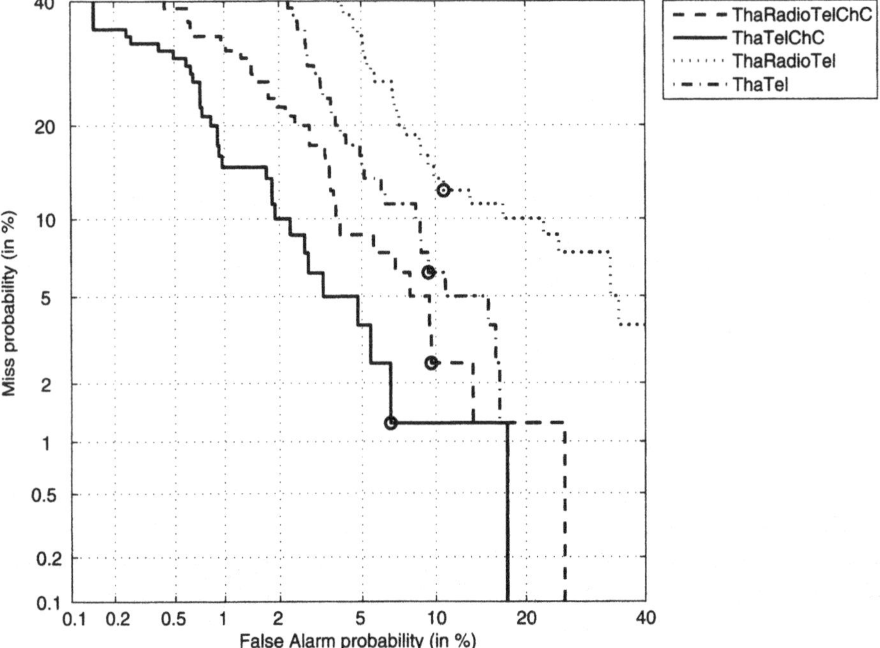

Fig. 3. Results achieved with both systems with and without channel compensation for Thai only. The det curves marked as ThaiTel and ThaiTelChC represent the results achieved on the original telephone data (CTS) with and without eigenchannel adaptation respectively. The det curves marked as ThaiRadioTel and ThaiRadioTelChC represent the results achieved on the telephone data acquired from radio with and without eigenchannel adaptation respectively.

First, both systems were trained without channel compensation. Then, eigenchannel adaptation was applied, using a matrix containing 50 eigenchannels computed without radio data.

The third experiment was performed by adding several eigenchannels estimated using only Thai data (both, CTS and radio data) and concatenated to the former eigenchannel matrix. This approach decreased the error in the system where Thai was trained on radio data, although this decrease was not significant.

However the recognizer trained on radio data does not bring as good results as the recognizer trained on CTS data, the results show (in comparison to the average DCF) that in case of language there are no available data for, radio data can be used.

5 Conclusions and Future Work

We introduced a simple but promising approach of acquiring telephone data for LID. Experiments with Thai language modeled using standard telephone data and telephone data acquired from broadcast were performed. Although the system trained on this data did not outperform the one trained on CTS, the degradation was not critical and we consider it a viable option for scenarios with very little or no data for a given language. Future investigation on more languages should be performed. Our main goal will be to obtain the radio data for all 14 languages and to train LID system on these data. We will also investigate into channel compensation techniques in cases several sources (CTS, broadcast-telephone, read speech) are mixed.

References

1. Řezníček, I.: Audiovisual recording system. Diploma thesis, Brno University of Technology FIT (2007)
2. Matějka, P., Burget, L., Schwarz, P., Černocký, J.: Nist language recognition evaluation 2005. In: Proceedings of NIST LRE 2005, pp. 1–37 (2006)
3. Burget, L., Matějka, P., Schwarz, P., Glembek, O., Černocký, J.: Analysis of feature extraction and channel compensation in gmm speaker recognition system. IEEE Transactions on Audio, Speech, and Language Processing 15(7), 1979–1986 (2007)
4. Reynolds, D.A., Quatieri, T.F., Dunn, R.B.: Speaker verification using adapted gaussian mixture models. Digital Signal Processing 10(1–3), 19–41 (2000)
5. Matějka, P., Burget, L., Schwarz, P., Černocký, J.: Brno university of technology system for nist 2005 language recognition evaluation. In: Proc. NIST LRE 2005 Workshop, San Juan, Puerto Rico, June 2006, pp. 57–64 (2005)
6. Torres-Carrasquillo, P.A., Singer, E., Kohler, M.A., Greene, R.J., Reynolds, D.A., Deller, J.R.: Approaches to language identification using gaussian mixture models and shifted delta cepstral features. In: Proc. 7th International Conference on Spoken Language Processing, Denver, Colorado, USA (September 2002)
7. Cohen, J., Kamm, T., Andreou, A.G.: Vocal tract normalization in speech recognition: Compensating for systematic speaker variability. The Journal of the Acoustical Society of America 97(5), 3246–3247 (1995)
8. Reynolds, D.A.: Comparison of background normalization methods for text-independent speaker verification. In: Proc. Eurospeech, Rhodes, Greece, September 1997, pp. 963–966 (1997)
9. Brummer, N.: Spescom DataVoice NIST 2004 system description. In: Proc. NIST Speaker Recognition Evaluation 2004, Toledo, Spain (June 2004)
10. The 2007 NIST Language Recognition Evaluation Plan (LRE 2007), http://www.nist.gov/speech/tests/lang/2007/LRE07EvalPlan-v8b.pdf

Multilingual Weighted Codebooks for Non-native Speech Recognition

Martin Raab[1,2], Rainer Gruhn[1,3], and Elmar Nöth[2]

[1] Harman Becker Automotive Systems, Speech Dialog Systems, Ulm, Germany
mraab@harmanbecker.com
http://www.harmanbecker.de
[2] University of Erlangen, Dept. of Pattern Recognition, Erlangen, Germany
[3] University of Ulm, Dept. of Information Technology, Ulm, Germany

Abstract. In many embedded systems commands and other words in the user's main language must be recognized with maximum accuracy, but it should be possible to use foreign names as they frequently occur in music titles or city names. Example systems with constrained resources are navigation systems, mobile phones and MP3 players.

Speech recognizers on embedded systems are typically semi-continuous speech recognizers based on vector quantization. Recently we introduced Multilingual Weighted Codebooks (MWCs) for such systems. Our previous work shows significant improvements for the recognition of multiple native languages. However, open questions remained regarding the performance on non-native speech.

We evaluate on four different non-native accents of English, and our MWCs produce always significantly better results than a native English codebook. Our best result is a 4.4% absolute word accuracy improvement. Further experiments with non-native accented speech give interesting insights in the attributes of non-native speech in general.

Keywords: Multilingual, codebook, semi-continuous, non-native.

1 Introduction

For speech recognition in embedded systems, multilinguality is a great challenge. Commands and other words in the user's main language must be recognized with maximum possible accuracy, but also words in other languages should be recognized well, e.g. for music titles. Therefore a system is needed that performs as well as possible for all languages under the constraint of keeping monolingual performance in the main language. Earlier approaches [1,2,3] did not address the constraint of conserving the main language performance when traversing from a mono- to a multi-lingual system.

An additional problem is that human users are uttering names in foreign languages. In most cases such pronunciations will differ significantly from native pronunciations of the same name. These deviations in non-native speech are well known to degrade the performance of speech recognizers severely.

We showed that MWCs can outperform traditional codebook generation methods for native speech in [4]. Results in this paper show that a similar conclusion can be

P. Sojka et al. (Eds.): TSD 2008, LNAI 5246, pp. 485–492, 2008.
© Springer-Verlag Berlin Heidelberg 2008

drawn for non-native speech. In addition, our results show that it is better to use more languages than the spoken language and the native language of the speaker for an MWC for non-native speech.

For completeness, the motivation and the idea of MWCs is briefly recapitulated. The motivation is based on the fact that traditional vector quantization algorithms like the LBG [5] are not optimal for our scenario. The aim of the LBG is to find a limited number of Gaussian prototypes in the feature space that cover the training data as well as possible. For the multilingual scenario, either only main language training data or data from all languages can be used with the LBG. In the first case the codebook is only optimized for the main language, not considering the performance on the additional languages. In the second case the codebook is optimized for all languages without prioritizing the main language.

Therefore the idea of MWCs is to generate a codebook that optimizes performance on all languages, with the constraint that the main language performance is more important than the performance on other languages. The first step is the construction of a codebook for each language. For this soft vector quantization based on the LBG approach is used. From these initial codebooks a new codebook is created. As this new codebook is based on codebooks from many languages, it is called multilingual, and as the influence of each original codebook can be adjusted, it is called weighted.

The remainder of this paper is organized as follows: Section 2 describes the baseline architecture to train recognizers for multiple languages. Section 3 explains how MWCs are constructed from initial codebooks. Section 4 describes our experimental setup. The results are given in Section 5. Finally, a conclusion is drawn.

2 Baseline System

We start with a well trained monolingual semi-continuous HMM speech recognizer. While keeping the codebook which was generated with the main language constant, the following is done for each additional language:

- Add all additional language HMMs to the recognizer,
- Train these additional HMMs with training data from the corresponding language, not changing the codebook.

Finally, we have a system with trained HMMs for all languages. As we explained in the introduction the main language codebook is not optimal for our scenario.

3 Extended System

To improve the performance on the additional languages, the monolingual codebook is replaced by a MWC. The MWC is basically the main language codebook plus some additional Gaussians. Figure 1 depicts an example for the extension of a codebook to cover an additional language. From left to right one iteration of the generation of MWCs is represented.

The picture to the left shows the initial situation. The Xs are mean vectors from the main language codebook, and the area that is roughly covered by them is indicated by the dotted line. Additionally, the numbered Os are mean vectors from the second language codebook. Supposing that both Xs and Os are optimal for the language they were created for, it is clear that the second language contains sound patterns that are not typical for the first language (Os 1, 2 and 3).

The middle picture shows the distance calculation. For each of the second language codebook vectors, the nearest neighbor among the main language Gaussians is determined. These nearest neighbor connections are indicated by the dotted lines. Our previous experiments showed that using the Mahalanobis distance produces the best results [4].

The right picture presents the outcome of one iteration. From each of the nearest neighbor connections, the largest one (O number 2) was chosen as this is obviously the mean vector which causes the largest vector quantization error. Thus, the Gaussian O number 2 was added to the main language codebook.

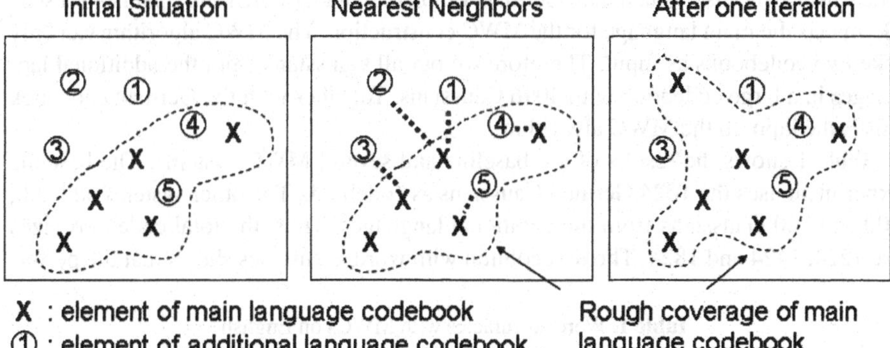

Fig. 1. Basic Idea of Multilingual Weighted Codebooks

4 Experimental Setup

Our semi-continuous speech recognizer uses 11 MFCCs with their first and second derivatives per frame and LDA for feature space transformation. All recognizers are trained on 200 hours of Speecon data [6]. The HMMs are context dependent.

We used five languages for the training of our recognizer (US English, French, German, Spanish and Italian). For each training language a codebook with 1024 Gaussians was created by the LBG algorithm. In Section 5.1 some results on native speech are given. The native test sets consist of city names. The noise conditions are uniform within each language, but not across languages.

For the non-native tests in Section 5.2 the HIWIRE data [7] is used. We test on the clean speech test which is provided with the data (50% of the HIWIRE data). Our results are lower than other published results, which we believe is due to the fact that we trained on Speecon, which contains some background noise. On some tests on low noise HIWIRE speech our systems clearly outperformed results in the literature. The HIWIRE database contains English from French, Spanish, Italian and Greek speakers. The adaptation data is not used.

To reduce the number of experiments, only results with German as the main user interaction language are presented.

5 Results

In this results section we want to give a complete overview of the benefits of MWCs for multilingual embedded systems. Therefore, both results of MWC systems on native and on non-native speech are presented. In addition, we present some further results that are necessary to answer some questions raised by the results on native and non-native speech in a plausible manner.

5.1 MWCs on Native Speech

The performance is evaluated on German, English, Italian, French and Spanish test sets. German is the main language for the MWC construction. The MWC algorithm can only take two codebooks as input. Therefore we put all Gaussians from the additional languages in a large codebook with 4096 Gaussians. Together with the German codebook this is the input to the MWC algorithm.

Table 1 shows the results of the baseline and several MWC systems. The baseline experiment uses the 1024 German Gaussians as codebook. The other systems add 200, 400 and 800 Gaussians from the additional languages. Thus, the total codebook sizes are 1224, 1424 and 1824. The first column with word accuracies shows that the performance on the German test set varies insignificantly. This is what we expected, as the LBG produces already an optimal codebook for German. Thus, the extensions to the codebook can not improve performance on German, but they also do not hurt.

Table 1. Word Accuracies with MWCs on English

Total Gaussians	Added Gaussians	German	English	Italian	French	Spanish
1024	0	84.1	65.6	85.2	68.7	88.3
1224	200	83.8	68.4	88.3	69.0	90.2
1424	400	84.0	70.9	87.9	71.3	**91.5**
1824	800	**84.3**	**72.0**	**89.7**	**72.9**	91.0

However, for the additional languages significant improvements are achieved. In general, the performance increase is correlated to the amount of Gaussians we add to the codebook, meaning the more we add, the better the performance. The differences between the different test sets are less relevant, as they are due to different noise conditions and on different languages.

It is also important to verify how the MWC algorithm performs compared to a common codebook that is trained with data from all languages. Therefore we build a codebook with 1424 Gaussians with the LBG algorithm. The results of this system are presented in Table 2, the baseline is the 1424 MWC system. For the additional languages, the codebook created by the LBG outperforms our new algorithm if performance over all four languages is compared. However, the performance on the additional

Table 2. Comparison to multilingual Codebook created only with LBG

Codebook	German	English	Italian	French	Spanish
1424 MWC	**84.0**	**70.9**	87.9	71.3	**91.5**
1424 LBG	80.8	70.5	**90.6**	**72.2**	91.4

languages was only our second priority aim. On German, the main language, our MWC significantly outperforms the LBG approach. Thus the MWC systems are better for multilingual systems that have one main interaction language.

5.2 MWCs on Non-native Speech

Table 3 shows the performance on non-native English with Spanish, French, Italian and Greek accent. The first column shows for reference the performance on native English. Again, German is the main language for the multilingual systems. Thus there are two systems that can be regarded as baseline. First, the system with an English codebook, as we want to improve performance compared to a system that would usually be used to recognize non-native English. On the other hand, a system with only the German codebook can also be regarded as baseline, as all our MWC systems base on the German codebook. The MWC systems are the same as in Section 5.1.

Table 3. Word Accuracies with MWCs on HIWIRE

Codebook	US City	Hiwire SP	Hiwire FR	Hiwire IT	Hiwire GR
English 1024	**75.5**	82.5	83.9	81.6	83.1
German 1024	65.6	85.4	**86.9**	82.8	85.5
5ling_1224	68.6	86.2	86.6	84.6	85.3
5ling_1424	68.4	86.4	86.7	**85.7**	85.8
5ling_1824	70.9	**86.9**	86.2	84.2	**86.3**

A completely unanticipated result is the performance of the two baseline systems on non-native speech. While for native English a native English codebook is significantly better, the system using the German codebook is in all cases better for the recognition of non-native English. This shows how strong non-native speech differs from native English.

But we also have to analyze the results for each of our baselines separately. When the focus is on the performance relative to the system with the English codebook, the MWCs steadily improve the performance. The larger the codebook is, the better the performance. The MWC system with 1824 Gaussians shows significant improvements for all accents. Thus we have shown that MWCs are a very valuable tool for the recognition of non-native speech.

However, when the focus is on the performance relative to a system with German codebook, the improvements on the non-native accents are rather moderate. To verify all the results we also evaluate each test set in a phoneme loop and align the recognition phonemes via dynamic programming to the reference phoneme string. As the HIWIRE

Table 4. Phoneme Loop Accuracies on HIWIRE

Codebook	Hiwire SP	Hiwire FR	Hiwire IT	Hiwire GR
English 1024	−1.1	8.1	−0.9	−6.4
German 1024	3.8	10.5	3.3	0.1
5ling_1224	5.9	12.8	5.1	1.0
5ling_1424	5.5	12.9	5.0	0.6
5ling_1824	**6.6**	**13.1**	**6.1**	**1.8**

corpus is not phonetically transcribed, the canonical phoneme sequence of each word in the reference is used as the reference phoneme string.

The results in Table 4 confirm the previous results, and the effect that the German codebook produces better performance than an English codebook is even more articulated. Obviously, all phoneme accuracies are in a very low range due to high insertion rates (the phone correctness rates for the classifiers are about 50%). These high insertion rates are due to two reasons. First, phoneme recognition on non-native speech is just a difficult problem. Second, we usually apply phonotactical models to keep insertion rates lower. However, for the experiments in Table 4, we did not want to influence the recognition rates by additional rules.

In the next section we analyze why a German codebook performs better than an English codebook and present a plausible explanation.

5.3 Native Language Codebooks on Non-native Speech

This set of experiments shows results of systems with codebooks that are built purely on native data of each language, each with 1024 Gaussians. The HMMs are trained with native English speech. The results show that for the native US City test, the US codebook is by far the best. Yet for non-native speech, the US English codebook performs worse than other codebooks. The strongest contrast can be observed for the Spanish case. The Spanish codebook loses almost 20% word accuracy on the native English test. On the other hand, the Spanish codebook performs better than the English codebook for English with Spanish accent.

These results show how important it is to have a good codebook, and how strong non-native English differs from native English. The German and Italian codebook outperform the English codebook significantly in all cases. A possible explanation is that a codebook performs better the more phonemes the corresponding language has. In our

Table 5. Word Accuracies with native language codebooks

Codebook	US City	Hiwire SP	Hiwire FR	Hiwire IT	Hiwire GR
German	65.6	85.4	**86.9**	82.8	85.5
Italian	62.8	**87.1**	84.3	**86.2**	**86.5**
Spanish	56.1	85.5	79.6	80.0	82.9
French	64.2	86.7	86.0	83.0	84.6
US English	**75.5**	82.5	83.9	81.6	83.1

phoneme sets, which are based on standard SAMPA [8] German has 59 Phonemes, Italian 50, English 46, French 37 and Spanish 29.

Apart from this unexpected aspect of non-native speech, the results also provide evidence for the common notion that non-native speakers use sounds of their mother tongue. The codebook built on the mother tongue of the speaker always performs very good, even if the language itself has fewer phonemes. The Spanish codebook system achieves 85.5% on the Spanish accented English, and the Italian codebook system achieves by far the best performance on the Italian accented speech.

6 Conclusion

This paper evaluates MWCs, a technique for improved vector quantization that has recently been introduced for multilingual scenarios. We showed that an MWC increases performance significantly on native speech of several languages while keeping monolingual performance for the main language. Furthermore, we were able to show that MWCs show significantly better performance on non-native English than a native English codebook.

The fact that a German codebook performed better than an English codebook on non-native English was a surprising aspect of our results. In Section 5.3 we showed that the performance of a codebook on non-native English is related to the amount of phonemes of the language that is used for building the codebook. In other words, building codebooks on phonetically rich languages is better for non-native speech recognition.

This gives an interesting insight into the attributes of non-native speech, being of interest for all work on non-native speech. It seems, that the sounds produced by non-native speakers are not only from their native language or the language they want to speak. Additionally, when they fail to produce foreign sounds they produce sounds that are typical for humans in general and easy to articulate.

We are aware that we have only evaluated several accents of English, but if we trust the conclusion above, we can generalize our results further. As MWCs can cover typical human sounds of all languages, not limited to the phonetical richness of one language, the conclusion would be that the MWC algorithm can produce optimal codebooks for non-native speech of all languages.

Acknowledgments

We thank the HIWIRE project from the EC 6th Framework IST Programme for the non-native test data [7].

References

1. Koehler, J.: Multilingual phone models for vocabulary-independent speech recognition tasks. Speech Communication Journal 35, 21–30 (2001)
2. Gruhn, R., Markov, K., Nakamura, S.: A statistical lexicon for non-native speech recognition. In: Proc. Interspeech, Jeju Island, Korea, pp. 1497–1500 (2004)

3. Schultz, T., Waibel, A.: Language-independent and language-adaptive acoustic modeling for speech recognition. Speech Communication 35, 31–51 (2001)
4. Raab, M., Gruhn, R., Noeth, E.: Multilingual weighted codebooks. In: Proc. ICASSP, Las Vegas, USA (2008)
5. Linde, Y., Buzo, A., Gray, R.: An algorithm for vector quantization design. IEEE Transactions on Communications 28, 84–95 (1980)
6. Iskra, D., Grosskopf, B., Marasek, K., van den Huevel, H., Diehl, F., Kiessling, A.: Speecon speech databases for consumer devices: Database specification and validation. In: Proc. LREC (2002)
7. Segura, J., et al.: The HIWIRE database, a noisy and non-native English speech corpus for cockpit communication (2007), http://www.hiwire.org/
8. Wells, J.: SAMPA (2008), http://www.phon.ucl.ac.uk/home/sampa/index.html

Prosodic Phrases and Semantic Accents in Speech Corpus for Czech TTS Synthesis*

Jan Romportl

[1] SpeechTech, s.r.o., Morseova 5, Plzeň, Czech Republic
[2] Department of Cybernetics, Faculty of Applied Sciences
University of West Bohemia, Univerzitní 8, Plzeň, Czech Republic
jan.romportl@speechtech.cz

Abstract. We describe a statistical method for assignment of prosodic phrases and semantic accents in read speech data. The method is based on statistical evaluation of listening test data by a maximum-likelihood approach with parameters estimated by an EM algorithm. We also present linguistically relevant quantitative results about the prosodic phrase and semantic accent distribution in 250 Czech sentences.

1 Introduction

The aim of this paper is to describe a method which is able to objectively assign (annotate) prosodic phrases and semantic accents in read speech data and present results of such an assignment in the Czech language. The concept of *prosodic phrase* – as understood here – basically corresponds to a traditional view or to what is meant by the term "phonemic clause" (or "discourse segment") in Czech literature [1], i.e. such a phonetic unit which constitutes perception of the *rhythmical* qualities in language. A prosodic phrase is mainly delimited by acoustical features of its boundaries and it can also contain an "intonation peak". However, as [1] discusses, there is no empirical evidence supporting any stronger assumption about the intonation peak presence/absence or their number in a Czech utterance.

We therefore assume solely that a speaker may *emphasise any number* of words by acoustic means to express (perhaps even unintentionally) their prominence in comparison with other words. The acoustic prominence of a word can deliver various kinds of information – from this point of view we can observe the acoustic prominence even on words where a phrase end is acoustically realised. However, such a prominence has a different function: we want to find words whose acoustical prominence has an emphasising function in terms of semantics or pragmatics. We have decided to call such a phenomenon a *semantic accent*.

Prosodic annotation within the ToBI framework and its objectiveness in terms of quantitative inter-transcriber reliability (together with extensive references on the topic) is presented for example in [2]. Such annotation, however, is very time-consuming and requires highly trained annotators. Moreover, in conformity with [3] we prefer labelling

* Support for this work was provided by the Ministry of Education of the Czech Republic, projects 2C06020 and LC536.

P. Sojka et al. (Eds.): TSD 2008, LNAI 5246, pp. 493–500, 2008.
© Springer-Verlag Berlin Heidelberg 2008

of what an annotator hears rather than the shape of F0. Similar task has been described in [4] where the authors conclude with the statement that it is also reliable in some way to work with naive listeners.

As it will become evident further in our paper, phrases and semantic accents in Czech speech often lack any clear and reliable acoustic cues, therefore the number of naive listeners annotating our data must be relatively very high to generate a robust statistics. A similar strategy can be found in [5] where 74 students took part in listening tests on prominence and boundary perception, albeit our goal (as well as test protocol and nature of the speech data) here is different – we primarily want to find out *where* in our speech data the intonation phenomena *are*, whereas *how* they are *perceived* is secondary for us.

2 Prosodic Phrases, Semantic Accents and Speech Synthesis

Performance of a unit selection TTS synthesis approach strongly depends on the way how the basic speech units (e.g diphones) in its database are parameterised in terms of their higher-level features (i.e. linguistical – as opposed to acoustical, which are obviously very important as well). Such a parameterisation should be compact and should try to cluster the speech units according to some repetitive suprasegmental patterns present in the speech data. This way usability of each unit can be as broad as possible while disallowing it to be used in such contexts where its specific properties would cause unwanted speech disruptions.

A discussion over the need of a *consistent* and reliable annotation of phrase boundaries together with problems in achieving it can be found in [6]. Moreover, the suprasegmental acoustic characteristics of strings of the basic speech units in emphasised words (i.e. the words bearing the semantic accent) are different from those which cover non-emphasised words. It is therefore important to have such units parameterised so that these differences are present in this parameterisation, hence the semantic accents must be designated in the speech corpus too.

The idea of the whole process is following: prosodic phrase boundaries and semantic accents are manually designated in a reasonable sub-part of the whole (presumably very large) real speech database so that there is *agreement* as high as possible *among many independent listeners*. The phrase boundaries and semantic accents (their model respectively) obtained this way are considered to be the "real" ones in the sense of "objectiveness", no matter our subjective opinion. In the second phase (not covered by this paper) a machine classifier trained on these data can automatically extend the phrase boundary and semantic accent designation to the rest of the speech database, without being "confused" by inconsistencies in training data subjectively annotated by a single person.

3 Inter-subjective Annotation Process

The inter-subjective agreement on the phrase boundary and semantic accent annotation has been achieved by a statistical model applied on data acquired by two independent listening tests.

3.1 Listening Tests

The listening tests were organised on the client-server basis using specially developed web application. We have used our speech corpus [7] designed as the source dataset for our text-to-speech system ARTIC. A test layout description and technical issues together with information about the source corpus and test participants are in [6].

The first listening test is described in detail together with its evaluation also in [6] and we will not discuss it here. It is relevant for us now that its result are 100 sentences from the aforementioned corpus with labelled prosodic phrase boundaries. No semantic accents were labelled in this test and the participants had taken part in it actually with no knowledge about the phenomenon of semantic accent.

The second listening test (which was carried out 3 months after the first test and we will analyse it in the present paper) consisted of two parts (further in the text denoted as Part 1 and Part 2). Part 1 was aimed at finding the semantic accents in the sentences where the prosodic phrase deployment is already given: the same sentences as in the first listening test have been used again and the participants had been instructed to listen to these sentences very carefully and subsequently designate words where they perceived the semantic accent. The textual form of the sentences was displayed together with the a priori prosodic phrase deployment acquired from the first test. The participants had to accept this phrasing and bring into line their semantic accent assignment with it. This part has also served as a "tutorial" for Part 2 because the participants could learn this way what is statistically considered as phrase boundaries.

Part 2 was actually a combination of Part 1 and the first test: we have selected another 150 sentences from our corpus and the participants were again instructed to listen to the sentence recordings and designate the semantic accents. Moreover, in this part the task was also to designate words where the participants *are sure* there is a phrase boundary and words where they feel there *might be* a phrase boundary (i.e. these two cases were distinguished). It means that for every word in each of these 150 sentences the test comprises three options: a) this word is emphasised; b) after this word there certainly is the phrase boundary; c) after this word there might be the phrase boundary; the options *b* and *c* are obviously mutually exclusive.

We have eventually received correctly finished electronic answer sheets from 99 participants. It is worth mentioning that the first listening test described in [6] was finished by 103 participants from which 46 took part also in the second test and finished it (i.e. they have already had previous experience with phrase boundary assignment).

3.2 Statistical Evaluation

The goal of the listening test was to find places in the given sentences where we can make inter-subjective agreement on phrase boundary and semantic accent occurrences. The resulting phrase and semantic accent deployment is then to be treated as an objective basis for any further research. We can transform the problem of such a deployment based on many independent observations into more abstract and formal level:

Let X be a random process defined as $X = \{X_t \mid t \in T\}$, where $T = \{1, 2, \dots n\}$ is a set of time points respective to the ordinal numbering of words in the test sentences (i.e. the first word in the first sentence has $t = 1$, the second word in the first sentence has $t = 2$, and so on), and X_t are random variables which hold $X_t = 1$ iff the t-th word finishes a prosodic phrase, and $X_t = 0$ otherwise. Exactly the same can be done for the semantic accents, such a random process is analogical to X and will be denoted as Y. We assume that the random processes X and Y are mutually independent.

Now let the test participants be numbered by the set $J = \{1, 2, \ldots m\}$, i.e. the first participant has $j = 1$, the last one has $j = m$. We can define m random processes $O^{(1)}, \ldots O^{(m)}$ representing the participants' responses (observations) such that $O^{(j)} = \{O_t^{(j)} \mid t \in T\}$, where t has the same meaning as for the process X, and $O_t^{(j)}$ are random variables which hold $O_t^{(j)} = 1$ iff the j-th participant asserts that the t-th word finishes a prosodic phrase, and $O_t^{(j)} = 0$ iff the j-th participant *does not* assert that the t-th word finishes a prosodic phrase.

Our goal can now be re-formulated as follows: knowing the observations $O^{(1)}, \ldots O^{(m)}$ we want to estimate the hidden trajectory of the process X which best satisfies the given observations.

This can be analogically defined for the process Y with the only difference that the observations refer to the semantic accents and are based on the whole set of 250 sentences, whereas X describes only the subset of 150 sentences additionally chosen for the second listening test. For the sake of lucidity we will speak further in the text only about the process X assuming that everything which holds for it, holds analogically also for the process Y. It is supported by the fact that the two variants of the answers on the phrase boundary presence/absence (i.e. "boundary for sure" and "boundary maybe") were treated equally – this was based on the assumption that if the "statistically relevant" number of participants think that there *might be* the phrase boundary at the given place, it *really is* there. The reason for allowing two levels of certainty from the participants' side was mainly due to the experience that if a listener is really not sure, he answers randomly – and this can be avoided by the "maybe" variant. The difference between these two variants is utilised in the participants' agreement calculation (see Section 4). We have decided not to use such two variants for the semantic accents because the semantic accent is defined more vaguely and a positive answer about its presence is actually most often an opinion like "this word might be emphasised".

The aforementioned goal can be transformed into the problem of finding the most likely model parameters given the observed data – a *maximum likelihood* approach. The relations between the unknown "real" boundary and a participant's assumption is expressed by the probabilities:

$$P(O_t^{(j)} = 1 | X_t = 1) = r_X^{(j)} \tag{1}$$

$$P(O_t^{(j)} = 0 | X_t = 1) = 1 - r_X^{(j)} \tag{2}$$

$$P(O_t^{(j)} = 0 | X_t = 0) = f_X^{(j)} \tag{3}$$

$$P(O_t^{(j)} = 1 | X_t = 0) = 1 - f_X^{(j)} \tag{4}$$

We further presuppose that X is a stationary process with the alternative probability distribution, thus:

$$X \sim A(p) \tag{5}$$

where $\forall h, i \in T : p_h = p_i = p$. In this point we really intentionally pretend that we do not know anything about phrasing behaviour so that all words have equal probability of bearing a phrase boundary (phrase lengths, lexical, syntactical, semantical or any other factors are excluded on account of the methodological constraints).

Through the Equations 1–5 we have postulated the structure of the probabilistic model of our problem and now we can see that it has the unknown parameters $r^{(j)}$, $f^{(j)}$ and p which we will further collectively denote as Θ.

The goal is to find the most likely parameters Θ^* given the observation $O = [O^{(1)}, \ldots, O^{(m)}]$, i.e. maximise the likelihood function

$$L(\Theta) = P(O|\Theta) \tag{6}$$

$$\Theta^* = \arg\max_{\Theta} L(\Theta) \tag{7}$$

There is not an analytical solution to the Equation 7 and therefore we have decided to estimate the parameters by an expectation-maximisation (EM) algorithm. The EM algorithm is proved not to decrease the likelihood function in any iteration. However, it tends to converge to a local maximum, hence the initial parameters must be chosen reasonably and perturbed in more experiments.

We have set the initial parameters Θ_0 heuristically and equal for both the estimation of X and of Y : $p = 0.5$, $r_t^{(j)} = 0.7$ and $f_t^{(j)} = 0.9$ for all j and t. The parameters converged in both cases already after 10 iterations of the EM algorithm to Θ^*. The parameter of the alternative distribution converges to $p_X^* = 0.8470$, i.e. $\forall t : P(X_t = 0) = 0.8470$, and $p_Y^* = 0.9783$, i.e. $\forall t : P(Y_t = 0) = 0.9783$. The values $r^{*(j)}$ and $f^{*(j)}$ are generally different for all j but still their brief characterisation can be found in Table 1. We have used the Baum-Welch algorithm simplified to suit the needs of this problem. Instead of explicit maximisation of $L(\Theta)$ the algorithm maximises $P(X|O)$ by iterative gradient changes of the parameters Θ – this process ensures growth of $L(\Theta)$.

The probability that the t-th word bears a phrase boundary given the observations $O_t = [O_t^{(1)}, \ldots O_t^{(m)}]$ is

$$P(X_t = 1|O_t) = \frac{\prod_{j \in J} P(O_t^{(j)}|X_t = 1) \cdot P(X_t = 1)}{P(O_t)} \tag{8}$$

and therefore we can formulate the decision criterion as

$$X_t = 1 \iff P(X_t = 1|O_t) > 1 - P(X_t = 1|O_t) \tag{9}$$

and since $P(O_t)$ is constant for the given t, we can omit it and compute only the numerator from the Equation 8. The same criterion holds also for the process Y.

4 Results and Conclusions

After having formally decided which words bear the phrase boundaries and semantic accents using the method described in the previous section, we can formulate some assertions about phrasing and emphasising in the Czech language. These assertions are based on the quantitative evaluation of the acquired data from various points of

Table 1. Probabilities of two kinds of errors the participants have done in placing the phrase boundaries (X) and semantic accents (Y), as estimated by the EM algorithm. These probabilities refer to the Equations 2 and 4.

	$X : P(0\|1)$	$X : P(1\|0)$	$Y : P(0\|1)$	$Y : P(1\|0)$
average	0.125	0.054	0.380	0.159
st. dev.	0.099	0.049	0.191	0.096
min	0.000	0.004	0.036	0.013
max	0.421	0.351	0.864	0.400

view and have rather linguistically-theoretical value. On the contrary, the labelled data themselves have great practical value for our TTS system development.

Pause presence is one of the most important acoustic features signalling the phrase boundary, therefore it is useful to classify all phrase boundary occurrences into two classes: the boundary without a pause (B1) and the boundary with a pause (B2). Table 2 comprises an overview of phrase boundary type frequencies together with similar information on the semantic accents (SA). The frequencies do not include phrase breaks at the sentence ends – it means that only "intra-sentential" boundaries have been considered. This table also shows the inter-participant agreement for both boundary types, given as the relative number of the participants who have placed these boundaries. We can see that if the phrase boundary is followed by a pause, then in average 97 % of the participants agree on the phrase boundary assertion. Considering also the standard deviation it is clear that most of the cases with a pause are agreed on by more than 90 % of the participants. This is in contrast with 72 % average agreement on the non-paused phrase boundaries but even such a number is still a very good indicator that the phrase boundaries without a pause are well recognised too. The minimum agreement in the class B1 is less than 50 % – in such a case the EM algorithm has "decided" to place the boundary because the votes from the more "credible" participants (i.e. with higher $r_X^{(j)}$ and lower $1 - f_X^{(j)}$) have higher weight. There are 23 phrase boundaries with the agreement value less than 50 %. The average agreement on the semantic accents is much lower than on the phrase boundaries – this is in conformity with much higher error probabilities (see Table 1), which point out that the semantic accent is a less clear language phenomenon than the prosodic phrase (we will see that a similar conclusion implies also from the labelling agreement among the participants).

Table 2. Boundary type (B1 – without a pause, B2 – with a pause) and semantic accent (SA) frequencies together with agreement among the participants on phrase boundary and semantic accent placement

	frequency	agreement average	st. dev.	min	max
B1	201 (39.11 %)	72 %	17 %	43 %	100 %
B2	313 (60.89 %)	97 %	4 %	75 %	100 %
SA	227	49 %	10 %	83 %	36 %

Table 3. Distribution of the sentence lengths given as the number of phrases in a sentence and the phrase lengths given as the number of words in a phrase

sentence length (in phrases)	frequency		
	whole test	part 1	part 2
1	4 (1.6 %)	0 (0.00 %)	4 (2.67 %)
2	89 (35.6 %)	22 (22.00 %)	67 (44.67 %)
3	78 (31.2 %)	35 (35.00 %)	43 (28.67 %)
4	56 (22.4 %)	27 (27.00 %)	29 (19.33 %)
5	14 (5.6 %)	11 (11.00 %)	3 (2.00 %)
6	9 (3.6 %)	5 (5.00 %)	4 (2.67 %)
phrase length (in words)	**frequency**		
	whole test	part 1	part 2
1	58 (7.59 %)	35 (10.23 %)	23 (5.45 %)
2	172 (22.51 %)	93 (27.19 %)	79 (18.72 %)
3	224 (29.32 %)	101 (29.53 %)	123 (29.15 %)
4	145 (18.98 %)	55 (16.08 %)	90 (21.33 %)
5	97 (12.7 %)	47 (13.74 %)	50 (11.85 %)
6	42 (5.5 %)	7 (2.05 %)	35 (8.29 %)
7	15 (1.96 %)	2 (0.58 %)	13 (3.08 %)
8	8 (1.05 %)	2 (0.58 %)	6 (1.42 %)
9	3 (0.39 %)	0 (0 %)	3 (0.71 %)

Another important factor describing the phrase distribution is length of the phrases, viewed either as the number of phrases in a sentence or the number of words in a phrase. The distribution of these lengths is in Table 3. Since the phrase deployment in Part 1 of the test has been acquired without regards to the semantic accents, whereas the phrases in Part 2 have been deployed concurrently with the semantic accents, we give comparison of the phrase lengths in both parts separately as well. We can simply state that the fact whether the phrases are assessed with or without regard to the semantic accents *influences* the resulting phrase assessment. This is most apparent in case of short phrases from Part 1 – they are often replaced in Part 2 by longer phrases with the semantic accent elsewhere than on the last word.

The latter statement is supported by data shown in Table 4. According to this table there is in Part 2 a significant decrease of the number of phrases with the semantic accent on the last word in comparison with Part 1. We can make one more important conclusion from this table: it proves the hypothesis that there can be at the most one semantic accent in a phrase. The test participants had no prior information about expected or allowed numbers of the semantic accents in the phrases. Still the resulting statistically underlain deployment places no more than one semantic accent into any phrase, no matter whether the semantic accents have been assessed concurrently with the phrase boundaries or separately afterwards.

We can also present calculation of the overall inter-participant agreements A_X (for the phrase boundaries) and A_Y (for the semantic accents) given as the average of mutual agreement of all possible pairs of the participants. The mutual agreement of two participants is calculated as a quotient s/n where s is the number of word tokens where both participants had the same answer ("same answer" for the phrase boundary assignment

Table 4. Distribution of the number of semantic accents (SA) in a phrase and the number of phrases with the semantic accent placed on the last word

	frequency (number of phrases)		
number of SA in phrase	whole test	part 1	part 2
0	537 (70.29 %)	224 (65.50 %)	313 (74.17 %)
1	227 (29.71 %)	118 (34.50 %)	109 (25.83 %)
SA on last word	181 (23.69 %)	100 (29.24 %)	81 (19.19 %)

means also the situation when one participant answers "maybe boundary" and the other can then answer anything) and n is the total number of tokens where they answered. We have obtained $A_X = 0.95$ and $A_Y = 0.76$. Another criterion representing the annotation reliability based on the inter-participant agreement is the Fleiss' kappa measure. Again, we have calculated it separately for the phrase boundary labels and the semantic accents. The overall Fleiss' kappa values for 99 % confidence level (i.e. $a = 0.01$) are $\kappa_X = 0.6636$ and $\kappa_Y = 0.1283$. Especially the value for the phrases (κ_X), which means that the agreement is much above chance, is very similar to what [5] shows. Even in spite of κ_Y being much smaller ([5] also reports κ smaller for prominences) we can expect the EM algorithm to produce reasonable labelling of the semantic accents as well.

Considering these values we can conclude the paper with a statement that we have obtained a reliable phrase annotation which has not been (to our knowledge) available for Czech so far and the correctness can be supported by the fact that the inter-participant reliability measures are in conformity with results of other researchers.

References

1. Palková, Z.: Rytmická výstavba prozaického textu (with English resume: The rhythmical potential of prose). Academia, Prague (1974)
2. Yoon, T.-J., Chavarría, S., Cole, J., Hasegawa-Johnson, M.: Intertranscriber reliability of prosodic labeling on telephone conversation using ToBI. In: Proc. Interspeech, Jeju, Korea, pp. 2729–2732 (2004)
3. Wightman, C.W.: ToBI or not ToBI. In: Proc. Speech Prosody, pp. 25–29. Aix-en-Provence, France (2002)
4. Buhmann, J., Caspers, J., van Heuven, V.J., Hoekstra, H., Martens, J.-P., Swerts, M.: Annotation of prominent words, prosodic boundaries and segmental lengthening by non-expert transcribers in the Spoken Dutch Corpus. In: Proc. LREC, Canary Islands, Spain, pp. 779–785 (2002)
5. Mo, Y., Cole, J., Lee, E.-K.: Naïve listeners' prominence and boundary perception. In: Proc. Speech Prosody, Campinas, Brazil, pp. 735–738 (2008)
6. Romportl, J.: Statistical evaluation of prosodic phrases in the Czech language. In: Proc. Speech Prosody, Campinas, Brazil, pp. 755–758 (2008)
7. Matoušek, J., Romportl, J.: Recording and annotation of speech corpus for Czech unit selection speech synthesis. In: Matoušek, V., Mautner, P. (eds.) TSD 2007. LNCS (LNAI), vol. 4629, pp. 326–333. Springer, Heidelberg (2007)

Making Speech Technologies Available in (Serviko) Romani Language

Milan Rusko[1], Sakhia Darjaa[1], Marián Trnka[1], Viliam Zeman[2], and Juraj Glovňa[3]

[1] Institute of Informatics of the Slovak Academy of Sciences, Dubravska cesta 9, 845 07 Bratislava, Slovakia
[2] Office of the Plenipotentiary of the Government of the Slovak Republic for Romanies, Cukrová 14, 813 39 Bratislava, Slovakia
[3] University of Constantinus, the Philosopher in Nitra, Štefánikova 64, 949 07 Nitra, Slovakia
{milan.rusko,utrrsach,trnka}@savba.sk, zeman@vlada.gov.sk, jglovna@ukf.sk

Abstract. The language of Romanies seems not to be commercially interesting for big companies. Not only the majority of people from Roma community are very poor, but the language itself is very difficult to work on, because it is extremely rich in local dialects and the standardized form that would be accepted by majority of Romanies does not exist. The Romani language belongs to the "digitally endangered languages". The paper gives a short description of Romani language in Slovakia. An effort to design basic tools needed to start using Romani speech and language in computer technologies is presented. As the authors are familiar with speech synthesis, they have chosen building several types of speech synthesizers in Romani as a pilot project. The paper shortly summarizes some facts on Romani orthography, phonetics, and prosody. The design of text corpus, diphone set, and speech database is described. The application part of the paper presents Romani synthesizers – both diphone and unit-selection, some of which are bilingual (Romani-Slovak). The demo of the synthesis can be tried on the authors' web-page.

Keywords: Romani language, speech technologies, Romani speech synthesis.

Motto: "Sa tu man šaj kames, kana dumakeri čhib nadžanes? Sar saj prindžarav tiro, jilo kana tiro lav hin gadžikano?" (How could you want me if you do not know my language? How could I know your heart, if I do not understand the Gadžo (non-Roma) words?) – fragment of a poem by J. Berky-Ĺuborecký

1 Anglelav – Introduction

Digitally endangered languages are languages used by people who are too few in number or too poor to make them attractive to commercial software developers. This means that native speakers of these languages end up having two barriers to climb to access computers – first, they have to learn English; then they have to learn IT skills. Their

P. Sojka et al. (Eds.): TSD 2008, LNAI 5246, pp. 501–508, 2008.
© Springer-Verlag Berlin Heidelberg 2008

native language is marginalised, and becomes digitally endangered. [2] From this point of view the language of Romanies (Romani čhib) surely belongs to the digitally endangered languages.

In Romani the expression “Rom” designates people of Roma origin. Realistic estimations report that the number of Romani speakers in Europe is approx. 4.6 million. The total number of Roma in Europe amounts to 6.6 million people. [3]

Until recently Romani was only talking language, without a written norm. During the last decades an attempt to create a written norm started in different countries and in several cases the written form was codified.

2 Serviko Romani čhib – The Language of Eastern Slovak Romanies

The language of European Romanies belongs to the group of Indian languages. Their language was always influenced by their habitat, where they stayed for certain time. Persian, Greek, Armenian, or even Slavonic words can still be found in this language. Slovak Romanies have in their language many words which were adopted from standard Slovak and even more expressions “borrowed” from local Slovak dialects. The language of Romanies living on the south of Slovakia, Romani language uses many expressions coming from Hungarian.

During the population census in 2001, 89 920 people in Slovakia have claimed that their nationality is Roma. Nevertheless the real number of the members of Roma community is estimated to be as big as 380 000. [4]

In Slovakia – as opposed to Czech Republic the Romani language (as a group of varieties) does not seem to disappear, although many of the dialects and the local varieties (especially in Western, Central and Southern Slovakia) are endangered or close to extinction. But mainly in Eastern Slovakia in some socially isolated localities a gradual change from traditional bilingualism to monolingualism in Romani (!) – or to a radical lowering of competence in Slovak (in generations which grew up to a productive age after the revolution in 1989) can be observed. This is in relation to the growth of unemployment and disintegration of social nets. [5]

The Eastern Slovak (Serviko) Romani is one of the three main dialects that are spoken by Romanies in Slovakia. It is spoken by approximately 80–85% of Roma population in Slovakia therefore Eastern Slovak Romani dialect was chosen for the basis for grammar, lexicon and phraseology of the codified language.

The term Servika comes from the words “Serbika”, “Serbos”, “Serbija”. The Roma in Slovakia were so called because they came from Serbia.

3 Irišagos – Orthography

The reserchers from the ROMBASE project say: “In view of the academic level, a type of norm has developed in recent years. Though never conventionalised among Romani linguists, a consensus seems to prevail on the use of wedge-accents, as employed in south-western Slavic alphabets, to indicate palato-alveolars (š, ž, č) and the use of -h to indicate distinctive aspiration on voiceless stops and affricates (ph, th, kh, čh). The marking of the palatisation with postponed j or with accent is one of the few open points of this convention for the written language." [6]

In Slovakia the Romani orthography was codified in 1971 and recodified in nineties. The codified form comes out from the orthographical rules of Slovak. Basically the Slovak alphabet was adopted including diacritical marks, which are however used according to different rules. The most recent publication available on the topic [4] gives us an overview on the grammar and phonetic rules of Romani language. In Romani a phoneme is always written with the same corresponding grapheme. This rule rule is consistently followed with the following phonemes: a, b, c, č, čh, d, e, f, g, h, i, j, k, kh, l, m, n, o, p, ph, r, s, š, t, th, u, v, z, ž. Palatalized d,t,n,l are written with wedge-accents even before vowels i and e.

There is a general rule is in Romani, that at the end of words the loose of voicedness and aspiration occurs. In the written form however the graphemes corresponding to the aspired and voiced phonemes are preserved (jakh [jak] – jakha [jakha] (eye-eyes)).

Combinations of vowels in the foreign words (in Slovak ia, ie, iu) are written as ija, ije, iju (geografija, gimnazijum). Romani does not use y.

4 Pheniben – Pronunciation

The phoneme set used in codified Serviko Romani is very similar to that of standard Slovak. The only Romani specific phonemes that do not exist in Slovak are ?h, kh, ph a th, which are pronounced with a slight aspiration. Without the aspiration the words have essentially different meaning, e.g.: pherel (draw, pump) – perel (fall), khoro (jar) – koro (blind). [4]

A problematic issue, significant also for prosody is the problem of vowel quantity. Long vowels are not marked by acute accent – orthographic form of the text therefore does not mark their position. But the position of the accent which has a strong influence on the vowel quantity is very important for the speech naturalness and intelligibility.

The first phase of ph, th, čh, kh phonemes is voiceless plosive, and so they are not coarticulated with their left neighbour. Therefore from the point of view of the left context they behave like Slovak p, t, č, k.

The rule that voicedness is lost at the end of words holds also for Slovak and therefore does not cause any difference in pronunciation between the languages.

5 Vakeribno Melodija – Intonation

Accent is generally placed on the pre-final syllable in Serviko Romani, which also holds for Eastern Slovak dialects. However, too many exceptions exist from this basic rule, so we had to use a lexicon of accentuation exceptions. (As our rule-based intonation model was built for Standard Slovak, which has accents always on the first syllable, we had to revise the whole model and change the rules of accentuation. The rules for phoneme lengths prediction, which are based mainly on the mean value of the phoneme length remained unchanged.)

6 Irišago Skidžipen – Text Corpus

The number of available texts in Romani is absolutely insufficient. To have at least some amount of texts to start with, we used the archive of the only Slovak Romani newspaper "Romano nevo žil" from the years 2003 to 2008 and unpublished Romani fairy tales by V. Zeman. In a whole we got from these two sources only 510 kB of texts.

7 Vakeribno Skidžipen – Speech Corpus

The situation with available speech recordings in Romani is even worse than with texts. In the first stage the only source of recorded Romani were ten audiocassettes of field research realized in Roma settlements by the students of Institute of Romological Studies in Nitra. These served us for preliminary listening studies of phonetics and prosody. Nevertheless these files contained speech influenced by local geographical dialect. The only recordings of the "standard" Romani were available from the Slovak TV Roma magazine "So vakeres?". From listening to these recordings of dialogues between educated, trained speakers – TV moderators – and common Romanians, one get a feeling, that the educated speakers (especially the younger of them) are much influenced by Slovak and the typical Romani intonation, accentuation, phoneme quality and speech timbre, has nearly disappeared from their speech. The speech signal of the recordings is moreover polluted with background music and noise and therefore practically unusable for speech technology applications.

For Speech synthesis database we have chosen an experienced Romani speaker from Eastern Slovakia, who is also a theoretical expert in Romani language. We have recorded a set of 500 sentences chosen from the text database (see paragraph 6). The sentences were chosen to cover the basic phoneme inventory, and the most common diphones. The speech was recorded in studio with a RODE K2 large diaphragm microphone and MOTU TRAVELER digital mixing device with Firewire IEEE 1394 PC interface. The sampling frequency was 96 kHz and resolution 24 bit. Prior to the use in unit-selection synthesizer the signal was down-sampled to 22 kHz, 16 bit format. The annotation was done by forced alignment using EHMM phoneme recognizer and our phoneme boundary correction routines. The annotation consists of pitch marks, phoneme boundaries, syllable boundaries, word boundaries, phrase boundaries and sentence boundaries markers. The phoneme level information also includes initial, central and final pitch, energy and length in milliseconds.

A similar database with the same speaker is also being built in Slovak.

8 Čhibakro modulatoris – Speech synthesizer

8.1 Diphone Synthesizer

For the first experiments we decided to port the existing Slovak synthesizer [1] to Romani language. This synthesizer is based on concatenation of small elements of a prerecorded speech signal, mainly diphones. An algorithm similar to TD-PSOLA [6] is used for concatenation.

To prepare a diphone database we had to define a set of words that contain the Romani specific diphones that are not included in the diphone set of the Slovak speech synthesizer. For the baseline version a set of diphones containing aspired phonemes – ph, kh, čh, th was sufficient. (The full version of the table, containing Romani words needed to record the whole database for a new Romani diphone synthesizer voice is behind the scope of this paper, and can be found on our webpage.)

Table 1 shows Romani words containing aspired consonants which were recorded by a native Romani speaker and used for phonetic analysis. We tried to cover combinations of vowels and aspired consonants from both sides.

Table 1. Examples of Romani words containing aspired consonants

čh	$t\int^{h}$	lačharela ačhavel čhamenger	e čhercheňa bilačhe prečhinel	vičhinel dičhiben	o čhohano ľočhipena	čhuvaleskro odučharel
kh	k^{h}	khamoro naarakheha	te khelel jekhetane	dikhipena lokhiben	o khosno polokhe	te khuvel mukhavkerel
ph	p^{h}	phabaj zaphenel	phenel barephikeskero	phirel priphandel	phosavel dophenel	phuv phurikano
th	t^{h}	tharel sathemeskro	themeskero prethovel	ithiskero prithovibe	te thovel odothar	thudeskero thuvalel

While studying the acoustic properties of the recorded aspired consonants, we found an easy way to transform Slovak triphones to Romani aspired diphones, which is explained on Figure 1.

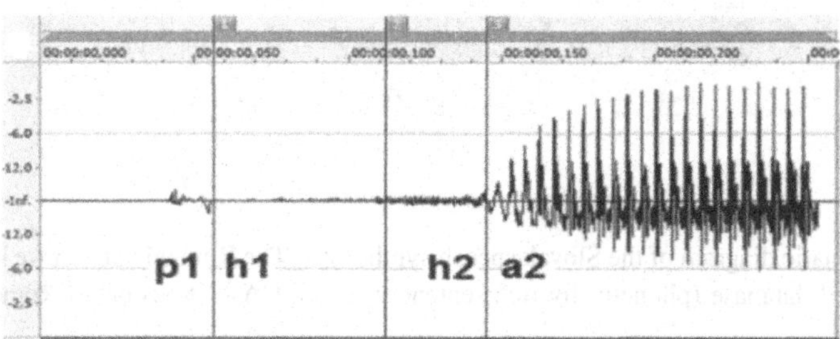

Fig. 1. Oscillogram of two diphones from the original Slovak inventory: p1-h1 and h2-a2, creating a p-h-a triphone. To get Romani diphone ph-a, the h1 section of the Slovak p-h-a triphone must be omitted.

Grapheme to phoneme conversion is based on a sophisticated set of rules supplemented by a pronunciation vocabulary and a list of exceptions. The Romani version of this block was created by changing Slovak rules according to Romani pronunciation rules mentioned in paragraph 4. The same grapheme to phoneme conversion routine was also used in the unit-selection synthesizer.

8.2 Unit-Selection Synthesizer

To create a Unit selection synthesizer we used the engine presented by Rusko [1]. This synthesizer does not calculate the joint and cost functions. It merely relies on phonetical – phonological pre-selection of elements (mainly syllables). The main features of the selection are phonological context, pitch and phoneme length.

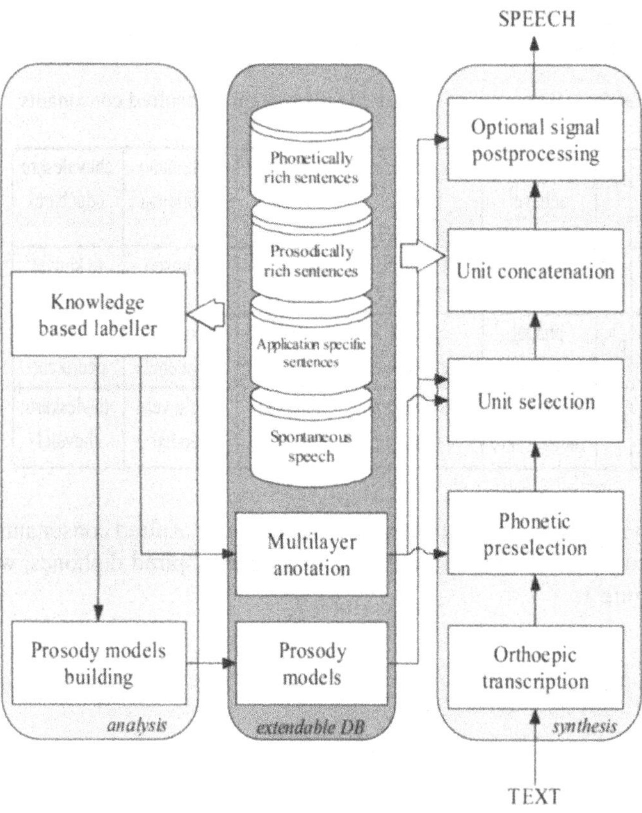

Fig. 2. Schematic diagram of the Slovak speech synthesizer. The Romani version we have created uses a limited database (phonetically rich sentences) and a CART trees based Romani prosody model.

8.3 Multilinguality

In 2004 we made a Hungarian version of our synthesizer to broaden the rank of possible users by the Hungarian speaking fellow-citizens. The Serviko Romani synthesizer represents our recent attempt for further broadening the multilingual capability of the Kempelen family of synthesizers.

Both versions (male and female) of Romani diphone synthesizer we ported from the Slovak synthesizer, use only the original Slovak diphones and the diphones derived from them. Thus a combination of Slovak and Romani synthesizer naturally gives a uniform

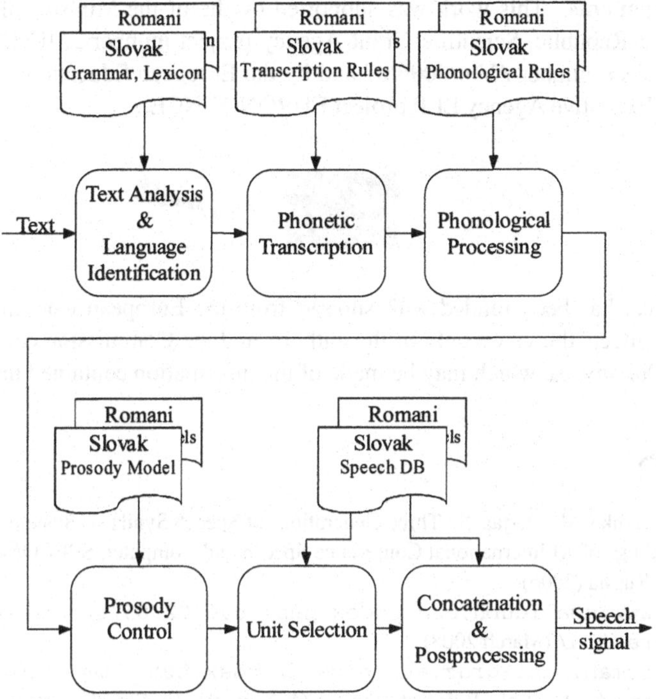

Fig. 3. Schematic diagram of the bilingual Slovak-Romani synthesizer

voice bilingual synthesizer. The language identification routine is based on Slovak and Romani lexicons.

The same procedure is being applied to the unit selection synthesizer (Figure 3). At present experiments the Slovak and Romani synthesizer use different voices, but our next step will be building a uniform-voice bilingual unit-selection synthesizer from bi-lingual speaker database.

9 Conclusion

With relatively small intervention in the text preprocessing unit, grapheme to phoneme rules, phoneme inventory and intonation model of the Slovak speech synthesizer we managed to create two versions of the basic level diphone speech synthesizer in Serviko Romani. The quality of the speech is adequate to the type of the synthesizer. The segmental quality is not perfect, but is acceptable, the main shortcommings being robotic and buzzy speech quality and very simple prosody model.

The unit-selection synthesizer with CART trees [7] based prosody model gives much better naturalness and allows for more advanced experiments in prosody modelling.

Anyway the aim of this work was not to create a perfect speech synthesizer, but to study the new language, find its main peculiarities, and prepare the basic speech processing background needed for further development of much more comprehensive speech technology applications like pedagogical tools and information systems in Romani.

Acknowledgements. This work was supported by the of the Ministry of Education of the Slovak Republic, Scientific Grant Agency project number 2/0138/08 Applied Research project number AV 4/0006/07 and by the European Education, Audiovisual and Culture Executive Agency LLP project EURONOUNCE.

This project has been funded with support from the European Commission. This publication reflects the views only of the authors, and the Commission cannot be held responsible for any use which may be made of the information contained therein.

References

1. Rusko, M., Trnka, M., Darjaa, S.: Three Generations of Speech Synthesis Systems in Slovakia. In: Proceedings of XI International Conference Speech and Computer, SPECOM 2006. Sankt Peterburg, Russia (2006)
2. http://www.mealldubh.org/index.php/2006/02/05/strength-in-confederation/ (March 2008)
3. http://romani.uni-graz.at/rombase/index.html (March 2008)
4. Hübschmannová, M., et al.: Rules of Romani Orthography (in Slovak). State Paedagogical Institute, Bratislava (2006)
5. Elšík, V.: personal communication (2007)
6. Hamon, C., Moulines, E., Charpentier, F.: A diphone synthesis system based on time-domain prosodic manipulations of speech. In: Proc. Int. Conf. Acoustics, Speech and Signal Processing, Glasgow, UK, p. 238 (1989)
7. Breiman, L., Friedman, J., Stone, C., Olshen, R.: Classification and Regression Trees. Chapman Hall, New York (1984)

Emulating Temporal Receptive Fields of Higher Level Auditory Neurons for ASR

Garimella S.V.S. Sivaram and Hynek Hermansky

IDIAP Research Institute, Martigny
Swiss Federal Institute of Technology at Lausanne (EPFL), Switzerland
{sgarimel,hynek}@idiap.ch

Abstract. This paper proposes modifications to the Multi-resolution RASTA (MRASTA) feature extraction technique for the automatic speech recognition (ASR). By emulating asymmetries of the temporal receptive field (TRF) profiles of higher level auditory neurons, we obtain more than 11.4% relative improvement in word error rate on OGI-Digits database. Experiments on TIMIT database confirm that proposed modifications are indeed useful.

Keywords: Feature extraction, auditory neurons and speech recognition.

1 Introduction

MRASTA ([2]) technique extracts features by filtering the temporal trajectory of each critical band energy of speech by a bank of finite impulse response (FIR) filters. Thus

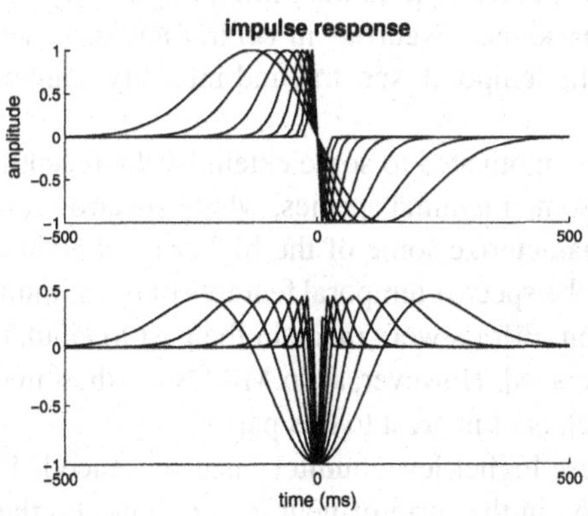

Fig. 1. Normalized impulse responses of the MRASTA filters, $\sigma = 8 - 130$ ms

each feature represents the convolution of the corresponding input critical band trajectory with the impulse response of a filter. Note that impulse response of each FIR filter is symmetric (even or odd) around the center as shown in the Figure 1.

In this paper, we propose modifications to these impulse responses, motivated by the asymmetries of the TRFs of the higher level auditory neurons, as shown in the Figure 2.

P. Sojka et al. (Eds.): TSD 2008, LNAI 5246, pp. 509–516, 2008.
© Springer-Verlag Berlin Heidelberg 2008

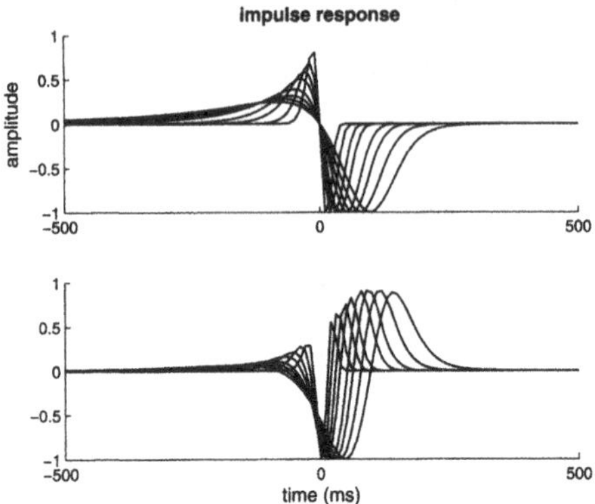

Fig. 2. Normalized impulse responses of the asymmetric filters, $m = -140$

The rest of the paper is organized as follows. The motivation for this work is presented in the Section 2. In Section 3, we give an overview of the MRASTA feature extraction technique and describe our proposed technique to emulate asymmetries of the TRF profiles. Then we discuss experimental results in Section 4. Finally we conclude in Section 5.

2 Motivation

The peripheral auditory system encodes the acoustic waveform into a neural code in the auditory nerve. This neural code is then interpreted by the central auditory pathways to identify various sounds. Neurons in central auditory stations are sensitive to dynamic variations in the temporal, spectral and intensity composition of the sensory stimulus.

MRASTA approach is motivated to some extent by the recent findings ([4] and [5]) in brain physiology of some mammal species, where spectro-temporal receptive fields (STRFs) are used to characterize some of the higher level auditory neurons. STRF, a linear model, describes the spectro-temporal features of the stimulus (speech) that most likely activate the neuron. Efforts were made in the past to emulate these STRFs using multiple 2-D Gabor filters [8]. However, as in MRASTA, their method did not emulate asymmetry in time which is of interest to this paper.

It is believed that these higher level auditory neurons encode information pertained to the speech recognition in the form of neural firing rate. Furthermore, it is possible to predict the neural firing rate of a neuron due to an arbitrary stimulus (speech) by convolving (2-D) the corresponding STRF with the input spectrogram of speech as given by the Equation 1 ([7]).

$$r_{pre}(t) = \sum_{i=1}^{nf} \int h_i(\tau)\, S_i(t - \tau)\, d\tau \qquad (1)$$

where $r_{pre}(t)$ – predicted firing rate,
nf – number of critical bands,
$h_{\{i\}}(t)$ – STRF,
$h_i(t)$ – temporal receptive field of ith frequency channel (critical band),
$S_i(t)$ – i^{th} critical band trajectory of speech.

One can think of this 2-D convolution as several 1-D convolutions at various critical band trajectories of speech and temporal receptive field (TRF) profiles of the STRF, and subsequent summation of all such convolutions. The TRF profile is obtained by slicing through the STRF at a particular frequency. Additionally, we note that these profiles ($h_i(t)$) are not symmetric ([6]) for higher level auditory neurons. MRASTA feature extraction technique fails to emulate these asymmetries as each of its filter has a symmetric impulse response. This observation motivates us to study the effect of using asymmetric filters in MRASTA feature extraction technique.

3 Feature Extraction

3.1 MRASTA Overview

Detailed description of this technique can be found in [2]. In this section, we describe only the FIR filter bank.

Energy in each critical band is extracted from 25 ms windowed speech for every 10 ms as described in [1]. Features are extracted for each frame (10ms) by filtering each of the 15 temporal trajectories of critical band spectral energies (OGI-Digits database) by a bank of 16 FIR filters (shown in the Figure 1). Thus the total number of features per frame are $15 \times 16 = 240$. Typically, three tap FIR filter with impulse response $\{-1, 0, 1\}$ is used for computing the first frequency derivatives ($16 \times 13 = 208$ features). Dimensionality is further increased by appending these frequency derivatives to the features described above ($240 + 208 = 448$ features). The schematic of this feature extraction technique is shown in the Figure 3.

In MRASTA, impulse response of each filter in the FIR filter bank is a discrete version of either first or second analytic derivative of the Gaussian function and is given by Equation 2 or 3.

$$g1[x] \propto -\frac{x}{\sigma^2} \exp\left(-\frac{x^2}{2\sigma^2}\right) \quad (2)$$

$$g2[x] \propto \left(\frac{x^2}{\sigma^4} - \frac{1}{\sigma^2}\right) \exp\left(-\frac{x^2}{2\sigma^2}\right) \quad (3)$$

where x is time, $x \in (-500, 500)$ ms with the step of 10 ms; standard deviation σ determines the effective width of the Gaussian. Filters with low σ values have finer temporal resolution whereas high σ filters cover wider temporal context and yield smoother trajectories. The impulse response of each filter is shown in the Figure 1 (total eight different σ values are used). Length of all filters is fixed at 101 frames, corresponding to roughly 1000 ms in time.

Figure 4 shows the impulse, magnitude and phase responses of few MRASTA filters for $\sigma = 40$ ms. Note that each filter has a zero-phase phase response in the passband as

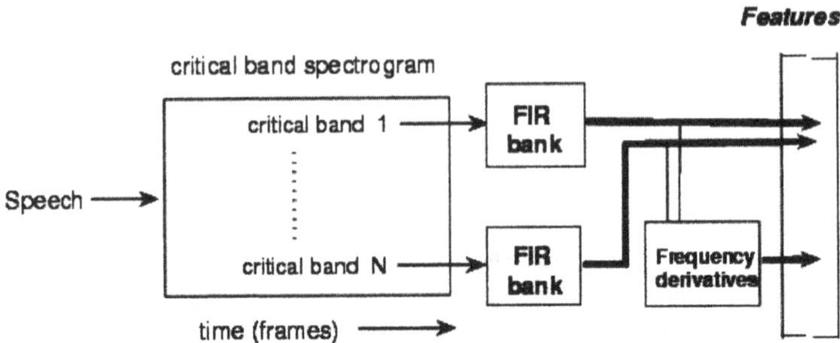

Fig. 3. Schematic of the feature extraction

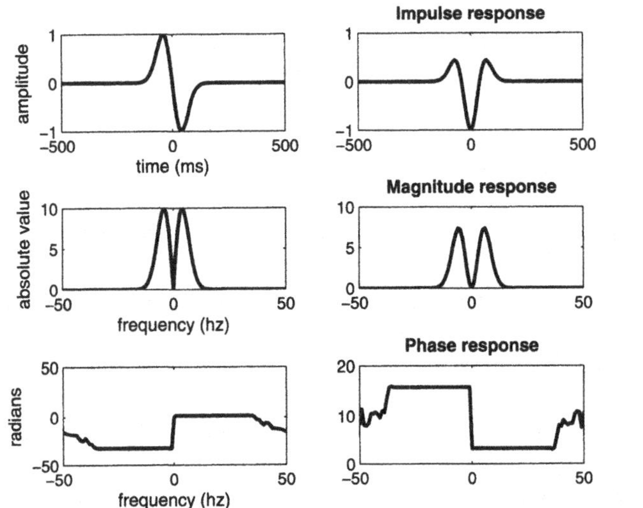

Fig. 4. Impulse, magnitude and phase responses of MRASTA filters ($\sigma = 40$ ms), left column: first Gaussian derivative, right column: second Gaussian derivative

the corresponding impulse response is symmetric (even or odd) around the center. Since interval between the frames is 10 ms, the highest frequency (modulation) component is 50 Hz as shown in the Figure 4. Therefore one can view this MRASTA technique as performing multiple filtering in modulation spectral domain of speech. Modulation spectral domain is the Fourier domain of the temporal trajectory of a critical band energy.

3.2 Asymmetric Filters (Proposed Technique)

To fit the observed temporal asymmetry of the TRF profile, [6] uses a modified Gabor function. Their idea is to first skew the time axis and then to fit a symmetric Gabor function. Generalized version of their Gabor envelope is given by the Equation 4.

$$g(x) = \exp\left(-\frac{\left(a\ \tan^{-1}(bx) - m\right)^2}{2c^2}\right) \tag{4}$$

$$x_{peak} = \frac{\tan\left(\frac{m}{a}\right)}{b}, \; when \; |m| < \left|\frac{a\pi}{2}\right| \tag{5}$$

The envelope (Equation 4) shows asymmetry about its peak point for non zero values of m and the degree of asymmetry increases with absolute value of m as shown in the Figure 5. The value of x for which envelope reaches its peak is given by the Equation 5. Note from the Equation 4 that $g(x)$ is an even function of x when $m = 0$. The first and second derivatives of the above envelope are given by the Equations 6 and 7 respectively.

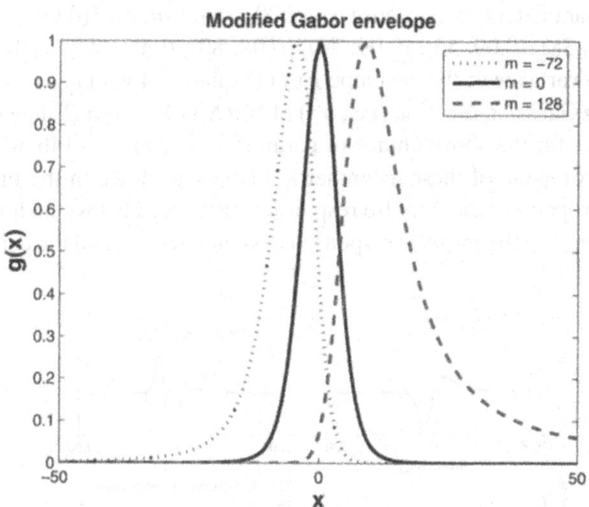

Fig. 5. Modified Gabor envelopes for $m = -72,\ 0 \; and \; 128$, $a = 600/\pi$, $b = 0.09$ and $c = 55$

$$g'(x) = -\frac{a\,b\,\exp\left(-\frac{\left(a\,\tan^{-1}(bx)-m\right)^2}{2c^2}\right)\left(a\,\tan^{-1}(bx)-m\right)}{c^2\left(1+b^2x^2\right)} \tag{6}$$

$$g''(x) = \frac{2\,a\,b^3\,\exp\left(-\frac{\left(a\,\tan^{-1}(bx)-m\right)^2}{2c^2}\right)\left(a\,\tan^{-1}(bx)-m\right)x}{c^2\left(1+b^2x^2\right)^2} + \frac{a^2\,b^2\,\exp\left(-\frac{\left(a\,\tan^{-1}(bx)-m\right)^2}{2c^2}\right)\left(a\,\tan^{-1}(bx)-m\right)^2}{c^4\left(1+b^2x^2\right)^2} - \frac{a^2\,b^2\,\exp\left(-\frac{\left(a\,\tan^{-1}(bx)-m\right)^2}{2c^2}\right)}{c^2\left(1+b^2x^2\right)^2} \tag{7}$$

The impulse responses of the asymmetric filters are obtained from these derivatives as per the Equations 8 and 9.

$$g1\,[x] = g'\left(\frac{x}{10} + x_{peak}\right) \tag{8}$$

$$g2\,[x] = g''\left(\frac{x}{10} + x_{peak}\right) \tag{9}$$

where x is time, $x \in (-500,\ 500)$ ms with the step of 10 ms; x_{peak} is given by the Equation 5. Furthermore, these impulse responses are symmetric for $m = 0$. We choose a set of parameters (not unique) $a = 600/\pi$ and (b, c)={(0.09, 13), (0.09, 20), (0.09, 29), (0.09, 38), (0.09, 55), (0.09, 70), (0.08, 80), (0.07, 90)} such that for each combination, the variance of the envelope $g\,(x)$ (Equation 4 with $m = 0$) matches that of the underlying Gaussian function (i.e., σ^2) of MRASTA. Figure 2 shows asymmetric impulse responses for the above choice of parameters and $m = -140$. Magnitude and phase responses of some of these asymmetric filters are shown in the Figure 6. Note that zero mean property of the impulse response is preserved but we no longer have the zero-phase response as the impulse response is asymmetric around the center.

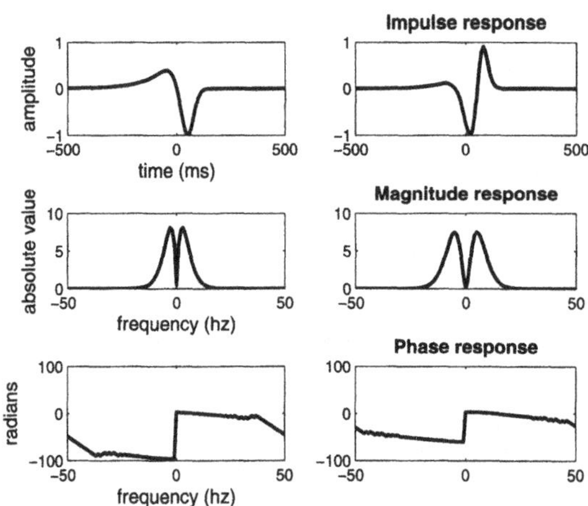

Fig. 6. Impulse, magnitude and phase responses of asymmetric filters ($a = 600/\pi$, $(b, c) =$ (0.09, 55) and $m = -140$), left column: first derivative (Equation 8), right column: second derivative (Equation 9)

Features are extracted from speech by using these asymmetric filters for different values of m. The section below describes the ASR experiments conducted on different databases and lists the performances of the proposed approach and the baseline MRASTA technique.

4 Experiments

Initial set of experiments consists of small vocabulary continuous digit recognition (OGI Digits database). Recognized words are eleven (0 – 9 and *zero*) digits in 28 pronunciation variants. Features are extracted from speech every 10 ms as described in Section 3. Multi-layer perceptron feed forward neural net (MLP) with 1800 hidden nodes is trained on the whole Stories database plus training part of Numbers95 database to estimate posterior probabilities of 29 English phonemes. Around 10% of the data is used for cross-validation. Log and Karhunen Loeve (KL) transforms are applied on these features in order to convert them into features appropriate for a conventional HMM recognizer ([3]). The HMM based recognizer, trained on training part of Numbers95 database, is used for classification.

The WER of **baseline** MRASTA features on OGI-Digits database is **3.5%**. Figure 7 shows the WER of proposed features for different values of the parameter m (values of other parameters are as in Section 3.2). Observe from the figure that the best WER of about **3.1%** corresponds to the parameter value $m = -140$. – *a relative improvement in WER of over* 11.4% *on OGI-Digits database.* The impulse responses of the asymmetric filters corresponding to the optimal parameter are shown in the Figure 2.

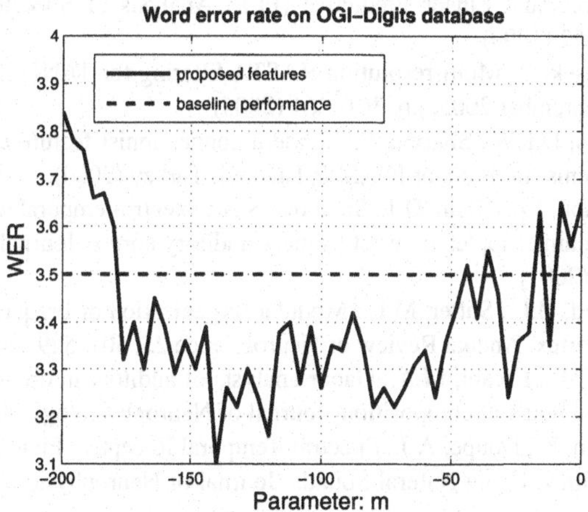

Fig. 7. Word error rate as a function of parameter m on OGI-Digits database, optimal $m = -140$

In order to test the effectiveness of the proposed features on a different database, phoneme classification experiments are conducted on TIMIT. MLP with 1000 hidden nodes is trained to convert input speech features into posterior probabilities of phoneme classes and decisions are made based on these probabilities (Viterbi decoding). Phoneme error rate (PER) is used as a measure to evaluate performance of the features. The PER of the baseline MRASTA features is **36.9%** while that of the proposed features ($m = -140$) is **35.7%**. *Thus the proposed features yield a relative improvement of about* 3.2% *over the baseline features on TIMIT database.* The above results indicate that asymmetry in filter shapes is indeed desired for speech recognition task.

5 Conclusions

Modifications, motivated by the asymmetries of the TRF profiles of higher level auditory neurons, to the MRASTA feature extraction technique has been proposed and tested for an ASR task. Results from the experiments on different databases seem to be promising, suggesting that careful emulation of STRFs of higher level auditory neurons would lead to better performance. With the proposed approach, we obtained more than 11.4% relative improvement in performance on OGI-Digits database. The proposed features also performed better when tested on different (TIMIT) database.

Acknowledgments

This work was supported by the European Union (EU) under the integrated project DIRAC, Detection and Identification of Rare Audio-visual Cues, contract number FP6-IST-027787, and by DARPA GALE program.

References

1. Hermansky, H.: Perceptual Linear Predictive (PLP) Analysis of Speech. J. Acoust. Soc. Am. 87(4), 1738–1752 (1990)
2. Hermansky, H., Fousek, P.: Multi-resolution RASTA filtering for TANDEM-based ASR. In: INTERSPEECH, September 2005, pp. 361–364 (2005)
3. Hermansky, H., Ellis, D.P.W., Sharma, S.: Tandem connectionist feature extraction for conventional HMMsystems. In: Proc. of ICASSP, Istanbul, Turkey (2000)
4. Depireux, D.A., Simon, J.Z., Klein, D.J., Shamma, S.A.: Spectro-temporal response field characterization with dynamic ripples in ferret primary auditory cortex. Journal of Neurophysiology 85, 1220–1234 (2001)
5. Schreiner, C.E., Read, H.L., Sutter, M.L.: Modular Organization of Frequency Integration in Primary Auditory Cortex. Annual Review of Neuroscience 23, 501–529 (2000)
6. Qiu, A., Schreiner, C.E., Escabi, M.A.: Gabor analysis of auditory midbrain receptive fields: Spectro-temporal and binaural composition. Journal of Neurophysiology 90 (2003)
7. Theunissen, F.E., Sen, K., Doupe, A.J.: Spectral-Temporal Receptive Fields of Nonlinear Auditory Neurons Obtained Using Natural Sounds. Journal of Neurophysiology 20, 2315–2331 (2000)
8. Kleinschmidt, M., Gelbart, D.: Improving Word Accuracy with Gabor Feature Extraction. In: Proc. of ICSLP, Colorado, USA (2002)

A Framework for Language-Independent Analysis and Prosodic Feature Annotation of Text Corpora

Dimitris Spiliotopoulos, Georgios Petasis, and Georgios Kouroupetroglou

Department of Informatics and Telecommunications
National and Kapodistrian University of Athens
Panepistimiopolis, Ilisia, GR-15784, Athens, Greece
dspiliot@di.uoa.gr, petasis@iit.demokritos.gr,
koupe@di.uoa.gr

Abstract. Concept-to-Speech systems include Natural Language Generators that produce linguistically enriched text descriptions which can lead to significantly improved quality of speech synthesis. There are cases, however, where either the generator modules produce pieces of non-analyzed, non-annotated plain text, or such modules are not available at all. Moreover, the language analysis is restricted by the usually limited domain coverage of the generator due to its embedded grammar. This work reports on a language-independent framework basis, linguistic resources and language analysis procedures (word/sentence identification, part-of-speech, prosodic feature annotation) for text annotation/processing for plain or enriched text corpora. It aims to produce an automated XML-annotated enriched prosodic markup for English and Greek texts, for improved synthetic speech. The markup includes information for both training the synthesizer and for actual input for synthesising. Depending on the domain and target, different methods may be used for automatic classification of entities (words, phrases, sentences) to one or more preset categories such as "emphatic event", "new/old information", "second argument to verb", "proper noun phrase", etc. The prosodic features are classified according to the analysis of the speech-specific characteristics for their role in prosody modelling and passed through to the synthesizer via an extended SOLE-ML description. Evaluation results show that using selectable hybrid methods for part-of-speech tagging high accuracy is achieved. Annotation of a large generated text corpus containing 50% enriched text and 50% canned plain text produces a fully annotated uniform SOLE-ML output containing all prosodic features found in the initial enriched source. Furthermore, additional automatically-derived prosodic feature annotation and speech synthesis related values are assigned, such as word-placement in sentences and phrases, previous and next word entity relations, emphatic phrases containing proper nouns, and more.

1 Introduction

Text annotation is a procedure where certain meta-information gets identified and associated with the entities in a text corpus. Such information is commonly used in computational linguistics for language analysis, speech processing, natural language processing, speech synthesis, and other areas. The type of information that is analyzed

P. Sojka et al. (Eds.): TSD 2008, LNAI 5246, pp. 517–524, 2008.
© Springer-Verlag Berlin Heidelberg 2008

and associated to text units may span the linguistic analysis tree (grammatical, syntactic, morphological, semantic, pragmatic, phonological, phonetic), as well as include any other description that may be of use.

Speech synthesizers traditionally perform a part-of-speech analysis and build the syntactic tree of the text in order to assign prosody [1]. General purpose Text-to-Speech (TtS) systems use certain language processing subsystems, such as sentence segmentation and part-of-speech tagging, for the analysis of the written text input. Depending on the actual system, such analysis may suffer from inherent statistical error accuracy that may be due to the design and implementation of the respective modules or language ambiguity. However, TtS systems may employ language analysis modules that are designed for high accuracy in specific thematic domains for which they seem to perform adequately. The respective accuracy when used for generic or other thematic domains may fall under unacceptable levels. Additionally, the language processing modules embedded in TtS systems are not usually designed to identify and extract higher-level linguistic information, such as semantic or pragmatic factors, that may be used to aid speech synthesis.

Concept-to-Speech (CtS) systems seem to provide an ideal means of text analysis. The Natural Language Generator (NLG) component of the CtS systems produces processed and annotated text as input for the speech synthesis module [2]. The NLG output text is generated as error-free syntactically annotated text exhibiting full disambiguation. In addition, further linguistic information may be generated providing considerable aid to guide synthesis. CtS systems, as a result, utilize the linguistic features from the natural language generation phase in order to produce significantly improved synthesized speech [3]. One of the major drawbacks of CtS systems is that the NLGs are designed to operate in specific thematic domains, and thus restricted to limited domain text generation. To make the things more complicated, the text output may not always be generated by the system. There may exist chunks of plain unprocessed text (canned) designed to be included in the output. These include groups of words, phrases or whole sentences that contain language that is too complicated for the NLG to fully generate. Such example is the MPIRO corpus [4] where more than 40% of the text descriptions of a museum exhibit domain is canned text. A linguistic analysis of that portion of text can provide a fully analyzed, uniformly annotated corpus, an essential and important benefit for speech synthesis.

Previous works that have explored natural language generated texts show that linguistically enriched annotated text input to a speech synthesizer can lead to improved naturalness of speech output [5,6]. Generation of tones and prosodic phrasing from high level linguistic input produces better prosody than plain texts do [7]. When such input can be provided, the language processing from the TtS system can be superseded.

In this work, a language-independent framework for language analysis and semantic annotation is presented. The aim is to produce uniform enriched text description, similar to the one generated by the natural language generation component of a CtS system, starting from plain or partially annotated text whether that may come from a natural language generator or a plain text document. This framework has been used successfully for the design, implementation and evaluation of a methodology for automatic annotation of large domain-dependent Greek text corpora.

This work reports on the set of linguistic features and information that needs to be considered and the description of the workflow and key modules that are employed for enriched text annotation for English and Greek text. Furthermore, the nature of the text analysis and prosodic feature incorporation are explored for focus prominence calculation for synthetic speech.

2 Enriched Text Annotation

TtS systems generally accept plain (or "raw") text as input, using specialized algorithms to internally generate the needed natural language data prior to synthesis. However, the algorithms that are usually implemented for such tasks are not powerful enough to broadly identify additional information about several linguistic phenomena from the plain text form, thus limiting the depth of text analysis and the derived description. A valuable alternative is to use pre-processed annotated text as input to the speech synthesizer. Enriched text of that kind exhibits major advantage over plain text as it retains structural and discourse level information. Each of the above types of linguistic information is described by sets of features that can be used to generate improved prosody in speech synthesis. Depending on the domain as well as the type of text, different sets of features may be used for maximum improvement.

As an alternative to generated text, existing plain text can be adequately processed to derive annotated NLG-similar output, essentially gaining advantage for the prosody modelling stage in speech synthesis. In order to do that efficiently, automated analysis and annotation should be made available for the most language analysis stages. A breakdown of the identifiable distinct processes is:

- Word/Sentence identification and segmentation.
- Morphological analysis (part-of-speech tagging and noun-phrase identification).
- Calculation/annotation of prosodic features.
- Creation/export to appropriate XML format description for speech synthesis.

As described in the following paragraphs, fully automated analysis can be achieved for all processes. The enriched linguistic annotation needs to be exported to a well-tested and reliable standard markup, such as XML. All the above processes have been implemented through the utilisation of the Ellogon Language Engineering Platform [8] platform and implemented the speech-oriented natural language analysis and annotation components [9].

As shown in Figure 1, the input may be either fully or partially annotated text (e.g. from a Natural Language Generator) or plain unformatted text. Information from the enriched input is extracted and used for the annotation of the plain text. The prosodic feature annotation assigns prosodically important values for the calculation of intonation focus for higher quality speech synthesis.

3 Morphological Analysis

Pre-processing mainly includes word and sentence identification, as well as part-of-speech (POS) tagging. For English texts, a POS tagger based on machine learning

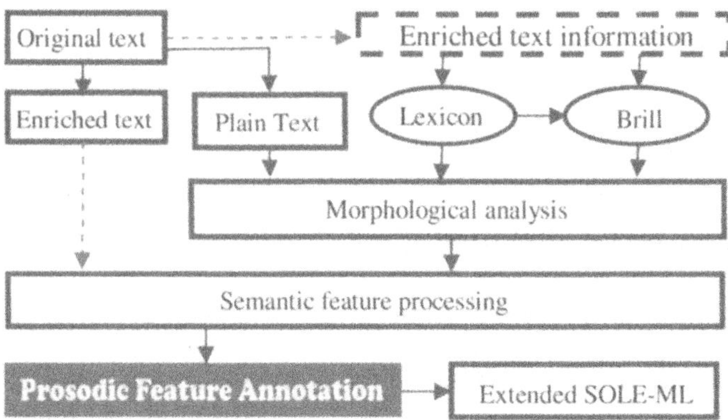

Fig. 1. The annotation workflow

is used, while for Greek texts a combination of lexicon-based and machine learning analysis is preferred. Word and sentence identification are performed by a rule-based component (HTokenizer) that presents an accuracy that approaches 100% for both languages. For part-of-speech tagging, the implementation is based on Transformation-based Error-driven learning [10] and provides models for English with an accuracy that approaches 97% measured as average of several accuracy measurements performed on various thematic domains.

For Greek, the common approach for most embedded systems is the use of Lexicon-based POS taggers. This approach is used by most speech synthesisers and yields accuracy between 75–85% depending on the domain of the text corpora. This low accuracy in most cases hinders poor final prosody prediction. This is due to Greek being an inflectional language with vast vocabulary that cannot be covered by lexicons. In order to increase the accuracy of POS tagging when processing documents in the Greek language, we used a hybrid approach, a combination of a lexicon-based POS tagger and a rule-based (Brill) POS tagging component. Two morphological lexicons for the Greek language have been combined in order to build a lexicon-based POS tagger with the highest possible coverage. The first lexicon is a large-scale morphological lexicon for the Greek language, developed exclusively for the system [11]. The lexicon consists of ~60,000 lemmas that correspond to ~710,000 different word forms (Greek is an inflexional language). The second lexicon is property of the Speech Group, University of Athens, used in the DEMOSTHeNES speech composer [12] and contains ~60,000 lemmas, which correspond to ~650,000 word forms. Both lexicons yield a word form identification span of ~880,000.

The hybrid approach was applied to the full generated corpus using two different ways in order to examine and evaluate the best procedure:

In the first approach, the built-in POS tagger and the lexicon-based POS tagger are both applied independently. Depending on the actual corpus and relative precision of the lexicon and HBrill modules, a word can be set to be assigned a value by either tagger (or both). The default state is that if a word contained in any of the two lexicons and thus is assigned a POS category by the lexicon-based tagger, this categorization becomes the final POS of the word, ignoring any categorization performed by the machine learning

POS tagger. On the other hand, if a word is not found in any of the two lexicons, the categorization presented by the built-in POS tagger is assigned. The machine learning based POS tagger uses an extension of the Penn Tree Bank tagset, which contains additional information regarding number and gender of words [13]. This approach achieved an accuracy of 95%.

The second approach sees that the Lexicon-based component is always followed be the machine learning POS tagger. Initial values are extrapolated from the lexicons and fed as initial states for the machine learning algorithm which provides the final value. In the case of partially annotated texts, the values of the pre-annotated word tokens were used for initial values in similar word forms since they were 100% correct coming from the natural language generator. This approach yielded total accuracy >97% for plain and >98% for partially annotated Greek texts and was the preferred choice.

4 Prosodic Feature Annotation

Previous research shows that higher-level linguistic information such as semantic features can be used to improve prosody modelling for speech synthesis [6]. This is because part-of-speech and phrase type information alone cannot always infer certain intonational focus points since those are not only affected by syntax but also by semantic and pragmatic factors [14]. For prosody modelling in speech synthesis, these factors can be used for calculation, deduction and verification of focus prominence and are accounted for by enriching the text corpus accordingly.

In our corpus, the plain text was annotated using the hybrid part-of-speech technique. Then, the results were validated and updated using the part-of-speech information from the enriched corpus. The benefit is twofold, the values are checked with the correct ones from the enriched text (if such is available for a lexical item) and key items are assigned specific values where appropriate. After that, certain semantic factors are calculated and added to the meta-information pool

Figure 2 shows an example of how semantic factors such as *newness (new or old infromation), contrast, explicit emphasis, first or second argument to verb* may be used for determining intonational focus prominence.

The intonational focus is assigned in a scale of three, strong focus '3', normal focus '2', and weak focus '1'. The features in bold are the ones computed from the information provided by the enriched portion of the text. Although *newness* is a key factor for strong intonational focus, certain validation checks in the algorithm make sure that only the proper lexical items are assigned. Validation factors are proper-noun and second-argument-to-verb (arg2) as well as explicit factors such as *emphasis* and *contrast*. As a result, strong focus '3' is assigned when validation factors arg2 and/or proper-noun exist for a new information (e.g., #1-2) while old information (e.g. #8-9) gets weak focus, as shown below:

Strong focus prominence: newness_TRUE (validation=passed)
Normal focus prominence: newness_FALSE (validation=passed)
Weak focus prominence: newness_TRUE (validation=failed)
No focus prominence: newness_FALSE (validation=failed)

However, it can be seen that explicit factors elevate the focus prominence, clearly providing explicit emphatic events as in the case of *splachnoscopy* (# 8) where if it were not for the *emphasis* factor it would have been assigned weak focus since it is an already given piece of information.

This exhibit is an [amphora] created during the [archaic period]. It dates from the early fifth century before [Christ.] [It was found in [Beotea] but it was made in [Athens]] It depicts a warrior performing [splachnoscopy] before leaving for battle. [Splachnoscopy] is the study of animal entrails, through which people tried to predict the future. It was one of the most common divination methods used in the [archaic period]. This [amphora] was painted by [the painter of Kleofrades] and was decorated with the [red figure] technique.

#	Lexical item	Focus	Prosodic features
1	amphora	3	[newness_TRUE, arg2]
2	archaic period	3	[newness_TRUE, arg2]
3	Christ	2	[newness_FALSE, proper-noun]
4	It was ... in Athens	-	[contrast]
5	Beotea	3	[newness_TRUE, arg2, ***proper-noun***]
6	Athens	3	[newness_TRUE, arg2, ***proper-noun***]
7	splachnoscopy	3	[newness_TRUE, arg2, ***emphasis***]
8	Splachnoscopy	3	[newness_FALSE, arg1, ***emphasis***]
9	archaic period	1	[newness_FALSE, arg2]
10	amphora	1	[newness_FALSE, arg2]
11	the painter of Kleofrades	3	[newness_TRUE, arg2, ***proper-noun***]
12	red figure	3	[newness_TRUE, arg2, ***proper-noun***]

Fig. 2. Focus prominence identification from semantic factors

Contrast is a rather generalized annotation that was implemented as a rule in the process and was initiated due to the fact that domain contained several instances of similarly NLG-derived phrases. The rule applies to both Greek and English text and elevates the main verb(s) and the conjunction to a mid-level emphasis, thus assigning explicitly a normal focus prominence marker (not showing in Figure 2).

From the above, it is obvious that the precision of part-of-speech identification is quite important since certain lexical items are validated for their assigned focus prominence using the part-of-speech information against the identified prosodic features.

5 SOLE-ML Description

The enriched text meta-information is encoded using an open XML schema. It is an extension (to cater for the semantic/prosodic description) of the SOLE-ML description [15], and was originally built as an annotation scheme for CtS synthesis, used as markup for the enriched text output of the ILEX generator [2]. It has been successfully used in earlier works, a well-tested means of representing enriched linguistic information, and is now standard input of the DEMOSTHeNES speech composer. The automatic extraction to the extended XML description based on SOLE-ML encodes all prosodic features. Figure 3 shows the XML output for the sentence *"It was found in Beotea but it was made in Athens."* from the text paragraph shown in Figure 2.

```
<utterance>
<relation name="Word" structure-type="list">
<wordlist>
<w id="w20">It</w>
<w id="w21">was</w>
<w id="w22">found</w>
<w id="w23">in</w>
<w id="w24">Beotea</w>
<w id="w25">but</w>
<w id="w26">it</w>
<w id="w27">was</w>
<w id="w28">made</w>
<w id="w29">in</w>
<w id="w30" punct=".">Athens</w>
</wordlist>
</relation>
<relation name="Group" structure-type="list">
</relation>
<relation name="Syntax" structure-type="tree">
<elem phrase-type="S">
<elem phrase-type="prosody" contrast>
<elem lex-cat="PRONOUN" href="w#id(w20)"/elem>
<elem phrase-type="prosody" mid-emphasis-verb>
<elem lex-cat="VERB" href="w#id(w21)"/elem>
<elem lex-cat="VERB" href="w#id(w22)"/elem>
</elem>
<elem lex-cat="PREPOS" href="w#id(w23)"/elem>
<elem phrase-type="prosody" newness="true", arg2, proper-noun>
<elem lex-cat="NOUN" href="w#id(w24)"/elem>
</elem>
<elem phrase-type="prosody" mid-emphasis-conj>
<elem lex-cat="CONJNCT" href="w#id(w25)"/elem>
</elem>
<elem lex-cat="PRONOUN" href="w#id(w26)"/elem>
<elem phrase-type="prosody" mid-emphasis-verb>
<elem lex-cat="VERB" href="w#id(w27)"/elem>
<elem lex-cat="VERB" href="wl#id(w28)"/elem>
</elem>
<elem lex-cat="PREPOS" href="w#id(w29)"/elem>
<elem phrase-type="prosody" newness="true", arg2, proper-noun >
<elem lex-cat="NOUN" href="w#id(w30)"/elem>
</elem>
</elem>
</elem>
</relation>
</utterance>
```

Fig. 3. The XML description

A wordlist of all tokens (words) and punctuation values takes up the first part (*<wordlist>*), followed by the syntax tree, prosodic features, and other high-level information (<relation>). This is the input for the speech synthesizer.

6 Evaluation and Discussion

The proposed framework utilises the meta-information contained in enriched automatically generated texts in order to compute and annotate both the enriched and the plain text with prosodic features that aid focus prominence in synthetic speech. The uniformly annotated target text contains enough elements to aid focus prominence using the modified speech synthesizer for Greek or an equivalent for English. An evaluation of the performance of the hybrid morphological analysis methods was performed for both Greek and English texts, shown in Table 1.

These results include the prime importance validation factor in our approach *proper-noun,* while exclude all other features that are calculated later in the process.

Enriched text annotation using naturally generated meta-information for a specific domain greatly enhances the intonational focus prominence predictors of a speech synthesizer. A strong indication of focus based on the new or already given information validated by the type of the lexical item works exceptionally well for domain-dependent corpora where the prosodic features can be more easily calculated automatically. This leads to enhanced input for speech synthesis, while bypassing all internal language analysis modules of the synthesizer, results on improved prosody prediction.

Table 1. Plain text part-of-speech annotation

Corpus (plain text)		Lexicon	Brill	Hybrid
English	precision	0.90	0.97	0.98
	recall	0.77	0.92	0.98
Greek	precision	0.88	0.94	0.98
	recall	0.75	0.84	0.92

Acknowledgements. The work described in this paper has been funded by the European Social Fund and Greek National Resources under the RHETOR project of the Information Society programme, Hellenic General Secretariat of Research and Technology.

References

1. Taylor, P., Black, A., Caley, R.: The architecture of the festival speech synthesis system. In: Proc. 3rd ESCA Workshop on Speech Synthesis, Australia, pp. 147–151 (1998)
2. O'Donnel, M., Mellish, C., Oberlander, J., Knott, A.: ILEX: An architecture for a dynamic hypertext generation system. Natural Language Engineering 7(3), 225–250 (2001)
3. Hitzeman, J., Black, A., Taylor, P., Mellish, C., Oberlander, J.: On the Use of Automatically Generated Discourse-Level Information in a Concept-to-Speech Synthesis System. In: Proc. 5th Int. Conf. on Spoken Language Generation (ICSLP), pp. 2763–2768 (1998)
4. Isard, A., Oberlander, J., Androutsopoulos, I., Matheson, C.: Speaking the Users' Languages. IEEE Intelligent Systems 18(1), 40–45 (2003)
5. Pan, S., McKeown, K., Hirschberg, J.: Exploring features from natural language generation for prosody modeling. Computer Speech and Language 16, 457–490 (2002)
6. Xydas, G., Spiliotopoulos, D., Kouroupetroglou, G.: Modeling Improved Prosody Generation from High-Level Linguistically Annotated Corpora. IEICE Trans. of Inf. and Syst., Special Section on Corpus-Based Speech Technologies E88-D(3), 510–518 (2005)
7. Black, A., Taylor, P.: Assigning intonation elements and prosodic phrasing for English speech synthesis from high level linguistic input. In: Proc. 3rd Int. Conf. on Spoken Language Processing, Yokohama, Japan, pp. 715–718 (1994)
8. Petasis, G., Karkaletsis, V., Paliouras, G., Androutsopoulos, I., Spyropoulos, C.D.: Ellogon: A New Text Engineering Platform. In: Proc. 3rd Int. Conf. on Language Resources and Evaluation (LREC 2002), Las Palmas, Canary Islands, Spain, May 2002, pp. 72–78 (2002)
9. Ellogon Language Engineering Platform, Speech tools add-ons, `http://www.ellogon.org/speech/`
10. Brill, E.: Transformation-Based Error-Driven Learning and Natural Language Processing: A Case Study in Part of Speech Tagging. Computational Linguistics 21, 543–565 (1995)
11. Petasis, G., Karkaletsis, V., Farmakiotou, D., Androutsopoulos, I., Spyropoulos, C.D.: A Greek Morphological Lexicon and its Exploitation by Natural Language Processing Applications. In: Manolopoulos, Y., Evripidou, S., Kakas, A.C. (eds.) PCI 2001. LNCS, vol. 2563, Springer, Heidelberg (2003)
12. Xydas, G., Kouroupetroglou, G.: The DEMOSTHeNES Speech Composer. In: Proc. 4th ISCA Workshop on Speech Synthesis, Perthshire, Scotland, pp. 167–172 (2001)
13. Petasis, G., Paliouras, G., Karkaletsis, V., Spyropoulos, C.D., Androutsopoulos, I.: Resolving Part-of-Speech Ambiguity in the Greek Language Using Learning Techniques. In: Fakotakis, et al. (eds.) Machine Learning in Human Language Technology, pp. 29–34 (1999)
14. Bolinger, D.: Intonation and its Uses: Melody in grammar and discourse. Edward Arnold, London (1989)
15. Hitzeman, J., Black, A., Mellish, C., Oberlander, J., Poesio, M., Taylor, P.: An annotation scheme for Concept-to-Speech synthesis. In: Proc. 7th European Workshop on Natural Language Generation, Toulouse France, pp. 59–66 (1999)

Quantification of Segmentation and F0 Errors and Their Effect on Emotion Recognition

Stefan Steidl, Anton Batliner, Elmar Nöth, and Joachim Hornegger*

Friedrich-Alexander-Universität Erlangen-Nürnberg, Lehrstuhl für Mustererkennung, Martensstraße 3, D-91058 Erlangen, Germany
stefan.steidl@informatik.uni-erlangen.de

Abstract. Prosodic features modelling pitch, energy, and duration play a major role in speech emotion recognition. Our word level features, especially duration and pitch features, rely on correct word segmentation and F0 extraction. For the FAU Aibo Emotion Corpus, the automatic segmentation of a forced alignment of the spoken word sequence and the automatically extracted F0 values have been manually corrected. Frequencies of different types of segmentation and F0errors are given and their influence on emotion recognition using different groups of prosodic features is evaluated. The classification results show that the impact of these errors on emotion recognition is small.

1 Introduction

Different types of features have been proposed in speech emotion recognition. In this paper, we focus on prosodic features, which have been proven to effectively discriminate emotional states and are widely used in this field. They model pitch, loudness, and accentuation as well as temporal aspects within suprasegmental units like words or whole utterances. The acoustic correlates are the fundamental frequency F0, the short-term signal energy, and durations of words, syllables, pauses, etc. A vast number of F0 extraction algorithms has been developed. For a comparative evaluation see [1]. Nevertheless, all of them are erroneous to some degree. F0 features are heavily affected by extraction errors; especially octave errors (doubled or halved F0 values) change the F0 extrema and the F0 range significantly. But other features like the slope and the error of the regression line are affected, too. Durations of words or subunits are obtained by a forced alignment of the spoken word sequence to the audio signal. A wrong start and end frame leads to a wrong duration of the word. Furthermore, our data is labelled and classified on word level for which the segmentation is needed as well. In this paper, different types of segmentation and F0 errors are identified and their frequency of occurrence in the FAU Aibo Emotion Corpus, a corpus of spontaneous children's speech in various realistic emotional and emotion-related states, is given. For this reason, the automatic segmentation of the forced alignment and the automatically extracted F0 values using ESPS have been manually corrected. The impact on emotion recognition is

* This work was partially funded by the European Commission (IST programme) in the framework of the PF-STAR project under Grant IST-2001-37599 and the NoE HUMAINE under Grant IST-2002-507422. The responsibility for the content lies with the authors.

P. Sojka et al. (Eds.): TSD 2008, LNAI 5246, pp. 525–534, 2008.
© Springer-Verlag Berlin Heidelberg 2008

evaluated by comparing the classification performance which results from features calculated with the corrected version with the classification results obtained by features based on the automatic version.

2 The FAU Aibo Emotion Corpus

For this study, the German FAU Aibo Emotion Corpus is used. Here, only a brief description of the corpus is given. More details can be found in [2] and papers quoted therein. The corpus contains speech recordings of 51 children (age 10–13, 21 male, 30 female) of two different schools who were communicating with Sony's pet robot Aibo. The children were led to believe that Aibo was responding to their commands, but the robot was actually being remote controlled by a human operator who caused Aibo to perform a fixed, predetermined sequence of actions. The children were given different tasks like directing Aibo to certain places or through a parcours. To evoke emotions, they were put slightly under time pressure by telling them to direct Aibo as fast as possible through the parcours. At certain predefined situations in the course of the experiment, Aibo did not obey to evoke anger. The task to let Aibo dance was supposed to induce joy. In some tasks, up to three feeding dishes are placed on the carpet. The children were told that one of them contains poison and that they have to make sure that Aibo does not go to this cup under any circumstances. Nevertheless, Aibo approaches exactly this cup in order to elicit slight forms of fear or panic. About 9.2 hours of speech – larger pauses have been removed – have been collected. The recordings of each child have been segmented automatically into smaller 'turns' using a pause threshold of 1 s. Five labellers (advanced students of linguistics) listened to the turns in sequential order and annotated each word independently of each other as neutral, which is the default, or belonging to one of ten other classes of emotion-related user states. These categories have been chosen in advance by inspection of the data. Actually, much of the data (48,401 words in total) is neutral. Other states are quite rare (sparse data problem). Hence, a subset of 6,070 words has been selected containing an almost balanced set of the four classes ***Angry*** (1,557 words), ***Motherese*** (1,223 words), ***Emphatic*** (1,645 words), and ***Neutral*** (1,645 words). The category ***Angry*** subsumes different but closely related forms of negative attitude like *slight anger*, *touchy/irritated* and *reprimanding*.

3 Manual Correction of the Word Segmentation

A segmentation is necessary to calculate our prosodic features on the word level. If the word boundaries are incorrect, frames outside the word might be considered for the calculation of the energy and F0 features while frames inside the word may be missing. Nevertheless, the impact on the energy and F0 features is supposed to be small since the mean, the extrema, etc. will not change significantly. In contrast, the duration features rely heavily on a high accuracy of the determination of the word, syllable, and phoneme durations which are given by the segmentation. Hence, segmentation errors might have very well an impact on the subsequent emotion recognition. In order to find out how large the influence actually is, the word boundaries have been manually corrected for

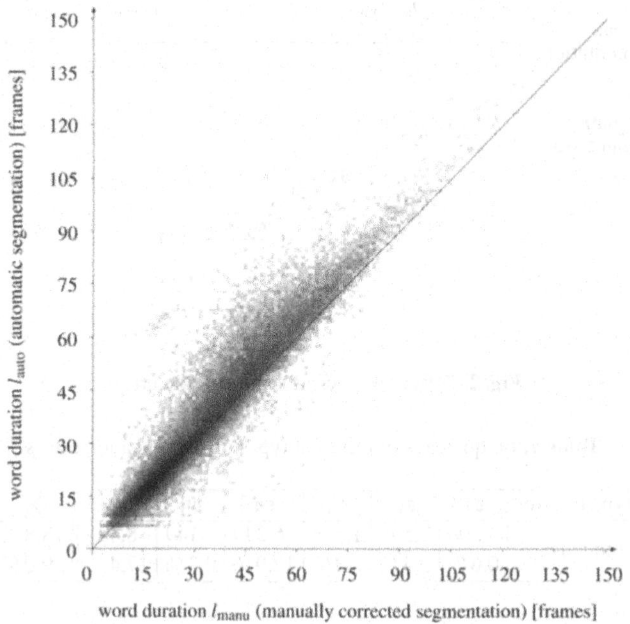

Fig. 1. Comparison of the manually corrected word segmentation with the automatic segmentation of the forced alignment. The histogram frequencies are displayed on a logarithmic gray scale.

the complete FAU Aibo Emotion Corpus on the basis of the automatic segmentation obtained by a forced alignment of the spoken word sequence using our own speech recognition system ISADORA [3]. Yet, the exact word boundaries are hard to identify. This is especially true for the end of the word due to reverberation, although a close-talk microphone has been used. The word durations l_{manu} of the manually corrected segmentation and the durations l_{auto} obtained by the forced alignment of the spoken word chain correlate highly (correlation of 0.93). On average, the word in the automatic segmentation is 36.8 frames (frame shift of 10 ms) long – 3.4 frames longer than the average word in the manually corrected segmentation. As this is a systematic error of the aligner which avoids small pauses between words, the impact on the prosodic features is supposed to be small. The two-dimensional histogram in Fig. 1 shows the frequencies of pairs (l_{manu}, l_{auto}) on a logarithmic gray scale. On average, a word in the forced alignment begins 1.5 frames too early and ends 2.0 frames too late.

In order to have a closer look at the occurring segmentation errors, they are categorised into six groups which are illustrated in Fig. 2. Errors of type $\mathfrak{s}3$ and $\mathfrak{s}6$ indicate that the automatically segmented word is either too short ($\mathfrak{s}3$) or too long ($\mathfrak{s}6$). Automatically segmented words that are shifted slightly to the left or to the right on the time axis, i. e. words that begin and end too early or too late, respectively, but where the automatic and the manual segmentation do overlap to some degree, are of type $\mathfrak{s}2$ ($\mathfrak{s}4$). In the case of no overlap between the automatic and manual segmentation, the words are of type $\mathfrak{s}1$ or $\mathfrak{s}5$ depending on whether the automatically segmented word

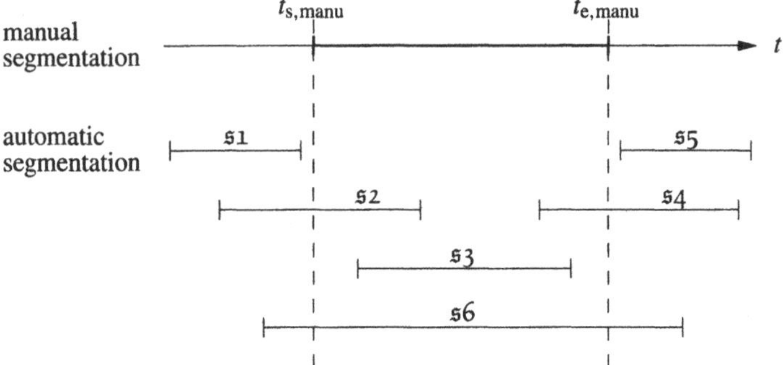

Fig. 2. Different types of segmentation errors

Table 1. Frequencies of different types of segmentation errors

type of error	s1	s2	s3	s4	s5	s6	$\sum$
frequency	300	2,598	1,133	6,249	149	8,427	18,856
	0.6 %	5.4 %	2.3 %	12.9 %	0.3 %	17.4 %	39.0 %

appears before (s1) or after (s5) the manually segmented one. The frequencies of the different error types are given in Table 1. Deviations of at most three frames at both word boundaries are tolerated. Using this threshold, the segmentation of 39.0 % of all words in the corpus is incorrect. In most cases (17.4 %) the automatically segmented words are too long (error type s6), due to the systematic error of the aligner mentioned above. In about 1 % of the cases, the words are completely misplaced by the automatic alignment (types s1 and s5). The average duration of these misplaced words is 264 ms compared to 334 ms of the average word.

4 Manual F0 Correction

F0 features model the rough course of the fundamental frequency. Especially features like the values of the extrema or the range of the F0 values, but also the regression line and the regression error are directly influenced by F0 extraction errors. It is an open question to what extent these errors influence the performance of the emotion recognition system. For this reason, the F0 values of 3,996 turns – the turns that contain amongst others the 6,070 words of the reduced data set described above – have been manually corrected by the second author. As a reference baseline, the F0 has been calculated automatically using the freely available and well established F0 algorithm of the popular *Entropic Signal Processing System* (ESPS) toolkit [4] which is often used for benchmarking. Due to the large amount of data ($> 10^6$ frames), it is impossible to manually determine the length of each period. Hence, the focus is set on the manual correction of obvious errors like voiced/unvoiced errors, octave jumps, or other gross errors. Besides real errors of the pitch extraction algorithm, there are irregularities in the speech production which actually change the fundamental frequency of the signal and can be perceived as suprasegmental irritations modulated onto the pitch contour, but which are not perceived as jumps up or down [5,6]. Since the manual correction is geared to human perception, a better term instead of 'correction' would be 'smoothed and adjusted to human perception'. We use the term *laryngealisation* for various types of irregular voiced stretches of speech. In [6], five types of laryngealisations have been established: glottalisation, diplophonia, damping, subharmonic, and aperiodicity. The manual correction mostly dealt with the following phenomena:

(1) octave jumps: the ESPS F0 has been corrected by one octave jump up, in some rare cases also two octave jumps up, or one octave jump down. This concerns rather smooth F0 curves which had to be transposed. In most cases, it is a matter of irregular phonation where the extraction algorithm modelled pitch rather 'close to the signal' instead of 'close to perception'. In a few cases, however, no clear sign of laryngealisation can be observed. Sometimes, the context and/or the perception had to decide whether an octave jump had to be corrected or not. If the whole word is laryngealised and the impression is low pitch throughout, then laryngealisation is not modulated onto pitch and the F0 values were kept unchanged.

(2) smoothing at irregularities: the ESPS curve is not smooth but irregular due to laryngealisations or voiceless parts which ESPS wrongly classified as voiced. Here, often the F0 values between the context to left and the context to the right were interpolated in order to result in a smoothed curve. In case of voiceless parts, the F0 values were set to zero.

(3) other phenomenalike irregularities at transitions which are not necessarily due to irregular phonation: smoothing at transitions is admittedly a bit delicate – when should it be done if the phenomenon is well known, e. g. in the case of higher F0 values after voiceless consonants. Sometimes, the context and/or the perception had to decide whether an octave jump had to be corrected or not. A typical problem is a hiatus, i. e. the sequence of one word ending in a vowel followed by, e. g., "Aibo". The perception is rather no pitch movement but 'something' modulated onto the pitch curve. In these cases, various F0 extraction errors can occur: the F0 values may be set to zero, i. e. the segment is classified as voiceless, octave jumps up or down may occur, the F0 values may be fully irregular, or values from low to higher may occur. Here, the F0 was sometimes interpolated, sometimes doubled, or sometimes not corrected (in the case from 'low to higher'). Sometimes, clear criteria for the one or the other solution could not be found, at least not with a reasonable effort. In voiced-unvoiced-voiced sequences within a word, e. g. in the word "Aibo", the plosive sometimes was set to voiceless even if voiced would have been possible – F0 postprocessing sometimes interpolates in such cases anyway. In some rare cases, it had to be 'educated guessing' and was not really based on strong criteria.

Fig. 3 shows an example with F0correction: below, the time signal, in the middle, the spectrogram, and above, the F0 values per frame (frame shift of 10 ms). Manually corrected F0 values are displayed with gray, filled circles. The colour of the background is set to gray if ESPS and manually corrected F0 values differ. The first part (the `[a]` in `[aI]`) of `/Aibo/` is clearly laryngealised: first glottalisation, then diplophonia, and in the last irregular part, aperiodicity. The intervocalic plosive `[b]` was set to voiceless

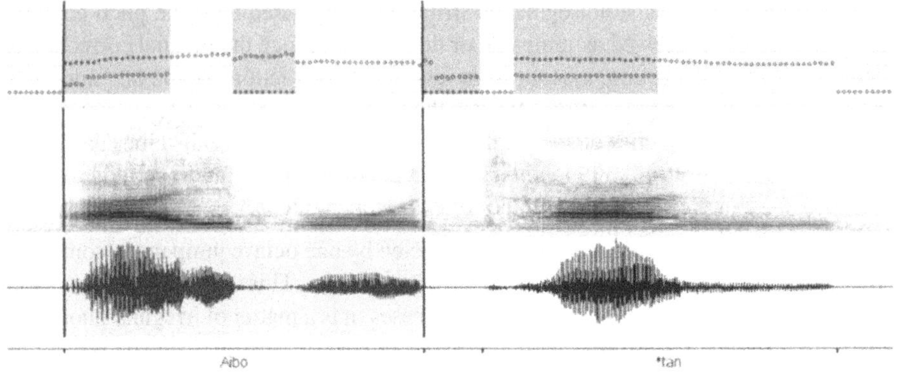

Fig. 3. Manual F0 correction for the utterance "Aibo, *tanz" (*Aibo, *dance*). The '*' marks word fragments.

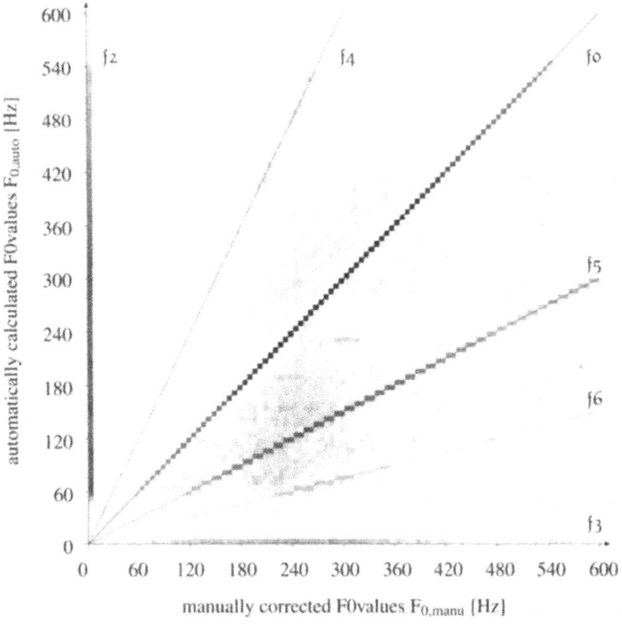

Fig. 4. Comparison of the automatically calculated and the manually corrected F0 values. The frequencies in the histogram are displayed on a logarithmic gray scale. The straight lines represent identical values (f0) and the error types f2 to f6. Errors of type f1 and f7 are located between these lines.

(note that this is regular in south German dialects). Without using the 'magnifying glass' to scale up the time signal, the `[a]` in `/tanz/` does not display clear signs of irregular phonation.

Table 2. Description of different F0 error types

type	short description	long description
f0	identical	F0 value calculated by ESPS is not changed by the manual correction
f1	minor error	deviation of the ESPS F0 value from the manually corrected F0 value is less than 10 %
f2	voiced error	ESPS calculates a F0 value for a frame which is considered to be unvoiced by the manual correction
f3	unvoiced error	a frame which is considered to be voiced by the manual correction is marked as unvoiced by ESPS
f4	octave error ↑	ESPS F0 value is twice the manually corrected F0 value with a tolerance of 10 %
f5	octave error ↓	ESPS F0 value is half the manually corrected F0 value with a tolerance of 10 %
f6	octave error ↓↓	ESPS F0 value is one fourth of the manually corrected F0 value with a tolerance of 10 %
f7	other gross error	deviation of the ESPS F0 value of more than 10 % but not one of the octave jumps mentioned above

To illustrate which types of F0 extraction errors occur how often, a two-dimensional histogram of the pairs ($F_{0,manu}$, $F_{0,auto}$) is given in Fig. 4. The frequencies of these pairs are displayed on a logarithmic gray scale in order to make less frequent errors visible as well. Cases where both ESPS and the human corrector decided for voiceless, i. e. pairs (0, 0), are discarded in the histogram due to their very high frequency. The histogram shows that F0 extraction errors can be categorised into various types of errors. They are denominated with f1 to f7 as defined in Table 2. Since only obvious errors have been corrected, the F0 values of most frames (94.3 %, s. Table 3) have been kept unchanged resulting in the dark diagonal (f0) in the histogram. Voiced errors (f2), i. e. F0 values which are considered to be voiceless by the human corrector and voiced by ESPS, result in the vertical straight line. Unvoiced errors (f3), i. e. F0values which are wrongly considered to be voiceless by ESPS, yield the horizontal straight line. Three more straight lines result from one (f5) or two (f6) octave jumps down (one half and one

Table 3. Frequencies of the different F0 error types evaluated on the whole turn or only within words

type of error	evaluation: whole turn		only within words	
f0 identical	1,050,450	94.3 %	574,485	93.7 %
f1 minor errors	455	0.0 %	452	0.1 %
f2 voiced errors	32,774	2.9 %	8,804	1.4 %
f3 unvoiced errors	1,884	0.2 %	1,877	0.3 %
f4 octave errors ↑	247	0.0 %	239	0.0 %
f5 octave errors ↓	23,718	2.1 %	23,498	3.8 %
f6 octave errors ↓↓	375	0.0 %	364	0.1 %
f7 other gross errors	3,634	0.3 %	3,559	0.6 %

Table 4. Classification results (average recall) for different types of prosodic features

features	manual correction no	manual correction yes
all features (PCA: 125 → 95)	61.3 %	61.4 %
F0 features without position (26)	46.5 %	47.8 %
energy features without position (31)	55.4 %	56.4 %
duration features (17)	52.7 %	52.8 %
duration and position of energy/F0 (30)	54.6 %	53.8 %
pause features (8)	34.9 %	34.1 %
POS features (30)	52.4 %	

fourth of the manually corrected F0 value, respectively) or one octave jump up (f(4), ESPS F0 value is twice the manually corrected one). Other gross F0 errors are located between these lines.

The frequencies of the different error types are given in Table 3. Numbers are given for the evaluation on the whole turns and for the evaluation only within words. As our prosodic features are word based, F0 values outside words (45 % of all frames) are irrelevant for our feature extraction. The comparison reveals that – as expected – almost all F0errors occur within words. Only voiced errors appear mostly outside words (73 %). The table also lists a few minor errors defined as deviations of less than 10 %. As explained above, minor errors were not in the focus of our manual correction. Anyway, state-of-the-art F0 features only model the rough course of the fundamental frequency. Thus, minor errors are highly unlikely to influence the emotion recognition.

5 Prosodic Features

We use a set of 95 relevant prosodic features modelling duration, energy and F0. The latter two groups of features model the course of the energy and the F0, respectively, within a certain context. Additionally, 30 linguistic features (part-of-speech features) are used. The context can be chosen from two words before and two words after the actual word. Thus we model, so to speak, a 'prosodic five-gram'. A full account of the prosodic features is beyond the scope of this paper; details are given in [7].

6 Experimental Results

For our 4-class classification problem, cf. Section 2, we use artificial neural networks (ANN), implemented within the software package SNNS [8]. A leave-one-speaker-out procedure is employed using 40 speakers for training, 10 speakers for validation and the remaining one for testing in each of the 51 runs. Classification results are obtained for the whole data set and are speaker-independent. In the training and the validation set, the samples of less frequent classes are upsampled to get a balanced set. The features are mapped onto a range from -1 to $+1$ and decorrelated using principal component analysis (PCA). In each run, the topology of the net and two parameters of the training algorithm (weight decay and random seed for initialisation of the network parameters) are optimised on the validation set. The ANNs consist of the input layer containing one node for each feature, one hidden layer of a varying number of nodes, and one output layer of four nodes – one for each class. In Table 4, the classification results are given in terms of the unweighted average recall over all four classes.

Six sets of feature groups are evaluated: The first set contains all prosodic features. In order to reduce the computational costs in the leave-one-speaker-out procedure, the number of features is reduced from 125 to 95 features using PCA. The second and the third set contain 26 F0 features and 31 energy features only, respectively. Features describing the position of the F0 and energy extrema are excluded since position features model the duration between the extremum and the reference point. Hence, they are regarded as duration features. The fourth and the fifth set are duration features. In the latter, the position features of the F0/energy extrema are included. The sixth set contains eight features describing filled and unfilled pauses between words. The last set contains 30 part-of-speech features. For all subsets as well as for the combination of them, the changes caused by segmentation and F0errors are not significant. The generally lower relevance of F0features, in comparison with energy and duration features, is in line with other studies, cf. [2] and [9].

7 Discussion and Concluding Remarks

Our corpus has been labelled and classified on the word level which is quite unique in research of emotion. This approach takes into account that emotion-related states can change rather quickly – even within utterances. Nevertheless, the opportunity to merge words into larger units like chunks or turns and to map emotion labels from word level onto these larger units is still possible. This approach has been pursued in [2]. One might as well argue that the influence of the described errors on emotion recognition also depends on the choice of features. In [2], results from our initiative CEICES are presented where F0 and duration features of the participating research institutes are combined covering a plethora of state-of-the-art prosodic features. Units of analysis were syntactically/semantically meaningful 'chunks' with 2.9 words per chunk on the average. The classification results confirm as well the low impact of F0errors on emotion recognition. Note that matters were different for the two-class problem prominence where erroneous F0 values yielded significantly lower classification performance. However, most important features were different: for corrected values, features modelling the slope (regression) were more important whereas for automatically extracted features, the more robust mean came to the fore. Thus we should not conclude that automatic extraction is generally good enough and does not contribute to classification errors: the automatic segmentation is based on word recognition which is only 'perfect' in forced alignment. In a fully automatic speech recognition system, wrong word recognition can yield wrong segmentation as well, and this in turn might very well contribute to wrong linguistic features for bag-of-words or part-of-speech classes. F0 errors might not be detrimental only if the feature vector models both specific and more general aspects.

With [2] and the present study it has been shown – to our knowledge, for the first time – that the impact of erroneous automatic extraction of pitch and segmentation on emotion recognition might be negligible, even if their frequency is not (some 4% octave errors within words, and almost 40% incorrect segmentation on the word level). This outcome makes it more likely that the difference in recognition relevance, observed for different feature groups, is not due to some 'surface phenomena' such as the extent of erroneous extraction.

References

1. de Cheveigné, A., Kawahara, H.: Comparative Evaluation of F0 estimation algorithms. In: Proc. Eurospeech 2001, Aalborg, Denmark, pp. 2451–2454.
2. Batliner, A., Steidl, S., Schuller, B., Seppi, D., Vogt, T., Devillers, L., Vidrascu, L., Amir, N., Kessous, L., Aharonson, V.: The Impact of F0 Extraction Errors on the Classification of Prominence and Emotion. In: Proc. ICPhS 2007, Saarbrücken, Germany, pp. 2201–2204 (2007)
3. Stemmer, G.: Modeling Variability in Speech Recognition. Logos Verlag, Berlin (2005)
4. Talkin, D.: A robust algorithm for pitch tracking (RAPT). In: Kleijn, W.B., Paliwal, K.K. (eds.) Speech coding and synthesis, pp. 495–518. Elsevier Science, Amsterdam (1995)
5. Batliner, A., Steidl, S., Nöth, E.: Laryngealizations and Emotions: How Many Babushkas? In: Proc. of International Workshop on Paralinguistic Speech – between Models and Data (ParaLing 2007), DFKI, Saarbrücken, Germany, pp. 17–22 (2007)
6. Batliner, A., Burger, S., Kießling, A.: MÜSLI: A Classification Scheme For Laryngealizations. In: House, D., Touati, P. (eds.) Proc. of an ESCA Workshop on Prosody, Lund University, Lund, Sweden, pp. 176–179 (1993)
7. Batliner, A., Fischer, K., Huber, R., Spilker, J., Nöth, E.: How to find trouble in communication. Speech Communication 40, 117–143 (2003)
8. Zell, A., Mache, N., Sommer, T., Korb, T.: The SNNS Neural Network Simulator. In: Radig, B. (ed.) Proc. of Mustererkennung 1991, 13. DAGM-Symposium, München, Germany, Informatik-Fachberichte, vol. 290, pp. 454–461. Springer, Heidelberg (1991)
9. Kochanski, G., Grabe, E., Coleman, J., Rosner, B.: Loudness predicts Prominence. Fundamental Frequency lends little. JASA 11, 1038–1054 (2005)

Deep Syntactic Analysis and Rule Based Accentuation in Text-to-Speech Synthesis

Antti Suni and Martti Vainio

Department of Speech Sciences, University of Helsinki, Finland

Abstract. With the emergence of the HMM-synthesis paradigm, producing natural, expressive prosody has become viable in speech synthesis. This paper describes the development of rule-based prominence prediction model for Finnish Text-to-Speech system, based on deep syntactic analysis and discourse structure.

Keywords: Speech synthesis, prosodic prominence, syntactic analysis, Finnish.

1 Introduction

Linguistically based accent prediction was a vital topic in Text-to-Speech synthesis (TTS)research in the 1980's and 1990's. Syntax and information structure (IS) were central in the field (see, for example, [1] and [2]). Since then, however, large annotated speech corpora have become available, and the success of statistical learning methods using shallow features have shifted the focus of research to the use various information theoretic features and data-driven algorithms. While the performance of the statistical methods using e.g. N-grams and part-of-speech sequences is indeed good, the drawback of this shift is that it is hard to gain insight into the causes behind accentuation patterns – theoretical research on information and discourse structure and TTS research no longer support each other.

Assigning acceptable accentuation and consequently believable prominence relations in TTS has been a problem ever since it became a viable field of research. The most simple means of assigning accents has usually been based on the so called FUNCTION vs. CONTENT word dichotomy. Although only 10% of Finnish running text consists of function words, (as opposed to approximately 50% in English), most Finnish TTS systems are still based on this simple lexical lookup and some systems have no word class analysis at all. In some cases, a statistical intonation model can separate the closed class words based on the surface forms and the resulting speech can still be acceptable in prosodic quality [3].

As such, pitch accenting in Finnish can not be well modeled by a simple accented-unaccented dichotomy and tests based on word accent prediction accuracy are not comparable with tests performed on systems from other languages. Therefore, a more subtle model for signaling the prominence relations within and between utterances is called for. Developing such a system requires reliable syntactic analysis which is necessarily deeper than mere part of speech tagging.

We have recently developed a new Finnish speech synthesis system, which utilizes a full syntactic parser based on Functional Dependency Grammar (FDG) [4]. The dependency parser analyzes sentences into part-of-speech tags, morpho-syntactic functions

P. Sojka et al. (Eds.): TSD 2008, LNAI 5246, pp. 535–542, 2008.
© Springer-Verlag Berlin Heidelberg 2008

and syntactic dependency functions (or roles); subjects, verbs, and objects as well as modifiers etc. These dependency functions can further be used to infer noun phrases. The relatively deep syntactic analysis allows for writing accentuation rules on a fairly high level and the separation of symbolic and signal based prosody can be made in a sensible place with regard to the TTS process – provided that the signal generation component of the system can produce meaningful pitch contours and segmental durations from a relatively high level input. The prosody and signal generation component of the system described in this paper is based on a Hidden Markov Model (HMM) -based synthesis system [5], which has been implemented for Finnish. We have previously shown that the system can reproduce the intended prominence patterns with good accuracy from automatically tagged corpus [6].

2 The Rules

In this chapter, we will describe the relevant rules of our current prominence prediction model, highlighting some concepts which might pose problems to non-linguistic accent prediction, especially regarding cues on information structure.

The rules are organized by context size, ranging from word level to sentence level, each rule modifying the prominence value predicted by previous rules. In this fashion the effect of each rule can be observed easily, and the possibility of unwanted rule interactions can be mostly avoided. The rules presented here are based, in addition to various TTS and IS research sources, on authors' introspection, especially concerning the generalizability of the observed phenomena in the limited Finnish speech corpora. Clearly, when dealing with unrestricted text, all rules are coarse generalizations and of heuristic nature. Application of any proper theoretical framework for information and discourse structure is also difficult due to the uncertainties and limitations of available information.

For brevity, we will use 'accented – deaccented' -terms with the meaning: produced with increased or decreased prominence. In rule examples, dec(w) means decreasing and inc(w) increasing the prominence of the word, while set(w, x) means setting the prominence of the word to x. Prominence itself can have four levels ranging from deaccented zero to emphatic(3).

2.1 Word Level

Part-of-Speech. We set the initial prominence values for words according to the generalization that content words are accented and function words are deaccented. Verbs are assigned lower prominence than other content words, according to their special status.

```
if PoS(w) in [N, Adj, Num, Adv]:  set(w,2)
else if PoS(w) in [V, Pron]: set(w,1)
else set(w,0)
```

The part-of-speech classes are not very good categories for our purposes, for example being an adverb gives little indication of the words accentability. With the aid of frequency lists and grammar [7], we have added some subcategories by listing common words whose default accentuation differs from the main category. These include for example various determiner types considered as simply pronouns by the parser, generic temporal and particle-like adverb, deverbals and generic nouns. We have listed about 300 generic or semantically light nouns and these cover approximately 25% of all noun instances in language, so sub-categorization has potentially a large effect on prediction. Recent experiments on utilizing *Accent Ratio* in accent prediction [8] support this claim. Of the listed nouns, some are clearly anaphoric like *situation, thing* and thus deaccented like pronouns, whereas the majority are typically deaccented only within a context of more specific material.

```
if sPoS(w) in [very generic noun, deverbal, particle Adv,...]: dec(w)
if sPoS(w) in [accentable Pron,...]: inc(w)
```

Givenness. It is generally believed that in discourse, new information tends to be accented and old information deaccented. However, it has been surprisingly difficult to confirm this view in corpus studies, see for example [9]. In our model, prominence of given words depends largely on the context as demonstrated by subsequent rules.

Our discourse structure model follows the method presented in [1]. Two levels of givenness are separated, local focus (given1) and global focus (given2). Local focus, containing the currently salient items in discourse, is modeled by a queue of lemmas of few previous words. The global focus contains the accessible items, modeled by a list of lemmas of all previously mentioned nouns in the paragraph.

2.2 Constituent Level

Noun Phrases. Accentuation of complex NPs is a difficult problem in Finnish. There is no agreed rhythmical preference for left or right-sidedness and equal prominence of both modifiers and the head is also normal. We try to approach the accentuation of NPs by simple information structural means, assigning less prominence to less informative parts of the phrase. The informativeness is defined by a heuristic ranking as exemplified below. Pronouns and words in local focus are considered the least informative and numerals and some quantifiers the most informative.

pronoun, given1 < generic N < demonstrative < given2 < N, Adj < Q, Num

```
for i in (0, length(NP)):
   if info(NP[i]) < info(NP[i-1]) or info(NP[i]) < info(NP[i+1]):
      dec(NP[i])
```

PEKAN KISSA, PEKKA's CAT
MOLEMMAT kissat, BOTH cats
VIISI senttiä, FIVE cents
minun KISSANI, my CAT
Pekan(given1) AUTO(generic), Pekka's CAR
HIENO kaupunki(generic), NICE town
PEKAN voittanut MIES(generic)
PEKKA won MAN (man who defeated Pekka)

This method produces acceptable prominence patterns for simple NPs above, but there are also many exceptions and the presented method without any structural analysis is decidedly inadequate. For English research on the subject, with deeper structural approach, see for example [10].

Focus Particles and Intensifiers. There is a rich selection of focus particles in Finnish that are used to bring an element to focus in sentence. This is potentially very useful for TTS as these focused elements can generally be given extra emphasis and thus bring liveliness to the synthetic speech.

The word to be emphasized usually follows the focus particle directly, but exceptions are frequent. The probable target is predicted using the previously assigned prominence values of the following words. The rule is conservative, as producing erroneous emphasis is a larger mistake than no emphasis at all.

```
if NP following focus particle:
  emphasize the most prominent word in NP
else:
  emphasize the closest most prominent word
  within 3 words of the focus particle if
  the prominence of that word > 1
```

Juha voisi myös(foc) korjata TUOLIN.
Juha voisi myös(foc) KORJATA tuolin(given1).
Juha could also fix the chair.

For intensifiers like *very, too, extremely* we have a rule that accents the intensifier and deaccents the intensified. Which of the items is accented seems to vary a lot though, perhaps depending on the expectedness of the intensified.

```
if intensifier(w) and PoS(w+1) in [Adv, Adj]:
   inc(w), dec(w+1)
```

2.3 Sentence Level

The default pattern of Finnish sentence stress is having the most prominent elements in the beginning(topic) and in the end(focus) of the sentence, while the middle part is less prominent. If the information structure does not agree with the default word order, the word order can be adjusted to adhere to the default pattern. However, there are enough exceptions and details to justify a more comprehensive model.

There is a large body of linguistic work on the subject of information structure and accent assignment for English, most notably Steedman's theory [11]. Unfortunately, the models typically require definite knowledge on phrase structure and information status of words and do not seem to adopt well to heuristic treatment. In any case, our sentence model roughly follows Steedman's division of clauses into thematic and rhematic parts, which further consist background and focus parts. The division into theme and rheme is initially made before the predicate, and the status of the predicate is handled separately.

Theme. The prominence of the theme depends largely on the thematic continuity of the discourse. If the topic changes, the theme element is often realized with high prominence, even if the said element is mentioned in the previous sentence, opposing the*deaccent given* generalization. On the other hand, new words can represent topic continuity and be deaccented, for example in case of synonyms or inferable content. However, tracking the topic would need a more sophisticated discourse model than our local and global focus stack, and we simply assume a shift of topic between paragraphs and no shift between sentence-internal clauses, except in the presence of words like *but, on the other hand,* suggesting contrast. We assume that at most one constituent can function as a theme focus and, following the Finnish tendency of placing the most prominent elements on peripheries, the first constituent if any, is the focus. On the following rules, xp corresponds to a constituent of undetermined type. Conjunctions and adverb particles are not considered constituents.

```
c = current_clause()
xp = first_xp()
w = get_most_prominent_word(xp)
if acc(w) < 2:
  if pos_in_paragraph(c) = 1: inc(w)
  if pos_in_sentence(c) > 1 and
     type(c) not in [CONTRAST, MAIN]:
          foreach w in xp: if w > 1: dec(w)
```

If pre-verbal part contains two constituents, the second one is suppressed. This is especially true regarding Finnish topicalization phenomenon. Any constituent independent of its syntactic role, can be moved to the front of the sentence, thus often functioning as a focus of the whole sentence [7]:

KAKUN Juha söi. (It was the) CAKE John ate.
EILEN Juha meni KEMIIN. YESTERDAY John went to KEMI.

```
if num_xp_before_verb > 1:
  if max_acc(first_xp) < 2: inc_max(first_xp)
  dec_all(second_xp)
```

Rheme. The order of words in free-word-order languages like Finnish is actually quite strict. Deviations from the default order have specific meaning regarding the information structure of the sentence, and has prosodic consequences [12].

In determining the accentuation of the post-verbal part of the sentence, we use aspects of a Czech model, *Topic-Focus Articulation* (TFA) [13]. In TFA, *Systemic Ordering* (SO), a language specific default order of arguments and modifiers is used to determine the background (topic) and focus of the sentence. To our knowledge, SO in Finnish has not been researched, but at least the following relations seem likely:

subject < object < location < manner < source < goal

In the following, we present our simplified version of the rule for background / focus determination, in verbal form:

```
If an element A precedes element B in text, but B precedes A in SO,
element A belongs to background. If the text order does follow SO,
the elements' background / focus status is ambiguous regarding
word order. In that case, all elements followed by more informative
elements belong to background and others in focus.  If all elements
after the predicate are just mentioned or very generic,
focus the predicate.
```

Here, informativeness is defined in a similar way as in the case of noun phrases, except that whole phrases are considered instead of individual words. When the background and focus status of elements has been determined, the prominence of the words is modified accordingly, suppressing the background elements.

```
if background(w) and acc(w) > 1: dec(w)
if focus(xp) and max_acc(xp) < 2: inc_max(xp)
```

The rule with the prototype Finnish SO produces the following acceptable patterns.

JUHA osti TELKKARIN TURUSTA.
JOHN bought a TV(object) from TURKU(source).

JUHA osti Turusta KALAN.
JOHN bought from Turku(source) a FISH(object).

JUHA matkusti Helsingistä Turkuun JUNALLA.
JOHN went from Helsinki(source) to Turku(goal) by TRAIN(manner). *JUHA OSTI sen. JOHN BOUGHT it(generic).*

In principle, TFA provides a sound method for handling over-accentuation, compared to more arbitrary rhythmical rules which are sometimes used. Unfortunately, the usefulness of the TFA method in our system is limited, due to somewhat weak performance of the FDG parser in providing function tags, and lack of research on Finnish SO.

2.4 Combining the Rules – Example

To illustrate how the rules work together we provide an example in Table 1. This is from the middle of a text about child discipline. By applying the prediction rules sequentially from part-of-speech to focus particles, we get the prominence pattern of the right-most column of the table, approximated in English below.

At THIS moment the child is already SO confused about the advice that he starts CRYING.

3 Discussion

In this paper, we have presented a selection of rules for Finnish prominence prediction, attempting to infer some information structural aspects of unrestricted text with the aid of automatic syntactic analysis. Although many aspects of the rules still need refinement

Table 1. Prominence prediction example

eng	word	PoS	sPoS	NP	Theme	Rheme	Focus part.
This	Tässä	1	1	1	2	2	2
moment	vaiheessa	2	1	0	0	0	0
child	lapsi	2	2	2	1	1	1
is	on	0	0	0	0	0	0
already	jo	2	1	1	1	1	0
so	niin	2	2	2	2	2	3
confused	sekaisin	2	2	2	2	2	1
advice-about	neuvoista	2	2	2	2	1	1
that	että	0	0	0	0	0	0
starts	alkaa	0	0	0	0	0	0
crying	itkemään	1	1	1	1	2	2

and many phenomena lack treatment, the current level of detail is certainly unsurpassed in Finnish accentuation modeling.

Many questions still remain unanswered and are subject to further research. Firstly, while informally we are very pleased with the performance of our model in our current TTS system, a proper evaluation of the rules and comparison against other methods is undone, due to lack of large annotated corpora with paragraph sized utterances. This question also concerns the accuracy of the syntactic parser; the rules depend heavily on the quality of the analysis. Independent of the numerical performance of the model, however, the rules function as an important step in research of Finnish sentence stress. The model provides testable claims that can readily be used as hypotheses in speech production and listening experiments.

Acknowledgments. This work has been supported by Grant NO. 107606 from the Academy of Finland to M. Vainio.

References

1. Hirschberg, J.: Pitch accent in context: predicting intonational prominence from text. Artif. Intell. 63(1-2), 305–340 (1993)
2. Prevost, S., Steedman, M.: Specifying intonation from context for speech synthesis. Speech Communication 15(1), 139–153 (1994)
3. Vainio, M.: Artificial Neural Network Based Prosody Models for Finnish Text-to-Speech Synthesis, ser. Publications of the Department of Phonetics, University of Helsinki. Yliopistopaino, 43 (2001)
4. Tapanainen, P.: Parsing in Two Frameworks: Finite-state and Functional Dependency Grammar. University of Helsinki, Dept. of General Linguistics (1999)
5. Tokuda, K., Yoshimura, T., Masuko, T., Kobayashi, T., Kitamura, T.: Speech parameter generation algorithms for HMM-based speechsynthesis. In: Proceedings of Int. Conf. of Acoustics, Speech, and Signal Processing, ICASSP 2000, pp. 1315–1318 (2000)
6. Vainio, M., Suni, A., Sirjola, P.: Accent and prominence in Finnish speech synthesis. In: Proceedings of the 10th International Conference on Speech and Computer (Specom 2005), University of Patras, Greece, October 2005, pp. 309–312 (2005)

7. Hakulinen, A., Vilkuna, M., Korhonen, R., Koivisto, V., Heinonen, T.R., Alho, I.: Iso suomen kielioppi. Suomalaisen Kirjallisuuden Seura (2004)
8. Brenier, J.M., Nenkova, A., Kothari, A., Whitton, L., Beaver, D., Jurafsky, D.: The (Non)Utility of Linguistic Features for Predicting Prominence in Spontaneous Speech. In: Proceedings of the IEEE / ACL 2006 Workshop on Spoken Language Technology. The Stanford Natural Language Processing Group (2006)
9. Sityaev, D.: The relationship between accentuation and information status of discourse referents: A corpus-based study. In: UCL Working Papers in Linguistics, vol. 12 (2000)
10. Zacharski, R.: Generation of accent in nominally premodified noun phrases. In: Proceedings of the 14th conference on Computational linguistics, Morristown, NJ, USA, pp. 253–259. Association for Computational Linguistics (1992)
11. Steedman, M.: Information structure and the syntax-phonology interface. Linguistic Inquiry 31(4), 649–689 (2000)
12. Vainio, M., Jarvikivi, J.: Focus in production: Tonal shape, intensity and word order. The Journal of the Acoustical Society of America 121(2), EL55–EL61 (2007), `http://link.aip.org/link/?JAS/121/EL55/1`
13. Hajičová, E., Sgall, P., Skoumalová, H.: Identifying topic and focus by an automatic procedure. In: EACL 93P, EACL 93L, pp. 178–182 (1993)

Error Prediction-Based Semi-automatic Segmentation of Speech Databases

Marcin Szymański[1,2] and Stefan Grocholewski[2]

[1] Adam Mickiewicz University Foundation, Laboratory of Speech and Language Technology
ul. Rubież 46, 61–612 Poznań, Poland
[2] Poznan University of Technology, Institute of Computing Science
ul. Piotrowo 2, 60–965 Poznań, Poland
mszymanski@cs.put.poznan.pl

Abstract. The manual segmentation of speech databases still outperforms the automatic segmentation algorithms and, at the same time, the quality of resulting synthetic voice depends on the accuracy of the phonetic segmentation. In this paper we describe a *semi-automatic* speech segmentation procedure, in which a human expert manually allocates the selected boundaries *prior* to the automatic segmentation of the rest of the corpus. Segmentation error predictor is designed, estimated and then used to generate a sequence of manual annotations done by an expert. The obtained error response curves are significantly better than random segmentation strategies. The results are presented for two different Polish corpora.

1 Introduction

In the process of constructing speech recognition and synthesis systems it is essential that the proper set of prerecorded utterances is available. Moreover, it should additionally contain such precise information as the sequence of phoneme labels and subsequent unit durations. In the case of recognition systems the accuracy of phoneme boundaries is not crucial, however, the errors in segmentation seriously affect the quality of obtained *synthesis* system.

The manual segmentation of speech is a very labor-intensive process. Moreover, it is prone to inconsistencies and should be performed by an expert (usually a phonetician). The simplest idea is to implement an algorithm which will perform this task automatically. Here, in contrast to the phoneme recognition problem, we assume that the phone sequence is already determined. Obviously, the obtained automatic boundary points will not be faultless.

1.1 Automatic Segmentation Method

The basic algorithmic solution of the segmentation is to run a HMM recognizer in *forced alignment* mode. The segmentation can now be considered a special case of recognition, where the word- and model-net are the simple concatenation of units corresponding to the imposed phonetic transcription of an utterance.

P. Sojka et al. (Eds.): TSD 2008, LNAI 5246, pp. 543–550, 2008.
© Springer-Verlag Berlin Heidelberg 2008

Since the state transition probability in standard HMM is represented by one constant value, the state duration have an implicit geometric probability density that most probably is inadequate as the duration model. For this reason, the observed phoneme duration is estimated, expanding the problem to the segment-model one [1].

The state-chain based methods return the results discretized according to the given frame-rate (usually 5 or 10 ms) and it was observed that they tend to make systematic errors for certain transition types. For that reason a boundary refinement stage was proposed [2]. It combines boundary-specific error estimates, several acoustic observation distributions and phoneme duration distribution in a single fine-tuning algorithm.

The method mentioned above was not optimized in this research, i.e. it is treated as an input to the following approach.

1.2 Objectives

In this work we test the approach in which a human expert performs the manual segmentation of selected transition cases and the rest of boundaries are calculated automatically. This is a major difference compared to the common approach, where the automatic aligner results are reviewed and manually corrected by the human expert. In the presented method the manual annotations can also be based on the automatic segmentation but the *human labor is not required for every boundary in the corpus.* The automatic stage, which *always follows* the manual one, does not change any boundary points entered by an expert.

The basic goal is to find a strategy of performing the manual segmentation so that the maximum error reduction is obtained compared to the expert labor required. It is clear that there can be many "optimal" solutions on different levels of allowed labor or required accuracy.

Table 1. Phoneme clusters used in this paper. This implies 10×10 partition of the boundary population.

Symbol	Phonemes	Symbol	Phonemes
S	sil sp	VLD	o u a õ ẽ
CVX	j w l	VU	i I e
CVN	m n n' N	CLAF	ts ts' tS s s' S
CV	dz dz' dZ r v z z' Z	CLP	p t k
CVP	b d g	CLF2	f x

In [3] it was quested for manual annotation strategy, i.e. a *sequence* of recommended human annotations, based on a simple clustering of Polish phonetic alphabet into several classes (which implied about 100 different transition classes). The analysis was done separately for two error measures. The example strategy yielded 75.4% of gross errors reduction while requiring 13.6% of the database to be manually segmented. The alphabet partition is presented in Table 1.

Problem space analysis for a cluster-based strategy was also done in [3]. In addition, it was observed that the most problematic transitions were those from speech to silence as well as transitions between two phones belonging to the same phoneme cluster (inter-vowel boundaries, in particular).

In this paper we develop a boundary confidence measure, which tries to predict the potential segmentation error related to each boundary. The predicted error is used to suggest the next manual annotation that should be entered by an expert. The objective of this semi-automatic approach is to reduce the trade-off between the human labor and the segmentation accuracy.

Authors of this research are aware that some unit-selection speech synthesis systems perform overlapping of the concatenated units, which can make the accurate corpus segmentation less important. For good quality speech, however, the database cannot contain major segmentation errors. Since the results show that satisfactory segmentation still cannot be obtained using the automatic algorithms only, some human expert contribution is required.

The rest of this paper is organized as follows: Section 2 introduces a synthetic error measure, which is to be predicted by the regression model; Section 3 presents the conditional features used to build an error predictor; in Section 4 the semi-automatic segmentation procedure is further discussed; in Section 5 we present the experimental results. The paper is concluded in Section 6.

2 Error Measures

There are two basic error measures used in our experiments: number of gross errors and the root mean square error. The gross error (GrE) occurs when an automatically located boundary passes beyond the adjacent manually labeled segments [4]. This kind of error is considered the most harmful for the speech synthesis applications.

The RMS error is defined as: $RMSE = \sqrt{\frac{1}{N}\sum_{i=1}^{N}\delta_i^2}$, where N is the number of boundaries in the corpus and $\delta_i = b_i - b_i^*$ is the ith boundary error, where b_i is the automatically yielded transition point and b_i^* is a reference annotation.

We think that both measures have their drawbacks:

- GrE is a boolean feature, which means that even a serious segmentation error, nonetheless not a gross error as such, will have a value $false$ for that particular boundary,
- if a boundary is automatically segmented with a gross error, the actual displacement (measured by $RMSE$) is of little importance as far as the unit selection synthesis is regarded.

For those reasons we propose a synthetic error measure, which is a segment duration-normalized gross error, SDNGE:

$$SDNGE = \min(1, \max(\tag{1}$$
$$|\delta_i|/short_i, \tag{2}$$
$$|\delta_i|/(long_i + \delta_i))) , \tag{3}$$

where $short_i$ is the duration of the reference segment that was shortened due to the segmentation error (left- or right-hand, depending on a sign of δ_i), $long_i$ is the duration of a segment that was made too long as a result of an error. The formula has the value 1 for every $|\delta_i| \geq short_i$ (that is, for a gross error), but greater than 0 for every fine error (ex. 2); the expression (3) measures the *contamination* of a segment on an opposite direction compared to the segmentation error.

It should be noted that the final results will be presented in terms of the two baseline measures (GrE and $RMSE$), i.e. $SDNGE$ is only used for building the error predictor.

3 Error Prediction Features

The squared error measure presented in Section 2 ($SDNGE^2$) is used as a predicted attribute in the regression tree (CART, [5]) training. The conditional attributes are:

- quad-phone context of the boundary, plus sound type, articulation place and manner and voicing features for each of the four phonemes (20 columns in total),
- five boolean attributes representing equality of the above features between two central phonemes,
- $\max(LMan, RMan)$ and $\min(LMan, RMan)$, where LMan and RMan are the distances to the nearest present manually placed boundaries in respective directions (the distances to start or end of the audio file if no human annotations are present),
- log probability density of the boundary having a given MFCC and filter-bank energy observations in a given phonetic context (2 columns),
- log likelihood of the left- and right-hand segments having observed durations (2 columns)

The example set was generated based on the segmentation of the training part of the corpus. CART was trained using wagon with a minimum of 100 examples in a node.

4 Semi-automatic Segmentation Procedure

In each step of the semi-automatic segmentation, the CART is evaluated for every boundary (except those already manually fixed) and the *one* with the highest predicted error is "revealed" based on the test set reference, which simulates the manual annotation. Some of the "strategies" presented in Section 5 use confidence measures that require the corpus to be segmented fully automatically before the first error prediction can be made. This process is iterated.

The automatic segmentation algorithm, based on Segment Models [1] (plus a fine tuning method [2]), was modified to consider *enforced* segmentation points. Every potential phoneme model transition that violates any human annotation is heavily penalized during both the segment-model decoding and the refinement procedure, thus ensuring the resulting segmentation reflects the manual annotation as closely as possible.[1] It may be worth noticing that inserting one boundary manually may cause shifting other boundaries, which are yielded automatically (this may be caused by the

[1] The minimum phoneme duration in segment-model algorithm is 30 ms. This may cause small deviations from the enforced manual segmentation.

previous boundary location being impossible in context of the neighboring manual annotation or be the effect of segment duration models).

In the results in Section 5 it is assumed that the affected utterance is re-segmented automatically after every human intervention. Whether in practice an expert would follow this assumption or enter several most recommended boundaries once the given audio file has been opened may be the matter of the expert's working style.

5 Experimental Results

The presented approach was tested in two different setups. First, we used a part of Polish Corpora [6] database. It consists of a total of 5 hours of speech, inside 28 folders of 180 separate sentences each, coming from 24 different speakers. Hence, we deal with a speaker-independent segmentation.[2] Those tests were performed in 7-fold cross-validation.

Second, prediction models trained on Corpora were tested on a 2-hour single-speaker database, Waldemar, containing 2700 sentences. No cross-validation was performed this time.

Table 2. Strategy qualities in terms of the mean area under the graph (± a single standard deviation)

Strategy	Measure	Corpora (cross-val.)		Corpora/Waldemar	
		%GrE Area	RMSE Ar. [ms]	%GrE Ar.	RMSE Area
random	—	.0638 ± .0346	9.213 ± 1.325	.069 ± .005	12.37 ± .09
10x10	$RMSE$	.0149 ± .0055	6.667 ± 0.578	.0662	10.92
10x10	GrE	.0058 ± .0030	8.114 ± 0.824	.0738	11.06
10x10	$SDNGE$	.0085 ± .0027	6.881 ± 0.494	.0703	10.92
CART	$SDNGE$ $(\mu_{C(b_i)})$	.0066 ± .0034	6.692 ± 0.525	.0342	8.892
CART	$SDNGE$ $(\mu - \frac{\sigma}{3})$	.0063 ± .0031	6.702 ± 0.544	.0364	8.883
CART	$SDNGE$ $(\mu + \frac{\sigma}{4})$	.0067 ± .0035	6.709 ± 0.524	.0339	8.682
CART	$SDNGE$ $(\mu + \frac{\sigma}{2})$	.0067 ± .0035	6.720 ± 0.523	.0346	8.737
CART	$SDNGE$w/Toler $(\mu_{C(b_i)})$	.0051 ± .0026	6.463 ± 0.497	.0316	8.552
CART	$SDNGE$w/Toler $(\mu - \frac{\sigma}{3})$	.0048 ± .0026	6.877 ± 0.590	.0362	9.158
CART	$SDNGE$w/Toler $(\mu + \frac{\sigma}{4})$	.0052 ± .0028	6.514 ± 0.498	.0314	8.599
CART	$SDNGE$w/Toler $(\mu + \frac{\sigma}{2})$	.0052 ± .0028	6.542 ± 0.501	.0327	8.611
greedy	$RMSE$	.0043 ± .0029	3.253 ± 0.182	.0005	3.671
greedy	GrE	.0001 ± .0001	—	.0001	—

In both setups the following strategies were generated:

- random strategy (repeated 4 times in case of the Corpora-Waldemar setup);
- three partition-based strategies – GrE, $RMSE$ and $SDNGE$ were estimated separately for a simple 10x10 transition clustering (see p.1.2, Table 1);

[2] It may be noted that the database was specifically designed to contain as many different diphones as possible. As some of its sentences are not uttered on a daily basis, this might have influenced the statistical models used in this work.

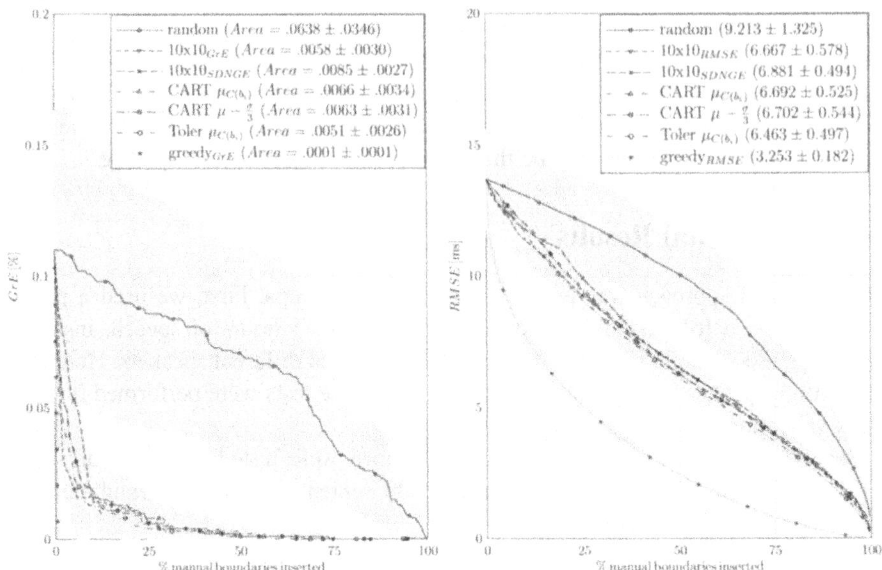

Fig. 1. Cross-validation on Corpora: left – percent of gross errors in function of required percentage workload; right – RMS error in function of required workload. (Pale areas denote single standard deviations. Graph areas in parentheses).

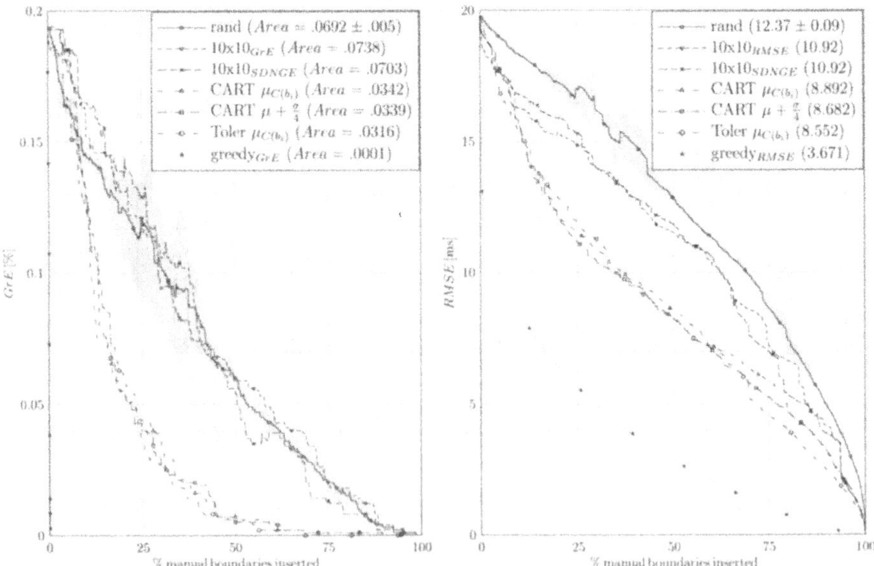

Fig. 2. Corpora models tested on Waldemar: left – percent of gross errors in function of required workload; right – root mean square error in function of required percentage workload. (Pale area denotes single standard deviation for the random strategy. Graph areas in parentheses).

- four error predictor-based strategies – in the standard CART-based strategy, the mean error found in the leaf cluster of the regression tree ($\mu_{C(b_i)}$) is assumed as a predicted error; the other three use the formulas $\mu_{C(b_i)} + \sigma_{C(b_i)}/4$, $\mu_{C(b_i)} + \sigma_{C(b_i)}/2$ and $\mu_{C(b_i)} - \sigma_{C(b_i)}/3$; this can be interpreted as testing two risk-averse and one risk-seeking approaches, respectively;
- two greedy strategies (performed separately for GrE and $RMSE$) – the largest segmentation error was removed based on a test-set reference; naturally, this is unrealistic and was done only to show a potential of a *perfect* predictor.[3]

Additionally, 8 ms tolerance was experimentally introduced into the SDNGE formula (Section 2) for the predictor-based procedures – any segmentation within the tolerance was treated as perfect during CART training.

The quality of each strategy is measured in terms of the *area* under the respective error graph: the lower the value, the greater error *reduction* can be obtained in the same number of manual actions throughout the strategy. The results are summarized in Table 2.

Figure 1 presents the response curves (error as a function of a number of manually entered boundaries) for Corpora cross-validation setup. The results for Corpora models tested on Waldemar are presented in Figure 2.

In general, similar strategies give very similar results. It can be observed that although the CART method is insignificantly different from the 10x10 partition based strategies for the Corpora cross-validation, it proved more robust than the latter when tested on Waldemar. Also, no partition method is better than any CART on *both* error measures.

When 8 ms tolerance is used for building CART, the error measures tend to reduce slightly. For using speaker-independent models on different corpora, a small "risk aversion" ($\mu_{C(b_i)} + \sigma_{C(b_i)}/4$) is recommended during error prediction.

6 Conclusions and Future Works

This paper introduces an error-prediction approach to semi-automatic speech segmentation. It was shown that it is possible, e.g., to obtain 61% reduction of the number of gross errors and 32% reduction of RMS segmentation error across an example database by manually locating 15% of the boundaries (using a predictor trained on a different corpus; see Figure 2).

While the obtained error response curves are significantly better than random segmentation strategies, effort still has to be made, in particular to eliminate *any* gross errors in a lower number of manual annotations. For this reason, an error prediction feature (see Section 3) based on a multi-hypothesis HMM-based segmentation has to be developed.

The distances to the nearest present manual annotations are already used as confidence features (Section 3). However, the predictor was trained on files segmented fully-automatically (distances to the start or end of the utterance were used). It is planned to retrain it using randomly placed forced transitions inside sentences.

[3] Note those are not necessarily equivalent to *optimal* strategies.

It was also observed that speech-to-silence (sentence final) transitions were the first suggested by the predictor in virtually every utterance. Since large $RMSE$ was also found for sentence-initial boundaries, this strongly suggests the need for a better speech-silence detector.

As a number of continuous features is used, we may also test the incorporation of the linear regression into the error prediction training.

In the presented work, no model *adaptation* based on a manually entered knowledge was made during a realization of the strategy. This would be very problematic technically in this experiment, but it is possible for an expert to retrain the acoustic and prediction models after the annotation of a part of a large corpus.

References

1. Ostendorf, M., Digalakis, V., Kimball, O.: From HMM's to Segment Models: A unified view of stochastic modeling for speech recognition. IEEE Trans. on Speech and Audio Proc. 4(5), 360–378 (1996)
2. Szymański, M., Grocholewski, S.: Post-processing of automatic segmentation of speech using dynamic programming. In: Proc. 9th International Conference on Text, Speech and Dialogue (2006)
3. Szymański, M., Grocholewski, S.: Semi-automatic segmentation of speech: manual segmentation strategy; problem space analysis. In: Proc. CORES 2005, Wroclaw (2005)
4. Kvale, K.: Segmentation and labelling of speech. Ph.D. thesis, Institutt for Teleteknikk, Trondheim (1993)
5. Breiman, L., Friedman, J.H., Olshen, R.A., Stone, C.J.: Classification and Regression Trees. Wadsworth International Group, Belmont (1984)
6. Grocholewski, S.: Corpora speech database for Polish diphones. In: Eurospeech 1997, pp. 1735–1738 (1997)

Part IV

Dialogue

"**Dialogue**: a discussion between two or more people or groups, especially one directed towards exploration of a particular subject or resolution of a problem: *interfaith dialogue*."

NODE (The New Oxford Dictionary of English), Oxford, OUP, 1998, page 509.

Cognitive and Emotional Interaction

Amel Achour, Jeanne Villaneau, and Dominique Duhaut

Valoria, Université de Bretagne Sud, Lorient-Vannes
Centre de recherche de Saint Maudé, 56100 Lorient, France
{amel.achour, jeanne.villaneau, dominique.duhaut}@univ-ubs.fr
http://www.univ-ubs.fr

Abstract. The ANR project EmotiRob aims at conceiving and realizing a companion robot which interacts emotionally with fragile children. However, the project MAPH which is an extension of EmotiRob tries to extend the cognitive abilities of the robot to implement a linguistic interaction with the child. For this, we studied a children corpus and got semantic links that could exist between each pair of words. This corpus elaborated by D. Bassano has been used to evaluate language development among children under five. Using this corpus, we tried to make a taxonomy in accordance with the conceptual world of children and tested its validity. Using the taxonomy and the semantic properties that we attributed to the corpus words, we defined rapprochement coefficients between words in order to generate new sentences, answer the child questions and play with him. As a perspective for this, we envisage to make the robot able of enriching its vocabulary, and to define new learning patterns basing on its reactions.

Keywords: Cognitive interaction, emotion, corpus, taxonomy, semantic links, child conceptual world.

1 Introduction

A new important field of study in robotics is the domain of companion robots which execute complex tasks and offer behavior enrichment through their interaction with human beings. The French project, EmotiRob, supported by the ANR (National Agency of Research), belongs to this research domain and aims at conceiving and realizing a "reactive" autonomous soft toy robot, which can interact emotionally with children weakened by disease, and bring them some comfort. Previous experiments have already shown the contribution of companion robots in this type of situation [6,5].

The research presented here corresponds to a part of the MAPH project (Active Media for the Handicap) which is related to the EmotiRob project. It aims at extending the robot's reaction capacities so that it could maintain a natural-language "conversation" with the child. The purpose of this work, indeed, is to build a linguistic and cognitive interaction module between the child and the robot by the generation of new subjects and sentences, and by the implementation of well-targeted games.

A recognition/comprehension module of the child's words supplies the cognitive interaction module inputs; this module of recognition/comprehension is the subject of another study within the EmotiRob project. The recognition of the child words is done using DRAGON software which is a commercial product developed by NUANCE. The

P. Sojka et al. (Eds.): TSD 2008, LNAI 5246, pp. 553–560, 2008.
© Springer-Verlag Berlin Heidelberg 2008

sentences generated by the robot depend on the emotional states of both robot and child. The basic tool of the cognitive interaction module is a corpus established by means of the DLPF tool which was realized by D. Bassano, F. Labrell, C. Champaud, F. Lemétayer and P. Bonnet [1]. This tool is an instrument intended to estimate the development of production language of French children whose age is between two and five. This corpus counts a little less than 1500 words including nouns, verbs, adjectives, adverbs, onomatopoeias and common expressions, articles and pronouns among which we find all the common vocabulary that could be said by a 4-year-old child. The problem which arises is then how to model the conceptual world of a very young child. We thus thought of conducting some surveys in elementary classes in order to find how children at an early age see their surroundings. By studying the semantic relations which could exist between the different words of the corpus, we established a classification of the words or rather taxonomy according to not only objective but also emotional properties.

In order to give our system robustness, we plan for corpus enrichment, as well as the addition of new knowledge. So that, the system must be capable of semantically connecting the new words added to those already in its base of knowledge and to draw up their lists of properties.

In the following section, we will give a small outline on what has already been done in the field of language treatment and robotics dedicated to children. Then, we will detail our approach and the work that we have done. First of all, we are going to describe the taxonomy which we created then the method used in the calculation of semantic links between the corpus words. Then, we will describe how we used this taxonomy in the generation of sentences and speech as well as games between child and robot, and how we plan to continue this work.

2 Related Works

Carrying on a "natural" conversation with a machine on a non constrained subject seems to be very difficult and even impossible [8] as we cannot model world knowledge right now. By restricting the field of the conversation and choosing a well-defined subject, the use of natural language in order to carry a "dialogue" becomes feasible but still very difficult. Some existing systems of human-machine dialogue that we can quote here are COALA [1] which is a system for documentary assistance in a town media library, CMU Communicator [2] which is realized by Carnegie Mellon University, and serves as a tourist guide. Problems that come up against the conception and the realization of human-machine dialogue systems are essentially oral language recognition and understanding [7], real time conversation constraint and finally, speaking with the machine must not require learning from the user [9]. Regarding our project, we are implementing a generic human-machine dialogue system dedicated to young children. We are dealing with vocabulary covering the child's entire surroundings. This is quite problematic as we have a non-restricted conversation domain. However, limiting the users of our system to young children makes the vocabulary we are interested in quite restricted. Moreover, despite the fact that we conceived different types of interactions, they are still well

[1] http://www-ic2.univ-lemans.fr/~lemeunie/these/node35.html

[2] http://www.speech.cs.cmu.edu/Communicator/

targeted. Under these conditions, producing a dialogue between the child and the robot is conceivable. In our case, we are working with a limited corpus of children whose age is under five, in order to carry on cognitive interaction with the robot. This interaction should also depend on the child's perception of the world and his/her emotional state. In spite of the important interest of emotions in generating a realistic dialogue [3,4], this concerned only dialogues between adults and ignored children. As far as corpus study and computing of semantic links between words are concerned, the works that we found are generally statistic methods applied to large corpus of words. The purpose of such systems is essentially improving the automatic extraction of knowledge, as well as documentary search. As we have already said, our objective is to imagine an ontology of the world as it is viewed by children not only regarding its cognitive aspects, but also regarding its emotional ones. We also aim at being in accordance with their way of thinking and their perception of things. That is why we tried to validate our research with children and to verify if we have satisfied these constraints. From the beginning, we opted for the development of a prototype of the system which will evolve throughout our work in order to satisfy our various specifications and requirements. In the following section, we are going to describe the approach used during this work.

3 Classification of the Corpus Words and Creation of the Taxonomy

3.1 Creation of the Taxonomy

In order to calculate the coefficients of semantic link between pairs of words, we began by classifying all the words of the corpus in various classes according to their senses. At the first level we find the verbs, the adjectives and the common nouns. Each of these three classes was afterward divided into several categories, which gave place to numerous sub-categories with a more specific sense of words, and so on. The figure below shows a small outline of the taxonomy that we obtained. The detail of the taxonomy is visible on the site of EmotiRob[3].

3.2 Validation of the Taxonomy with the Children

To validate the word taxonomy that we created, we thought of making a questionnaire and having children, between 5 and 7 years old, fill it out. According to Piaget [2] the children of lower age tend to fantasize and to say anything when they do not know the answer to the question or even when they are not sure of the answer. It is what Piaget calls the "whateverism" phenomenon. To implement the various questions and parts of the questionnaire, we used the tests method, which is often employed on the study of the childish beliefs. This method requires two essential conditions: the first one demands that questions should be the same for all the subjects and asked in the same circumstances. The second condition requires that all the answers must be reported to the same evaluation scale. This method has the advantage of giving useful information statically speaking; however, the risk of falsifying the child spirit orientation remains

[3] http://www-valoria.univ-ubs.fr/emotirob/

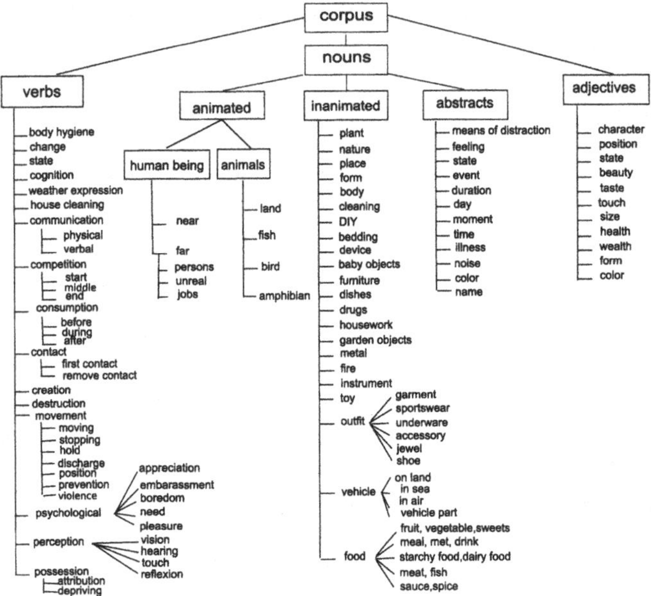

Fig. 1. Taxonomy of children words

important. To remedy this problem, it is necessary to vary the questions and to make counter-suggestions. Thus, in our questionnaire we opted for various types of questions such as questions with multiple choices, tables to fill out, as well as searching for adequate solutions among a set of possible ones.

This questionnaire essentially concerns the animated beings that surround the children: human or animal, and with whom they can have more or less emotional links. The first part of the questionnaire was dedicated to human beings. At first, we asked the children to distinguish the characters we find around us from those who exist only in tales. And second, we tried to find the different primary feelings which evoke these characters in their minds.

The second part was dedicated to animals and aimed at verifying their belonging to the different categories established in the taxonomy. The questionnaire was filled in by first-grade pupils from "Sainte Marie" elementary school in Lorient twice. After the counting of the answers, we noticed that, in the majority of the cases, the results are in accordance with the taxonomy and answer, indeed, our expectations. Nevertheless, certain results surprised us and made us modify the taxonomy. For example, according to 92% of the questioned children, "king" and "prince" characters do not exist in the real world and are only in tales and stories we tell them. "Father Christmas" belongs to the set of imaginary characters for only the half of the children, the other half consider him as a real person. 80% of children are "happy" when they see a "magician" and consider him as a "kind person". To many children, almost half, an "ostrich" and a "penguin" are not birds, and nearly 70% of them think that a "whale" is a fish.

4 Calculation of Rapprochement Coefficients

According to the taxonomy, we tried to measure the semantic links which exist between the words. For that purpose, we defined rapprochement coefficients between pairs of words. In the next part, we describe these rapprochement coefficients, as well as the method used for calculation.

4.1 Rapprochement between Two Common Nouns

The set of common nouns was divided into three big classes, the class of animate, inanimate and abstracts. Each of these classes was afterward divided in several sub-categories, and so on. Here is the formula used to calculate the rapprochement Rapproch $(N1, N2)$ between two common nouns $N1$ and $N2$:

$$\text{Rapproch}(N1, N2) = \frac{C1 * R1(N1, N2) + C2 * R2(N1, N2)}{C1 + C2} \quad (1)$$

As can be noticed, the rapprochement coefficient between two common nouns is a weighted average between two coefficients, the first of which, $R1(N1, N2)$, calculates the rapprochement between both words in the taxonomy, whereas the second evaluates their rapprochement regarding to their common properties number. We distinguished two types of properties: affective properties and objective ones. Each property was balanced with a weight measuring its importance in defining a certain set of words. $R2$ is then the weighted average of an affective rapprochement *Raff(N1,N2)* weighted by an affective coefficient Qa, and an objective rapprochement $\text{Robj}(N1, N2)$ weighted by an objective coefficient Qo.

$$\text{Raff}(N1, N2) = \frac{\text{nbr\_prop\_aff\_com}(N1, N2)}{\max(\text{nbr\_prop\_aff}(N1), \text{nbr\_prop\_aff}(N2))} \quad (2)$$

$$\text{Robj}(N1, N2) = \frac{\text{nbr\_prop\_obj\_com}(N1, N2)}{\max(\text{nbr\_prop\_obj}(N1), \text{nbr\_prop\_obj}(N2))} \quad (3)$$

$$R2(N1, N2) = \frac{Qa * \text{Raff}(N1, N2) + Qo * \text{Robj}(N1, N2)}{Qa + Qo} \quad (4)$$

The rapprochement coefficients we obtained depend on the Qa and Qo that we chose. For instance, "ladybug" and "louse" will be semantically close if Qo is bigger than Qa. Otherwise, they will be distant.

4.2 Rapprochement between Two Verbs

To describe the semantic rapprochement between any two verbs, we defined two types of coefficients. The first one calculates the rapprochement between both verbs with regard to their place in the taxonomy. The second coefficient is calculated according to their respective types (intransitive, transitive or double transitive) and according to the taxonomic rapprochement between their subjects and respective complements. So, the rapprochement coefficient between two verbs is a weighted average of both coefficients calculated previously.

4.3 Rapprochement between Two Adjectives

The calculation of the rapprochement coefficient between two adjectives bases on the preliminary calculation of two rapprochement coefficients. The first one is calculated with regard to the properties described by adjectives (size, shape, taste, etc.). So, it represents their rapprochement in the taxonomy. As for the second coefficient, it is calculated according to the types of subjects that can be applied to both adjectives. The rapprochement coefficient between two adjectives is then a weighted average of these two coefficients.

4.4 Rapprochement between Verb and Noun, Adjective and Noun

We defined the rapprochement coefficient between noun and verb to measure the applicability of some verbs to a particular noun. For example, we can say that the verb "to bore" applies perfectly to an animate subject whereas if the same verb can be applied to a "chair", that is possible only in an artistic or a funny context. Also for adjectives, the rapprochement coefficient between noun and adjective measures rather the use of an adjective with a certain name in a particular speech context.

5 Sentence Generation and Maintenance of the Child-Robot Speech

The sentence generation model that we adopted works with simple input sentences such as (subject, adjective), (subject, intransitive verb), (subject, transitive verb, complement) or (subject, verb double transitive, complement1, complement2). At the present time, we have limited our choice to one type of sentence which is the affirmative sentences. Later, we intend to work on the acts of language and to introduce the interrogative, imperative sentences, etc.

The generation module takes several parameters into account which describe the emotional state of the child, as well as the humor of the robot. First of all, it analyzes the input sentence so as to find the context that the speech is about. Secondly, it looks for words which represent a certain value of the semantic rapprochement coefficients with the words composing the input sentence. Several cases are possible, according to the speech context (realistic, funny, artistic, etc.) and according to the emotional states of both the child and the robot. The robot will then be able to answer the child either by keeping the same subject of discussion or a similar one, or by approaching another completely different subject.

5.1 Child-Robot Interactive Games

Because of the playful and psychological interest of games for children, we considered it very interesting to introduce a game module in order to distract the child and maximize his comfort. As an example of game, we thought of a simple play on words where the second opponent should formulate a sentence which begins with the last word pronounced by the first opponent or to keep some words of the input sentence and to change the others while ensuring a coherent meaning to the sentence. In another game, the child asks the robot to guess something from certain properties that he would give him as indications. For that purpose, we defined a certain number of prototypes related to what we think belong to the conceptual world of children. Each of these prototypes is described by a certain number of properties. An example of such a prototype is "pet" which can be described by the following list of properties, present in our taxonomy: (it lives in the house, it is sweet, you play with it, etc.) or also the prototype "bird" which can be defined as: (it has feathers, it has a beak, it lays eggs, it has wings, it flies, etc.).

We have now implemented a classic riddle game in which the child begins by choosing an animal and on the other hand, the robot tries to guess it by asking a series of questions. The questions asked essentially concern the classification of animals described in the taxonomy. The robot adopts a strategy concerning the choice of the questions which essentially depends on the child's answers and which can change during the game. However, what increases the interest of the game is that the robot can know if the child made a mistake in answering one of his questions or if he cheated on purpose. By checking all the answers supplied by the child, the robot can thus protest if the child gave him a bad answer. As the robot will be endowed with an internal humor, its reaction can be declined according to several modes: annoyed, amused or simply neutral.

6 Conclusion and Perspectives

The system that we have developed allows both cognitive and emotional communication and interaction between the child and the robot. It aims at the entertainment of the child by putting him in front of a companion capable of oral interaction and endowed with cognitive and linguistic capacities. Our future work will now focus on finding other types of interactions and new series of games. As we have already said, one of the capacities with which we want to endow our system is to be able to enrich its vocabulary: new words can be added to the taxonomy and be bound semantically and in an automatic way to what already exists in the system knowledge database. We are currently thinking about the way we will allow the system to evolve. We also intend to define learning frameworks; according to the answers the system is going to produce, to adapt it to our expectations and to those of the children. The evaluation of our work is rather delicate because it involves child psychology. We thus have to foresee evaluation methods which allow for the measuring of the quality of the interaction between the robot and the child and the comfort brought to him. More simply, we thought of elaborating a questionnaire which should be filled in by voluntary adults and children in order to test their reactions face to face with the robot, and to see what they think of games and of the quality of the robot reactions.

Acknowledgments. We thank the region of Brittany for the interest and the support which it brings to the project. We also wish to thank Dominique Bassano for her collaboration by giving us access to her child corpus.

References

1. Bassano, D., labrell, F., Champaud, C., Lemétayer, F., Bonnet, P.: Le DLPF: un nouvel outil pour lévaluation du développement du langage de production en français. Enfance 2, 171–208 (2005)
2. Piaget, J.: La représentation du monde chez l'enfant. Presses universitaires de France (1947)
3. Keltner, D., Haidt, J.: Social functions of emotions. In: Mayne, T., Bonanno, G.A. (eds.) Emotions: Current issues and future directions, pp. 192–213. Guilford Press, New York (2001)
4. Adam, C., Evrard, F.: Donner des émotions aux agents conversationnels. In: Workshop Francophone sur les Agents Conversationnels Animés, Grenoble, France, pp. 135–144 (2005)
5. Shibata, T.: An overview of human interactive robots for psychological enrichment. Proceedings of the IEEE 92(11), 1749–1758 (2004)
6. Wada, K., Shibata, T., Saito, T., Tanie, K.: Effects of Robot-Assisted activity for elderly people and nurse at day service center. Proceedings of the IEEE 92(11), 1780–1788 (2004)
7. Glass, J.: Challenges for spoken Dialogue Systems. In: IEEE ASRU Workshop, KeyStone, Colorado, USA, pp. 39–61 (1999)
8. Wilks, Y., Catizone, R.: Human-Computer Conversation. In: Encyclopedia of Microcomputers, Dekker, New York (2000)
9. Gazdar, G.: The handling of natural language. In: The simulation of Human intelligence, Donald Broadbent edn., pp. 150–177. Blackwell, Oxford (1993)

Architecture Model and Tools for Perceptual Dialog Systems

Jan Cuřín and Jan Kleindienst

IBM Czech Republic, Voice Technologies and Systems,
V Parku 4/2294, Praha 4, Czech Republic
{jan_curin, jankle}@cz.ibm.com

Abstract. In this paper, we present an architecture model for context-aware dialog-based services. We consider the term "context" in a broader meaning including presence and location of humans and objects, human behavior, human-to-human or human-to-computer interaction, activities of daily living, etc. We expect that the surrounding environment from which context is gathered is a "smart environment", i.e. a space (such as office, house, or public area) equipped with different sets of sensors, including audio and visual perception. Designing the underlying perceptual systems is a non-trivial task which involves interdisciplinary effort dealing with the integration of voice and image recognition technologies, situation modeling middleware, and context-aware interfaces into a robust and self-manageable software framework. To support fast development and tuning of dialog-based services, we introduce a simulation framework compliant with the proposed architecture. The framework is capable of gathering information from a broad set of sensors, of an event-based abstraction of such information, interaction with an integrated dialog system, and of a virtual representation of a smart environment in schematic 2D or realistic 3D projections. We elaborate on use-cases of architecture model referring the two projects where our system was successfully deployed.

Keywords: Architecture, simulation, dialog systems, multi-modal fusion, perceptual systems, visualization, smart spaces, situation modeling.

1 Introduction

Research in the field of conversational and dialog systems has a long tradition starting in 1966 with Weizenbaum's Eliza [1]. More recently, spoken dialog systems [2] are used in smart phones [3], navigation systems, embedded devices for cars, etc.

This paper does not explore techniques or implementation details of a dialog system, but rather we propose a software architecture model for dealing with various types of contextual information gathered from the surrounding environment that can help the dialog management system provide better, more effective, and non-obtrusive services to the user. The proposed architecture model is based on several years of experience gained in two large HCI (human-computer interaction) projects: one dealing with building intelligent meeting room services, the other with domotic perceptual services for elderly. We introduce a software framework capable both of collecting data from a set of sensors

P. Sojka et al. (Eds.): TSD 2008, LNAI 5246, pp. 561–568, 2008.
© Springer-Verlag Berlin Heidelberg 2008

and perceptual components (complying with a specific interface), and of simulating the perceptual environment based on supplied scenarios or recordings. In addition, the system provides an infrastructure for plugging-in software components (called situation machines) capable of aggregation, filtering and abstraction of events received from sensors. We refer to an existing dialog management system – CIMA platform – integrated within the proposed framework.

During the initial phase, we have identified two reliable pervasive computing systems which may serve as context acquisition and modeling platforms for multimodal services: UbiREAL [4] and Context Toolkit [5]. But in contrast to these systems, which integrate contextual information directly from various sensors, we needed a system which relies on information provided by more complex perceptual components, i.e. more complex context-acquisition components such as person trackers and speech recognizers. A body tracker might be at once capable of detecting location, heading and posture of persons, identifying them, and tracking subjects of their interest. Such scenarios led us to separate the perceptual components layer from the layer that deals with higher abstraction – the situation modeling. Defining and modeling situations based on a wide range of context-acquisition components was not supported by other environments such as UbiREAL and Context Toolkit, so we decided to implement a new framework, called SitCom which we describe in Section 3. Section 4 presents several uses cases implemented in this framework taking advantage of the integration of perceptual technologies with a dialog management system.

2 Reference Architecture Model

The construction of contextual services demands an interdisciplinary effort because it has to integrate three related parts: environment sensing, situation modeling, and application logic. Thus, even a simple system built for an intelligent room requires at least three different developer roles, each with a different set of skills:

- **Perceptual Technology Providers** supply sensing components such as person trackers, sound and speech recognizers, activity detectors, etc. that can see, hear, and feel the environment. The skills needed here include signal processing, pattern recognition, statistical modeling, etc.
- **Context Model Builders** create models that synthesize the flood of information acquired by the sensing layer into semantically higher-level information suitable to user services. Here, the skills are directed toward probabilistic modeling, inferencing, logic, etc.
- **Dialog System Designers** construct the context-aware dialog system by using the abstracted information from the context model (the situation modeling layer). Needed skills comprise application logic, dialog management, and user interface design; multi-modal user interfaces typically require yet another set of roles and skills.

Upon starting, the dialog system designer does not, typically, have access to a fully-equipped room with sensors, nor does she have the technology for detecting people and meetings. But for an initial prototype of such a service, pre-recorded input data may suffice.

One of the key features of our approach is an effective separation of the effort of different development groups facilitating easy integration. This resulted in the layered architecture design – *Reference Architecture Model* [6]. The *Reference Architecture Model* provides a collection of structuring principles, specifications and *Application Programming Interfaces* (APIs) that govern the assemblage of components into highly distributed and heterogeneous interoperable systems. See the schema in Figure 1.

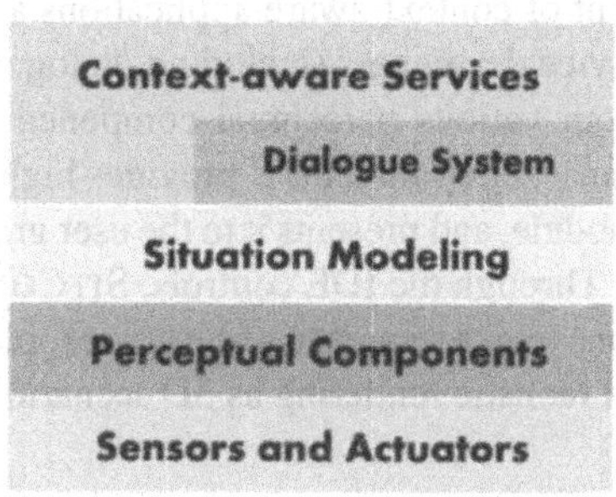

Fig. 1. Layers of the Reference Architecture Model

Each level derives new information from the abstraction of the information processed and data flow characteristics such as latency and bandwidth, based on the functional requirements from the system design phase.

The components at the **Logical Sensors and Actuators** layer are either sensors feeding their data to the upper layers, or various actuators, such as output devices receiving data or steering mechanisms receiving control instructions. The aim of this layer is to organize the sensors and actuators into classes with well-defined and well-described functionality, to make them available remotely over a network and to standardize their data access.

The **Perceptual Components** layer extracts meaningful events from continuous streams of video, audio and other sensor signals. Perceptual Components provide interpretation of data streams coming from various logical sensors and provide a collection of well-defined APIs that allow replacing one technology engine for another. These components still process information in a linear fashion, from sensors to higher level semantics.

The **Situation Modeling** layer is the place where the situation context received from audio and video sensors is processed and modeled. The context information acquired by the components at this layer helps services to respond better to varying user activities and environmental changes.

The **Dialog-based Context-aware Services** are implemented as applications generated from scripted dialog strategies in a modular spoken language framework. The dialog-driven components at this level are responsible for communication with the user and for presenting the appropriate information at appropriate time-spatial interaction spots, for which they utilize the contextual information available from the situation modeling layer.

3 The Framework

The framework used for the actual implementation of our perceptual dialog-based services consists of two key components: a situation modeling tool – SitCom – and a conversation management platform – CIMA.

Situation Composer for Smart Environments (SitCom) is a simulation tool and runtime for the development of context-aware applications and services [7]. SITCOM supports context-aware service developers (including dialog system designers) by collecting the information from various perceptual components (body trackers, speech recognition engines) in a smart environment. It provides higher (semantic) abstraction in the situation modeling module, and presents it to the user graphically in schematic 2D or realistic 3D projections. Through the IDE controls, SITCOM also facilitates the capture and creation of situations (e.g. a sequence of several people meeting in a conference room) and their subsequent realistic rendering as 3D scenarios. See the screen-shots of SITCOM GUI in Figure 3.

Conversational Interaction Management Architecture (CIMA) is a development platform and runtime for creating multimodal applications. It integrates an advanced speech recognition module, where the recognition domain is controlled by language models and/or real-time generated grammars, and a concatenative text-to-speech (TTS) module with a graphical user interface. The application dialog logic is encoded in SCXML (State-Chart XML) format[1].

4 Use Cases – Dialog Scenarios

4.1 CHIL Connector

Let us describe the use of our framework on a CHIL[2] Connector scenario proposed and exploited in [8]. The Connector service is responsible for detecting acceptable interruptions (phone call, SMS, targeted audio, etc.) of a particular person in a "smart" office.

Let us assume that our smart office is equipped with multiple cameras and microphones on the *sensor* level. The audio and video data are streamed into the following *perceptual components* (as depicted in Figure 2):

- **Body Tracker** is a video-based tracker providing 3D coordinates of the head centroid for each person in the room;
- **Facial Features Tracker** is a video-based face visibility detector, providing nose visibility for each participant from each camera;
- **Automatic Speech Recognition** is providing speech transcription for each participant in the room.

[1] SCXML, see W3C specification at `http://www.w3.org/2005/07/scxml`

[2] CHIL (Computers in the Human Interaction Loop) – an integrated project (IP 506909) under the EC's Sixth Framework Programme finished in October 2007.

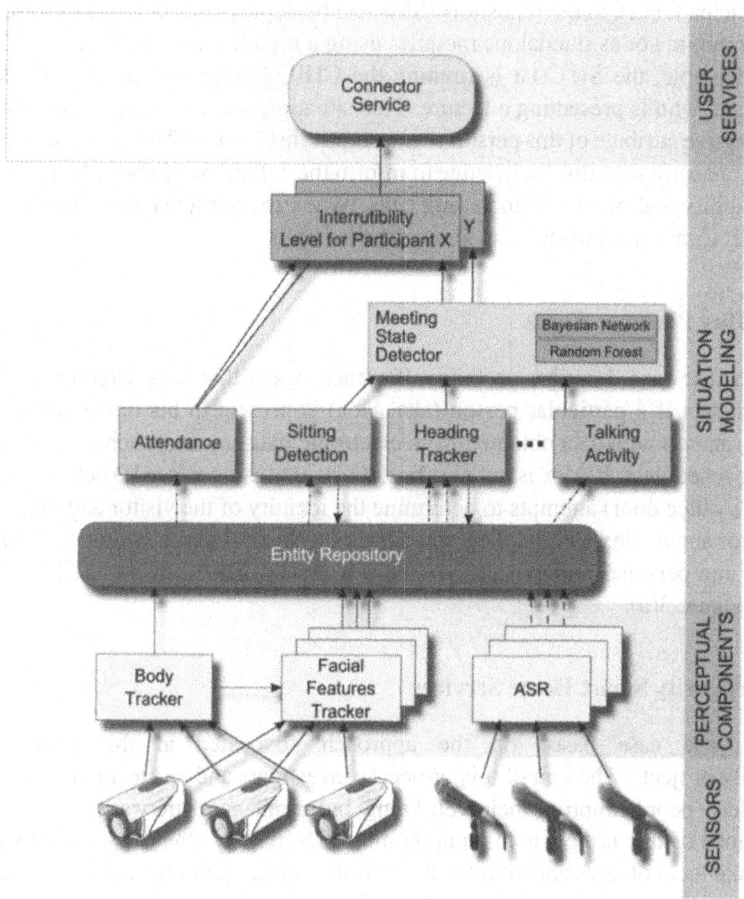

Fig. 2. Data flow in the CHIL connector scenario

As a plug-in to the architecture outlined in Section 2, we have designed a situation model that recognized meeting-related activities (ongoing meeting, meeting break, number of participants, etc.) and thus supports the context-aware dialog behavior of the Connector service.

As we have shown in Section 2, the conceptual goal of the situation model is to extract higher semantic information from the stream of environment perceptions. Technically, this is achieved through *situation machines* (SM), where each SM is implemented as a finite state graph. The state of the SM is determined by a state of entities in the entity repository that mirror real objects in the environment and by current states of other SMs. For this use case of the meeting state recognition task, we used a SM with a statistical classification module trained off-line.

Some situation machines can update attributes of entities in the entity repository. For example the *Heading Tracker SM* updates the *Heading* attribute of person entities, whereas the *Sitting Detection SM* propagates the observed sitting or standing positions of a participant back to the repository. SMs can be deployed as plug-ins that are part of the Java runtime or as standalone modules using a remote API.

For example, the SITCOM is running the CHIL Connector service for a specific participant, who is presenting a lecture. This situation is detected by a SM, which sets the respective attribute of this person's "interruptibility" to low. The dialog system of the CHIL Connector uses this knowledge to inform the calling party about the participant's unavailability and blocks all incoming calls. When the participant becomes available, the CHIL Connector informs him about diverted calls.

4.2 Office Presence Agent

A similar use case describes an Office Presence Agent that uses perceptual technologies to detect if a particular person (aka Alex) is present in his office and combines the information with data obtained from electronic data sources, such as calendars and meeting schedulers. If Alex is out of office, the agent (using a face ID detector mounted at Alex's office door) attempts to determine the identity of the visitor and then informs the visitor about Alex's availability choosing a level of details according to hes relation to Alex and personal preferences. See the screenshot of the SitCom GUI for this use case in Figure 3(a).

4.3 NetCarity Smart Home Services

Another use case based on the approach advocated in this paper is the NetCarity project[3]. The aim of this project is investigate and test technologies that can help elderly people improve their well-being, independence, safety and health at home.

The idea of this project is to equip homes of NetCarity clients by a variable set of perceptual technologies and connect them to the central point providing various types of services. The most important activity of the central service point, called *NetCarity Server* (NCS), is to "keep an eye" on the clients health and safety exploiting different perceptual technologies, e.g. an acoustic event detector, body tracker, etc. In addition, NetCarity can provide services to improve activities of daily living for the elderly person. Examples of such additional services are chatting or videoconferencing with family members or friends, on-line health care service, ordering of food and other goods, etc.

Currently the following three dialog-based services are being investigated and developed:

- **Morning Wellness** is a health checking procedure done regularly in the morning by the NetCarity client. The system drives the elderly through several steps (such as weight and blood pressure measurements) and allows her to initialize an audio or video call with a NetCarity assistant or a medical doctor. A multi-modal dialog system is available to the client providing a speech enabled remote control interface.
- **Door Access Service** is a service connected to a camera mounted close to the entrance of the NetCarity client's flat or home. It informs the client about a visitor

[3] NetCarity – an ongoing integrated project supported by the EC under the Sixth Framework Programme (IST-2006-045508).

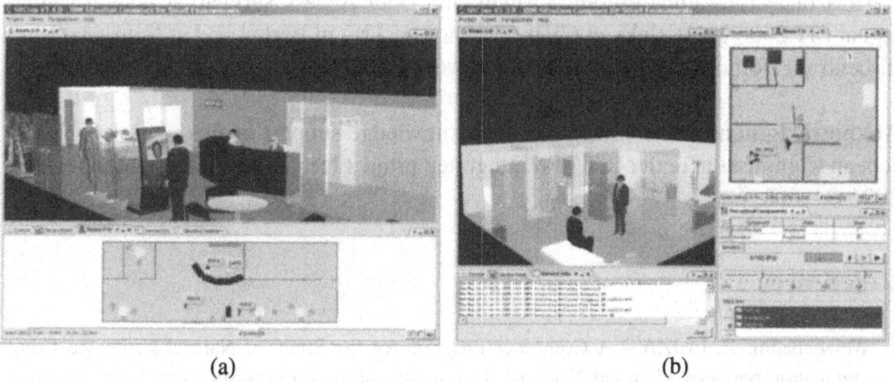

(a) (b)

Fig. 3. Screen-shots of (a) Office Presence Agent kiosk at the reception and (b) Virtual smart home for the NetCarity project

providing both video and audio stream from the camera. The client can decide to open the door by herself or consult a Netcarity operator to decide whether it is safe to let the visitor in.

- **Interactive JukeBox** is an entertaining service that allows the client to use voice command to play her favorite music, movie, or TV channel. The user can either say directly the name of a song or album she wants to hear or she may choose a song by a genre or an interpret. The dialog system interacts when it needs to disambiguate the user's choice or when it is asked to provide some information about a particular song or movie.

The project is currently under development. The virtual smart home created in our framework (see the the screenshot in Figure 3(b)) allows the dialog system designers to build and tune dialog strategies before the real environment becomes available.

5 Conclusion

This paper introduces an architecture model and tooling for perceptual dialog systems, where the dialog takes advantage of the underlying situation context – information gathered from smart room sensors. Designing the perceptual systems is a non-trivial task. To support fast development and tuning of dialog based services, we have introduced the SITCOM tool compliant with the proposed architecture and featuring support for 3D simulation and visualization, pluggable modules (situation machines) for context acquisition, integration flexibility (deployed in 10+ third-party systems), and implementation stability (4+ years of development).

The SITCOM environment was successfully applied during the whole development cycle of the CHIL project and it is now helping to bootstrap the NetCarity system. The integration of SITCOM with our dialog management system brought new possibilities in creating context-aware application using a broader context gathered from the surrounding environment and thus allowed to provide more effective non-obtrusive services to the users. We have presented several uses cases, and demonstrated some of them (the

CHIL Connector, Office Presence Agent, and Door Access Service) at several venues e.g. at IST 2006 in Helsinki, at CHIL Technology Day in Berlin, and at Nationaal Kenniscentrum Domotica exhibition in Eindhoven in 2007.

Acknowledgments. We would like to acknowledge support of this work by the European Commission under IST FP6 integrated project NetCarity, contract number IST-2006-045508.

References

1. Weizenbaum, J.: ELIZA – A Computer Program for the Study of Natural Language Communication between Man and Machine. Communications of the Association for Computing Machinery 9, 36–45 (1966)
2. McTear, M.F.: Spoken Dialogue Technology: Towards the Conversational User Interface. Springer, Heidelberg (2004)
3. Turunen, M., Hakulinen, J.: Spoken and multimodal communication systems in mobile settings. In: Esposito, A., Faúndez-Zanuy, M., Keller, E., Marinaro, M. (eds.) COST 2102 Workshop. LNCS, vol. 4775, pp. 227–241. Springer, Heidelberg (2007)
4. Nishikawa, H., Yamamoto, S., Tamai, M., Nishigaki, K., Kitani, T., Shibata, N., Yasumoto, K., Ito, M.: UbiREAL: Realistic smartspace simulator for systematic testing. In: Dourish, P., Friday, A. (eds.) UbiComp 2006. LNCS, vol. 4206, pp. 459–476. Springer, Heidelberg (2006)
5. Dey, A., Salber, D., Abowd, G.: A conceptual framework and a toolkit for supporting the rapid prototyping of context-aware applications. Human-Computer Interaction (HCI) Journal 16, 97–166 (2001)
6. Kleindienst, J., Cuřín, J., Fleury, P.: Reference architecture for multi-modal perceptual systems: Tooling for application development. In: Proceedings of 3rd IET International Conference on Intelligent Environments (IE 2007), Ulm, Germany (2007)
7. Fleury, P., Cuřín, J., Kleindienst, J.: SitCom – development platform for multimodal perceptual services. In: Mařík, V., Vyatkin, V., Colombo, A.W. (eds.) HoloMAS 2007. LNCS (LNAI), vol. 4659, pp. 104–113. Springer, Heidelberg (2007)
8. Danninger, M., Robles, E., Takayama, L., Wang, Q., Kluge, T., Nass, C., Stiefelhagen, R.: The connector service – predicting availability in mobile contexts. In: Proc. of the 3rd Joint Workshop on Multimodal Interaction and Related Machine Learning Algorithms (MLMI), Washington DC, US (2006)

The Generation of Emotional Expressions for a Text-Based Dialogue Agent

Siska Fitrianie[1] and Leon J.M. Rothkrantz[1,2]

[1] Man-Machine Interaction, Delft University of Technology
[2] Netherlands Defence Academy
{s.fitrianie,l.j.m.rothkrantz}@tudelft.nl
http://mmi.tudelft.nl

Abstract. Emotion influences the choice of facial expression. In a dialogue the emotional state is co-determined by the events that happen during a dialogue. To enable rich, human like expressivity of a dialogue agent, the facial displays should show a correct expression of the state of the agent in the dialogue. This paper reports about our study in building knowledge on how to appropriately express emotions in face to face communication. We have analyzed the appearance of facial expressions and corresponding dialogue-text (in balloons) of characters of selected cartoon illustrations. From the facial expressions and dialogue-text, we have extracted independently the emotional state and the communicative function. We also collected emotion words from the dialogue-text. The emotional states (label) and the emotion words are represented along two dimensions "arousal" and "valence". Here, the relationship between facial expressions and text were explored. The final goal of this research is to develop emotional-display rules for a text-based dialogue agent.

Keywords: Emotion, the relationship of facial expression and text, human-like qualities of a dialogue agent.

1 Introduction

The human face is the primary channel to express emotion. The instantaneous emotional state is directly linked to the displayed facial expression [6]. Facial expressions occur synchronizly to one's own speech or to the speech of others [7,3]. Human face-to-face dialogue involves both language and nonverbal behavior. Seeing faces, interpreting their expression, understanding the linguistics contents of speech are all part of human communication.

Multimodal user interfaces are interfaces with multiple channels that act on multiple modalities. Dialogue is supported by multiple coordinated activities of various cognitive levels. As a result communication becomes highly flexible and robust, so that failure of one channel is recovered by another channel and a message in one channel can be explained by another channel. Many researchers showed that the capability of the agent communicating with humans using both verbal and nonverbal behaviors will make the interaction more intimate and human-like [16,17]. Using facial displays as means to communicate have been found to provide natural and compelling dialogue

P. Sojka et al. (Eds.): TSD 2008, LNAI 5246, pp. 569–576, 2008.
© Springer-Verlag Berlin Heidelberg 2008

agents [11,17]. The realization of a dialogue agent that can perceive and express emotions can enhance an effective and more natural communication. Directly displaying the emotion expressions from a dialogue-text, however, provides challenges on several levels.

Firstly, emotional linguistic content consists of entities of complexity and ambiguity such as syntax, semantics and emotions. Most research in text-based emotion analysis describe how words with an affective meaning are being used within a sentence e.g. [15,13,8,9]. Such a model fails to describe emotion expressed by phrases requiring complex phrase/sentence-level analyses, since words are interrelated and influence each other's affect-related interpretation.

Secondly, along with the role emotional signal, other functions of facial displays should also be taken into account: (a) communicative function, indicating the mental state of speaker and listener e.g. alignment, acknowledgement [19] and (b) interpersonal functions, showing how one feels/thinks about the other and the relation to one self [13,8]. Finally, facial expressions for emotions are certainly not always trustful [5]. Facial expressions do not always correspond to felt emotions: they can be faked (showing an expression of an unfelt emotion), masked (masking a felt emotion by an unfelt emotion), superposed (showing a mixed of emotions), inhibited (masking the expression of emotion with neutral expression), suppressed (de-intensifying the expression of an emotion) or exaggerated (intensifying the expression of an emotion).

This work aims at studying the role of facial expressions to convey emotions in a dialogue. The results will be applied on our developed dialogue agent. For this purpose, we have analyzed selected dialogues of some characters in a number of fragments from some selected cartoon illustrations. We used this data set because of the following reasons: (1) the emotion expressions in cartoons are associated with particular human emotions, (2) each character shows emotion with largely expressive facial displays; therefore the emotion can be observed easily and (3) the data includes the context why an emotion occurs and why a particular facial expression is displayed. From the facial expressions and text, we have extracted independently the emotional state and the communicative function. We also collected emotion words from the dialogue-text. The emotional states and the emotion words are represented along two dimensions "arousal" and "valence". Here, the relationship between facial expressions and text was explored.

The structure of this paper is as follows. In the following section, we start with related work. Next, the description of the study using cartoon illustrations and our analysis are presented. Further, our developed dialogue agent is presented. Finally, we conclude the paper with a discussion on future work.

2 Related Work

Currently, text-based emotion analysis is approached mostly as a text-classification problem. A textual unit is classified as expressing positive or negative (or pleasant and unpleasant) feelings, e.g. word sentiment analysis [15], propagation towards an object or event [13], text orientation measure [9]. Based on [13], Fitrianie & Rothkrantz found 5 types of emotional intention in language [8]: (a) emotionally active toward an object, (b) emotionally directed by an object, (c) emotions that provoked by an object,

(d) emotions that experienced towards an object and (e) appraisal toward an object. This analyzing the emotional aspect of language needs a large-scale affective lexicon resource database. The largest database – DAL [22], we found so far, which contains 8742 words in a 2D circumplex model. Real-world commonsense knowledge was proposed by Liu et al. [12] for understanding the underlying semantics of language.

A study using cartoons found that the accuration of text-based emotion annotation was twice higher when correspondence facial expressions were referred [18]. To alleviate the difficulties of directly displaying human faces from speech, the BEAT system takes annotated text to be spoken by an animated figure as input, and outputs appropriate synchronized nonverbal behaviors and synthesized speech in a form that can be sent to a number of different animation systems [2]. A rule based approach has been used to govern the emotion expressions by (a) Wojdel and Rothkrantz [23] and connect their facial expression modulator to a dictionary of predefined facial expressions, (b) Prendinger et al. [16] with considerations to social conventions (power and distance) and user personality, and (c) Reflexive Agent [1] for analyzing emotional nature factors (i.e. valence, social acceptance, emotion of the addressee) and scenario factors (i.e. personality, goals, type of relationship, type of interaction). Both [16] and [1] applied the rules to decide the intensity of emotions or even totally inhibit it.

3 Corpus-Based Knowledge Engineering

The purpose of our experiment was to find out: (1) what kind of emotional expressions are shown most often in a dialogue, (2) what kind of emotion words appear in the dialogue-text, and (3) whether the emotion expressions are in correspondence (parallel) to the emotion loading of the dialogue-text.

3.1 Annotation Method

To collect data for the analysis we prepared 5 cartoon illustrations consisting of 27 dialogue fragments (277 dialogues –460 facial expressions and 357 text balloons). We selected dialogs between 2 characters in diverse situations which evoke various emotional states. 9 groups of 2 students were asked to annotate the data. One member of every couple annotated only the facial expressions, the other only the dialogue-text. The experiment was performed as follows:

- The participants with the facial expression corpus were asked to: (1) identify emotion expressions, (2) label the identified expressions using emotion words and (3) identify the function of the expression in the dialogue.
- The participants with the dialogue-text data were asked to: (1) collect emotion words in the dialogue-text, (2) label the dialogue-text using emotion words and (3) identify the function of the expression in the dialogue.

The communication function category consists of question, acknowledgement, statement, clarification and confirmation [19]. The agreement rates between raters of the facial expressions were 78% and of the dialogue-text were 73%.

3.2 Analysis Results

From the experimental results, we found that the participants used a large range of emotion words (54 words) to label the emotion types of the characters in the story: 45 from the facial expression corpus and 32 from the text corpus. To formulate coherent results, we decided to group the emotion words into 8 octants based on [4] (Table 1). We found that joyful (18%), angry (14%), unpleasantly-surprised (13%) and (pleasantly-) surprised (12%) are shown most often.

To study which emotion words correlate to a certain emotion, we selected 207 dialogue-lines (consisting of 357 dialogue-texts), in which the characters show their facial expression and prompt text. From the results, we found that only about 23% of the dialogue-text using emotion words like "great" and "bad" (adjective) or "hate" and "love" (verb). Other dialogue-text (about 41%) needs our common sense to correlate them with a certain emotion, e.g. "how funny! It really works" for amused and "this time he's going flat on his beak!" for indignant. Some dialogues (16%) use textual-prosody to show the character's emotion, e.g. "grrrr" (indignant), "gloek" (flabbergasted), and "snap!" (hostile). The rests of the dialogue-text (about 19%) do not contain any emotion words and are considered as neutral, while the facial expressions show a certain emotion.

The emotion labels of the facial expressions are not always correspondence with the dialogue-text. As shown in Table 2, negative emotions are more often masked or inhibited, while the positive emotions are always conveyed. In the story, the characters mask their negative emotions with positive emotions to pretend their actual feeling (usually bad intention to others). A fake joyful emotion is often used to mask negative emotions. The characters suppressed or even inhibit their negative emotions if they need to be polite or if they are feeling shameful. Only 1% of neutral emotions are expressed as neutral. The characters use facial displays to complete or even (in some cases) exaggerate the expression of an emotion. Superposed emotions are shown for surprised with joyful or anger, however, most annotators label them as surprised and shocked.

We used the collected emotion words in 83 dialogue-texts to study whether the facial expressions correlate with the use of emotion words in the text. For each character i and each dialogue-text j, the words $w_{1..n}$ are plotted on two-dimensional "arousal" and "valence" using the following equation:

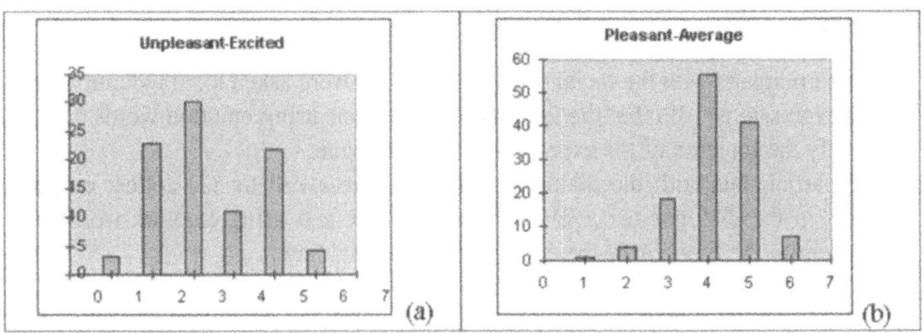

Fig. 1. Histogram for elapsed time: (a) unpleasant-excited and (b) pleasant-average

Table 1. 54 emotion labels found in the corpus (in 8 octants based on [4])

No.	Valence-Arousal	Facial expression labels (43 words)	Dialogue-text labels (34 words)
1.	Neutral-Excited	{curious, 2}, {amazed, 7}, {concentrated/ confident, 7}, {eager/ excited, 11}	{enthusiastic, 18}, {amazed, 3}
2.	Pleasant-Excited	{inspired, 5}, {desiring, 3}, {in love, 4}	{in love, 4}, {inspired, 5}
3.	Pleasant-Average	{surprised, 57}, {happy/ cheerful, 83}, {arrogant/ amused, 15}, {teasing/ admiring, 11}	{teasing, 3}, {surprised, 31}, {cocky/ conceited/ amused, 15}, {good-humored/ delighted/ cheerful, 77}
4.	Pleasant-Calm	{satisfied/pleased, 5}	{militant/pleased, 9}
5.	Neutral-Calm	{hope/calm/normal, 23}	{neutral/hope, 61}
6.	Unpleasant-Calm	{sighing, 2}, {bored, 2}, {sad/distressed, 23}	{sighing, 2}, {sad, 11}
7.	Unpleasant-Average	{jealous, 5}, {disappointed, 9}, {wicked/ mischief, 27}, {confused/ dumbfounded/ scared, 17}	{wicked/ mischief/ roguish, 16}, {confused/ fear/ bewildered, 9}
8.	Unpleasant-Excited	{alarmed, 3}, {hostile, 2}, {greedy, 2}, {hate/ disgusted, 11}, {irritated/ disturbed, 11}, {pleading/ frustrated, 1}, {angry/ taken-back, 66}, {shocked/ unpleasant-surprised, 59}	{shocked, 31}, {hostile, 1}, {annoyed, 5}, {greedy, 4}, {desperate/ pleading, 3}, {angry/ fed-up, 49}

Table 2. Correspondance of 357 emotion labels of associated text and facial expressions

Octant#		**Facial expression corpus**							
		1	**2**	**3**	**4**	**5**	**6**	**7**	**8**
Dialogue-text corpus	**1**	11		10					
	2		9						
	3			126					
	4		1	3	5				
	5	3	5	19	6	7		6	13
	6						12	1	
	7					2		25	
	8			7	1	3		13	69

$$v_{ij} = 1/n\left(v_{w_1} + v_{w_2} + \ldots + v_{w_n}\right) \text{ and } a_{ij} = 1/n\left(a_{w_1} + a_{w_2} + \ldots + a_{w_n}\right)$$

The valence $v\,[-1\ldots 1]$ and arousal $a\,[-1\ldots 1]$ scores of a word w are obtained from DAL [22]. The equation measures the text emotional-orientation. Then, we divided the circumplex into the 8 octants using the following rules: (1) $-0.2 \leq v \leq 0.2$ is neutral, $v > 0.2$ is pleasant, $v < -0.2$ is unpleasant and (2) $-0.2 \leq a \leq 0.2$ is average, $a > 0.2$ is excited, $v < -0.2$ is calm. Finally, for each character and each dialogue line, we mapped it with the corresponding facial expression corpus to find whether they

Table 3. Correspondance of 83 emotion labels of associated text and facial expressions

Octant#		Facial expression corpus							
		1	**2**	**3**	**4**	**5**	**6**	**7**	**8**
Dialogue-text corpus	**1**	3		3					
	2		4						
	3	3		27					
	4			1	1				
	5					1			
	6					2	5	1	
	7							8	
	8	2		3	1			3	14

correspond to each other. Table 3 shows similar composition as Table 2. In most cases, the characters show their felt emotion.

In the final experiment, we analyzed the distribution of the elapsed frames of facial expressions. For this purpose, we plotted appropriate histograms for each octant with interval 1 frame (dialogue-line). From the histograms (see example in Figure 1), we noticed that emotion expressions mostly appear for 2–3 frames. However, 40% of unpleasant-excited and 90% of pleasant-average could appear until 4–6 frames. The pleasant-average histogram also contained a tail that comprises of the longest observed expression duration. We attributed this to the dual role of the expression joyful and anger as both short emotion signal and mood indicator. In this study, we also found that the changing emotions between octants mostly occurs (92%) when the characters acknowledge a new event or object. The resulted emotion is directed or provoked by the event or object.

4 My_Eliza: A Multimodal Dialogue Agent

We applied our experimental results on our developed dialogue agent, My_Eliza [10]. My_Eliza is a multimodal question-answering system based on the famous Eliza program [21]. Our system recognizes emotion from natural language. It shows a facial expression for each sentence typed by the user. Subsequently, it gives a text reply together with an appropriate facial expression to convey emotional content. The system generates its responses using an extended-XML script, AIML (Artificial Intelligent Mark-up Language [20]). AIML provides specifications for pattern input matching and reply generation. Besides matching user input patterns, its transformation rules allow user-system conversation focusing on a certain topic and generating responses based on previous conversation.

The system analyzes the emotions loading of text by classifying emotion words and applying a set of rules to select a corresponding emotion expression. Emotion thermometers are used to observe the changing of emotions in a dialogue and to decide the intensity of the displayed emotion. The current version of the system analyzes both input and output text to distinguish the communicative function (based on [19]) and the emotional intention (based on [8]). If it acknowledges a new object or event or if the dialogue topic changes, the system will consider whether this supports or violates its

User:	My name is Siska.	My_Eliza
My_Eliza	Nice name! I am Eliza.	
User	Hello Eliza.	My_Eliza
My_Eliza	Enough, what's the problem?	

User:	My computer just crashed.	My_Eliza
My_Eliza	Sorry to hear that, what will you do then?	
User	I do not know.	My_Eliza
My_Eliza	I do not know anything about computer either.	

Fig. 2. Example of a fragment between a user and My_Eliza

goals/principles. A new emotional state from different octant may be selected and all thermometers are refreshed. Figure 2 shows example of a dialogue.

5 Conclusion and Discussion

Using cartoon illustrations, the role of facial expressions to convey emotions in a dialogue has been studied. The results have been applied on a text-based dialogue agent. These experimental results indicate that text-based emotion expression generation depends mostly on the context, since emotion words do not always appear in text. Knowing the intention of the speaker/listener, the communicative function of the dialogue state, and a large-scale of common sense knowledge can enhance the accuracy of the analysis results. Furthermore, we found in most cases the characters convey felt-emotions, although in some cases negative emotions are masked or inhibited to be polite or hide personal intentions to others. The results also show that mood indicator expressions were shown much longer in dialogues. The changing of the valence sign mostly occurs when a new event or object is acknowledged. The resulted emotion is usually directed or provoked by the event or object. The next step will include emotion analysis using other discourse information, such as moods, personality characteristics, anaphoric information, and background contexts of story. Future work is necessary to evaluate the knowledge applied in the developed system.

Acknowledgments. The research reported here is part of the Interactive Collaborative Information Systems (ICIS) project, supported by the Dutch Ministry of Economic Affairs, grant nr: BSIK03024.

References

1. De Carolis, B., Pelachaud, C., Poggi, I., De Rosis, F.: Behavior Planning for a Reflexive Agent. In: Proc. of IJCAI 2001, Portugal, pp. 1059–1066 (2001)
2. Cassell, J., Vilhjálmsson, H., Bickmore, T.: BEAT – The Behavior Expression Animation Toolkit. In: Proc of SIGGRAPH 2001, Los Angeles, CA, pp. 477–486 (2001)
3. Condon, W., Osgton, W.: Speech and Body Motion Synchrony of the Speaker-Hearer. In: Horton, D., Jenkins, J. (eds.) The perception of Language, pp. 150–184. Academic Press, London (1971)

4. Desmet, P.: Designing Emotion, Doctoral Dissertation, TU Delft (2002)
5. Ekman, P.: Telling Lies: Clues to Deceit in the Marketplace, Marriage, and Politics. WW Norton, New York (2001)
6. Ekman, P.: Basic Emotions. In: Dalgleish, T., Power, M. (eds.) Handbook of Cognition and Emotion, UK. John Wiley and Sons Ltd, Chichester (1999)
7. Ekman, P.: About brows: Emotional and Conversational Signals. In: Human Ethology: Claims and limits of a new discipline: contributions to the Colloquium, Cambridge, pp. 169–248 (1979)
8. Fitrianie, S., Rothkrantz, L.J.M.: An Automated Text-Based Synthetic Face with Emotions for Web Lectures. CCAI 23(1–4), 89–98 (2007)
9. Fitrianie, S., Rothkrantz, L.J.M.: An Automated Crisis Online Dispatcher. Int. Journal of Emergency Management, Research Method (to appear, 2008)
10. Fitrianie, S., Wiggers, P., Rothkrantz, L.J.M.: A Multimodal Eliza Using Natural Language Processing and Emotion Recognition. In: Matoušek, V., Mautner, P. (eds.) TSD 2003. LNCS (LNAI), vol. 2807, pp. 394–399. Springer, Heidelberg (2003)
11. Kiesler, S., Sproull, L.: Social Human Computer Interaction, Human Values and Design of Computer Technology, pp. 191–199. CSLI Pub., Standford (1997)
12. Liu, H., Lieberman, H., Selker, T.: A Model of Textual Affect Sensing using Real World Knowledge, Technical Report, MIT Media Laboratory, Cambridge (2003)
13. Mulder, M., Nijholt, A., den Uyl, M., Terpstra, P.: A Lexical Grammatical Implementation of Affect. In: Sojka, P., Kopeček, I., Pala, K. (eds.) TSD 2004. LNCS (LNAI), vol. 3206, pp. 171–178. Springer, Heidelberg (2004)
14. Niewiadomski, R., Pelachaud, C.: Model of Facial Expressions Management for an Embodied Conversational Agent. In: Paiva, A.C.R., Prada, R., Picard, R.W. (eds.) ACII 2007. LNCS, vol. 4738, pp. 12–23. Springer, Heidelberg (2007)
15. Pang, B., Lee, L.: A Sentimental Education: Sentiment Analysis Using Subjectivity Summarization Based on Minimum Cuts. In: ACL 2004, Spain, pp. 271–278 (2004)
16. Predinger, H., Ishizuka, M.: Social Role Awareness in Animated Agents. In: Proc. of AGENT 2001, pp. 270–276. ACM, New York (2001)
17. Schiano, D.J., Erlich, S.M., Rahardja, K., Sheridan, K.: Face to Interface: Facial Affect in (Hu)man and Machine. In: Proc. of ACM CHI 2000, pp. 193–200 (2000)
18. Tokuhisa, M., Murakami, J., Ikehara, S.: Construction and Evaluation of Text-Dialog Corpus with Emotion Tags Focusing on Facial Expression in Comic. In: Gabrys, B., Howlett, R.J., Jain, L.C. (eds.) KES 2006. LNCS (LNAI), vol. 4253, pp. 715–724. Springer, Heidelberg (2006)
19. van Vark, R.J., Vreught, J.P., Rothkrantz, L.J.M.: Classification of Public Transport Information Dialogues Using an Information-Based Coding Scheme. In: Dialogue Processing in Spoken Language Systems Workshop, pp. 55–69. Springer, Heidelberg (1997)
20. Wallace, R.: Alicebot (1995), `http://www.Alicebot.org`
21. Weizenbaum, J.: ELIZA – A Computer Program for the Study of Natural Language Communication between Man and Machine. Comm. of ACM 9(1), 36–45 (1966)
22. Whissell, C.M.: The Dictionary of Affect and Language. In: Plutchik, R., Kellerman, H. (eds.) Emotion: Theory, Research, and Experience, vol. 4, pp. 113–131 (1989)
23. Wojdel, A., Rothkrantz, L.J.M.: Parametric Generation of Facial Expressions Based on Facs. Computer Graphics forum 4(24), 1–15 (2005)

Intelligent Voice Navigation of Spreadsheets
An Empirical Evaluation

Derek Flood, Kevin Mc Daid, Fergal Mc Caffery, and Brian Bishop

Dundalk Institute of Technology, Dundalk, Co. Louth, Ireland
derek.flood@dkit.ie, kevin.mcdaid@dkit.ie,
fergal.mccaffery@dkit.ie, brian.bishop@dkit.ie

Abstract. Interaction with software systems has become second nature to most computer users, however when voice recognition is introduced, this simple procedure becomes quite complex. To reduce this complexity for spreadsheet users, the authors have developed an intelligent voice navigation system called iVoice. This paper outlines the iVoice system and details an experiment that was conducted to determine the efficiency of iVoice when compared to a leading voice recognition technology.

1 Introduction

Technological advances and a desire for mobility have substantially reduced the size of complex systems, calling into question the practicality of traditional input mechanisms. New mechanisms, such as voice recognition, are emerging that will allow for easier control of these new mobile systems.

Voice recognition technology will not only benefit users of mobile systems. Using voice recognition technology sufferers of RSI (Repetitive Strain Injury) can more comfortably control computer applications, which would otherwise require extensive use of the keyboard and mouse [1]. One such application is Microsoft Excel, the most common spreadsheet package. Spreadsheets are used for a wide variety of tasks from record keeping to financial statements and although spreadsheets are a ubiquitous software tool, in recent years their reliability has been shown to be very poor. Section 2 overviews the spreadsheet technology and ways to address this absence of reliability.

Voice control of software applications, such as spreadsheets, has been proven to be difficult [2]. Some software domains, as outlined in Section 3, have been adapted to take advantage of the unique features offered by voice recognition technology. With the aim of improving the control of spreadsheets through voice recognition, the authors have developed an intelligent navigation system called iVoice, as outlined in Section 4.

An experiment, detailed in Section 5, was conducted to evaluate iVoice against a state of the art voice recognition technology. Section 6 summarizes the results of this experiment and Section 7 concludes this paper.

2 Speech Technology

Spreadsheets can be used to help companies make multi-million euro decisions on a daily basis. Within the financial district of London, spreadsheets have been described as the primary front line tool of analysis [3].

P. Sojka et al. (Eds.): TSD 2008, LNAI 5246, pp. 577–584, 2008.
© Springer-Verlag Berlin Heidelberg 2008

Despite the importance of the decisions being made based on the information contained in these spreadsheets, there is very little done in practice to ensure their quality. Two independant studies [4,5] have revealed that over 90% of spreadsheets contain errors. These errors can range in severity from simple spelling errors to complicated formula errors.

A range of techniques have been proposed to improve the quality of spreadsheets. Some of these initiatives focus on the importance of development methodologies similar to existing lifecycle processes in place for software development. One such technique is Test Driven Development [6]. This method requires users to write test cases before they develop the spreadsheet. In this way developers are forced to build the solution in small steps and to consider the design at the outset. They will also have a series of tests in place that will ensure the integrity of the spreadsheet as development proceeds.

3 Voice Recognition Technology

Voice recognition technology has been primarily used for creating and modifying text documents [7]. Although vendors of Voice Recognition Technology claim it can be faster than traditional keyboard and mouse input, the author has found no evidence to support this claim. A recent study [8] found that through voice recognition technology, users could quickly achieve input rates of 150 WPM with an accuracy of around 90%.

With the increased interest in voice recognition technology, new applications have emerged across multiple domains. Begel [9] has allowed software developers to create java applications by speaking in a natural way. As the developer speaks, the system, Spoken Java, will generate the java source code.

There are many voice recognition engines available; the best of which is Dragon NaturallySpeaking(DrNS), which boasts an accuracy of 99%. The most basic version, Standard Edition, features support for Microsoft® Word and Microsoft® Internet Explorer. The Preferred Edition also includes support for Microsoft® Excel. Special editions for the legal and medical domains also exist [7].

4 iVoice

In an earlier experiment [2], three experienced spreadsheet users were asked to audit a spreadsheet, seeded with errors, using voice recognition technology. Although none of the participants had any prior experience with voice recognition, they had sufficient experience with spreadsheets to complete the task easily. Their performance was compared to another study [10] in which 13 professional spreadsheet users were asked to audit the same spreadsheet using a keyboard and mouse. Both studies recorded the behaviour of each participant through logging of cell selection activity via the T-CAT [11] (Time-stamped Cell Activity Tracker) tool. It was found that voice recognition participants found 14% less of the seeded errors despite taking twice as long.

A number of elements, namely editing formulas, entering data and navigation were looked at to try to identify key differences in the behaviour and performance of the groups. The results showed that the voice control users struggled in all of these aspects. Given that navigation is the most fundamental of spreadsheet activities it was decided to explore the development of new technologies that could improve the efficiency of voice-controlled navigation of spreadsheets.

The resulting iVoice system, which integrates with DrNS, provides support for three particular actions, navigation to referenced cells, automatic navigation of a range of similar cells, and navigation directly to the next non-blank cell. Each of the three features is explained and subsequently investigated through an experiment.

4.1 Navigation to Referenced Cells

When looking at the navigational behaviour of all participants in [2], it was observed that upon initial entry to a cell containing a cell reference users would navigate to this cell in order to ensure that the reference was correct. Traditional voice recognition software requires users to navigate to the target worksheet through all intermediary worksheets. iVoice allows users to skip the intermediate worksheets bringing them directly to the remotely referenced cell.

By assigning each referenced cell in a formula a unique colour, users can move directly to the required cell through the command *Jump <colour>*, where *<colour>* is the colour of the desired destination cell. A *Jump Back* command is also provided to enable users to move back to the original cell. It is hypothesised that the time to reach a referenced sheet will, based upon the number of commands, decrease using the iVoice technology for non-adjacent worksheets. The colours and the names used by the system can be displayed and hidden at any time through the *Show Colours* and *Hide Colours* commands.

4.2 Scan Command

The second component of the iVoice system allows users to navigate through a list of semantically similar cells. Semantically similar cells are those whose contents are of a similar structure and purpose [12]. Spreadsheets are in general composed of regions of such cells. It has been observed that users will examine these regions sequentially, spending on average between 0.33 seconds and 1.5 seconds on each cell.

iVoice supports this activity through the provision of a scan command which automatically moves to the next cell in the chosen direction after one second. This delay allows users the chance to review the contents of the cell before moving on. To initiate this command users say *Scan <direction>* where *<direction>* is the direction they wish to scan, be it left, right, up or down. It is hoped that this feature will reduce the time to perform this task, as it requires a single voice command rather than a series of such commands.

4.3 Jump Blank Cells

The third component of the iVoice system allows users to skip over blank cells. By saying *"Jump <direction>"* where <direction> is the way the user wants to move. The system moves directly to the next non blank cell in that direction. It is felt that this command is more efficient and natural than dictating the associated keyboard shortcut as is currently required in DrNS.

5 Experiment

A quantitative experiment was designed to compare the iVoice system with DrNS. This was followed by a qualitative study, where participants took part in a structured interviewed to establish their view of the iVoice technology.

The quantitative experiment asked six experienced spreadsheet users to highlight, through the *Mark Error* command, as many errors as they could in two spreadsheets through voice recognition technology. This highlighting could be later removed, if required, through the *Unmark Error* command. To randomise the experiment participants were split randomly into two groups of three. Group 1 first audited Spreadsheet 1 using DrNS and then Spreadsheet 2 using iVoice. Group 2 reversed the use of the technologies. The cell selection behaviour of all participants was recorded with the T-CAT tool.

Before the trial commenced participants were asked to configure the voice recognition software. This took approximately ten minutes to complete, after which participants were introduced to the voice-control systems through a navigation exercise which asked them to move about a sample spreadsheet. When participants felt they had mastered the navigation commands, the trial commenced with all participants auditing Spreadsheet 1. No time limit was set for the task allowing participants to finish when they believed they could find no more errors.

The first spreadsheet audited, Spreadsheet 1, was comprised of three worksheets, *Wages*, *Expenses*, and *2007 Department Spending*. The spreadsheet calculates the total expenditure for each of three departments in a company. The *Wages* sheet details all employees wages and which department they belong to. The *Expenses* worksheet is used to detail different expenses to the company and what department these expenses should be assigned to. The final worksheet totals all costs and apportions company-wide expenses to each department based on the number of employees in that department.

When the first spreadsheet had been completed participants were given a break to facilitate the changing of the technology. Participants then received training in the second technology and subsequently audited the second spreadsheet with the second technology. Again, no time limit was set allowing participants to finish when they judged the exercise was complete.

The second spreadsheet, Spreadsheet 2, also contained three worksheets, *Opening Stock*, *Purchases* and *Sales and Profit*. This spreadsheet was used to calculate the profit made on each of 18 products over a given period. The *Opening Stock* worksheet detailed the quantity and value of each of the products at the start of the period. The *Purchases* worksheet detailed the purchases that were made during the period, and the *Sales and Profit* worksheet detailed the sales and closing stock of each product. This worksheet also uses the costs from the *Opening Stock* and *Purchases* worksheets to calculate the profit for the period.

6 Analysis of Results

Before examining the individual features, the overall performance and behaviour of participants using each technology was examined. The results obtained by the T-CAT tool allowed for a detailed analysis of the experiment.

The measures used for performance were spreadsheet coverage and errors found. Evidence has been found to suggest that there is a relationship between the number of cells evaluated and the number of errors found [10]. For the purpose of this experiment coverage was defined as the percentage of the spreadsheet that was actually reviewed by participants where a cell was considered to be reviewed if a participant spent more than 0.3 seconds on it. Only cells that contain numerical data or formula were considered as all other cells could be reviewed without being entered.

Table 1. Cell Coverage

	Spreadsheet 1	Spreadsheet 2
iVoice	87.5	44.6
DrNS	53.6	37.1

Table 1 shows that iVoice users covered a higher percentage of the cells than users of DrNS. In Spreadsheet 1 users who used iVoice covered 87.5% in contrast to the 53.6% covered by those using DrNS. For Spreadsheet 2, using iVoice, participants covered 44.6% of the spreadsheet whereas participants using DrNS covered 37.1%.

Table 2. Time spent auditing spreadsheet in minutes

	Spreadsheet 1	Spreadsheet 2
Group 1	28.4	21.5
Group 2	26.8	20.0

It is important to establish that any difference in performance is not due to a difference in time spent by the groups on the tasks. Table 2 shows the average time each group spent auditing each spreadsheet. It was found that Group 1 spent on average 90 seconds more on each spreadsheet than those in Group 2, regardless of the technology that was employed. The difference is small and so it was concluded that this effect is insignificant.

Table 3. Overall Performance Per Spreadsheet

	Spreadsheet 1	Spreadsheet 2
iVoice	69.4%	61.1%
DrNS	63.9%	57.4%

Table 3 shows the average percentage of errors that were found by each group. It was found that users of iVoice found between 3% and 6% more than users of DrNS. On Spreadsheet 1 those using iVoice found 69.4% of the errors whereas those using DrNS found 63.9%. For Spreadsheet 2 those using iVoice found 61.1% of the errors whereas those using DrNS found 57.4%. While the sample sizes are insufficient to establish a statistically significant difference these values indicate that iVoice leads to better coverage and performance than Dragon NaturallySpeaking. The results obtained through a more detailed analysis of each of the iVoice features is now presented.

6.1 Navigate to Referenced Cells

The time to navigate to referenced cells is measured from when the participants leave the source cell until they enter the destination cell, omitting the time spent in both the source and destination cell. The average values are quoted only for participants who performed this action three or more times. This feature was used by participants between 1 and 10 times during the experiment.

While using DrNS it was found that users spent on average 4.1 seconds moving from a cell that contains a reference to the referenced worksheet, passing through one intermediate worksheet. Participants needed to spend on average a further 2.7 seconds returning to the original cell. These results indicate that through the use of iVoice users can save approximately seven seconds checking one remote reference, as the iVoice system can bring users directly to such a reference in one command and back to the original cell through a second command. As the number of intermediate worksheets increases it is expected that the savings quoted above would also increase.

6.2 Scan Command

The scan command was evaluated by measuring the average time participants spent in each cell of a scanned region. Each evaluated region contained a minimum of three cells. The first cell in each region was discarded as it is believed users spend longer reviewing this cell and including this time would distort the results. For the analysis, each participant must have scanned at least three regions. Only one participant failed to scan three such regions. This feature was used between 2 and 25 times by participants during the trial.

Table 4. Overall Performance Per Spreadsheet

	Spreadsheet 1	Spreadsheet 2
iVoice	0.93	0.97
DrNS	2.62	2.92

Table 4 shows average time participants spent on each cell while scanning through a region. It was found that when users used the iVoice function they were able to spend the expected 1 second on each cell in a scanned region. When using DrNS it was found that they spent on average 2.5 to 3.0 seconds on each cell, almost three times as long.

6.3 Jump Blank Cells

In order to evaluate the effectiveness of the "Jump Direction" command, the time at which users left a cell to the time they entered the next non-blank cell was measured. It was found that participants using Dragon NaturallySpeaking spent approximately 1.3 seconds on average performing this action. With iVoice the equivalent time would be zero as they are brought directly form one cell to the next non-blank cell. Although not all participants used this feature, a number of participants used it extensively, with one participant using it 45 times.

6.4 Discussion with Participants

Upon completion of the quantitative trial, a structured interview was conducted to find out participants opinions of the technologies they had used and their prior level of experience with both spreadsheets and voice recognition technology.

It was found that most of the six participants preferred the iVoice navigation system to Dragon NaturallySpeakings own in-built navigation system. The participants remarked that the iVoice commands made tasks like moving to remote references easier and also mentioned that it was easier to concentrate on auditing the spreadsheet while using iVoice.

7 Conclusions

This paper details a controlled experiment that suggestes the performance of spreadsheet auditors using voice recognition technology can be improved through the use of an intelligent navigation system. The experiment compared the performance of six experienced spreadsheet users using two technologies, Dragon NaturallySpeaking, a leading voice recognition software and iVoice, an intelligent navigation system. The experiment showed that participants using the iVoice system found a higher percentage of errors than those using Dragon NaturallySpeaking. iVoice simplifies navigation of a spreadsheet through three features; scanning through a range of cells, navigating to references off screen and moving over blank cells.

A number of enhancements were identified through a qualitative evaluation of the technology. At present the scan function stops for one second on each cell. In certain situations this time was found to be unsuitable and by allowing users alter this time the scan function could become more efficient. Other features which support the debugging process were also mentioned. One such feature would enable the scan function to automatically stop on cells which are semantically different from the preceding cells. This may be an indication of an error.

Acknowledgement. This work is supported by the Irish Research Council for Science, Engineering and Technology.

References

1. The repetitive strain injury association rsi facts and figures, `http://rsi.websitehosting-services.co.uk/facts_&_figures.pdf`
2. Flood, D., McDaid, K.: Voice-controlled auditing of spreadsheets. In: Proc. of the Conf. of the European Spreadsheet Risks Interest Group (2007)
3. Croll, G.: The importance and criticality of spreadsheets in the city of london. In: Proc. of the Conf. of the European Spreadsheet Risks Interest Group (2005)
4. Powell, S.G., Baker, K.R., Lawson, B.: Errors in operational spreadsheets (2007), `http://mba.tuck.dartmouth.edu/spreadsheet/product_pubs.html`
5. Panko, R.R.: What we know about spreadsheet errors. J. End User Comput. 10, 15–21 (1998)
6. McDaid, K., Rust, A., Bishop, B.: Test-driven development: can it work for spreadsheets? In: WEUSE 2008: Proceedings of the 4th international workshop on End-user software engineering, pp. 25–29. ACM, New York (2008)

7. NaturallySpeaking (2007), `http://www.nuance.com/naturallyspeaking`
8. Bailey, G.A.: Achieving speed and accuracy with speech recognition software. In: Americas Conference on Information Systems (2006)
9. Begel, A.: Spoken language support for software development. In: VLHCC 2004: Proceedings of the 2004 IEEE Symposium on Visual Languages – Human Centric Computing, pp. 271–272. IEEE Computer Society, Los Alamitos (2004)
10. Bishop, B., McDaid, K.: An empirical study of end-user behaviour in spreadsheet debugging. In: The 3rd Annual Work-In-Progress Meeting Of Psychology of Programming Interest Group (2007)
11. Bishop, B., McDaid, K.: Spreadsheet debugging behaviour of expert and novice end-users. In: WEUSE 2008: Proceedings of the 4th international workshop on End-user software engineering, pp. 56–60. ACM, New York (2008)
12. Clermont, M.: A toolkit for scalable spreadsheet visualization. In: Proc. of the Conference of the European Spreadsheet Risks Interest Group (2004)

A Semi-automatic Wizard of Oz Technique for Let'sFly Spoken Dialogue System

Alexey Karpov, Andrey Ronzhin, and Anastasia Leontyeva

St. Petersburg Institute for Informatics and Automation of RAS
SPIIRAS, 39, 14th line, St. Petersburg, Russia
{karpov,ronzhin,an_leo}@iias.spb.su
http://www.spiiras.nw.ru/speech

Abstract. The paper presents Let'sFly spoken dialogue system intended for natural human-computer interaction via telephone lines in travel planning domain. The system uses ASR, keyword spotting and TTS methods for continuous Russian speech and a dialogue manager with mixed-initiative strategy. Semi-automatic Wizard of Oz technique used for collecting speech data and real dialogues is described. Semi-automatic model is a tradeoff between a fully automatic spoken dialogue system and an interaction with a hidden human-operator working like a computer system. The experimental data obtained with this technique are discussed in the paper.

Keywords: Spoken dialogue systems, automatic speech recognition, Wizard of Oz study, spoken language understanding, call-centers.

1 Introduction

Development of spoken dialogue systems and automated call-centers with Automatic Speech Recognition (ASR) and and Text-To-Speech (TTS) is outstanding field within the framework of speech and language technologies [1]. Currently there exist several successful examples of such systems in order to obtain bus, train or flight schedule information by telephone, for instance American systems Let's Go! [2] and DARPA Communicator [3], or French system LIMSI ARISE [4], etc. It is known that communicative schemes of one language are not well appropriate to another. Therefore, in spite of the fact that in the travel planning domain there are well defined automated systems for other languages, we should begin all over again and study communicative schemes used in Russian speech taking into account experience of other teams.

The inflective nature of the Russian language leads to reach morphology and complex structure of word formation, thus a recognition vocabulary has to be increased in several orders because of many prefixes, suffixes and flexions. Moreover the order of words in Russian sentences is not defined by strict grammatical constructions and uses almost free grammar. It much complicates statistical language models and grammars as well as essentially decreases their efficiency.

Authors of the paper jointly with NewVoice telecommunication company are involved in the process of development of Let'sFly automated call-center within the framework of the EU INTAS innovative project. The call-center is intended to

P. Sojka et al. (Eds.): TSD 2008, LNAI 5246, pp. 585–592, 2008.
© Springer-Verlag Berlin Heidelberg 2008

automate flight ticket reservations of the "Rossiya" Russian Airlines company `http://www.rossiya-airlines.com`. By Let'sFly spoken dialogue system users will be able to obtain two kinds of information: (1) flight schedule; (2) fare for one-way or round-trip tickets.

2 Spoken Dialogue System Realization

This main section gives description of the most important steps completed on the way of development of automated call-center with a natural spoken dialogue: speech and dialogue corpus collection, development of the system's architecture, dialogue model, ASR and keyword spotter for Russian.

2.1 Initial Data Collection

The work was started from collection and expert analysis of the conversational dialogues between clients and a human-operator of the Rossiya service agency. This research was started in early 2007. The corpus of speech dialogues has been received using the simulation of human-computer interaction where operator's responses were degraded by a vocoder and telephone calls were recorded in parallel.

Totally 216 dialogues were obtained: 102 were made by women and 114 by men (all of them are native Russian speakers). All the dialogues are well balanced on age and sex. The corpus contains 4 hours 20 minutes of continuous speech. The recorded dialogues were labeled in terms of dialogue acts and frames of phrases. One of the main objectives of the research was analysis of typical Russian spoken sentences made by users in real dialogues. We were interested also what types of utterances prevail in the dialogues.

It was discovered that phrases "Mne nuzhno" (I need) and "Ja by hotel" (I would like) are the most frequent in user's requests. These are fragments of polite spoken forms. Words "Da" (Yes) and "Net" (No) are very often too and occupied 18.9 + 7.9 = 26.8% of utterances (or 2.9 + 1.2 = 4.1% of words) of the whole corpus. These words are used by users in confirmation procedures. The successful dialogues last in range 30–200 seconds having 12 turns on average. Thus, it is possible to tell, that the users are ready for productive and long dialogue (up to 4 minutes).

An analysis of user's phrases has shown that the first request has highest recognition complexity. It is necessary to note, that in analyzed dialogues the majority of the first user's utterances possesses high correlation of such word-combinations as "Good evening", "I would like", "Could you", etc. These word-combinations are cultural components of the request. We noted also that users quickly understood that they speak with a human but not with a computer. The user also knows that he/she needs to get information as soon as possible. As the result, user's speech was saturated by polite forms.

The speech data collected were used to train acoustical models (context dependent phonemes) of ASR of Let'sFly.

2.2 System's Architecture

The first version of Let'sFly call-center was constructed and installed in the middle of 2007. The general architecture of the system is presented in Figure 1. It contains modules for Russian speech recognition and synthesis and a dialogue manager that controls dialogue flow, organizes Wizard of Oz support as well as Internet access to Rossiya flight database located at the remote server. Let'sFly system was installed on a standard Pentium-IV computer connected to analogue Public Switched Telephone Network via a 3COM telecommunication board and uses Telephone API as software telephone interface. Users can apply ordinary or mobile phones in order to get voice access to the flight schedule information. The speech signal coming through the telecommunication board is sampled with 8 KHz and the energy-based voice activity detector corresponding to G.711 standard is used. The detected speech signal is saved into a unique PCM wave file after that SIRIUS ASR engine [5] for Russian is applied to recognize the captured signal. Both modeling and recognition of phonemes and words are based on HMMs. For the speech parameterization MFCCs with 1st and 2nd derivatives are used. In order to reduce the influence of channel noises two methods for speech enhancement are applied: (1) band-pass filter 200–3500 Hz to reject some low and high-frequency channel noises; (2) Cepstral Mean Subtraction [6] to remove channel distortion from the signal. The result of recognition is presented as N-best list which is processed by the dialogue manager.

Wizard of Oz technique [7] with an attraction of a human-operator is applied in a semi-automatic mode, it means that an operator may correct a recognition result only if ASR has made recognition mistakes several times for the same system's query and the user has explicitly confirmed it with the phrase "Net" (No). In classical Wizard of Oz technique confirmations are not necessary. When the user of Let'sFly confirms the best recognition hypothesis, then the dialogue manager generates the text query and sends it via Internet to the flight server which runs the search script. For searching a flight schedule on concrete user's date, two weeks time period is used in the text query. The beginning date is calculated as user's date subtracting 7 days. It is connected with the fact that some flights occur odd/even days or once a week only. But considering two

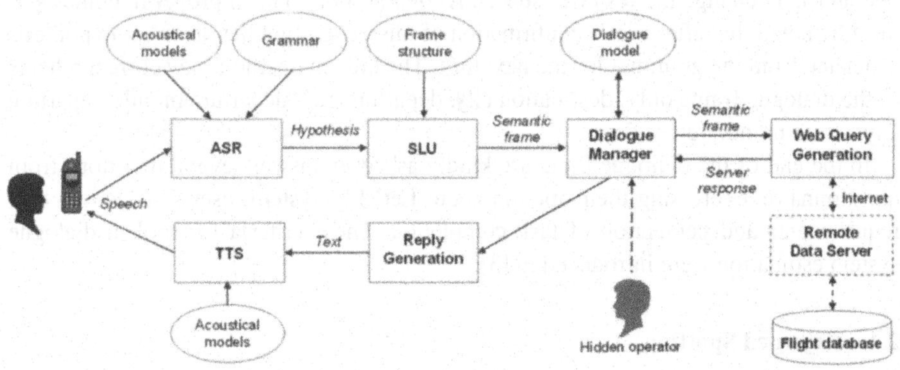

Fig. 1. General architecture of the spoken dialogue system

weeks period the systems can propose alternative flights both before and after the date needed.

The result of the query from the flight database is parsed by the dialogue manager which generates response and sends it to the TTS module for synthesis. TTS synthesizer of Russian speech was developed by Belarusian United Institute of Informatics Problems within the frameworks of the cooperative EU INTAS project. The speech synthesis is based on Allophone and Multi-Allophone Natural Waves method of speech signal concatenation [8]. Male synthetic voice is used to generate dynamically speech response in the Let'sFly system.

2.3 Dialogue Model

The dialogue manager was developed to assist a client in interaction with the system. The dialogue model uses mixed-initiative strategy and combines a finite-state automaton with a frame-based approach that provides a tradeoff between naturalness of the communication and high dialogue completion rate. The general diagram of spoken dialogue is presented in Figure 2. It shows the fields of the frame that should be filled in to complete a dialogue. Dialogue model relies on a combination of analysis of initial data collected by human-human interaction and our own intuition.

Each call to the system starts with a welcome message that prompts the user to say his/her request. The user receives an initiative in the beginning of the conversation when he/she has to answer system's question "How may I help you?" and the system waits for the user's answer. It was discovered that answers to this question follow in very free form therefore it is impossible to recognize it with a low WER using a grammar or a statistical language model, at least at this stage of the research. So in order to extract maximum of useful information from the first user's utterance the keyword spotting approach was developed to search destination city names in an utterance (see the next Section). After the first voice input the initiative is transmitted to the system which leads the dialogue to fill in empty fields of the query frame. Each recognition result is given to the user in the speech synthesized form and the user is asked to confirm it explicitly. If user says "No" confirming recognition mistake 2(3) times on the same question the system automatically switches to Wizard of Oz mode and a human-operator may change the result of speech recognition or confirm proper hypothesis. It must be noted that after user's confirmation of misrecognition the incorrect hypothesis is deleted from the grammar for the next turn. The hidden operator may correct 6 fields of the dialogue frame only: destination city, departure city, departure month, departure day, return month, return day.

In the end of the dialogue users are kindly asked to answer several questions from questionnaire, expressing their opinion about Let'sFly system, user satisfaction, task requirements and perception of task completion. These criteria for spoken dialogue system estimation were introduced in [3].

2.4 Keyword Spotting

An analysis of the collected human-human dialogues has shown that the first phrase has highest variability of the grammar. It was discovered that 82% first phrases in the

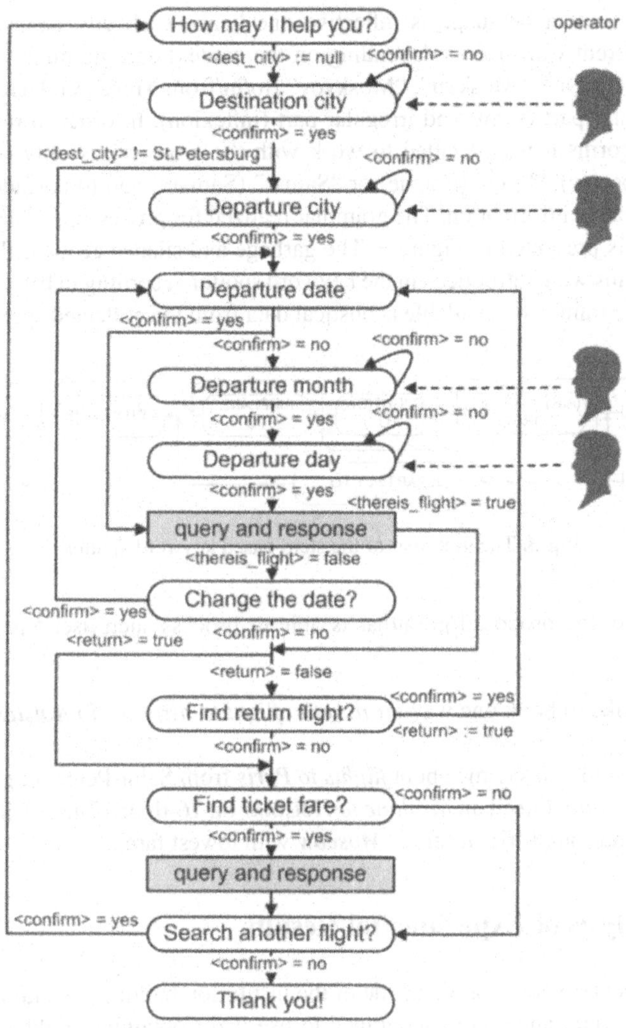

Fig. 2. Diagram of the spoken dialogue model

dialogues contain a destination city; 23% contain a departure date; 3% contain both a date of departure and a return date; 18% phrases do not contain any useful information. Polite forms of first phrases were observed in 96% utterances. Users may employ a wide range of phrases to refer to a destination city. However it was found that in 93% cases the destination city follows after the words "flight to", "ticket to", "airplane to", "ticket to flight to" or "fly to" (or their plural versions and combinations). So we decided to use keyword spotting method for searching destination cities only. 30 most popular cities (of more than 300 items of cities vocabulary) were selected from the Rossiya flight database, the flights to these cities cover over 90% regular flights of the air company. These cities are large Russian or European cities mainly. It must be noted that the Rossiya schedule server is able to handle with direct flights only.

Moreover Russian language is inflective one hence each city name can be presented in different word-forms depending on the context and grammar (for instance, “Moskva”, “Moskve”, “Moskvu”, “Moskvoy” (to/in/from Moscow). Here each word-form has regular part (stem) and irregular part (inflexion). In order to reduce the list of key word-forms it was decided to work with the word stems only (for instance, “Moskv” (Moscow), “Prag” (Prague) or “Samar” (Samara) and to consider inflexions as non-informative information. The grammar realized for processing of first phrases in the dialogues is presented in Figure 3. The garbage and silence acoustical models and vocabulary items were introduced in the keyword spotter according to [9]. HMMs of the keywords were trained by available acoustical data from the collected speech corpus.

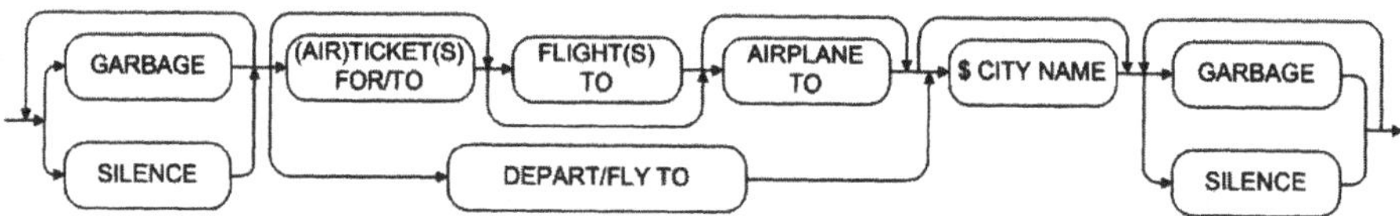

Fig. 3. The grammar of the stem-based keyword spotter

For instance, the proposed grammar is able to process such user’s phrases as the following:

- “I would like to book one way air *ticket to flight to airplane to* ***Amsterdam*** on the next week”
- “Please, could you say me about *flights to* ***Paris*** from Saint-Petersburg”
- “Good morning, I need an *airplane to* ***Helsinki*** on 16-th or 17-th April”
- “Please, find round-trip *tickets to* ***Moscow*** with lowest fare”

3 An Analysis of Experimental Results

Let’sFly call-center was installed inside of the institution building in autumn 2007 and all employees could and were encouraged to use it for obtaining flight schedule and fares when they need it. Totally 118 real dialogues were collected up to 2008 with semi-automatic WoZ technique and then processed manually. Most of the calls were really made to find out information needed, but some of them were made by persons with the purpose to “play” with the automatic call-center or to cause misunderstanding of the systems. Totally all the recorded dialogues contain 2489 phrases, so each dialogue has about 21 turns on average. 108 times an operator had to correct recognition results with the semi-automatic WoZ. The dialogue completion rate was 78% taking into account that all the calls in which the system did not give any information to the user were labeled as failures. Posterior satisfaction estimates of successful users were rather positive.

Figure 4 shows the distribution of dialogue length in turns, it can be seen that there are two maximums in the distribution. Some users started the dialogue but stopped it before fifth turn because they were not satisfied with the quality of work of speech recognition or speech synthesis. The minimal dialogue length to get flight schedule is

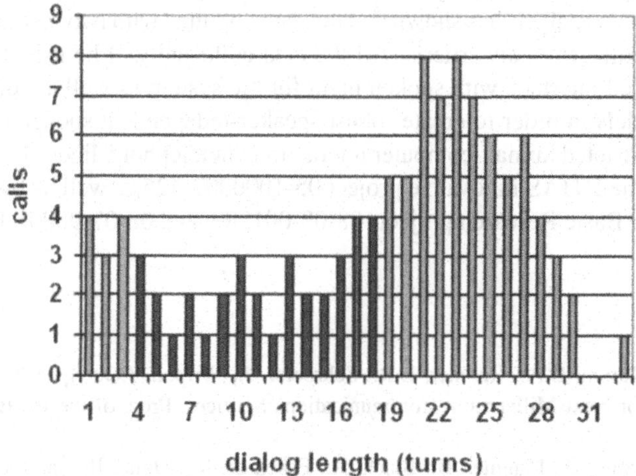

Fig. 4. Distribution of dialogue length (in turns)

8 turns, so the second maximum is located in the zone of successful dialogues. Speech recognition was more or less acceptable but far from perfect. WER was 14% for the short utterances (without first phrases) however about 55% of the phrases contained explicit confirmations only, which had 1.5% WER. The recognition vocabulary of the Let'sFly system contains about 500 items including city names, countries, dates, ticket class as well as speech artifacts, noise and garbage models (breath, sounded pauses, clicking lips or mouth), which are arisen in real dialogues. All the cities are divided into two classes: (1) cities with direct flights of Rossiya company (320 cities); (2) large world cities without direct connection from St. Petersburg (130 cities).

Stem-based keyword spotting method has shown 14% false alarms and 32% false rejections using the proposed grammar and 47% false alarms and 29% false rejections without the grammar. Keyword spotter based on full word-forms and the grammar has demonstrated worse results in false rejections (41%) than stem-based one. Durations of the first phrases in the dialogues were shorter than at human-human conversation (1.6 s. versus 2.3 s. in average). It was surprisingly that many users started to speak with a polite form "Hello" or "Good morning". It was found also that informative words were not so frequent in the first phrases in comparison with the training dialogue corpus: 66% queries contain a destination city; 12% contain a departure date; 2% phrases contain both a date of departure and a return date; 34% phrases do not contain any useful information.

4 Conclusions

The paper presented the first steps of development of the spoken dialogue system named Let'sFly which is aimed for getting flight schedule and fare information from the Rossiya Russian Airlines on-line database. Two methods of dialogue collection were applied and compared: human-human conversation which emulates HCI and Wizard of

Oz methodology that mixes fully automatic HCI with that attracting a hidden human-operator. The data analysis has shown that the data obtained with two methods are rather different according to many criteria. All the data collected will be used for designing a more advanced interface with spoken input for the system as well as for training the acoustical models in order to create robust speaker-independent spoken dialogue system with unrestricted human-computer interaction via telephone lines. This research is supported by the INTAS innovative project 05-1000007-426 as well as by the Russian Foundation for Basic Research projects 08-08-00128-a and 08-07-90002-Bel-a.

References

1. Cox, R.V., Kamm, C.A., Rabiner, L.R., Schroeter, J., Wilpon, J.G.: Speech and Language Processing for Next-Millennium Communications Services. Proc. of the IEEE 88(8), 1314–1337 (2000)
2. Raux, A., Bohus, D., Langner, B., Black, A., Eskenazi, M.: Doing Research on a Deployed Spoken Dialog System: One Year of Let's Go! Experience. In: 9th International Conference on Spoken Language Processing ICSLP, Pittsburgh, USA (2006)
3. Walker, M.A., Rudnicky, A.I., Prasad, R., Aberdeen, J., Bratt, E.O., Garofolo, J.S., Hastie, H., Le, A.N., Pellom, B., Potamianos, A., Passonneau, R., Roukos, S., Sanders, G.A., Seneff, S., Stallard, D.: DARPA communicator: cross-system results for the 2001 evaluation. In: 7-th International Conference on Spoken Language Processing ICSLP, Denver, USA, pp. 269–272 (2002)
4. Lamel, L., Rosset, S., Gauvain, J.L., Bennacef, S., Garnier-Rizet, M., Prouts, B.: The LIMSI ARISE System. Speech Communication 31(4), 339–354 (2000)
5. Ronzhin, A.L., Karpov, A.A.: Russian Voice Interface. Pattern Recognition and Image Analysis 17(2), 321–336 (2007)
6. Juang, B.H.: Speech recognition in adverse environments. Computer Speech and Language 5, 275–294 (1991)
7. Dahlback, N., Jonsson, A., Ahrenberg, L.: Wizard of Oz Studies - Why and How. Knowledge Based Systems 6(4), 258–266 (1993)
8. Hoffmann, R., Jokisch, O., Lobanov, B., Tsirulnik, L., Shpilewsky, E., Piurkowska, B., Ronzhin, A., Karpov, A.: Slavonic TTS and SST Conversion for Let's Fly Dialogue System. In: 12-th International Conference on Speech and Computer SPECOM 2007, Moscow, Russia, pp. 729–733 (2007)
9. Caminero, J., de la Torre, C., Villarrubia, L., Martin, C., Hernandez, L.: On-line garbage modeling with discriminant analysis for utterance verification. In: 4th International Conference on Spoken Language Processing ICSLP, Philadelphia, USA, pp. 2111–2114 (1996)

Developing a Dialogue System

How to Grant a Customer's Directive?

Mare Koit, Olga Gerassimenko, Riina Kasterpalu, Andriela Rääbis, and Krista Strandson

University of Tartu, J. Liivi 2, 50409 Tartu, Estonia
{mare.koit,olga.gerassimenko,riina.kasterpalu, andriela.raabis,krista.strandson}@ut.ee
http://www.cl.ut.ee

Abstract. Estonian phone calls to travel agencies are analyzed with the further aim to develop a dialogue system. The analysis is based on the Estonian Dialogue Corpus. Customers' initial requests introduce a topic or check the competencies of the official, and they have to be adjusted in a following information-sharing sub-dialogue. Information is given briefly, using short sentences or phrases. A collaborative travel agent offers substituting information or action if s/he is unable to fulfill the customer's request. A dialogue system is being developed which gives travel information in Estonian. Ready-made sentence patterns are used in the current version for granting the users' requests. Our study will help to make the grants more natural.

Keywords: Dialogue system, corpus analysis, request, grant, Estonian.

1 Introduction

There are many dialogue systems (DS) which interact with a user in natural language giving information or helping to perform practical tasks [1,2]. Dialogue corpora have been collected and used for studying human-human spoken communication in order to model it in a DS (Switchboard, Verbmobil, BNC, etc.).

Our goal is to develop a DS that performs the role of an information agent interacting with a user in Estonian. For this reason, we are studying Estonian human-human institutional calls in order to explain how a customer makes his/her requests, how the requests are further processed in collaboration of participants, linguistic means used by participants in conversation, the general structure of institutional calls, etc. Our further aim is to implement a DS which follows norms and rules of human-human communication. In a previous paper [3], customers' requests were analyzed in Estonian spoken human-human dialogues with the aim to find out the linguistic cues that can be used for their automatic recognition. In this paper, we will study the officials' grants to the customers' directive acts.

The paper is organized as follows. In Section 2 we give an overview of our empirical material – the Estonian Dialogue Corpus and the dialogue act typology used for annotation of the corpus. Section 3 is dedicated to the corpus analysis. Different types of customers' directives and agents' reactions are considered. Section 4 discusses some

P. Sojka et al. (Eds.): TSD 2008, LNAI 5246, pp. 593–600, 2008.
© Springer-Verlag Berlin Heidelberg 2008

results of the corpus analysis which will be implemented in an experimental DS. In Section 5 we will make conclusions.

2 Corpus and Dialogue Act Typology

Our current study is based on the Estonian Dialogue Corpus (EDiC) [4]. The corpus contains over 1,100 authentic human-human spoken dialogues, including about 1,000 phone calls. Dialogue acts are annotated in the corpus. A DAMSL-like typology of dialogue acts is used for annotation [5]. Our typology is influenced by the conversation analysis (CA) [6] that focuses on the techniques actually used in social interaction. According to CA, some dialogue acts form adjacency pairs (AP) where producing the first pair act makes the second one relevant (e. g. a request requires a grant). In naturally occurring talk some violations of the norm are possible: the grant may come in overlap with a request end or after a clarifying insertion sequence. Still, the second pair part remains relevant if it is not produced in the next turn. The computer as a dialogue participant must be able to differentiate the first part of an AP (which is expecting a response) from acts that do not require particular responsive actions (non-AP acts).

Acts are divided into nine groups in our typology. The overall number of dialogue acts is 127 (90 AP acts, 37 non-AP acts). Names of dialogue acts consist of two parts separated by a colon: 1) the first two letters give abbreviation of the name of an act-group, e.g. DI – directives, QU – questions, AI – additional information acts. The third letter is used only for AP acts – the first (F) or the second (S) part of an AP act; 2) full name of the act, e.g. DIF: Request, DIS: Giving information, AI: Specifying.

In this paper, we will consider directive APs. Their possible first parts are request, proposal and offer in our typology. Request expresses author's need or intent to get information or to release an action of the partner. Giving information or missing information is a suitable reaction to information request (Ex 1, C – customer, A – travel agent; transcription of CA is used in examples). Agreement or refusal to do an action is a suitable reaction to action request.

(1)
C: *sooviks 'Norrasse sõita.* DIF: REQUEST
I would like to travel to Norway.
A: *ää=ee* DIS: OTHER
uh
C: *'jõu-lude ajal.* AI: SPECIFICATION
in Advent.
A: *ei, kahjuks meie ei=tee sinna sellel=ajal 'reise.* DIS: MISSING INFORMATION
no, unfortunately, we do not organize trips there at that time

Proposal and offer (to do an action) expect an agreement or refusal in response (Ex 2). In the first case action will originate from partner (proposal: *call me later*) and in the second case from author (offer: *I'll call you later*).

(2)
C: *.hhh õhh tundub väga põnev 'reis, nii et ma kindlasti $ 'mõtlen asja üle veel. $* DIF: OFFER
.hhh uh, it seems to be a very exciting trip, so I certainly $ will consider it. $
A: *jah, hästi.* DIS: AGREEMENT
OK, fine

For this paper, 36 calls to travel agencies (total 12,104 tokens) were selected from the EDiC. All the analyzed dialogues are information calls – a customer's goal is to get information, s/he is asking for some preliminary travel information and does not book a certain trip.

3 Analysis

3.1 Customer's Goal

In all the analyzed dialogues, a customer's goal is to get information. S/he gets requested information only in 12 cases out of 36. The typical reason of failure is that shared knowledge is missing – a customer does not have previous knowledge about the fields of activity of the agency (e.g. s/he asks how to travel to England but the agency offers only trips inside Estonia). In three dialogues, there are no more vacancies available for the requested group-trip.

Calling a travel agency, customers use one dialogue act for setting up the initial goal in 22 dialogues, and two acts (utterances) in one turn in 5 cases (mostly a request together with specifying information). In the remaining 9 cases, a response of the travel agent (continuer) follows to the customer's request which signals that the agent is waiting for adjustment of the initial request. After that, the customer asks a certain question or adds specifying information to his/her request (Ex 1). This can be considered as a collaborative behavior because information comes to the partner step by step which makes it easier to understand.

3.2 Directive Adjacency Pairs

Let us consider here the directive APs where the first part is produced by a customer and the second one by a travel agent. There are 39 first and 44 second parts of such APs in our analyzed sub-corpus. However, some of the first parts remain without the second part and some of them get more than one second part.

Request and offer are the actual first parts of directive APs in our analyzed sub-corpus (32 and 5 cases, respectively). There are no APs that would begin with a customer's proposal. In two cases, a customer's dialogue act is annotated as DIF: Other (Ex 3).

(3)
C: *.hh ee ma 'sooviksin natuke informatsiooni selle 'Lapimaa reisi kohta.* DIF: REQUEST
.hh uh I would like to get some information about the trip to Lapland.

A: *jaa?* VR: NEUTRAL CONTINUER
yes
C: *[.hh et=õ]* DIF: OTHER
[.hh so=uh]
A: *[millise]=mis 'kuupäeva kohta.* QUF: WH-QUESTION + ADF: ADJUSTING THE CONDITIONS OF ANSWER
[what]=what's the date.

In five cases, a request or an offer is simultaneously the second part of a previous AP (agreement as a request, limited agreement as an offer, Ex 4).

(4)
A: *.h aga: aga ma saan teile anda 'Hermann reiside telefoninumbri.=* DIF: OFFER
.h bu:t I can give you the phone number of Herman travel.=
C: *=jaa* DIS: AGREEMENT + DIF: REQUEST
=yes
C: *see oleks 'hea.* DIS: AGREEMENT + DIF: REQUEST
it would be nice.

Request. Request is the customers' most frequent first part of a directive AP. 15 requests out of 28 introduce a topic in the beginning of a call (Ex 5). (The remaining calls begin with questions that will not be considered here.) The remaining requests arise during a dialogue. In four cases, a customer's agreement with an agent's offer is simultaneously a request (Ex 4).

(5)
C: *ma küsiksin nende odavate lennuki'piletite kohta.* DIF: REQUEST
I'd (like to) ask about the low-price plane tickets.
A: *jaa?* VR: NEUTRAL CONTINUER
yes?
C: *kas need tuleb osta kohe 'välja.* QUF: CLOSED YES-NO QUESTION
whether they have to be bought off immediately.
(1.0)
A: *reeglina 'küll.* QUS: YES
as a rule, yes.

It is significant that a customer often starts a call with a general request which serves as a topic introduction. The actual request will be detailed in a following information-sharing sub-dialogue. This is why a customer's initial request as the first part of an AP remains without the second part. Still, s/he gets the answer to his/her question(s).

The typical response to a request is giving information. Phone numbers or web addresses are given in several short utterances which are separated with a customer's continuer (Ex 6). Behavior of both parties is reciprocally cooperative. By dividing information into small pieces a travel agent is collaborative as it would be difficult for a customer to record and to remember long sequences. Continuers between information giving utterances are signs of successfully received information.

(6)
A: *.hh see=on kolm null 'üks,* DIS: GIVING INFORMATION
.hh it=is three zero one,
C: *jaa?* VR: NEUTRAL CONTINUER
yes?
A: *'neli neli neli.* DIS: GIVING INFORMATION
four four four.

Sometimes a postponement precedes giving information (Ex 7).

(7)
A: *Austria?* DIS: POSTPONEMENT
ma kirjutan omale üles 'autoga? DIS: POSTPONEMENT
Austria? I'll write it down for myself, by car?
mhmh VR: NEUTRAL ACKNOWLEDGEMENT
uh huh
.hh eahh noh üsõnaga me müüme Austrias: 'enamvähem 'kõiki suusa'kuurorte. DIS: GIVING INFORMATION
uh, in short, well we sell more or less all the ski resorts in Austria.

If the requested information is missing in a data base the customer is informed immediately. Still, a collaborative agent mostly proposes substituting information (another agency, another trip, Ex 8).

(8)
C: *.hh olen uvitatud reisidest Skandi'naaviamaadesse=h.* DIF: REQUEST
.hh I'm interested in trips to Scandinavian countries=h.
A: *e='jaah?* VR: NEUTRAL ACKNOWLEDGEMENT
uh=yes?
.hh et asi on 'selles=et=e vähemalt 'praegu: me=ei 'korralda neid. DIS: MISSING INFORMATION
.hh the fact is that at least at the moment we do not organize them.
(0.5)
C: *ahah* VR: NEUTRAL CHANGE OF STATE
I see
(.)
A: *et=sis=meil='on 'üks reis 'Lapimaale kui=te=oleksite 'huvitatud.* DIF: OFFER
but=we=have a trip to Lapland if=you=are interested in.

There are three cases where the agent refuses to fulfill the customer's request. However, all the refusals are softened with the agent's promise to fulfill the request later.

Offer. All the offers are made by customers in the end of a dialogue. An agreement follows immediately (Ex 2).

Distance between the Two Parts of an AP. As it was said before, sometimes the first part of an AP remains without the second part. Mostly a customer's request used to introduce a topic at the beginning of a call does not get a grant. There are 15 such dialogues in our sub-corpus. However, in a following information-sharing sub-dialogue the request will be detailed by adjusting questions, and answers will be given to the questions.

In other 14 cases, the second part immediately follows the first part. If the first part is a request (9 cases) then the answer comes immediately if information is missing, or the answering is postponed in order to search a data base, or the request is directed to another, a more competent travel agent. If the first part is an offer (5 cases) then agreement follows immediately.

In the remaining cases, an information-sharing sub-dialogue follows a request. During a sub-dialogue, the names, ages and number of travellers, etc. will be detailed, i.e. information will be collected that is needed in order to grant the request.

How the Second Parts of Directive APs Are Expressed in Estonian? Our analysis has shown that preferable reactions (giving information, agreement to do an action) are typically shorter than non-preferable (missing information, refusal). Giving information is mostly expressed by a simple sentence. When giving a phone number or web address, a sequence of noun phrases is used (Ex 6). Such cases form in total 90%.

Some fixed expressions are used for postponement (*üks hetk palun* /a moment please, *üks minut* /a minute). Sometimes, postponement consists of a comment of the agent's current action (*vaatame* / let's look, *proovin ühendada* / I'll try to connect, *kirjutan üles* / I'll write it down) or of repetition of the partner's previous turn (* *oli kahekümn=neljandast.* * / it was since twenty fourth).

Agreement is expressed by positive response particle *jah* (yes), *hea* (good), *hästi* (fine, OK) or a verb phrase in imperative which acclaims the action offered by the customer (*mõelge järgi.* /(please) think it over, *teeme nii.* / let's do so).

Missing information is formulated as a negative sentence (Ex 1) or as a declarative sentence which offers substituting information.

Refusal is formulated as a declarative sentence remising to another agent (*kohe ma suunan Teid Moonikale* / just a moment, I'll direct you to Moonika), or as an interrogative sentence which is simultaneously an adjusting question (*saab kuskile 'tagasi helistada.* / is it possible to call back anywhere).

4 Discussion

A simple DS is being implemented by Margus Treumuth which interacts with a user in Estonian and gives information about fligths leaving from the Tallinn Airport [3]. The user has to put in his/her question as an Estonian text, the DS's output is optional (written or spoken). A finite state automaton is used for dialogue management. Recognition of user's dialogue acts is based on cue words and expressions found in utterances. Morphological analyzer of Estonian and text-to-speech synthesizer are integrated into DS. Currently, the DS is able to answer only questions about departing times of fligths. Speech recognition will be integrated in future.

The following lessons are learnt from the corpus analysis.

- An institutional call has certain typical structure. It starts with a conventional beginning where participants greet each other and introduce themselves. Then the main part follows where a customer makes a request and will get information. Information-sharing sub-dialogues are used to adjust the request. Similarly, information-sharing sub-dialogues possibly follow the grant if the received information needs to be adjusted. An offer can be made by a participant and acknowledged by the partner if the request was not fulfilled. The dialogue ends with a conventional part where the participants thank each other and take leave themselves (sometimes, not always).

- A collaborative agent gives substituting information or proposes substituting action if s/he is unable to grant the customer's request.
- It is typical for calls to travel agencies that a customer makes a general request which only introduces a topic, and his/her actual need has to be explained in the following information-sharing sub-dialogue.
- When giving information, short sentences are preferred. Pauses have to be made in order to enable the partner to record (write down) the information. The partner's feedback signalizes that information is successfully received and s/he is ready to continue.
- Some limited sets of expressions are used by people for greeting, leave-taking, agreeing, refusing etc. It has to be taken into account when modelling a travel agent.

The typical structure of a call to a travel agency can be represented by the following formal grammar (cf. [3]).

```
dialogue ::= conventional-beginning (request
        (information-sharing)* grant
        (information-sharing)*)+
        (offer   agreement)* conventional-ending
grant ::=  giving-information | missing-information
       | [limited-]agreement | postponement
       | refusal | other
```

Still, postponement is needed for a human agent to seek a data base and is not needed for a DS that will find data quickly.

5 Conclusion and Future Work

Estonian calls to travel agencies were analysed with the further aim to develop a DS. A customer's initial request sets up a goal which will be achieved in collaboration with an agent. We investigated directive adjacency pairs where the first part sets an expectation of a certain second part. The possible first parts are request, proposal, and offer. The second parts are giving information, missing information, agreement, and refusal. We concentrated on the APs initiated by customers who call a travel agency, and especially on the second parts of APs produced by travel agents. The most frequent APs are request – giving information and offer – agreement. Information is given by travel agents in an effective way, using short sentences or phrases. It is typical for calls to travel agencies that the first request has to be detailed in an information-sharing sub-dialogue. If information is missing then the second part of an AP comes immediately.

Our study was done on the Estonian Dialogue Corpus where dialogue acts are annotated.

A simple DS is implemented which gives information about flights leaving from the Tallinn Airport. Our future work concerns implementation of results of the current analysis in the DS.

Acknowledgment. This work is supported by the Estonian Science Foundation (grant No 7503) and the Estonian Ministry of Research and Education (National Programme of Language Technology).

References

1. McTear, M.F.: Spoken Dialogue Technology. Toward the Conversational User Interface. Springer, Berlin (2004)
2. Minker, W., Bennacef, S.: Speech and Human-Machine Dialog. Kluwer Academic Publishers, Boston (2004)
3. Koit, M., Valdisoo, M., Gerassimenko, O., Hennoste, T., Kasterpalu, R., Rääbis, A., Strandson, K.: Processing of Requests in Estonian Institutional Dialogues: Corpus Analysis. In: Matousek, V., Mautner, P. (eds.) Text, Speech and Dialogue, Proceedings, pp. 621–628. Springer, Berlin (2006)
4. Estonian Dialogue Corpus, `http://lepo.it.da.ut.ee/~koit/Dialoog/EDiC.html`
5. Gerassimenko, O., Hennoste, T., Koit, M., Rääbis, A., Strandson, K., Valdisoo, M., Vutt, E.: Annotated Dialogue Corpus as a Language Resource: An Experience of Building the Estonian Dialogue Corpus. In: The First Baltic Conference: Human Language Technologies, Riga, Latvia. The Baltic Perspective, pp. 150–155 (2004)
6. Schegloff, E.: The routine as achievement. Human Studies 9, 111–152 (1986)

Dialogue-Based Processing of Graphics and Graphical Ontologies

Ivan Kopeček and Radek Ošlejšek

Faculty of Informatics, Masaryk University
Botanická 68a, CZ-602 00 Brno, Czech Republic
kopecek@fi.muni.cz, oslejsek@fi.muni.cz

Abstract. Dialogue-based processing of graphics forms a framework for enabling visually impaired people to access computer graphics. The presented approach integrates current technologies of dialogue systems, computer graphics and ontologies for the purpose of overcoming the obstacles and problems that the development of the system used for the dialogue processing of graphics entails. In this paper we focus on problems related to creating graphical ontologies and exploiting them in the annotation of graphical objects as well as in the development of dialogue strategies. We also present an illustrative example.

1 Introduction

Thanks to the possibility of using natural language, the dialogue-based processing of graphics targets the applications that are convenient for inexperienced users or the handicapped, i.e. those who cannot use any graphical interface. This approach can make graphics (at least partly) accessible especially to visually impaired people. Up to now, the accessibility of graphics for the blind and visually impaired has been mostly connected with the use of tactile devices (see e.g. [1,2,3,4]) or use of sound to help explore graphical objects (see e.g. [5,6]). Recently, the SVG graphical format has been utilized to permit vector graphics [7,8,9] to access object-oriented graphical information. We also use SVG format for processing raster graphics, exploiting its ability to represent text as text strings allowing the graphics to be fully searched for content and its support for a rich set of geometrical primitives. These capabilities provide the user with desired pieces of information in combinations of verbal and non-verbal form.

In what follows we show that supporting the graphics processing by graphical ontologies helps the user to efficiently annotate the graphics, prevents chaos in terminology and keeps the semantic hierarchy consistent. Simultaneously, graphical ontologies combine the information about the picture with the information about the objects that appear in the picture and their description in the real world, which makes the annotator's work easier by automatically supplying them with the attributes of described objects. Moreover, graphical ontologies can be used for enhancing the dialogue strategies in the process of retrieving the information about the graphics as well as in the process of generating graphical objects.

P. Sojka et al. (Eds.): TSD 2008, LNAI 5246, pp. 601–608, 2008.
© Springer-Verlag Berlin Heidelberg 2008

2 GATE Dialogue System

GATE (= Graphics Accessible To Everyone) is an ongoing project aiming to achieve the following goals. First, development of utilities deployed for easy picture annotation. Second, provision of blind users with support for exploring ("viewing") pictures. And finally, development of a system utilized for generating images by means of dialogue and enabling the blind to create some limited form of computer graphics. The project is currently being developed at the the Faculty of Informatics, Masaryk University Brno, Czech Republic. As regards the *GATE* system, let us briefly describe its basic modules and principles.

The *ANNOTATOR* module supports image navigation and is closely connected to the graphical ontology. Its basic task is to inform the user about the graphical content in a non-visual way. The *GATE* system provides two basic ways doing so – verbally and by means of sound. There are two basic tools supporting this communication: *What-Where Language* and *Recursive Navigation Grid*. *What-Where Language*, *WWL*, is a simple fragment of English. Each sentence of this language has the form of WHAT is WHERE or WHERE is WHAT. It enables the user to ask simple questions about the objects in the scene and their position (e.g. "Where is the tower?", "What is in the middle?", "What is in the background?").

Recursive Navigation Grid [10,11], *RNG*, represents the navigation backbone of the system, dividing the picture space into nine identical rectangular sectors analogously to the layout of numerical keys 1-9 of the numerical keyboard. Each sector is subdivided in the same way recursively. This enables the user to investigate a point or region with demanded precision and to carry out "zooming".

Verbal Information Module, *VIM*, controls the verbal part of the dialogue including the *WWL* communication. Possible misunderstandings in the communication are solved by *VIM* by invoking dialogue repairing strategies.

Two basic strategies of retrieving information are supported, being represented by *GUIDE* and *EXPLORER* modules. The task of *GUIDE* is to provide verbal information, exploiting both the pieces of information obtained by tagging the picture and the pieces of information gained directly from the picture format. The module provides *EXPLORER* with relevant information and cooperates with *VIM* and *RNG*.

The communication of *EXPLORER* is not primarily verbal, but analogue. It is controlled by means of mouse, digitizer, or numerical keyboard. The output sound information is also primarily non-verbal. The *RNG* module is exploited for navigation. The pieces of information that are related to the place, object or rectangle pointed to are both verbal and non-verbal. This allows the user to perform a quick dynamic exploration of the non-annotated details of the picture. The information about the explored color is provided by a procedure which is based on a sound representation of colors. The basic idea assumes the sound information to be a combination of special sounds assigned to the primary colors of a suitable color model (based on the RGB color model). The boundary of the picture and its vicinity is signalized by a special sound.

3 Graphical Ontology

In general, the ontology represents a knowledge base, which is composed of a domain and individuals. Graphical ontology describes global properties representing important

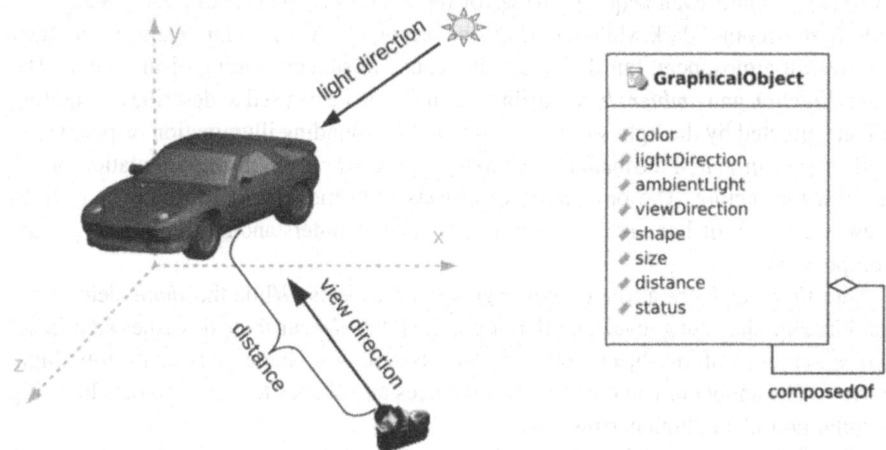

Fig. 1. Visual properties in 3D (left) and the top-level class of the graphical ontology (right)

visual aspects of graphical objects and contributing to the semantic description of graphical information.

Basic classification divides graphical objects into classes according to the graphical type, e.g. photography, vector graphics, cartoon, map, graph, etc. In this paper, we concentrate on 3D color photography, i.e. a color photography of three dimensional scene. The reason is that it represents a widespread type of graphical objects, which is still inaccessible to the blind and, at the same time, complex enough for the presented approach to be easily applicable to most of the other graphical types.

In the graphical ontology which is developed within the *GATE* project, the individuals are created during the annotation process through classification concrete graphical objects in the ontology domain. The domain is specified by *OWL Lite* [12] and predefines visual attributes, which are available in all classes of the ontology. These global attributes cover visual characteristics that affect visual perception and inform users about visual aspects of the scene in a non-visual way, e.g. verbally. Most of these attributes are similar to the features describing the energy distribution in 3D scenes (Figure 1). However, they are adjusted for 2D pictures as well as for the approximate form required by ontologies.

The *color* attribute enables the annotator to assign a dominant color to the object in a picture. Many types of objects have some typical dominant color, e.g. snow is white, grass is green, etc. The ontology can restrict the domain of the *color* attribute within specific subclasses. For example, for *Animal* class there are no color restrictions while the *Swan* class restricts the color to white. This knowledge is fixed in the ontology as a fact and the annotator has neither a reason for nor way of changing it. On the contrary, the *Cat* class restricts the color domain to all colors except the shades of blue, green, red, etc. In this case, the annotator is restricted to the given extent of shades and therefore unable to describe a blue cat.

The *color* attribute determines the basic color in standard light conditions. However, the real colors in the picture can vary depending on specific light conditions. For example, a white swan acquires shades of red at sunset, one half of a face exposed to side-light becomes dark while the other one bright, etc. These color changes can affect the overall atmosphere, but they may also cause misunderstanding of the scene. The *light direction* and *ambient light* attributes can therefore be used to describe the lighting effects affected by the light source position and surrounding illumination respectively.

The *view direction* attribute is used to determine the approximate orientation of objects in the picture. The orientation of objects with meaningful attributes, e.g. front view, rear view or left view, is often important for understanding the overall picture composition.

The *shape* and *size* attributes cover geometric aspects. While the *shape* determines a noticeable shape, e.g. oval, angular or crooked, the *size* attribute describes significant size divergence of an object from the obvious state, e.g. huge, big, small, tiny, high, narrow. The annotator can combine several sizes and shapes to define objects like "big angular house" or "high narrow tower".

The *distance* attribute describes a relative distance between the object and the viewer using terms such as foreground, background, middle part. This attribute indicates the approximate depth of information in the scene. The lower-right image in Figure 2 shows an example of the annotated scene with three depth regions marked.

The attributes discussed so far represent features that are common to all the semantic categories in the graphical ontology. However, a concrete category (class) can have its own additional visual characteristics distinguishing its individuals (graphical objects). For example, a swan looks different in different situations, e.g. when swimming, flying or walking, and in these situations it may also find itself in different environments (water, sky, ground), buildings differ in their type, etc. The *status* attribute is used to describe these individual properties.

A graphical object can be composed by means of manipulating graphical content and creating more complex objects and scenes by combining more primitive sub-objects. For example, a human being can be composed of a head, body, legs and arms. The head can be composed of hair, eyes, nose, ears and mouth. This composition structure forms an information tree, into which the objects are organized. The upper classes in the tree represent the more general objects while the components represent their details. The graphical ontology therefore includes the *composedOf* relationship, which can be exploited for the information filtering as well as efficient navigation during the picture annotation and exploration, as discussed in the following section.

4 The Role of Graphical Ontologies in Dialogue Processing of Graphics

Within the *GATE* project, graphical ontologies are closely connected with the dialogue communication of the following three primary tasks – picture annotation, picture investigation and picture generation.

The extent to which annotated SVG images are accessible to the blind user depends both on the quality of the annotation and the supporting software. In relation to picture annotation, the graphical ontology:

- provides support for multilinguality of the annotation,
- unifies the annotation processes and prevents chaos in terminology,
- keeps the semantic hierarchy of the annotation process consistent with other annotated graphical objects.

Further, it navigates the annotator and makes their work easier by allowing the graphics recognition algorithms to be exploited in all the situations where it is possible. For instance, if graphics recognition process recognizes a graphical object as a swimming swan, it annotates the object as "swan" and fills in the status attribute field with "swimming". The human annotator only checks the correctness. On the other hand, even if graphics recognition process fails to recognize the object, the "status" attribute always displays its values, allowing the annotator to define it just by clicking on the right option.

At the beginning of the picture investigation process, the user is provided with short basic information about the picture. The technical information, like the number of pixels and colors, compression type, etc., are obtained automatically from the data stored in the format of the picture. Then the user starts the exploration by using the picture information tree structure, into which the data describing the picture are organized, navigation by keywords, querying by means of *WWL*, and the use of the *EXPLORER* module.

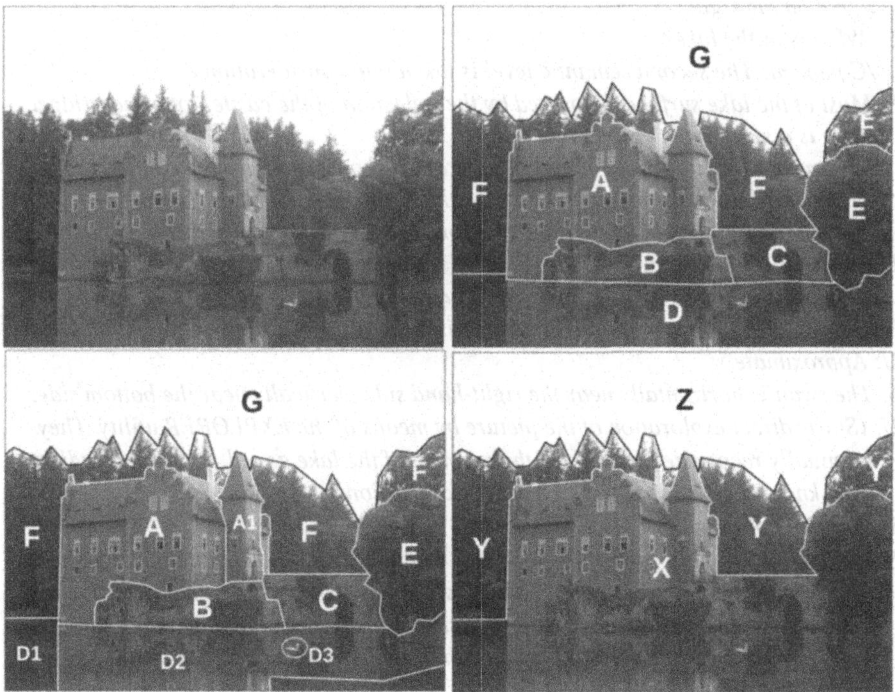

Fig. 2. The original photo (upper-left), the tagged first (upper-right) and second (lower-left) semantic levels, and the tagged foreground (X), central part (Y) and background (Z)

In the following example a blind user explores a color photo (photo of Cervena Lhota castle in south Bohemia – see Figure 2). The characters in the pictures stand for: A: the castle, B: a rocky island, C: an entrance bridge, D: a surface of the lake, E: a tree, F: a forest in the background, G: a clear sky, A1: an entrance tower of the castle, D1: a surface of the lake, D2: the castle, rock and bridge reflected on the lake surface, D3: a swan on the lake.

The following dialogue between a blind user U and the system S was recorded during the WOZ simulation. It illustrates the exploration strategies and the role of graphical ontology (originally, the conversation was recorded in Czech):

U: (Enters the system by clicking on the picture.)
S: This is a color photo of Cervena Lhota castle in south Bohemia. For more information, choose Description, Technical information, use keywords or What-Where language.
U: Technical information.
S: This is a color photograph in full color. The size is tree megapixels.
...
U: What is in the foreground?
/Comment: The user ask a typical WWL question. The X area and the first semantic level are taken into consideration./
S: In the foreground, there is a lake, a rocky island, a bridge, a castle and a tree on the right.
U: What is in the lake?
/Comment: The second semantic level is taken into consideration./
S: Most of the lake surface is covered by the reflection of the castle, rock and bridge. There is a swan swimming in the lake.
/Comment: This information, based on graphical ontology, is inferred from the status attribute ''swimming'' of the graphical object denoted as ''swan'', the location of the object, and the relation lake --> water --> swim./
U: Where is the swan?
/Comment: Another type of WWL question./
S: Do you wish to get an approximate localization or use Recursive Navigation Grid?
U: Approximate.
S: The swan is horizontally near the right-hand side, vertically near the bottom side.
U: (Starts direct exploration of the picture by means of the EXPLORER utility. They eventually recognize the color of the segment of the lake as red, which contradicts their knowledge and ask the system for explanation.)
U: Explain color.
S: The red color of this part of the lake may be caused by a reflection of the red castle.
/Comment: The ''Explain color'' query is an important tool that facilitates the blind user picture recognition, especially by those that are born blind and therefore have no experience with the visual phenomena like mirroring, shadow, perspective, etc./

Graphical ontology also plays an important role in describing an object that we have in mind, i.e. the first stage of the dialogue communication that takes place during the dialogue generation of graphics. The crucial problem is that of determining the point where the system considers the graphical object to be sufficiently described. It is also here that the graphical ontology supports the well-balanced decision based on the information about both the relevant attributes and the typical attribute values. For instance, knowing that the typical value for the "color" attribute of the graphical object "swan" is "white", the system does not demand the information about the color.

5 Conclusions and Future Work

Supporting dialogue systems by means of graphical ontologies seems to be a promising approach in a wide range of applications targeting processing graphics. It is capable of helping the blind and visually impaired to explore, to a certain extent, the photographs and other types of graphics. This renders making artistic paintings accessible to these people an interesting and challenging task. Within our project, we are collaborating with the Department of Graphical Design to create a Web gallery for blind students. Making maps accessible to the blind represents another attractive challenge. The same applies to the use of the annotated SVG format in electronic textbooks.

In order to enhance the efficiency of creating graphical ontologies, we want to focus on the issue of exploiting the methods of automatic recognition of graphical objects and detection of shape in connection with a similarity search performed in large graphical databases. Another objective is testing the developed modules for the purpose of optimizing dialogue strategies.

Acknowledgment. The authors are grateful to the students and staff of the Support Centre for Students with Special Needs of Masaryk University for their advice, support and collaboration. This work has been supported by the Czech Science Foundation, Czech Republic, under Contract No. GA 201/07/0881. and Contract No. GA 201/06/P247.

References

1. Edman, P.K.: Tactile Graphics. American Foundation for the Blind, New York (1992)
2. Kaczmarek, K.: Electrotactile display for computer graphics to blind. Research Report 5-R01-EY10019-08. University of Wisconsin (2004)
3. Kurze, M.: Tdraw: a computer-based tactile drawing tool for blind people. In: Proceedings of the second annual ACM conference on Assistive technologies, Vancouver, pp. 131–138 (1996)
4. Satoshi, I.: Computer graphics for the blind. ACM SIGCAPH Newsletter, pp. 16–21 (1996)
5. Daunys, G., Lauruska, V.: Maps sonification system using digitiser for visually impaired children. In: Miesenberger, K., Klaus, J., Zagler, W., Karshmer, A.I. (eds.) ICCHP 2006. LNCS, vol. 4061, pp. 12–15. Springer, Heidelberg (2006)
6. Matta, S., Rudolph, H., Kumar, D.K.: Auditory eyes: Representing visual information in sound and tactile cues. In: 13th European Signal Processing Conference, Antalya (2005)
7. Bulatov, V., Gardner, J.: Making graphics accessible. In: 3rd Annual Conference on Scalable Vector Graphics, Tokyo (2004)
8. Fredj, Z.B., Duce, D.: Grassml: Accessible smart schematic diagrams for all. In: Theory and Practice of Computer Graphics, pp. 49–55. IEEE Computer Society Press, Los Alamitos (2003)

9. Mathis, R.M.: Constraint scalable vector graphics, accessibility and the semantic web. In: SoutheastCon Proceedings, pp. 588–593. IEEE Computer Society Press, Los Alamitos (2005)
10. Kopeček, I., Ošlejšek, R.: The blind and creating computer graphics. In: Second IASTED International Conference on Computational Intelligence, Zurich, pp. 343–348 (2006)
11. Kamel, H.M., Landay, J.A.: Sketching images eyes-free: a grid-based dynamic drawing tool for the blind. In: Fifth international ACM conf. on Assistive technologies, pp. 33–40 (2002)
12. Lacy, L.W.: Owl: Representing Information Using the Web Ontology Language. Trafford Publishing (2005)

An Issue-Based Approach to Information Search Modelling: Analysis of a Human Dialog Corpus

Alain Loisel, Jean-Philippe Kotowicz, and Nathalie Chaignaud

INSA de Rouen – LITIS EA 4108, 76131 Mont-Saint-Aignan, France
{alain.loisel,jean-philippe.kotowicz,
nathalie.chaignaud}@insa-rouen.fr

Abstract. We aim at improving the health information search engine CISMeF, by including a conversational agent that interacts with the user in natural language. To study the cognitive processes involved during information search, a bottom-up methodology was adopted. An experiment has been set up to obtain human dialogs related to such searches. In this article, the emphasis lays on the analysis of these human dialogs.

1 Introduction

Most information search systems are based on mere keyword matching and do not try to understand the user's intention. To adapt itself to the user, the system has to help the user to describe what he is looking for, by considering the human discursive strategies, the human variability of language and the search context.

Our aim is to design a conversational agent that interacts with the user in natural language. Each *move* in the human-computer dialog is translated into *items* that improve the whole query. The system is based on cooperative strategies: it evaluates the current state of the search; and suggests assistance and choices. It determines with the user the keywords to propose to the database. It is able to test several queries in parallel. Moreover, if necessary, it expands the user's intention by suggesting new keywords. Our target application is CISMeF, an index of French-language health resources, which assists health professionals and patients in their search for information available on the Internet. The terms used come from the MeSH (medical subject headings): meta-term (medical specialties), keyword, subheading (symptoms, treatments, etc.) and resource type. We have set up an experimentation to obtain human spoken dialogs between an expert and users dealing with medical information search. These dialogs have been analyzed to extract their discursive structure and their linguistic features.

2 Dialog Systems

2.1 The Questions Under Discussion Theory

The QUD model [1] is a conventional theory of dialog, which is based on formal semantics to solve ellipsis. Comparing with work on *questions* like [2], originality of QUD comes from the use of a structured conversational board, which is led by the semantics

P. Sojka et al. (Eds.): TSD 2008, LNAI 5246, pp. 609–616, 2008.
© Springer-Verlag Berlin Heidelberg 2008

of questions in context. QUD aims at analysing precisely the properties of issues and at specifying how the questions under discussion complete the conversational board: what can be asserted or asked at the moment? Which short answer can be made to the discussed questions? By reusing concepts of the dialog games theory [3], Ginzburg comprehends dialog as moves in games.

2.2 Issue-Based Dialog and GoDIS

Based on QUD, GoDIS [4] is a computational model, which uses a simplified version of the semantics of the questions. There exist three types of questions: complete questions ($?\lambda\{\}.P$), partial questions ($?\lambda x.P(x)$) or questions with a choice list ($?set(P_1(x), P_2(y), \ldots, P_n(z))$). Issues are integrated into a structure of dialog plan comprising static plans, which are sequences of abstract actions. These plans are called *actionPlan*. Plans in GoDIS represent both the task and dialog. GoDIS uses the *IS (Information State)* (like a conversational board). This IS is complex and is structured into two parts:

Private:	*Shared*:
`Agenda`: action queue	`Com`: set of propositions
`Plan`: planConstruct stack	`Qud`: question stack
`Bel`: set of propositions	`Issues`: question stack
`NIM`: dialog act queue	`Actions`: action stack
	`Previous moves`: dialog act queue
	`Last utterance`: (`enunciator`: participant) (`moves`: set of dialog acts)

The private part represents the mental state of the agent and the shared part defines the conversational board, which memorize shared information between the two interlocutors. Updating rules and selection rules can change the IS.

The use of semantic relations enables to determine the possible answers to a question. According to the initial dialog act, the possible reactive dialog acts are given. The main asset of GoDIS is that the semantic relations introduce the notion of *accommodation*, which brings flexibility to the user. Accommodation effects enable to discuss questions in a non-trivial way. Like QUD, GoDIS uses accommodation effects and it can handle several fields of the IS.

The *issue* can be considered as a specialization of the dialog games [5]. However, as pointed out in [6], intentional relations and informational relations can be specified between dialog games. But GoDIS defines only the dominance relation between issues and it is not recursive. The different issues follows one another in an implicit way (relation of implicit sequence). Since there is no precedence constraint, GoDIS allows to accommodate any dialog plan at anytime. This model can only manage simple dialogs, where each plan is a simple sequence of actions or questions. Moreover, it does not accept dialogs, which digress from the task (for example, to give an explanation to the user), except for the *common ground* establishment (information shared during the dialog). This deficiency is solved in [7] by the introduction of the notion of dialog strategy. This strategy is chosen according to the best adjustment of the moves to the goals:

- The *cooperative strategy* adjusts the goal to the user's intention. In GoDIS, This is a dynamic modification of planConstruct, keeping the goal unchanged.
- The *constructive strategy* temporarily leaves the current goal for a new one. In GoDIS, that corresponds to leaving the stack of planConstruct and choosing a new goal dynamically. Digressions are thus introduced and the dialog manager does not follow the sequence of actions linked with the task.

3 Corpus Collection

As mentioned above, we have set up an experimentation to obtain human spoken dialogs between an expert and users dealing with medical information search. We have hand-analyzed these dialogs and we have shown that the concepts of GoDIS enable to describe a natural interaction between the subjects for the task.

The users were voluntary members of the LITIS laboratory (secretary, Ph.D. students, researchers and teachers), who wanted to obtain answers about medical inquiries. The experts were two members of our project, trained to CISMeF and its terminology. The experimentation took place as follows: one expert and one user were facing a computer using the advanced search interface of the system and recording all the queries along with their answers in a log. The expert was in charge of conducting the search by conversing with the user and verbalizing each action. The experimentation ended when relevant documents were given to the user or when it seemed that no answer existed in the system. A textual corpus has been constituted from the transcription of the 21 spoken dialogs.

4 Corpus Analysis

4.1 Linking Sub-dialogs with the Task

We can draw from the analysis of the corpus, an ideal global structure, that breaks them down into several sub-dialogs linked with the task (see Figure 1). A dialog always begins with an *opening sub-dialog*, following a *choice sub-dialog*, which generally leads to a main *search sub-dialog*. The latter comprises several search sequences, each corresponding to the solving of an inquiry, and finally the dialog finishes with a *closing sub-dialog*. The *search sub-dialog* is twofold: a *query-building sub-dialog*, which fills in a current query, and a *testing sub-dialog* to test this query. During the *search sub-dialog*, the current query is always improved. The *query-building sub-dialog* in turn is composed of two sub-dialogs: a *free-building sub-dialog*, which aims at understanding the type of inquiry (comings and goings between the user and the expert) and a *query sub-dialog* where the queries are built in cooperation with the user. The queries are processed in a *query-execution sub-dialog* and the documents are given to the user in an *evaluation sub-dialog*. This list is described in a *document-describing sub-dialog* and one or more documents can be opened and inspected precisely in a *document-selection sub-dialog*. At anytime, this sub-dialog can be interrupted by precision inquiries. The user inquiry can be also a question about a definition (*definition sub-dialog*) or about an explanation of the system (*explanation sub-dialog*).

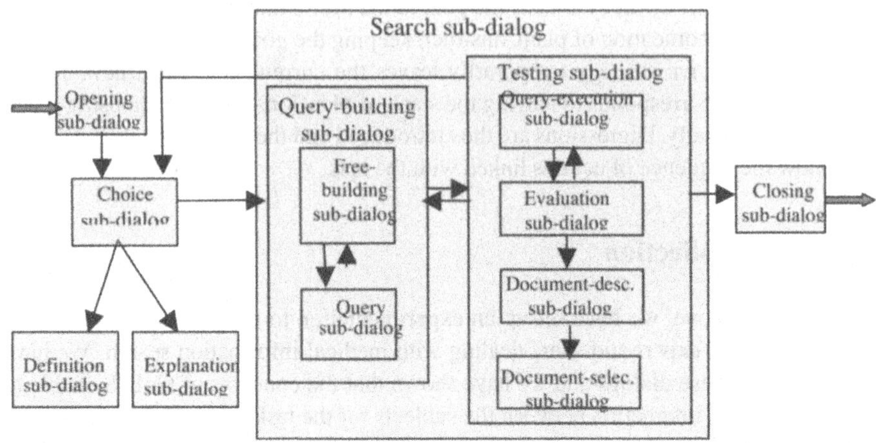

Fig. 1. Links between sub-dialogs

These sub-dialogs can be interpreted in terms of Discourse Segment Purpose (DSP) [8]. In this theory, discourse structure is composed of three components: the structure of the sequence of utterances (the *linguistic structure*), a structure of purposes (the *intentional structure* which captures the discourse-relevant purposes), and the state of focus of attention (called the *attentional state*). There exist two types of relations between discourse segments: *dominance* and *satisfaction-precedence*.

In Figure 1, embedded sub-dialogs show the dominance relations between them whereas arrows represent natural transitions deducted from the task, which do not violate the satisfaction-precedence relations. These sub-dialogs are opportunistic. This structure of "ideal dialog" is always interrupted by one of the interlocutors. Some sub-dialogs are optional and all of them can be left explicitly or implicitly. Incident dialogs can be opened to update the common ground. The use of classical planning is not appropriate to model human dialogs.

After this task-oriented analysis, a list of dialog acts has been built according to linguistic features found in the corpus. This taxonomy comes from [9] and has been adapted to our corpus [10].

4.2 Questions about the Task

Dialog acts are classified according to the different phases of the task. To each sub-dialog corresponds a set of issues, which make the task progress. The corpus has been analysed in terms of issues with a formal representation. An issue can be parsed into several contiguous (or not) sequences of dialog acts.

For example, the following dialog with its translation, corresponding to a *query-building sub-dialog*, shows a proposition from the user of a new keyword:

- *user: I would like to propose "arthroscopy"*
  expert: Yes, "arthroscopy"
  `Suggest(addKeyWord(arthroscopy.mc))`
  `icm:sem*pos(addKeyWord(arthroscopy.mc))`

The evaluation of ?interesting(SetD) on a given document set SetD answers the implicit question: "Are the results interesting ?"

- *expert: that is not satisfactory*
  `Inform(¬interesting(SetD))`
- *user: this type of documents is interesting*
  `Inform(interesting(SetD))`

4.3 Instantiation of Issues

Several instantiations of issues are in general possible. In fact, a question immediately followed by a answer is only one example of interaction among all of them. To handle this, we have defined consistency relations between issues.

Examples drawn from our corpus can be interpreted in several sequences of dialog acts for just one question-answer.

- *expert: do you want to know the evolution, the diagnosis or the treatments?*
  user: the diagnosis

At the beginning, the question is under discussion in the fields Qud and Issues. At the receipt of the answer, the question is deleted from Qud, but not from Issues [11,4] and thus remains under discussion after being answered, allowing the interlocutor to rectify his answer with *reaccommodation* effects. Moreover, if the interlocutor does not answer the question, the question is also deleted from Qud. Some questions accept several answers, thus the first one is chosen but the question is still under discussion.

- *expert: Can you be a little more precise?*
```
Com:
Qud: ? λx.precisions(x)
Issues: ? λx.precisions(x)
LastMove: RequestInfo(? λx.precisions(x))
```
- *user: What do we have to do in order to donate organs?*
  user: Do we have to do medical examination?

```
Com: precisions("process","organ donor"),precisions("examination")
Qud:
Issues: ? λx.precisions(x)
LastMove: Answer("examination")
```

Two answers to this question are acceptable. The use of connectors like "also" and "more" specifies that a new answer is expected. The following example shows that the interlocutor answers himself his own question.

- *expert: does an other subheading interest us? There is prevention and control. We can also try "diagnosis of knuckle problems"*
```
RequestInfo(? λM.addSubheading(M))
Answer(addSubheading(prevention.qu))
Suggest(addSubheading(diagnosis.qu))
```

GoDIS considers only one acceptable and resolving answer for each question. For us, the two answers are acceptable and resolving.

- *user: perhaps, it is "osteo"*
  expert: Yes, osteopathy, thanks! oh no, it does not exist in the index.
  user: rheumatologist, rheumatology
  expert: Ok, we will try rheumatology. I run the query.

In this example, the first answer is resolving but the user can propose a second answer. It is a correction to the first answer, which has to be cancelled. The reaccommodation mechanism of GoDIS can be used to model this dialog.

In our corpus, there are numerous dialog acts of assertions, information and suggestions. These utterances have been associated to issues, even when they are not enunciated in an interrogative form. We also observed cooperative strategies when the interlocutor answers his own question.

- *expert: How to translate it? Maybe therapeutic.*

The user is free to contest this answer and to propose some others. The expert can answer to an unasked question hereby seeking approval from the user.

4.4 Links between Issues

We analysed how to link the different issues into a sub-dialog. In our corpus, we look for the relation between issues by differentiating interactionnal subordinate relations (the second issue is an explanation, a correction or a reformulation to place the first one in the common ground) from interactionnal coordinate relations (greetings, thanks, etc.) [6]. Moreover, there are intentionnal relations between the issues and the task. For example, the query-building action action(queryBuilding)) dominates the question about keywords adding (?λM.addKeyWord(M)). And there is a satisfaction-precedence relation between action(queryBuilding) and action(queryExecution).

4.5 Transitions between the Sub-dialogs

Once the sub-dialogs and their relations with the questions are identified, the sequences of the sub-dialogs have to be studied. The dialog game theory [3] describes explicitly the transitions between two games.

- *expert: Do you have another inquiry?*
  user: No, that is all!

The transition to a new sub-dialog is often expressed by the use of connectors like "so", "firstly", etc., corresponding to the first issue of the sub-dialog.

- *expert: ok! I type firstly the initial inquiry. So, that is a problem with food?*
  user: yes
  expert: ok! So, we try thematic access

We can consider that the issue of the last utterance noted InformIntent() is added in Qud by fact accommodation.

As already said, the actions linked to the task are in dominance relation or in satisfaction-precedence relation. The same holds for the sub-dialogs, which explains that transitions between sub-dialogs can be implicit. Sub-dialog accommodation (called *plan accommodation* in GoDIS) enables to avoid some sequences.

- *expert: I begin the search with parasomnia. Ok! Nothing!*
  user: So, I think we are finished

The user proposes to end the dialog after a failure: transition from the document-describing sub-dialog to the closing sub-dialog. The user's utterance is accommodated because satisfaction-precedence relations are satisfied.

Finally, constructive strategies advise assistance and explanation sequences in the dialog corresponding to embedded dialog games. When the collaborative sequence is ended, the previous context is recovered.

- *expert: Hence insomnia. That is not a keyword. That is strange!*
  ```
  Suggest(?addKeyWord(Insomnia familial.mc))
  Inform(¬terminology(insomnia familial.mc))
  Com:(¬terminology(insomnia.mc))
  Qud: ?Cause(¬terminology(insomnia.mc))
  ```
- *user: Perhaps, it is included in sleeping disorders*
  ```
  Suggest(Cause(¬terminology(insomnia.mc),(included(insomnia.mc,
  "trouble sleeping"))))
  Com:(¬terminology(insomnia.mc),Cause(¬terminology(insomnia.mc),
  (included(insomnia.mc,"trouble sleeping"))))
  ```

5 Discussion

Applications using GoDIS are relatively simple (e.g. the control of a multimedia reader). We think that GoDIS can be used for more complex applications with a task model and an interaction model, provided new notions are introduced:

- Several answers for a single question: in GoDIS, once a solving answer is given, the question is deleted from Qud. Whereas in our application, to know if the answer is solving or not, we have to test it in CISMeF and analyse the results. If they are relevant, the question stays in Qud.
- The satisfaction-precedence relations: accommodation in GoDIS brings flexibility in the dialog. However, the dialog must remind consistent. Satisfaction-precedence relations have to be considered, which forbid issue accommodation if it is inconsistent with the task.
- The cooperative and constructive strategies: in the document search, the finding of relevant information enables either to improve the current plan (collaborative strategies) or to leave it (constructive strategies). Thus, the dialog manager (not the user) accommodates new issues.
- Dialog games: accommodation effects proposed in [3] combine consistency between issues and issue semantics.

6 Conclusion

We adopted an experimental approach to design a human-computer dialog system for medical information search. We collected and analyzed a rich textual corpus from which the building of a common ground and accommodation effects on the user have been emphasized. Dialogs can be divided into sub-dialogs directly linked to the task. We are designing a software agent able to converse with a user and find information. This agent is composed of:

- The language model, which performs a lexical and syntactical analysis, a pragmatic analysis and a semantic analysis (identification of CISMeF terms).
- The dialog model, which comprises the dialog manager and the sentence generator based on incomplete sentences.
- The task model, with the CISMeF interface. It includes also a query builder from the recognized terms and a query interpreter to refine the queries.
- This conversational agent is under development in Prolog [10].

References

1. Ginzburg, J.: Interrogatives: Questions, facts and dialogue. The Handbook of Contemporary Semantic Theory 5, 359–423 (1996)
2. Groenendijik, J., Stokhof, M.: Studies on the semantics of questions and the pragmatics of answers. Ph.D. thesis, University of Amsterdam (1984)
3. Maudet, N., Moore, D.J.: Dialogue games as dialogue models for interacting with, and via, computers. Journal of Informal Logic 21, 219–243 (2001)
4. Larsson, S.: Issue-based Dialogue Management. Ph.D. thesis, Goteborg U (2002)
5. Hulstijn, J.: Dialogue Models for Inquiry and Transaction. Ph.D. thesis, Twente U (2000)
6. Beveridge, M., Milward, D.: Ontologies and the structure of dialogue. In: 8th Workshop on the Semantics and Pragmatics of Dialogue, Barcelona, 69–76 (2000)
7. Caelen, J.: Strategies of dialogue. In: Speech Technology and Human-Computer Dialogue Conference, Bucarest, pp. 27–42 (2003)
8. Grosz, B., Sidner, C.L.: Attention, intentions, and the structure of discourse. Computational Linguistics 12, 175–204 (1986)
9. Weisser, M.: Spaacy: A tool for annotating dialogue. International Journal of Corpus Linguistics 8 (2003)
10. Loisel, A., Chaignaud, N., Kotowicz, J.P.: Modeling human interaction to design a human-computer dialog system. In: International Conference on Enterprise Information Systems, Barcelona (to appear, 2008)
11. Cooper, R., Larsson, S.: Accommodation and reaccommodation in dialogue. Presuppositions and Discourse (2003)

Two-Level Fusion to Improve Emotion Classification in Spoken Dialogue Systems*

Ramón López-Cózar[1], Zoraida Callejas[1], Martin Kroul[2], Jan Nouza[2], and Jan Silovský[2]

[1] Dept. of Languages and Computer Systems, University of Granada, Spain
{rlopezc,zoraida}@ugr.es
[2] Institute of Information Technology and Electronics
Technical University of Liberec, Czech Republic
{martin.kroul,jan.nouza,jan.silovsky}@tul.cz

Abstract. This paper proposes a technique to enhance emotion classification in spoken dialogue systems by means of two fusion modules. The first combines emotion predictions generated by a set of classifiers that deal with different kinds of information about each sentence uttered by the user. To do this, the module employs several fusion methods that produce other predictions about the emotional state of the user. The predictions are the input to the second fusion module, where they are combined to deduce the user's emotional state. Experiments have been carried out considering two emotion categories ('Non-negative' and 'Negative') and classifiers that deal with prosodic, acoustic, lexical and dialogue acts information. The results show that the first fusion module significantly increases the classification rates of a baseline and the classifiers working separately, as has been observed previously in the literature. The novelty of the technique is the inclusion of the second fusion module, which enhances classification rate by 2.25% absolute.

1 Introduction

Research on automatic classification of user emotion aims to design methods to make computers interact more naturally with humans, by making them take into account not only the information provided by the users, but also the way in which it is provided. Given the importance of emotions in human communication, many attempts have been made to include some kind of emotion recognition in computers. Many studies have considered acoustic/prosodic features only, for example pitch, loudness, energy contours or speaking rate [1]. Several authors have used a variety of methods based on pattern classification to carry out this task [2,3], for example, maximum likelihood Bayes classification, linear discriminant classification, kernel regression, k-nearest neighbourhood, neural networks, support vector machines or decision trees. Some studies have also made use of hybrid or ensemble methods [4].

* This work has been funded by the Spanish project HADA TIN2007-64718, and the grant no. 1QS108040569 of the Agency of Sciences of the Czech Republic.

P. Sojka et al. (Eds.): TSD 2008, LNAI 5246, pp. 617–624, 2008.
© Springer-Verlag Berlin Heidelberg 2008

2 The Proposed Technique

The technique presented in this paper considers that a set of classifiers $\Omega = C_1, C_2, \ldots, C_m$ receives as input feature vectors f related to each user utterance. Employing a classification algorithm, each classifier generates one emotion prediction. This prediction is a vector of pairs $(h_i, p_i), i = 1 \ldots S$, where h_i is an emotion category, p_i is the probability of the utterance belonging to h_i according to the classification algorithm, and S is the number of emotion categories considered. The predictions are the input to the Fusion-0 module, which employs n fusion methods called F_{0i}, $i = 1 \ldots n$, to generate other predictions. These predictions are vectors of pairs $(h_{0j,k}, p_{0j,k})$, $j = 1 \ldots n$, $k = 1 \ldots S$, where $h_{0j,k}$ is an emotion category, and $p_{0j,k}$ is the probability of the utterance belonging to $h_{0j,k}$ according to the fusion method F_{0j}. The Fusion-1 module receives the predictions provided by Fusion-0 and generates the pair (h_F, p_F), where h_F is the emotion category with highest probability, p_F. This emotion category is determined employing a fusion method called F_F, and represents the user's emotional state deduced by the technique. The best combination of fusion methods to be used in Fusion-0 $(F_{01}, F_{02}, \ldots, F_{0j}, 1 \leq j \leq n)$ and the best fusion method to be used in Fusion-1 (F_F) are experimentally determined.

3 Individual Classifiers

In this section we describe the 4 classifiers used in the experiments to deal with prosodic, acoustic, lexical and dialogue acts information to deduce the user's emotional state.

3.1 Prosodic Classifier

We used global statistics of pitch and energy and features derived from the duration of voiced/unvoiced segments to create one n-dimensional feature vector per utterance, which was the input to the prosodic classifier. After carrying out experiments to find the appropriate feature set, we decided to use the following 11 features: pitch mean, minimum and maximum, pitch derivatives mean, mean and variance of absolute values of pitch derivatives, energy maximum, mean of absolute value of energy derivatives, correlation of pitch and energy derivatives, average length of voiced segments, and duration of longest monotonous segment. The classifier employed gender-dependent Gaussian Mixture Models (GMMs) to represent emotion categories. The likelihood for the n-dimensional feature vector (x) representing the input utterance, given an emotion category λ, was defined as a weighted linear combination of Q unimodal Gaussian densities $P_l(x)$:

$$P(x|\lambda) = \sum_{l=1}^{Q} w_l P_l(x) \tag{1}$$

The density function $P_l(x)$ was defined as:

$$P_l(x) = \frac{1}{\sqrt{2\pi^n det \sum_l}} exp(-\frac{1}{2}(x - \mu_l)' \sum_l^{-1} (x - \mu l) \tag{2}$$

where the μ_l's are mean vectors and the $\sum_l$'s covariance matrices. The emotion category deduced by the classifier, h, was decided according to the following expression:

$$h = \arg\max_S P(x|\lambda^S) \tag{3}$$

where λ^S represents the models of the 2 emotion categories considered ('Non-negative' and 'Negative'), and the max function is computed employing the EM (Expectation-Maximization) algorithm. To compute the probabilities p_i for the emotion prediction of the classifier we used the following expression:

$$p_i = \frac{\beta_i}{\sum_{k=1}^{2} \beta_k} \tag{4}$$

where β_i is the log-likelihood of h_i, and the β_k's are the log-likelihoods of the 2 emotion categories considered.

3.2 Acoustic Classifier

The emotion patterns of the input utterances were modelled by gender-dependent GMMs, as with the prosodic classifier, but each input utterance was represented employing a sequence of feature vectors $x = x_1, \cdots, x_T$ instead of one n-dimensional vector. We assumed mutual independence of the feature vectors in x, and computed the log-likelihood for an emotion category λ as follows:

$$p(x|\lambda) = \sum_{t=1}^{T} log P(x_t|\lambda) \tag{5}$$

The emotion category deduced by the classifier, h, was decided employing Equation 4, whereas Equation 5 was used to compute the probabilities for the prediction, i.e. for the vector of pairs (h_i, p_i).

3.3 Lexical Classifier

We observed in previous experiments that words that were acoustically similar to others or that were considerably affected by the users' accents were more likely to be misrecognised, which usually caused negative emotional states of the users. Hence, the goal of the lexical classifier was to deduce the emotion category for each input utterance from the emotional information associated with the words in the utterance. To do this we employed the concept of "emotional salience" [5]. The emotional salience of a word for a given emotion category can be defined as the mutual information between the word and the emotion category. Let W be a sentence (speech recognition result) comprised of a sequence of n words: $W = w_1 w_2 \ldots w_n$, and E a set of emotion categories, a weighted linear combination of Q unimodal Gaussian densities $E = \{e_1, e_2, \ldots, e_S\}$. The emotional salience of the word w_i for an emotion category e_j is defined as:

$$salience(w_i, e_j) = P(e_j|w_i) \cdot mutual\_information(w_i, e_j) \tag{6}$$

To carry out the emotion classification at the sentence level, we considered that each word in a sentence is independent of the rest, and mapped the sentence W to one emotion categories in E. To do this, we computed an activation value a_k for each emotion category as follows:

$$a_k = \sum_{m=1}^{n} I_m w_{mk} + w_k \tag{7}$$

where $k = 1 \cdots S$, n is the number of words in W, I_m represents an indicator that has the value 1 if w_k is a salient word for the emotion category (i.e. $salience(w_i, e_j) \neq 0$) and the value 0 otherwise; w_{mk} is the connection weight between the word and the emotion category, and w_k represents bias. The connection weight was defined as: $w_{mk} = mutual\_information(w_m, e_k)$, whereas the bias was computed as: $w_k = log P(e_k)$. Finally, the deduced emotion category, h, was the one with highest activation value a_k:

$$h = \arg\max_k (a_k) \tag{8}$$

To compute the probabilities p_i for the emotion prediction, we used the following expression:

$$p_i = \frac{a_i}{\sum_{j=1}^{2} a_j} \tag{9}$$

where a_i represents the activation value of h_i, and the a_j's are the activation values of the 2 emotion categories considered.

3.4 Dialogue Acts Classifier

By inspecting the dialogue corpus employed for these experiments (to be discussed in Section 4.1) we observed that users of a spoken dialogue system got tired or angry if the system generated the same prompt reiteratively (i.e. repeated the same dialogue act) to try to get a particular data item. Hence, the goal of the dialogue acts classifier was to predict these negative emotional states by detecting successive repetitions of the same system's prompt types, e.g. prompts to get the telephone number. The emotion category of a user's dialogue turn, E_n, was the one which maximises the posterior probability given a sequence of the most recent system prompts:

$$E_n = \arg\max_k P(E_k | DA_{n-(2L-1)}, \ldots, DA_{n-7}, DA_{n-5}, DA_{n-3}, DA_{n-1}) \tag{10}$$

where the prompt sequence was represented by a sequence of dialogue acts (DA_i's) and L is the length of the sequence, i.e. the number of system's dialogue turns in the sequence. Note that if $L = 1$ then the decision about E_n depends only on the previous system prompt.

4 Experiments

The goal of the experiments was to test the proposed technique by employing the two emotion categories 'Non-negative' and 'Negative', four classifiers: prosodic classifier

(C1), acoustic classifier (C2), lexical classifier (C3) and dialogue acts classifier (C4), and three fusion methods: Average of Probabilities (AP), Multiplication of Probabilities (MP) [6] and Unweighted Vote (UV) [4].

The UV method combined the predictions by counting the number of classifiers (if used in Fusion-0) or fusion methods (if used in Fusion-1) that considered an emotion category h_i as the most likely for the input utterance. The probability p_i for h_i using this classifier was computed as follows:

$$P(h_i|X,Y) = \frac{Vh_i}{\sum_{j=1}^{2} Vh_j} \tag{11}$$

where X and Y denote the 2 emotion categories, Vh_i is the number of votes for h_i, and the Vh_j's are the number of votes for the 2 emotion categories. The experiments were carried out by taking as input a set of labelled dialogues in a test corpus. Each dialogue was processed to locate within it, from the beginning to the end: i) each prompt of an experimental dialogue system called Saplen [7], ii) the voice samples file that contains the user's response to the prompt, and iii) the result provided by the system's speech recogniser (sentence in text format). The prompt type was used to create a sequence of dialogue acts of length L, which was the input to the dialogue acts classifier. The voice samples file was the input to the prosodic and acoustic classifiers, and the speech recognition result was the input to the lexical classifier. In this way, the technique was tested just as if it had been used during the system's interaction with the users. The performance of each classifier was stored in a log file that was used for posterior performance analysis.

4.1 Speech Database

The speech database used in the experiments was constructed from a corpus of 440 telephone-based dialogues between University students and the Saplen system. Each dialogue was stored in a log file in text format that included each system prompt, the type of prompt, the name of the voice samples file (utterance) that stored the user response to the prompt, and the speech recognition result for the utterance. The orthographic transcriptions of the user responses were included manually in the log files after the collection of the corpus. The dialogue corpus contains 7,923 utterances, 50.3% of which were recorded by male users, and the remaining by female users. To train and test the classifiers we divided the dialogue corpus into two disjunct corpora: one for training (5,938 utterances corresponding to 75% of the dialogues) and the other for testing (1,985 utterances corresponding to the 25% remaining dialogues). The division was made in such a way that both sets contained utterances representative of the 18 different utterance types in the dialogue corpus: product orders, telephone numbers, postal codes, addresses, etc. The utterances were annotated by 4 labellers (2 male and 2 female). The order of the utterances was randomly chosen to avoid influencing the labellers by the situation in the dialogues, thus minimising the effect of discourse context. The labellers assigned one tag to each utterance, either 'Neutral', 'Tired' or 'Angry', according to the perceived emotional state of the user. To carry out the experiments we assigned one tag to each utterance according to the majority opinion of the labellers. We found

Table 1. Performance of Fusion-0 (results in %)

Fusion method	Classifiers	Classification rate
	Aco+Pro	84.15
	Lex+Pro	85.04
	DA+Pro	90.49
AP	Aco+Lex+Pro	89.20
	Aco+DA+Pro	90.24
	DA+Lex+Pro	90.02
	Aco+DA+Lex+Pro	**90.49**
	Average	88.66
	Aco+Pro	84.15
	Lex+Pro	85.16
	DA+Pro	91.49
MP	Aco+Lex+Pro	89.17
	Aco+DA+Pro	91.33
	DA+Lex+Pro	90.06
	Aco+DA+Lex+Pro	**92.23**
	Average	89.08
	Aco+Pro	88.64
	Lex+Pro	86.40
	DA+Pro	88.20
UV	Aco+Lex+Pro	88.76
	Aco+DA+Pro	88.91
	DA+Lex+Pro	88.47
	Aco+DA+Lex+Pro	**89.04**
	Average	88.35

that 81% of the utterances were annotated with the tag 'Neutral', 9.5% with the tag 'Tired' and 9.4% with the tag 'Angry'. This shows that the experimental database was clearly unbalanced in terms of emotion categories.

4.2 Experimental Results

We wish to employ the proposed technique to enable our experimental spoken dialogue system in transferring the call to a human operator if the user starts interacting in a negative emotional state because of system malfunction. Taking into account this purpose, making a distinction between utterances tagged as 'Tired' or 'Angry' is irrelevant for us, as both emotional states suggest that the call should be transferred. Hence, in the experiments we collapsed these two emotion categories into one generic category called 'Negative', whereas the 'Neutral' category was treated as 'Non-negative'. Given that 81% of the utterances in the speech database were annotated as 'Neutral', for comparison purposes we considered a baseline that classified each input utterance as belonging to this emotion category.

Table 1 sets out the average results obtained for Fusion-0. As can be observed, MP was the best fusion method, with average classification rate of 89.08%. The best classification rate (92.23%) was obtained employing the 4 classifiers, in which case Fusion-0

Table 2. Performance of Fusion-1 (results in %)

Fusion methods in Fusion-0	Fusion method in Fusion 1		
	AP	MP	UV
AP+MP	93.68	**94.48**	93.53
AP+UV	93.20	93.23	93.20
MP+UV	93.44	94.38	93.20
AP+MP+UV	93.23	94.36	93.17
Average	93.40	94.11	93.28

outperformed the baseline (81%) by 11.23%. Analysis of the performance log file of Fusion-0 shows that when AP was used the classification rates were 95.75% for 'Non-negative' and 85.37% for 'Negative', whereas when MP was used the rates were 95.93% and 88.91% respectively. This shows that MP outperformed AP in predicting both emotion categories.

Table 2 shows the results obtained when Fusion-1 was used to combine the predictions of Fusion-0. The 3 fusion methods were tested in Fusion-1, with Fusion-0 employing several combinations of these methods: AP+MP, AP+UV, MP+UV, AP+MP+UV. In all cases Fusion-0 used the four classifiers as this is the configuration that provides the highest classification accuracy according to Table 1. The best performance of Fusion-1 was attained employing MP (94.48%). This fact corroborates a conclusion of [6], namely that when independent feature spaces are available, the classifiers' outcomes should be multiplied to attain optimal results. Hence, in accordance with our experimental results, this conclusion is not only applicable to classifiers but also to the independent predictions generated by fusion methods.

5 Conclusions and Future Work

Comparison of the MP rates in Table 1 and Table 2 shows that the improvement attained by employing Fusion-1 is 2.25% absolute, from 92.23% to 94.48%. This benefit is achieved because Fusion-1 combines emotion predictions obtained by employing different fusion methods, and thus it benefits from their respective advantages. AP was good at obtaining less error-sensitive classifications, and was especially useful in case of very highly correlated feature spaces (prosodic and acoustic). MP was good at attaining maximum gain from the independent feature spaces (speech signal, speech recognition errors, and dialogue context), and was especially useful when the classifiers made small errors. UV was good as well at enhancing the classification rate, but not as much as AP and MP. The sources we have dealt with in the experiments (prosodic, acoustic, lexical, and dialogue acts) are those most commonly employed in previous studies. Future work will include testing the proposed technique by employing additional information sources not considered in this study, such as speaking style, subject and problem identification, and non-verbal cues. We also plan to investigate the use of weights in the fusion processes.

References

1. Bänziger, T., Scherer, K.R.: The role of intonation in emotional expressions. Speech communication 46, 252–267 (2005)
2. Ai, H., Litman, D., Forbes-Riley, K., Rotaru, M., Tetreault, J., Purandare, A.: Using systems and user performance features to improve emotion detection in spoken tutoring dialogs. In: Proc. of Interspeech 2006-ICSLP, Pittsburgh, USA, pp. 797–800 (2006)
3. Devillers, L., Scherer, K.: Real-life emotions detection with lexical and paralinguistic cues on human-human call center dialogs. In: Proc. of Interspeech 2006-ICSLP, Pittsburgh, USA, pp. 801–804 (2006)
4. Morrison, D., Wang, R., Silva, L.C.D.: Ensemble methods for spoken emotion recognition in call-centers. Speech communication 49, 98–112 (2007)
5. Lee, C.M., Narayanan, S.S.: Toward detecting emotions in spoken dialogs. IEEE Transactions on Speech and Audio Processing 13, 293–303 (2005)
6. Tax, D., Breukelen, M.V., Duin, R., Kittler, J.: Combining multiple classifiers by averaging or multiplying. Pattern Recognition 33, 1475–1485 (2001)
7. López-Cózar, R., Callejas, Z.: Combining language models in the input interface of a spoken dialogue system. Computer Speech and Language 20, 420–440 (2005)

Automatic Semantic Annotation of Polish Dialogue Corpus

Agnieszka Mykowiecka, Małgorzata Marciniak, and Katarzyna Głowińska

Institute of Computer Science, Polish Academy of Sciences,
J. K. Ordona 21, 01-237 Warsaw, Poland
{agn,mm}@ipipan.waw.pl, k.glowinska@gazeta.pl

Abstract. In the paper we present a method of automatic annotation of transliterated spontaneous human-human dialogues on the level of domain attributes. It has been used for the preparation of an annotated corpus of dialogs within LUNA project. We describe the domain ontology, process of manual creation of rules, annotation schema and evaluation.

1 Introduction

The main goal of the LUNA[1] (spoken **L**anguage **UN**derstanding in multilingu**A**l communication systems) project is to create an effective, multilingual spoken language understanding (SLU) module. As preparation of such a module requires domain dependent knowledge, collection of adequately annotated training corpora for all selected languages (French, Italian and Polish) is a significant part of the project. This task requires solving two main problems: selecting appropriate tagsets; establishing reliable and quick annotation procedures. In this paper we will address both issues at the semantic (attribute) level. While morphological tagsets are widely elaborated and standardized, semantic tags are much less universal and have to be designed for a particular domain. Annotating procedures always have to be adjusted to the task in hand and resources available. It should also be taken into account that transcribed speech is different than written text – speakers turns rarely contain full sentences and many fragments are grammatically incorrect. In the paper we present our approach to automatic semantic annotation of Polish data.

A corpus of Polish dialogues[2] was collected from the Warsaw City Transportation information center where people can get information on: routes between given points in the city, time schedules of public transport, expected trip duration, stops, tickets and reduced-fares. 500 dialogues were chosen for the corpus which consists of directories containing an acoustic signal of the dialogue, its transliteration and a set of XML files with its annotation on different levels.

Recorded dialogues were manually annotated on an acoustic level with help of Transcriber [1]. Then, for each dialogue a file representing information about speaker turns was automatically created. The next processing step concerned a morphosyntactic annotation. For each dialog a separate file describes information on POS of each word,

[1] This work is supported by LUNA (IST 033549) project.

[2] For detailed description of the collection procedure see [6].

P. Sojka et al. (Eds.): TSD 2008, LNAI 5246, pp. 625–632, 2008.
© Springer-Verlag Berlin Heidelberg 2008

its base form, and morphological characteristic. Morphological analysis was done automatically by AMOR [8], but disambiguation of forms was generally done manually. On the next annotation level, dialogues were segmented into elementary syntactic chunks by a program developed within the project.

In the paper we describe the next stage of corpora preparation: an automatic annotation of Polish dialogs on the level of domain attributes. The principles of annotations on this level were similar to the semantic annotation of Media corpus [2]. But as the chosen domain is complex a probability of changes within a domain model was very high. In such a case, relying on manual annotation did not seem to be adequate – apart from standard problems with achieving homogeneity between annotators it would require repetition of work already done. To avoid these problems, we decided to use a rule-based Information Extraction method, postulated in [5]. But in the case of our approach, we decided not to use a semi-automatic method of annotation but a fully automatic one instead. The set of rules was incrementally created and after positive validation used to annotate the corpus. Human annotators were used to check the results and to perform the final correction of annotation.

2 Domain Description

The starting point for any semantic annotation should be a clear definition of a chosen domain (see [3]). In this area there are still too few resources which can be reused, so nearly always a new specialized domain model has to be elaborated. Nowadays, a widely accepted method of representing domain semantics, is through defining its ontology. Although universality of this approach is in practice much more difficult then it is sometimes claimed, see [7], it is still the best existing solution. For the purpose of describing semantics of dialogs concerning Warsaw public transportation, we developed a new small OWL-DL ontology.

- different transportation lines (buses, trams, local trains and metro),
- town topology in the aspects needed for traveling (stops, important buildings, street names);
- time descriptions, timetables;
- prices and information connected with reduced-fares;
- types of requests.

Figure 1 shows the top of a tree of classes. Apart from the typology, an ontology allows for describing class properties, but this part is not represented here. The ontology was created on the bases of the selected part of data (the training set described in Section 5). This ontology was translated into attribute-value schema consistent with LUNA guidelines.

The domain model contains 134 concepts actually found within the data. Most concepts have only a few possible values but a few of them have many, eg. bus number, time description, street name etc.

TRANSPORT
    TRANSP_MEAN (bus, tram, metro, skm)
    TRANSP_LINE (bus_line, tram_line, metro_line, skm_line)
    TIMETIBLE
    LOCALIZATION
        LOCALIZATION_ABS: AREA, BLD, CROSSROADS, STREET, TOWN, TD)
        LOCALIZATION_REL
    PLACE: PLACE_IN_TOWN, TOWN
    ROUTE_FEATURES
        ROUTE_DIRECTION (GoStraight, Turn (Left, Right), GoBack)
        ROUTE_ELEM
    ROUTE
    PASSAGE
    ROAD_DISTURBANCE (DETOUR, TRAFFIC_JAM, ...)
    FARE (FREE, NORMAL, DISCOUNT (48, 50))
    DISCOUNT_TITLE (ID_CARD, STUDENT_CARD, ...)
    TRIP
    CONNECTION_QUALITY_FEATURES (Direct, Best, Longest)
    PERSON_FEATURE (AGE, EDUC)

TIME
    TIME_POINT_DESC:
        ABSOLUTE: AT_HOUR, AFTER_HOUR, AROUND_HOUR, BEFORE_HOUR
        REL: IN_X_MINUTES
    TIME_SPAN: TIME_SPAN_MIN
    REPETITIVE_MOMENTS: TIME_PERIODICS (every 5 min), EVERY_HOUR (5 after)

DIALOG_INTERACTIONS
    QUESTION: Q_WHAT_TIME, Q_WHAT, Q_CONF, ...
    REACTION: POSITIVE, NEGATIVE

Fig. 1. Fragment of the class topology

3 Annotation Rules

On the basis of the training set of dialogs a set of rules which identify concepts was defined. As some concepts are built on the basis of others (eg. recognition of source street depends on recognition of a street name), rules were divided into three sets. Rules in the second set can use results of rules from the first one and the third set can use the results of both the first and second. A rule schema is given in Figure 2. Each rule consists of three parts: attribute name, attribute value and recognition pattern (separated by a tab character). Comments are preceded by the % character.
The interpretation of the particular fields is as follows:

- AttrName is an attribute name which is a string containing letters and underscores beginning with '$' – it is a concept name and at the same time a variable which can be used by rules from a higher group;

- AttrValue is a value assigned to a recognized concept – it can be a sequence of strings or values recognized by a search pattern where $n refers to the *n*th element of the pattern (including optional ones);
- RealizationSchema is a sequence of conditions imposed on subsequent elements of a pattern; they may address a word form, its lemma, morphological annotation or chunk to which it belongs (i.e. it can use values from files prepared at the previous stages of annotation); a pattern can also contain a concept name already recognized.

AttrName	AttrValue	#RealizationSchema
$Name	StringExpression	#ConditionOnWord1+ConditionOnWord2+...

Fig. 2. Rule schema

Figure 3 shows three rules. The first one describes a question, the second and third describe places. The first rule recognizes Q_WHAT question on the basis of two words *jak* 'how' and a finite form of a verb *jechać* 'to go'. The value of the concept is a string *Route*. The second rule describes location being a street recognized by the second pattern element. The third one, differing only in case value of the second pattern element, recognizes a street as being a goal. (In both cases $2 represents a value of a $STREET variable.)

```
$Q_WHAT "Route" #CC(lemma=jak)+VV(lemma=jechać)    %'how does it run'
$LOCATION_STR  $2  #PreP(lemma=na)+ $STREET(morph=loc)
               %na Wilanowskiej 'at Wilanowska'
$GOAL_STR  $2 #PreP(lemma=na)+$STREET(morph=acc)
               %na Wilanowską 'to Wilanowska'
```

Fig. 3. Examples of semantic rules

The first set of rules contains 345 rules recognizing 40 concepts, the second one contains 346 rules for 41 concepts and the third one 259 rules for 85 concepts (in this calculation some concepts are counted two or even three times). The greater number of concepts in the third set is connected with the existence of many localization and time representing attributes.

In a rule, exactly one order of elements is defined, and assumes correctness of morphological values of words. But as it is spoken language we encounter many irregularities.

- Change of order:
  jakie tam autobusy są? jakie są autobusy tam? ('what buses there are there?'). In such case we can define as many rules as there are possible permutations of phrase elements.
- Inflectional errors:
  w kierunku Marki ('in the direction of Marki') – *Marki* is in nominative case but should be in genitive. To allow for the recognizing understandable but syntactically

improper phrases, we put minimal inflectional restrictions into the rules. In the example we only require that a word after 'kierunku' belongs to the proper semantic class (TOWN or TOWN_DISTRICT).

- Interjections:
  - *o której odjeżdża 510 autobus z przystanku nie wiem Świętokrzyska?* ('what time does the bus number 510 depart from the bus stop – I don't know – Świętokrzyska?')
  - *przystanek w stronę tego Żoliborza* ('bus stop to – you know – Żoliborz').

  In such cases concepts are not well recognized (in the first case we have street Świętokrzyska instead of a stop of that name, in the second we have separate concepts STOP and TOWN_DISTRICT). We plan to test the solution allowing a small group of words to occur in any place within recognized sequences.
- Elliptical phrases:
  dwanaście po ('twelve past [e.g., seven]') is a description of time. The interpretation of the phrase (after known hour or after every hour) is usually clear to both actors of a dialog but this information is not easy to infer. In this case we leave the specification incomplete.
- Ambiguity of words or word sequences:
  tram numbers and minutes (from 1 to 36), bus numbers (400, 500) and distance to walk to reach the target (400, 500 meters), bus number (100) and expressions like *płacić 100%* ('to pay 100%'), *na 100 procent* ('for sure'). If the context is well defined (tram number 25, 400 meters to walk), we disambiguate the phrase. In other cases we use the concept tag NUMB without taking a decision what this number represents.

The process of rules (and domain model) creation was incremental. First, on the basis of 20 dialogs from the training set an initial set of concepts and rules was defined. It was tested on the same set of dialogs and several randomly selected additional dialogs from the training set. The process was repeated three times and took an expert about 140 hours. During this time all dialogs from the training set were checked at least in fragments. The rules were not fully adjusted to the text – if something could not be plausibly inferred from the near context, it was left not fully disambiguated, e.g. the word *between* can introduce both time and location items. If this could not be stated, the LOC_TIME_REL concept was assigned. Such ambiguities can be solved on the higher semantic (i.e. frame) level.

4 Annotation Process

Semantic annotation consists in assigning attributes' names to phrases from dialogs. In our approach this process is rule based and it is performed in several stages. First, concepts are assigned on the basis of special vocabularies which serve two purposes:

- to assign concepts to words which are important for the domain and should be recognized even without any context (e.g *przystanek* 'stop'),
- to recognize proper names and assign them their type and lemmatized form.

In the next step, the semantic annotation process uses rules described in Section 3.

4.1 Proper Names Recognition

Dialogues concerning town transportation system contain a lot of proper names. Although their recognition is simplified because in our data they are the only words which begin with capital letters. The following problems have to be solved:

- names lemmatization: *Grójeckiej, Grójecką* are forms of *Grójecka* (street);
- type recognition: for proper dialogue understanding it is important whether a speaker talks about a town district *Wola* or a building *Fort Wola*;
- names separation: as in our corpus there are no commas, we have to find other ways of recognizing that *Józefów Otwock* is a sequence of two names.

As the current main goal was to prepare a well annotated corpus, we prepared a lexicon which helped us to solve all three problems mentioned. A list of most common names was prepared on the basis of different sources, and their elements were added to the morphological dictionary used. On its basis, a list of dictionary pairs was prepared: sequences of base forms of words and real base forms. So it is possible to get from e.g. *Jeziorkiem Czerniakowskim* via *Jeziorko Czerniakowski* the form *Jeziorko Czerniakowskie*. The list also includes type names to which all names belong. The list contains names of streets, buildings, towns near Warsaw and town areas. In future we plan to test pattern matching and learning techniques on the prepared data.

4.2 Rule-Based Concept Annotation

After proper names recognition the defined sets of rules are applied. The procedure follows the standard shallow parsing approach and is done in three cascaded steps – one for each rule set. For each level, a program (a Perl script) is obtained automatically on the basis of a file with rules definitions. For any dialog word, these rules are applied sequentially and the one which covers the longest string is chosen. Searching for concepts starts from the next word.

For example, for the phrase *z ulicy Grójeckiej* ('from Grójecka street'), the annotation process is carried out as follows:

- first the lemma for the form 'Grójeckiej' (i.e. 'Grójecki') is found in the street names lexicon. On the basis of its contents the variable $STREET_N of the value 'Grójecka' is assigned to the word 'Grójeckiej',
- the following rule from the first rules set is applied:
  $STREET $2 #Nc(lemma=ulica)+$STREET_N,
- then, the following rule from the second rules set is used:
  $STREET_STR $2 #PreP(lemma=z)+$STREET(morph=gen).

4.3 Annotation Example

Figure 4 shows a fragment of an annotated dialog. It contains one question concerning trip duration between two points. Square brackets inserted into the text show which strings were recognized.[3]

[3] In this example all text is annotated but it is not frequently the case.

... [jak długo jedzie] [autobus linii sto pięǳiesiąt siedem] [z ulicy Grójeckiej] [przy Bitwy Warszawskiej] [na Plac Wilsona]

... how long does the bus 157 go from Grójecka street close to Bitwy Warszawskiej to Wilson Square.

<c id="7" span="word_11..word_13" attribute="Q_HOW_LONG" value="RideTime" />
<c id="8" span="word_14..word_18" attribute="BUS" value="sto pięćdziesiąt siedem" />
<c id="9" span="word_19..word_21" attribute="SOURCE_STR" value="Grójecka" />
<c id="9" span="word_22..word_24" attribute="LOC_STR" value="Bitwy Warszawskiej" />
<c id="10" span="word_25..word_27" attribute="GOAL_STR" value="Plac Wilsona" />

Fig. 4. Semantic annotation

5 Evaluation and Conclusions

At the moment, the entire corpus consists of 465 dialogs which have 13772 turns and 76560 word forms occurrences. Dialogs are divided into five thematic groups concerning: question about the particular lines, questions about routes to given places, request concerning times of departure, stops localization and reduced-fares. In the described experiment we used a training set of 146 dialogs (from all groups) for rules creation and incremental improving, and 26 dialogues as a test set. Quantitative characteristic of these sets is given in Table 1. The test set was annotated automatically and then manually corrected. Manual correction for a dialogue took on average 14 minutes including making notes and learning a dedicated program. This time will be probably shorter for next dialogues group. Manual annotation of MEDIA dialogs took on average 2 hours [4]. The two obtained annotations were next automatically compared. Table 2 contains the results of the evaluation. It should be noted here that corrected files include all information which can be inferred from the context by a human speaker and which can be difficult to cover by rules. The achieved 20.43% attribute/value error rate justified the adopted strategy. In [9] the annotation of Italian corpus using incremental statistical learning approach is described. They achieved 53,6% attribute/value error rate in the eight turns, for a training set of 140 dialogs.

The manual preparation of rules seems to be a right solution for complicated domains where a lot different concepts should be recognized and no training data is available. In comparison to hand annotation, the described procedure allows for better standardization of the results. All decisions are consistent which makes changes in annotation schema relatively easy, as long as a new schema can be expressed by using the same

Table 1. Data characteristics

	training set	test set
number of dialogues	146	26
turns	3313	726
word forms	19847	4409
different word forms	2255	934
different lemmas	1404	642
chunks	17762	3892

Table 2. Results of attribute/value error rate (A/V ER)

concepts	training set	test set	test set corrected	A/V ER
concept types	132	90	88	-
all concepts	6040	1374	1229	20.43
AT_HOUR	341	57	54	8.77
all location types	512	146	138	11.56
LOCATION_STR	111	33	33	3.03
GOAL_STR	111	19	8	5.00
BUS	355	94	86	12.77
TRAM	21	24	24	58.62
STOP_DESCRIPTION	235	43	33	34.29

type of rules. After corrections made on the basis of the presented evaluation, the described rule set will be used to prepare the first corpus of Polish spontaneous dialogs and for annotation of man-machine dialogues corpus which is being collected at the moment. The prepared corpora will be also used for testing learning approaches to concept tagging.

References

1. Barras, C., Geoffrois, E., Wu, Z., Liberman, M.: Transcriber: a free tool for segmenting, labeling and transcribing speech. In: First International Conference on Language Resources and Evaluation (1998)
2. Bonneau-Maynard, H., Rosset, S., Ayache, C., Kuhn, A., Mostefa, D.: Semantic annotation of the MEDIA corpus for spoken dialog. ISCA Interspeech, 3457–3460 (2005)
3. Bonneau-Maynard, H., Rosset, S.: A semantic representation for spoken dialog. Eurospeech, Geneva (2003)
4. Duvert, F., Meurs, M.-J., Servan, Ch., Béchet, F., Lefèvre, F., de Mori, R.: From concepts to interpretations: semantic composition process on the MEDIA corpus. In: Proc. of Intelligent Information Systems, Zakopane (in press, 2008)
5. Erdmann, M., Maedche, A., Schnurr, H., Staab, S.: From manual to semi-automatic semantic annotation: About ontology-based text annotation tools. In: Proceedings of the COLING 2000 Workshop on Semantic Annotation and Intelligent Content (2000)
6. Mykowiecka, A., Marasek, K., Marciniak, M., Gubrynowicz, R., Rabiega-Wiśniewska, J.: Annotation of Polish spoken dialogs in LUNA project. In: LTC 2007 (2007)
7. Paslaru-Bontas, E.: A Contextual Approach to Ontology Reuse Methodology, Methods and Tools for the Semantic Web. Ph.D. thesis, Fachbereich Mathematik u. Informatik, Freie Universität Berlin (2007)
8. Rabiega-Wiśniewska, J., Rudolf, M.: Towards a bi-modular automatic analyzer of large Polish corpora. In: Kosta, R., Błaszczak, J., Frasek, J., Geist, L., Żygis, M. (eds.) nvestigations into Formal Slavic Linguistics. FDSL IV (2003)
9. Raymond, Ch., Rodriguez, K.J., Riccardi, G.: Active annotation in the LUNA Italian corpus of spontaneous dialogues. LREC (2008)

Evaluation of the Slovak Spoken Dialogue System Based on ITU-T

Stanislav Ondáš, Jozef Juhár, and Anton Čižmár

Technical University of Košice, Faculty of Electrical Engineering and Informatics, Letná 9, 042 00 Košice, Slovakia
{stanislav.ondas, jozef.juhar, anton.cizmar}@tuke.sk

Abstract. The development of the Slovak spoken dialogue system started in year 2006. The developed system is publicly available as a trial system and provides weather forecast and timetables information services. One of the very important questions is how to evaluate quality of such a system. A new method for quality assessment of the spoken dialogue system is proposed. The designed method is based on ITU-T P.851 recommendation. Three types of questionnaires were prepared – A, B and C. The questionnaires serve for obtaining information about user's background, completed interactions with the system and about overall impression of the system and its services. Scenarios, methodology of coding, classifying and rating of collected data were also developed. There are also six classes of quality for representation of system's features. Introduced method was used for evaluation of the dialogue system and timetables information service. This paper also summarizes the results of performed experiment.

Keywords: Spoken dialogue system, subjective evaluation, quality of service, questionnaire.

1 Introduction

Spoken dialogue systems (SDS) are nowadays widely used in several domains. This fact brings a need of evaluation, comparison and categorization of dialogue systems and their services. The quality of the interaction with a telephone-based speech service can be addressed from two separate points-of view. System developers are concerned of system/system's modules performance. From user's point-of view, the perceived quality or overall opinion are the most important aspects. [5] Subjective measures are the only way how to find out the user's opinion of the system. Objective measures, such as performance, do not have a direct link to the user satisfaction [3].

While extensive consideration has been given to the definition and measurement of efficiency and effectiveness in user interactions with speech systems [7], comparatively little emphasis has been given to measurement of subjective satisfaction [1]. There are few projects, which have focused on this domain.

The first one is the SASSI (Subjective Assesment of Speech-System Interface) methodology, which considers "validity" and "reliability" aspects of the questionnaires as a most important. The questionnaires in this project are prepared in iterative design process, in which, at first, a pool of attitude statements is designed and in several iteration only a relevant set of statements is selected. An initial 50 item questionnaire was

P. Sojka et al. (Eds.): TSD 2008, LNAI 5246, pp. 633–640, 2008.
© Springer-Verlag Berlin Heidelberg 2008

designed. Each attitude statement was rated according to a seven point scale. Authors identify six factors or quality aspects: perceived system response accuracy, likeability, cognitive demand, annoyance, habitability and speed [7].

SERVQUAL method for subjective quality evaluation was adopted from the area of marketing applications and is suitable also for subjective evaluation of dialogue systems. Authors in [3] view the spoken dialogue system as a service, which is provided to the users. This method is based on two principles: Service quality can be divided into dimensions, and measured as a difference of expectations and perceptions [4]. SERVQUAL method defines five service quality dimensions: tangibles, reliability, responsiveness, assurance and empathy. SERVQUAL method uses 22 items questionnaires. Respondents assess how the reality meets their expectations.

The subjective evaluation methods providing information about the quality of telephone services is also described in the ITU Recommendation P.851 [5]. The evaluation methods described in this recommendation address different aspects of quality from a user's point of view, as are the usability of the service, the communication efficiency, task and service efficiency, user satisfaction, perceived speech input and output quality, the system's cooperativity, and so on [5]. Described methods are based on laboratory experiments in which subjects interact with the spoken dialogue system in order to perform a pre-defined, realistic task. Then they fill a set of questionnaires, which reflected your opinion of assessed system and services. Recommendation contains also a set of questions (items) related to described aspects of quality and the examples of test scenarios.

The motivation for proposing a new subjective evaluation method is:

- Perceived quality of services provided by the Slovak dialogue system has never been adequately evaluated.
- There is a need of a simple quality evaluation method, which will be able to give usable and understandable results.
- There is a need to identify the most important aspects, which influence overall impression of the spoken dialogue system and its services.

The following chapters introduce the Slovak spoken dialogue system, its architecture, components and provided services (Section 2), the developed method for quality evaluation (Section 3) and Section 4 summarizes results of performed questionnaire experiment.

2 The Slovak Spoken Dialogue System

The development of the Slovak spoken dialogue system started in year 2003 [6,8]. Since 2006 it is publicly available as a trial system interacting with users in Slovak language. The solution of the system is based on Galaxy infrastructure [11] and consists of six servers and Galaxy hub process (Figure 1).

Telephony server connects whole system into the telephony network. It supports analog telephony cards based on springware architecture (Dialogic D120/41JCT-LSEuro). **Automatic Speech Recognition** (ASR) **server** is hidden Markov model – based speech recognizer, which was built by the *Application Toolkit for HTK* (ATK). Context dependent (triphone) acoustic models were trained on SpeechDat-SK and MobilDat-Sk

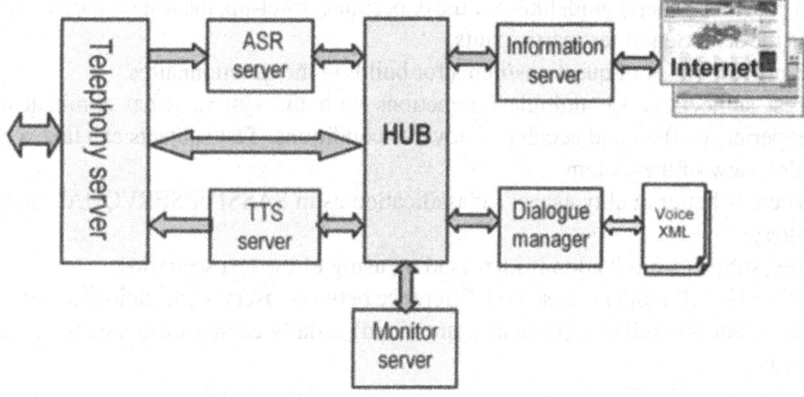

Fig. 1. Architecture of the Slovak spoken dialogue system

databases [12] in a training procedure compatible with "refrec". The Slovak **Text-to-Speech** (TTS) **server** is based on concatenation of diphones, which uses similar algorithm to the Time Domain Pitch Synchronous Overlap and Add (TD-PSOLA) algorithm [13]. **Dialogue manager** is a brain of the system, which manages all system actions. It uses VoiceXML 1.0 language for writing of dialogues. It also follows W3C Speech Interface Framework specifications [14]. **Information server** connects whole SDS to the internet. It is responsible for retrieving information from web. **Monitor server** makes automatic evaluation of system and its services. It is based on collecting of a set of interaction parameters [10].

In more detail the Slovak spoken dialogue system is described in [8].

2.1 Telephone-Based Pilot Applications

Since 2006 the dialogue system is providing the weather forecast and train timetable information services in the Slovak language. In 2007 also bus timetable service was started. Weather forecast service provides information about weather for more than 80 district towns in Slovakia. Timetable services provide information about train/bus domestic connections in Slovakia. The information in these services is obtained from appropriate web locations in real time [9].

Voice dialogues in provided services are designed as a system-directed and they are written in VoiceXML 1.0 language. Services are available through public telephony network (PSTN, GSM) and through VoIP (Skype) in the common dialog. The structure of the service's subdialogues is shown in [16].

3 A New Method for Quality Evaluation of the Slovak Spoken Dialogue System and Services

ITU Recommendation P.851 is adopted as a good background for proposed method. There are following reasons for this selection:

- It provides general guidelines for the experimental set-up, the test scenarios, as well as the selection of test participants.
- It provides a set of questions/items for building the questionnaires.
- Test subjects make multiple interactions with the system, what eliminate first-experience effect and accidental adverse conditions. Test subjects can take a complex view of the system.
- There is better quality aspect's classification as in SASSI or SERVQUAL methodology.
- Test subject's motivation is increased by using of the test scenarios.
- SERVQUAL model is based on difference between users' expectations and perceptions, but test subjects (primarily untrained) usually cannot formulate your expectations.

The designed method is questionnaire-based and it is appointed to realization of subjective evaluation experiments for obtaining of user judgments and opinions of the spoken dialogue system's and service's quality. In this method a set of questionnaires, test scenarios and rating scales are defined. Also there are defined both, new categorization of quality aspects and classes of quality for rating of this categories.

3.1 Questionnaires and Test Scenarios for Evaluation Experiment

The questionnaire form of evaluation was selected as the most suitable form of obtaining information about perceived quality of dialogue system and its services. From a set of questions/items, defined in ITU T.Rec P.851, three types of questionnaires were prepared:

- **Questionnaire A** – contains items related to user's background, their knowledge about domain and system. The items in this questionnaire were modified for testing of the Slovak timetable information service. It contains 12 items.
- **Questionnaire B** – comprises 17 items related to individual interaction with the system.
- **Questionnaire C** – contains 14 closed and 3 open items related to the overall impression of the system and provided services.

Table 1 contains item's numbers according ITU T.Rec P.851, which was adopted in to the designed questionnaires.

For obtaining a complex view on quality aspects it is necessary to interact with the system more than once. Also the sufficient motivation on the side of the test subjects is required. Because of these facts, a set of test scenarios were designed. Each user (test subject) makes two calls on the system's telephone number. First, the test subject fills in the questionnaire A. Then he makes a call on the one of the telephone numbers of

Table 1. Items used to building of questionnaires B and C

Type of questionnaire	Item's numbers
Questionnaire B	overall impression, 1, 2, 4–9, 11, 15–17, 19–22
Questionnaire C	1, 3, 4, 6–8, 10, 11, 13–21

the Slovak SDS and in spoken interaction realizes given "common scenario". After the interaction he fills in the questionnaire B (B1), in which he assesses the prior interaction. Then he makes a second call and carries out given "individual scenario". He assesses the prior interaction in questionnaire B (B2). At last the test subject fills in the questionnaire C, in which assesses both interactions and his overall impression of the system and its service.

3.2 Methodology of Questionnaire's Assessment

The assessment of questionnaires consists of two operations – coding and categorization. The coding is a substitution of data by symbols, which will be used in statistical methods. The rating scale, which can hold values from 0 to 1, was designed for coding of questionnaire's items. The "0" value represents the worst (the lowest) level of the quality (property). Conversely, the "1" value represents the best (the highest) level of the quality.

The items/questions in the questionnaires were coded with five types of scales:

- the growing 7-level scale,
- the decreased 7-level scale,
- 5-level Likert scale,
- backward 5-level Likert scale,
- 5-level centered scale (the highest value in the middle).

All scales were coded within the $< 0, 1 >$ interval.

There was designed categorization on six aspects of quality (categories). The first four categories are the same as in ITU T. recommendation. The last two categories were created for obtaining direct information about user's satisfaction and about usability of the system and its services. The designed categories are the following:

Information obtained from the user; communication with the system; system's behavior; dialog; user satisfaction; usability.

Each category is characterized by a set of questions from questionnaires B (1 and 2) and C. One question can be assigned to several categories. Then the categories are rated by one of the class of quality. There were designed six classes of quality, from A to FX, as is shown in Table 2.

For each category the class of the quality is evaluated as an arithmetic mean of all item's responses for the given category (their score, assigned during coding) in percentage.

Table 2. The classes of quality

Class of quality	Range (%)
A	100–91
B	90–81
C	80–71
D	70–61
E	60–51
FX	51–0

4 Evaluation Experiment Based on the Proposed Method

4.1 Conditions of the Experiment

The evaluation experiment was carried out on 26 test subjects (students). They made 52 interactions with the system, in which they complete ten individual scenarios and one common scenario. They filled in 104 questionnaires of A, B and C type. The calls were made through PSTN network in two acoustic environments – in the office (the silent environment) and in the laboratory with twelve students (noisy environment).

The main aim of the experiment was to obtain understandable and usable information about perceived quality of the Slovak spoken dialogue system and its timetable information service. Testing of the proposed method for quality evaluation was the next reason for conducting this experiment.

4.2 Preliminary Results of the Experiment

The **A** questionnaire gives us the following information: Students (the test subjects) are about 22 years old in average. More than 72% of them travel by bus or train weekly. We can say that they are the experts in the analyzed domain. Almost all users (97%) obtain information about bus/train connections from the internet; on the second place is obtaining of information directly on the bus or train station (next to 30%). For nearly 70% of test subjects this was the first experience with such a system.

According to the proposed method, the judgments of questionnaire's items are divided in to 6 categories. Scores, calculated across the items in each category, determine the class of the quality for given dimensions. This mean, that the overall category's score is calculated as an arithmetic mean of item's scores obtained in coding process. Table 3. contains scores and the classes of quality for all categories.

Table 3. Evaluated classes of quality for the Slovak spoken dialogue system and the Timetables information service

N The category	Score	The class of quality
1. Information obtained from the user	*79.4*	C
2. Communication with the system	*65.9*	D
3. System's behaviour	*78.5*	C
4. Dialog	*71*	C
5. User satisfaction	*64.5*	D
6. Usability	*62.3*	D

Table 3. shows, that quality of the evaluated system and its services is rated in the middle of the range of classes. The first category acquires the highest score (79.4). The lowest score is for sixth category – usability (only 62.3) and user satisfaction (64.5). It means, that there are still some reasons, for which, the users do not use evaluated system or service. For example, the last item in questionnaire B shows, that, the users judge the quality of synthesized speech as very poor (0.40).

Earlier evaluation experiment based on collecting of interaction parameters [15] was described in [16]. In this experiment 69 real users interact with timetable information service. The successfulness of interaction was about 60%. We define the term successfulness as a parameter, which represents information, whether the user obtains required information or not. Categories 1., 5, and 6. are related to mentioned parameter.

5 Conclusions

The proposed paper introduces a new simple method for subjective evaluation of the spoken dialogue system and its services, which is based on ITU T. recommendation. The introduced method consists of a set of test scenarios, questionnaires and rating scales. There are also defined a new categorization and rating-approach based on classes.

The paper also describes the performed evaluation experiment, in which 26 test subjects interact with the Slovak spoken dialogue system and briefly describes the main results of executed experiment.

In the near future, we intend to verify validity and reliability of obtained user judgments and to compare results of questionnaire-based experiment with results obtained by objective evaluation methods based on collecting of interaction parameters. We intend to execute both types of experiments together.

References

1. Hone, K.S., Graham, R.: Subjective assessment of speech-system interface usability. In: EUROSPEECH-2001, pp. 2083–2086 (2001)
2. Walker, M.A., Litman, D.J., Kamm, C.A., Abella, A.: PARADISE – A Framework for Evaluating Spoken Dialogue Agents. In: Proceedings of ACL/EACL 35th Annual Meeting of the Association for Computational Linguistics, pp. 271–280. Morgan Kaufmann, San Francisco (1997)
3. Hartikainen, M., Salonen, E., Turunen, M.: Subjective evaluation of spoken dialogue systems using SERQUAL method. In: ICSLP 2004. International conference on spoken language processing ICSLP, 2004 (2004)
4. Parasuraman, A., Zeithaml, V.A., Berry, L.L.: SERVQUAL – A multiple-item scale for measuring consumer perceptions of service quality. Journal of Retailing 64(1) (1988)
5. Möller, S.: Evaluating telephone-based interactive systems. In: ASIDE-2005, Aalborg, Denmark, November 10–11, 2005, p. 42 (2005)
6. Juhár, J., et al.: Voice Operated Information System in Slovak. Computing and Informatics 26, 577–603 (2007)
7. ITU-T Rec. P.851, Subjective Quality Evaluation of Telephone Services Based on Spoken Dialogue Systems, International Telecommunication Union, Geneva (2003), www.itu.int/rec/T-REC-P.851-200311-I/en
8. Juhár, J., et al.: Development of Slovak GALAXY/VoiceXML based spoken language dialogue system to retrieve information from the internet. In: INTERSPEECH-2006, paper 2056-Mon2FoP.10 (2006)
9. Juhár, J., et al.: Speech interface to the bus and city buses timetable information. In: RTT 2007: Research in Telecommunication Technology, 8th International Conference, Žilina-Liptovský Ján, September 10–12, 2007, pages 4. Žilina, ŽU (2007) ISBN 978-80-8070-735-4

10. Ondáš, S., Juhár, J., Val'o, L'.:: Evaluation tool for spoken dialogue system and its services. In: RTT 2007: Research in Telecomunication Technology 2007. 8th International Conference, Žilina-Liptovský Ján, September 10–12, 2007, pages 4. Žilina, ŽU (2007) ISBN 978-80-8070-735-4
11. Galaxy communicator website, `http://communicator.sourceforge.net/`
12. Rusko, M., Trnka, M., Darjaa, S.: MobilDat-SK – A Mobile Telephone Extension to the SpeechDat-E SK Telephone Speech Database in Slovak. In: SPEECOM 2006, Sankt Peterburg, Russia (July 2006)
13. Rusko, M., et al.: Unit-selection speech synthesis in Slovak. In: Noise and vibration in practice: Proceedings of the 9th International Acoustic Conference, Slovenská technická univerzita Strojnícka fakulta, pp. 85–90 (2004) ISBN 80-227-1901-3
14. W3C Voice Browser Activity website, `http://www.w3.org/Voice/`
15. Möller, S.: Quality of Telephone-Based Spoken Dialogue Systems 2005. Springer, Boston (2004)
16. Ondáš, S., Juhár, J.: Automatic evaluation of Slovak spoken language dialogue system. In: ECMS 2007 & Doctoral School: 8th international workshop on Electronics, Control, Modelling, Measurement and Signals, May 21–23, 2007, p. 20. Technical University of Liberec, Liberec (2007) ISBN 978-80-7372-202-9

Shedding Light on a Troublesome Issue in NLIDBS

Word Economy in Query Formulation

Rodolfo Pazos[1,2], René Santaolalaya S.[1], Juan C. Rojas P.[1], and Joaquín Pérez O.[1]

[1]Centro Nacional de Investigación y Desarrollo Tecnológico (CENIDET)
Departamento de Ciencias Computacionales
AP 5-164, Cuernavaca 62490, México
[2]Instituto Tecnológico de Ciudad Madero, Cd. Madero, México
{pazos,rene,jcrojasp06c,jperez}@cenidet.edu.mx

Abstract. A natural language interface to databases (NLIDB) without help mechanisms that permit clarifying queries is prone to incorrect query translation. In this paper we draw attention to a problem in NLIDBs that has been overlooked and has not been dealt with systematically: word economy; i.e., the omission of words when expressing a query in natural language (NL). In order to get an idea of the magnitude of this problem, we conducted experiments on EnglishQuery when applied to a corpora of economized-wording queries. The results show that the percentage of correctly answered queries is 18%, which is substantially lower than those obtained with corpora of regular queries (53%–83%). In this paper we describe a typification of problems found in economized-wording queries, which has been used to implement domain-independent dialog processes for an NLIDB in Spanish. The incorporation of dialog processes in an NLIDB permits users to clarify queries in NL, thus improving the percentage of correctly answered queries. This paper presents the tests of a dialog manager that deals with four types of query problems, which permits to improve the percentage of correctly answered queries from 60% to 91%. Due to the generality of our approach, we claim that it can be applied to other domain-dependent or domain-independent NLIDBs, as well as other NLs such as English, French, Italian, etc.

Keywords: Natural Language (NL), Natural language interface to databases (NLIDB), Dialog manager.

1 Introduction

NLIDBs were proposed as a solution to the problem of accessing information in a simple way, allowing ideally any type of users, mainly inexperienced ones, to retrieve information from a database (DB) using natural language (NL). However, the NLIDBs promise has not been completely fulfilled: there still exist problems which have not been fully or adequately solved by existing NLIDBs. Some of the main issues are: query translation from NL to a DB query language (usually SQL) and domain independence.

In this work we propose a new approach for the translation process, in which we cope with **one of the most difficult problems: word economy**. To this end, we analyzed the structure of NL queries and found several patterns of recurrent problems. We noticed

P. Sojka et al. (Eds.): TSD 2008, LNAI 5246, pp. 641–648, 2008.
© Springer-Verlag Berlin Heidelberg 2008

that one of the most frequent phenomena (besides anaphora, ellipsis and cataphora) in query formulation is the omission or economy of words when expressing a query in NL.

Word economy can be illustrated by the following query from a hypothetical Registrar's DB: "list grades for John Smith". Though, most MS students in Computer Science can translate this query into SQL given the DB schema, it may prove difficult to be answered correctly by a domain-independent NLIDB, as it is shown by the experiment results presented in Subsection 4.1, which yielded a surprising result: EnglishQuery can answer correctly 18.1% of this type of queries, which reveals the magnitude of this issue. The problem is that the query does not specify the following information: the table (*courses*) where course grades are stored, the table (*student*) and the column (*name*) where student names are stored. If the query specified this information then it would read as follows: "list grades of the *courses* taken by the *student* whose *name* is John Smith", and it would be easily answered by any NLIDB. Unfortunately, most human beings do not formulate queries this way, and therefore, NLIDBs have to live with this problem.

Word economy may cause an incorrect understanding of a query and, therefore, an incorrect translation of it. For solving this problem, we propose a typification of query problems that helps identify problems in NL queries, which permits to devise dialogs for determining omitted words, thus improving the effectiveness of query translation.

For testing our approach we used a domain-independent NLIDB in Spanish (described in [1]) and developed dialog processes corresponding to the problems identified in the typification. The dialog processes permit the user to clarify the query, thus improving the percentage of correctly answered queries.

2 Related Works

Many works on NLIDBs have been developed since the 60's, which can be classified according to several characteristics. For the purposes of this work we consider two characteristics: domain independence and user-interface dialog.

Some of the most important NLIDBs that are domain dependent and do not include dialog are: VILIB [2], Kid [3]. In this type of interfaces the percentage of correctly answered queries is high (69.4–96.2% [4]), mainly because they are limited to one domain. PLANES [5] is an example of a domain-dependent NLIDB that includes dialog.

In domain-independent interfaces that do not include dialog, the success percentage is usually lower than that of domain-dependent interfaces (89.1–92.5% [4]). The most important of this type of interfaces are: EnglishQuery [6], PRECISE [7], ELF [8], SQL-HAL [9], and MASQUE/SQL [10]. The main domain-independent interfaces that include dialog are: Rendezvous [11], STEP [12], TAMIC [13], CoBase [14], CLARE [15], LOQUI [16], and Inbase [17], and the most important commercial NLIDBs with these characteristics are: IBM's LanguageAccess [18] (discontinued in 1992), BBN's PARLANCE [19], and Natural Language [20]. The success percentage for these interfaces is usually larger (84–95% for STEP [21]) than that of the previous type, since they include a dialog manager that allows users to clarify their queries.

Though all of the NLIDBs implementers have faced the word-economy problem and they have tried to solve it, in a thorough survey of the specialized literature on NLIDBs we could not find evidence that word economy has been identified as a major problem nor that it has been dealt with systematically.

3 Typification of Problems in Queries

For the typification of query problems we first analyzed two corpora related to two databases: Northwind (198 queries in Spanish) and Pubs (70 queries in Spanish), where

Table 1. Typification of problems in queries

Cases	Description
1	**Explicit columns and tables**
1.1	Queries that include Select and Where clauses and explicitly include column and table names (therefore, no clarification of information is needed).
Example: What is the home phone of the employee whose territory identifier is 30346?	
1.2	Queries that include only a Select clause and explicitly include column and table names (therefore, no clarification of information is needed).
Example: Show the name of employees.	
2	**Univocal implicit column in the Select clause**
2.1	Queries that include a table name *without specifying which of its columns is(are) requested.*
Example: To which warehouse does the following address belong 679 Carson St.?	
3	**Univocal implicit column in the Where clause**
3.1	Queries that include a table name and a value in the search condition *without specifying the table column related to the value.*
Example: Select the discount for warehouse 8042.	
4	**Multivocal implicit column in the Select clause**
4.1	Queries that include a column name *without specifying one of the several tables to which it may be related.*
Example: What is the name of the contact that has the order date 07/30/1996?	
5	**Multivocal implicit column in the Where clause**
5.1	Queries that include a value in the search condition *without specifying the column nor the table to which it may be related* (therefore, the value may be related to any column of any table).
Example: Show the hire date of Margaret Peacock.	
5.2	Queries that include the name of a column *without specifying one of the several tables to which it may be related.*
Example: Show the name of the company whose identifier is 1.	
6	**Inexistent column**
6.1	Queries that include a *column name that can not be related to any database table.*
Example: I want the type of food whose category is Grains/Cereals.	
7	**Inexistent table**
7.1	Queries that include the name of a *table that does not belong to the database.*
Example: Show the address of the Antonio Moreno taco shop	

the queries of both corpora were formulated by students and are detailed in [22]. The analysis aimed at trying to find problem patterns which are domain independent. The rationale of the analysis involves the following principles:

1. Each reasonably formulated NL query that can be translated into an SQL statement, must always include a phrase that maps to the Select clause of the SQL statement, and usually includes a phrase that maps to the Where clause of the SQL statement. In our analysis such phases will be called *Select clause* and *Where clause*.
2. Word economy in query formulation only involves two general cases: omission of column names and omission of table names.
3. The situation in which a column or table name is omitted differs, depending if it occurs in the Select clause or the Where clause.
4. An omitted column (or table) name may refer either to one column (or table) or several columns (or tables). The first situation is referred to as *univocal* and second *multivocal*.
5. Though any SQL query always includes a From clause, it can always be ignored for our purposes, since all the table names included in the From clause are always included in the Select and Where clauses, so the From clause can be automatically generated from the Select and Where clauses.

Considering the previous analysis, we have obtained the typification of query problems summarized in Table 1. It is important to note that according to this typification a given query may involve more than one problem, and therefore it may be classified in two or more types. We are currently applying our typification to a corpus for the ATIS database (2,884 benchmark queries in English [23]) to make sure that our typification is complete.

In Sections 3.1 and 3.2 algorithms for two dialog processes are briefly described for the proposed typification (Table 1). Prior to execution of the dialog algorithms the query is processed by the NL-to-SQL translator, which involves three phases: identification of the Select and Where clauses in the query, identification of the tables and columns referred to by query words, and construction of a semantic graph [1]. The dialog processes are devised so as not to "guess" what the user is looking for, but to obtain the correct information omitted through user-interface interaction.

3.1 Dialog Process for Case 3.1

The dialog process devised for queries that include a table name and a value in the search condition without specifying the table column related to the value, proceeds as follows:

1. Identify the name of the table and the data type (string, integer, date, etc.) of the value supplied in the search condition.
2. Look into the DB for the columns of the table identified, whose data types match the data type of the value supplied.
3. Display to the user a dialog window that shows a question and a list of the names of the columns found in the previous step, from which the user has to select a list item.
4. Modify the original query adding the omitted information.
5. Execute the modified query.

3.2 Dialog Process for Case 4.1

The dialog process developed for queries that include a column name without specifying one of the several tables to which it may be related, is the following:

1. Identify the name of the column.
2. Look into the DB for tables that contain the column specified in the query.
3. Display to the user a dialog window that shows a question and the names of the tables that contain the column specified in the query, from which the user has to select a list item.
4. Modify the original query adding the omitted information.
5. Execute the modified query.

4 Experimental Results

4.1 Assessing NLIDB Performance with Economized-Wording Queries

Initially, we conducted an experiment on the commercial interface EnglishQuery (EQ) in order to find out how it fared with a corpus of problematic queries: a set of 86 queries were selected (out of 198 queries) from the Northwind DB corpus (detailed in [22]), which involve the problems described in Table 1.

To carry out the test, the queries of the Northwind corpus were translated into English. Then, since the behavior of EQ depends on the customization needed to adapt it to a given database, we decided to use as many customizations as possible. Unfortunately, the customization of EQ (like most NLIDBs) requires some expertise, so we could only get five customizers: one expert (a professor whose Ph.D. dissertation dealt with NLIDBs) and four MS students in Computer Science. A corpus subset consisting of 12 representative queries was used for customization. Finally, after customization, the 86 queries were input to EQ, and the results obtained are shown in Table 2.

Table 2. Results from the execution of queries involving four problem types using EQ

Customizer	**Query Results**	**Types of Problems**				**Total**	**%**
		3.1	**4.1**	**5.1**	**7.1**		
Professor	Correct	8	10	1	0	19	**22**
	Incorrect/Unanswered	29	33	2	3	67	74
Student1	Correct	6	8	1	0	15	**17**
	Incorrect/Unanswered	31	35	2	3	71	83
Student2	Correct	6	9	2	0	17	**20**
	Incorrect/Unanswered	31	34	1	3	69	80
Student3	Correct	7	8	1	0	16	**19**
	Incorrect/Unanswered	30	35	2	3	70	81
Student4	Correct	1	9	1	0	11	**13**
	Incorrect/Unanswered	36	34	2	3	75	87
Average	Correct	5.6	8.8	1.2	0	15.6	**18.1**
	Incorrect/Unanswered	31.4	34.2	1.8	3	70.4	81.9

Table 2 shows that the average success rate of EQ was 18%, which compared to the results reported by other researchers (53%–83% [7], 58% [24]) imply a performance reduction of 35–65%, which can be attributed to the difficulty of our corpus queries.

4.2 Assessing Performance Improvement with a Dialog Manager

For this experiment our NLIDB [1] was provided with dialog processes for four types of query problems related to the typification in Table 1: 3.1, 4.1, 5.1, and 7.1. Subsequently, the 86 queries from the Northwind DB corpus were input first to the NLIDB without dialog manager and afterwards with dialog manager. Table 3 shows the results obtained, which show a 31% increase in the number of correctly answered queries by the NLIDB with dialog manager.

Table 3. Results for four types of problems in queries from the Northwind DB corpus

NLIDB	Query Results	Types of Problems				Total	%
		3.1	4.1	5.1	7.1		
Without dialog	Correct	22	29	1	0	52	**60**
	Incorrect/Unanswered	15	14	2	3	34	40
	Total	37	43	3	3	86	100
With dialog	Correct	35	37	3	3	78	**91**
	Incorrect/Unanswered	2	6	0	0	8	9
	Total	37	43	3	3	86	100
Success rate improvement:							**31**

5 Conclusions and Future Work

This work is a continuation of a domain-independent NLIDB in Spanish, which has an undemanding customization process [1]. This paper describes a domain-independent dialog manager based on a typification of query problems, which constitutes an approach entirely different to previous ones, and due to its generality, it can be applied to other domain-dependent or domain-independent NLIDBs, as well as other NLs such as English, French, Italian, etc.

The proposed typification of query problems has two important features: **domain independence**; notice that the typification of problems presented in Table 1 is not related in any way to some specific database; and **completeness**; the principles (1–4) described in Section 3 comprise all the situations found in the corpora for the Northwind and Pubs databases and the situations found so far in the corpus for the ATIS database.

The results of the experiments carried out lead to the following conclusions: the success rate of EnglishQuery undergoes a reduction of 35–65% when applied to a corpus of economized-wording queries, which shows the difficulty posed by the word-economy problem; and the use of a dialog manager can improve the success rate of an NLIDB from 60 to 91%.

Up to now we have implemented dialogs for four types of problems (3.1, 4.1, 5.1 and 7.1). Since the results obtained have been encouraging, we are devising and implementing dialog processes for the rest of the problems.

References

1. Pazos, R., Pérez, J., González, B., Gelbukh, A., Sidorov, G., Rodríguez, M.: A Domain Independent Natural Language Interface to Databases Capable of Processing Complex Queries. In: Gelbukh, A., de Albornoz, Á., Terashima-Marín, H. (eds.) MICAI 2005. LNCS (LNAI), vol. 3789, pp. 833–842. Springer, Heidelberg (2005)
2. VILIB Virtual Library (1999), www.islp.uni-koeln.de/aktuell/vilib/
3. Chae, J., Lee, S.: Frame-based Decomposition Method for Korean Language Query Processing. Computer Processing of Oriental Languages (1998)
4. Popescu, A.: Modern Natural Language Interfaces to Databases: Composing Statistical Parsing with Semantic Tractability, University of Washington (2004)
5. Waltz, D.: An English Language Question Answering System for a Large Relational Database. Communications of the ACM (1978)
6. Microsoft TechNet., ch. 32- English Query Best Practices (2008), www.microsoft.com/technet/prodtechnol/sql/2000/reskit/part9/c3261.mspx?mfr=true
7. Popescu, A., Etzioni, O., Kautz, H.: Towards a Theory of Natural Language Interfaces to Databases. In: Proc. IUI-2003, Miami, USA (2003)
8. ELF Software, ELF Software Documentation Series (2002), www.elfsoft.com/help/accelf/Overview.htm
9. SQL-HAL, www.csse.monash.edu.au/hons/projects/2000/Supun.Ruwanpura/
10. Androutsopoulus, I., Ritchie, G., Thanish, P.: MASQUE/SQL, An Efficient and Portables Language Query Interface for Relational Databases, Department of Artificial Intelligence, University of Edinburgh (1993)
11. Minock, M.: A STEP Towards Realizing Codd's Vision of Rendezvous with the Casual User. In: Proc. 33rd International Conference on Very Large Databases (VLDB-2007), Demonstration Session, Vienna, Austria (2007)
12. Minock, M.: Natural Language Access to Relational Databases through STEP. Technical report, Department of Computer Science, Umea University (2004)
13. Bagnasco, C., Bresciani, P., Magnini, B., Strapparava, C.: Natural Language Interpretation for Public Administrations Database Querying in the TAMIC Demonstrator. In: The Proc. Second International Workshop on Applications of Natural Language to Information Systems (1996)
14. Chu, W., Yang, H., Chiang, K., Minock, M., Chow, G., Larson, C.: Cobase – A Scalable and Extensible Cooperative Information System. Journal of Intelligent Information System 6, 253–259 (1996)
15. Alshawi, H., Carter, D., Crouch, R., Pulman, S.: CLARE: A Contextual Reasoning and Cooperative Response Framework for the Core Language Engine. Technical report CRC-028 (1994)
16. Binot, J., Debille, L., Sedlock, D., Vandecapelle, B.: Natural Language Interfaces: A New Philosophy, SunExpert, Magazine (1991)
17. Boldasov, M., Sokolova, G.E.: QGen – Generation Module for the Register Restricted InBASE System. In: Computational Linguistics and Intelligent Text Processing, 4th International Conference, vol. 2588, pp. 465–476 (2003)
18. Ott, N.: Aspects of the Automatic Generation of SQL Statements in a Natural Language Query Interface. Information Systems 17(2), 147–159 (1992)
19. Bates, M.: Rapid Porting of the Parlance Natural Language Interface. In: Proc. Workshop on Speech and Natural Language, pp. 83–88 (1989)
20. Manferdelli, J.: Natural Language Inc., Sun Technology (1989)

21. Naslund, A., Olofsson, P.: Authoring Semantic Grammars in a Web Environment, MS thesis in Computer science, Umea University, Sweden (2007)
22. Gonzalez, B.J.J.: Traductor de Lenguaje Natural Español a SQL para un Sistema de Consultas a Bases de Datos. Ph.D. dissertation, Depto. de Ciencias Computacionales, Centro Nacional de Investigación y Desarrollo Tecnológico, Cuernavaca, Mexico (2005)
23. DARPA Air Travel Information System (ATIS0), `www.ldc.upenn.edu/Catalog/readme_files/atis/sdtd/trn_prmp.html`
24. Bhootra, R.: Natural Language Interfaces: Comparing English Language Front End and English Query. MS thesis, Virginia Commonwealth University (2004)

Dialogue Based Text Editing

Jaromír Plhák

Faculty of Informatics, Masaryk University
Botanická 68a, CZ-602 00 Brno, Czech Republic
xplhak@mail.muni.cz

Abstract. This paper presents the basic principles for text editing by means of dialogue. First, the usage of the text division algorithm is discussed as well as its enhancement. Then, the dialogue text processing interface which co-operates with the voice synthesizer is described. We propose basic functions, formulate the most notable problems and suggest and discuss their possible solutions.

1 Introduction

Editing text is one of the general purposes of using personal computers. Many different text editors focused on every type of text have been developed for many years and they offer a great functionality according the users' requirements.

Nowadays, a specific software (e.g. screen readers [1]) is used by the visually impaired users for editing text with the common text editors. Orientation of these users in text might be complicated, while they are not able to recognize the paragraph indenting, quickly overview the context in the scientific or mathematical text, etc.

Some commercial editors aspire to satisfy requirements of visually impaired users and implement features that make text processing easier for them (e.g. [2]). Other projects focus on implementing voice recognition and speech synthesis inside current popular text editors, allowing them to use some simple commands for text editing (select last sentence, change font,...) (e.g. [3]). However, all these solutions are not primarily oriented on the visually impaired users, therefore the majority of functions are completely useless for them.

In the first part of the paper, the most common algorithms for the text segmentation are presented. We designed modifications of the algorithms and these are discussed at the end of the following section. In the second part, the dialogue interface is presented as well as the most notable challenges.

2 Text Preprocessing

First, we analyze the input text in order to achieve effective processing of divided text into small units. The analysis consists of the language detection and the text structure identification, where the language detection is necessary for an accurate evaluation of the text division algorithms. The identified structure is considered to be suitable for further linguistic processing. Otherwise, the data are converted into plain text by removing the pictures and the other non-textual objects.

P. Sojka et al. (Eds.): TSD 2008, LNAI 5246, pp. 649–655, 2008.
© Springer-Verlag Berlin Heidelberg 2008

In the second step, the text is processed using general knowledge about natural language. Various sub-tasks for natural language processing (NLP) are specified in [4], only processes of word tokenization, sentence segmentation and paragraph boundaries detection are discussed.

2.1 Boundaries Detection

Text segmentation into the smaller blocks like articles (chapters), paragraphs, sentences and words is far from trivial while the low error rate is required. In the following, we discuss the determination of the boundaries for mentioned language units in a bottom-up manner.

Tokenization is the process of splitting the text into discrete units, each usually corresponds to a word in text. The problem of tokenization is mostly related to the words boundary detection, what seems to be an easy task. Nevertheless, we cannot establish the whitespaces as the only delimiters, but have to consider the punctuation as well as the words with the punctuated suffixes (*"won't", "I'm"*) and the hyphenated words (*"self-made", "well-aimed"*) [5]. Moreover, some languages like Chinese or Japanese do not contain spaces between words.

Special attention is dedicated to the sentence segmentation, while finding the sentence boundaries includes many practical problems. The efficiency of the sentence segmentation depends on the text complexity and on the number of abbreviations and proper names, especially those possibly starting with both lower and upper case letters (*"Mr. Brown has brown hair."*). In most cases the segmentation is relatively simple. We are searching for periods, exclamation and question marks that usually indicate a sentence boundary. However, in some cases a period does not signal the end of the sentence, because it denotes a decimal point or is a part of the abbreviation [6]. In the Czech language, period should also stand behind ordinal number (*"2. ledna"*) or be a part of ellipsis. This problem might be solved by defining the list of abbreviation. Nevertheless, the list cannot be complete, because the abbreviations do not form a close set. Finally, abbreviation can be the last token in a sentence, implying that a period acts at the same time as a part of abbreviation and as the end of sentence (*"This happened at 3 p.m."*) [6]. On the other hand, we can find periods, exclamation and question marks as a part of proper names (*"Yahoo!"*).

Basic approaches to sentence boundary detection are as follows:

- **"Period-space-capital letter" algorithm** – marks every period (question, exclamation mark) followed by at least single whitespace and a capital letter as a sentences' separator [6].
- **Machine learning (ML)** – systems based on ML algorithms implements decision trees and neural networks. These systems need some small lexicon and the training corpus for the algorithm adaptation for a particular language.
  In [7], Palmer and Hearst describe the SATZ system. This system handles English, German and French languages extremely fast with the relatively low error rate.
- **Regular expressions** – some rules for the sentence boundary detection are specified by regular expressions [8].
- **Maximum entropy model** – a system based on the maximum entropy principle [9] requests only a list of abbreviations induced from the training data. Therefore it is

very flexible and it can be applied to various languages. The algorithm uses a set of potential sentence's separators (candidates) and processes them using prefixes and suffixes.

- **Probability model** – The probability system [10] uses a statistical model of likely abbreviations, sentence-initial words and the words that precede and follow the numbers. The abbreviation recognition is enhanced by suffix analysis.

The paragraph boundary detection coincides with sentence segmentation. Once we process all sentences, we conclude, that the paragraphs' separators are (considering the format of the text):

- End of sentence + NewLine character (NLC) + one or more tabulator characters.
- End of sentence + NLC + NLC.
- End of sentence + NLC – in special cases, when the text is formatted without the indention.
- Other – for example a tagged text (<p>).

Similarly, the articles' (chapters') separators should be inserted in front of the one-sentence long paragraph surrounded by a NLC.

2.2 Text Representation

We consider two possible internal representations of the preprocessed input text in the system. In the first case, the text is marked using particular tags. We distinguish two different approaches in tag assigning – part-of-speech (PoS) tagging and unique tag insertion between language units. Implementation of the PoS tagging of natural language text into the algorithm is necessary (some algorithms mentioned earlier already implement PoS analysis). Some markup languages for PoS tagging like The Prague Markup Language [11] have been already specified.

Another possibility is to insert unique marks between individual language units. This approach seems to be more convenient as the tagging and dividing words do not increase the efficiency of further text processing. The advantage of the markup approach is that the algorithm for finding boundaries is applied only once (then it is applied to while the text is edited). Manual correction of misplaced tags should be possible even for visually impaired users in the alternative editing mode. Problems corresponding to the tags' misplacement are discussed in the following section.

The implementation of the marking algorithm is not the only possible solution for the text preprocessing. We can also process the text using a minimal modifications to the structure, while the original specifications of the text document preserve. The modified text division algorithm for positioning in the text will be implemented in the dialogue interface. However, this approach disables the language segmentation errors to be corrected. On the other hand, the number of the input text transformations is reduced, which implies lower probability of the transformation error appearance.

2.3 Implementation Difficulties

Quality of the input text is very important factor for the text preprocessing. Many electronic formats like PDF, (X)HTML, XML, or RTF documents are accepted and processed. But, for example, the text received from Optical Character Recognition (OCR) may contain many extraneous and incorrect characters. Some other documents consists only of the upper-case (lower-case) text, or contain incorrectly recognized punctuation or capitalization. These errors degrade the efficiency of the algorithms based on the capitalization rules.

3 Dialogue Interface

The approach to dialogue based editing is adopted from the well-known functions and principles that are applied in the common editors. However, we do not adjust some features used in the common editors since either they manipulate with visual appearance of letters (e.g. changing fonts) or they are extremely complex to be converted into a dialogue format (e.g. spellchecking). While discussing the most relevant functions, we encounter with many remarkable difficulties that are analyzed in this section.

3.1 Basic Functions

The first step in composing the dialogue editor is detailed definition of basic functions. Some of the following functions, that are converted to the dialogue representation, are inspired in the common editors. Others are particularly defined for the dialogue based editor (e.g. "read text").

- Create a new text.
- Select the part of the text by assigning:
  - one language unit (paragraph, sentence, ...);
  - precise text to be selected;
  - "from" (string) and "to" (string);
  - "from" (string);
  - "to" (string).
- Edit the selected text:
  - delete;
  - copy;
  - cut = delete and copy;
  - paste (copied or cut string);
  - replace (string);
  - move (another location).
- Replace the text:
  - behind "cursor";
  - what (string), replace with (string).
- Find the parts of the text;
  - find (string);
  - find next (string).
- Text formatting:
  - create the paragraph;
  - create new article (chapter).
- "Undo" and "Redo" functions.
- Read the text from the:

- current position;
- specific language unit (sequential number);
- precise text (string).

When the user inserts only one of the "from" or "to" strings while selecting a part of the text, the current "cursor" position is considered as the second coordinate.

3.2 Principles

In the initial phase of the dialogue, the user is requested to enter the name of the file to be edited. If the file with the corresponding name does not exist, completely new one is created. Otherwise, the text file is loaded and processed using the segmentation algorithm. After the users are informed whether the file is created or edited, they choose one of the possible actions, specified in the previous section.

The users enter the text using the keyboard or input it verbally. Output is operated either by the terminal or by the suitable voice synthesizer (e.g. Demosthenes [12]).

3.3 Notable Problems

Many notable problems appear while defining the structure and dialogue strategies. Their description and proposed solutions are discussed in this section.

Navigation. The selection of convenient dialogue strategy for navigation in the text has a significant impact on efficiency of the text processing. In the first proposed strategy we assume that the segmentation algorithm is applied, whilst another is based on the string searching.

1. The dialogue interface allows the users to read, select or edit the entire block of the text, that is equal to one language unit (paragraph, sentence, word). Typical user commands are, for instance:
   - *Read third paragraph.*
   - *Start reading the text from second sentence in third paragraph.*
   - *Select first two paragraphs in first article.*
   - *Delete fifth and sixth sentence in this paragraph.*

   The user is allowed to interrupt the text interpretation at any time. Then, the current position is stored. This position can be referred by following commands, for instance:
   - *Read the next word.*
   - *Select this sentence.*
   - *Edit the previous sentence.*
   - *Tell me the sequence number of this paragraph.*

   However, the efficiency of this approach is critically dependent on the error rate of the segmentation algorithm.
2. This approach applies the unique string searching for position determination. It allows the users to select a part of the text crossing more language units. The users remember some parts of the text from the reading and utilize this knowledge for the text selection:

- *Start reading from "She went to the kitchen".*
- *Select the text from "She went" to "knife".*
- *Delete the text from "Sunday" to "beautiful weather".*
- *Read the text from next word to "never ends".*

String Searching. String searching algorithm in the dialogue editing enables the users to either select the part of the text or navigate themselves in the text. We can easily handle the situations, when the particular string is not found or unique string in the text is matched. On the other hand, handling the multiple matches is the interesting challenge. There are three possible approaches at this problem:

1. The user is informed about multiple matches, and the dialogue changes the initiative back to the user (to specify the context or to give other command).
2. We specify the property that allows matches to be sorted. The first item in the sorted queue is chosen (e.g. the nearest match, the shortest match, ...).
3. The dialogue sends all matches to output including their context. The user is able to read them subsequently or at once (sorted or unsorted).

Selection of Arbitrary Part. This problem is closely related to the previous, only the string length metric is considered. In this situation, when exactly one of the initial and terminal strings is uniquely matched, the shortest string is offered first.

4 Conclusions and Future Work

In this paper the basic ideas of dialogue based text editing are described. General problems like navigation in the text occurred while defining the functions and principles of the dialogue interface. We discussed these problems and offer some possible solutions as well. Adaptation of algorithm for text segmentation is considered as the suitable solution.

In the future work, we would like to focus on partial implementation of dialogue based editor and its testing by visually impaired users, obtaining more information for assigning the direction of future implementation.

References

1. Screen Reader – Assistive Technology for Blind and Visually Impaired People, `http://www.screenreader.co.uk/`
2. EditPad Pro – Mighty Fine Text Editor, `http://www.editpadpro.com/`
3. Nuance – Dragon NaturallySpeaking 9, `http://www.nuance.com/naturallyspeaking/`
4. Ben-Dov, M., Feldman, R.: Text Mining and Information Extraction. In: Data Mining and Knowledge Discovery Handbook, pp. 801–831. Springer, Heidelberg (2005)
5. Williams, K.: A framework for text categorization. Ph.D. Thesis, University of Sydney (2002), `http://www.softlab.ntua.gr`
6. Mikheev, A.: Periods, capitalized words, etc. Computational linguistic 28(3), 289–318 (2002)

7. Palmer, D.D., Hearst, M.A.: Adaptive Multilingual Sentence Boundary Disambiguation. Computational Linguistic 23(2), 241–267 (1997)
8. Grefenstette, G., Tapanainen, P.: What is a word, what is a sentence. Problems of tokenization. In: The 3rd International Conference on Computational Lexicography, Budapest, pp. 79–87 (1994)
9. Reynar, J., Ratnaparkhi, A.: A maximum entropy approach to identifying sentence boundaries. In: Proceedings of the Fifth Conference on Applied Natural Language Processing, Washington, DC, pp. 16–19 (1997)
10. Smith, H.: Unsupervised Learning of Period Disambiguation for Tokenisation. Internal Report, IMS, University of Stuttgart (2000)
11. The Prague Markup Language (Version 1.1), `http://ufal.mff.cuni.cz/jazz/PML/doc/`
12. Kopeček, I.: Speech synthesis based on the composed syllable segments. In: Proceedings of TSD 1998, Brno, pp. 247–250 (1998)

SSML for Arabic Language

Noor Shaker[1], Mohamed Abou-Zleikha[1], and Oumayma Al Dakkak[2]

[1] Department of Artificial Intelligence, University of Damascus, Damascus, Syria
noor.shaker@gmail.com, mhd_az@hotmail.com
[2] HIAST P.o.Box 31983, Damascus, Syria
odakkak@hiast.edu.sy

Abstract. This paper introduces SSML for using with Arabic language. SSML is part of a larger set of markup specifications for voice browsers developed through the open processes of the W3C. The essential role of the markup language is to give authors of synthesizable content a standard way to control aspects of speech output such as pronunciation, volume, pitch, rate, etc. across different synthesis-capable platforms. We study SSML, the validity to extend SSML for Arabic language by building Arabic SSML project for parsing SSML document and extracting the speech output.

Keywords: Speech synthesis, SSML markup language, Arabic text-to-speech system.

1 Introduction

SSML (Speech Synthesis Markup Language) is one of the standards that have been developed by Voice Browser Working Group to enable access to the Web using spoken interaction; it is designed to provide a rich, XML-based markup language for assisting the generation of synthetic speech in Web and other applications.

In the objective of building a complete Text-to-Speech system of standard spoken Arabic with a high speech quality, based on SSML, we've built our Arabic SSML system.

The input to this system is an XML document, containing the vocalized Arabic text enclosed in SSML tags. An expert system based on TOPH (Orthographic-Phonetic Transliteration) language [1,2] and [3] transcripts the text into phonetic codes, then MBROLA diphones [4] are used to generate speech. In fact, MBROLA permits the control of some prosodic features such as fundamental frequency F0 and duration. It enabled us to construct our prosodic models and test it. In what follows, we discuss automatic speech generation in our Arabic SSML.

2 Arabic SSML

To achieve our goals, we adopted the same steps as introduced in the W3C [7] see Figure 1. They are: XML parse, Structure analysis, Text normalization, Text-to-phoneme conversion, Prosody analysis, and Waveform production.

P. Sojka et al. (Eds.): TSD 2008, LNAI 5246, pp. 657–664, 2008.
© Springer-Verlag Berlin Heidelberg 2008

Arabic SSML is composed of various modules that represent each stage, and has the capability of parsing SSML documents.

When dealing with Arabic Language there are several specific points to be taken into account, which are not handled in SSML: the choice of the correct pronunciation of numbers and plural nouns because they depend on the context and this feature is not supported in SSML. In fact, if we want to say "nine pens" or "nine sheets" in Arabic, the pronunciation of the word "nine" changes according to the gender of the following word. This pronunciation changes again if the word is definite, and with it the two words permute. Similar and more complicated cases arise for other numbers.

In addition, in Arabic, plural can be of two types: dual or trial and above, the morphology and the pronunciation change according to the plural and to the part of speech of the word (subject, object, adverb...)

To handle the above problems and similar ones, we customized some tags by adding some attributes that provide the context and allow us to choose the correct word to be pronounced (gender, number, definite or not, POS...).

3 Structure of Arabic SSML System

The architecture is designed by specifying the data format serving as input and/or output for each module. This design enables us to easily extend the system to other languages and to change entire modules without affecting the others.

Our Arabic SSML system architecture is described in Figure 1.

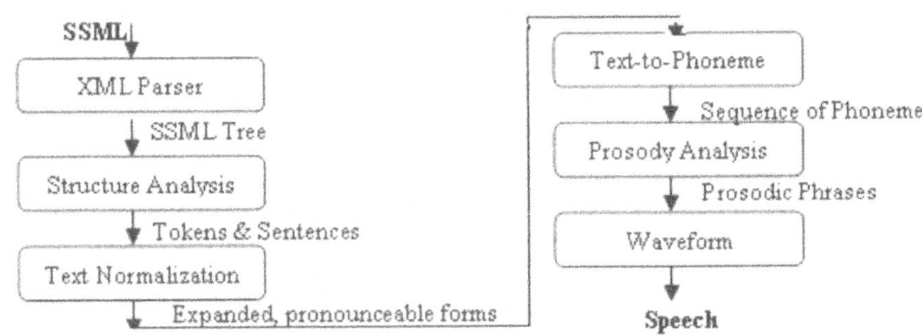

Fig. 1. The architecture of the Arabic SSML system

It contains an XML parser, an Analysis Module, and a Text Normalization Module; the output of this Module constitutes the input to the speech synthesizer containing the Text to phoneme Module, a Prosody generator and a speech waveform generator. We describe briefly, each of the system components.

3.1 XML Parser

The input markup language is translated by this module into a tree. We convert this extracted tree into another more detailed tree, upon which subsequent modules will operate. The tree we build consists of nodes, each node corresponds to a tag in the SSML document.

For example: if we have the following SSML document:
<speack> <p>
الساعة الآن
</p> <say-as interpret-as="Time" format ="hh:mm">10:15</say-as>
</speack>

We build the corresponding XML tree, and from it, we build another one that contains all the details to be added by each module. In this stage the tree will be as in Figure 2.

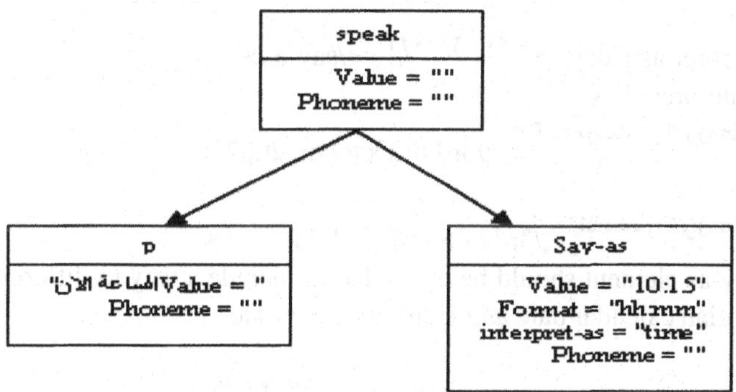

Fig. 2. The extracted tree

3.2 Structure Analysis

The structure of a document influences the way in which a document should be read. The input text needs to be segmented into sentences and words, relying on blanks and punctuation marks. For example, there are common speaking patterns associated with paragraphs and sentences.

The **p** and **s** elements, defined in SSML, which stand for Paragraph and Sentence respectively, explicitly indicate document structures that affect the speech output.

3.3 Text Normalization

Numbers, special symbols, abbreviations, have to be conveniently expanded to normalize the text into a standard format. Text normalization is an automated process that performs a conversion of the written form (orthographic form) into the spoken form.

By the end of this step, the text to be spoken has been converted completely into tokens.

In this module, we process the following SSML elements [8,5]:

<sub> element The substitution element indicates that the text in the alias attribute value replaces the contained text for pronunciation.

For example, ل.س may be spoken as "ليرة سورية" /Syrian pound/, etc.

_{ل.س}

<say-as> element This element allows the author to indicate information on the type of text contained within the element, and to specify the level of detail to render the contained text. The say-as element has three attributes:

- **interpret-as:** always required, indicates the content type of the contained text. interpret-as value: date, time, telephone, characters, cardinal, and ordinal.
- **format:** optional attribute, its legal value depends on the value of the interpret-as attribute. For instance: an Interpret as Time may have the format: hh:mm:ss or hh:mm followed by am. or pm., etc.
- **Detail:** optional attribute, indicates the level of detail to be rendered.

For example:

- date

  <say-as interpret-as="date"> ٢٠٠٢/١٠/٣ </say-as>
  Should be interpreted as
  "الثالث من تشرين الأول عام ألفين و اثنين" ;/3rd of October, 2002/
  or
  "الثالث من شوال عام ألفين و اثنين" ;/3rd of Shawaal, 02 /
  Here, the say-as element should be expanded to include a way to differentiate between Gregorian calendar date and Hejri calendar date.
- Time

  <say-as interpret-as="time" format="hms24"> ٩:٢١:٣٠ </say-as>
  Should be interpreted as
  "التاسعة و واحدٌ و عشرون دقيقة و ثلاثون ثانية صباحاً"
  or "التاسعة وواحدٍ وعشرين دقيقة و ثلاثين ثانية صباحاً" /9:21:30 am /.
  In fact, Arabic has diacretizations which are often omitted short vowels in the written language. These short vowels change according to the part of speech (POS) of the words. Other words change phonemes depending on their POS like male plural which formed from singular by adding /uun/ if the word is a subject or /iin/ if the word is an object or is preceded by a preposition [10]. In the above examples the actual spoken word ("ثلاثين","ثلاثون") /30/ is as male plural, depends on the context, which should be provided to determine the correct pronunciation of the word.
- Telephone number
  <say-as interpret-as="telephone" format="963"> ٠٩٨٥٩٥٩١ </say-as>
  This is a telephone number in Syria (country code is "963"). As a mobile phone number, it should be interpreted as
  "تسعة ستة ثلاثة صفر تسعة ثمانية خمسة تسعة خمسة تسعة واحد"
- Cardinal number
  <say-as interpret-as="cardinal" detail=","> ٢٣٤ </say-as>
  Should be interpreted as
  "مئتين و أربعٍ و ثلاثين" or "مئتان و أربعٌ وثلاثون" ; /234 Two hundred.../
  Here again, the correct words to be spoken depend on the context which must be provided. It depends on the POS of the word, its gender, its number, etc. the same note applies to ordinal numbers.

- Ordinal number
  <say-as interpret-as="ordinal">٣٢</say-as>
  This example will likely be spoken as ;/32/
  "الثانية و الثلاثون" Or "الثاني و الثلاثون" or "الثانية و الثلاثين" Or "الثاني و الثلاثين"
  In Arabic, the actual spoken numbers must cope with Arabic numbers' rules which depend on implied context which must be provided. The words indicating the numbers can be themselves male or female. In some cases they take the gender of the numbered objects, in others they take the opposite.

  In our implementation of the system we took into consideration all these rules and incorporated them in the system, and added the proper attributes in order to get the appropriate output speech.

3.4 Text-to-Phoneme Conversion

Once the synthesizer processor has determined the set of words to be spoken, it must derive pronunciations for each word. Word pronunciations may be conveniently described as sequences of phonemes.

In this module, we process the following SSML element:

<phoneme> element This element allows a phonemic sequence to be provided for any word sequence. All other words in the text are converted to its corresponding set of phonemes using TOPH after its adaptation to Arabic Language.

For example
<phoneme alphabet="ipa" ph="ibda;"> إبدأ </phoneme>

3.5 Prosody Analysis

Prosody is the set of features of speech output that includes pitch, timing, pausing, speaking rate, and emphasis on words, etc. Producing human-like prosody is important for making speech sound natural, and for correctly conveying the meaning of the spoken language. The prosody module adds prosody information for the phoneme according to what the author indicates in SSML tags.

The system generates automatic prosody for all other sentences where the prosody is not given explicitly. This prosody relies only on the punctuation to give the type of the sentence: Exclamation if it ends by the exclamation point "!", Interrogative if it ends by a question mark "?". Continuous affirmation if ended by comma "," and Affirmation (Long) if ended by a point "." [2]

This module processes the following SSML elements:

<emphasis> element This element implies speaking the text with emphasis. The attributes are:

- **level:** the optional level attribute indicates the strength of emphasis to be applied. Defined values are "strong'", "moderate", "none" and "reduced". The default level is "moderate". For example
  <emphasis level = "strong"> ما </emphasis> أروع عملك هذا! ;/What marvelous is your work! /

<break> element The break element is an empty element that controls the pausing or other prosodic boundaries between words. The attributes on this element are:

- **strength:** optional attribute having one of the following values: "none", "x-weak", "weak", "medium" (default value), "strong", or "x-strong". This attribute is used to indicate the strength of the prosodic break in the speech output.
- **time:** optional attribute indicating the duration of a pause to be inserted in the output in seconds or milliseconds.

If a break element is used without strength or time attributes, a break will be produced according to the type of the sentence.

For example
رجاءً اضغط الرمز واحد أو انتظر سماع الصوت <break time="3s"/> ;/Press one or wait till you hear the tone/

<prosody> element The prosody element permits control of the pitch, speaking rate and volume of the speech output. These attributes are all optional. They are the following [3]:

- **pitch:** the baseline pitch for the contained text, Legal values are: a number followed by "Hz", a relative change or "x-low", "low", "medium", "high", "x-high", or "default".
- **contour:** sets the actual pitch contour for the contained text.
  <prosody contour="0%, +20Hz" (10%, +30%) (40,+10Hz)> صباح الخير </prosody> ;/good morning/
- **range:** the pitch range (variability) for the contained text. Legal values are: a number followed by "Hz", a relative change or "x-low", "low", "medium", "high", "x-high", or "default".
- **rate:** a change in the speaking rate for the contained text. Legal values are: a relative change or "x-slow", "slow", "medium", "fast", "x-fast", or "default".
  سعر القطعة <prosody rate = "-25%"> 100 ل.س</prosody> ;/The price is 32SP/
- **duration:** a value in seconds or milliseconds for the desired time to take to read the element contents.
- **volume:** the volume for the contained text in the range 0.0 to 100.0 specifying a value of zero is equivalent to specifying "silent". Legal values are: number, a relative change or "silent", "x-soft", "soft", "medium", "loud", "x-loud", or "default". The default value is 100.0.

<voice> element This element is a production element. It requests a change in the voice speaker. Attributes are: age, variant, name, gender, xml:lang (language). They are all optional.

3.6 Waveform Production

This module uses the phonemes and the generated parameters from the prosodic information to produce the audio waveform. The format for the output is compatible with MBROLA since it's the tool we use to produce the speech output [4].

4 Implementation

An interface has been designed and implemented, which allows the user to easily investigate parts of the system's architecture tree, each intermediate processing result can serve as input, and any subsequent processing result can be output.

Individual processing steps can be carried out, allowing the user to understand the function of each module, or to investigate the source of an error. We followed the design architecture described in MARY [6].

For text-to-phoneme and prosody, we used the system described in [3] and incorporated it in our interface. Our System has been developed and tested using Java programming language. The system handles the following SSML tags: speak, p, s, sub, say as, phoneme, emphasize, break, prosody; including all their attributes.

5 Future Perspectives

1. It would be useful if SSML contains elements to modify prosody parameters to express emotions such as (happiness, sadness, anger, surprise, fear).
2. Some text type like messages and SMS contains symbols like and it would be useful if SSML contains elements to express such symbols.
3. Extend SSML to create audio versions of the mathematical expressions. That may include an extension to say-as element to handle math formats like $\sqrt{a+b/d}$.

6 Conclusion

An automated tool has been developed for Arabic SSML. An overview of the processing components of the system has been given. It has been described how a system-internal tree-based data representation, can be used to make partial processing results available outside the system. The advantages of this design architecture are:

1. All intermediate processing results can be made visible.
2. These intermediate results can be modified and fed back as input into the system for future improvements.
3. Each module can be easily replaced by another one in case they have the same input and output data type.

These features are very helpful for teaching purposes, for non-technical users and for research and development of TTS synthesis.

References

1. Ghneim, N., Habash, H.: Text-to-Phonemes in Arabic. Damascus University Journal for the Basic Sciences 19(1) (2003)
2. Al-Dakkak, O., Ghneim, N., Abou Zliekha, M., Al-Moubayed, S.: Prosodic Feature Introduction and Emotion Incorporation in an Arabic TTS.In: ICTTA (2006)

3. Al-Dakkak, O., Ghneim, N., Abou Zliekha, M., Al-Moubayed, S.: Prosodic Feature Introduction and Emotion Incorporation in an Arabic TTS. In: IEEE Int. Conf. on Information and Communication Technologies.In: ICTTA. Damascus-SYRIA (2006)
4. Dutoit, T., Pagel, V., Pierret, N., Bataille, F., van der Vrecken, O.: The MBROLA project: towards a set of high quality speech synthesizers free of use for non commercial purposes. In: ICSLP 1996. Proceedings. Fourth International Conference (1996)
5. Bonardo, D., Baggia, P.: (Loquendo) SSML 1.0: an XML-based language to improve TTS rendering (2005)
6. Schröder, M., Trouvain, J.: The German Text-to-Speech Synthesis System MARY: A Tool for Research. Development and Teaching (2003)
7. ScanSoft Speech Synthesis Markup Language (SSML) Version 1.0, W3C Recommendation 7 September (2004), http://www.w3.org/TR/speech-synthesis/
8. SSML 1.0 say-as attribute values W3C Working Group Note 26 May (2005), http://www.w3.org/TR/2005/NOTE-ssml-sayas-20050526

Author Index

Lecture Notes in Artificial Intelligence (LNAI)

Vol. 5253: D. Dochev, M. Pistore, P. Traverso (Eds.), Artificial Intelligence: Methodology, Systems, and Applications. XII, 416 pages. 2008.

Vol. 5246: P. Sojka, A. Horák, I. Kopeček, K. Pala (Eds.), Text, Speech and Dialogue. XVIII, 667 pages. 2008.

Vol. 5239: K.-D. Althoff, R. Bergmann, M. Minor, A. Hanft (Eds.), Advances in Case-Based Reasoning. XIV, 632 pages. 2008.

Vol. 5227: D.-S. Huang, D.C. Wunsch II, D.S. Levine, K.-H. Jo (Eds.), Advanced Intelligent Computing Theories and Applications. XXVII, 1251 pages. 2008.

Vol. 5221: B. Nordström, A. Ranta (Eds.), Advances in Natural Language Processing. XII, 512 pages. 2008.

Vol. 5212: W. Daelemans, B. Goethals, K. Morik (Eds.), Machine Learning and Knowledge Discovery in Databases, Part II. XXIII, 698 pages. 2008.

Vol. 5211: W. Daelemans, B. Goethals, K. Morik (Eds.), Machine Learning and Knowledge Discovery in Databases, Part I. XXIV, 692 pages. 2008.

Vol. 5208: H. Prendinger, J. Lester, M. Ishizuka (Eds.), Intelligent Virtual Agents. XVII, 557 pages. 2008.

Vol. 5195: A. Armando, P. Baumgartner, G. Dowek (Eds.), Automated Reasoning. XII, 556 pages. 2008.

Vol. 5194: F. Železný, N. Lavrač (Eds.), Inductive Logic Programming. X, 349 pages. 2008.

Vol. 5190: A. Teixeira, V.L.S. de Lima, L.C. de Oliveira, P. Quaresma (Eds.), Computational Processing of the Portuguese Language. XIV, 278 pages. 2008.

Vol. 5180: M. Klusch, M. Pechoucek, A. Polleres (Eds.), Cooperative Information Agents XII. IX, 321 pages. 2008.

Vol. 5179: I. Lovrek, R.J. Howlett, L.C. Jain (Eds.), Knowledge-Based Intelligent Information and Engineering Systems, Part III. XXXVI, 817 pages. 2008.

Vol. 5178: I. Lovrek, R.J. Howlett, L.C. Jain (Eds.), Knowledge-Based Intelligent Information and Engineering Systems, Part II. XXXVIII, 1043 pages. 2008.

Vol. 5177: I. Lovrek, R.J. Howlett, L.C. Jain (Eds.), Knowledge-Based Intelligent Information and Engineering Systems, Part I. LVI, 781 pages. 2008.

Vol. 5144: S. Autexier, J. Campbell, J. Rubio, V. Sorge, M. Suzuki, F. Wiedijk (Eds.), Intelligent Computer Mathematics. XIV, 600 pages. 2008.

Vol. 5118: M. Dastani, A. El Fallah Seghrouchni, J. Leite, P. Torroni (Eds.), Languages, Methodologies and Development Tools for Multi-Agent Systems. X, 279 pages. 2008.

Vol. 5113: P. Eklund, O. Haemmerlé (Eds.), Conceptual Structures: Knowledge Visualization and Reasoning. X, 311 pages. 2008.

Vol. 5110: W. Hodges, R. de Queiroz (Eds.), Logic, Language, Information and Computation. VIII, 313 pages. 2008.

Vol. 5108: P. Perner, O. Salvetti (Eds.), Advances in Mass Data Analysis of Images and Signals in Medicine, Biotechnology, Chemistry and Food Industry. X, 173 pages. 2008.

Vol. 5097: L. Rutkowski, R. Tadeusiewicz, L.A. Zadeh, J.M. Zurada (Eds.), Artificial Intelligence and Soft Computing – ICAISC 2008. XVI, 1269 pages. 2008.

Vol. 5078: E. André, L. Dybkjær, W. Minker, H. Neumann, R. Pieraccini, M. Weber (Eds.), Perception in Multimodal Dialogue Systems. X, 311 pages. 2008.

Vol. 5077: P. Perner (Ed.), Advances in Data Mining. XI, 428 pages. 2008.

Vol. 5076: R. van der Meyden, L. van der Torre (Eds.), Deontic Logic in Computer Science. X, 279 pages. 2008.

Vol. 5064: L. Prevost, S. Marinai, F. Schwenker (Eds.), Artificial Neural Networks in Pattern Recognition. IX, 318 pages. 2008.

Vol. 5049: D. Weyns, S.A. Brueckner, Y. Demazeau (Eds.), Engineering Environment-Mediated Multi-Agent Systems. X, 297 pages. 2008.

Vol. 5043: N. Jamali, P. Scerri, T. Sugawara (Eds.), Massively Multi-Agent Technology. XII, 191 pages. 2008.

Vol. 5040: M. Asada, J.C.T. Hallam, J.-A. Meyer, J. Tani (Eds.), From Animals to Animats 10. XIII, 530 pages. 2008.

Vol. 5032: S. Bergler (Ed.), Advances in Artificial Intelligence. XI, 382 pages. 2008.

Vol. 5027: N.T. Nguyen, L. Borzemski, A. Grzech, M. Ali (Eds.), New Frontiers in Applied Artificial Intelligence. XVIII, 879 pages. 2008.

Vol. 5012: T. Washio, E. Suzuki, K.M. Ting, A. Inokuchi (Eds.), Advances in Knowledge Discovery and Data Mining. XXIV, 1102 pages. 2008.

Vol. 5009: G. Wang, T. Li, J.W. Grzymala-Busse, D. Miao, A. Skowron, Y. Yao (Eds.), Rough Sets and Knowledge Technology. XVIII, 765 pages. 2008.

Vol. 5003: L. Antunes, M. Paolucci, E. Norling (Eds.), Multi-Agent-Based Simulation VIII. IX, 141 pages. 2008.

Vol. 5001: U. Visser, F. Ribeiro, T. Ohashi, F. Dellaert (Eds.), RoboCup 2007: Robot Soccer World Cup XI. XIV, 566 pages. 2008.

Vol. 4999: L. Maicher, L.M. Garshol (Eds.), Scaling Topic Maps. XI, 253 pages. 2008.

Vol. 4994: A. An, S. Matwin, Z.W. Raś, D. Ślęzak (Eds.), Foundations of Intelligent Systems. XVII, 653 pages. 2008.

Vol. 4953: N.T. Nguyen, G.S. Jo, R.J. Howlett, L.C. Jain (Eds.), Agent and Multi-Agent Systems: Technologies and Applications. XX, 909 pages. 2008.

Vol. 4946: I. Rahwan, S. Parsons, C. Reed (Eds.), Argumentation in Multi-Agent Systems. X, 235 pages. 2008.

Vol. 4944: Z.W. Raś, S. Tsumoto, D.A. Zighed (Eds.), Mining Complex Data. X, 265 pages. 2008.

Vol. 4938: T. Tokunaga, A. Ortega (Eds.), Large-Scale Knowledge Resources. IX, 367 pages. 2008.

Vol. 4933: R. Medina, S. Obiedkov (Eds.), Formal Concept Analysis. XII, 325 pages. 2008.

Vol. 4930: I. Wachsmuth, G. Knoblich (Eds.), Modeling Communication with Robots and Virtual Humans. X, 337 pages. 2008.

Vol. 4929: M. Helmert, Understanding Planning Tasks. XIV, 270 pages. 2008.

Vol. 4924: D. Riaño (Ed.), Knowledge Management for Health Care Procedures. X, 161 pages. 2008.

Vol. 4923: S.B. Yahia, E.M. Nguifo, R. Belohlavek (Eds.), Concept Lattices and Their Applications. XII, 283 pages. 2008.

Vol. 4914: K. Satoh, A. Inokuchi, K. Nagao, T. Kawamura (Eds.), New Frontiers in Artificial Intelligence. X, 404 pages. 2008.

Vol. 4911: L. De Raedt, P. Frasconi, K. Kersting, S. Muggleton (Eds.), Probabilistic Inductive Logic Programming. VIII, 341 pages. 2008.

Vol. 4908: M. Dastani, A. El Fallah Seghrouchni, A. Ricci, M. Winikoff (Eds.), Programming Multi-Agent Systems. XII, 267 pages. 2008.

Vol. 4898: M. Kolp, B. Henderson-Sellers, H. Mouratidis, A. Garcia, A.K. Ghose, P. Bresciani (Eds.), Agent-Oriented Information Systems IV. X, 292 pages. 2008.

Vol. 4897: M. Baldoni, T.C. Son, M.B. van Riemsdijk, M. Winikoff (Eds.), Declarative Agent Languages and Technologies V. X, 245 pages. 2008.

Vol. 4894: H. Blockeel, J. Ramon, J. Shavlik, P. Tadepalli (Eds.), Inductive Logic Programming. XI, 307 pages. 2008.

Vol. 4885: M. Chetouani, A. Hussain, B. Gas, M. Milgram, J.-L. Zarader (Eds.), Advances in Nonlinear Speech Processing. XI, 284 pages. 2007.

Vol. 4884: P. Casanovas, G. Sartor, N. Casellas, R. Rubino (Eds.), Computable Models of the Law. XI, 341 pages. 2008.

Vol. 4874: J. Neves, M.F. Santos, J.M. Machado (Eds.), Progress in Artificial Intelligence. XVIII, 704 pages. 2007.

Vol. 4870: J.S. Sichman, J. Padget, S. Ossowski, P. Noriega (Eds.), Coordination, Organizations, Institutions, and Norms in Agent Systems III. XII, 331 pages. 2008.

Vol. 4869: F. Botana, T. Recio (Eds.), Automated Deduction in Geometry. X, 213 pages. 2007.

Vol. 4865: K. Tuyls, A. Nowe, Z. Guessoum, D. Kudenko (Eds.), Adaptive Agents and Multi-Agent Systems III. VIII, 255 pages. 2008.

Vol. 4850: M. Lungarella, F. Iida, J.C. Bongard, R. Pfeifer (Eds.), 50 Years of Artificial Intelligence. X, 399 pages. 2007.

Vol. 4845: N. Zhong, J. Liu, Y. Yao, J. Wu, S. Lu, K. Li (Eds.), Web Intelligence Meets Brain Informatics. XI, 516 pages. 2007.

Vol. 4840: L. Paletta, E. Rome (Eds.), Attention in Cognitive Systems. XI, 497 pages. 2007.

Vol. 4830: M.A. Orgun, J. Thornton (Eds.), AI 2007: Advances in Artificial Intelligence. XIX, 841 pages. 2007.

Vol. 4828: M. Randall, H.A. Abbass, J. Wiles (Eds.), Progress in Artificial Life. XII, 402 pages. 2007.

Vol. 4827: A. Gelbukh, Á.F. Kuri Morales (Eds.), MICAI 2007: Advances in Artificial Intelligence. XXIV, 1234 pages. 2007.

Vol. 4826: P. Perner, O. Salvetti (Eds.), Advances in Mass Data Analysis of Signals and Images in Medicine, Biotechnology and Chemistry. X, 183 pages. 2007.

Vol. 4819: T. Washio, Z.-H. Zhou, J.Z. Huang, X. Hu, J. Li, C. Xie, J. He, D. Zou, K.-C. Li, M.M. Freire (Eds.), Emerging Technologies in Knowledge Discovery and Data Mining. XIV, 675 pages. 2007.

Vol. 4811: O. Nasraoui, M. Spiliopoulou, J. Srivastava, B. Mobasher, B. Masand (Eds.), Advances in Web Mining and Web Usage Analysis. XII, 247 pages. 2007.

Vol. 4798: Z. Zhang, J.H. Siekmann (Eds.), Knowledge Science, Engineering and Management. XVI, 669 pages. 2007.

Vol. 4795: F. Schilder, G. Katz, J. Pustejovsky (Eds.), Annotating, Extracting and Reasoning about Time and Events. VII, 141 pages. 2007.

Vol. 4790: N. Dershowitz, A. Voronkov (Eds.), Logic for Programming, Artificial Intelligence, and Reasoning. XIII, 562 pages. 2007.

Vol. 4788: D. Borrajo, L. Castillo, J.M. Corchado (Eds.), Current Topics in Artificial Intelligence. XI, 280 pages. 2007.

Vol. 4775: A. Esposito, M. Faundez-Zanuy, E. Keller, M. Marinaro (Eds.), Verbal and Nonverbal Communication Behaviours. XII, 325 pages. 2007.

Vol. 4772: H. Prade, V.S. Subrahmanian (Eds.), Scalable Uncertainty Management. X, 277 pages. 2007.

Vol. 4766: N. Maudet, S. Parsons, I. Rahwan (Eds.), Argumentation in Multi-Agent Systems. XII, 211 pages. 2007.

Vol. 4760: E. Rome, J. Hertzberg, G. Dorffner (Eds.), Towards Affordance-Based Robot Control. IX, 211 pages. 2008.

Vol. 4755: V. Corruble, M. Takeda, E. Suzuki (Eds.), Discovery Science. XI, 298 pages. 2007.

Vol. 4754: M. Hutter, R.A. Servedio, E. Takimoto (Eds.), Algorithmic Learning Theory. XI, 403 pages. 2007.

Vol. 4737: B. Berendt, A. Hotho, D. Mladenic, G. Semeraro (Eds.), From Web to Social Web: Discovering and Deploying User and Content Profiles. XI, 161 pages. 2007.